17-32

School of chemica[l]

98448 ~ 094 이승욱.

- Analogy in transport phenomena

	momentum (mv)	heat	mass
driving force	vel. grad	temp. grad	conc. grad
rate equation	$\tau = -\mu \cdot \frac{dV}{dy}$	$\frac{Q}{A} = -K \cdot \frac{dt}{dy}$	$J = -D \cdot \frac{dC}{dy}$
transport properties	μ: viscosity	K: thermal conductivity	D: diffusivity
	K	μ	D
for perfect gas	$\frac{1}{3}\lambda K \bar{C} N$	$\frac{1}{3}\lambda \bar{C} m N$	$\frac{1}{3}\lambda \bar{C}$
	J/k·m·s	kg/m·s	m^2/s

$$\frac{K}{\rho \cdot C_p} = \alpha \quad = \quad \frac{\mu}{\rho} = \nu \quad = [m^2/s] \quad //$$

12/18 (木)
14:30 – ~~[illegible]~~ 17:00 〈mass 기말고사 : 숙제〉

① 맥케이 & ② 버드

Conversion factors relating common U.S. and SI units

U.S. unit	Equivalent value in SI units
angstrom	0.1 nm
atmosphere (standard)	101.325 kPa
Btu	1.055056 kJ
Btu/(lb·°F) (heat capacity)	4.186 8 kJ/(kg·K)
Btu/h	0.293 971 1 W
Btu/ft^2	11.356 53 kJ/m^2
Btu/(ft^2·h·°F) (heat transfer coefficient)	5.678 263 W/m^2·K
Btu/(ft^2·h) (heat flux)	3.154 591 W/m^2
Btu/(ft·h·°F) (thermal conductivity)	1.730 735 W/m·K
calorie	4.1868 J
cal/(g·°C) (heat capacity)	4186 J/kg·K
cal/g mol	4186 J/kmol
cal/(g mol)(K)	4186 J/kmol·k
centipoise (absolute viscosity)	1.0 mPa·s
centistoke (kinematic viscosity)	1.0×10^{-6} m^2/s
t (°F)	$(t + 459.67)/(1.8)$ K
T(°R)	$T/(1.8)$ K
dyne	10.0 μN
foot	0.3048 m
ft^2	$9.290\,304 \times 10^{-2}$ m^2
ft^3	$2.831\,685 \times 10^{-2}$ m^3
gallon (U.S. liquid)	$3.785\,412 \times 10^{-3}$ m^3
horsepower (550 ft·lb$_f$/s)	745.699 9 W
inch	2.54×10^{-2} m
in. Hg (60°F) (inches mercury pressure)	3.376 85 kPa
in. H_2O (60°F) (inches water pressure)	0.248 84 kPa
kg$_f$ (kilogram force)	9.806 65 N
mass transfer coefficient	0.004450 kmol/m^3·s
mass transfer flux	0.001356 kmol/m^2·s
mile	1 609.344 m
mmHg (0°C) (millimeters mercury pressure)	0.133 322 kPa
poise (absolute viscosity)	0.1 Pa·s
lb$_f$ (pounds force)	4.448 222 N
lb (pounds mass—avoirdupois)	0.453 592 4 kg
lb/in^2 (pounds per square inch pressure)	6.894 757 kPa
stoke (kinematic viscosity)	1.0×10^{-4} m^2/s

1 Pa = 1 N/m^2

$Bi = \frac{hr_i}{k} \ll 1.$

Analogy.

	Heat		Mass
Bi	$\frac{h \cdot L}{k_s}$ (p317 ex21-1)		$\frac{k_{\bar{x}} \cdot L}{D_{AB}}$
Fo	$\frac{\alpha \cdot \theta}{L^2}$		$\frac{D_{AB} \cdot \theta}{L^2}$
Gr	$\frac{g \beta \cdot \Delta t \cdot L^3}{\nu^2}$		$\frac{g\, \varphi \cdot \Delta x_{A} \cdot L^3}{\nu^2}$. $\varphi = -\frac{1}{\rho} \left(\frac{\partial \rho}{\partial x_A}\right)_{P.T.}$
Nu	$\frac{h \cdot L}{k}$	Sh	$\frac{k_{\bar{x}} \cdot L}{\hat{\rho} \cdot D_{AB}}$ or $\frac{k_c \cdot L}{D_{AB}}$ (= Sherwood #)
Pr	$\frac{\nu}{\alpha} = \frac{C_p \cdot \mu}{k}$	Sc	$\frac{\nu}{D_{AB}}$ ∃ $Le = \frac{Sc}{Pr} = \frac{\alpha}{D_{AB}}$ ∟(Lewis #).

Pr≒1, Prandtl mixing length.

$\nu_e = \alpha_e = D_{AB}.$

Pe	Re·Pr	Re·Sc
Ra	Gr·Pr < 10^9: laminar, > 10^9: turbulent	Gr·Sc

∟ Natural convection 관련.

$D_{eq} = 4 \times r_H - \left(\frac{\text{Cross sectional Area}}{\text{Wetted parameter}}\right)$

$t_b = \frac{\int_A u t\, dA}{u_b A}$ ∃ p347. . $h = \frac{k}{t_0 - t_s} \left(\frac{dt}{dy}\right)_{y=0}$

- entry length.

$\frac{L_e}{D} \doteqdot 0.05 \cdot Re \cdot Pr$ (laminar

$\frac{L_e}{D} \doteqdot 40 \sim 100$

MOMENTUM, HEAT, AND MASS TRANSFER

BUILDING THE LITERATURE OF A PROFESSION

Fifteen prominent chemical engineers first met in New York more than 50 years ago to plan a continuing literature for their rapidly growing profession. From industry came such pioneer practitioners as Leo H. Baekeland, Arthur D. Little, Charles L. Reese, John V. N. Dorr, M. C. Whitaker, and R. S. McBride. From the universities came such eminent educators as William H. Walker, Alfred H. White, D. D. Jackson, J. H. James, Warren K. Lewis, and Harry A. Curtis. H. C. Parmelee, then editor of *Chemical and Metallurgical Engineering*, served as chairman and was joined subsequently by S. D. Kirkpatrick as consulting editor.

After several meetings, this committee submitted its report to the McGraw-Hill Book Company in September 1925. In the report were detailed specifications for a correlated series of more than a dozen texts and reference books which have since become the McGraw-Hill Series in Chemical Engineering and which became the cornerstone of the chemical engineering curriculum.

From this beginning there has evolved a series of texts surpassing by far the scope and longevity envisioned by the founding Editorial Board. The McGraw-Hill Series in Chemical Engineering stands as a unique historical record of the development of chemical engineering education and practice. In the series one finds the milestones of the subject's evolution: industrial chemistry, stoichiometry, unit operations and processes, thermodynamics, kinetics, and transfer operations.

Chemical engineering is a dynamic profession, and its literature continues to evolve. McGraw-Hill and its consulting editors remain committed to a publishing policy that will serve, and indeed lead, the needs of the chemical engineering profession during the years to come.

The Series

BAILEY AND OLLIS: Biochemical Engineering Fundamentals
BENNETT AND MYERS: Momentum, Heat, and Mass Transfer
BEVERIDGE AND SCHECHTER: Optimization: Theory and Practice
CARBERRY: Chemical and Catalytic Reaction Engineering
CHURCHILL: The Interpretation and Use of Rate Data—The Rate Concept
CLARKE AND DAVIDSON: Manual for Process Engineering Calculations
COUGHANOWR AND KOPPEL: Process Systems Analysis and Control
DANCKWERTS: Gas Liquid Reactions
FINLAYSON: Nonlinear Analysis in Chemical Engineering
GATES, KATZER, AND SCHUIT: Chemistry of Catalytic Processes
HARRIOTT: Process Control
HOLLAND: Fundamentals of Multicomponent Distillation
JOHNSON: Automatic Process Control
JOHNSTONE AND THRING: Pilot Plants, Models, and Scale-up Methods in Chemical Engineering
KATZ, CORNELL, KOBAYASHI, POETTMANN, VARY, ELLENBAAS, AND WEINAUG: Handbook of Natural Gas Engineering
KING: Separation Processes
KLINZING: Gas-Solid Transport
KNUDSEN AND KATZ: Fluid Dynamics and Heat Transfer
LAPIDUS: Digital Computation for Chemical Engineers
LUYBEN: Process Modeling, Simulation, and Control for Chemical Engineers
MCCABE AND SMITH, J. C.: Unit Operations of Chemical Engineering
MICKLEY, SHERWOOD, AND REED: Applied Mathematics in Chemical Engineering
NELSON: Petroleum Refinery Engineering
PERRY AND CHILTON (EDITORS): Chemical Engineers' Handbook
PETERS: Elementary Chemical Engineering
PETERS AND TIMMERHAUS: Plant Design and Economics for Chemical Engineers
PROBSTEIN AND HICKS: Synthetic Fuels
RAY: Advanced Process Control
REED AND GUBBINS: Applied Statistical Mechanics
REID, PRAUSNITZ, AND SHERWOOD: The Properties of Gases and Liquids
RESNICK: Process Analysis and Design for Chemical Engineers
SATTERFIELD: Heterogeneous Catalysis in Practice
SHERWOOD, PIGFORD, AND WILKE: Mass Transfer
SLATTERY: Momentum, Energy, and Mass Transfer in Continua
SMITH, B. D.: Design of Equilibrium Stage Processes
SMITH, J. M.: Chemical Engineering Kinetics
SMITH, J. M., AND VAN NESS: Introduction to Chemical Engineering Thermodynamics
THOMPSON AND CECKLER: Introduction to Chemical Engineering
TREYBAL: Mass Transfer Operations
VAN NESS AND ABBOTT: Classical Thermodynamics of Nonelectrolyte Solutions: With Applications to Phase Equilibria
VAN WINKLE: Distillation
VOLK: Applied Statistics for Engineers
WALAS: Reaction Kinetics for Chemical Engineers
WEI, RUSSELL, AND SWARTZLANDER: The Structure of the Chemical Processing Industries
WHITWELL AND TONER: Conservation of Mass and Energy

Momentum, Heat, and Mass Transfer

THIRD EDITION

C. O. BENNETT

Professor of Chemical Engineering
University of Connecticut
Storrs

J. E. MYERS

Dean of the College of Engineering
University of California
Santa Barbara

McGRAW-HILL BOOK COMPANY

Auckland Bogotá Guatemala Hamburg Lisbon
London Madrid Mexico New Delhi Panama Paris
San Juan São Paulo Singapore Sydney Tokyo

MOMENTUM, HEAT, AND MASS TRANSFER
INTERNATIONAL EDITION 1983

7 8 9 10 SLP 20 9 8 7 6 5 4 3 2 1

This book was set in Times Roman..
The editors were Julienne V. Brown and Madelaine Eichberg.
The production supervisor was Phil Galea.

Library of Congress Cataloging in Publication Data

Bennett, C. O. (Carroll O.)
Momentum, heat, and mass transfer.

(McGraw-Hill chemical engineering series)
Includes index.
I. Momentum transfer. 2. Chemical engineering.
I. Myers, J. E. (John Earle), date.
II. Title. III. Series.
TP156.T7B46 1982 660.2'842 81-8401
ISBN 0-07-004671-9 AACR2

When ordering this title use ISBN 0-07-066180-4

Printed in Singapore

CONTENTS

	Preface to the Third Edition	ix
	Preface to the Second Edition	xi
1	Introduction	1
Part 1	**Fluid Dynamics**	
2	Introduction to Fluid Behavior	17
3	The Overall Mass Balance	30
4	The Overall Energy Balance	41
5	The Overall Momentum Balance	64
6	Flow Measurement	75
7	The Differential Mass Balance	90
8	The Differential Energy Balance	97
9	The Differential Momentum Balance	103
10	Some Solutions of the Equations of Motion	116
11	Boundary-Layer Flow	135
12	Velocity Distribution and Drag with Turbulent Flow	151
13	Dimensional Analysis with Applications in Fluid Dynamics	179
14	Some Design Equations for the Flow of Incompressible Fluids	197
15	Filtration	224

Part 2 Heat Transfer

16	Introduction to Heat Transfer	249
17	Conduction and Thermal Conductivity	252
18	Steady-State Heat Conduction	260
19	Unsteady-State Heat Conduction	274
20	Numerical, Graphical, and Analog Methods in the Analysis of Heat Conduction	293
21	Convective Heat-Transfer Coefficients	310
22	Heat Transfer with Laminar Flow	334
23	Heat Transfer with Turbulent Flow	355
24	Some Design Equations for Convective Heat Transfer	378
25	Boiling and Condensation	406
26	Radiant Heat Transfer	426
27	Heat-Exchange Equipment	461

Part 3 Mass Transfer

28	Introduction to Mass Transfer	491
29	Molecular Diffusion and Diffusivity	495
30	Diffusion in Binary Mixtures	503
31	Convective Mass-Transfer Coefficients	529
32	Mass Transfer with Laminar Flow	550
33	Mass Transfer with Turbulent Flow	560
34	Some Design Equations for Convective Mass Transfer	577
35	Continuous Contacting of Immiscible Phases	598
36	Simultaneous Momentum, Heat, and Mass Transfer	624
37	Separation by Equilibrium Stages; Immiscible Phases	657
38	Contacting of Partially Miscible Phases	680
39	Distillation of Binary Mixtures	700
40	Multicomponent Separations	752
	Appendixes	781
	Index	821

PREFACE TO THE THIRD EDITION

The third edition of this book was prepared to assist instructors in transport processes who are making partial use of SI units in their classes. Although some favor a sharp break with past systems of engineering units, we think there are good reasons why students should be educated using both the traditional engineering units and SI units. The quantities pounds, mass, and feet are embedded in our culture and will probably remain so for some time to come. It is sufficiently difficult for most students to learn the mathematical formulation of problems in engineering without adding to their burdens by expressing results entirely in unfamiliar units. It is likely that all engineering freshmen have seen a pressure gauge and that they will have some concept of the magnitude of a pound-force and a square inch. Few if any will have seen a gauge reading in pascals, and it is not likely that many will for some time to come. We think it is important for engineering students to be expected to get numerical answers to their problems as soon as possible when starting the study of transport processes so they can begin to acquire the habit of critically evaluating their answers. This is best done if the answers are expressed in units that have some meaning to them.

A second reason for making only a partial conversion of this book to SI is that most sources of engineering data still contain results expressed in the traditional engineering units. Thus for some time to come engineers will have to be able to move readily from one system of units to the other. We would be doing our students no favor if they were not able to handle both systems.

In the derivations in this book we have chosen to continue using g_c, the venerable dimensional constant of U.S. chemical engineering calculations. If students learn that in SI it has a value of unity and no dimensions, it should cause no harm in that system while it continues to be essential in the other. In most chapters the first example problem has been left in engineering units and the second and succeeding alternative examples converted to SI units. Approximately half of all numerical problems in fluid mechanics and heat transfer have been stated in SI units. However, the majority of problems in mass transfer have been left unaltered, chiefly because the changes would have been trivial. A number of quantities in the mass-transfer section have been left in non-SI metric units simply because these are the units in which the data are found.

Few other changes have been made in this edition. Abbreviations of *all* units of measure, however, whether used in the traditional system *or* the SI system, have been updated to conform with current symbols recommended by the American National Standards Institute (ANSI). This means that the former sec is now s, for example, while the former hr is now h. Additionally, some updating of references has been done and a few sections have been revised to reflect more modern interests or new approaches to solving certain problems. On the whole, interest in the approach to teaching transport processes used in this book has remained strong. Editions have now been published in five foreign countries and in three foreign languages. We continue to believe that a single unified study of the fundamental physical and mathematical bases of transport processes carried through to the application of theory to solve real engineering problems is the best way to educate engineers.

C. O. BENNETT
J. E. MYERS

PREFACE TO THE SECOND EDITION

This book is written as an undergraduate textbook for the study of momentum, heat, and mass transfer. The work is not a complete treatise, and we make no claim to cover all aspects, either theoretical or practical, of the broad subjects discussed. Instead, we have attempted to present the most important parts of the theory and applications in a manner at once rigorous and uncluttered by excessive detail.

A book covering such a wide range of subjects can serve only as an introduction. A student who wishes to learn more will go on to the specialized texts in fluid dynamics, heat transfer, mass transfer, and the unit operations. However, the engineering student is now required to study so many topics that it seems to us essential to provide a unified introduction to the transport processes which can be covered in a year's study. We have attempted to discuss most of the important parts of the transport processes and yet provide enough detail to permit the reader to use the material given for the solution of typical problems.

We introduce the unit operations in this book as applications of the theory and show the bases of the correlations which the engineer has used in the past in the hope that he, himself, can provide new correlations in the future as he needs them. The book is not written for only the best students or for only the students who will go to graduate school. It is written for all students who will ultimately do work in which there can be no substitute for a fundamental understanding of the transport processes. In spite of the organization of the subject matter on the basis of theory, we shall expect the student using our book to be able to apply scientific principles to concrete situations, to be able to find data in a handbook, to understand engineering terminology, and to be able to obtain the numerical solution of a problem.

There are already in existence many volumes containing long descriptions and numerous pictures of filters, evaporators, and other process equipment. Although we have used sufficient drawings and pictures to illustrate our subject matter, we refer the reader to Perry's *Chemical Engineers' Handbook* for details of equipment and empirical correlations.

As the various fields of engineering become better understood, the differences between the disciplines diminish. Although the authors' professional experience has been in chemical engineering, we hope this book will be studied in courses intended to introduce engineers in all fields to momentum, heat, and mass transfer. The treatment of fluid dynamics and heat transfer presents the fundamental concepts and laws which all engineers should know. The study of mass transfer has, in the past, been limited almost entirely to chemical engineers. Recently, however, engineers in other fields have shown greatly increased interest in mass transfer. It is our opinion that these engineers will profit from a presentation which gives mass transfer equal standing with momentum and heat transfer.

It is assumed that our readers have a basic knowledge of thermodynamics, calculus, and ordinary differential equations. Vector notation is not used in the main text, but the important equations are developed in compact notation in the appendixes. Thus the teacher or reader can choose for himself the manner of presentation of the basic equations.

The entire subject matter and a selection of the problems of this book can be covered in about 125 class hours, i.e., two 4-semester-hour courses. These courses can extend over the junior year if sufficient thermodynamics and physical chemistry courses are available beforehand. If the subject matter is to be covered in two 3-hour courses, as has been done in our experience, a number of chapters can either be treated qualitatively or omitted without loss of continuity; for instance, in Part 1, "Fluid Dynamics," those chapters on the differential balances and turbulent velocity distributions. Various choices of subject matter will occur to the qualified teacher, and we do not presume to advise him at length on the topics which will be most appropriate for his students.

This book, in its present form, is the result of many years of interaction with colleagues and students too numerous to list. As they read this edition, we hope they will enjoy the evidence that we often have decided to take their advice.

C. O. BENNETT
J. E. MYERS

1
INTRODUCTION

Engineers are employed in many industries in which raw materials are converted to products by chemical reactions and physical changes. Some industries depending upon chemical changes are the manufacture of sulfuric acid from sulfide ores, the production of ammonia, and the conversion of petroleum into a variety of petrochemicals. In other industries, such as the production of sugar from sugar beets or the extraction of oil from corn, physical changes predominate. Most industrial operations consist of a sequence of physical and chemical transformations, and this sequence is called a *process*.

The work of many engineers in industry involves the development of processes and the design and operation of plants. *Process development* is a term used by engineers to describe the search for optimum equipment and conditions for the process. The steps constituting the process are usually known; combining these steps so that the process is profitable when carried out in large-scale equipment is called "developing" a process. The job of chemical engineers in design is usually to determine certain general characteristics and dimensions of the equipment. For example, they might specify the height, diameter, number of plates, and method of control of a distillation column. The thickness of the column walls and the dimensions of the foundation are likely to be specified by mechanical engineers, and the details of the control systems are normally in the hands of electrical engineers. The function of the engineer in plant operation is not only to supervise routine production, but to improve plant efficiency by changes within the existing process. Engineers can perform their functions best if they understand the principles behind the chemical and physical changes of the process.

Chemical industries operated long before the profession of chemical engineering was recognized. The technology of each industry was regarded as a special branch of knowledge, and the people who did the jobs now done by chemical engineers were trained as chemists, mechanical engineers, and technologists. Early courses in chemical engineering were based on the study of industrial technology. These courses were greatly changed by the introduction of the concept of the unit operations. The similarity of the physical changes occurring in widely differing industries led to the study of the many steps common to the industries; these became known as the *unit* operations. For example, it was recognized that the evaporation of a liquid from a solution followed the same principles whether it occurred in a process for manufacturing sugar or fertilizer. Thus evaporation became one of the earliest recognized unit operations. Many other steps achieved the status of unit operations, including fluid flow, heat transfer, humidification, drying, distillation, gas absorption, extraction, crushing and grinding, crystallization, filtration, and mixing.

As the unit operations became better understood, it was apparent that they were not distinct entities. Filtration was obviously a special case of fluid flow; evaporation was a form of heat transfer; extraction and gas absorption involved mass transfer. Drying and distillation were recognized as operations in which both heat and mass transfer were important. The unit operations came to be regarded, therefore, as special cases or combinations of heat transfer, mass transfer, and fluid flow. The latter are referred to by engineers as the *transport* phenomena and are the basis of the unit operations. Any fundamental study of unit operations becomes ultimately a study of these transport phenomena.

Most of the important unit operations are concerned with the behavior of fluids in process equipment. The chemical reactor is usually at the heart of the process, and here the engineer may utilize simultaneously the principles of fluid mechanics, heat transfer, and mass transfer, as well as chemical kinetics and thermodynamics. In the preparation of reactants and in the separation of products, the unit operations of filtration, leaching, absorption, extraction, and distillation are important. Underlying every step of the process are the principles of fluid flow and heat transfer; the fluid must be transported, and its temperature must be controlled. In a chemical process, where composition is a variable, the principles of mass transfer are essential for the design of separation or reaction equipment.

The transport phenomena, which are basic in the study of the unit operations, also underlie the problems in fluid flow and heat transfer which are solved by mechanical and aeronautical engineers. These problems traditionally have not been complicated by chemical reactions and have involved relatively simple geometry, such as flow in a pipe or around a body of revolution. Because of the complications which arise from simultaneous heat and mass transfer in a chemically reacting system and from the complicated geometry of interphase contacting devices, the empirical approach has been fruitful for chemical engineers. Until recently the basic theory of fluid dynamics and heat transfer has been studied mostly by physicists and mechanical and aeronautical engineers. On the other hand, chemical engineers have made significant contributions in mass transfer. However, in recent years other engineers have become more interested in mass transfer, and chemical engineers have found many applications of the theory of fluid dynamics and heat transfer. The practical importance of theory has been greatly enhanced by digital computers. These devices have made possible the application of theory to complicated situations for which it was formerly necessary to be satisfied with empirical methods.

From the above discussion it is clear that as various types of engineers advance in the study of their fields, the differences which have caused the traditional specialization diminish. The study of the principles of the transport phenomena becomes a central topic for every engineer, no matter what the field of specialization. Although many of the examples which illustrate the principles presented in this book are taken from applications in chemical engineering, only Chap. 15 "Filtration," and Chap. 39, "Distillation of Binary Mixtures," concern topics which may be considered special to the chemical engineer.

The more fundamental approach to problems in the transport phenomena and unit operations means that the mathematical analysis of these problems takes on added importance. In addition to its use in design calculations, the computer is now being used to control the operating variables in a plant. In order to put the computer into this application and to give it instructions to run the process so that it operates at an economic optimum, a mathematical model of the process is usually necessary. This model depends on a more detailed understanding and on a more mathematical analysis of the process than was formerly necessary.

In a computer-controlled plant the operating conditions may be changed rather frequently in response to changes in raw materials, catalyst activity, equipment scaling, weather conditions, or market demand. It is therefore necessary for the engineer to understand the unsteady-state behavior of a process, and the mathematics is made even more difficult. This subject is usually called *process dynamics.*

For the reasons given above, the reader will find that the analytical, mathematical approach is stressed in this book in the hope of furnishing a firm foundation for subsequent work in more advanced engineering problems.

NOTATION, UNITS, AND DIMENSIONS

The table at the end of this chapter gives an explanation of most of the symbols in this book. Certain symbols may be explained more fully in the text, and some symbols used only a few times are omitted from the table.

In most discussions of mechanics, it is customary to define the so-called fundamental units with reference to Newton's law,

$$F = Ma \tag{1-1}$$

This procedure leads to the following relations:

Centimeter-gram-second (cgs) system:

$$1 \text{ dyn} = \frac{(1 \text{ g})(1 \text{ cm})}{\text{s}^2} \tag{1-2}$$

Meter-kilogram-second (mks) system:

$$1 \text{ N} = \frac{(1 \text{ kg})(1 \text{ m})}{\text{s}^2} \tag{1-3}$$

British mass system:

$$1 \text{ pdl} = \frac{(1 \text{ lb})(1 \text{ ft})}{\text{s}^2} \tag{1-4}$$

In the systems illustrated, the units of force are derived from the units of mass, length, and time, which are usually considered as fundamental in mechanics. In terms of dimensions,[1] we speak of an $[M]$, $[L]$, $[\theta]$ system, and force then has the dimensions of $[ML/\theta^2]$.

On the other hand, in the so-called American engineering system, the pound force is taken as a fundamental unit, so that we have

$$1 \text{ lb}_f = \frac{(1 \text{ slug})(1 \text{ ft})}{\text{s}^2} \tag{1-5}$$

The pound force is defined as (0.4536)(9.807), or 4.448 N, that is, 4.448 (kg)(m)/s^2. In this $[F]$, $[L]$, $[\theta]$ system a slug is equal to 1 (lb$_f$)(s^2)/ft, and mass has the dimensions of $[F\theta^2/L]$.

The traditional practice in chemical engineering in the United States has been to use four fundamental units and dimensions rather than three, namely, the second, foot, pound, and pound force. The pound force is defined as the force exerted on one pound (mass) which gives it an acceleration of 32.1740 ft/s^2. This

[1] The expression $\mu = [M/L\theta]$, for example, means that viscosity has the dimensions of $[M/L\theta]$.

acceleration corresponds to the acceleration in the standard gravitational field. With this choice of units, it is necessary to use a dimensional constant in Newton's second law,

$$F = \frac{Ma}{g_c} \tag{1-6}$$

so that

$$1\ \text{lb}_f = \frac{(1\ \text{lb})(32.17\ \text{ft})/\text{s}^2}{g_c} \tag{1-7}$$

Therefore the constant g_c is

$$g_c = 32.17\ (\text{lb})(\text{ft})/(\text{lb}_f)(\text{s}^2) \tag{1-8}$$

The pound force defined in this $[F]$, $[M]$, $[L]$, $[\theta]$ system has a definite fixed value and does not depend on the local value of g as do the pound weight or the kilogram weight. The value of g_c has, of course, been chosen so that one pound mass exerts a force of one pound force when exposed to the standard gravitational field of the earth. Much of this book is written in terms of this system, which we shall call the *USChE system.*

The use of this system is in accord with most industrial measuring devices. It is rare indeed to find a set of "weights" in the United States calibrated in slugs or a pressure gauge in England which reads in poundals per square inch. As a matter of fact, even in countries in which the metric system is used, the bar (1 bar $= 10^5$ N/m^2 $= 0.9869$ atm) is used and so is the kg$_f$/cm^2 ($= 0.9628$ atm). A gauge indicating pascals (N/m^2) is rare indeed.

There is now a trend throughout the world to adopt the major features of the SI (*système international*) version of the metric system.[1] The units are chosen mostly from the mks system rather than the cgs system. This rationalization is most advanced in electrical units, but it will take some time to permeate the process industries. Thus we retain the English-based units in the Table of Notation, but a competent engineer must be prepared to use metric units with equal facility.

In order to be applicable to any consistent set of units, an equation must be dimensionally homogeneous; i.e., all the terms must have the same dimensions. In the $[F]$, $[M]$, $[L]$, $[\theta]$ system, the dimensional constant g_c is necessary in Newton's law to ensure dimensional homogeneity. Its numerical value depends on the system of units; for instance, it might equal 9.81 (kg)(m)/(kg$_f$)(s^2). Any numbers which appear in dimensionally homogeneous algebraic equations are dimensionless. Equation (1-6) would no longer be dimensionally homogeneous if it were written as $32.2F = Ma$. This equation does not hold for any units other than those given in

[1] J. W. Mullin, *AIChE J.*, **18**:222 (1972); J. Y. Oldshue, *Chem. Eng. Progr.*, **73**(8):135 (1977).

Eq. (1-8). On the other hand, in the $[M]$, $[L]$, $[\theta]$ system, Eq. (1-1) is dimensionally homogeneous. By defining one less fundamental dimension and letting F have the dimensions of Ma, g_c has a value of 1.0 with no dimensions.

We have so far discussed only mechanical systems. If thermal effects are important, we use a thermal energy dimension $[H]$ and a temperature dimension $[T]$. Thus the complete set of fundamental dimensions used in this book is $[F]$, $[M]$, $[L]$, $[\theta]$, $[H]$, $[T]$. Of course, this number could be reduced, not only with regard to the mechanical quantities, as already discussed, but also with regard to $[H]$, which has the dimensions $[FL]$. Indeed, the SI unit joule is used for both thermal and mechanical energy.

The symbols V, H, etc., refer to intensive properties, i.e., ft^3/lb or Btu/lb. The extensive properties are designated by a superscript, $\check{V} = MV$, $\check{H} = MH$, total volume or total enthalpy. We also use a superscript to denote molar quantities: $\tilde{V}$ or $\tilde{H}$, molar volume, in ft^3/lb mol, or molar enthalpy, in Btu/lb mol. The superscript is also used to distinguish between G, mass flux, in lb/(h)(ft^2), and $\tilde{G}$, molar flux, in lb mol/(h)(ft^2); x_A, mass fraction, and $\tilde{x}_A$, mole fraction; and a number of similar symbols.

The dimensions of the variables used can be inferred from the units given in the table of notation. For instance, the viscosity μ in the table has the units of lb/(ft)(h) and thus the dimensions of $[M]/[L][\theta]$. In the cgs system this becomes g/(cm)(s) or poises; in the mks system, kg/(m)(s), usually expressed as N·s/m^2 or Pa·s in SI units. Numerical conversions can be made with the help of the tables on the inside of the covers of this book.

Example 1-1 We shall learn that the viscosity μ is related to the shear stress τ and the velocity gradient du/dy by the equation

$$\mu = \frac{g_c \tau}{du/dy}$$

Show that this equation is dimensionally homogeneous in the USChE system and also in SI units.

Substitution of units into the right-hand side of the equation gives

$$\frac{(\text{lb})(\text{ft})}{(\text{lb}_f)(\text{s}^2)} \times \frac{(\text{lb}_f)/(\text{ft}^2)}{\text{ft}/(\text{s})(\text{ft})}$$

or lb/(s)(ft), the units of viscosity. In SI, g_c becomes (kg)(m)/(N)(s^2), and the definition [Eq. (1-3)] of a newton means that g_c is the dimensionless number 1.

Thus μ is

$$\frac{1 \times (\text{N/m}^2)}{(\text{m})/[(\text{s})(\text{m})]} = \frac{\text{s} \cdot \text{N}}{\text{m}^2}$$

or the equivalent Pa · s or kg/s · m.[1] ////

It is interesting to notice that use of g_c is implied even in many practical applications of the metric system. If a service station attendant checks your tires in Europe, the pressure is measured not in pascals (SI or N/m^2) or even bars (10^6 dyn/cm^2) but as "1800 g." What is meant is g force/cm^2; in this system of units, g_c is 980 (g)(cm)/(g force)(s^2).

SI units are now standard in publications of the American Institute of Chemical Engineers, and engineers will eventually find that they must work in this system. However, the change is being made so slowly that for many years they will still be confronted with gauges reading psi (lb$_f$/in^2), flow meters indicating gal/min, and data given in Btu/lb. For these reasons both the USChE and SI units are used in this book. However, it should be emphasized that practically all the equations of this book are dimensionally homogeneous. Any consistent system of units can be used, in particular, cgs, SI, and USChE; for the first two, unity is g_c, and for the third, it is the value given in Eq. (1-8).

A feature of SI is the attempt to persuade people to use the units needed in dimensionally consistent equations as the everyday units. However, we expect bars, liters, kilowatts, and other terms to continue in use. As is true for any system, the engineer must realize that these are not the units which go into equations. A bar is 10^5 Pa, just as a psi is 144 lb$_f$/ft^2. Of course, in this day of hand-held calculators, both conversions are equally easy to make. While we wait for interest rates expressed in s^{-1}, we need to be able to work with non-SI units like the year. In SI steam tables, Haywood[2] uses kN/m^2, MN/m^2, and kJ/kg, whereas Irvine and Hartnett[3] use bars and kJ/kg. Although mol (gram mole) is the SI unit, we shall use kmol (kilogram mole). This unit is consistent with molecular weights expressed as kg/kmol (or g/mol). The gas constant is 8314 J/kmol · K.

[1] Please note that, throughout this book, traditional units of the form (s)(m) will be written s · m when SI units are involved.

[2] R. W. Haywood, "Thermodynamic Tables in SI (metric) Units," Cambridge University Press, London, England, 1972.

[3] T. F. Irvine, Jr. and James P. Hartnett, "Steam and Air Tables in SI Units," Hemisphere Publishing Corp., Washington, 1976.

TABLE OF NOTATION

Symbol	Explanation	Typical units	
		USChE	SI
A	Area; cross-sectional area	ft^2	m^2
$\mathcal{A}$	Absorption factor, $\tilde{L}/m\tilde{G}$	dimensionless	...
A_i	Interfacial area	ft^2	m^2
B	Bottom product or raffinate flow rate	lb/h	kg/s
C	Sonic velocity	ft/s	m/s
C_D	Drag coefficient	dimensionless	...
C_D^o	Value of C_D in absence of bulk flow	dimensionless	...
C_o	Orifice coefficient defined by Eq. (6-13)	dimensionless	...
C_p	Specific heat at constant pressure	Btu/(lb)(°F)	J/kg · K
C_{pH}	Humid heat defined by Eq. (36-46)	Btu/(lb)(°F)	J/kg · K
C_r	Rotameter coefficient defined by Eq. (6-16)	dimensionless	...
C_v	Specific heat at constant volume	Btu/(lb)(°F)	J/kg · K
	Venturi coefficient defined by Eq. (6-11)	dimensionless	...
D	Diameter	ft	m
	Distillate or extract flow rate	lb/h	kg/s
D_{AB}	Diffusivity of A through B	ft^2/h	m^2/s
D_{ABe}	Eddy diffusivity	ft^2/h	m^2/s
D_{ABT}	Thermal diffusivity	lb/(h)(ft)	kg/s · m
$D/D\theta$	Substantial derivative operator	s^{-1}	s^{-1}
E	Total energy defined by Eq. (4-2)	(ft)(lb_f)/lb	J/kg
E_G	Murphree V-phase stage efficiency	dimensionless	...
E_{GP}	Murphree V-phase point efficiency	dimensionless	...
E_o	Overall efficiency	dimensionless	...
Eu	Euler number, $g_c p/u^2\rho$	dimensionless	...
F	Feed rate	lb/h	kg/s
$\check{F}$	Force	lb_f	N
$\check{F}_d$	Drag force	lb_f	N
$\check{F}_g$	Force of gravity	lb_f	N
$\check{F}_{xd}$	Force in x direction due to drag at control surface	lb_f	N
$\check{F}_{xp}$	Force in x direction due to pressure difference	lb_f	N
$\check{F}_x$, $\check{F}_y$, $\check{F}_z$	Components of force vector	lb_f	N
F_{12}, $\bar{F}_{12}$, $\mathscr{F}_{12}$	View factors for radiant-heat transfer	dimensionless	...
Fr	Froude number, u^2/gL	dimensionless	...
G	Mass velocity of V phase	lb/(h)(ft^2)	kg/s · m^2
G_B	Mass velocity of component B	lb/(h)(ft^2)	kg/s · m^2
Gr	Grashof number, $gL^3\beta\,\Delta t/\nu^2$	dimensionless	...
H	Enthalpy	Btu/lb	J/kg
H_A, H_x, H_y	Enthalpy of component A, L phase, or V phase	Btu/lb	J/kg
H_G, H_L	Height of individual transfer unit	ft	m
H_{OG}, H_{OL}	Height of overall transfer unit	ft	m
H_{tx}, H_{ty}	Height of transfer unit for heat transfer	ft	m
I	Intensity of turbulence	dimensionless	...
	Intensity of radiant energy	Btu/(h)(ft^2)	W/m^2
$\tilde{I}_A$	Flux of A with respect to $\tilde{u}$	lb mol/(h)(ft^2)	kmol/s · m^2
J	Mechanical equivalent of heat	778 (ft)(lb_f)/Btu	unity
J_A	Flux of A with respect to u	lb/(h)(ft^2)	kg/s · m^2
K	Consistency index	[(lb)(ft)/s^2]s^n	kg m · s^{n-2}
	Constant in Eq. (12-40)	dimensionless	...

TABLE OF NOTATION *(continued)*

Symbol	Explanation	Typical units USChE	Typical units SI
K_A	Equilibrium vaporization ratio, $\tilde{y}_A/\tilde{x}_A$	dimensionless	...
K_{ABT}	Thermal-diffusion ratio	dimensionless	...
K_ρ, K_x, K_y, $K_{\tilde{x}}$, $K_{\tilde{y}}$	Overall mass-transfer coefficients		
L	Mass velocity of L phase	lb/(h)(ft^2)	kg/s · m^2
	Length; thickness; scale of turbulence	ft	m
L_{AB}	Phenomenological coefficient defined by Eq. (29-5)	lb mol/(ft)(h)(Btu)	kmol/m · s · J
L_C	Mass velocity of C	lb/(h)(ft^2)	kg/s · m^2
L_e	Entrance length	ft	m
L_f	Cake thickness	ft	m
Le	Lewis number, **Sc**/**Pr** $= k/\rho D_{AB} C_p$	dimensionless	...
M	Mass	lb	kg
M_c	Mass of filter cake	lb	kg
M_A, M_B, M_m	Molecular weight	lb/lb mol	kg/kmol
Ma	Mach number, u/C	dimensionless	...
N	Number of stages; number of tubes	dimensionless	...
	Flux with respect to fixed axes	lb/(h)(ft^2)	kg/s · m^2
N_A	Flux of A with respect to fixed axes	lb/(h)(ft^2)	kg/s · m^2
N_p	Number of particles	dimensionless	...
N_1	Number of tube rows	dimensionless	...
Nu	Nusselt number, hL/k	dimensionless	...
Nu$_a$	Nusselt number, $h_a L/k$	dimensionless	...
Nu$_m$	Mean Nusselt number $h_m L/k$	dimensionless	...
Nu$_x$	Local Nusselt number $h_x x/k$	dimensionless	...
O	Flow rate of L phase	lb/h	kg/s
O_C	Flow rate of component C	lb/h	kg/s
P	Upstream pressure	lb$_f$/ft^2	N/m^2(Pa)
	Power	(ft)(lb$_f$)/s	W
$\check{P}_x$, $\check{P}_y$, $\check{P}_z$	Components of momentum vector	(lb)(ft)/s	kg · m/s
Pe	Peclet number, **Re Pr** $= C_p \rho u L/k$	dimensionless	...
Pr	Prandtl number, $C_p \mu/k$	dimensionless	...
Q	Heat absorbed per unit mass	Btu/lb	J/kg
R	Drying rate	lb/(h)(ft^2)	kg/s · m^2
	Heat-transfer resistance	(°F)(h)/Btu	K · s/J
	Fractional recovery of pressure loss	dimensionless	...
R_i	Rate of generation of i	lb/h	kg/s
R_m	Resistance to flow of filter medium	ft^{-1}	m^{-1}
R_x, R_y, R_z	Components of resultant force vector	lb$_f$	N
$R(y)$	Correlation coefficient	dimensionless	...
Re	Reynolds number, $Du_b \rho/\mu$	dimensionless	...
Re$_L$	Reynolds number, $Lu_o \rho/\mu$	dimensionless	...
Re$_p$	Reynolds number, $Du_{bs} \rho/\mu(1 - \varepsilon)$	dimensionless	...
Re$_x$	Reynolds number, $xu_o \rho/\mu$	dimensionless	...
S	Entropy	Btu/(lb)(°R)	J/kg · K
$\mathcal{S}$	Stripping factor, $m\tilde{G}/\tilde{L}$	dimensionless	...
S_p	Surface area of a particle	ft^2	m^2
S_v	Specific surface, S_p/v_p	ft^{-1}	m^{-1}
Sc	Schmidt number, $\mu/\rho D_{AB}$	dimensionless	...
Sh	Sherwood number, $k_p L/D_{AB}$	dimensionless	...
St	Stanton number, $h/C_p u \rho$	dimensionless	...
T	Absolute temperature	°R	K
T_o	Stagnation temperature	°R	K

TABLE OF NOTATION *(continued)*

Symbol	Explanation	Typical units	
		USChE	SI
U	Internal energy	Btu/lb	J/kg
	Overall heat-transfer coefficient	Btu/(h)(ft^2)(°F)	W/m$^2\cdot$K
V	Specific volume	ft^3/lb	m^3/kg
	Flow rate of V phase	lb/h	kg/s
V_B	Flow rate of component B	lb/h	kg/s
V_f	Filtrate volume	ft^3	m^3
$\check{V}_f$	Rotameter float volume	ft^3	m^3
W	Work done by unit mass	(ft)(lb$_f$)/lb	J/kg
	Flux of radiant energy	Btu/(h)(ft^2)	W/m^2
$W_{B\lambda}$	Flux of radiant energy in interval λ to $\lambda + d\lambda$ divided by $d\lambda$	Btu/(h)(ft^3)	W/m^3
W_s	Shaft work per unit mass	(ft)(lb$_f$)/lb	J/kg
$\dot{W}$	Rate of work done	(ft)(lb$_f$)/h	W
$\dot{W}_s$	Rate of shaft work	(ft)(lb$_f$)/h	W
We	Weber number, $\rho u^2 L/\sigma$	dimensionless	...
X, Y, Z	Components of body force	lb$_f$/lb	N/kg
X_A	Mass ratio of A in L phase	dimensionless	...
X_c	Critical moisture content	dimensionless	...
X_e	Equilibrium moisture content	dimensionless	...
Y	Correction to Δt_{lm}	dimensionless	...
Y_A	Mass ratio of A in V phase	dimensionless	...
a	Interfacial area per unit volume	ft^2/ft^3	m^2/m^3
a_A	Activity of component A	dimensionless	...
a_x, a_y, a_z	Components of acceleration a	ft/s^2	m/s^2
e	Effective height of roughness	ft	m
f	Friction factor, $g_c D l w_f / 2 L u_b^2$	dimensionless	...
f_p	Friction factor defined by Eq. (14-21)	dimensionless	...
g	Acceleration of gravity	ft/s^2	m/s^2
g_c	Constant	32.17 (lb)(ft)/(lb$_f$)(s^2)	unity
h	Individual heat-transfer coefficient	Btu/(h)(ft^2)(°F)	W/m$^2\cdot$K
h_a	Value of h based on arithmetic mean Δt	Btu/(h)(ft^2)(°F)	W/m$^2\cdot$K
h_d	Fouling coefficient	Btu/(h)(ft^2)(°F)	W/m$^2\cdot$K
h_{lm}	Value of h based on logarithmic mean Δt	Btu/(h)(ft^2)(°F)	W/m$^2\cdot$K
h_m	Mean value of h (constant t_s)	Btu/(h)(ft^2)(°F)	W/m$^2\cdot$K
h_r	Radiant-heat-transfer coefficient	Btu/(h)(ft^2)(°F)	W/m$^2\cdot$K
h_x	Local value of h	Btu/(h)(ft^2)(°F)	W/m$^2\cdot$K
h^o	Value of h in absence of bulk flow	Btu/(h)(ft^2)(°F)	W/m$^2\cdot$K
$\tilde{h}_R, \tilde{h}_n$	Enthalpy	Btu/lb mol	J/kmol
j_H	Colburn j factor for heat transfer	dimensionless	...
j_M	Colburn j factor for mass transfer	dimensionless	...
k	Thermal conductivity	Btu/(h)(ft)(°F)	W/m$\cdot$K
	Ratio of specific heats, C_p/C_v	dimensionless	...
k_x, k_y	Individual mass-transfer coefficient based on Δx_A or Δy_A	lb/(h)(ft^2)	kg/s$\cdot$m^2
$k_{\tilde{x}}, k_{\tilde{y}}$	Individual mass-transfer coefficient based on $\Delta \tilde{x}_A$ or $\Delta \tilde{y}_A$	lb mol/(h)(ft^2)	kmol/s$\cdot$m^2
k_ρ	Individual mass-transfer coefficient based on $\Delta \rho_A$	ft/h	m/s
k_x^o, k_y^o	Value of k_x or k_y in absence of bulk flow		
k_ρ^o	Value of k_ρ in absence of bulk flow	lb/(h)(ft^2)	kg/s$\cdot$m^2
k'_ρ	Mass-transfer coefficient defined by Eq.	ft/h	m/s
	(36-7)	ft/h	m/s

TABLE OF NOTATION *(continued)*

Symbol	Explanation	Typical units: USChE	Typical units: SI
ℓ	Prandtl mixing length	ft	m
lw	Friction, or lost work	$(ft)(lb_f)/lb$	J/kg
lw_f	lw due to pipe friction alone	$(ft)(lb_f)/lb$	J/kg
m	Mass ratio of wet to dry filter cake; stage number; constant in equilibrium relation $\tilde{y}_A = m\tilde{x}_A$	dimensionless	...
n	Flow-behavior index; stage number	dimensionless	...
n_G, n_L, n_{OG}, n_{OL}	Number of transfer units	dimensionless	...
p	Total pressure	lb_f/ft^2	Pa
p_A	Vapor pressure of A	lb_f/ft^2	Pa
$\bar{p}_A$	Partial pressure of A	lb_f/ft^2	Pa
p_i	Pressure at filter medium	lb_f/ft^2	Pa
p_o	Stagnation pressure; free-stream pressure	lb_f/ft^2	Pa
p_s	Mechanical stress on filter cake	lb_f/ft^2	Pa
q	Rate of heat transfer	Btu/h	W
	Volumetric rate of flow	ft^3/h	m^3/s
	Moles of liquid produced on feed plate per mole of feed	dimensionless	...
r	Radius	ft	m
r_H	Hydraulic radius	ft	m
r_A, r_i	Rate of generation of A or i	$lb/(h)(ft^3)$	$kg/s \cdot m^3$
s	Mass-fraction solid in slurry	dimensionless	...
	Fractional rate of surface renewal	s^{-1}	s^{-1}
t	Temperature	°F	°C
t_{as}	Adiabatic saturation temperature	°F	°C
t_b	Bulk, or mixing-cup, temperature	°F	°C
t_m	Temperature of fluid medium	°F	°C
t_o, t_∞	Temperature outside boundary layer	°F	°C
$t_{s\ell}$	Surface temperature	°F	°C
t_{sv}	Temperature of saturated vapor	°F	°C
t_s	Temperature of saturated liquid	°F	°C
t_{wb}	Wet-bulb temperature	°F	°C
u	Velocity; magnitude of velocity vector	ft/s	m/s
u_b	Bulk velocity	ft/s	m/s
u_{br}	Bulk velocity at rotameter annulus	ft/s	m/s
u_{bs}	Superficial velocity	ft/s	m/s
u_o	Velocity outside boundary layer	ft/s	m/s
u_x, u_y, u_z	Components of velocity vector	ft/s	m/s
u_{ys}	Velocity normal to surface at surface	ft/s	m/s
u'_x, u'_y, u'_z	Components of fluctuating velocity	ft/s	m/s
$\bar{u}_x, \bar{u}_y, \bar{u}_z$	Components of time mean velocity	ft/s	m/s
u^*	Friction velocity	ft/s	m/s
u^+	u/u^*	dimensionless	...
v_p	Volume of a particle	ft^3	m^3
w	Mass-flow rate	lb/h	kg/s
x, y, z	Cartesian-coordinate distances	ft	m
x	Distance from leading edge	ft	m
x_A	Mass fraction A, especially in L phase	dimensionless	...
x_e	Condensate film thickness	ft	m
y	Distance from surface	ft	m
y_A	Mass fraction A, especially in V phase	dimensionless	...

TABLE OF NOTATION (*continued*)

Symbol	Explanation	Typical units: USChE	SI
y^+	yu^*/ν	dimensionless	...
z	Height of tower; height of manometer fluid	ft	m
α	Angle between velocity vector and outward normal to control surface	rad	rad
	Thermal diffusivity, $k/\rho C_p$	ft^2/h	m^2/s
	Specific cake resistance	ft/lb	m/kg
	Absorptivity; relative volatility	dimensionless	...
α_{AB}	Relative volatility of A to B	dimensionless	...
α_e	Eddy thermal diffusivity	ft^2/h	m^2/s
β	Angle between x direction and downward-directed vertical	rad	rad
	Ratio of diameters	dimensionless	...
	Coefficient of thermal expansion	$(°R)^{-1}$	K^{-1}
Γ	Mass flow per unit of perimeter	lb/(h)(ft)	kg/s·m
γ	Rate of change of angle ϕ	rad/s	rad/s
γ_A	Activity coefficient of component A	dimensionless	...
Δ	Operator indicating final minus initial or outlet minus inlet, for example, $\Delta w = w_2 - w_1$	dimensionless	...
	Net flow toward end of cascade	lb/h	kg/s
δ	Boundary-layer thickness	ft	m
δ_c	Concentration boundary-layer thickness	ft	m
δ_m	Film thickness	ft	m
δ_{th}	Thermal boundary-layer thickness	ft	m
ε	Void fraction; emissivity	dimensionless	...
η	Dimensionless distance, $y\sqrt{u_o/\nu x}$	dimensionless	...
η_f	Fin efficiency	dimensionless	...
η_p, η_t	Efficiency of pump or turbine	dimensionless	...
θ	Time	s	s
	Angle in polar coordinates	rad	rad
Λ	Kármán number, $\mathbf{Re}\sqrt{f}$	dimensionless	...
λ	Latent heat	Btu/lb	J/kg
	Wavelength	ft	m
μ	Viscosity	lb/(ft)(h)	kg/s·m
μ_A	Chemical potential of A	Btu/lb mol	J/kmol
ν	Kinematic viscosity	ft^2/h	m^2/s
ν_e	Eddy kinematic viscosity	ft^2/h	m^2/s
ρ	Density	lb/ft^3	kg/m^3
	Reflectivity	dimensionless	...
ρ_A	Concentration of A	lb/ft^3	kg/m^3
ρ_{Am}	Concentration of A in fluid medium	lb/ft^3	kg/m^3
ρ_{Ao}	Concentration of A outside boundary layer	lb/ft^3	kg/m^3
ρ_{As}	Concentration of A at surface	lb/ft^3	kg/m^3
ρ_s	Density of solid	lb/ft^3	kg/m^3
σ	Surface tension	lb_f/ft	N/m
	Stefan-Boltzmann constant	$Btu/(h)(ft^2)(°R)^4$	$W/m^2 \cdot K^4$
τ	Transmissivity	dimensionless	...
	Shear stress	lb_f/ft^2	Pa
τ_s	Shear stress at surface	lb_f/ft^2	Pa
$\bar{\tau}^t$	Total stress	lb_f/ft^2	Pa

TABLE OF NOTATION *(continued)*

Symbol	Explanation	Typical units	
		USChE	SI
$\bar{\tau}^r$	Reynolds stress	lb_f/ft^2	Pa
Φ	Dissipation function	Btu/ft^3	J/m^3
ϕ	Angle defined in Fig. 9-4; angular variable in spherical coordinates	rad	rad
	Potential function	ft^2/s	m^2/s
ψ	Stream function	ft^2/s	m^2/s
Ω	Potential energy	$(ft)(lb_f)/lb$	J/kg
ω	Angular velocity	rad/s	rad/s

PART ONE

Fluid Dynamics

2

INTRODUCTION TO FLUID BEHAVIOR

Many problems which arise in the consideration of processes can be solved by viewing the process figuratively from outside its physical enclosure. The changes within the enclosure are measured in terms of the properties of the inlet and outlet streams and the exchanges of energy, in the form of heat and work, between the enclosure and its surroundings. This approach immediately calls to mind the idea of a system in thermodynamics and the application of the first law of thermodynamics thereto. What we shall call the *overall energy balance* is merely the first law of thermodynamics applied to a general situation.

The same remarks can be made about the simpler case—the overall mass balance—and its relation to the usual ideas of a material balance encountered in the study of stoichiometry. The overall force balance, involving fluxes of momentum rather than energy or mass, is perhaps less familiar to the student.

We use the term *overall* to describe these balances because of our viewpoint outside the enclosure; the details of what happens inside do not enter into the analysis. However, in many situations we do wish to consider the details of the process occurring inside the enclosure, and to achieve this end it is useful to write similar balances for a small or differential element of volume. These differential balances may then be integrated for situations where we know some of the characteristic properties of a fluid, such as the viscosity of a newtonian liquid. By this method of analysis we obtain a detailed picture of what happens inside the process enclosure. For example, the velocity distribution may be determined by a differential analysis, whereas average inlet and outlet velocities are all that can be considered in an overall balance.

The overall balances can, in general, be obtained by the integration of the differential balances; conversely, the latter can be obtained by the reduction of the enclosure considered to a differential volume. For the sake of clarity, and to avoid certain mathematical complications, the two types of balance will be derived independently in this book; we shall start with the overall balance because of its relative simplicity and its value in the solution of many important problems.

In spite of the fact that the overall mass, energy, and momentum balances can be derived and used without the extensive consideration of the details of fluid behavior, we shall find it useful to have a few preliminary ideas about the nature of a fluid and how it flows.

Viscosity of Newtonian Fluids

A fluid is distinguished from a solid in this discussion of viscosity by its behavior when subjected to an applied stress. Whereas an elastic solid deforms by an amount proportional to the applied stress, a fluid under similar circumstances continues to deform—i.e., flow—at a velocity which increases with increasing stress. These ideas are clarified by a quantitative definition of viscosity. In Fig. 2-1 a fluid in laminar motion is contained between two infinite parallel plates.

If the upper plate moves with a constant velocity relative to the lower one, a steady-state variation of velocity in the fluid between the plates is eventually achieved. For newtonian fluids, the shear stress τ (the applied force per unit area of plate needed to maintain the constant velocity) is proportional to Δu and inversely proportional to Δy.

$$\tau = \mu \frac{\Delta u}{\Delta y} \tag{2-1}$$

where μ is by definition the viscosity. Since μ as defined by Eq. (2-1) is really an average viscosity over the distance Δy and the curve of u versus y will not in general be a straight line, the relationship is made exact by letting Δy approach zero and using the definition of the derivative to obtain

$$\tau = \mu \frac{du}{dy} \tag{2-2}$$

The defining expression for μ is thus a differential equation, as are many other basic relations of physics.

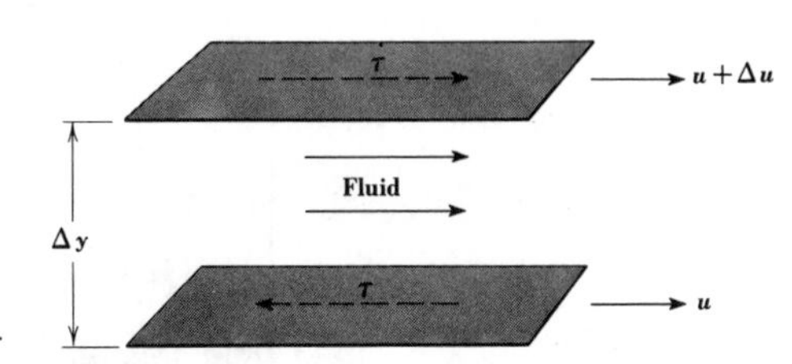

FIGURE 2-1
Fluid shear between parallel plates.

The units for the viscosity given by Eq. (2-2) are (dyn)(s)/cm^2 or (lb$_f$)(s)/ft^2. The cgs unit, alternatively expressed as g/(cm)(s), is called the *poise*, and the usual unit appearing in tables is the *centipoise*, equal to $\frac{1}{100}$ poise. In the system of units commonly employed by engineers, viscosity is given in lb/(ft)(s), and the defining equation is written

$$\tau = \frac{\mu}{g_c}\frac{du}{dy} \tag{2-3}$$

We shall follow this custom. Centipoises are converted to lb/(ft)(s) by multiplying by 6.72×10^{-4}, and to lb/(ft)(h) by multiplying by 2.42. To get Pa · s or kg/m · s, multiply centipoises by 10^{-3}.

If the viscosity of a fluid is divided by its density, a useful quantity, the kinematic viscosity, is obtained. It is written as

$$\nu = \frac{\mu}{\rho} \tag{2-4}$$

where the cgs units are the stoke and the centistoke.

Laminar and Turbulent Flow

It is an empirical fact that a fluid flowing in a small tube or at low velocity does so by the mechanism of laminar flow, also called viscous, or streamline, flow. The layers of fluid slide over each other with no macroscopic mixing, and the velocity in macroscopic steady flow is constant at any point. At higher velocities flow becomes turbulent; there is mixing by eddy motion between the layers, and even in overall steady flow the velocity at a point fluctuates about some mean value. The existence of these two types of flow was demonstrated by Reynolds,[1] who performed the experiment illustrated in Fig. 2-2.

If different sizes of circular pipe and different fluids are used, it is found that laminar flow generally exists when the dimensionless ratio $Du_b\rho/\mu$ is less than 2100. This ratio, which is called the *Reynolds number*, is extremely useful and will

[1] O. Reynolds, *Trans. Roy. Soc. London*, **A174**:935 (1883).

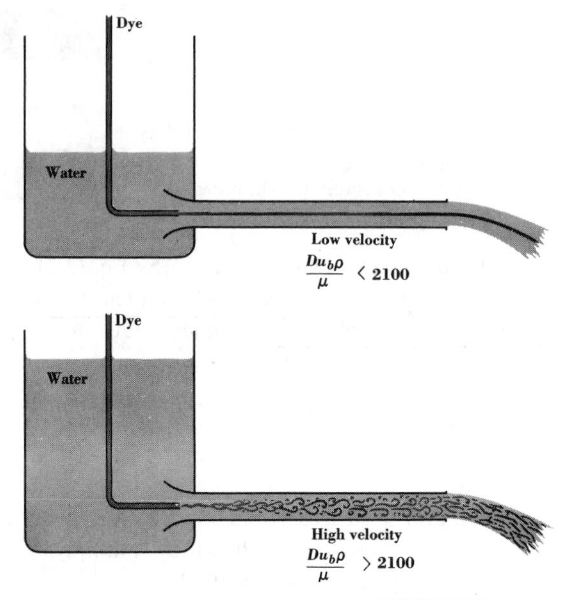

FIGURE 2-2
Reynolds' experiment.

often be encountered in our analyses; it will be explained further in the discussion of dimensional analysis in Chap. 13. The magnitude of the Reynolds number is independent of the system of units employed. Laminar flow can exist at Reynolds numbers above 2100 in very smooth pipes, but the flow is unstable, and small disturbances may cause a transition to turbulent flow. Of two fluids flowing in identical tubes at a given velocity, the one having the lower kinematic viscosity will be the more susceptible to the development of turbulence. Deviations from straight-line flow are damped by the viscous character of the fluid, and the inertia of a deviating element is proportional to its density. Thus fluids with low viscosity and high density tend to support turbulence. The Reynolds number expressed in terms of the kinematic viscosity becomes Du_b/ν. It is therefore a measure of the ratio of the inertia effect to the viscous effect and of the likelihood of the development of turbulence. It is interesting to note that at room temperature ν is 1.0 cSt for water and 15.0 cSt for air at atmospheric pressure.

A remark on the structure of fluids should also be made here before writing expressions involving a differential volume of a fluid. The element really contains an assembly of individual molecules. We shall choose, however, to regard the fluid as a continuum, so that the differential volume will be treated as a homogeneous sample rather than as a collection of molecules and empty spaces between the molecules. This simplification permits us to write differential equations to characterize and analyze fluid behavior.

For liquids, and for gases at all but such low pressure that the mean free path of a molecule is of the order of the tube diameter, we assume that there is no slip at the wall. The velocity then increases from zero at the wall to a maximum at the center of a tube. The curve of u versus y (distance from the wall) is referred to as the *velocity distribution* or *profile*. At a sufficiently large distance from the tube entrance, the velocity profile assumes a constant shape; the flow is said to be developed. The velocity distribution for developed flow depends on whether the flow is turbulent or laminar.

Momentum Flux

Let us now observe that Eq. (2-2) defines a flux of momentum. The shear stress τ can be written in the units (kg)(m/s)/(m^2)(s), that is, momentum per unit area-second. The argument can be put into more concrete terms by considering the interaction of two adjacent layers of a gas in which there is a velocity gradient $du/dy \neq 0$. The random motions of the molecules in the faster-moving layer send some of them into the slower-moving layer, where they collide with the slower-moving molecules and tend to speed them up. In similar fashion the slower layer acts to retard the faster one. (Note that "slow" and "fast" refer to the velocities of the layers, not to the random molecular velocities which determine the temperature.) This exchange of molecules produces a transfer of momentum parallel to the velocity gradient, and a certain force per unit area is required to overcome the drag between the layers and maintain the velocity gradient.

For a newtonian fluid the viscosity is a property of the fluid and thus a function of only the state (pressure, temperature, and composition) of the fluid, so that a plot of τ against du/dy gives a straight line of slope μ. If the flow in a channel as illustrated in Fig. 2-1 is laminar, so that all the layers of fluid slide over each other in the same direction, Eq. (2-3) can be integrated over a finite Δy, provided that τ is known as a function of y. In this way we shall obtain many useful results for problems with laminar flow.

In turbulent flow the velocity at any point is a function of time, and in terms of this instantaneous unsteady-state velocity, Eq. (2-3) still holds; after all, μ is a property of the fluid. In addition to the exchange of molecules between layers, macroscopic amounts of fluid are also exchanged, and momentum is transferred by this mixing process. For these reasons the ordinary viscosity is often called the molecular viscosity, whereas the viscosity which would be measured using Eq. (2-1)

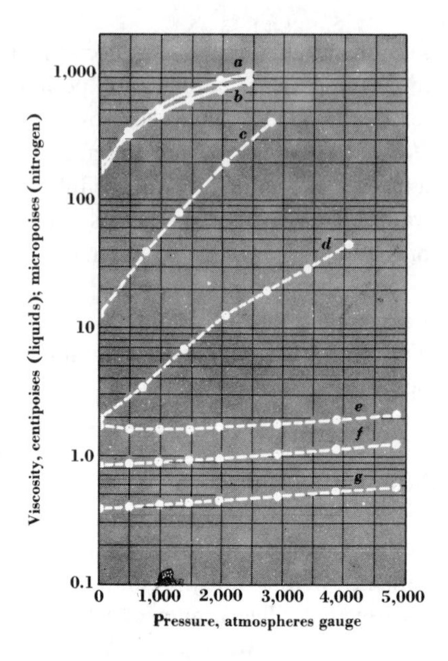

FIGURE 2-3
Viscosities of 9,*n*-octyl heptadecane, water, and nitrogen.
a Nitrogen, 25°C (4).
b Nitrogen, 75°C (4).
c *n*-octyl heptadecane, 20°C (1).
d *n*-octyl heptadecane, 99°C (1).
e Water, 0°C (2,3).
f Water, 20°C (2,3).
g Water, 75°C (2,3).
(1) *Pressure-Viscosity Report, Am. Soc. Mech. Eng., New York, 1953.*
(2) *P. W. Bridgman, "The Physics of High Pressure," G. Bell & Sons, Ltd., London, 1950.*
(3) *J. H. Keenan and F. G. Keyes, "Thermodynamic Properties of Steam," John Wiley & Sons, Inc., New York, 1936.*
(4) *F. Lazarre and B. Vodar, Proc. Conf. Thermo. and Transport Properties of Fluids, Inst. Mech. Engrs., London, 1958.*

(u signifies now a time mean velocity) for turbulent flow in the channel of Fig. 2-1 is the sum of the molecular viscosity and a quantity called the eddy viscosity. The eddy viscosity attains values much larger than the molecular viscosity. It is not a property of the fluid, since it is a function of the extent of mixing (turbulence) in the fluid; it varies from point to point. Turbulent flow will be considered in detail in Chap. 12.

The student should know in general that the viscosity of a gas increases with increasing temperature; the viscosity of a liquid, which is much larger than that of the same substance in its vapor state at the same temperature, decreases with increasing temperature. The viscosity of an ideal gas is independent of pressure, but the viscosities of real gases and liquids usually increase with pressure. Figure 2-3 gives the viscosities of a few substances and the effects of pressure and temperature. More data appear in the appendix, Tables A-8–A-10. If no experimental viscosity data are available, there are several methods available by which this property can be estimated from the fundamental physical constants for a fluid. Refer to Bird et al.[1] for an introduction to this subject.

[1] R. B. Bird, W. E. Stewart, and E. N. Lightfoot, "Transport Phenomena," John Wiley & Sons, Inc., New York, 1960.

Rheology of Nonnewtonian Fluids[2]

The science of the deformation and flow of materials is often called *rheology*: an important branch of this subject concerns the behavior of nonnewtonian fluids. For a newtonian fluid there is a linear relation between the shear stress τ and the velocity gradient du/dy. (Many authors refer to the velocity gradient as the rate of shear.) For nonnewtonian fluids, the relation is not linear: the character of the relation is used to classify the several types of nonnewtonian fluid. The stress rates of shear curves for some of these types are shown in Fig. 2-4, a logarithmic plot of τ versus du/dy.

There is a large range of rate of shear over which the curves for many nonnewtonian fluids on Fig. 2-4 are essentially straight lines. It is therefore convenient to write[3]

$$\tau = \frac{K}{g_c}\left(\frac{du}{dy}\right)^n \tag{2-5}$$

This equation does not apply for precise work over a large range of du/dy; however, it is useful for many engineering applications. For newtonian fluids, which include all gases and most liquids of low molecular weight, n is unity and K is then the same as μ. The quantity n is often called the *flow-behavior* index, and K is the *consistency* index. We shall now discuss several types of nonnewtonian fluids.

Bingham-plastic fluids These fluids require the application of a stress τ_p to cause any deformation. At stresses below τ_p they behave like solids, and at stresses above τ_p a plot of τ versus du/dy is linear, so that

$$\tau = \tau_p + \frac{\mu_p}{g_c}\frac{du}{dy} \qquad \tau \geqq \tau_p \tag{2-6}$$

where μ_p is a constant analogous to the viscosity of a newtonian fluid. This is the simplest kind of nonnewtonian behavior. It is exhibited by fluids such as drilling muds and suspensions of regular, granular solids which take on a rigid, three-dimensional structure when at rest.

Pseudoplastic fluids Most nonnewtonian fluids fall into this group; they are typified by the curve in Fig. 2-4 having a slope between zero and unity. The slope often approaches unity at very high or very low values of the rate of shear; that is, the fluids become more newtonian.

[2] A. B. Metzner, "Non-Newtonian Technology," *Adv. Chem. Eng.*, **1**:77–153 (1956); J. M. McKelvey, "Polymer Processing," John Wiley & Sons, Inc., New York, 1962; A. H. P. Skelland, "Non-Newtonian Flow and Heat Transfer," McGraw-Hill Book Company, New York, 1967; R. B. Bird, R. C. Armstrong, and O. Hassager, "Dynamics of Polymer Liquids," John Wiley & Sons, Inc., New York, 1977.

[3] This equation often is replaced in the literature by $du/dy = Kg_c\tau^n$.

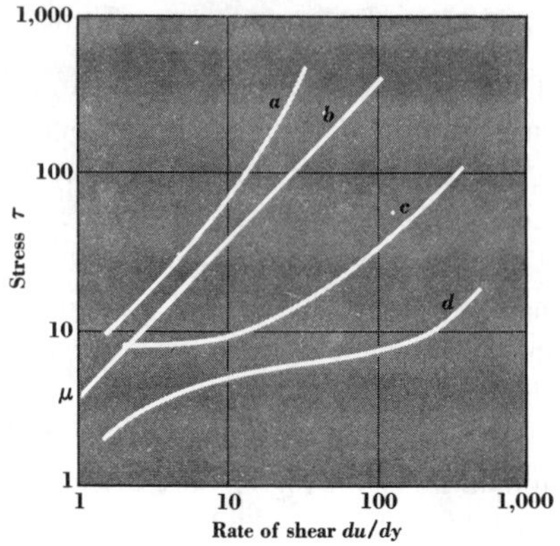

FIGURE 2-4
Shear behavior of fluids.
a Dilatant.
b Newtonian.
c Bingham-plastic.
d Pseudoplastic.

If Eq. (2-5) is put in the form

$$\tau = \frac{K \mid du/dy \mid^{n-1}}{g_c} \frac{du}{dy} \tag{2-7}$$

the term $K \mid du/dy \mid^{n-1}$ is sometimes called the *apparent viscosity*. This quantity decreases with increasing shear rate for pseudoplastic fluids, for which n is less than 1.0. Solutions of polymers and other large elongated molecules behave in this way, as well as colloidal or ordinary suspensions of asymmetric particles. These molecules or particles are looked upon as entangled at low shear rates, but more nearly aligned with each other at higher rates, decreasing the apparent viscosity. At very low shear rates the effect of entanglement is less important, whereas at very high shear rates the particles are aligned so that the entanglement is small. This explains the shape of the curve for pseudoplastic fluids in Fig. 2-4; newtonian behavior (a slope of unity) is approached at very low and at very high shear rates.

Dilatant fluids The curve in Fig. 2-4 typifying dilatant fluids has a slope greater than unity; n is greater than 1.0. A dilatant fluid is often thought of as containing only enough liquid to fill the void between particles at rest or at very low shear rates; for these cases the fluid is almost newtonian. As the particles move over each other more rapidly, they need more room; the fluid as a whole dilates. Since there is no longer enough liquid to fill the larger void space, the apparent viscosity increases. This behavior is in accord with Eq. (2-7) for $n > 1.0$. Suspensions of starch, potassium silicate, and sand are examples of dilatant fluids.

Time-dependent nonnewtonian fluids For these fluids, the shear stress is a function of time at constant shear rate. This type of behavior is complicated, and little progress has been made toward its analytical representation.

For a rheopectic fluid the shear stress increases with time at constant shear rate; for a thixotropic fluid it decreases with time. Many paints are thixotropic, a characteristic which aids their application with either a brush or a spray nozzle and retards their subsequent flow when applied to a vertical surface.

Elastic solids are characterized by Hooke's law, which states that stress is proportional to strain, or extent of deformation. For a newtonian fluid, stress is proportional to the rate of strain or deformation. Many viscous fluids, for example molten polymers, exhibit some elasticity, a property usually associated with solids. Such fluids are called *viscoelastic*; their resistance to deformation is proportional to the usual viscous effect plus an added elastic effect, which is a function of time. When the rate of strain of a viscoelastic fluid is suddenly increased, there is a relaxation time during which the stress changes from its original value to a new steady-state value.

Most nonnewtonian fluids have apparent viscosities which are relatively high compared with the viscosity of water.

Laminar Flow in a Circular Tube

In Chap. 9 we shall derive the general partial differential equations which can be used to obtain velocity distributions for two- and three-dimensional flow systems; the viscous stress defined by Eq. (2-3) is an important part of the analysis. However, to show an immediate application of the definition of viscosity to a simple and important case, let us analyze the one-dimensional, steady, laminar flow of an incompressible newtonian fluid in a horizontal, circular tube of constant diameter in order to find the velocity distribution.

This problem will be solved by making a force balance on the cylindrical element of fluid shown in Fig. 2-5. Since the element is not accelerated, the pressure force acting on its ends just balances the drag force acting on its cylindrical surface

$$\pi r^2 \, \Delta p = -2\pi r L \tau \tag{2-8}$$

so that the shear stress varies with radius according to

$$\tau = \frac{-\Delta p}{2L} r \tag{2-9}$$

At the wall this equation becomes

$$\tau_s = \frac{-\Delta p D}{4L} \tag{2-10}$$

These two equations, unlike those to follow, apply to both laminar and turbulent flow, provided the time mean pressure and velocity are used for the latter. The

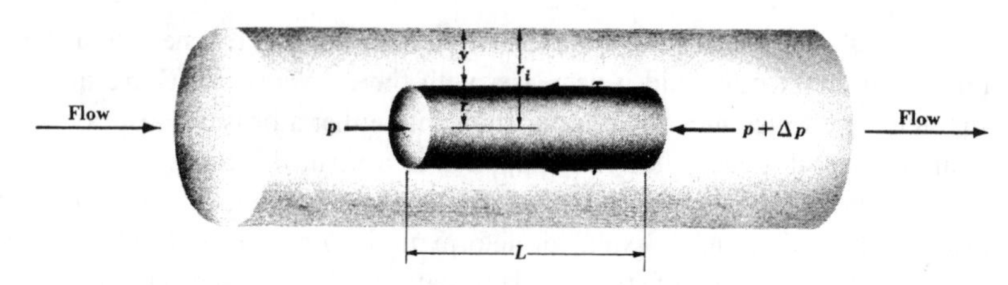

FIGURE 2-5
Forces acting on an element of fluid flowing in a circular tube.

relations do not depend on the constitutive equation for the fluid [Eqs. (2-3) and (2-7) are such equations]. For laminar flow of a newtonian fluid, the velocity distribution is found by substituting

$$\tau = \frac{\mu}{g_c}\left(-\frac{du}{dr}\right) \tag{2-11}$$

into Eq. (2-9). This is Eq. (2-3) modified with a minus sign because du/dr is negative. After rearrangement and the introduction of integral signs, the substitution gives

$$\frac{-\Delta p g_c}{2\mu L}\int_0^r r\,dr = -\int_{u_{\max}}^u du \tag{2-12}$$

The integrated result is

$$u = u_{\max} + \frac{\Delta p g_c}{4\mu L} r^2 \tag{2-13}$$

Since $u = 0$ at $r = r_i$ from the assumption of no slip at the wall, it follows that

$$u_{\max} = \frac{-\Delta p g_c r_i^{\,2}}{4\mu L} \tag{2-14}$$

and thus we obtain

$$u = u_{\max}\left[1 - \left(\frac{r}{r_i}\right)^2\right] \tag{2-15}$$

This equation shows that for ordinary laminar flow there is a parabolic variation of the velocity from zero at the wall to $u_{\max}$ at the center of a tube. Since a fluid near the entrance to a pipe may be accelerated, Eq. (2-15) does not apply in that region, but rather downstream where we characterize the flow as fully developed.

Example 2-1 Find the relation between velocity and radial position for the laminar flow of a nonnewtonian fluid through a circular pipe at steady state. The stress rate of shear relation is given by

$$\tau = \frac{K}{g_c}\left(-\frac{du}{dr}\right)^n \tag{1}$$

This is Eq. (2-5), except for the minus sign, which is necessary in this case.

Equation (1) is substituted into Eq. (2-9) to yield

$$\frac{-\Delta p g_c}{2KL} r = \left(-\frac{du}{dr}\right)^n \tag{2}$$

Both sides of this equation are now raised to the $1/n$ power, and the equation is rearranged to give

$$\left(\frac{-\Delta p g_c}{2KL}\right)^{1/n} \int_r^{r_i} r^{1/n}\, dr = -\int_u^0 du \tag{3}$$

We find upon integration

$$u = \frac{n}{n+1}\left(\frac{-\Delta p g_c}{2KL}\right)^{1/n} (r_i)^{(n+1)/n}\left[1 - \left(\frac{r}{r_i}\right)^{(n+1)/n}\right] \tag{4}$$

and since $u = u_{\max}$ at $r = 0$,

$$u = u_{\max}\left[1 - \left(\frac{r}{r_i}\right)^{(n+1)/n}\right] \tag{5}$$

This result, if compared with Eq. (2-15), shows that the velocity profile for the laminar flow of a nonnewtonian fluid can be greatly different from that of a newtonian fluid. For pseudoplastic fluids, a relatively flat profile is obtained, and in the limit of an infinitely pseudoplastic fluid ($n = 0$) there is rodlike flow. For dilatant fluids, the profile is sharper, and for the limiting case of an infinitely dilatant fluid ($n = \infty$), the velocity is a linear function of the radius. The velocity profile is conical. ////

Fluxes of Momentum, Mass, and Energy

Earlier in this chapter we showed that the shear stress τ can be regarded as a flux of momentum. For an incompressible fluid we can rewrite Eq. (2-3) as[1]

$$g_c \tau = -\nu \frac{d(u\rho)}{dy} \tag{2-16}$$

[1] The minus sign in this equation has been introduced to emphasize the analogy to heat and mass transfer. According to convention, the sign is positive, as in Eq. (2-3) and elsewhere in this book.

This equation is of the form: flux of momentum [(lb)(ft/s)/(ft^2)(s)] equals a conductivity or diffusivity (ft^2/s) times a gradient of momentum concentration [(lb)(ft/s)/(ft^3)(ft)]. We insert the minus sign because the momentum is transferred toward the region of lower concentration of momentum (velocity). The kinematic viscosity is equivalent to a diffusivity of momentum. Equations in which flux equals diffusivity times concentration gradient are often called phenomenological equations; they represent an empirical method of correlating observed phenomena. Similar equations can be written for fluxes of mass, energy, electricity, and other entities.

The equation for the transfer of mass is written

$$J_A = -D_{AB} \frac{d\rho_A}{dy} \qquad (2\text{-}17)$$

This relation applies to a binary mixture of A and B; J_A is the flux of A, D_{AB} is the diffusivity of A through the mixture of A and B, and ρ_A is the density (concentration) of A. The diffusivity, like the kinematic viscosity, has the units square feet per second. Component A flows from a region of high concentration of A to one of lower concentration. The flux is measured in lb A/(s)(ft^2) across an area perpendicular to the direction of the concentration gradient, expressed as (lb A/ft^3)/ft.

The equation for the flux of energy is written

$$\frac{q}{A} = -k \frac{dt}{dy} \qquad (2\text{-}18)$$

where k is the thermal conductivity. For a fluid of constant density and specific heat, this equation may be put into the form of Eqs. (2-16) and (2-17):

$$\frac{q}{A} = -\alpha \frac{d(C_p \rho t)}{dy} \qquad (2\text{-}19)$$

where $\alpha = k/C_p\rho$. This coefficient is called the *thermal diffusivity*, and like ν and D_{AB}, its units are square feet per second. The energy flux is expressed as Btu/(s) × (ft^2), and the gradient of concentration of energy as (Btu/ft^3)/ft. The energy flux is measured over an area perpendicular to the temperature gradient, and the energy flows from a region of high concentration of energy (temperature) to one of lower concentration.

In Chaps. 3 to 5 we shall see many similarities among the overall balances of mass, energy, and momentum. In the preceding paragraphs we have seen a rather striking similarity among the equations for the fluxes of these quantities. After considering some applications of the overall balances derived in those chapters, we shall study the differential balances of mass, energy, and momentum in Chaps. 7 to 9. In doing so we shall use the equations for the fluxes and the definitions of the various diffusivities, and we shall continue to find similarities among the three transport phenomena. In the present chapter we have discussed only viscosity and momentum transfer in detail. More extensive treatments of thermal conductivity and molecular diffusivity are deferred to the later parts of the book, which deal specifically with heat transfer and with mass transfer.

PROBLEMS

2-1 Derive the relations analogous to Eqs. (2-14) and (2-15) for the velocity distribution and for the pressure drop in laminar, newtonian flow through a slit of height y_o and infinite width.

2-2 Repeat Prob. 2-1 for a pseudoplastic fluid which follows the power law, Eq. (2-5), with $n = 0.5$.

2-3 Consider a newtonian fluid in laminar flow down a plane surface of infinite width. Find the velocity distribution within the liquid layer and the velocity at the free surface.

2-4 What pressure drop is necessary to make water at 20°C flow through a 3-m-long tube of 10-mm ID at a centerline velocity of 0.07 m/s?

2-5 What pressure drop in psi is needed to make air at 68°F and 1 atm flow through the tube of Prob. 2-4 at 3 in/s?

2-6 The velocity measured 5 mm from the wall of a 1-cm-ID tube is 10 cm/s for water flowing in the tube in laminar flow at 10°C. What is the rate of change of pressure along the tube in torr/cm?

2-7 Ethyl alcohol at 68°F is flowing through a tube of $\frac{1}{8}$-in ID. The pressure drop over 1 ft of length is measured as 0.25 in of mercury. What is the velocity of the water at the center of the tube?

2-8 Water at 30°C is being forced through a 3-mm-ID capillary at a centerline velocity of 1 cm/s. Calculate the shear stress at the wall in dyn/cm^2.

2-9 Make a plot of u/u_{max} as a function of r/r_i for: (a) a pseudoplastic fluid with $n = 0.5$; (b) a dilatant fluid of $n = 2.5$.

2-10 A laboratory experiment involves water at 20°C flowing through a 1-mm ID capillary tube. We want to triple the fluid *velocity* inside the tube by using a larger tube of the same length; the pressure drop stays constant. Flow is laminar in both cases. What ID tube should we use, and what will be the ratio of the new mass flow rate to the old one?

2-11 A pitot tube, which measures the fluid velocity at a certain radius in a tube, indicates that the velocity of water flowing at 35°C at 3 mm from the center of a 10-mm ID tube is 0.1 m/s. Calculate the pressure drop along the tube in Pa/m; the flow is laminar.

2-12 Air is flowing in laminar flow through a certain capillary of 0.5 mm ID at a volumetric flow rate of 10 cm^3/min. If the tube is replaced by one of the same length but only 0.2 mm ID, what volumetric flow rate would be expected? The pressure gradient along the tube is the same in both cases.

3

THE OVERALL MASS BALANCE

INTRODUCTION

As already mentioned, we shall develop rather general equations for the basic balances. The electronic computer makes practical the omission of many simplifying assumptions which were formerly necessary to make problems tractable. To begin the discussion of the overall mass balance, however, let us first consider a simple geometry and then later find the equations which apply in general.

Figure 3-1 represents a tank to which mass is added at a rate w_1 and withdrawn at w_2. A simple application of conservation of mass leads to the relation that the rate of mass flow out minus the rate of mass flow in plus the rate of mass accumulation equals zero. The equation which expresses this relation is

$$w_2 - w_1 + \frac{dM}{d\theta} = 0 \qquad (3\text{-}1)$$

The region over which the balance is made is called the *control volume.*

Example 3-1 A cylindrical tank has a cross-sectional area of 4 ft^2 and is filled with water to a depth of 6 ft. A valve is opened at the bottom of the tank. The flow rate of water out decreases as the depth of the water goes down according to the formula

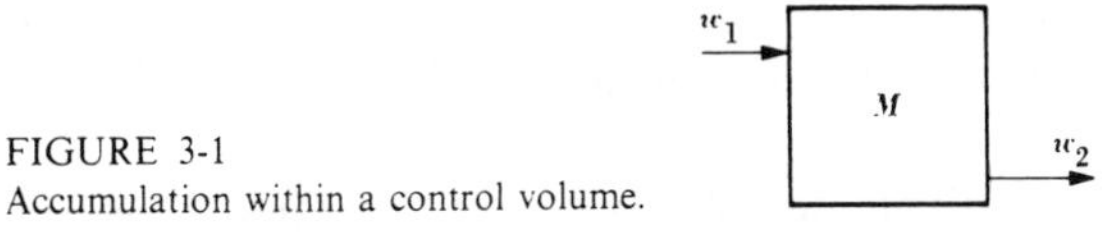

FIGURE 3-1
Accumulation within a control volume.

$$w = 20\sqrt{z} \tag{1}$$

where w = flow rate out, lb/min

z = depth of water in the tank, ft

How long will it take for the depth to fall to 2 ft?

For this application of Eq. (3-1), w_1 is zero and w_2 is given as w. We also know that

$$M = 4\rho z \tag{2}$$

so that Eq. (3-1) becomes

$$w + \frac{dM}{d\theta} = 0 \tag{3}$$

and

$$20\sqrt{z} + 4\rho\frac{dz}{d\theta} = 0 \tag{4}$$

Then we have

$$-\int_6^2 \frac{dz}{\sqrt{z}} = \frac{5}{\rho}\int_0^\theta d\theta \tag{5}$$

and integration gives $\theta = 26.0$ min.

Incidentally, note that the mass balance [Eq. (3-1)] is dimensionally homogeneous but the special Eq. (1) is not.

If there are several nonreacting components involved in the balance, conservation of mass applies to each one;

$$w_{i_2} - w_{i_1} + \frac{dM_i}{d\theta} = 0 \tag{3-2}$$

For each stream, the mass fraction x_i is

$$x_i = \frac{w_i}{w} \tag{3-3}$$

so that

$$w_2 x_{i_2} - w_1 x_{i_1} + \frac{dM_i}{d\theta} = 0 \tag{3-4}$$

Summing over i gives Eq. (3-1) again; for N components there are $N - 1$ independent overall mass balances.

Example 3-2 Water is flowing into a well-stirred tank at 150 kg/h, and salt (NaCl) is being added at 30 kg/h. The resulting solution is leaving the tank at 120 kg/h; because of the effective stirring, the concentration of the outlet solution is the same as that within the tank. There is 100 kg of fresh water in the tank at the start of the operation, and the rates of input and output remain constant thereafter. Calculate the outlet concentration (mass fraction salt) after 1 h.

Equations (3-2) and (3-1) must be solved simultaneously. Consider first a balance on the salt (component A).

$$w_{A2} - w_{A1} + \frac{dM_A}{d\theta} = 0 \tag{1}$$

Let x_A represent the mass fraction salt in the solution leaving the tank. Then we have

$$w_2 x_A - w_{A1} + \frac{d(Mx_A)}{d\theta} = 0 \tag{2}$$

and

$$120x_A - 30 + M\frac{dx_A}{d\theta} + x_A\frac{dM}{d\theta} = 0 \tag{3}$$

This equation is simplified, using the balance on the total streams,

$$w_2 - w_1 + \frac{dM}{d\theta} = 0 \tag{4}$$

and we get

$$120 - (150 + 30) + \frac{dM}{d\theta} = 0 \tag{5}$$

This gives the rate of accumulation $dM/d\theta$ as 60 kg/h. The inlet is made up of two streams, the water and the salt, so $w_1 = 150 + 30$. Equation (5) can also be integrated to give

$$M = 60\theta + M_o \tag{6}$$

where M_o is 100 kg for this problem. These results are substituted into Eq. (3) so that it contains only θ and x_A as variables.

$$120x_A - 30 + (60\theta + 100)\frac{dx_A}{d\theta} + 60x_A = 0 \tag{7}$$

The variables are separated, resulting in

$$\int_0^{\theta} \frac{d\theta}{60\theta + 100} = -\int_0^{x_A} \frac{dx_A}{180x_A - 30} \tag{8}$$

which, upon integration and rearrangement, yields

$$x_A = \frac{1}{6}\left[1 - \left(\frac{10}{6\theta + 10}\right)^3\right] \tag{9}$$

For $\theta = 1$ h, $x_A = 0.126$ mass fraction NaCl. Equation (9) shows that as θ becomes very large, x_A tends toward $\frac{1}{6}$, as common sense would indicate. ////

If the components can react chemically, a generation term must be added to the balances. Also, it is usually more convenient to work in molar units. The appropriate component balances are then

$$\tilde{w}_2 \tilde{x}_{i_2} - \tilde{w}_1 \tilde{x}_{i_1} + \frac{d\tilde{M}_i}{d\theta} = \tilde{R}_i \tag{3-5}$$

Summing over N components yields

$$\tilde{w}_2 - \tilde{w}_1 + \frac{d\tilde{M}}{d\theta} = \sum_i^N \tilde{R}_i \tag{3-6}$$

Chemical reaction within the control volume may change the total number of moles. The $\tilde{R}_i$ are interrelated by the stoichiometry of the reaction, which we write in general form, using B_i to denote the symbols for the species,

$$\sum_i^N b_i B_i = 0 \tag{3-7}$$

where $b_i > 0$ for products and $b_i < 0$ for reactants. We can now relate any one of the $\tilde{R}_i$ to a reference rate, say $\tilde{R}_A$ by the equation[1]

$$\tilde{R}_i = b_i \frac{\tilde{R}_A}{b_A} \tag{3-8}$$

and so the summation term in Eq. (3-6) is

$$\sum_{i=A}^N \tilde{R}_i = \frac{\tilde{R}_A}{b_A} \sum_i^N b_i \tag{3-9}$$

These ideas are best understood by an application to a particular reaction.

[1] Note that $\tilde{R}_A/b_A$ is positive by definition; $\tilde{R}_i > 0$ for products, $\tilde{R}_i < 0$ for reactants.

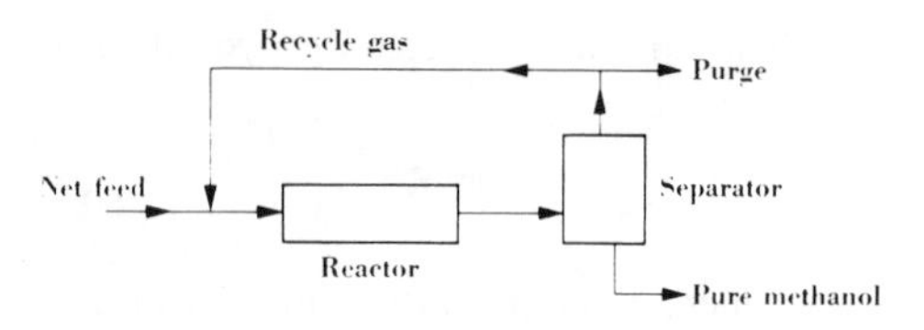

FIGURE 3-2
Flow sheet for methanol process.

Example 3-3 Methanol is made by the reaction

$$CO + 2H_2 \rightleftarrows CH_3OH \tag{1}$$

which occurs when a gaseous mixture of the reactants is passed over a $ZnO - Cr_2O_3$ catalyst at 200 atm and 375°C. Figure 3-2 shows a simplified flow sheet. The net feed to the process is 66 mol percent H_2, 33 mol percent CO, and 1 mol percent Ar. All the methanol is removed from the reactor product in the separator, and there are no side reactions. The purge rate is 1 mol/10 mol of net feed. Calculate the composition of the recycle gas and the ratio of moles of methanol produced to moles of net feed.

The mass balance can be applied over various control volumes in such a problem, but it is usually effective to start with a balance around the whole process. We designate the streams as $\tilde{w}_1$, net feed; $\tilde{w}_2$, purge; and $\tilde{w}_3$, methanol product. Then Eq. (3-6) becomes, at steady state,

$$\tilde{w}_2 + \tilde{w}_3 - \tilde{w}_1 = \tilde{R}_{MeOH} + \tilde{R}_{CO} + \tilde{R}_{H_2} \tag{2}$$

The chemical equation, written in the form of Eq. (3-7), is

$$CH_3OH - CO - 2H_2 = 0 \tag{3}$$

so that $b_{MeOH} = 1$, $b_{CO} = -1$, and $b_{H_2} = -2$. (Also, $b_{Ar} = 0$). Choosing methanol as the reference (A), Eq. (3-9) gives for the right-hand side of Eq. (2):

$$\sum_i^N \tilde{R}_i = -2\tilde{R}_{MeOH} \tag{4}$$

The application of the balance Eq. (3-5) to methanol gives

$$\tilde{w}_3 = \tilde{R}_{MeOH} \tag{5}$$

since $\tilde{x}_{MeOH1}$, $\tilde{x}_{MeOH2}$, and $dM_{MeOH}/d\theta$ are zero. Equation (2) now becomes

$$\tilde{w}_2 - \tilde{w}_1 = -3\tilde{R}_{MeOH} \tag{6}$$

Since $\tilde{w}_2 = 0.1\tilde{w}_1$, we find that $\tilde{R}_{MeOH}/\tilde{w}_1 = 0.30$.

The composition of the recycle gas, which is the same as that of the purge, is deduced from a balance on argon:

$$\tilde{w}_2 \tilde{x}_{Ar_2} - 0.01\tilde{w}_1 = 0$$

and $\tilde{x}_{Ar_2} = 0.10$. Since the CO and H_2 are always present in their stoichiometric ratio, the recycle gas is 10 mol percent argon, 60 mol percent H_2, and 30 mol percent CO. To compute the flow rate of the recycle, we would need to know the degree of conversion across the reactor.

The mastery of the systematic application of mass balances to chemical processes requires considerable practice.[1]

GENERAL EQUATIONS

We now apply the principle of conservation of mass to a hypothetical enclosure fixed in space; we call the space enclosed the *control volume* and its boundary the *control surface*. As in the simpler geometries, the rate of accumulation of mass within the volume is equal to the rate of mass flow in minus the rate of mass flow out, the equation expressing the principle can be written almost immediately. Defining α as the angle between a line normal to the surface and directed outward and a line representing the direction of the velocity at the point considered, the net outward flow across the surface becomes

$$\iint_A u\rho \cos \alpha \, dA \qquad (3\text{-}10)$$

Figure 3-3 illustrates this assertion.

Where the velocity u is parallel to the surface, $\cos \alpha$ is zero; where u is directed inward, α is greater than $\pi/2$ and $\cos \alpha$ is negative. Hence it can be seen that the integration over the whole surface does indeed give the net outflow of mass. The quantity $u\rho$ has the units of lb/(h)(ft^2) and is called a *flux*. It is also called the mass velocity G.

The integral (3-10), the units of which are lb/h, must equal the negative of the rate of accumulation of mass in the control volume. This statement is true at any instant, even for the unsteady-state process. Since in general the density is a function of position within the volume, we write the total accumulation rate as

$$\frac{d}{d\theta} \iiint_{\check{V}} \rho \, d\check{V} = \frac{dM}{d\theta} \qquad (3\text{-}11)$$

[1] See, for example, D. M. Himmelblau, "Basic Principles and Calculations in Chemical Engineering," Prentice-Hall, Inc., Englewood Cliffs, N.J., 1967; and J. C. Whitwell and R. K. Toner, "Conservation of Mass and Energy," McGraw-Hill Book Company, New York, 1973.

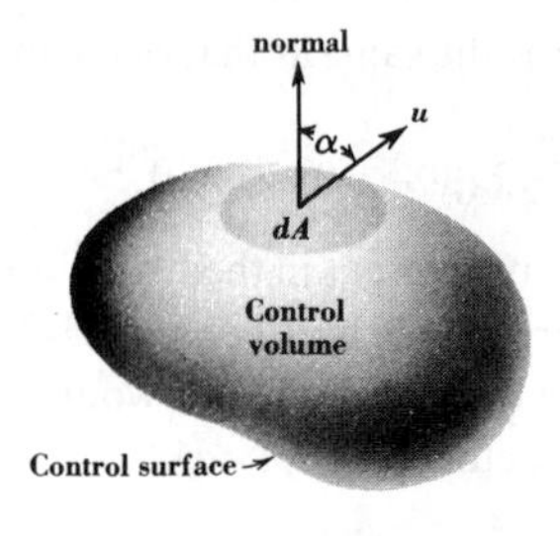

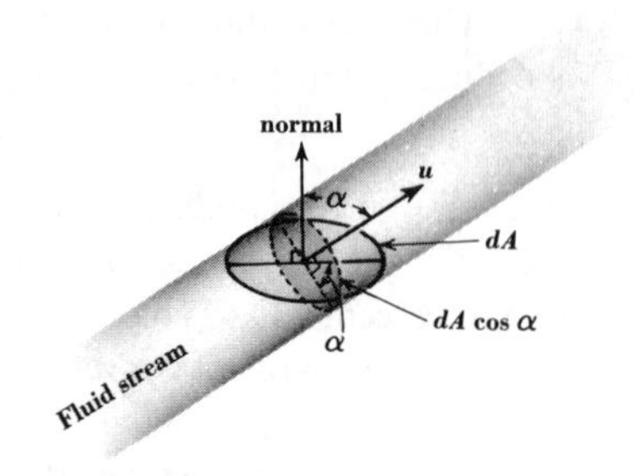

FIGURE 3-3
Flow through differential area on control surface.

These two expressions (3-10) and (3-11) are then combined to give the overall mass balance in general form

$$\iint_A u\rho \cos\alpha \, dA + \frac{d}{d\theta}\iiint_{\check{V}} \rho \, d\check{V} = 0 \qquad (3\text{-}12)$$

The application of Eq. (3-12) is clarified if it is written for the most usual situation encountered, for which all the flow inward is normal to an area A_1 and all the flow out is normal to an area A_2, as shown in Fig. 3-4. The flow is parallel to the other parts of the control surface. For these conditions

$$u_{b2}\rho_2 A_2 - u_{b1}\rho_1 A_1 + \frac{dM}{d\theta} = 0 \qquad (3\text{-}13)$$

The quantity u_b is called the *bulk*, or average, velocity and is defined by

$$u_b = \frac{1}{A}\iint_A u \, dA \qquad (3\text{-}14)$$

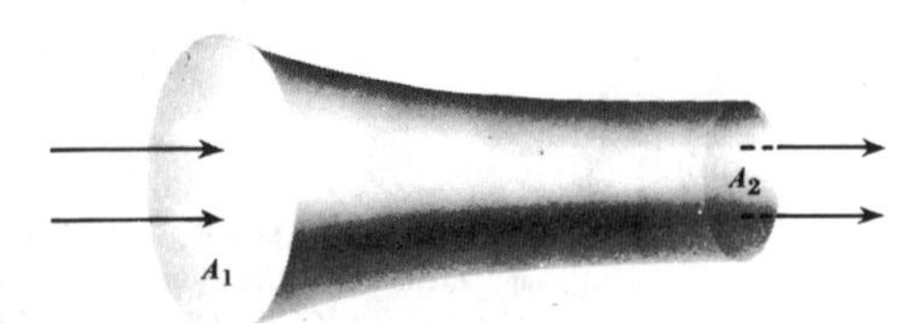

FIGURE 3-4
Simplified control volume.

for a surface over which u is normal to A. The density is assumed constant over the area of integration. We define w, the mass rate of flow, by

$$w = u_b \rho A \tag{3-15}$$

so that

$$\Delta w + \frac{dM}{d\theta} = 0 \tag{3-16}$$

Equation (3-16) can be integrated immediately if w_1 and w_2 are given as functions of θ. Often one or more of the flow rates is constant; if both flow rates are equal and constant, the process is at steady state and we have

$$\frac{dM}{d\theta} = 0 \tag{3-17}$$

These equations are readily extended to apply to a component, i, in a multicomponent system. For the case illustrated by Fig. 3-4, we have

$$\Delta w_i - R_i + \frac{dM_i}{d\theta} = 0 \tag{3-18}$$

where R_i is the rate of generation of i by chemical reaction within the volume in pounds per hour, and there are no significant fluxes by diffusion at the inlet or outlet [diffusion terms could easily be added to Eq. (3-18)]. Equation (3-18) can be written for each constituent; the sum of these equations will give Eq. (3-16), because

$$\sum_{i=1}^{n} R_i = 0 \tag{3-19}$$

for n components in a system.

Although some of the equations which have been presented here may seem rather obvious, the method of reasoning used will be helpful in the derivations of the energy and momentum balances, where the results are not so easy to obtain.

Example 3-4 In order to show how the variation of velocity across a part of the control surface is taken into account, let us calculate the rate of flow of 98 percent sulfuric acid at 0°C in a 2-in, schedule-40 pipe. The velocity is measured at the center of the pipe by a pitot tube and is 1 m/s. The flow is laminar, and the variation of the velocity across the pipe radius is given for laminar flow by

$$u = u_{\max}\left(1 - \frac{r^2}{r_i^2}\right) \tag{2-15}$$

Since $w = u_b \rho A$, the problem is to find u_b from

$$u_b = \frac{1}{A} \iint_A u \, dA \tag{3-14}$$

In cartesian coordinates dA would be expressed by $dx\,dy$, but for flow through a circular pipe polar coordinates are more appropriate, so we use $dA = r\,dr\,d\theta$ (θ is of course the angle in polar coordinates, not the time). Equation (2-15) is substituted into Eq. (3-14) to get

$$u_b = \frac{u_{max}}{\pi r_i^2} \int_0^{2\pi} \int_0^{r_i} \left(1 - \frac{r^2}{r_i^2}\right) r\,dr\,d\theta \tag{1}$$

The variable θ does not appear in Eq. (2-15) because there is axial symmetry. The evaluation of the integral of Eq. (1) gives the simple result

$$u_b = \frac{u_{max}}{2} \tag{2}$$

and so we obtain

$$w = \frac{u_{max} \rho \pi r_i^2}{2} = \frac{u_{max} \rho \pi D^2}{8} \tag{3}$$

The ID of the pipe D is found from Appendix Table A-5A as 52.5 mm; the density of 98 percent sulfuric acid at 0°C is 1.8567 g/cm^3 (Perry, p. 3-80).[1] Therefore

$$w = \frac{(1.0)(1856.7)(3.14)(5.25 \times 10^{-2})^2}{8}$$

$$= 2.01 \text{ kg/s}$$

The calculation could have been shortened by reading the inside sectional area from Appendix Table A-5A as 0.2165×10^{-2} m^2.

Let us also calculate the Reynolds number; the bulk velocity should be used.

$$\mathbf{Re} = \frac{D u_b \rho}{\mu} = \frac{(5.25 \times 10^{-2})(0.5)(1856.7)}{47.0 \times 10^{-3}}$$

$$= 1037$$

The viscosity (47.0×10^{-3} Pa·s) is found from Appendix Fig. A-3. One should always check calculations for dimensional consistency.

[1] Perry and Chilton, "Chemical Engineers' Handbook," 5th ed., McGraw-Hill Book Company, New York, 1973. Future mention of Perry, without footnote, will refer to this handbook.

Since the Reynolds number is less than 2100, it is verified that the flow is laminar.

PROBLEMS

3-1 One hundred pound moles per hour of a mixture containing 0.35 mole fraction toluene and 0.65 mole fraction benzene is fed continuously to a distillation process. Two product streams are obtained, one containing 0.99 mole fraction benzene and one containing 5 percent of the entering benzene. Find the rates of flow (pound moles per hour) of both streams and the composition of the toluene-rich stream. There is no accumulation within the system.

3-2 A batch distillation unit is charged with 150 kg mol of a liquid mixture containing 60 mol percent benzene and 40 mol percent toluene. The composition of the vapor leaving the still is related to the composition of the liquid remaining by the equation $\tilde{y}_A = \alpha\tilde{x}_A/[1 + (\alpha - 1)\tilde{x}_A]$, in which $\tilde{y}_A$ and $\tilde{x}_A$ represent the mole fraction of benzene in the vapor and liquid, respectively, and α, the relative volatility, is a constant, 2.57. If the distillation is continued until only 30 kg mol of liquid remains in the still, what is the composition of the collected distillate?

3-3 At a Reynolds number of 110,000, the velocity profile is given by

$$u = u_{\text{max}}\left(\frac{r_i - r}{r_i}\right)^{1/7}$$

for turbulent flow in a smooth, circular tube. Find the value of u_b/u_{max} for this situation. If a variable y is substituted for $r_i - r$, the integration will be simplified.

3-4 The velocity profile developed in a layer of liquid flowing in laminar flow down a vertical surface can be represented by the equation

$$u_y = \frac{\rho g}{\mu}\left(Lx - \frac{x^2}{2}\right)$$

where L = thickness of liquid layer
u_y = downward velocity at a distance x from wall

(*a*) Prove that the average velocity of the liquid is two-thirds the velocity at the free surface.

(*b*) Find the thickness of a layer of water at 60°F flowing down a vertical surface 3 ft wide at the rate of 1 gal/min.

3-5 Steam enters a long section of 3-in, schedule-40 steel pipe at 200 psia, 600°F, and a bulk velocity of 10 ft/s. At a point downstream the pressure is 140 psia and the temperature is 590°F; what is the bulk velocity at this point? Calculate also the Reynolds numbers at both the upstream and the downstream points. Use data from the steam tables and assume steady flow.

3-6 In an agitated tank in a part of a simplified process for making H_3PO_4, 10,000 lb/h of $Ca_3(PO_4)_2$ in a suspension in water is reacted continuously with a stoichiometric amount of 94 mass percent H_2SO_4. Sufficient water is added with the ground phosphate rock to give 40 mass percent phosphoric acid when the process operates at steady state. The phosphoric acid solution and the $CaSO_4 \cdot 2H_2O$ (gypsum) formed are removed from the agitator at a constant rate, so that the total mass in the tank remains constant in all parts of this problem.

If the operation is started when the tank contains 10,000 lb of 20 mass percent phosphoric acid solution, what is the concentration in the tank after 1 h?

3-7 Water flowing between parallel plates has a velocity profile described by the equation

$$u = u_{max}\left[1 - \left(\frac{y}{y_o}\right)^2\right]$$

A sketch of the system is shown in Fig. 10-1. There is no flow or variation of any properties in the z direction. Use the velocity profile equation to derive a numerical value for the ratio u_b/u_{max} for this system.

3-8 A vacuum pump operates so that it pumps gas at a constant volumetric rate based on the inlet conditions. For a pumping speed of 10 L/min, how long will it take to pump down a 100-L tank of air from 1 atm abs to 0.01 atm abs? The process is slow enough so that the gas in the tank remains at constant temperature.

3-9 A tank containing 4 m^3 of 95 mass percent ethyl alcohol solution in water operates at steady state with a continuous flow in and out of 6×10^{-3} m^3/s of 95 mass percent alcohol (density, 804 kg/m^3). The alcohol flow is suddenly stopped and replaced by a flow of 6×10^{-3} m^3/s of pure water. If the total mass of material in the tank remains constant and is well mixed, how long will it take for the mass percent alcohol in the tank to fall to 5 percent?

3-10 Formaldehyde is made by the partial oxidation of methanol:

$$CH_3OH + \frac{1}{2}O_2 \rightarrow CH_2O + H_2O$$

The gas mixture fed to the reactor in Fig. 3-5 contains 8 mol percent methanol and 10 mol percent oxygen (stream 4).

The methanol is completely converted to formaldehyde in the reactor, which contains a bed of $Fe_2O_3 \cdot MoO_3$ catalyst particles.

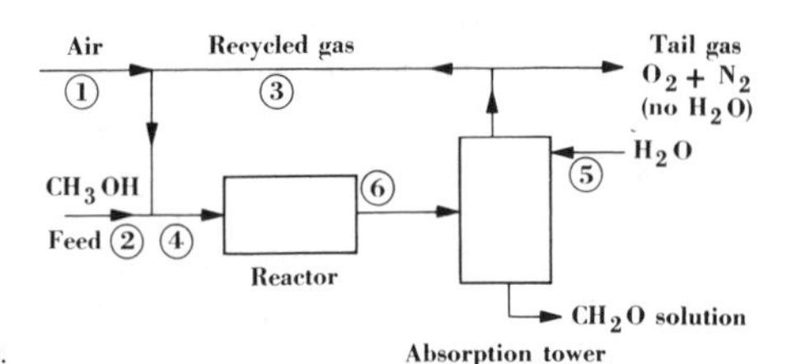

FIGURE 3-5
Formaldehyde process.

Calculate the composition (mole fraction) and flow rate (kmol/s) of streams 1, 2, 3, 4, and 5 needed to make 30 kg/s of 37 mass percent formaldehyde solution.

3-11 Synthetic ethyl alcohol is being made by the hydration of ethylene:

$$CH_2{=}CH_2 + H_2O \rightleftarrows C_2H_5OH$$

as follows:

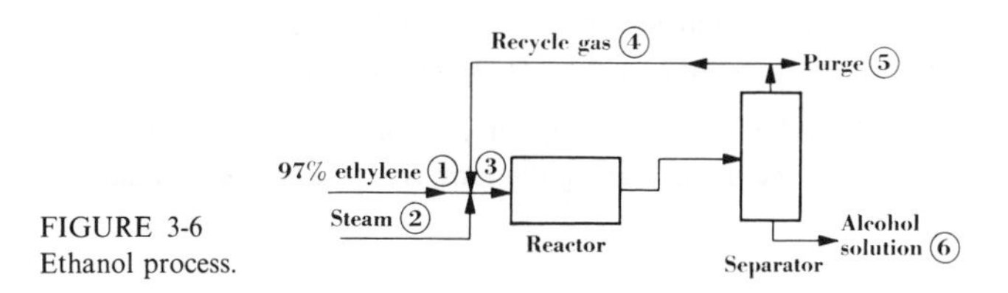

FIGURE 3-6
Ethanol process.

The feed gas to the plant is 97 mol percent ethylene and 3 percent methane; the latter is inert in the process. For proper reactor operation, the feed to the reactor (3) must be

	$\tilde{y}_i$
C_2H_4	0.50
H_2O	0.30
CH_4	0.20

The conversion of ethylene in the reactor is 4 percent.

How many moles of purge gas are there per mole of net feed of 97 percent ethylene?

3-12 Find the relation between u_{max} and u_b, and also the expression for the pressure drop for a nonnewtonian fluid following Eq. (2-5) for laminar flow in a circular pipe.

3-13 Repeat Prob. 3-12 for a Bingham-plastic fluid.

3-14 A small pump is used to pump ambient air into a 100-L tank. It operates so that the constant inlet volumetric rate is 10 L/min. How long will it take to raise the pressure in the tank to 10 atm? The process is slow enough so that the gas in the tank remains at constant temperature.

3-15 A gas-phase stirred vessel has a volume of 25 mL. Pure helium flows to and from the vessel at 3 mL/s. At time zero the He feed is suddenly switched to argon; all else remains the same. How long does it take for the mole fraction of helium in the vessel to fall to 0.50?

3-16 A well-mixed tank is fed water at 10 gal/min and the water leaves the tank at the same rate via an overflow when the tank holds 100 gal. At time zero a dye is added to the feed stream at a rate such that the feed to the tank contains 100 ppm (parts per million, mass) of dye. How long will it take for the dye concentration in the water leaving the tank to reach 99 ppm?

3-17 A liquid stream under pressure is expanded in a steady-state flow process through a valve into a tank. In the process, part of the liquid vaporizes (flashes) and leaves through a line at the top of the tank; the liquid leaves from the bottom of the tank. The feed is 60 mol percent butane and 40 mol percent pentane. Equilibrium between the gas and liquid exists in the tank, and this requires that

$$\tilde{y} = 2.5\tilde{x}/[1 + 1.5\tilde{x}]$$

where $\tilde{y}$ = mole fraction butane in the vapor
$\tilde{x}$ = mole fraction butane in the liquid

For a feed rate of 100 mol/h, calculate the fraction of the feed which leaves as liquid when the liquid leaving the tank is 40 mol percent butane.

4

THE OVERALL ENERGY BALANCE

An overall or macroscopic energy balance will now be obtained by the application of the principle of the conservation of energy to a control volume fixed in space, in much the same way as the principle of the conservation of mass was applied in the previous chapter. The first law of thermodynamics may be written as

$$\Delta E = Q - W \tag{4-1}$$

where

$$E = U + \frac{u^2}{2g_c} + \frac{gz}{g_c} \tag{4-2}$$

The terms on the right-hand side of this equation represent, respectively, the internal energy, the kinetic energy, and the potential energy of 1 lb of fluid. In Eq. (4-1) Q is the heat absorbed per pound of fluid, and W is the work of all kinds done per pound of fluid upon the surroundings. In computation, each term must, of course, be expressed in the same units—usually Btu/lb or (ft)(lb_f)/lb. If it is, the mechanical equivalent of heat J need not appear in our equations. By considering the flow out of, the flow into, and the accumulation within the control volume, we write

$$\iint_A u\rho(\cos\alpha)E\,dA + \frac{d}{d\theta}\iiint_{\check{V}} E\rho\,d\check{V} = q - \dot{W} \tag{4-3}$$

where q and $\dot{W}$, as well as the two integrals, are expressed in energy per unit time.

The energy E is a function of x, y, z, and θ; so we may write

$$dE = \frac{\partial E}{\partial x}dx + \frac{\partial E}{\partial y}dy + \frac{\partial E}{\partial z}dz + \frac{\partial E}{\partial \theta}d\theta \tag{4-4}$$

The rate of change of E with respect to time is

$$\frac{dE}{d\theta} = \frac{\partial E}{\partial x}\frac{dx}{d\theta} + \frac{\partial E}{\partial y}\frac{dy}{d\theta} + \frac{\partial E}{\partial z}\frac{dz}{d\theta} + \frac{\partial E}{\partial \theta} \tag{4-5}$$

The physical meaning of this derivative can be understood if we visualize an instrument for reading E that can be moved in the fluid. For example, often the potential and kinetic energies are small compared to the internal energy; in this case, for an ideal gas or an incompressible liquid, we can consider the energies in Eq. (4-5) to be represented approximately by temperature. Thus if a thermocouple is moved about in the temperature field of a fluid described by $\partial T/\partial x$, $\partial T/\partial y$, $\partial T/\partial z$, and $\partial T/\partial\theta$, the rate of change of the reading $dT/d\theta$ is given by

$$\frac{dT}{d\theta} = \frac{\partial T}{\partial x}\frac{dx}{d\theta} + \frac{\partial T}{\partial y}\frac{dy}{d\theta} + \frac{\partial T}{\partial z}\frac{dz}{d\theta} + \frac{\partial T}{\partial \theta} \tag{4-6}$$

The quantities $dx/d\theta$, $dy/d\theta$, and $dz/d\theta$ are the components of the velocity of the thermocouple. We refer to $dE/d\theta$ or $dT/d\theta$ as a *total derivative*.

If the thermocouple is fixed in space, $dT/d\theta = \partial T/\partial\theta$. If the thermocouple floats along with the flow, or "follows" the flow, then its velocity is identical with the fluid velocity and we have

$$\frac{DT}{D\theta} = \frac{\partial T}{\partial x}u_x + \frac{\partial T}{\partial y}u_y + \frac{\partial T}{\partial z}u_z + \frac{\partial T}{\partial \theta} \tag{4-7}$$

The derivative defined by Eq. (4-7) is written as $DT/D\theta$ or $DE/D\theta$, to distinguish it from the total derivative $dT/d\theta$ or $dE/d\theta$. This new derivative is called the *substantial*, or *material*, derivative, or sometimes a *Lagrangian* derivative. The quantity $DE/D\theta$ represents the rate of change of the energy of a constant mass of fluid as it moves; the volume (density) of this mass of fluid may change with time as it moves.

We can now see that Eq. (4-1) can be written for an arbitrary, constant mass, with respect to an observer following the fluid, as

$$\frac{D\check{E}}{D\theta} = q - \dot{W} \tag{4-8}$$

In view of Eq. (4-3), there results

$$\frac{D\check{E}}{D\theta} = \iint_A E\rho u(\cos \alpha)\, dA + \frac{\partial \check{E}}{\partial \theta} \tag{4-9}$$

In this equation the volume integral of Eq. (4-3) has been replaced by its equivalent, the rate of change of the total energy in the control volume. The left-hand side of this equation is the rate of change of the energy of a moving, constant mass M of fluid. This substantial derivative equals the sum of the flux of energy across the fixed control surface enclosing the mass at that instant plus the rate of change of the energy inside the control surface at that instant also. This principle applies also to any other continuous function of space and time such as thermodynamic state functions, velocity, and momentum. Equation (4-9) is often called the *Reynolds transport* equation.

It is usual to divide the work $\dot{W}$ into several categories, as follows:

$$\dot{W} = \dot{W}_s + \iint_A u\rho p V(\cos \alpha)\, dA + \iint_{A_s} u_s p(\cos \beta)\, dA_s \tag{4-10}$$

The term $\dot{W}_s$ is the shaft work. It is purely mechanical and must be directly identifiable with a turning shaft which crosses the control surface. In a steady-state problem it can also represent work transferred by a reciprocating motion that results in a constant average control volume.

The first integral in Eq. (4-10) is the net work done by the fluid as it flows into and out of the control volume. We assume that there are no viscous forces over the part of the control surface through which fluid flows. Then as one pound of fluid flows out at a point where the pressure is p, it displaces a volume V in the surroundings. Since this work is given by $\int_0^V p\, dV$, the work per pound of fluid is pV.

The second integral term represents the work done by a noncyclic movement of a solid part of the control surface at a speed u_s in a direction inclined at an angle β from the outward-directed normal to the element of solid surface dA_s. For a batch (nonflow) system, u and $\dot{W}_s$ are zero; so Eq. (4-10) becomes equivalent to $W = \int p\, dV$, as it should.

Recalling that the enthalpy is defined by

$$H = U + pV \tag{4-11}$$

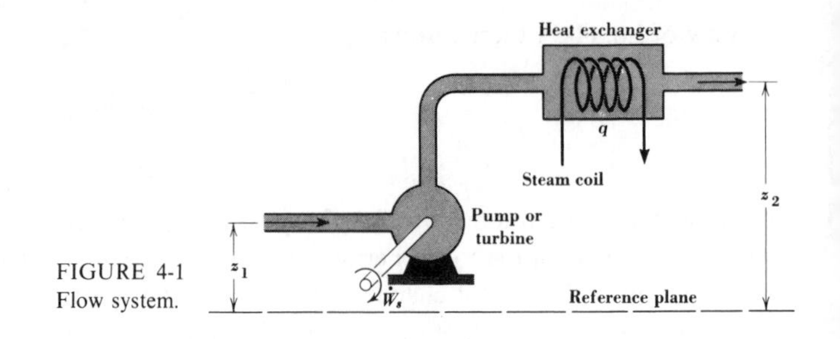

FIGURE 4-1
Flow system.

we combine Eqs. (4-3), (4-10), and (4-11) to obtain the overall energy balance:

$$\iint_{A_s} u_s p(\cos \beta)\, dA_s + \iint_A u\rho \cos \alpha \left(\frac{u^2}{2g_c} + \frac{gz}{g_c} + H \right) dA + \frac{\partial \check{E}}{\partial \theta} = q - \dot{W}_s \tag{4-12}$$

For the simple type of control surface defined in Fig. 4-1, Eq. (4-12) reduces to (where $u_s = 0$)

$$\frac{w_2(u^3)_{\text{av},2}}{2g_c u_{b2}} + \frac{w_2 g(uz)_{\text{av},2}}{g_c u_{b2}} + \frac{w_2(uH)_{\text{av},2}}{u_{b2}} - \frac{w_1(u^3)_{\text{av},1}}{2g_c u_{b1}} - \frac{w_1 g(uz)_{\text{av},1}}{g_c u_{b1}} - \frac{w_1(uH)_{\text{av},1}}{u_{b1}} + \frac{\partial \check{E}}{\partial \theta} = q - \dot{W}_S \tag{4-13}$$

In writing this equation in terms of averages across the flow areas, we have used the mean-value theorem of calculus, which says that there is an average value of an arbitrary continuous function Z given by

$$Z_{\text{av}} = \frac{1}{A} \iint_A Z\, dA \tag{4-14}$$

Thus we can write, for example,

$$\left(\frac{u^3 \rho \cos \alpha}{2g_c} \right)_{\text{av}} = \frac{1}{A} \iint_A \frac{u^3 \rho \cos \alpha}{2g_c}\, dA \tag{4-15}$$

By taking ρ as constant and α as zero over the flow areas, there results

$$(u^3)_{\text{av}} = \frac{1}{A} \iint_A u^3\, dA \tag{4-16}$$

Since

$$w = u_b \rho A$$

the final expression for the integral becomes

$$\frac{w(u^3)_{av}}{2g_c u_b}$$

These simple averaging procedures can also be used to define the average temperature of a flowing stream. A stream with a profile of temperature and velocity over the inlet flow area enters a control volume and passes adiabatically and at steady state through a mixing section from which it emerges through a second area equal to the first. Over this second area the temperature and velocity are constant at T_b and u_b; the temperature defined in this way is often called the *mixing-cup* temperature. It is a measure of the effective temperature of a stream of nonuniform temperature (in this case, the inlet stream). An application of energy balance in the form of Eq. (4-12) to this situation gives, for $u_s = 0$, $u^2/2g_c$ and gz/g_c negligible, and $\partial \check{E}/\partial\theta$, q, and $\dot{W}_s = 0$,

$$\iint_A u\rho(\cos\alpha)H\,dA = 0 \qquad (4\text{-}17)$$

and, for ρ constant and $\cos\alpha = 1$,

$$u_b H_b = \frac{1}{A}\iint_A uH\,dA \qquad (4\text{-}18)$$

If H is proportional to T, we have for the mixing-cup temperature

$$T_b = \frac{1}{u_b A}\iint_A uT\,dA \qquad (4\text{-}19)$$

It is usually necessary to solve Eq. (4-13) simultaneously with the overall mass balance and the equation of state of the fluid. In the simplified case studied here, the overall mass balance is given by Eq. (3-16).

Certain simplified forms of Eqs. (4-12) and (4-13) are so important that they will be given here. Let us write Eq. (4-13) as

$$\frac{1}{2g_c}\Delta\frac{w(u^3)_{av}}{u_b} + \frac{g}{g_c}\Delta\frac{w(uz)_{av}}{u_b} + \Delta\frac{w(uH)_{av}}{u_b} + \frac{\partial \check{E}}{\partial\theta} = q - \dot{W}_s \qquad (4\text{-}20)$$

When input equals output, Δw equals zero; in this case $q = wQ$ and $\dot{W}_s = wW_s$. We obtain from Eq. (4-20), for $\Delta w = 0$,

$$\frac{1}{2g_c}\Delta\frac{(u^3)_{av}}{u_b} + \frac{g}{g_c}\Delta\frac{(uz)_{av}}{u_b} + \Delta\frac{(uH)_{av}}{u_b} + \frac{M}{w}\frac{\partial E}{\partial\theta} = Q - W_s \qquad (4\text{-}21)$$

If there is no accumulation of energy, $\partial E/\partial \theta$ is zero. If there are, in addition, negligible variations of velocity, height, and temperature over each of the two areas, there results the familiar general energy balance which appears in texts on thermodynamics:

$$\frac{\Delta u_b^{\ 2}}{2g_c} + \frac{g}{g_c}\Delta z + \Delta H = Q - W_s \qquad \text{(4-22)}$$

It should be emphasized that Q or q represents all the heat transferred across the control surface, including that by conduction through the fluid. Problems involving reacting mixtures can be solved by Eq. (4-12), using for the value of H or U its thermochemical value; this procedure is illustrated in Example 4-4.

No attempt will be made here to present all the applications of the overall energy balance, although a number of these applications are illustrated in the exercises and problems which follow. For instance, in a process involving a chemical reaction the ΔH term is usually so large that kinetic and potential energy terms are negligible. Conversely, in flow through a nozzle, the kinetic energy term is important, and in the production of hydroelectric power, the potential energy term is significant.

In many practical problems the variations in u or H across an inlet or outlet stream can be ignored. For example, the kinetic energy term becomes important only at high velocities, but the velocity across a stream becomes more nearly uniform at higher Reynolds numbers. However, an awareness of the effect of these variations is important in applying the energy balance to each new situation; in certain cases there may be significant variations of temperature or velocity over a cross section. The best way to obtain an appreciation of these matters is to study the following examples and solve some numerical problems such as those given later in this chapter.

Example 4-1 In laminar flow the effects of the variations of velocity and temperature across a circular pipe may be significant. As an illustration, let us calculate the value of the kinetic energy term of Eq. (4-21), $(u^3)_{\text{av}}/2g_c u_b$, in terms of u_b. For rodlike flow, the result is $u_b^{\ 2}/2g_c$; for other velocity distributions, constants other than 2 appear in the denominator.

We substitute the equation for u as a function of r, [Eq. (2-15)], into Eq. (4-16) to obtain

$$(u^3)_{\text{av}} = \frac{u_{\max}^{\ 3}}{\pi r_i^{\ 2}} \int_0^{2\pi} \int_0^{r_i} \left(1 - \frac{r^2}{r_i^{\ 2}}\right)^3 r\, dr\, d\theta \qquad (1)$$

This equation is similar to Eq. (1) of Example 3-4. Evaluation of the integral yields

$$(u^3)_{av} = \frac{(u_{max})^3}{4} \tag{2}$$

and so

$$\frac{(u^3)_{av}}{2g_c u_b} = \frac{(u_{max})^3}{8g_c u_b} \tag{3}$$

Since $u_b = u_{max}/2$, the final result is

$$\frac{(u^3)_{av}}{2g_c u_b} = \frac{u_b{}^2}{g_c} \tag{4}$$

As was remarked earlier, in laminar flow the kinetic energy term is usually negligible. However, the analogous calculation of $(uH)_{av}/u_b$ gives a result which is frequently important, as in the case of heat transfer to a fluid in laminar flow. A computation of this sort is required in Prob. 4-7. ////

Example 4-2 A dilute solution at 70°F is added to a well-stirred tank at the rate of 180 lb/h. A heating coil having an area of 10.0 ft^2 is located in the tank and contains steam condensing at 300°F. The heated liquid leaves at 120 lb/h and at the temperature of the solution in the tank, maintained uniform by effective agitation. There is 500 lb of solution at 100°F in the tank at the start of the operation. Calculate the outlet temperature after 1 h.

The heat flux to the liquid is known empirically to be proportional to the temperature difference between the steam and the liquid in the tank. This observation is expressed mathematically by

$$\frac{q}{A} = U(t_s - t) \tag{1}$$

where U is the constant of proportionality, called the *overall heat-transfer coefficient*. This quantity is, in general, a function of the properties of the two fluids involved, their flow patterns, and the shape and material of the coil. To simplify this problem, these factors are assumed to have a negligible effect, and U is taken as 70 Btu/(h)(ft^2)(°F).

The overall energy balance in the form of Eq. (4-20) is used for the solution. In most problems where there are large temperature changes combined with moderate velocities (i.e., less than 100 ft/s), the potential and kinetic energy terms are negligible; such is the case in this problem. Since there are no turning shafts or moving piston rods, $\dot{W}_s$ is zero also (work of stirring is negligible). Equation (4-20) then becomes

$$\Delta \frac{w(uH)_{\text{av}}}{u_b} + \frac{d\check{E}}{d\theta} = q \tag{2}$$

For the liquid considered here, essentially water, $H = C_p(t - t_o)$, where t_o is a constant reference temperature, and C_p is the specific heat of water. It will be assumed that the temperatures of the inlet and outlet streams given are the mixing-cup, or bulk, temperatures, so that

$$\Delta \frac{w(uH)_{\text{av}}}{u_b} = w_2 H_2 - w_1 H_1 = C_p(w_2 t - w_1 t_1) - C_p t_o(w_2 - w_1) \tag{3}$$

In addition, we have the approximate relation

$$\check{E} = \check{U} = \check{H} = C_p M(t - t_o) \tag{4}$$

If Eqs. (1) to (4) are substituted into Eq. (2), there results

$$C_p w_2 t - C_p w_1 t_1 - C_p t_o(w_2 - w_1) + MC_p \frac{dt}{d\theta} + C_p(t - t_o)\frac{dM}{d\theta} = UA(t_s - t) \tag{5}$$

The overall mass balance gives $dM/d\theta = 60$ and $M = 60\theta + 500$. t_o can be given any reasonable value; zero is the most convenient one to choose here. Making these substitutions in Eq. (5) and rearranging the results, one obtains

$$\frac{dt}{d\theta} + \frac{60C_p + w_2 C_p + UA}{C_p(60\theta + 500)} t = \frac{UAt_s + C_p w_1 t_1}{C_p(60\theta + 500)} \tag{6}$$

Upon inserting the numerical values $C_p = 1.0$, $w_2 = 120$, $U = 70$, $A = 10$, $w_1 = 180$, $t_s = 300$, and $t_1 = 70$, there results

$$\frac{dt}{d\theta} + \frac{880}{60\theta + 500} t = \frac{222{,}600}{60\theta + 500} \tag{7}$$

This is an ordinary, first-order, linear differential equation and is integrated to give

$$t = 253 + \frac{C}{(60\theta + 500)^{14.67}} \tag{8}$$

The constant of integration C is obtained from the requirement that $t = 100$ when $\theta = 0$, so the final equation is

$$t = 253 - \frac{6.07 \times 10^{+41}}{(60\theta + 500)^{14.67}} \tag{9}$$

For $\theta = 1$, t is 223°F; after a long period of operation, the temperature will reach a steady-state value of 253°F, if the tank does not overflow. ////

Example 4-3 The overall energy balance in the form of Eq. (4-22) can be used to estimate the temperature drop of a gas flowing through an adiabatic throttling valve. The kinetic and potential energy terms are negligible; there is no heat transfer or shaft work, so

$$\Delta H = 0 \tag{1}$$

For an ideal gas, $\Delta H = C_p \, \Delta t$ and there is no temperature drop.

As an example using a real gas, let us consider steam entering the valve at 100 psia and 340°F and leaving at 52 psia. The enthalpy upstream[1] is $H_1 = 1194.3$ Btu/lb.

Since ΔH must be zero, the temperature downstream must be such that $H_2 = 1194.3$ at $p_2 = 52$; this temperature is 320°F. ////

Example 4-4 In part of a new process for making sulfuric acid, sulfur dioxide and oxygen enter the process at 1 atm and 50°C. The gases react completely in passing over a catalyst, and the sulfur trioxide produced is cooled to 50°C. Calculate the amount of heat removed in the process for every pound mole of SO_2 reacted.

The steady-state form of the energy balance is to be used; all terms are negligible except $\Delta\tilde{H}$ and $\tilde{Q}$, so that

$$\Delta\tilde{H} = \tilde{Q} \tag{1}$$

$\Delta\tilde{H}$ is found from the principles of thermochemistry:

$$\Delta\tilde{H} = \Delta\tilde{H}_{f,\,SO_3} - \Delta\tilde{H}_{f,\,SO_2}$$

where $\Delta\tilde{H}_f$ is the enthalpy of formation. From Perry, p. 3-144,

$$\begin{aligned}\Delta\tilde{H} &= -94.39 - (-70.94)\\ &= -23.45 \text{ kcal/g mol}\\ &= \tilde{Q} = -9.82 \times 10^7 \text{ J/kmol}\end{aligned}$$

The enthalpies of formation are given at 25°C, but the effect of a change to 50°C is small and will be neglected here. Thus approximately 10^8 J must be dissipated from the process for every kilogram mole of SO_2 reacted. ////

Examples 4-3 and 4-4 are essentially problems in thermodynamics. They are presented here to emphasize the fact that a great variety of problems can be viewed as

[1] J. H. Keenan, F. G. Keyes, P. J. Hill, and J. G. Moore, "Thermodynamic Properties of Steam," John Wiley & Sons, Inc., New York, 1969.

applications of the overall energy balance. The velocity-distribution and unsteady-state problems, Examples 4-1 and 4-2, require a more general form of the overall energy balance than is usually encountered in beginning courses in physical chemistry or thermodynamics. We now turn to another example in order to introduce the important idea of the friction, or lost work, encountered in the flow of fluids.

Example 4-5 Water is flowing through an insulated, horizontal pipe at steady state. Since $W_s = 0$, $Q = 0$, $\Delta u_b^2/2g_c = 0$, and $\Delta zg/g_c = 0$, ΔH must also be zero. For an incompressible liquid, since $\Delta U = C_v \, \Delta t \cong C_p \, \Delta t$,

$$\Delta H = C_p \, \Delta t + \frac{\Delta p}{\rho} \tag{1}$$

Thus

$$C_p \, \Delta t = \frac{-\Delta p}{\rho} \tag{2}$$

Experiment shows that there is a pressure drop through the pipe, and Eq. (2) shows that this effect will cause a rise in the temperature of the fluid. For instance, if $\Delta p = 40$ kPa,

$$\Delta t = \frac{(4 \times 10^4)}{(1000)(4187)} = 0.0096°\text{C} \qquad ////$$

In the preceding example it has been shown that there can be a loss of pressure resulting in a decrease in the term $\Delta p/\rho$, with no corresponding gain in the other forms of mechanical energy, namely, $\Delta u_b^2/2g_c$, $\Delta zg/g_c$, and W_s. The gain in internal energy causes a scarcely perceptible temperature rise in the fluid; this mechanical energy is for all practical purposes "lost." The pressure loss is caused by frictional resistance to flow; the friction arises from the viscous nature of the fluid. In later chapters we shall see that this loss can be calculated from a knowledge of the properties of the viscous fluid, its velocity, and the geometry of the conduit. It is therefore convenient to write an energy balance in terms of this loss, variously referred to as the friction or the lost work. When this is done an equation is obtained involving only mechanical energy terms; it is called the *mechanical energy balance*. The mechanical energy balance for a system in an unsteady state is complicated; only the case of steady-state flow will be considered here.

The lost work will be defined by reference to the batch work done by a pound of fluid as it passes from the inlet to the outlet.

$$W' = \int_{V_1}^{V_2} p \, dV - lw \qquad lw > 0 \tag{4-23}$$

This work can be visualized as that which would be done by a small piston and cylinder enclosing the pound of fluid as it flows: it differs from the work of Eq. (4-1), which includes the kinetic and potential energy effects.

The first law of thermodynamics may be written for this batch process

$$\Delta U = Q - W \qquad (4\text{-}24)$$

The Q here is that of Eq. (4-1). Equations (4-23) and (4-24) are combined to give

$$\Delta U = Q - \int_{V_1}^{V_2} p\,dV + lw \qquad (4\text{-}25)$$

The definition of enthalpy, Eq. (4-11), is written as

$$\Delta H = \Delta U + \Delta pV \qquad (4\text{-}26)$$

and ΔpV is replaced, using

$$\Delta pV = \int_{V_1}^{V_2} p\,dV + \int_{p_1}^{p_2} V\,dp \qquad (4\text{-}27)$$

Equations (4-25) to (4-27) are combined and solved for ΔH:

$$\Delta H = Q + \int_{p_1}^{p_2} V\,dp + lw \qquad (4\text{-}28)$$

If Eq. (4-28) is substituted for ΔH in Eq. (4-22), we obtain the mechanical energy balance, so called because it contains only mechanical energy terms.

$$\frac{\Delta u_b^2}{2g_c} + \frac{g}{g_c}\Delta z + \int_{p_1}^{p_2} \frac{dp}{\rho} + lw + W_s = 0 \qquad (4\text{-}29)$$

The value of the integral depends on the path of the process and on the equation of state of the fluid. For an incompressible liquid the integral reduces to $\Delta p/\rho$. If, in addition, there is no shaft work or lost work (friction), Eq. (4-29) reduces to a form of Bernoulli's equation,

$$\frac{\Delta u_b^2}{2g_c} + \frac{g}{g_c}\Delta z + \frac{\Delta p}{\rho} = 0 \qquad (4\text{-}30)$$

Bernoulli's equation can also be obtained from a differential force balance, as will be shown in Chap. 10. Equation (4-29) is sometimes called the *extended Bernoulli equation.*

Equation (4-29) departs somewhat from our idea of an overall balance, which we postulated as containing only terms which can be evaluated from a knowledge of the state of the inlet and outlet streams and the exchanges of heat and work with the surroundings. Nevertheless, Eq. (4-29) reduces in many cases to one which depends only on the end states; Eq. (4-30) is an example.

Since the integral in Eq. (4-29) can be expressed in terms of the inlet and outlet conditions for incompressible fluids, lw can be calculated in these cases from a knowledge of the externally measurable terms. However, the value of lw depends on the detailed nature of the flow, so that its independent calculation requires the use of a differential balance and the knowledge of the properties of the viscous fluid.

These latter methods for the calculation of lw from the differential momentum balance are used for flow through conduits, but not for flow through a pump or a turbine. For such machines it is convenient to divide the loss into lw_f, the loss in the conduit, and lw_t, the loss in the turbine.

$$lw = lw_f + lw_t \qquad (4\text{-}31)$$

This equation applies to flow through a conduit including a turbine, for which, according to convention, $W_s > 0$. It is customary to take into account the loss in the turbine by defining an efficiency η_t by

$$\eta_t = \frac{W_s}{W_s + lw_t} \qquad W_s > 0 \qquad (4\text{-}32)$$

The shaft work W_s is the rate at which energy is transmitted by the turbine shaft. The sum $W_s + lw_t$ is the rate at which the fluid gives up its mechanical energy in the turbine.

If Eqs. (4-31) and (4-32) are incorporated into Eq. (4-29), there results

$$\frac{\Delta u_b{}^2}{2g_c} + \frac{g\,\Delta z}{g_c} + \int_{p_1}^{p_2} \frac{dp}{\rho} + lw_f + \frac{W_s}{\eta_t} = 0 \qquad (4\text{-}33)$$

For a pump, we let

$$lw = lw_f + lw_p \qquad (4\text{-}34)$$

Once again W_s is the rate at which energy is transmitted by the shaft. However, mechanical energy is absorbed by the fluid in the pump at the lower rate $W_s + lw_p$ because of the consumption of energy at the rate of lw_p to overcome friction. The terms W_s and lw_p will always have opposite signs when calculated for a pump. The efficiency for a pump is

$$\eta_p = \frac{W_s + lw_p}{W_s} \qquad W_s < 0 \qquad (4\text{-}35)$$

so that

$$\frac{\Delta u_b{}^2}{2g_c} + \frac{g\,\Delta z}{g_c} + \int_{p_1}^{p_2} \frac{dp}{\rho} + lw_f + \eta_p W_s = 0 \qquad (4\text{-}36)$$

The examples and problems which follow often involve the simultaneous use of the general energy balance and the mechanical energy balance. These equations will prove particularly useful after we have studied how to estimate lw_f for turbulent flow; for laminar flow we can use the results of the force balance leading to Eq. (2-14), rearranged as follows:

$$-\Delta p = \frac{4\mu L u_{max}}{g_c r_i^2} \qquad (4\text{-}37)$$

In Example 3-4 the overall mass balance was used to show that $u_b = u_{max}/2$. This result is substituted into Eq. (4-37) to give

$$-\Delta p = \frac{8\mu L u_b}{g_c r_i^2} = \frac{32\mu L u_b}{g_c D^2} \qquad (4\text{-}38)$$

This is the Hagen-Poiseuille equation for the pressure drop for steady-state, laminar flow in a horizontal, round pipe of constant cross section. From the mechanical energy balance, Eq. (4-36), it is seen that

$$lw_f = \frac{32\mu L u_b}{\rho g_c D^2} \qquad (4\text{-}39)$$

for laminar flow (**Re** < 2100). Although the frictional loss in Eq. (4-39) was derived for a horizontal pipe, the analysis could have been made for a vertical pipe and would have given the same value for lw_f. If the velocity profile in the pipe is not changed by the orientation of the pipe, the expression for lw_f is not affected.

Example 4-6 Calculate the power required and the pressure which should be developed by a pump of 70 percent efficiency in order to send 15 gal/min of 98 percent sulfuric acid at 68°F from a tank at atmospheric pressure through 1000 ft of 2-in, schedule-40 steel pipe to a tank at 10 psig, where the level is 10 ft above that in the lower tank.

We calculate the bulk velocity and the Reynolds number in order to be sure that Eq. (4-39) is applicable.

$$u_b = \frac{15}{(60)(7.48)(0.0233)} = 1.431 \text{ ft/s}$$

The cross-sectional area of the pipe, 0.0233 ft^2, is found in Appendix Table A-5. The specific gravity of 98 percent sulfuric acid is 1.836 (p. 3-80, Perry), and its viscosity is 26.0 cP (Appendix Fig. A-3). Thus

$$\mathbf{Re} = \frac{(2.067/12)(1.431)(1.836)(62.4)}{(0.000672)(26.0)} = 1615$$

The flow is laminar, so Eq. (4-39) gives

$$lw_f = \frac{(32)(0.000672)(26.0)(1000)(1.431)}{(1.836)(62.4)(32.2)(2.067/12)^2} = 7.31 \text{ (ft)(lb}_f\text{)/lb}$$

The work required is found from the mechanical energy balance, Eq. (4-36), which becomes

$$\frac{g}{g_c}\Delta z + \frac{\Delta p}{\rho} + lw_f + \eta_p W_s = 0$$

The kinetic energy term and the expansion and contraction losses are negligible for this laminar-flow problem. Substitution of numerical values yields

$$10 + \frac{(10)(144)}{(62.4)(1.836)} + 7.31 + 0.70W_s = 0$$

and $W_s = -42.7$ (ft)(lb$_f$)/lb.

The horsepower required is

$$\frac{(42.7)(3.83)}{550} = 0.298 \text{ hp}$$

for the flow rate of 3.83 lb/s; there are 550 (ft)(lb$_f$)/s in 1 hp.

The pressure rise across the pump is given by

$$\frac{\Delta p}{\rho} + \eta_p W_s = 0$$

and

$$\Delta p = -\frac{(0.70)(-42.7)(62.4)(1.836)}{144}$$

$$= 23.8 \text{ lb}_f/\text{in}^2$$

////

Example 4-7 Water is flowing from an elevated reservoir through a conduit to a turbine at a lower level and out of the turbine through a similar conduit. At a point in the conduit 100 m above the turbine the pressure is 200 kPa abs; at a point in the conduit 3 m below the turbine the pressure is 120 kPa abs. The water is flowing at 1000 kg/s, and the output at the shaft of the turbine is 8×10^5 W. If the efficiency of the turbine is known to be 90 percent, calculate the loss by friction in the conduit. If there were no heat transfer to the surroundings, how much would the water be heated in flowing through the conduit and turbine?

Let us start the solution of this problem by using Eq. (4-33). Since water can be treated as incompressible, the integral can be written as $\Delta p/\rho$; in addition, the inlet and outlet conduits are of the same diameter, so the kinetic-energy-difference term is negligible. (The latter statement does not invalidate the fact that the turbine itself operates on the principle of converting potential energy into kinetic energy.)

We write Eq. (4-33) as

$$\frac{g}{g_c}\Delta z + \frac{\Delta p}{\rho} + lw_f + \frac{\dot{W}_s}{w\eta_t} = 0 \qquad (1)$$

and then

$$(9.81)(-103) + \frac{(-0.8 \times 10^5)}{1000} + lw_f + \frac{8 \times 10^5}{(1000)(0.90)} = 0$$

Finally, we have

$$-1010 - 80 + lw_f + 889 = 0$$

$$lw_f = 201 \text{ J/kg}$$

The temperature rise of the water can be calculated from the overall energy balance,

$$\frac{g\,\Delta z}{g_c} + \Delta H = -W_s \qquad (2)$$

from which

$$\Delta H = -800 + 1010 = 210$$

$$C_p\,\Delta t + \frac{\Delta p}{\rho} = 210$$

$$C_p\,\Delta t + (-80) = 210$$

The temperature rise is, then,

$$\Delta t = \frac{210 + 80}{4187} = 0.0693°\text{C}$$

For steady flow the friction is, in general, given by

$$lw = \Delta U - Q + \int_{V_1}^{V_2} p\,dV \qquad (3)$$

which reduces, for an incompressible fluid, to

$$lw = C_p\,\Delta t - Q \qquad (4)$$

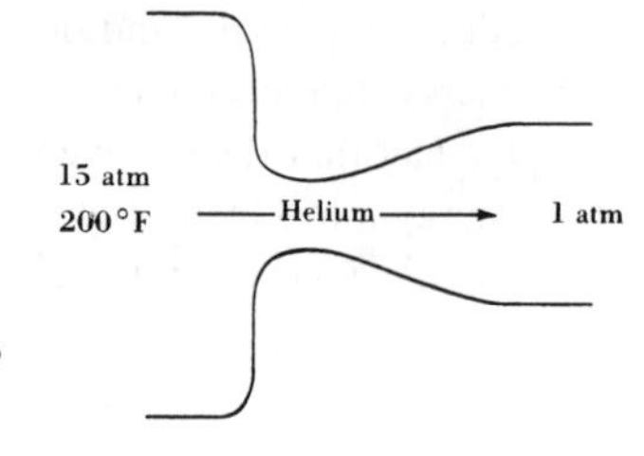

FIGURE 4-2
Converging-diverging nozzle. (Refer to Example 4-8.)

and for adiabatic flow of this fluid to

$$lw = C_p\ \Delta t$$
$$= 62.7\ (\text{ft})(\text{lb}_f)/\text{lb} \qquad (5)$$

This represents the total loss of mechanical energy in the conduit and in the turbine. ////

Example 4-8 Helium, which behaves like an ideal gas for the conditions of this problem, flows steadily at low velocity into a converging-diverging nozzle (Fig. 4-2) at a pressure of 15 atm abs and a temperature of 200°F. At a point near the end of the diverging section of the nozzle, the pressure has fallen to 1 atm abs. Calculate the velocity and temperature at this point.

Variations of velocity and temperature across the diameter of the nozzle are small; there is no exchange of heat or shaft work with the surroundings, and the nozzle is horizontal. The overall energy balance then becomes

$$\frac{\Delta u_b{}^2}{2g_c} + \Delta H = 0 \qquad (1)$$

For an ideal gas,

$$\Delta H = C_p\ \Delta t \qquad (2)$$

so that once Δt is known, the velocity is readily calculated from these equations.

The process is isentropic (reversible), as thermodynamics would indicate for a frictionless, isolated process. We can therefore calculate the downstream temperature by the formula

$$\frac{T_2}{T_1} = \left(\frac{p_1}{p_2}\right)^{(C_v - C_p)/C_p} \qquad (3)$$

which is valid for the isentropic expansion of an ideal gas.

For helium, $\tilde{C}_p$ is 5 and $\tilde{C}_v$ is 3, so

$$T_2 = (660)(15)^{-2/5} = 223°R$$
$$= -237°F$$

From Eq. (2)

$$\Delta H = \frac{(5)}{4}(-237 - 200) = -546 \text{ Btu/lb}$$

so that

$$\frac{u_{b2}^{\ 2}}{2g_c} = (546)(778)$$

and $u_{b2} = 5230$ ft/s.

The velocity obtained is much greater than the velocity of sound, and such nozzles are used to propel rockets and jet aircraft.

PROBLEMS

4-1 A moving body possesses kinetic energy equal to the amount of work required to take it from zero velocity to velocity u. By combining Newton's second law with the concept that work equals force times distance, show that this amount of work is $u^2/2g_c$ for a mass of 1 lb.

4-2 The potential energy change in a flow process is written $g\,\Delta z/g_c$, which can be shown to be the amount of work required to raise a mass of one pound a distance Δz ft. In most chemical engineering problems the variation in g is correctly assumed to be negligible, so an approximate value of 32.2 ft/s^2 is used which cancels numerically the proportionality constant g_c, having the value 32.2 (lb)(ft)/(lb$_f$)(s^2).

To emphasize the fact that g is not a constant and that the expression $g\,\Delta z/g_c$ does not represent the potential energy change in all situations, derive an expression for the potential energy change in raising an object of constant mass M a distance Δz above the earth's surface. The variation of g with the distance r from the center of the earth can be represented by the inverse-square law

$$g = 32.2\left(\frac{r_o}{r}\right)^2$$

where r_o is the earth's radius (3960 mi). Find the numerical answer for the case of $\Delta z = 500$ mi and compare it with the answer obtained by taking g as a constant at 32.2 ft/s^2.

4-3 In a chemical plant the water used in an absorption tower is pumped continuously from a pond through a well-insulated 3-in, schedule-40 pipe to a spray head at the top of the absorption tower. The temperature of the pond water is 15.5°C, and the vertical distance from the surface of the pond to the spray head is 30 m. The rate of pumping is 0.0126 m^3/s, and the power input to the pump is 10500 W. The water may be assumed to have a flat velocity profile at the pipe discharge. Determine the temperature of the water as it enters the spray head, where the pressure is 122 kPa abs.

4-4 Steam expands through an adiabatic converging nozzle. It enters the nozzle with a velocity of 20 ft/s at a pressure of 400 psia and a temperature of 600°F. At the throat of the nozzle the steam pressure is 216 psia and the temperature is 460°F. Find the velocity of the steam in the throat of the nozzle. The velocity profile may be assumed to be flat at all points.

4-5 A gas enters a horizontal 3-in, schedule-40 pipe at a constant rate of 0.2 kg/s. The gas is at a temperature of 20°C and a pressure of 116.5 kPa abs. It has a molecular weight of 29. The pipe is wrapped with an 80-kW electric-heating coil which is covered with a thick layer of insulation.

At the point where the gas is discharged, the pressure is 106.4 kPa abs. Find the temperature of the gas at this point. Assume that the gas behaves ideally and its heat capacity $C_p = 1000$ J/kg · K.

4-6 Water is to be drained from the bottom of an open cylindrical tank which has an ID of 10 ft. The initial depth of water in the tank is 4 ft. Estimate the time for all the water in the tank to drain through a 2-in opening if the process is frictionless and the discharge pressure is 1 atm.

4-7 A viscous oil flows at a constant rate (**Re** = 1000) out of a tank through a smoothly rounded entrance and into a length of 2-in, schedule-40 steel pipe. At the pipe entrance the velocity is essentially uniform at 0.9 m/s across the pipe diameter, and the temperature is 93°C. At a point downstream in the pipe the temperature at the center line of the pipe is 49°C, and at the pipe wall, 27°C. Calculate the rate of heat flow to the surroundings from the fluid, Btu/h, assuming a flat velocity profile and a temperature distribution across the pipe at the downstream point given by

$$t - t_s = (t_{\max} - t_s)\left[1 - \left(\frac{r}{r_i}\right)^2\right]$$

Compare this result with that obtained using the same temperature profile and the isothermal laminar velocity distribution of Eq. (2-15). The density of the oil is 865 kg/m^3, and its specific heat is 1883 J/kg · K.

4-8 A cross-country pipeline transports oil at the rate of 5000 bbl/d. The pressure of the oil leaving pumping station *A* is 250 psig. The pressure at the inlet to the next station *B* is 115 psig. Station *B* is at an elevation 57 ft higher than station *A*. Find the lost work per pound of oil and the corresponding power consumption (horsepower). Oil density, 48 lb/ft^3; 1 bbl = 42 gal.

4-9 (*a*) Write the overall mass and energy balances for a process involving two separate entering streams and one leaving stream.

(*b*) Simplify these equations for the common case in which the variations of velocity and temperature across each stream are negligible.

(*c*) Finally, write the equations for steady state.

(*d*) Generalize the results of part (*a*) to apply to the case of *m* streams in and *p* streams out. Use summation symbols to make the equations as compact as possible.

4-10 Water is taken from the bottom of a large tank, where the pressure is 446 kPa abs, and pumped to a nozzle discharging to the atmosphere 15 m above the bottom of the tank. The flow rate is 45 kg/s and the nozzle velocity is 21 m/s. If the efficiency of the pump is 75 percent and 7450 W is furnished to the pump shaft, calculate in (ft)(lb_f)/lb (*a*) the friction in the pump; (*b*) the friction in the rest of the system.

4-11 Water flows from the bottom of a large tank where the pressure is 100 psig through a pipe to a turbine which produces 5.82 hp. The pipe leading from the turbine is 60 ft below the bottom of the tank. In this pipe the pressure is 50 psig, the velocity 70 ft/s, and the flow rate 100 lb/s. If the friction loss in the system, excluding the turbine, is 40 (ft)(lb_f)/lb, find the efficiency of the turbine.

4-12 Oil is to be pumped at a rate of at least 6.3×10^{-5} m^3/s through a horizontal line with an equivalent length of 300 m. The pump to be used has under these conditions a maximum discharge pressure of 198 kPa abs. The line will discharge into an open tank containing oil to a depth of 6 m above the inlet. If the oil has a density of 800 kg/m^3 and a viscosity of 0.003 Pa · s, what size schedule-40 pipe should be recommended for the line?

4-13 A pipeline 20 mi long delivers 5000 bbl/d of petroleum. The pressure drop over the line is 500 psi. Find the new capacity of the pipeline if a parallel, identical line is laid along the last 12 mi of the line. The pressure drop is to remain 500 psi and the flow is laminar in all cases.

4-14 Glycerol is flowing at 60°C through a 30-m length of vertical 1-in, schedule-40 steel pipe. If the pressure at the lower end of the pipe is 138 kPa greater than at the upper end, what is the mass rate of flow of glycerol? What is the direction of flow?

4-15 A capillary tube (6 mm ID) is 300 m long and is attached to the bottom of a tank.

(*a*) When the outlet of the tube is 3 m below the level of the water in the tank, calculate the flow velocity in the tube. The fluid is water, for which the viscosity is 0.001 Pa · s.

(*b*) What is the Reynolds number of the fluid flowing in the tube?

4-16 Consider an insulated cylindrical tank having an ID of 2 ft filled to a depth of 2 ft with water at 80°F. Water at 180°F is now added at a constant rate of 100 lb/min until the liquid depth is 6 ft. The contents are always well mixed.

(*a*) How long does it take to fill the tank, and what is the final water temperature?

(*b*) To speed the heating a steam coil is added to the tank. It adds heat according to the formula

$$q = UA(t_{steam} - t)$$

where U = heat transfer coefficient, 50 Btu/(h)(ft^2)(°F)

A = area, 30 ft^2

t_{steam} = 230°F

t = variable water temperature.

What is the new final water temperature?

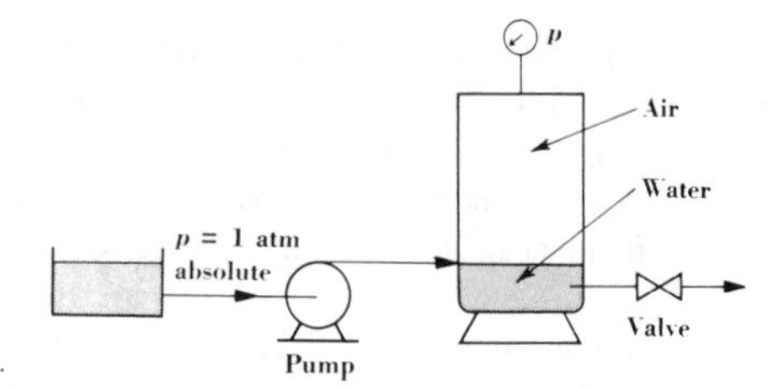

FIGURE 4-3
Pumping system for Prob. 4-18.

4-17 A centrifugal pump is tested with water at 70°F by measuring the flow rate and the corresponding pressure rise across the pump. The flow rate can be controlled by a valve downstream from the outlet pressure gauge. In this way the curve of Fig. 13-5 is obtained. For the pump that generated this curve, calculate the horsepower input at 2500 gal min.

4-18 A centrifugal pump transfers water from a large reservoir at atmospheric pressure to a surge tank, as shown in Fig. 4-3. When the water level in the tank is the same as that in the reservoir, the air pressure gauge reads 20 psig and the level is 1 ft above the bottom of the cylindrical tank. Neglect all friction in the lines. For the pump characteristics, use the curves of Fig. 13-5.

(*a*) What is the pump power (hp) consumption?

(*b*) The valve is adjusted until the level rises to 11 ft above the bottom of the tank. The total tank height is 18 ft. Assume the air stays at constant temperature as it is compressed. What is the pump power consumption now?

4-19 An ideal gas with a specific heat of 1000 J/kg · K flows through a horizontal tube in a constant temperature bath in which the liquid is held at 24°C. The gas enters the tube at a pressure of 446 kPa abs and leaves at 170.3 kPa abs. The temperature of the gas both entering and leaving is 24°C and the inlet velocity is 30 m/s. Find (*a*) the heat transfer rate and (*b*) the direction in which the heat is flowing.

4-20 A fountain discharges water vertically into the atmosphere from a circular opening of 1-in diameter. The velocity of the water as it leaves the opening is 30 ft/s. Assuming that the water has a flat velocity profile at all times as it rises and that there is negligible air drag or entrainment at the air-water interface, find (*a*) the height the water will rise and (*b*) the diameter of the rising jet at a distance 10 ft above the opening.

4-21 Water is pumped from one storage tank to a higher tank at a steady rate of 9.5×10^{-4} m^3/s. The difference in the elevations of the two water levels is 30 m. The tank which serves as a source is open to the atmosphere while the tank which receives the water has a pressure of 170.3 kPa abs in the vapor space above the water. Pressure gauges in the pipeline at the inlet and outlet of the pump read 34.5 kPa and 551.6 kPa, respectively. The power supplied by an electric motor to the pump shaft is 750 W. All piping is 1-in, schedule-40 steel pipe. Find pump efficiency and friction loss in the pipe per pound of water flowing.

4-22 In a particular flow system water is pumped at a steady rate from one tank to another at a different elevation. The pressure in the vapor space of one tank is 10 psig and in the other 20 psig. The water level in the tank at 10 psig is higher than that in the other tank by 30 ft. Determine the direction in which the water is flowing and give your reasons. Assume that frictional losses in the system are negligible.

4-23 A small city obtains a portion of its water supply from a deep well. The water (mean temperature 7°C) is pumped to a standpipe, whose level is held constant, 100 m above the ground level. The well is 75 m deep and the depth of water above the bottom of the well is estimated to be 6 m. Friction losses amount to 750 J kg through the entire length of the system of 5-in. schedule-40 pipe (see Appendix Table A-5A). The pump capacity is 1.6×10^{-2} m^3 s and the overall efficiency of the pumping apparatus is 83 percent. Calculate the daily pumping cost if electrical energy is available at 0.925¢ kWh.

4-24 The local fire department needs information on friction losses occurring between a water main and an open fire hydrant. At maximum main pressure (85 psig), the water discharge rate is 1620 gal min through a $2\frac{1}{2}$-in open hydrant butt. The water main, in which the fluid velocity is very small, is situated 8 ft below the hydrant discharge point. Determine the friction loss from the main to the discharge point. Assume atmospheric pressure is 15 psia.

4-25 The water level in an open supply tank is located 60 m above its discharge point. Water is discharged below at atmospheric pressure. Total friction losses are 570 J kg. What is the discharge velocity?

4-26 A two-bladed windmill has a diameter of 125 ft. What is the maximum theoretical power output in kilowatts from this machine at a uniform wind speed of 21 ft s? (1 kW = 0.948 Btu s.) State all assumptions.

4-27 Water from behind a dam flows at the rate of 5000 gal min through a turbine and discharges through a pipe into the atmosphere at a point 300 ft lower in elevation than the water level behind the dam. The turbine produces 200 hp and has an efficiency of 60 percent. Find the lost work due to friction in the line carrying the water. The bulk water velocity in the pipe to and from the turbine is 10 ft s.

4-28 A fluid with a flat velocity profile enters a tube of radius *R*. At some point downstream the fluid is considered to be in fully developed, laminar flow. Find the change in the total kinetic energy carried by the fluid between the entrance point and the point where developed flow exists.

5

THE OVERALL MOMENTUM BALANCE

The overall momentum balance proves useful for the analysis of the forces involved in fluid systems. Newton's second law can be written as a substantial derivative which applies to a mass M of moving fluid:

$$\frac{D(Mu_x)}{D\theta} = g_c \check{F}_x \tag{5-1}$$

This situation is inherently more complicated than those previously treated, because the force $\check{F}_x$ and the momentum Mu_x are vectors. In order to describe force and momentum, their direction must be specified as well as their magnitude; for mass and energy, only the magnitude is required. In this treatment we shall consider only the x components of the vectors.

A control volume is defined just as was done in deriving the mass and energy balances. The force acting on the volume is given by the rate of change of the momentum of the fluid passing through the volume at any instant; this quantity is in turn made up of the flux of momentum summed over the entire control surface and the rate of change of the momentum within the volume (recall that the control surface is fixed in space).

$$\frac{D(Mu_x)}{D\theta} = \iint_A u_x \rho u(\cos \alpha)\, dA + \frac{\partial}{\partial \theta} \iiint_V u_x \rho\, d\check{V} = g_c \check{F}_x \tag{5-2}$$

This equation is recognized as an application of Eq. (4-9), the Reynolds transport equation. Mu_x is the total x-directed momentum of the moving mass M enclosed at a particular instant in the control volume. The velocity u_x, which can be thought of as x-directed momentum per unit mass, replaces E in the term of Eq. (4-9), which sums the fluxes over the control surface. The volume integral represents the x-directed momentum of the fluid inside the control volume, Mu_x, or $\check{P}_x$. Thus, P_x equals u_x. There are equivalent equations for the y and z directions.

The force $\check{F}_x$ is composed of the sum of several forces. To avoid complication involving vectors and tensors, we consider here the integrals of these various forces as they act on the whole control volume. $\check{F}_{xg}$ is the x-directed force caused by gravity acting on the volume; it is zero if the x direction is horizontal. $\check{F}_{xp}$ is the x-directed force caused by the integral of the pressures acting on the surface of the control volume. Where the control surface cuts through a fluid, the pressure is taken directed inward perpendicular to the surface; the appropriate component must be considered. If a part of the control surface is solid and this wall is included inside the control surface, there is a contribution to $\check{F}_{xp}$ from the pressure on the outside of the wall; it is typically then atmospheric pressure.

Another important force is $\check{F}_{xd}$, the integrated x-directed drag, friction, or shear force. By this quantity we designate forces parallel to the control surface when it cuts between the fluid and a solid surface. In this simplified treatment this force is taken as parallel to the fluid flow only; of course, the appropriate component must be considered for areas not directed along the x axis.

The final force considered is $\check{R}_x$, which represents the x component of the resultant of the forces acting on the control volume at places where the control surface cuts through a solid. This occurs typically when a section of pipe and the fluid it contains are taken as the control volume.

Equation (5-2) is now rewritten in terms of these forces, and the volume integral is replaced by the derivative of the momentum of the mass within the control volume with respect to time.

$$\iint_A \frac{u_x \rho}{g_c} u(\cos \alpha)\, dA + \frac{1}{g_c} \frac{\partial \check{P}_x}{\partial \theta} = \check{R}_x + \check{F}_{xp} + \check{F}_{xd} + \check{F}_{xg} \tag{5-3}$$

If Eq. (5-3) is now applied to a section of conduit with its axis in the x direction as illustrated in Fig. 3-4, u and u_x become identical, ρA equals w/u_b, $\cos \alpha$ equals ± 1.0, and we have

$$\frac{1}{g_c} \Delta \left[\frac{w(u_x^2)_{av}}{u_b} \right] + \frac{1}{g_c} \frac{\partial \check{P}_x}{\partial \theta} = \check{R}_x + \check{F}_{xp} + \check{F}_{xd} + \check{F}_{xg} \tag{5-4}$$

where
$$(u_x^2)_{av} = \frac{1}{A} \iint_A u_x^2 \, dA \tag{5-5}$$

At steady state. Eq. (5-4) reduces to

$$\frac{w}{g_c} \Delta \frac{(u_x^2)_{av}}{u_b} = \check{R}_x + \check{F}_{xp} + \check{F}_{xd} + \check{F}_{xg} \tag{5-6}$$

Equation (5-4) or (5-6) can be combined with the mass balance and the equation of state to find the relation between the forces caused by a flowing fluid and its velocity and properties.

If there is no variation in velocity across the areas A_1 and A_2. Eq. (5-4) becomes

$$\frac{1}{g_c} \Delta(wu_x) + \frac{1}{g_c} \frac{\partial \check{P}_x}{\partial \theta} = \check{R}_x + \check{F}_{xp} + \check{F}_{xd} + \check{F}_{xg} \tag{5-7}$$

which applies whether the axis of the conduit is curved or not. If the conduit axis is curved. the analogous balance of y momentum must be solved in order to complete the analysis.

$$\frac{1}{g_c} \Delta(wu_y) + \frac{1}{g_c} \frac{\partial \check{P}_y}{\partial \theta} = \check{R}_y + \check{F}_{yp} + \check{F}_{yd} + \check{F}_{yg} \tag{5-8}$$

A problem involving a curved condit is illustrated in Example 5-3. Of course, in the most general case the z momentum must also be considered simultaneously.

In the analysis and design of rotating machinery such as turbines and rotary pumps, the momentum balance is conveniently written in terms of torque (moment) and angular momentum. We shall not consider this type of problem here; it is discussed in other texts, such as Hunsaker and Rightmire.[1] The study of the momentum balance will aid us in future work with the equations of motion and the boundary layer, as well as in the following examples and problems illustrating some immediate uses of the overall momentum balance.

Example 5-1 Water is flowing at 150 gal/min through a horizontal converging nozzle (Fig. 5-1). The upstream ID is 3 in, and downstream ID is 1 in. Calculate the resultant force on the nozzle when it discharges to the atmosphere. Consider that the nozzle is attached at its upstream end and that frictional forces are negligible.

[1] J. C. Hunsaker and B. G. Rightmire, "Engineering Applications of Fluid Mechanics," McGraw-Hill Book Company, New York, 1947. For a comprehensive treatment, see George F. Wislicenus, "Fluid Mechanics of Turbomachinery," 2 vols., Dover Publications, Inc., New York, 1965.

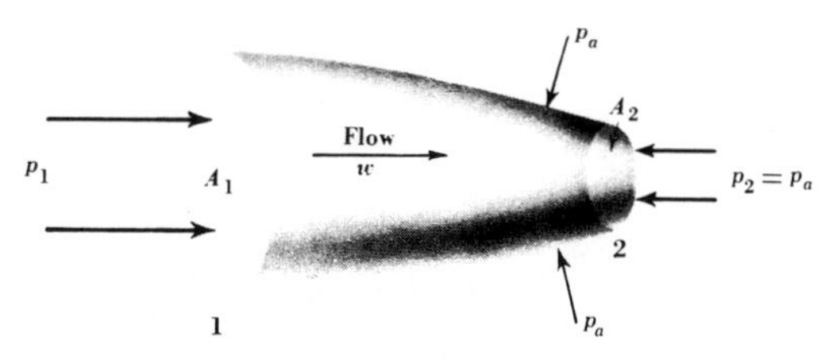

FIGURE 5-1
Flow through converging nozzle.

We must first compute the upstream pressure, or more specifically, Δp. This is accomplished by using Bernoulli's equation, Eq. (4-30), which neglects any friction in the flow.

$$\frac{\Delta u_b{}^2}{2g_c} + \frac{\Delta p}{\rho} = 0 \tag{1}$$

The velocities have been assumed uniform across the nozzle cross sections. From the given flow rate,

$$w = \frac{(150)(8.33)}{60} = 20.85 \text{ lb/s}$$

the velocities are calculated.

$$u_{b1} = \frac{w}{\rho A_1} = \frac{(20.85)(4)(144)}{(62.4)(3.14)(9)}$$

$$= 6.80 \text{ ft/s}$$

$$u_{b2} = (6.80)(\tfrac{3}{1})^2 = 61.2 \text{ ft/s}$$

From Eq. (1),

$$\Delta p = -\left[\frac{(61.2)^2 - (6.80)^2}{(2)(32.2)}\right](62.4)$$

$$= -3580 \text{ lb}_f/\text{ft}^2 = -24.9 \text{ psi}$$

The momentum balance, Eq. (5-6), is used to obtain $\check{R}_x$:

$$\frac{w}{g_c}\Delta u_x = \check{R}_x + \check{F}_{xp} \tag{2}$$

The solution of this problem requires that the control volume include the nozzle walls, so that $\check{R}_x$ enters into the force balance. At A_1 there is a force

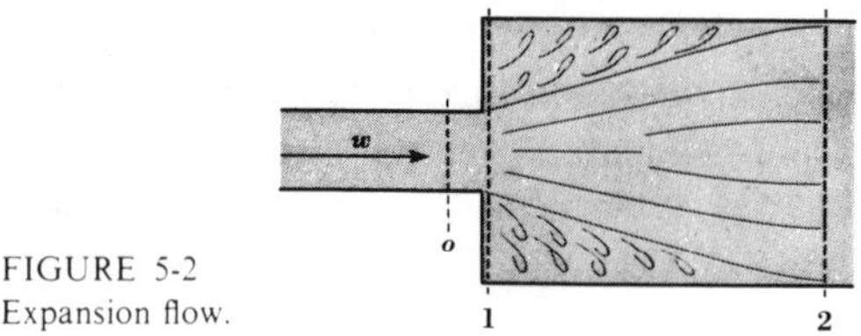

FIGURE 5-2
Expansion flow.

$p_1 A_1$ directed to the right in Fig. 5-1; at A_2, there is a force $A_2 p_2$ directed to the left. In addition, atmospheric pressure acting on the outside of the nozzle exerts a net force equivalent to $p_a(A_1 - A_2)$ directed to the left. Therefore the pressure force becomes

$$\check{F}_{xp} = p_1 A_1 - p_a(A_1 - A_2) - p_2 A_2$$

Because $p_2 = p_a$ we can write

$$\check{F}_{xp} = A_1(p_1 - p_a)$$

This pressure force is toward the right (positive x); the water is accelerated so that Δu_x is also positive. Thus, the signs of Δu_x and $\check{F}_{xp}$ in Eq. (2) are positive. Substitution of numerical values yields

$$\frac{(20.85)}{32.2}(61.2 - 6.80) = \check{R}_x + \frac{(3.14)(9)}{4}(24.9)$$

$$\check{R}_x = 35.2 - 175.8 = -140.6 \text{ lb}_f$$

Because the resultant force $\check{R}_x$ on the nozzle is negative, it is acting in the negative x direction, opposite to the direction of flow; the pipe wall is in tension since the nozzle is supported at its upstream end. ////

Example 5-2 When a fluid flows from a small pipe to one of larger diameter through a sudden expansion (Fig. 5-2), there is a mechanical energy loss caused by the resulting eddies. A gradual change in cross section, such as occurs in the nozzle of Example 5-1, is accompanied by very little lost work. Use the momentum balance to obtain an expression for *lw* for a sudden expansion for an incompressible fluid.

Let us consider the simple case for which the velocities in the two pipes do not vary across the diameter. Equation (5-6) becomes

$$\frac{w}{g_c}\Delta u_b = \check{F}_{xp} \tag{1}$$

The control volume is chosen so as not to include the pipe wall, so $\check{R}_x$ drops out. The boundaries of the control volume are planes 1 and 2. Flow through plane 1, however, occurs only over the area A_o. In addition, all the loss is assumed to be from eddies within the volume, making $\check{F}_{xd}$ equal to zero. Substituting $A_2(p_1 - p_2)$ for $\check{F}_{xp}$ gives

$$\frac{w}{g_c}\Delta u_b = A_2(p_1 - p_2) \tag{2}$$

It is assumed that p_1 and p_2 are uniform over the cross-sectional area; also, $p_o = p_1$.

Using the relations $w = u_{bo}\rho A_o$ and $u_{b2} = (A_o/A_2)u_{bo}$, this equation is changed to

$$\frac{u_{bo}^2}{g_c}\frac{A_o}{A_2}\left(1 - \frac{A_o}{A_2}\right) = \frac{\Delta p}{\rho} \tag{3}$$

The mechanical energy balance applied to the problem gives

$$\frac{u_{bo}^2}{2g_c}\left[1 - \left(\frac{A_o}{A_2}\right)^2\right] - lw = \frac{\Delta p}{\rho} \tag{4}$$

From the combination of these latter two equations we obtain the desired result:

$$lw = \frac{u_{bo}^2}{2g_c}\left(1 - \frac{A_o}{A_2}\right)^2 \tag{5}$$

In many situations the loss at the expansion is only a small part of the total loss and Eq. (5) is sufficiently accurate. If the expansion loss is a large fraction of the total loss, a more precise expression is desirable and can be obtained by using the actual upstream velocity distribution in the analysis (Prob. 5-5). ////

Example 5-3 Water is flowing through a horizontal, right-angle bend of gradual curvature, illustrated in Fig. 5-3. The velocity is 20 m/s, the pressure at the inlet is 240 kPa abs, and the ID of the pipe is 0.05 m. If friction can be neglected, calculate the magnitude and direction of the force required to hold the piece of pipe in equilibrium. The control volume includes the pipe walls. Neglect gravity.

To work this problem, the momentum balance is solved first for the x direction and then for the y direction. For the former,

$$\frac{w}{g_c}(u_{x2} - u_{x1}) = \check{R}_x + \check{F}_{xp} \tag{1}$$

The velocities are assumed constant across pipe cross sections. Since there is no x component of velocity at point 2, $u_{x2} = 0$. The x component of force

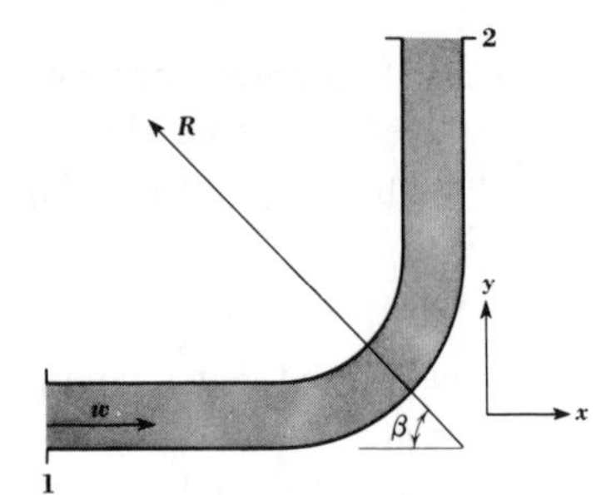

FIGURE 5-3
Flow through a curved pipe.

caused by pressure arises from p_1 over the area A_1 and atmospheric pressure p_a over the rest of the pipe, except A_2. The atmospheric pressure acts equally in all directions. Thus the net value of F_{xp} is $A_1(p_1 - p_a)$.

$$\check{R}_x = -A_1(p_1 - p_a) - \frac{w}{g_c} u_{x1}$$

$$= -\frac{\pi(0.05)^2(1.4 \times 10^5)}{4} - \frac{(20)(1000)(\pi)(0.05)^2(20)}{(1)(4)}$$

$$= -275 - 785 = 1060 \text{ N} \qquad (2)$$

By similar reasoning, we obtain

$$\check{R}_y = A_2(p_2 - p_a) + \frac{w}{g_c} u_{y2}$$

$$= 275 + 785 = 1060 \text{ N} \qquad (3)$$

The resultant force is directed as shown in Fig. 5-3, with $\beta = 45°$, and its magnitude is 1060/cos 45° = 1500 N. ////

PROBLEMS

5-1 Water is flowing through a horizontal diffuser consisting of a section of conduit for which the ID gradually increases from 0.05 to 0.10 m. The diffuser is constructed so that friction is negligible. Calculate the resultant force on the diffuser when the upstream pressure is 500 kPa abs and the flow rate is 0.0631 m^3/s.

5-2 Water flows through a horizontal 45° elbow of 2-in, schedule-40 steel pipe at 200 gal/min. If the upstream pressure is 50 psig and friction is neglected, calculate the resultant force on the elbow.

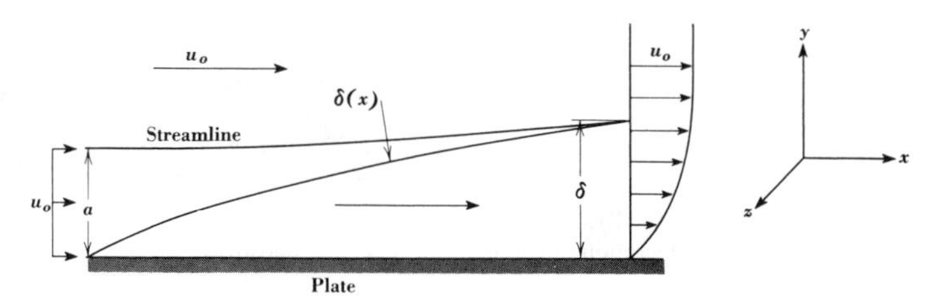

FIGURE 5-4
Boundary layer on flat plate.

5-3 An incompressible fluid flows past one side of a flat plate of unit width. The flow far from the plate is everywhere uniform at u_o, and the plate is oriented parallel to the general flow. The viscous fluid flowing near the plate is slowed by the layers nearer the plate, and there is no slip at the plate surface. At the leading edge a boundary layer develops which becomes thicker with distance downstream. In fact, the boundary layer is defined as the zone in which the velocity changes from zero at the plate to the uniform stream velocity u_o some distance from the plate. At any distance downstream the thickness is δ, and the velocity parallel to the plate is approximated by

$$\frac{u_x}{u_o} = \frac{3}{2}\left(\frac{y}{\delta}\right) - \frac{1}{2}\left(\frac{y}{\delta}\right)^3 \qquad 0 \le \frac{y}{\delta} < 1$$

$$\frac{u_x}{u_o} = 1.0 \qquad \frac{y}{\delta} \ge 1$$

Use the overall momentum balance to find the drag force on one side of the plate as a function of u_o, ρ, and δ. There is no change in pressure. Choose the control surface so there is flow in and out only in the x direction; Fig. 5-4 illustrates this problem. A streamline is a line (surface) across which no fluid flows; there is flow across the line $\delta(x)$ shown in the figure.

5-4 The ejector sketched in Fig. 5-5 is to be used to pump air to the atmosphere from a space at a pressure 500 Pa less than atmospheric. The pipe is 3-in, schedule-40 steel pipe, and the cross-sectional area of the nozzle is small enough to neglect in comparison with the cross section of the 3-in pipe. The injected air is pumped in at a pressure just sufficient to cause the air to leave the nozzle at 340 m/s, about sonic velocity. The density of the air at all points changes very little and can be taken as constant at 1.20 kg/m^3.

Find a relation between the flow rate of entrained air w_e and the flow rate of injected air w_i by writing the overall momentum balance between 1 and 2 in Fig. 5-5. The two streams are well mixed at 2 and the velocity is constant across the pipe at 2 and just before 1. The drag force $\check{F}_{xd}$ can be neglected. Calculate also the maximum flow rate of entrained air and the corresponding nozzle size.

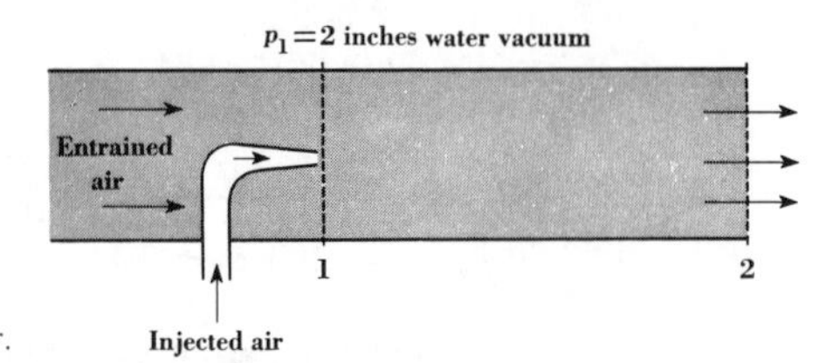

FIGURE 5-5
Operation of an air ejector.

5-5 An incompressible fluid flows from a small pipe, where it is in fully developed laminar flow, through a sudden expansion to a larger pipe, where the velocity is uniform. Derive an equation for lw, as was done in Example 5-2.
This problem[1] is sometimes considered with reference to the equation

$$\frac{\Delta u_b^2}{2g_c} + \frac{\Delta p}{\rho} + K_e \frac{u_{b1}^2}{2g_c} = 0 \tag{1}$$

Show that

$$K_e = 1 + \left(\frac{A_1}{A_2}\right)^2 - \frac{8}{3}\left(\frac{A_1}{A_2}\right) \tag{2}$$

This equation for K_e takes into account not only the lost-work term, but also the fact that $u_b^2/2g_c$ does not correctly represent the kinetic energy for laminar flow in the small pipe.

5-6 A stream of water at 15°C is discharged from a circular nozzle, travels in a horizontal direction, hits a flat wall oriented normal to the stream and falls directly to the ground. The nozzle has a diameter of 10 mm and the water leaves the nozzle with a flat profile at a velocity of 6 m/s. Find the force necessary to support the wall. Neglect air drag on the jet.

5-7 A straight, horizontal garden hose has an inside diameter of $\frac{1}{2}$ in and a nozzle diameter of 0.30 in. The inlet pressure to the hose is 10 psig when the water flow rate is 1 ft^3/min. The water leaving the nozzle discharges into the atmosphere. Find the lost work due to friction and the value of $\check{R}_x$ and the direction in which it acts relative to the direction of flow. Assume flat velocity profiles.

5-8 Jet boats are propelled by taking in water at a low velocity through a large opening, increasing the pressure on the water by a pump, and then discharging it at a high velocity through a small opening. A particular boat is being designed which must be capable of exerting a 900-N pull on a horizontal tow line in a static test. The water intake at the front of the boat is a 0.15-m diameter circular opening inclined at an angle of 85° to the vertical direction. The motor and the pump are of such size as to deliver 9.5×10^{-3} m^3/s through the circular outlet nozzle which discharges the water horizontally. Find (*a*) the size of the outlet nozzle and (*b*) the power requirement

[1] W. M. Kays, *Trans. Am. Soc. Mech. Eng.* **72**:1067 (1950). This article is concerned with the calculation of the pressure drop through a short heat exchanger, in which the expansion loss may be an appreciable fraction of the total loss.

if the pump and motor have an efficiency of 100 percent and there is negligible friction in the tubing or fittings.

5-9 Water at 60°F and 2 psig flows into a standard 2-in pipefitting at the rate of 104.5 gal/min. Calculate the resultant force necessary to support the fitting in each of the following examples. Assume friction is negligible in each case.

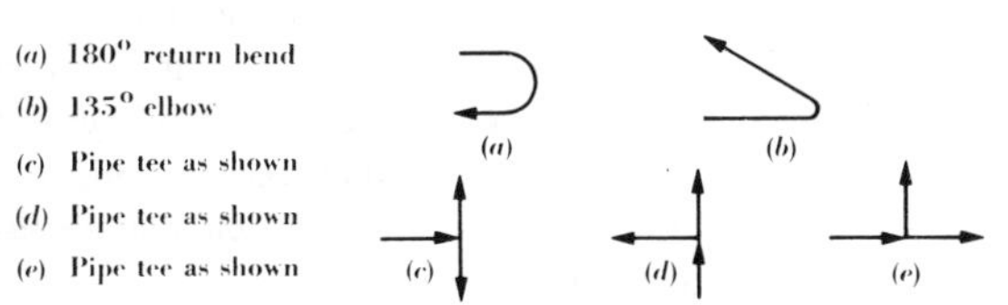

In cases (*c*), (*d*), and (*e*), assume streams are split into equal outlet flows.

5-10 Prove that a jet of water with a flat velocity profile discharged from a nozzle at ground level at an angle of 45° from vertical travels a greater horizontal distance before returning to its original elevation than a jet discharged at any other angle. Air drag must be assumed negligible.

5-11 Find the expression for the velocity distribution and pressure drop for a fluid in laminar flow in the annular space between two horizontal, concentric pipes. In this derivation the momentum balance must be applied to a control volume of differential magnitude, namely, an annular shell of thickness dr. Show that this analysis results in the differential equation

$$\frac{d}{dr}(r\tau) = \frac{r\,\Delta p}{L}$$

from which the desired relations are obtained by the substitution of Newton's law, Eq. (2-3), and two integrations.

5-12 A geothermal well produces 100 lb/s of a mixture of saturated steam and entrained water droplets, both phases being at 340°F and 118 psia. The mixture is 90 weight percent vapor and 10 weight percent liquid. Before the steam can be sent to a turbine, the liquid must be completely removed from the vapor in a separator. In this simple device the vapor phase changes direction from horizontal to vertical while the liquid impinges on the end of the separator and is drained off. Find the magnitude and direction of the external horizontal mechanical force necessary to support the separator. Assume steady-state operation, negligible gravitational effects, and no pressure or temperature drop in the separator.

5-13 A newtonian fluid flows through a circular pipe at a steady rate. The fluid has a flat velocity profile at the pipe inlet but by the time the fluid gets to the outlet the profile has changed to the standard parabolic profile. Does this change in velocity profile have any effect on the flow rate of momentum? Another way of asking this question is to ask if the amount of momentum flowing out of the pipe is equal to the amount flowing in. Derive the equation expressing the momentum change, if any, in terms of the usual variables.

5-14 A venturi has a 2-in-diameter circular inlet and outlet and a $\frac{3}{4}$-in-diameter circular throat. It is horizontal and assumed frictionless. Inlet and outlet pressures are 50 psig when water at 68°F is flowing through at a rate of 100 gal/min.

(*a*) Find the pressure at the throat.

(*b*) Find the R_x force at the throat and state whether the wall at the throat is in a state of axial tension (or compression) relative to the entire venturi.

(*c*) Find the magnitude and direction of the R_x force for the entire venturi.

5-15 Water at 60°F flows into a horizontal, 1-in pipe cross fitting with an inlet pressure of 10 psig. The inlet rate is 30 gal/min and the outlet rate for each opening is 10 gal/min. Assuming frictional losses to be negligible, find the resultant mechanical force necessary to support the fitting.

6
FLOW MEASUREMENT

In the preceding chapters the three overall balances have been developed, and they have enabled us to solve a number of problems. Before considering additional theory, the overall balances will be applied in this chapter to the analysis of the common types of flowmeter. The overall balances, together with some empirical observations, permit the practical solution of most flowmeter problems. Some important contributions to the understanding of flowmeter performance can also be obtained from the theory of flow of an ideal fluid, defined here as a fluid of zero viscosity.

Pitot Tube

Figure 6-1 shows the general appearance of a simple pitot tube. It consists of one tube with its opening normal to the direction of flow and a second tube in which the opening is parallel to the flow.

The velocity of flow is calculated from the difference between the pressure at the opening parallel to the flow, registering the static pressure, and the pressure in the impact tube, called the *stagnation pressure*. In Fig. 6-1 the difference between these two pressures is indicated by the difference in levels of the liquid in the manometer.

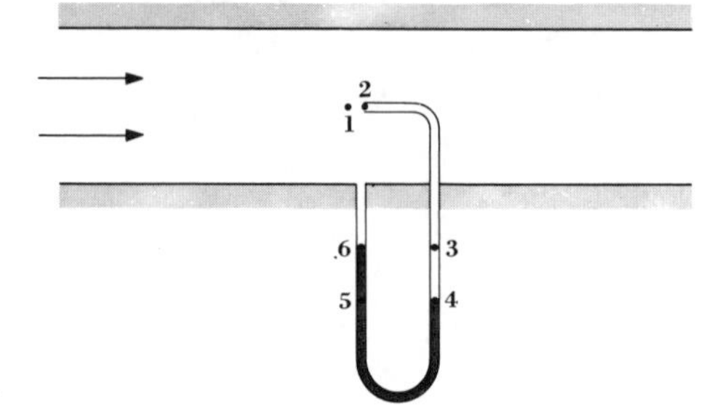

FIGURE 6-1
Schematic diagram of simple pitot tube.

Bernoulli's equation (4-30) can be written for the incompressible fluid at points 1 and 2. Since the velocity at 2 is zero,

$$\frac{u_1^2}{2g_c} = \frac{p_2 - p_1}{\rho} \tag{6-1}$$

The working equation for the pitot tube is usually written

$$u = C\sqrt{\frac{2g_c\,\Delta p}{\rho}} \tag{6-2}$$

The factor C is introduced to take into account deviations from Eq. (6-1): for most pitot tubes C is approximately unity, but for precise determinations of velocity the value should be determined by calibration of the instrument.

The value of Δp is related to the manometer reading by a simple hydrostatic analysis. This is approached most easily by applying Eq. (4-30) for the case of zero flow:

$$\Delta p = -\rho\,\Delta z\,\frac{g}{g_c} \tag{6-3}$$

Reference to Fig. (6-1) shows that

$$\begin{aligned}\Delta p = p_2 - p_1 &= (p_2 - p_3) + (p_3 - p_4) \\ &\quad + (p_4 - p_5) + (p_5 - p_6) + (p_6 - p_1)\end{aligned} \tag{6-4}$$

From Eq. (6-3) it is evident that two points at the same level joined by a continuous column of stagnant fluid are at the same pressure. Thus $p_4 - p_5$ is zero, and $p_2 - p_3$ equals $-(p_6 - p_1)$. The fluid between 6 and 1 is stagnant in the sense that there is no net flow in the vertical direction. These considerations reduce Eq. (6-4) to

$$\Delta p = (p_3 - p_4) + (p_5 - p_6) \tag{6-5}$$

By using Eq. (6-3) and the fact that $z_3 - z_4$ equals $-(z_5 - z_6)$, we obtain

$$\Delta p = \frac{g}{g_c}(\rho_m - \rho)(z_3 - z_4) \tag{6-6}$$

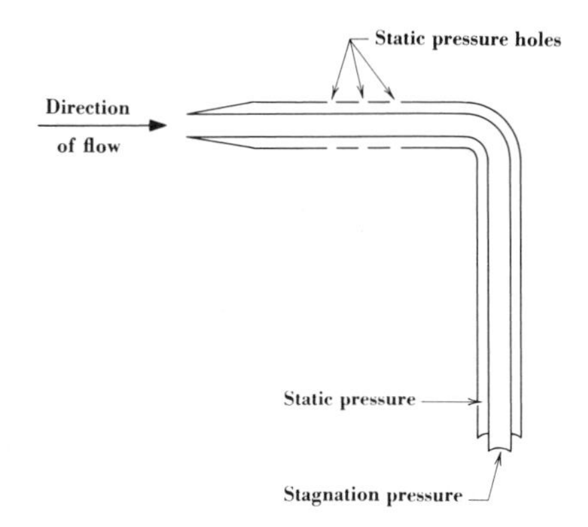

FIGURE 6-2
Schematic diagram of compact pitot tube.

where ρ_m is the density of the manometer fluid. If $z_3 - z_4$ is called Δh, Eqs. (6-6) and (6-2) can be combined to give

$$u = C\sqrt{\frac{2g(\rho_m - \rho)\,\Delta h}{\rho}} \tag{6-7}$$

There are many possible manometer configurations (Perry, 5th ed., p. 5-5), but they can all be analyzed by the procedure applied above.

In addition to the simple pitot tube which we have discussed, there exist many modifications. A compact design which makes use of concentric tubes is shown in Fig. 6-2. Static pressure holes parallel to the flow are drilled in the outer tube. This instrument can be mounted so that readings can be taken at various points across the pipe.

The pitot tube measures the point or local velocity; readings at successive points across a duct are used to obtain velocity profiles. The bulk velocity is obtained from the profile in a conduit by the evaluation, numerical or graphical if necessary, of Eq. (3-14).

This treatment applies to incompressible fluids and approximately to gases at moderate velocities for which the pressure change is less than 15 percent of the total pressure.

Venturi Meter

A venturi meter is shown in Fig. 6-3; the shape of the converging and diverging sections minimizes losses by eddy formation. Experimental measurements show that for $\mathbf{Re}_1 > 10{,}000$ the friction loss from 1 to 3 is about 10 percent of $(p_2 - p_1)/\rho$

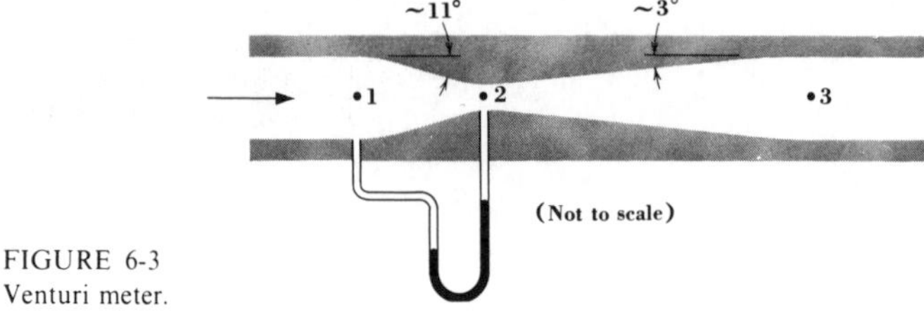

FIGURE 6-3
Venturi meter.

when the tube is horizontal. Referring to $(p_2 - p_1)/\rho$ as $\Delta p/\rho$, for incompressible fluids, we have

$$0.1\frac{\Delta p}{\rho} + lw_{13} = 0 \tag{6-8}$$

Assuming the loss from 1 to 2 is roughly half lw_{13}, the mechanical energy balance for a horizontal tube from 1 to 2 is

$$\frac{\Delta p}{\rho} + \frac{\Delta u_b{}^2}{2g_c} - 0.05\frac{\Delta p}{\rho} = 0 \tag{6-9}$$

Solution for u_{b2} gives

$$u_{b2} = 0.975\sqrt{\frac{2g_c(-\Delta p)}{\rho(1-\beta^4)}} \tag{6-10}$$

where β is the ratio of inside diameters D_2/D_1. The quantity $\sqrt{1/(1-\beta^4)}$ is called the *velocity-of-approach* factor; it is near unity for common values of β. For instance, for $\beta = 0.50$, $\sqrt{1/(1-\beta^4)} = 1.03$.

The coefficient 0.975 in Eq. (6-10) was obtained from the overall loss to show the relation between these two quantities. For precision, the coefficient is obtained from a direct measurement of the flow rate w by independent means, such as weighing the amount of liquid collected in a given time. Equation (6-10) is then written

$$u_{b2} = C_v\sqrt{\frac{2g_c(-\Delta p)}{\rho(1-\beta^4)}} \tag{6-11}$$

For $\mathbf{Re}_1 > 10{,}000$, C_v is 0.98; for smaller Reynolds numbers the coefficient decreases rapidly, as shown in Perry (3d ed.), fig. 53, p. 407. This effect is partially caused by the nonuniform velocity distribution across the diameter in laminar flow. Some results for C_v are given by Jorissen.[1]

[1] A. L. Jorissen, Discharge Coefficients of Herschel-type Venturi Tubes, *Trans. Am. Soc. Mech. Eng.*, **72**:1067 (1950), and **74**:905 (1952).

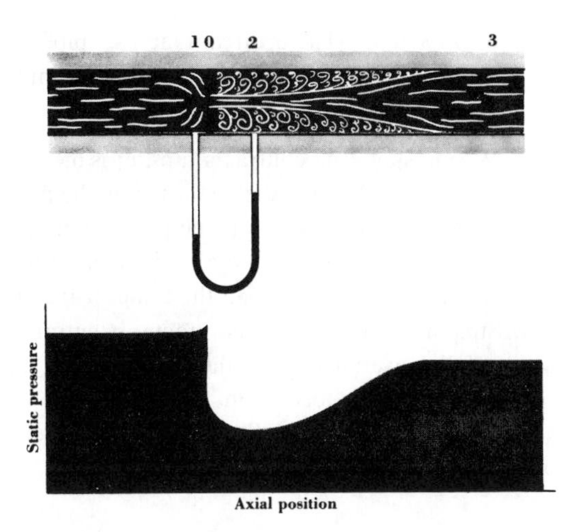

FIGURE 6-4
Orifice meter.

Orifice Meter

The orifice meter, illustrated in Fig. 6-4, operates on the same principle as the venturi meter but with some important differences. The orifice plate can easily be changed to accommodate widely different flow rates, whereas the throat diameter of a venturi is fixed, so that its range of flow rates is circumscribed by the practical limits of Δp. The orifice meter has a large permanent loss of pressure because of the presence of eddies on the downstream side of the plate; the shape of the venturi meter prevents the formation of these eddies and greatly reduces the permanent loss.

There are a number of customary positions for the pressure taps which lead to the manometer.

1 Plate taps Radial holes are drilled in the orifice plate and emerge on opposite sides of the plate near the pipe wall.

2 Corner taps Holes are drilled through the flange and emerge from the pipe wall in the corner adjacent to the orifice plate. These holes are normal to the holes for plate taps.

3 Flange taps The centers of the holes are located 1 in upstream and 1 in downstream from the faces of the plate.

4 Radius taps One pressure tap is 1 pipe diameter upstream from the orifice plate, and the other tap is $\frac{1}{2}$ pipe diameter downstream from the plate.

5 Vena contracta taps The upstream tap is usually 1 pipe diameter from the plate, and the downstream tap at the point of minimum pressure. This position is a function of the ratio of orifice and pipe diameters, as shown in Perry, p. 5-12.

6 *Pipe taps* The upstream tap is $2\frac{1}{2}$ pipe diameters from the orifice plate, and the downstream tap 8 pipe diameters from the plate. The downstream tap is at or beyond the point of maximum downstream pressure.

When using vena contracta taps, p_1 is measured far enough upstream to be in a region where the flow is little disturbed by the presence of the orifice; p_2 is measured at the vena contracta, which is the point of minimum static pressure and minimum cross section of the jet of fluid after it emerges from the orifice. This jet enters the relatively stagnant fluid behind the orifice plate, and as the jet expands to fill the pipe downstream from the vena contracta, it entrains some of the surrounding fluid. Most of the overall loss in the orifice arises from this viscous interaction and accompanying eddy formation.

Flow through the orifice is described in terms of Bernoulli's equation and an orifice coefficient C_o:

$$C_o^2 \frac{p_2 - p_1}{\rho} + \frac{u_{bo}^2 - u_{b1}^2}{2g_c} = 0 \qquad (6\text{-}12)$$

If β is defined as D_o/D_1, Eq. (6-12) becomes

$$u_{bo} = C_o \sqrt{\frac{2g_c(p_1 - p_2)}{\rho(1 - \beta^4)}} \qquad (6\text{-}13)$$

The discharge coefficient sometimes includes the velocity-of-approach factor $\sqrt{1/(1 - \beta^4)}$.

Empirically determined values of C_o for sharp-edged orifices with corner taps are given in Fig. 6-5 as a function of β and the Reynolds number through the orifice. The value of C_o is dependent on the location of the pressure taps. Tabulations of C_o for flange taps, vena contracta taps, and pipe taps are available in "Fluid Meters: Their Theory and Application," part I.[1]

The coefficient C_o departs from unity largely because the velocity u_{bo} differs from u_{b2}, and the lower pressure p_2 corresponds approximately to the velocity at the vena contracta, u_{b2}. Equations (6-12) and (6-13) are written in terms of u_{bo} because this velocity is conveniently calculated from the size of the orifice. Also included in C_o are, of course, the effects of friction, as they are for C_v. Above $\mathbf{Re}_o$ of about 10^4, C_o has a value of 0.61; at lower Reynolds numbers the orifice coefficient becomes a strong function of $\mathbf{Re}_o$, as shown by Fig. 6-5. This effect is partly caused by variations in the velocity across the pipe at point 1. Most orifices are designed so that $\mathbf{Re}_o > 10^4$.

[1] American Society of Mechanical Engineers, New York, 1959. A review of the state of the art for flowmeters is given by G. Thibessard, La Houille Blanche, no. 5, p. 441, 1969. See also, R. B. Dowdell, ed., "Flow; Its Measurement and Control in Science and Industry," vols. 1–3, Instrument Society of America, Pittsburgh, 1974.

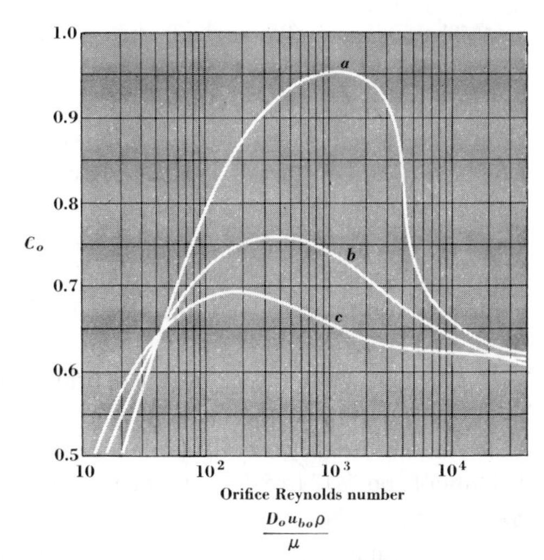

FIGURE 6-5
Orifice coefficient in Eq. (6-13) (Perry, p. 5-13).
a $D_o/D_1 = 0.80$.
b $D_o/D_1 = 0.60$.
c $D_o/D_1 = 0.20$.

In the design of piping systems it is important not to forget the permanent pressure drop caused by an orifice. The typical pressure profile shown in Fig. 6-4 indicates that $(p_1 - p_3)/(p_1 - p_2)$ may be a large fraction of $p_1 - p_2$, which is the indicated Δp associated with the orifice. We express this effect in terms of R, where

$$R = \frac{p_3 - p_2}{p_1 - p_2} \tag{6-14}$$

$$1 - R = \frac{p_1 - p_3}{p_1 - p_2} \tag{6-15}$$

The recovery fraction R and the permanent loss fraction $1 - R$ are given in Fig. 6-6 for a typical orifice for turbulent flow.

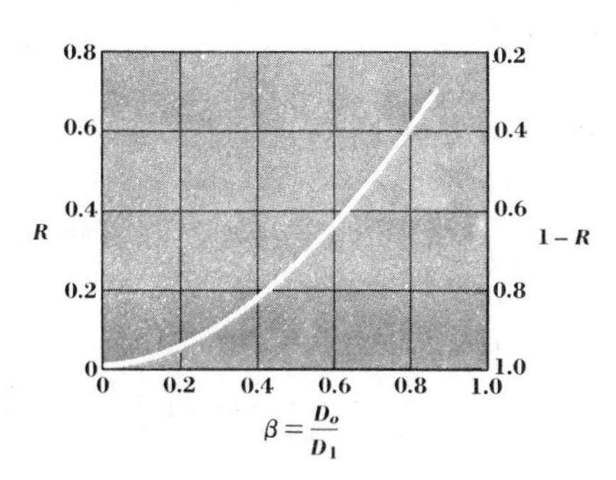

FIGURE 6-6
Overall pressure change for orifice.

The orifice discharge coefficient is significantly affected by flow disturbances which originate in valves, bends, and other fittings located upstream from the orifice. It is less affected by downstream disturbances. As a general rule, the meter should be placed 50 pipe diameters downstream and 10 pipe diameters upstream from any disturbances. It should be apparent that the necessary distances depend on the nature of the disturbance and in many cases are much less than the general specifications given above. Perry (p. 5-14) tabulates recommended locations for many specific cases. The upstream distance can often be reduced by placing straightening vanes in the pipe, and recommendations for the use of these vanes are also given by Perry.

Our purpose in this discussion of orifice meters has been to show some applications of the overall balances and to inquire briefly into the principles of flow through orifices. The extension of the formulas to compressible gases is discussed in Perry's Handbook, pp. 5-12 to 5-14. A much more detailed treatment of orifices is given by Stearns et al.[1] Not only are extensive tables of the discharge coefficient given, but much useful information is included about the location of the orifice with respect to upstream and downstream disturbances and about the details of orifice installation.

Example 6-1 Water at 68°F is flowing through an orifice arranged as in Fig. 6-4; the discharge at point 3 is directly to the atmosphere. The actual pipe ID is 1 in, and β is 0.6. The manometer contains an oil of specific gravity 1.10 and the reading for Δh is 1.50 in. Calculate the flow rate in gallons per minute and the gauge pressure at point 1 in inches of water.

We incorporate Δh into Eq. (6-13) to obtain

$$u_{bo} = C_o \sqrt{\frac{2g(\rho_m - \rho)\,\Delta h}{\rho(1 - \beta^4)}} \tag{1}$$

The velocity is at first estimated with $C_o = 0.61$.

$$u_{bo} = 0.61 \sqrt{\frac{2(32.2)(1.10 - 1.0)(0.125)}{1.0(1 - 0.13)}}$$

$$u_{bo} = 0.61(0.964) = 0.585 \text{ ft/s} \tag{2}$$

$$\mathbf{Re}_o = \frac{0.6(0.585)(62.4)}{12(6.72 \times 10^{-4})}$$

$$= 2.72 \times 10^3 \tag{3}$$

[1] R. F. Stearns, R. R. Johnson, R. M. Jackson, and C. A. Larson, "Flow Measurement with Orifice Meters," D. Van Nostrand Company, Inc., Princeton, N.J., 1951.

From Fig. 6-5 the value of C_o at this $\mathbf{Re}_o$ is about 0.69. Using this value to recalculate u_{bo} gives 0.66 ft/s and $\mathbf{Re}_o = 3.09 \times 10^3$. At this new $\mathbf{Re}_o$, C_o is 0.68, little changed from 0.69 because of the logarithmic abscissa of the figure. If desired, we can make this trial and error calculation more analytical by combining the elements of Eqs. (2) and (3) to give

$$\mathbf{Re}_o = 4.48 \times 10^3 C_o \tag{4}$$

The intersection of this equation and curve b of Fig. 6-5 gives the solution.

The above procedures give $u_{bo} = 0.66$ ft/s.
The flow rate in gal/min is

$$\begin{aligned} \text{gal/min} &= u_{bo} A_o (60)(7.48) \\ &= (0.66)(1.96 \times 10^{-3})(60)(7.48) \\ &= 0.580 \end{aligned}$$

The pressure drop over the manometer $(p_1 - p_2)$ corresponds to $[(1.10 - 1)/1](1.5) = 0.15$ in of water. Figure 6-6 shows that for $\beta = 0.6$ about 0.63 of this Δh is permanent loss, or 0.0945 in of water. ////

Rotameter

The rotameter is illustrated in Fig. 6-7. In the meters already studied, the area of the constriction remains constant and the pressure drop varies with flow rate; in the rotameter, the pressure drop remains nearly constant and the area of the constriction varies. The fluid flows vertically upward through the tapered rotameter tube, and the float comes to equilibrium at a point where the annular flow area is such that the

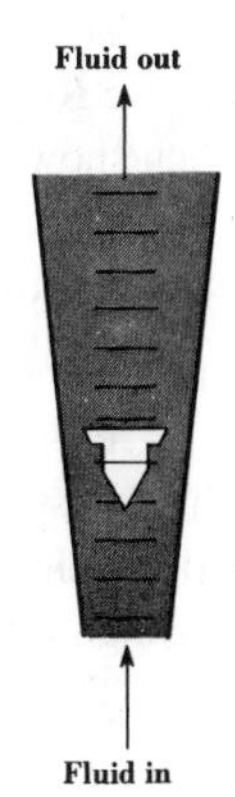

FIGURE 6-7
Operation of rotameter.

velocity increase has produced the necessary pressure difference. A higher flow rate causes the float to rise to a point where the annular area is larger.

Equation (6-13) is adapted to the rotameter by defining a coefficient C_r by

$$u_{br} = C_r \sqrt{\frac{2g_c(p_1 - p_2)}{\rho}} \qquad (6\text{-}16)$$

The factor $1/\sqrt{1 - \beta^4}$ is included in C_r; u_{br} is the velocity in the annular constriction. The pressure difference, $p_1 - p_2$, is found by making a force balance on the stationary float. The downward force caused by gravity is balanced by the upward buoyant force plus the force exerted by the pressure difference across the float caused by the velocity increase in the annular constriction. Thus we write

$$\rho_f \check{V}_f \frac{g}{g_c} = \rho \check{V}_f \frac{g}{g_c} + A_f(p_1 - p_2) \qquad (6\text{-}17)$$

and

$$p_1 - p_2 = \frac{\check{V}_f(\rho_f - \rho)g}{A_f g_c} \qquad (6\text{-}18)$$

where $\check{V}_f$ = volume of float

A_f = maximum cross-sectional area of float

The substitution of Eq. (6-18) into Eq. (6-16) gives

$$u_{br} = C_r \sqrt{\frac{2g\check{V}_f(\rho_f - \rho)}{A_f \rho}} \qquad (6\text{-}19)$$

The coefficient of discharge C_r is given in Fig. 6-8 as a function of the Reynolds number in the annulus and the type of float. The Reynolds number is defined as

$$\mathbf{Re} = \frac{(D - D_f)u_{br}\rho}{\mu} \qquad (6\text{-}20)$$

where D = diameter of tube at level of float

D_f = maximum diameter of float

Above **Re** = 10,000, C_r is constant for the floats shown in Fig. 6-8. Equation (6-19) then shows that u_{br} is constant when C_r is constant, for a given fluid and rotameter. The mass rate of flow or the upstream velocity is easily found from an overall mass balance.

$$w = u_{br}\rho A_r = u_{b1}\rho A_1 \qquad (6\text{-}21)$$

Flow around float a shown in Fig. 6-8 resembles flow through a venturi in that the cross-sectional area for flow changes gradually. As a consequence, C_r approaches C_v at high Reynolds numbers. Flow around float c resembles flow through an orifice, with the result that C_r approaches C_o at high **Re**.

The annular area A_r is a function of the float position. Although the flow through a rotameter can be estimated from the geometry of the float and tube and

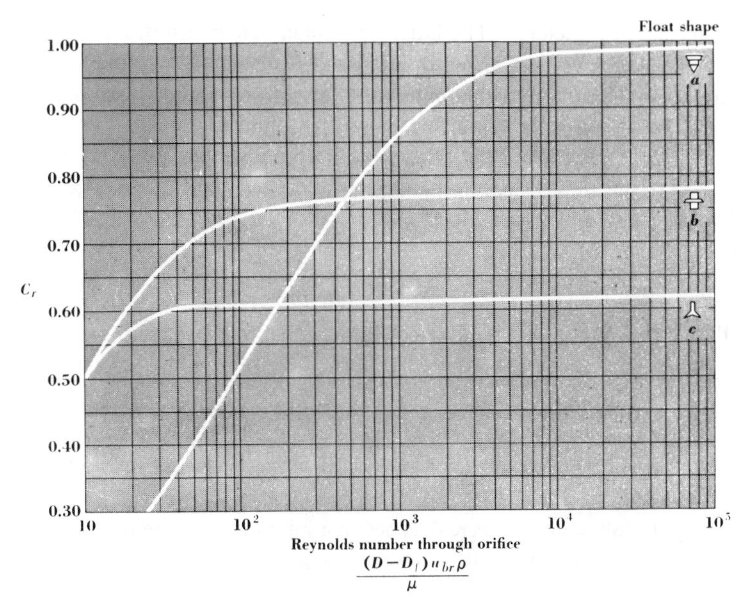

FIGURE 6-8
Rotameter discharge coefficients. (*From G. G. Brown et al., "Unit Operations," fig. 145, p. 158, John Wiley & Sons, Inc., New York, 1950.*)

the above equation, the manufacturer usually furnishes a reliable calibration. The above relations, particularly Eq. (6-19), permit the calibration for one fluid to be adapted for use with a different fluid.

Other Meters

An example of a nozzle is shown in Fig. 6-9. It has more overall pressure loss than a venturi but less than an orifice; some models can be installed in a pipe in the same manner as an orifice plate.

There are many other types of meters besides the foregoing differential-pressure meters. They include weirs and notches for flow in open channels and various kinds

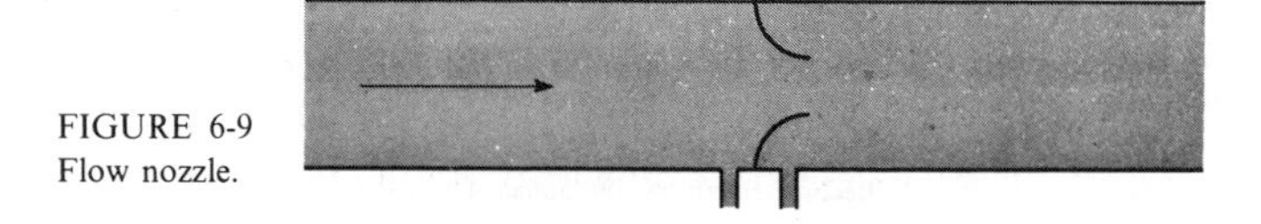

FIGURE 6-9
Flow nozzle.

of mechanical meters. The latter operate either on a positive-displacement principle, such as a wet test meter, or as a current meter, such as an anemometer. There are many other more specialized types; for pictures and discussion of their operation, refer, for example, to Perry, Sec. 5, and to Linford.[1]

Data on coefficients and operating and installation procedures are given in publications by the American Society of Mechanical Engineers[2] and the International Standards Association.[3]

PROBLEMS

6-1 A closed-U-tube manometer filled with mercury is attached to the underside of a line carrying water, as shown in Fig. 6-10. At a point directly above the closed-U-tube-manometer tap, the upstream tap of an inverted U tube is located. The inverted-U-tube manometer is filled with a liquid of specific gravity 0.5. What are p_1 and p_2 in psia?

6-2 The following data were obtained in a pitot-tube traverse of a 3-in, schedule-40 steel pipe in which water at 58°F was flowing:

Distance from wall, in	Manometer reading, in. carbon tetrachloride at 58°F
0.28	2.20
0.53	2.40
0.78	2.60
1.03	2.80
1.28	3.00
1.53	3.10
1.78	2.85
2.03	2.70
2.28	2.40
2.53	2.00
2.78	1.65

At the same time that these measurements were being taken, a series of samples of the flow were taken by discharging the total stream into a weigh tank. The discharge rate was found to be 1400 lb of water in 194.4 s. Find the pitot-tube coefficient.

6-3 Water at 15.5°C is flowing through a $3\frac{1}{2}$-in, schedule-40 steel pipe. The corner taps of a 25-mm circular, square-edged orifice in the pipe are connected to a manometer containing methyl benzoate (specific gravity 1.10). The difference in fluid levels in the manometer is 0.132 m. Find the flow rate in the pipe.

[1] A. Linford, "Flow Measurement and Meters," E. and F. N. Spoon Ltd., London, 1949.

[2] Am. Soc. Mech. Eng. Special Committee on Fluid Meters, "Fluid Meters," pts. 1 to 3, New York, 1931–1937 and 1959.

[3] Rules for Measuring the Flow of Fluids by Means of Nozzles and Orifice Plates, Preliminary Recommendations, *Int. Stand. Assoc. Bulls.* 9 and 12, December, 1935, and August, 1936.

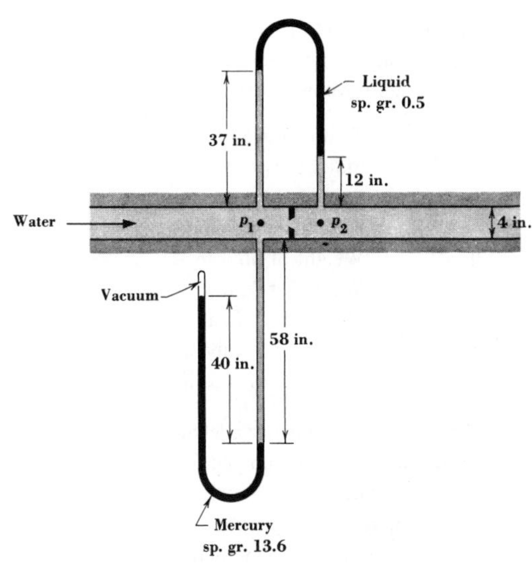

FIGURE 6-10
Orifice meter for Prob. 6-1.

6-4 During a laboratory experiment 4.5 ft^3 of air at an upstream pressure of 749.5 mmHg and a temperature of 84°F is passed through an orifice in 131 s. The orifice diameter is 0.25 in and is small compared with the pipe diameter. The pressure drop across the orifice is 5.6 in of water. Calculate the orifice coefficient.

6-5 Water at 15.5°C is flowing through a 3-in, schedule-40 steel pipe at the rather low rate of 0.09 kg/s. The liquid is metered by a sharp-edged orifice having a diameter of 0.058 m. It is later found necessary to meter acetone at 4.5°C in the same system at the same rate of flow, 0.09 kg/s. What will be the ratio of the pressure drop across the orifice using acetone to the pressure drop using water?

6-6 Air at 60°F is flowing through a 12-in-diameter circular duct into a spray drier at the rate of 300 ft^3/min. The flow rate is to be measured using an orifice meter to which is attached a water manometer. Since the pressure in the air line is only 2 psig and the air must reach the drier at a pressure of at least 1.8 psig, it is essential that the permanent head loss due to the orifice not exceed 0.2 psi. Find the maximum manometer differential (in inches of water) which can be obtained under these conditions.

6-7 Water at 21°C is to be pumped through a length of 2-in, schedule-40 steel pipe, and its flow rate is to be measured by an orifice meter or a venturi meter. The water enters the pump at atmospheric pressure and leaves the meter at 584 kPa abs at the same elevation as the pump entrance. At a flow rate of 6.3×10^{-3} m^3/s the difference in mercury levels in the attached manometer is to be 0.25 m. How much will the daily cost of electrical energy to run the pump be reduced by using the venturi meter instead of the orifice meter? The motor-pump combination has an efficiency of 70 percent. Electricity cost is 2 ¢/kWh.

6-8 A rotameter has the sight glass marked into 25 divisions. The float which is used is made of an unknown material, and its total volume is also unknown. Air at the rate of 3 $ft^3/122$ s is allowed to flow through the meter. The following data were taken:

Temperature of air	92°F
Static pressure	2.05 in water
Barometric pressure	29.49 inHg
Rotameter reading	130
Weight of float in air	4.62 g at 60°F
Weight of float in water	2.96 g at 60°F
Diameter of float	0.50 in
Diameter of tube at top	0.6482 in
Diameter of tube at bottom	0.5128 in

Calculate the rotameter coefficient.

6-9 Water at 15.5°C is pumped at a steady rate of flow through a pipeline 180 m long. Pressure gauges immediately upstream and downstream from the pump read 69 and 173 kPa, respectively. An orifice meter in the line with a 25-mm-diameter orifice ($C_o = 0.61$) shows a manometer differential of 52.3 mm of mercury. If the pipeline has an ID of 40.9 mm and the pump has an efficiency of 60 percent, find the power supplied to the pump.

6-10 Water at 60°F is flowing through a 3-in, schedule-40 steel pipe. The line contains both a venturi meter and an orifice meter, each with a throat diameter of 1 in. Each meter is connected to its own manometer containing carbon tetrachloride (specific gravity, 1.6) as a fluid. The manometer differential for the orifice meter is 18 in. What is the manometer differential for the venturi meter? The orifice coefficient $C_o = 0.61$ and the venturi coefficient $C_v = 0.98$. The meters, although operating simultaneously, may be assumed to be far enough apart so that each operates as though the other were not present.

6-11 An orifice is installed at the end of a pipe so that p_3 of Fig. 6-4 is atmospheric pressure. Water at 20°C is flowing through the orifice so that the reading on a mercury manometer connected across the taps is 0.15 m. For a pipe ID of 0.05 m and an orifice diameter of 0.03 m, calculate the flow rate and the pressure p_1.

6-12 Air is flowing through a 15-cm-ID duct at 40°C and 1.20 atm abs. If the bulk velocity of the air is 1 m/s, what size orifice should be used to give a (vertical) manometer reading of 1.0 cm of water across the orifice taps? For sufficient precision, an inclined manometer might be used.

If the air conditions are changed to 20°C and 2.00 atm abs, what bulk velocity will the same reading of 1 cm now indicate? What is the ratio of the new flow to the old on a mass basis?

6-13 A certain orifice and manometer setup has been calibrated with air at 20°C and 1 atm abs. Make a graph that will facilitate the use of the orifice for other gases; i.e., plot a correction factor C versus ρ/ρ_o where ρ_o is the density of air at the calibration conditions. The quantity C is the ratio of the mass flow of gas of density ρ to the reference flow of air of density ρ_o at the same Δh. Is your graph restricted to certain flow rates?

6-14 A rotameter is equipped with a float of type c of Fig. 6-8, of density 1500 kg/m^3. If the rotameter reads 100 when 6000 mL/h of water flow through it at 20°C, calculate, for the same reading of 100, (*a*) the flow of acetone at 10°C, (*b*) a 20 mass percent aqueous solution of K_2CO_3 at 60°C, and (*c*) 40 mass percent H_2SO_4 at 10°C.

6-15 Argon is flowing through a rotameter at 100 cm^3/min. What flow of helium will give the same reading on the rotameter? C_r is always 0.61. The float density is 2 g/cm^3.

6-16 A hydrocarbon stream with a density of 50 lb/ft^3 is flowing through a 4-in, schedule-40 steel pipe. The pipe contains an orifice plate with a circular opening 1.50 in in diameter having an orifice coefficient $C_o = 0.61$ for the conditions in which it is being used. The orifice taps are connected to a manometer containing mercury which shows a difference in levels of 12 in. Unfortunately some water has gotten into the manometer so, in addition to the 12-in differential in mercury, there is a 3-in column of water on top of the mercury in the high-pressure leg of the manometer. What is the flow rate of the hydrocarbon stream in gallons per minute?

6-17 A wind tunnel can be regarded as a venturi meter because it has tapered converging and diverging cross sections. If the inlet velocity is 13.0 ft/s (as measured by a hot-wire anemometer) and the manometer reading at the test section is 6.0 in of water, what is the venturi coefficient for the apparatus? The inlet pressure is exactly 1 atm and the temperature throughout the system can be assumed to be uniform at 70°F. Also assume flat velocity profiles and ideal gases.

6-18 An orifice meter is used to measure the flow of water in a 1-in, schedule-40 steel pipe. The pressure drop over the orifice for determining the velocity is found using a manometer containing blue oil (S.G. = 1.75). At a flow rate of 2 gal/min the manometer shows a differential height of 23 in. What is the size of the orifice?

6-19 In a laboratory experiment on calibrating flow meters, the orifice-meter coefficient C_o for water at 20°C was found to be 0.715. Presumably this value should have been 0.63 as indicated on Fig. 6-5. The discrepancy was explained as possibly resulting from an air bubble in one of the manometer lines. Assuming the correct value of this orifice coefficient to be $C_o = 0.63$, what is the vertical length of an air bubble sufficient to cause the error observed and on which side of the manometer would the air bubble be located (i.e., high-pressure side or low-pressure side)?

Data:

$u_o = 6.72$ ft/s, $D_o = 0.553$ in, $D_1 = 0.772$ in

sp. gr. of manometer fluid = 1.75, $\Delta h_{obs} = 16.2$ in

7

THE DIFFERENTIAL MASS BALANCE

The overall balances have permitted the solution of many problems, but to advance further in our study we must investigate in more detail what goes on inside a control volume. The overall balances are powerful because they do not require knowledge of details inside the control volume, but this very characteristic prevents their use to study these details. The derivations of the differential balances will be similar to those for finite balances, but now the element always cuts through a single phase. In use we integrate a differential balance over a phase, but only up to a phase boundary, where we must know the boundary conditions in order to complete the solution. On the contrary, a typical applied control volume for the overall balances included parts of a solid wall, and flow was restricted to certain areas only.

Single-Component System

A mass balance will now be made over a differential element of volume in much the same way as it was made over a finite volume in Chap. 3.

Figure 7-1 shows the differential element of volume; we shall use as before the

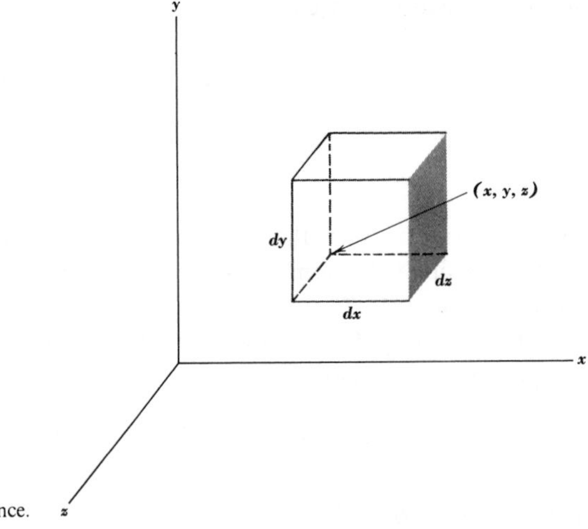

FIGURE 7-1
Differential element for mass balance.

fact that output minus input plus accumulation must equal zero. In the x direction, output less input by mass flow is given by

$$[u_x \rho + d(u_x \rho)]\, dy\, dz - u_x \rho\, dy\, dz \tag{7-1}$$

The second term is the input in the x direction through the face of area $dy\, dz$ located at a distance x from the plane $x = 0$; the first term is the output through the parallel face located at $x + dx$. The expression (7-1) reduces to

$$d(u_x \rho)\, dy\, dz \tag{7-2}$$

It will be convenient to express the differential change in $u_x \rho$ over the distance dx by

$$d(u_x \rho) = \frac{\partial(u_x \rho)}{\partial x}\, dx \tag{7-3}$$

As a result the output minus input in the x direction is

$$\frac{\partial(u_x \rho)}{\partial x}\, dx\, dy\, dz \tag{7-4}$$

Similar expressions can be written for flow in the y and z directions. The rate of accumulation in the element is

$$\frac{\partial \rho}{\partial \theta}\, dx\, dy\, dz \tag{7-5}$$

so that the mass balance becomes

$$\frac{\partial(u_x\rho)}{\partial x}dx\,dy\,dz + \frac{\partial(u_y\rho)}{\partial y}dy\,dx\,dz + \frac{\partial(u_z\rho)}{\partial z}dz\,dx\,dy + \frac{\partial\rho}{\partial\theta}dx\,dy\,dz = 0 \qquad (7\text{-}6)$$

The similarity between this equation and Eqs. (3-13) and (3-16) should be noticed. The first three terms of Eq. (7-6) are equivalent to $\Delta(u\rho A)$, that is, Δw, and the fourth term to $\partial M/\partial\theta$.

From Eq. (7-6) we now write the equation of continuity[1] for a single-component, unsteady-state system.

$$\frac{\partial(u_x\rho)}{\partial x} + \frac{\partial(u_y\rho)}{\partial y} + \frac{\partial(u_z\rho)}{\partial z} + \frac{\partial\rho}{\partial\theta} = 0 \qquad (7\text{-}7)$$

This equation may be expanded to give

$$\rho\left(\frac{\partial u_x}{\partial x} + \frac{\partial u_y}{\partial y} + \frac{\partial u_z}{\partial z}\right) + u_x\frac{\partial\rho}{\partial x} + u_y\frac{\partial\rho}{\partial y} + u_z\frac{\partial\rho}{\partial z} + \frac{\partial\rho}{\partial\theta} = 0 \qquad (7\text{-}8)$$

Because the density of the fluid ρ is a function of x, y, z, and θ; following the definition of the substantial derivative [see Eq. (4-7)], one may write

$$\frac{\partial u_x}{\partial x} + \frac{\partial u_y}{\partial y} + \frac{\partial u_z}{\partial z} + \frac{1}{\rho}\frac{D\rho}{D\theta} = 0 \qquad (7\text{-}9)$$

Let us now use the idea of the substantial derivative to try to give a physical significance to the first three terms of Eq. (7-9). Consider a fluid element of unit mass moving with the fluid; the volume and density of the element change with time. By differentiating the relation $\rho V = 1$ with respect to time, we get

$$\rho\frac{DV}{D\theta} + V\frac{D\rho}{D\theta} = 0 \qquad (7\text{-}10)$$

or

$$\frac{1}{V}\frac{DV}{D\theta} + \frac{1}{\rho}\frac{D\rho}{D\theta} = 0 \qquad (7\text{-}11)$$

A comparison of Eqs. (7-11) and (7-9) shows that $1/V(DV/D\theta)$ is equal to the sum of $\partial u_x/\partial x$, $\partial u_y/\partial y$, and $\partial u_z/\partial z$. The physical interpretation for this relationship is that the first term in Eq. (7-11), which represents the rate of volumetric strain or expansion of the element, is really the same as the sum of the first three terms of Eq.

[1] Although Eq. (7-7) is referred to as the equation of continuity in fluid mechanics, some engineers also give this name to the steady-state form of Eq. (3-16), that is, $\Delta(u\rho A) = 0$.

(7-9), which represent the three linear rates of strain. This sum is called the *divergence of the velocity vector.*

The analysis leading to Eq. (7-8), as well as that used in deriving the overall balances of Chaps. 3 to 5, is based on the consideration of the flow in and out of an element of fixed size, fixed in space, and containing a variable mass. This is often called the point of view of Euler. On the other hand, the analysis which leads to Eq. (7-11) is based on a fixed mass of fluid as it moves in space; the volume of the element may vary. This point of view is referred to as that of Lagrange. The Lagrangian analysis is convenient and will be used in the next two chapters to derive the differential energy and momentum balances.

The main use of the equation of continuity will be to simplify the differential heat and momentum equations, with which it must hold simultaneously in flow problems.

At steady state, Eq. (7-8) is easily modified by setting $\partial \rho / \partial \theta$ equal to zero. For the common case of an incompressible fluid, at steady or unsteady state, we have

$$\frac{\partial u_x}{\partial x} + \frac{\partial u_y}{\partial y} + \frac{\partial u_z}{\partial z} = 0 \qquad (7\text{-}12)$$

Example 7-1 When an incompressible viscous fluid flows over a flat surface, the velocity changes over a small boundary-layer region from zero at the surface to its free-stream velocity (see Fig. 5-4). Use the equation of continuity to show qualitatively the nature of the velocity components near the surface.

We assume that there is no velocity in the z direction (two-dimensional flow) and write Eq. (7-12) as

$$\frac{\partial u_x}{\partial x} + \frac{\partial u_y}{\partial y} = 0$$

where the x axis is along the plate in the direction of flow. Since the boundary layer thickens at greater distances along the plate, at a given small value of y the value of u_x must decrease from its free-stream velocity to a lower value inside the boundary layer. Thus, $\partial u_x/\partial x$ is negative and, by Eq. (1), $\partial u_y/\partial y$ is positive; within the boundary layer, therefore, there is a component of velocity away from the plate. ////

For this example the equation of continuity as developed in rectangular coordinates was convenient. For steady flow in a round tube, there is symmetry around the tube axis (z direction), and so all points at the same radius have the same velocity. Thus, if cylindrical coordinates are used, one independent variable (r) is needed instead of two (x and y). The equation of continuity in cylindrical and spherical coordinates is given in Table A-1.

Two-Component System

A differential mass balance can also be made for a component A in a binary mixture. In the overall balance of Chap. 3 we did not include the effect of diffusion; this time let us include it in the differential balance, for the equation thus obtained will prove very useful in the later study of mass transfer by diffusion.

The entry of component A into the differential element through face $dy\,dz$ is by two mechanisms. One is the contribution resulting from the general movement of the fluid, this flux being $u_x \rho_A$ lb/(h)(ft^2). The other is by molecular diffusion, which may be considered as a flux superimposed on the general movement of the fluid; this flux is written as J_{Ax} lb/(h)(ft^2). The total flux is the sum of the individual fluxes, $u_x \rho_A + J_{Ax}$. It follows that the total flux out of the opposite face of the element is $(u_x \rho_A + J_{Ax}) + d(u_x \rho_A + J_{Ax})$. The difference, which is output minus input in the x direction, is

$$d(u_x \rho_A + J_{Ax})\,dy\,dz = \left[\frac{\partial(u_x \rho_A)}{\partial x} + \frac{\partial J_{Ax}}{\partial x}\right] dx\,dy\,dz \qquad (7\text{-}13)$$

which is analogous to the expression (7-4).

It follows that the equation of continuity for component A in its three-dimensional form is then

$$\frac{\partial(u_x \rho_A)}{\partial x} + \frac{\partial(u_y \rho_A)}{\partial y} + \frac{\partial(u_z \rho_A)}{\partial z} + \frac{\partial \rho_A}{\partial \theta} + \frac{\partial J_{Ax}}{\partial x} + \frac{\partial J_{Ay}}{\partial y} + \frac{\partial J_{Az}}{\partial z} - r_A = 0 \qquad (7\text{-}14)$$

or

$$\rho_A\left(\frac{\partial u_x}{\partial x} + \frac{\partial u_y}{\partial y} + \frac{\partial u_z}{\partial z}\right) + \frac{D\rho_A}{D\theta} + \frac{\partial J_{Ax}}{\partial x} + \frac{\partial J_{Ay}}{\partial y} + \frac{\partial J_{Az}}{\partial z} - r_A = 0 \qquad (7\text{-}15)$$

The term r_A in these equations has been added, as was R_i in Eq. (3-18); it is the rate of generation of component A in lb/(h)(ft^3) within the element by chemical reaction. With the appropriate sign it may also represent the rate of consumption of component A. The term $\partial \rho_A/\partial \theta$ is the rate of accumulation of component A in lb/(h)(ft^3).

The fluxes J_A can arise in many ways, such as thermal diffusion, pressure diffusion, ionic diffusion, or ordinary molecular diffusion. For a case involving only binary molecular diffusion we usually write

$$J_{Ax} = -D_{AB}\frac{\partial \rho_A}{\partial x} \qquad (7\text{-}16)$$

Similar equations exist for the other two directions. If these equations are substituted into Eq. (7-14) and D_{AB} is constant, we obtain

$$\frac{\partial(u_x \rho_A)}{\partial x} + \frac{\partial(u_y \rho_A)}{\partial y} + \frac{\partial(u_z \rho_A)}{\partial z} + \frac{\partial \rho_A}{\partial \theta}$$

$$= D_{AB}\left(\frac{\partial^2 \rho_A}{\partial x^2} + \frac{\partial^2 \rho_A}{\partial y^2} + \frac{\partial^2 \rho_A}{\partial z^2}\right) + r_A \qquad (7\text{-}17)$$

The first three terms may be expanded to give

$$\rho_A\left(\frac{\partial u_x}{\partial x} + \frac{\partial u_y}{\partial y} + \frac{\partial u_z}{\partial z}\right) + u_x \frac{\partial \rho_A}{\partial x} + u_y \frac{\partial \rho_A}{\partial y} + u_z \frac{\partial \rho_A}{\partial z} + \frac{\partial \rho_A}{\partial \theta}$$

$$= D_{AB}\left(\frac{\partial^2 \rho_A}{\partial x^2} + \frac{\partial^2 \rho_A}{\partial y^2} + \frac{\partial^2 \rho_A}{\partial z^2}\right) + r_A \qquad (7\text{-}18)$$

If the density of the mixture is constant, $\partial u_x/\partial x + \partial u_y/\partial y + \partial u_z/\partial z$ equals zero, according to Eq. (7-12). Using the substantial derivative notation, Eq. (7-18) becomes

$$\frac{D\rho_A}{D\theta} = D_{AB}\left(\frac{\partial^2 \rho_A}{\partial x^2} + \frac{\partial^2 \rho_A}{\partial y^2} + \frac{\partial^2 \rho_A}{\partial z^2}\right) + r_A \qquad (7\text{-}19)$$

If diffusion and chemical reaction are important, ρ may be changing with distance. This effect can be particularly significant in a gaseous system, so that Eq. (7-19) may be an oversimplification. This situation is somewhat complicated, and so we shall reserve a detailed discussion of this matter for the part of the book which deals specifically with mass transfer. The study of mixtures of more than two components is rather difficult and will not be included. We defer until the chapters on mass transfer any discussion of the use of molar units in these equations. The differential mass balance for other coordinate systems appears in Table A-2.

PROBLEMS

7-1 Use a cylindrical shell of dimensions dz and dr to obtain the equation of continuity for flow in a circular pipe with axial symmetry. Assume ρ is constant. Could this equation be obtained from Eq. (7-8)?

7-2 Find the equation of continuity for component A in a binary mixture in a tubular reactor with axial symmetry. Use cylindrical coordinates. Assume ρ is constant.

7-3 Find the differential equation describing the concentration of component A (no reaction occurring) at any point in a spherical drop. The drop, which contains a binary mixture, is exposed to a medium of composition different from the drop, so that component A is diffusing at unsteady state from within the drop. Assume ρ is constant.

7-4 Hinze[1] studied the flow of suspensions of discrete solid particles in fluids and derived for such systems the general equations for continuity, energy, and momentum in cartesian coordinates. Derive the continuity equation using the following nomenclature:

u_{fx}, u_{fy}, u_{fz} = velocity components of fluid
u_{sx}, u_{sy}, u_{sz} = velocity components of solid particles
ε = volumetric fraction of solid
ρ_f, ρ_s = density of fluid and solid, respectively

All these quantities should be considered to be functions of x, y, z, θ, except for ρ_s.

7-5 A fluid flows into a circular tube through a bell-shaped entrance in such a way that its velocity distribution is uniform across the diameter of the tube at its entrance. From this point downstream, the velocity profile changes gradually to that for fully developed flow, such as a parabolic profile for laminar flow. Use the equation of continuity to show whether there is a radial component of velocity in the entrance region and in the fully developed region.

7-6 Flow from a porous structure into a vertical pipe (a well) is sometimes represented by $u_r = -c/r$; $u_\theta = u_z = 0$ in cylindrical coordinates. Show that the equation of continuity is satisfied.

7-7 Flow around a bluff body can be represented by

$$u_x = -\left(A + \frac{Cx}{x^2 + y^2}\right)$$

$$u_y = \frac{-Cy}{x^2 + y^2} \qquad u_z = 0$$

Show that the equation of continuity is satisfied.

7-8 Potential two-dimensional flow around a circular cylinder is given by

$$u_r = \cos\theta\left(\frac{C}{r^2} - D\right)$$

$$u_\theta = \sin\theta\left(\frac{C}{r^2} + D\right)$$

Is the equation of continuity satisfied?

7-9 A fluid is in steady, one-dimensional flow through a circular tube. A reactant A is disappearing according to the rate equation $r_A = -k\rho_A$. If we know that the term involving D_{AB} is small compared to the others, what is the sign of $\partial^2\rho_A/\partial z^2$, where z is the direction of flow?

[1] J. O. Hinze, *Appl. Sci. Res.*, **A11**:33 (1963).

8

THE DIFFERENTIAL ENERGY BALANCE

A differential energy balance could be written with reference to a constant-volume element fixed in space, as in the first part of Chap. 7. However, this time the point of view of Lagrange will be found simpler than that of Euler, so the element is chosen as a fixed mass moving with the fluid at the fluid velocity. The basis for any energy balance is the first law of thermodynamics, so we consider the internal energy change of the element as it flows along and exchanges heat and work with the surrounding fluid. This is the same point of view as that used in the discussion leading to Eqs. (4-24) and (4-25). Equation (4-25), written as the rate of change of internal energy of one pound of moving fluid, is

$$\frac{DU}{D\theta} = \frac{DQ}{D\theta} - p\frac{DV}{D\theta} + \frac{D(lw)}{D\theta} \tag{8-1}$$

The term $DQ/D\theta$ is the rate of heat flow to the surface of the pound of fluid from the surrounding stream with which it is moving. In order to analyze this term it is convenient to apply Eq. (8-1) to a cubic differential element of mass $\rho\, dx\, dy\, dz$. We define a loss term Φ, which equals $\rho D(lw)/D\theta$ and has units of Btu/(h)(ft^3). Hence we obtain the equation

$$\frac{DU}{D\theta}\rho\,dx\,dy\,dz = \rho\frac{DQ}{D\theta}dx\,dy\,dz - p\frac{DV}{D\theta}\rho\,dx\,dy\,dz + \Phi\,dx\,dy\,dz \tag{8-2}$$

It is now possible to relate $\rho(DQ/D\theta)\,dx\,dy\,dz$ to the heat fluxes across the faces of the cube. Since the heat term is positive for a net inflow, we write the inflow minus outflow as

$$\begin{aligned}\rho\frac{DQ}{D\theta}dx\,dy\,dz = \left(\frac{q}{A}\right)_x dy\,dz &- \left[\left(\frac{q}{A}\right)_x + d\left(\frac{q}{A}\right)_x\right]dy\,dz + \left(\frac{q}{A}\right)_y dx\,dz\\ &- \left[\left(\frac{q}{A}\right)_y + d\left(\frac{q}{A}\right)_y\right]dx\,dz + \left(\frac{q}{A}\right)_z dx\,dy\\ &- \left[\left(\frac{q}{A}\right)_z + d\left(\frac{q}{A}\right)_z\right]dx\,dy\end{aligned} \tag{8-3}$$

or

$$\rho\frac{DQ}{D\theta}dx\,dy\,dz = -\left[\frac{\partial(q/A)_x}{\partial x} + \frac{\partial(q/A)_y}{\partial y} + \frac{\partial(q/A)_z}{\partial z}\right]dx\,dy\,dz \tag{8-4}$$

By substituting Eq. (8-4) into Eq. (8-2) and rearranging, we obtain the differential energy balance in terms of the heat flux q/A.

$$\rho\frac{DU}{D\theta} + p\rho\frac{DV}{D\theta} = -\left[\frac{\partial(q/A)_x}{\partial x} + \frac{\partial(q/A)_y}{\partial y} + \frac{\partial(q/A)_z}{\partial z}\right] + \Phi \tag{8-5}$$

It is often convenient to replace the internal energy by the enthalpy H, where

$$H = U + pV \tag{8-6}$$

Taking the substantial derivative and multiplying by ρ, we get

$$\rho\frac{DH}{D\theta} = \rho\frac{DU}{D\theta} + p\rho\frac{DV}{D\theta} + \frac{Dp}{D\theta} \tag{8-7}$$

so that the differential energy balance (8-5) written in terms of the enthalpy is

$$\rho\frac{DH}{D\theta} - \frac{Dp}{D\theta} = -\left[\frac{\partial(q/A)_x}{\partial x} + \frac{\partial(q/A)_y}{\partial y} + \frac{\partial(q/A)_z}{\partial z}\right] + \Phi \tag{8-8}$$

The heat flux in laminar flow or for solids arises from heat flow by conduction, so that

$$\left(\frac{q}{A}\right)_x = -k\frac{\partial t}{\partial x} \tag{8-9}$$

Similar equations exist for the other two directions. If these equations are substituted into Eq. (8-8), they give, for constant k,

$$\rho \frac{DH}{D\theta} - \frac{Dp}{D\theta} = k\left(\frac{\partial^2 t}{\partial x^2} + \frac{\partial^2 t}{\partial y^2} + \frac{\partial^2 t}{\partial z^2}\right) + \Phi \qquad (8\text{-}10)$$

For viscous fluids Φ is the rate at which work is done per unit volume to change the shape of the element. A detailed discussion of the relation of this work to the shearing stresses and in turn to the viscosity and the deformation is given in more specialized texts.[1] It is related to the lost work of Chap. 4 in that it is the rate of work lost per unit volume because of friction within the fluid. It is also referred to as the *viscous dissipation term.* Although lw is essential in the mechanical energy balance for the calculation of pressure drop, Φ usually has a negligible effect on the temperature of the fluid and may be omitted from the energy balance except when it is used for highly viscous fluids or velocities of the order of the sonic velocity.

Starting from thermodynamics, it can be shown that for an incompressible fluid the following equation holds:

$$\rho \frac{DH}{D\theta} = \rho C_p \frac{Dt}{D\theta} + \frac{Dp}{D\theta}$$

This can be substituted in Eq. (8-10) to give

$$\frac{Dt}{D\theta} = u_x \frac{\partial t}{\partial x} + u_y \frac{\partial t}{\partial y} + u_z \frac{\partial t}{\partial z} + \frac{\partial t}{\partial \theta} = \frac{k}{\rho C_p}\left(\frac{\partial^2 t}{\partial x^2} + \frac{\partial^2 t}{\partial y^2} + \frac{\partial^2 t}{\partial z^2}\right) \qquad (8\text{-}11)$$

where Φ has been omitted. For a solid, the equation becomes

$$\frac{\partial t}{\partial \theta} = \frac{k}{\rho C_p}\left(\frac{\partial^2 t}{\partial x^2} + \frac{\partial^2 t}{\partial y^2} + \frac{\partial^2 t}{\partial z^2}\right) \qquad (8\text{-}12)$$

and its ultimate simplification for a solid at steady state is

$$\frac{\partial^2 t}{\partial x^2} + \frac{\partial^2 t}{\partial y^2} + \frac{\partial^2 t}{\partial z^2} = 0 \qquad (8\text{-}13)$$

This is Laplace's equation. The solution of the above equations for several important cases will be considered in detail in the section on heat transfer.

[1] H. Lamb, "Hydrodynamics," 6th ed., p. 579, Dover Publications, Inc., New York, 1945; H. Schlichting and Kestin, "Boundary Layer Theory," 6th ed., chap. 12, McGraw-Hill Book Company, New York, 1968.

If heat is being generated within the element of fluid by a chemical reaction, a simple way of modifying Eq. (8-11) is to add a term involving q_r, the rate of heat generation in Btu/(h)(ft^3).

$$\frac{Dt}{D\theta} = \frac{k}{\rho C_p}\left(\frac{\partial^2 t}{\partial x^2} + \frac{\partial^2 t}{\partial y^2} + \frac{\partial^2 t}{\partial z^2}\right) + \frac{q_r}{\rho C_p} \tag{8-14}$$

Equation (8-14), for an incompressible fluid, is very similar to Eq. (7-19), for mass transfer in a binary mixture.

$$\frac{D\rho_A}{D\theta} = D_{AB}\left(\frac{\partial^2 \rho_A}{\partial x^2} + \frac{\partial^2 \rho_A}{\partial y^2} + \frac{\partial^2 \rho_A}{\partial z^2}\right) + r_A \tag{7-19}$$

The analogy extends to $k/\rho C_p$, which has the same units as D_{AB}. Because of this similarity, $k/\rho C_p$ is often called the *thermal diffusivity*, as already mentioned at the end of Chap. 2. Having shown this analogy between mass and heat transfer, we shall proceed in the next chapter to develop similar equations for momentum transfer, for which similar analogies will become evident.

In presenting the basic differential equations for mass, heat, and momentum transfer in this portion of the book, we are leading up to the equations of motion of a fluid. These equations will then be used to solve a number of isothermal-fluid-flow problems. The equation of continuity for a single component must be simultaneously satisfied, as well as the equation of state of the fluid. As presented here, the equations increase in complexity in going from mass to heat to momentum transfer. Paradoxically, for fluid systems, the isothermal-momentum-transfer equation is the easiest to solve; for mass and heat transfer, the pertinent equations must in general be solved simultaneously with the momentum equation. The detailed applications of Eq. (7-17) and Eq. (8-10) will be presented later, after those of the equations of motion. The energy balance is given in Tables A-2 and A-3, in cylindrical and spherical coordinates.

Example 8-1 Problems in unsteady heat conduction can be analyzed in terms of Eq. (8-12). As an illustration let us consider heat transfer from the plane bottom of a low-temperature storage tank buried in the ground. The tank is initially empty and we assume that the temperature of the ground below it is constant at T_o. The tank is filled relatively quickly, so that the wall is suddenly cooled to T_1. The relation for T, as a function of y and θ, is desired. We consider points under the middle of the tank, so that temperature variation in a horizontal plane can be neglected. Equation (8-12) becomes

$$\frac{\partial T}{\partial \theta} = \alpha \frac{\partial^2 T}{\partial y^2} \tag{1}$$

where α is the thermal diffusivity $k/C_p\rho$. The boundary conditions are simplified by defining a dimensionless temperature Y:

$$Y = \frac{T - T_o}{T_1 - T_o} \qquad (2)$$

Then for $Y(\theta, y)$ we find

$$\begin{aligned} Y(0, y) &= 0 \\ Y(\theta, 0) &= 1 \qquad \theta > 0 \\ Y(\theta, \infty) &= 0 \qquad \theta > 0 \end{aligned}$$

The temperature disturbance, initially concentrated at $y = 0$, penetrates with time into the ground, considered infinite in extent; for a finite solid the problem would be complicated by the approach of the temperature wave to another boundary.

We shall learn in later chapters how to find a transformation of the variables to change Eq. (1) to an ordinary differential equation. For now we simply assert that this is achieved by the variable

$$\eta = \frac{y}{\sqrt{4\alpha\theta}} \qquad (3)$$

The reader should verify that this expression can be used to transform Eqs. (1) and (2) to

$$\frac{d^2Y}{d\eta^2} + 2\eta\frac{dY}{d\eta} = 0 \qquad (4)$$

with the boundary conditions on $Y(\eta)$

$$\begin{aligned} Y(0) &= 1 \\ Y(\infty) &= 0 \end{aligned}$$

The solution of this differential equation is

$$Y = \frac{T - T_o}{T_1 - T_o} = 1 - \operatorname{erf}\left(\frac{y}{\sqrt{4\alpha\theta}}\right) \qquad (5)$$

where erf (η) is the error function, a common tabulated function.[1] The method of solution of Eq. (5) is given in Example 19-1. ////

[1] M. Abramowitz and I. A. Stegun, "Handbook of Mathematical Functions," p. 310, Natl. Bur. Std., Washington, 1964.

PROBLEMS

8-1 Find the expression for the differential energy balance in cylindrical coordinates. Obtain first the equation analogous to Eq. (8-10) (k = const) and then its simplification for k, ρ, and C_p constant [analogous to Eq. (8-11)].

8-2 Find the equation analogous to Eq. (8-14) which would be convenient for a tubular reactor with axial symmetry and developed flow.

8-3 It is desired to study the time required to cool the center of a hot ball bearing to a certain level. Find the form of the energy balance which would be convenient for this unsteady-state heat-transfer problem.

9

THE DIFFERENTIAL MOMENTUM BALANCE

EQUATIONS IN TERMS OF THE SHEAR-STRESS COMPONENTS

We now proceed to the derivation of the equations of motion for a real fluid; these equations, which are based on Newton's second law, will permit the determination of the way the velocity varies with position. It will then be possible to find quantities such as the pressure drop in laminar flow. With certain modifications, the equations can also be used for turbulent flow, but this application will be discussed separately; in what follows the flow is considered laminar.

Newton's Second Law Applied to an Element of Fluid

The force-momentum balances apply in the three coordinate directions; the equation in the x direction is

$$\check{F}_x = \frac{Ma_x}{g_c} \qquad (9\text{-}1)$$

which can be written in terms of rate of change of momentum as

$$\check{F}_x = \frac{1}{g_c}\frac{d(Mu_x)}{d\theta} \tag{9-2}$$

We apply this equation to a fluid element of constant mass moving with velocity of the fluid. This is the point of view of Lagrange, so for the same reasons given in Chap. 8, the derivative in Eq. (9-2) is written as a substantial derivative.

$$\check{F}_x = \frac{M}{g_c}\frac{Du_x}{D\theta} \tag{9-3}$$

If the element of fluid is a differential mass $\rho\, dx\, dy\, dz$, Eq. (9-3) becomes

$$d\check{F}_x = \frac{1}{g_c}\rho\, dx\, dy\, dz \frac{Du_x}{D\theta} \tag{9-4}$$

The Body Force

The force $d\check{F}_x$ represents the resultant of all the forces acting in the x direction on the element. It is convenient to divide it into a body force X, expressed in lb_f/lb, and the forces arising from the mechanical stresses on the surface of the element. A body force acts on the element as a whole; the force on a charged body in an electric field is an example. For the differential element, the body force in the x direction is

$$d\check{F}_{xB} = X\rho\, dx\, dy\, dz \tag{9-5}$$

The only body force we consider is that caused by gravity. It can be written as

$$X = \frac{g}{g_c}\cos\beta = F_{xg} \tag{9-6}$$

in which β is the angle between the x direction and the direction in which gravity acts.

The Stresses on an Element of Fluid

Each of the six faces of the cube will be subjected to mechanical stresses arising from the adjacent fluid, and each of these stresses can in turn be resolved into three components, parallel to the three axes. Figure 9-1 shows the stress components acting in the x direction, and Fig. 9-2 illustrates the three stress components operating on a given face. These figures show the system used for the subscripts on τ. The first subscript indicates the face on which the stress is acting by giving the axis perpendicular to the face, and the second subscript gives the direction in which

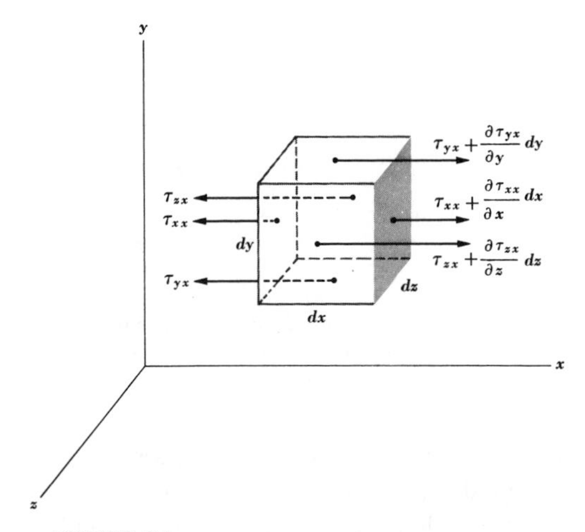

FIGURE 9-1
Element of fluid showing stresses acting in the x direction.

the stress acts. The stresses with the mixed subscripts, such as τ_{xy}, are shear stresses, tending to deform the element and change the angles between the faces. The convention for the positive direction of the shear stresses is illustrated in Fig. 9-3. The stresses with repeated subscripts, such as τ_{xx}, are normal stresses, which are closely related to the hydrostatic pressure. As is customary in mechanics, normal stresses are considered positive for tension. If the size of the elements illustrated by Figs. 9-1

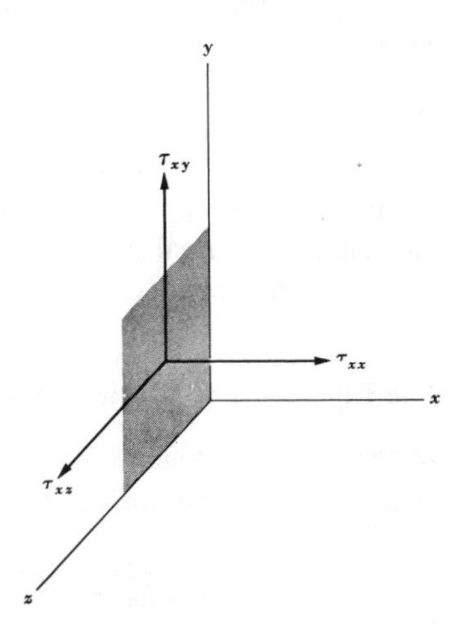

FIGURE 9-2
Stresses on a single face of fluid element.

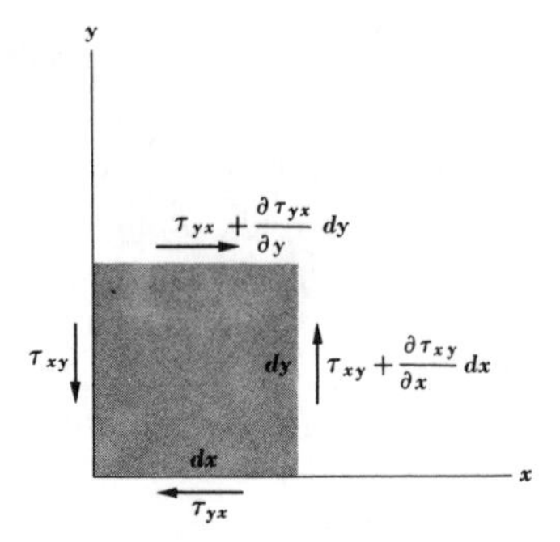

FIGURE 9-3
Shearing stresses causing moments about an axis parallel to the z axis.

and 9-3 approaches zero, the stresses on the opposite faces become identical in magnitude and opposite in sign. Consequently, the state of stress at a point can be completely described by the six components of shear stress and the three components of normal stress.

It can be shown that not all the shear-stress terms are independent by considering the possibility of rotation of the element. For example, rotation about an axis parallel to the z axis may be caused by a torque produced by the stresses shown in Fig. 9-3. From the principles of mechanics, the torque equals the moment of inertia multiplied by the angular acceleration. This can be shown[1] to lead to the equation

$$\tau_{xy} - \tau_{yx} = \rho \text{ (radius of gyration)}^2 \text{ (angular acceleration)}$$

If the size of the fluid element approaches zero, the radius of gyration goes to zero, so that τ_{xy} must equal τ_{yx}. Similar procedures show that $\tau_{xz} = \tau_{zx}$ and $\tau_{yz} = \tau_{zy}$.

The normal stresses are related to the pressure by the following arbitrary definition:

$$-p = \tfrac{1}{3}(\tau_{xx} + \tau_{yy} + \tau_{zz}) \qquad (9\text{-}7)$$

Since the normal stresses are positive for tension, and the pressure is positive for compression, the minus sign is necessary. If the velocity is not everywhere the same, the individual normal stresses will differ among themselves, and the absolute value of no one of them need equal the pressure.

The Equations of Motion in Terms of the Stresses

We refer again to Fig. 9-1 and write the net force in the x direction resulting from the variation of the appropriate stresses with position.

[1] J. G. Knudsen and D. L. Katz, "Fluid Dynamics and Heat Transfer," p. 31, McGraw-Hill Book Company, New York, 1958.

$$d\check{F}_{x,\,\text{stresses}} = \left(\tau_{xx} + \frac{\partial \tau_{xx}}{\partial x}\,dx\right) dy\,dz + \left(\tau_{yx} + \frac{\partial \tau_{yx}}{\partial y}\,dy\right) dx\,dz$$
$$+ \left(\tau_{zx} + \frac{\partial \tau_{zx}}{\partial z}\,dz\right) dx\,dy - \tau_{xx}\,dy\,dz - \tau_{yx}\,dx\,dz$$
$$- \tau_{zx}\,dx\,dy \qquad (9\text{-}8)$$

This equation simplifies to

$$d\check{F}_{x,\,\text{stresses}} = \left(\frac{\partial \tau_{xx}}{\partial x} + \frac{\partial \tau_{yx}}{\partial y} + \frac{\partial \tau_{zx}}{\partial z}\right) dx\,dy\,dz \qquad (9\text{-}9)$$

The force arising from the stresses, Eq. (9-9), and the body force, Eq. (9-5), are the only two forces acting in the x direction. The term $d\check{F}_x$ in Eq. (9-4) is the sum of the above two forces, so the combination of Eqs. (9-4), (9-5), (9-6), and (9-9) gives

$$\frac{\rho}{g_c}\frac{Du_x}{D\theta} = \rho\frac{g}{g_c}\cos\beta + \frac{\partial \tau_{xx}}{\partial x} + \frac{\partial \tau_{yx}}{\partial y} + \frac{\partial \tau_{zx}}{\partial z} \qquad (9\text{-}10)$$

Equation (9-10) is given in terms of the stresses; the reader will recall that these stresses may also be viewed as fluxes of x momentum. The term $(\partial\tau_{xx}/\partial x)(dx)(g_c)$ represents the net flux of x momentum out through the two planes of the cubical element perpendicular to the x axis; the term $(\partial\tau_{yx}/\partial y)(dy)(g_c)$ gives the net flux of x momentum out through the planes of the element perpendicular to the y axis; and $(\partial\tau_{zx}/\partial z)(dz)(g_c)$ has a similar meaning. The last three terms of Eq. (9-10) are comparable with three similar terms in Eq. (7-15) (fluxes of mass) and in Eq. (8-8) (fluxes of energy).

In addition to Eq. (9-10), there are similar equations for the y and z directions. Let us write all three equations for future reference.

$$\frac{\rho}{g_c}\frac{Du_x}{D\theta} = \rho X + \frac{\partial \tau_{xx}}{\partial x} + \frac{\partial \tau_{yx}}{\partial y} + \frac{\partial \tau_{zx}}{\partial z} \qquad (9\text{-}11)$$

$$\frac{\rho}{g_c}\frac{Du_y}{D\theta} = \rho Y + \frac{\partial \tau_{xy}}{\partial x} + \frac{\partial \tau_{yy}}{\partial y} + \frac{\partial \tau_{zy}}{\partial z} \qquad (9\text{-}12)$$

$$\frac{\rho}{g_c}\frac{Du_z}{D\theta} = \rho Z + \frac{\partial \tau_{xz}}{\partial x} + \frac{\partial \tau_{yz}}{\partial y} + \frac{\partial \tau_{zz}}{\partial z} \qquad (9\text{-}13)$$

The next step is to obtain an equation which can be used to calculate a velocity distribution or a pressure drop. To do this we relate the stresses to the viscosity and the rate of deformation of the element. This deformation rate will be expressed in terms of the derivatives of the velocity components of the flowing fluid. If the fluid is nonnewtonian, Eq. (9-10) holds, but modification would need to be made in what follows in relating the stresses and strain rates through the ordinary viscosity.

THE STRESSES IN RELATION TO THE VISCOSITY

The relations among the stresses and the various rates of deformation are quite complex for the three-dimensional case. In particular, an understanding of the role of the normal viscous stress requires considerable effort. The treatment given here is a bit roundabout in order to avoid using vectors and tensors; a derivation making use of cartesian tensors is given in the appendix. It is possible to obtain the equation to be given below by an extension of the stress-strain relation for solids;[1] however, we give rather a development which applies directly to fluids.[2]

Shear Stresses

Elastic solids deform according to Hooke's law, which states that stress is proportional to strain, that is, to the amount of deformation; but for newtonian fluids, the stress is proportional to the rate of deformation. The rate of deformation is sometimes called the *rate of shear*. The defining equation for viscosity in terms of rate of shear has been shown in Chap. 2 in a one-dimensional analysis to be

$$\tau = \frac{\mu}{g_c}\frac{du_x}{dy} \tag{9-14}$$

For more complicated deformations it is convenient to define the rate of deformation as the rate of change of the angle ϕ, shown in Fig. 9-4. For the simple one-dimensional case, the change in the angle ϕ in time $d\theta$ can be expressed in radians, as the arc $(du_x/dy)\,dy\,d\theta$ divided by the radius dy:

$$d\phi = -\frac{(du_x/dy)\,dy\,d\theta}{dy} \tag{9-15}$$

The rate of deformation, therefore, is

$$\frac{d\phi}{d\theta} = -\frac{du_x}{dy} \tag{9-16}$$

which is substituted in Eq. (9-14) to give

$$\tau = -\frac{\mu}{g_c}\frac{d\phi}{d\theta} \tag{9-17}$$

[1] J. W. Daily and D. R. F. Harleman, "Fluid Dynamics," Addison-Wesley Publishing Company, Inc., Reading, Mass., 1966.

[2] This treatment follows that of J. M. Kay, "An Introduction to Fluid Mechanics and Heat Transfer," appendix 4, Cambridge University Press, London, 1957.

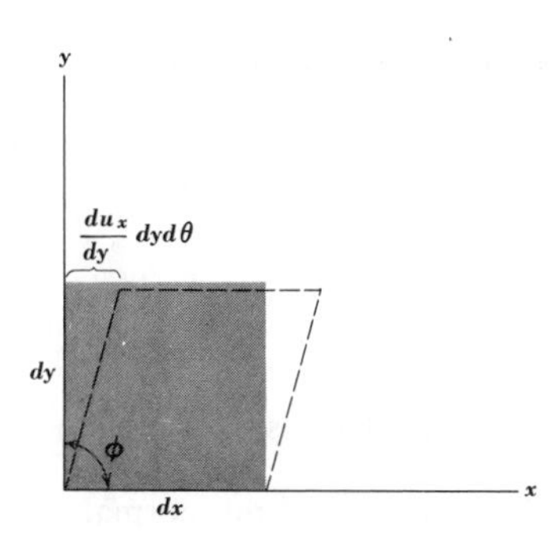

FIGURE 9-4
Deformation caused by a shear in a one-dimensional flow.

Equation (9-17) is now used in the analysis of the deformation of the element $dx\, dy\, dz$. Figure 9-5 shows the shear-stress components which act on the faces perpendicular to the xy plane to produce the illustrated deformation. The change in the angle ϕ consists of two portions, as shown in Fig. 9-5*b*: one part is $(-\partial u_x/\partial y)\, d\theta$, as derived by reasoning similar to that leading to Eq. (9-5); the other part is $(-\partial u_y/\partial x)\, d\theta$. Therefore we obtain

$$\frac{d\phi}{d\theta} = -\left(\frac{\partial u_x}{\partial y} + \frac{\partial u_y}{\partial x}\right) \tag{9-18}$$

Substitution of this into Eq. (9-17) gives

$$\tau_{xy} = \tau_{yx} = \frac{\mu}{g_c}\left(\frac{\partial u_x}{\partial y} + \frac{\partial u_y}{\partial x}\right) \tag{9-19}$$

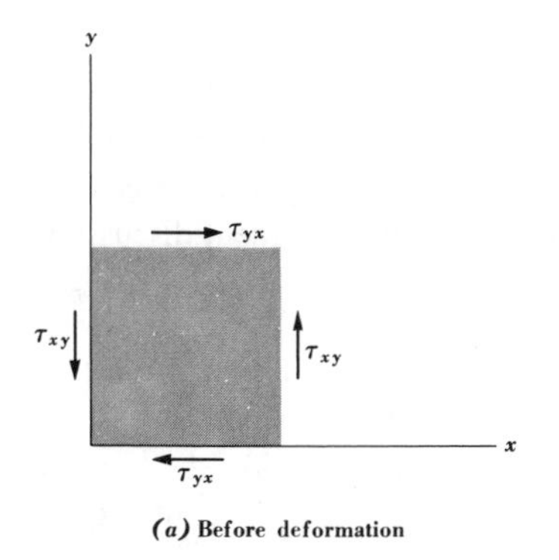

(*a*) Before deformation

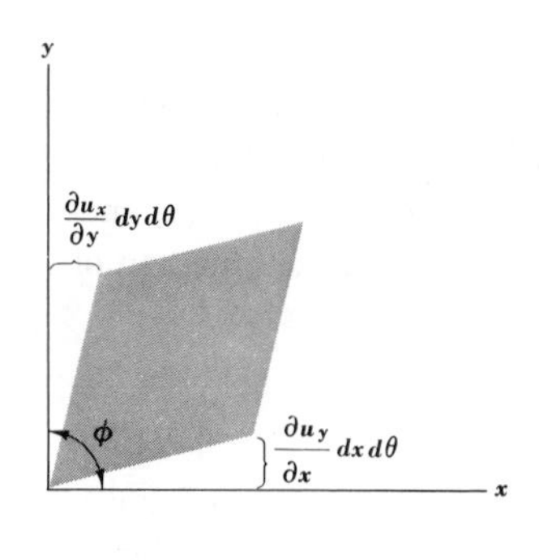

(*b*) After time $d\theta$

FIGURE 9-5
Deformation caused by shear stresses.

In a similar way, the other shear stresses are found to be

$$\tau_{yz} = \tau_{zy} = \frac{\mu}{g_c}\left(\frac{\partial u_z}{\partial y} + \frac{\partial u_y}{\partial z}\right) \tag{9-20}$$

$$\tau_{zx} = \tau_{xz} = \frac{\mu}{g_c}\left(\frac{\partial u_x}{\partial z} + \frac{\partial u_z}{\partial x}\right) \tag{9-21}$$

Normal Stresses

The problem of the normal stresses is somewhat more involved. First of all, by definition, we have

$$p = -\tfrac{1}{3}(\tau_{xx} + \tau_{yy} + \tau_{zz}) \tag{9-7}$$

For the specific case of a fluid at rest or in motion such that the velocity is everywhere the same, each normal stress is numerically equal to the pressure. In general, the normal stress is visualized as made up of a contribution from the pressure and a contribution from the viscous stresses which are associated with the linear deformation of the element in the direction of the normal stress. We state this idea mathematically for the x direction as

$$\tau_{xx} = -p + \sigma_x \tag{9-22}$$

We have already defined p in Eq. (9-7). This is combined with Eq. (9-22) as follows:

$$\sigma_x = \tau_{xx} + p \tag{9-23}$$

$$\sigma_x = \tau_{xx} - \tfrac{1}{3}(\tau_{xx} + \tau_{yy} + \tau_{zz}) \tag{9-24}$$

$$\sigma_x = \tfrac{2}{3}\tau_{xx} - \tfrac{1}{3}(\tau_{yy} + \tau_{zz}) \tag{9-25}$$

It will prove convenient to divide the normal viscous stress σ_x into two parts, so we arrange Eq. (9-25) into the form

$$\sigma_x = \tfrac{1}{3}(\tau_{xx} - \tau_{yy}) - \tfrac{1}{3}(\tau_{zz} - \tau_{xx}) \tag{9-26}$$

Similar equations can be derived for σ_y and σ_z. The quantity σ_x is substituted in Eq. (9-22) to give (9-27). Similar substitutions produce Eqs. (9-28) and (9-29). The designations (*a*) to (*d*) under the various groups in these three equations will be used as a convenient method of identification in subsequent discussions.

$$\tau_{xx} = \underset{(a)}{-p} + \underset{(b)}{\tfrac{1}{3}(\tau_{xx} - \tau_{yy})} - \underset{(c)}{\tfrac{1}{3}(\tau_{zz} - \tau_{xx})} \tag{9-27}$$

$$\tau_{yy} = \underset{(a)}{-p} - \underset{(b)}{\tfrac{1}{3}(\tau_{xx} - \tau_{yy})} + \underset{(d)}{\tfrac{1}{3}(\tau_{yy} - \tau_{zz})} \tag{9-28}$$

$$\tau_{zz} = \underset{(a)}{-p} + \underset{(c)}{\tfrac{1}{3}(\tau_{zz} - \tau_{xx})} - \underset{(d)}{\tfrac{1}{3}(\tau_{yy} - \tau_{zz})} \tag{9-29}$$

The effect of the normal stress τ_{xx} is to cause a deformation in the x direction. It will be assumed that the deformation rate in the x direction $\partial u_x/\partial x$ can be evaluated by adding the effects of the terms from the three equations for normal stress.

The contribution to $\partial u_x/\partial x$ by the pressure term is considered first; this is written as $(\partial u_x/\partial x)_a$. The variation of the pressure $Dp/D\theta$ causes a change in the density of the element; the rate of change is expressed by the equation of continuity:

$$-\frac{1}{\rho}\left(\frac{D\rho}{D\theta}\right) = \frac{\partial u_x}{\partial x} + \frac{\partial u_y}{\partial y} + \frac{\partial u_z}{\partial z} \tag{9-30}$$

The left-hand side of this equation is also equal to the rate of volumetric deformation (strain) $1/V(DV/D\theta)$, as was shown in Eq. (7-11). Since the pressure is equal in all directions, the contribution of pressure to the rate of deformation in the x direction is given by

$$\left(\frac{\partial u_x}{\partial x}\right)_a = \frac{1}{3}\frac{1}{V}\frac{DV}{D\theta} = \frac{1}{3}\left(\frac{\partial u_x}{\partial x} + \frac{\partial u_y}{\partial y} + \frac{\partial u_z}{\partial z}\right) \tag{9-31}$$

The other stress terms (b), (c), and (d) in Eqs. (9-27) to (9-29) will be examined in pairs chosen in such a way that the element can be considered as subjected to pure shear, a case which is easy to analyze. In pure shear an element experiencing a compression in one direction equal in magnitude to a tension in the perpendicular direction behaves as though it were subjected to pure shear stresses on planes inclined at 45° to the axes. There are no normal stresses on these planes, and the shear stresses are equal in magnitude to the applied normal stresses on the outside of the element.[1]

We choose first the terms labeled (b) in Eqs. (9-27) and (9-28); these stresses act as shown in Fig. 9-6 with positive stresses in the x direction (tension) and negative stresses in the y direction (compression). The shear stress τ_b is also proportional to the rate of angular deformation.

$$\tau_b = -\frac{\mu}{g_c}\frac{d\phi}{d\theta} = \tfrac{1}{3}(\tau_{xx} - \tau_{yy}) \tag{9-32}$$

The quantity $d\phi/d\theta$ is to be found in terms of $(\partial u_x/\partial x)_b$, the contribution to $\partial u_x/\partial x$ of the stress term under study. Since there are no normal stresses acting on the element of side h of Fig. 9-6, the sides of this element will not change length with deformation. Figure 9-6 shows that the angle $\phi/2$ can be expressed by the relation

$$\tan\frac{\phi}{2} = \frac{\lambda_y}{\lambda_x} \tag{9-33}$$

[1] These concepts are explained more fully in standard texts in mechanics. See, for example, S. Timoshenko, "Strength of Materials," 3rd ed., chap. 2, D. Van Nostrand Company, Inc., Princeton, N.J., 1955.

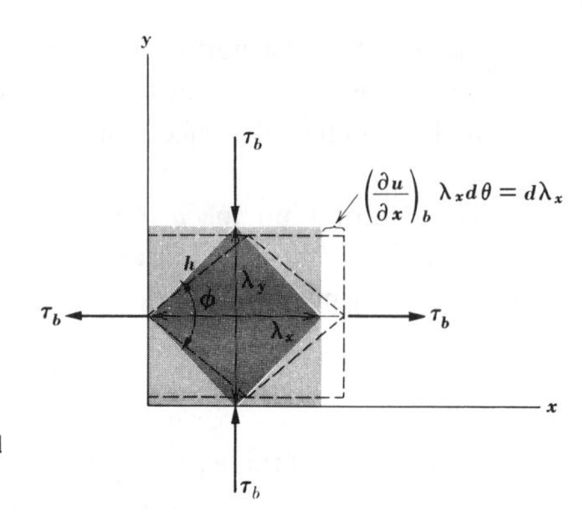

FIGURE 9-6
A contribution to the deformation caused by normal stresses (I).

The fact that the sides h are of constant length yields

$$\lambda_y{}^2 + \lambda_x{}^2 = \text{const} \tag{9-34}$$

and since λ_y approximately equals λ_x, it follows that

$$d\lambda_y = -d\lambda_x \tag{9-35}$$

We next write Eq. (9-33) in differential form.

$$\tfrac{1}{2}\sec^2\left(\frac{\phi}{2}\right)d\phi = \frac{\lambda_x\,d\lambda_y - \lambda_y\,d\lambda_x}{\lambda_x{}^2} \tag{9-36}$$

We substitute $-d\lambda_x$ for $d\lambda_y$ and use the fact that λ_x and λ_y differ only by a differential amount to obtain

$$\tfrac{1}{2}\sec^2\left(\frac{\phi}{2}\right)d\phi = \frac{-2d\lambda_x}{\lambda_x} \tag{9-37}$$

We then divide Eq. (9-37) by $d\theta$ and set $\phi = 90°$ to get

$$\frac{d\phi}{d\theta} = \frac{-2}{\lambda_x}\frac{d\lambda_x}{d\theta} \tag{9-38}$$

From Fig. 9-6 we notice that $d\lambda_x$ equals $(\partial u_x/\partial x)_b\,\lambda_x\,d\theta$, and there results

$$\frac{d\phi}{d\theta} = -2\left(\frac{\partial u_x}{\partial x}\right)_b \tag{9-39}$$

The desired stress is then found by combining Eqs. (9-32) and (9-39),

$$\tau_b = \frac{2\mu}{g_c}\left(\frac{\partial u_x}{\partial x}\right)_b \tag{9-40}$$

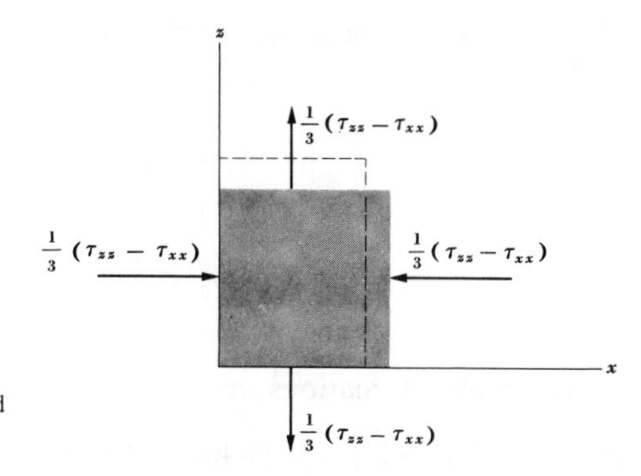

FIGURE 9-7
A contribution to the deformation caused by normal stresses (II).

and we obtain finally

$$\left(\frac{\partial u_x}{\partial x}\right)_b = \frac{\tau_b g_c}{2\mu} \tag{9-41}$$

or

$$\left(\frac{\partial u_x}{\partial x}\right)_b = \frac{g_c}{2\mu}\frac{\tau_{xx} - \tau_{yy}}{3} \tag{9-42}$$

We have now considered two of the three stress components in Eqs. (9-27) to (9-29). In the third case we consider the element again in pure shear, as shown in Fig. 9-7. By reasoning similar to that leading to $(\partial u_x/\partial x)_b$ we find

$$\left(\frac{\partial u_x}{\partial x}\right)_c = \frac{g_c}{2\mu}\frac{\tau_{xx} - \tau_{zz}}{3} \tag{9-43}$$

The remaining two stresses in Eqs. (9-27) to (9-29), labeled (*d*), do not cause any distortion in the x direction; that is, $(\partial u_x/\partial x)_d = 0$.

The sum of the deformation rates considered above gives the rate of deformation in the x direction caused by the normal stresses.

$$\frac{\partial u_x}{\partial x} = \frac{1}{3}\left(\frac{\partial u_x}{\partial x} + \frac{\partial u_y}{\partial y} + \frac{\partial u_z}{\partial z}\right) + \frac{g_c}{2\mu}\frac{\tau_{xx} - \tau_{yy}}{3} + \frac{g_c}{2\mu}\frac{\tau_{xx} - \tau_{zz}}{3} \tag{9-44}$$

or

$$\frac{\partial u_x}{\partial x} = \frac{1}{3}\left(\frac{\partial u_x}{\partial x} + \frac{\partial u_y}{\partial y} + \frac{\partial u_z}{\partial z}\right) + \frac{\tau_{xx} g_c}{2\mu} - \frac{g_c}{2\mu}\frac{\tau_{xx} + \tau_{yy} + \tau_{zz}}{3} \tag{9-45}$$

Using the definition of the pressure [Eq. (9-7)] and solving for τ_{xx}, we obtain

$$\tau_{xx} = -p + \frac{2\mu}{g_c}\frac{\partial u_x}{\partial x} - \frac{2}{3}\frac{\mu}{g_c}\left(\frac{\partial u_x}{\partial x} + \frac{\partial u_y}{\partial y} + \frac{\partial u_z}{\partial z}\right) \tag{9-46}$$

The equations for the stresses for the other two directions can be derived in a similar manner.

$$\tau_{yy} = -p + \frac{2\mu}{g_c}\frac{\partial u_y}{\partial y} - \frac{2}{3}\frac{\mu}{g_c}\left(\frac{\partial u_x}{\partial x} + \frac{\partial u_y}{\partial y} + \frac{\partial u_z}{\partial z}\right) \tag{9-47}$$

$$\tau_{zz} = -p + \frac{2\mu}{g_c}\frac{\partial u_z}{\partial z} - \frac{2}{3}\frac{\mu}{g_c}\left(\frac{\partial u_x}{\partial x} + \frac{\partial u_y}{\partial y} + \frac{\partial u_z}{\partial z}\right) \tag{9-48}$$

Navier-Stokes Equations

If Eqs. (9-19), (9-21), and (9-46) for the three stress components directed in the x direction are substituted into Eq. (9-11), we obtain the complete equation of motion for the x direction.

$$\frac{\rho}{g_c}\frac{Du_x}{D\theta} = \rho X - \frac{\partial p}{\partial x} + \frac{2\mu}{g_c}\frac{\partial^2 u_x}{\partial x^2} - \frac{2}{3}\frac{\mu}{g_c}\left(\frac{\partial^2 u_x}{\partial x^2} + \frac{\partial^2 u_y}{\partial x\,\partial y} + \frac{\partial^2 u_z}{\partial x\,\partial z}\right)$$
$$+ \frac{\mu}{g_c}\left(\frac{\partial^2 u_x}{\partial y^2} + \frac{\partial^2 u_y}{\partial y\,\partial x}\right) + \frac{\mu}{g_c}\left(\frac{\partial^2 u_x}{\partial z^2} + \frac{\partial^2 u_z}{\partial z\,\partial x}\right) \tag{9-49}$$

Rearranging, we have

$$\frac{\rho}{g_c}\frac{Du_x}{D\theta} = \rho X - \frac{\partial p}{\partial x} + \frac{\mu}{g_c}\left(\frac{\partial^2 u_x}{\partial x^2} + \frac{\partial^2 u_x}{\partial y^2} + \frac{\partial^2 u_x}{\partial z^2}\right)$$
$$+ \frac{1}{3}\frac{\mu}{g_c}\frac{\partial}{\partial x}\left(\frac{\partial u_x}{\partial x} + \frac{\partial u_y}{\partial y} + \frac{\partial u_z}{\partial z}\right) \tag{9-50}$$

Similar equations exist for the y and z directions. These three equations are called the *Navier-Stokes* equations, or the equations of motion; they are the basis of a large body of useful results obtained in fluid mechanics.

For incompressible flow, with which we shall deal for the most part, the equation of continuity gives

$$\frac{\partial u_x}{\partial x} + \frac{\partial u_y}{\partial y} + \frac{\partial u_z}{\partial z} = 0 \tag{9-51}$$

and so the three Navier-Stokes equations for this case are

$$u_x\frac{\partial u_x}{\partial x} + u_y\frac{\partial u_x}{\partial y} + u_z\frac{\partial u_x}{\partial z} + \frac{\partial u_x}{\partial \theta}$$
$$= g_c X - \frac{g_c}{\rho}\frac{\partial p}{\partial x} + \nu\left(\frac{\partial^2 u_x}{\partial x^2} + \frac{\partial^2 u_x}{\partial y^2} + \frac{\partial^2 u_x}{\partial z^2}\right) \tag{9-52}$$

$$u_x \frac{\partial u_y}{\partial x} + u_y \frac{\partial u_y}{\partial y} + u_z \frac{\partial u_y}{\partial z} + \frac{\partial u_y}{\partial \theta} = g_c Y - \frac{g_c}{\rho} \frac{\partial p}{\partial y} + \nu \left(\frac{\partial^2 u_y}{\partial x^2} + \frac{\partial^2 u_y}{\partial y^2} + \frac{\partial^2 u_y}{\partial z^2} \right) \tag{9-53}$$

$$u_x \frac{\partial u_z}{\partial x} + u_y \frac{\partial u_z}{\partial y} + u_z \frac{\partial u_z}{\partial z} + \frac{\partial u_z}{\partial \theta} = g_c Z - \frac{g_c}{\rho} \frac{\partial p}{\partial z} + \nu \left(\frac{\partial^2 u_z}{\partial x^2} + \frac{\partial^2 u_z}{\partial y^2} + \frac{\partial^2 u_z}{\partial z^2} \right) \tag{9-54}$$

These equations have a formidable appearance, which arises mostly from their being written in a general form applicable to a three-dimensional system lacking symmetry. Fortunately, many engineering problems can be solved by applying the equations in one- or two-dimensional form, as will be shown in the next chapters. As given above, Eqs. (9-52) to (9-54) are applicable to the laminar flow of a fluid of constant density and viscosity. The modifications required to adapt them for use with turbulent flow will be studied in Chap. 11. The equations of motion in other coordinate systems appear in Table A-4 of the appendix.

10

SOME SOLUTIONS OF THE EQUATIONS OF MOTION

In Chap. 2 we wrote a force-momentum balance for a finite, cylindrical element of fluid in laminar flow in a cylindrical pipe. This analysis led to the Hagen-Poiseuille equation. In Chap. 9 we wrote a force balance for a differential element of fluid in a general situation; the shape of the conduit or immersed body was not specified. The result of this analysis was a set of nonlinear partial differential equations, the Navier-Stokes equations. Because of the mathematical complexity of these equations, exact solutions have been found for only relatively simple cases, for which many terms can be set equal to zero. Sometimes the whole problem is reduced to the solution of one equation rather than several equations simultaneously. This simplification can be made for laminar flow in a circular pipe; later in this chapter we shall use this procedure to obtain the equation for the parabolic velocity distribution from the Navier-Stokes equations.

After considering a few exact solutions of the Navier-Stokes equations, we shall study some approximations which simplify the differential equations and make it possible to solve them analytically. Terms will be left out which are not exactly zero, but which are small in comparison with the terms retained. The approximations we shall consider are called *creeping flow*, *potential flow*, and (in Chap. 11) *boundary-layer flow*.

Laminar Parallel Flow between Flat Walls

One of the simplest types of fluid flow is the steady, laminar flow of an incompressible fluid between flat, parallel walls of infinite width, at a point far from the inlet or outlet of the channel. If the x axis is taken in the direction of flow, u_y and u_z are zero, and the equation of continuity gives

$$\frac{\partial u_x}{\partial x} = 0 \qquad (10\text{-}1)$$

Of the Navier-Stokes equations, we consider first (9-52); since $\partial u_x/\partial \theta = 0$, $u_y = 0$, $u_z = 0$, and $\partial u_x/\partial x = 0$, Eq. (9-52) reduces to

$$\frac{g_c}{\rho}\frac{\partial p}{\partial x} = g_c X + \nu\left(\frac{\partial^2 u_x}{\partial y^2} + \frac{\partial^2 u_x}{\partial z^2}\right) \qquad (10\text{-}2)$$

or

$$\frac{\partial p}{\partial x} = \rho X + \frac{\mu}{g_c}\left(\frac{\partial^2 u_x}{\partial y^2} + \frac{\partial^2 u_x}{\partial z^2}\right) \qquad (10\text{-}3)$$

If the channel is horizontal, the body force X is zero [Eq. (9-6)]. Figure 10-1 shows the orientation of the axes. The conduit extends indefinitely in the z direction, and its width is $2y_o$, as shown. There is therefore no change of u_x with z, and so $\partial^2 u_x/\partial z^2$ is zero, changing Eq. (10-3) to

$$\frac{\partial p}{\partial x} = \frac{\mu}{g_c}\frac{\partial^2 u_x}{\partial y^2} \qquad (10\text{-}4)$$

We now consider briefly the remaining two Navier-Stokes equations. In Eq. (9-54), the body force Z is zero, as well as all the other terms involving u_z, so that $\partial p/\partial z$ is zero. If $2y_o$ is presumed to be small, Y can also be taken as negligible, and Eq. (9-53) reduces to $\partial p/\partial y = 0$. Thus p is independent of y and z, and $\partial p/\partial x$ can be written dp/dx. Since u_x, and thus $\partial^2 u_x/\partial y^2$, is not a function of x, we can say finally that dp/dx is a constant everywhere in the parts of the conduit subject to the condition of this problem. Thus the final differential equation to be solved is

$$\frac{d^2 u_x}{dy^2} = \frac{g_c}{\mu}\frac{dp}{dx} = \text{const} \qquad (10\text{-}5)$$

The total pressure in the above equation for flow in a slit can be considered as the sum of the hydrostatic pressure (that pressure which would exist if there were no flow) and a quantity which is called the dynamic pressure. It is really the gradient of the dynamic pressure which causes flow; in a fluid at rest the total pressure gradient equals the hydrostatic pressure gradient, and the gradient of dynamic pressure is zero. If the above definition of the total pressure is inserted in the Navier-Stokes equations (9-52) to (9-54), it is found that the body-force terms disappear and that

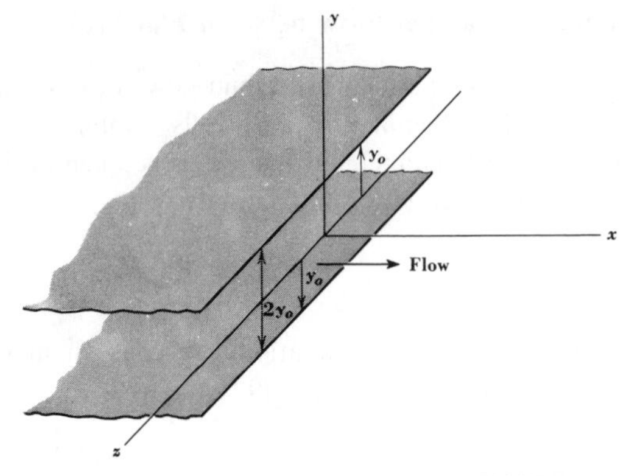

FIGURE 10-1
Flow in a slit.

the total pressure is replaced by the dynamic pressure; the other terms in the equations are unaffected. Thus it is rigorously correct that $\partial p_d/\partial y$ and $\partial p_d/\partial z$ equal zero, even for a nonhorizontal conduit, and the equation

$$\frac{d^2u_x}{dy^2} = \frac{g_c}{\mu}\frac{dp_d}{dx} = \text{const} \tag{10-6}$$

is true also for a nonhorizontal conduit. The original analysis for a horizontal conduit led to Eq. (10-5), which is not quite exact, since the total pressure at the bottom of the channel is slightly greater than at the top. However, there is no y gradient of dynamic pressure, and Eq. (10-6) is exact. In many texts on fluid dynamics the Navier-Stokes equations are written in terms of the dynamic pressure.

A first integration of Eq. (10-5) gives

$$\frac{du}{dy} = \frac{g_c}{\mu}\frac{dp}{dx}y \tag{10-7}$$

in which the constant of integration is zero because du/dy is zero at $y = 0$; the subscript on u has been dropped for convenience. A second integration yields

$$u = \frac{g_c}{2\mu}\frac{dp}{dx}(y^2 - y_o^{\,2}) \tag{10-8}$$

because u is zero at $y = y_o$. This equation is usually written as

$$u = -\frac{g_c}{2\mu}\frac{dp}{dx}(y_o^{\,2} - y^2) \tag{10-9}$$

since we have $dp/dx < 0$ and $y^2 < y_o^2$. The maximum velocity u_{max} occurs at $y = 0$. Therefore Eq. (10-9) can be written alternatively as

$$u = u_{max}\left[1 - \left(\frac{y}{y_o}\right)^2\right] \qquad (10\text{-}10)$$

This equation can be used in conjunction with the overall mass balance, as shown for several examples in Chap. 3, to give

$$\frac{dp}{dx} = -\frac{3\mu u_b}{g_c y_o^{\,2}} \qquad (10\text{-}11)$$

The above relations could have been found by starting from a force balance on an element of fluid defined to take advantage of the symmetry of this system: this derivation was the subject of Prob. 2-1. In the present discussion the equivalent result has been obtained from the Navier-Stokes equations, with attention to several theoretical details.

Laminar Flow in a Circular Pipe

This is another case of parallel flow, for which u_x, u_y, $\partial u_z/\partial z$, and $\partial u_z/\partial\theta$ are all zero for steady-state flow, far from the pipe inlet. The axes are oriented as shown in Fig. 10-2.

The Navier-Stokes equations for an incompressible fluid reduce to

$$\frac{dp}{dz} = \frac{\mu}{g_c}\left(\frac{\partial^2 u}{\partial y^2} + \frac{\partial^2 u}{\partial x^2}\right) \qquad (10\text{-}12)$$

In this equation, p may be interpreted as the dynamic pressure, particularly if the z axis is not horizontal, u replaces u_z, and dp/dz is a constant. These modifications follow from the same arguments used for the case of flow between flat walls.

In the present case u is a function of both y and x, so both second derivatives must be retained. The problem is made much easier to solve if cylindrical coordinates are used, for which

$$z = z \qquad (10\text{-}13)$$

$$x = r\cos\theta \qquad (10\text{-}14)$$

$$y = r\sin\theta \qquad (10\text{-}15)$$

as shown in Fig. 10-2. Conversely,

$$r = \sqrt{x^2 + y^2} \qquad (10\text{-}16)$$

and

$$\theta = \tan^{-1}\frac{y}{x} \qquad (10\text{-}17)$$

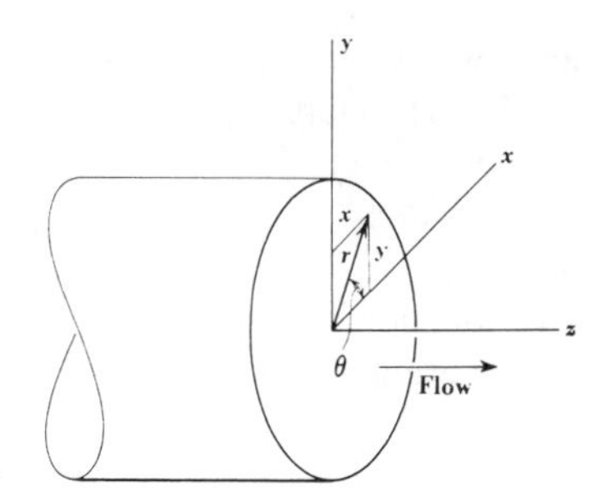

FIGURE 10-2
Flow in a circular tube.

These relations may be used to transform Eq. (10-12) into its equivalent in cylindrical coordinates.

$$\frac{\partial^2 u}{\partial r^2} + \frac{1}{r}\frac{\partial u}{\partial r} + \frac{1}{r^2}\frac{\partial^2 u}{\partial \theta^2} = \frac{g_c}{\mu}\frac{dp}{dz} \qquad (10\text{-}18)$$

The algebra involved in this transformation is further discussed in Appendix 1. Alternatively, the z component of the equation of motion in Table A-4 is used in the analysis. We obtain Eq. (10-18) by setting $\partial u_z/\partial \theta'$, u_r, u_θ, $\partial u_z/\partial \theta$, $\partial u_z/\partial z$, and X_z equal to zero.

Since we should understand the physical principles involved in the derivation of the Navier-Stokes equation from our study of Chap. 9, we can take advantage of Table A-4 to apply the equations to various problems without repeating the derivation each time. We choose the coordinate system so as to take advantage of the symmetry of a problem and thus reduce the number of independent space variables. Thus we have used cylindrical coordinates for a tube, and we would use spherical coordinates for flow around a sphere.

Returning to Eq. (10-18), we notice that the flow is symmetrical about the z axis, $\partial^2 u/\partial \theta^2$ is zero, and we obtain the following ordinary differential equation:

$$\frac{d^2 u}{dr^2} + \frac{1}{r}\frac{du}{dr} = \frac{g_c}{\mu}\frac{dp}{dz} = \text{const} \qquad (10\text{-}19)$$

or

$$\frac{1}{r}\frac{d}{dr}\left(r\frac{du}{dr}\right) = \frac{g_c}{\mu}\frac{dp}{dz} = \text{const} \qquad (10\text{-}20)$$

Equation (10-20) is integrated once, subject to the boundary condition that $du/dr = 0$ at $r = 0$. The result is then integrated again, subject to the condition that $u = 0$ at $r = r_i$, giving

$$u = -\frac{g_c}{4\mu}\frac{dp}{dz}(r_i^{\,2} - r^2) \qquad (10\text{-}21)$$

This result is analogous to Eq. (10-9) for flow through a slit and is easily converted to

$$u = u_{max}\left[1 - \left(\frac{r}{r_i}\right)^2\right] \qquad (10\text{-}22)$$

which is the same as Eq. (2-15). An integration over the pipe cross section gives

$$\frac{dp}{dz} = -\frac{8\mu u_b}{g_c r_i^2} \qquad (10\text{-}23)$$

Applications of (10-23), the Hagen-Poiseuille equation, have been given in Example 3-4 and Probs. 2-4 to 2-8.

Laminar Flow in an Annulus

Flow in the space between two concentric pipes is a common practical situation. For this case Eq. (10-20) still applies; an alternative derivation of this equation has been suggested in Prob. 5-11. The boundary condition for the first integration of Eq. (10-20) is $du/dr = 0$ at $r = r_{max}$, the radius at which the maximum velocity occurs. This gives

$$r\frac{du}{dr} = \frac{g_c}{2\mu}\frac{dp}{dz}(r^2 - r_{max}^2) \qquad (10\text{-}24)$$

The second integration is subject to the condition that u is zero at the inner boundary r_1, giving

$$u = \frac{g_c}{2\mu}\frac{dp}{dz}\left(\frac{r^2 - r_1^2}{2} - r_{max}^2 \ln\frac{r}{r_1}\right) \qquad (10\text{-}25)$$

However, at $r = r_2$, u also is zero, so an alternative expression is

$$u = \frac{g_c}{2\mu}\frac{dp}{dz}\left(\frac{r^2 - r_2^2}{2} - r_{max}^2 \ln\frac{r}{r_2}\right) \qquad (10\text{-}26)$$

These two equations permit r_{max} to be found in terms of r_1 and r_2:

$$r_{max} = \sqrt{\frac{r_2^2 - r_1^2}{2\ln(r_2/r_1)}} \qquad (10\text{-}27)$$

By an integration over the cross section of the annulus there results

$$\frac{dp}{dz} = -\frac{8\mu u_b}{g_c}\frac{1}{r_2^2 + r_1^2 - 2r_{max}^2} \qquad (10\text{-}28)$$

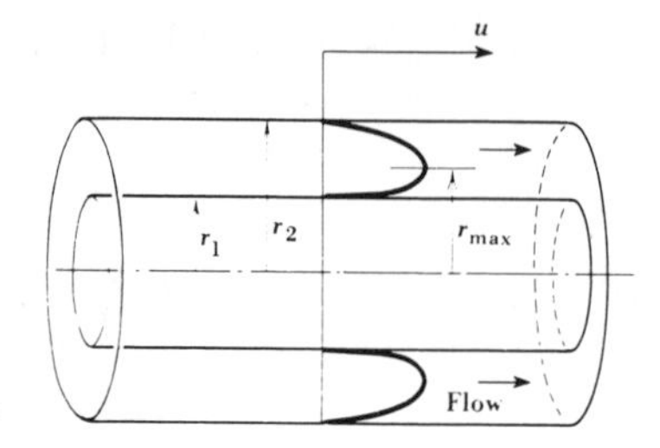

FIGURE 10-3
Velocity profiles for flow in an annulus.

The velocity profile predicted by Eq. (10-26) is shown in Fig. 10-3. For the limiting case of $r_1 = 0$, r_{max} becomes zero and Eqs. (10-26) and (10-28) reduce to the equivalent equations for the circular pipe, (10-21) and (10-23).

It must be remembered that these exact solutions of the equations of motion hold only for the isothermal, laminar, steady-state flow of an incompressible fluid far from the entrance of the channel. For a circular pipe **Re** must be less than 2100. The significance of Reynolds numbers for noncircular conduits will be discussed in Chap. 14.

No more exact solutions will be given here, but others are to be found in more specialized works.[1]

Creeping Flow

The term *creeping flow* is used to describe flow at very low velocity, or to be more precise, at very low Reynolds numbers. The principal interest in this type of flow stems from the fact that it applies for the fall of small particles through a fluid; it is the basis of Stokes' law, which is useful in solving problems of settling and sedimentation. Creeping flow also occurs in certain problems in lubrication.

For Reynolds numbers less than unity, the viscous forces involved in flow outweigh the inertia forces. In flow around a sphere, for example, a fluid particle changes direction and velocity in a complex way, and if the inertia effects connected with these changes were important, it would be necessary to retain all the terms in all three Navier-Stokes equations. It will be recalled, however, that the substantial derivative in the statement of Newton's second law (Eq. 9-4) is proportional to the force required to overcome the inertia of the fluid element. As a result, this term is

[1] J. G. Knudsen and D. L. Katz, "Fluid Dynamics and Heat Transfer," chap. 4, McGraw-Hill Book Company, New York, 1958; H. Schlichting and Kestin, "Boundary Layer Theory," 6th ed., chap. 5, McGraw-Hill Book Company, New York, 1968; R. B. Bird, W. E. Stewart, and E. N. Lightfoot, "Transport Phenomena," chap. 2, John Wiley & Sons, Inc., New York, 1960.

omitted from the equations of motion representing creeping flow. Results based on this assumption have been found to agree with experiment at **Re** < 1. Thus the equations of motion for creeping flow of an incompressible fluid are

$$\frac{\partial p}{\partial x} = \frac{\mu}{g_c}\left(\frac{\partial^2 u_x}{\partial x^2} + \frac{\partial^2 u_x}{\partial y^2} + \frac{\partial^2 u_x}{\partial z^2}\right) \qquad (10\text{-}29)$$

$$\frac{\partial p}{\partial y} = \frac{\mu}{g_c}\left(\frac{\partial^2 u_y}{\partial x^2} + \frac{\partial^2 u_y}{\partial y^2} + \frac{\partial^2 u_y}{\partial z^2}\right) \qquad (10\text{-}30)$$

$$\frac{\partial p}{\partial z} = \frac{\mu}{g_c}\left(\frac{\partial^2 u_z}{\partial x^2} + \frac{\partial^2 u_z}{\partial y^2} + \frac{\partial^2 u_z}{\partial z^2}\right) \qquad (10\text{-}31)$$

and the equation of continuity is

$$\frac{\partial u_x}{\partial x} + \frac{\partial u_y}{\partial y} + \frac{\partial u_z}{\partial z} = 0 \qquad (10\text{-}32)$$

At the surface of the sphere, the tangential and normal velocities must be zero. The problem can be reduced to one of only two independent variables by the use of spherical coordinates. The mathematics required to achieve the equations describing the velocity distribution is long and complicated, so we merely state that the problem has been solved.[1] From the resulting pressure distribution over the sphere it can be shown by integration over the whole surface that the form drag caused by the pressure distribution is

$$F_{d,\,\text{form}} = \frac{2\pi\mu r_o u_o}{g_c} \qquad (10\text{-}33)$$

where r_o = radius of sphere

u_o = velocity of parallel flow far from sphere (or velocity of travel of sphere through a quiescent fluid)

From the velocity distribution and the basic relation between viscosity and shear stress at the surface, an integration over the whole surface of the sphere gives the total force arising from viscous drag—the skin friction.

$$F_{d,\,\text{skin}} = \frac{4\pi\mu r_o u_o}{g_c} \qquad (10\text{-}34)$$

The total drag is then

$$F_d = \frac{6\pi\mu r_o u_o}{g_c} \qquad (10\text{-}35)$$

which is the Stokes equation. The drag force is proportional to the velocity.

[1] H. Lamb, "Hydrodynamics," 6th ed., pp. 597–598, Dover Publications, New York, 1945.

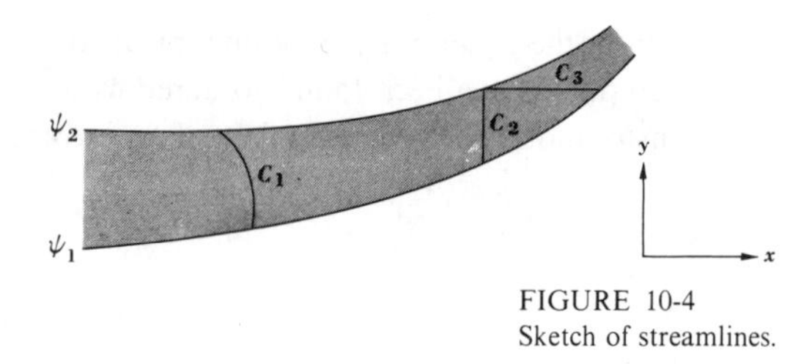

FIGURE 10-4
Sketch of streamlines.

Streamlines and the Stream Function

Another class of analytical solutions of the Navier-Stokes equation applies to flow of fluids for which viscous effects are not important; the Reynolds number is thus very large. A typical application of this theory is to flow around or past objects.

In order to describe these flow systems graphically it is common practice to use streamlines. We now consider their mathematical definition. To avoid unprofitable complications, only two-dimensional, incompressible steady flow will be discussed, i.e., cases where there is no variation of velocity in the z direction taken perpendicular to the plane of the paper. A streamline is then defined as a line across which no fluid flows.

It is useful to define a quantity which is called the *stream function* ψ in such a way that streamlines represent lines of constant ψ. The velocity components u_x and u_y can be related to the orientation of the streamlines by the following analysis.

Consider two lines of constant ψ as shown by Fig. 10-4. The mass rate of flow through the space having a cross section defined by the two streamlines and a unit distance in the direction perpendicular to the paper is constant across any line, say C_1, joining the two streamlines and is given by

$$w = \int_{C_1} u\rho \cos \alpha \, ds \qquad (10\text{-}36)$$

where s = distance along line C_1 joining streamlines

α = angle between velocity vector and outward-directed normal to ds

The stream function is related to the mass rate of flow by

$$w = (\psi_2 - \psi_1)\rho \qquad (10\text{-}37)$$

Equations (10-36) and (10-37) can also be written for a differential change in the stream function as

$$dw = \rho \, d\psi = u\rho \cos \alpha \, ds \qquad (10\text{-}38)$$

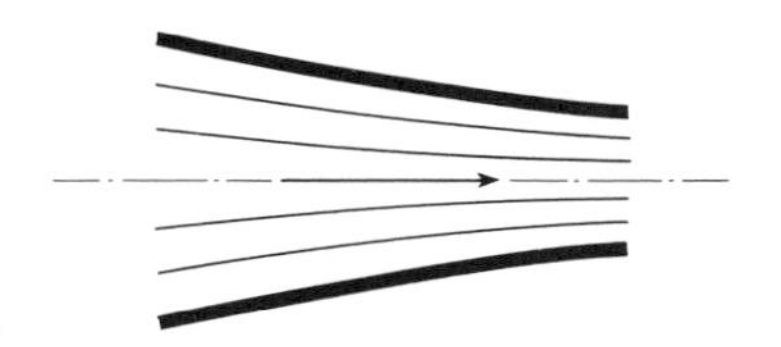

FIGURE 10-5
Streamlines in a nozzle.

The value of $\int \rho \, d\psi$ is the same for any path joining the streamlines. If the path C_2 is chosen, oriented in the y direction, then $ds = dy$ and $u \cos \alpha = u_x$, so that

$$\frac{\partial \psi}{\partial y} = u_x \qquad (10\text{-}39)$$

For path C_3, $ds = -dx$ and $u \cos \alpha = u_y$, giving

$$\frac{\partial \psi}{\partial x} = -u_y \qquad (10\text{-}40)$$

If ds is chosen so its normal coincides with the direction of flow, we obtain

$$\frac{d\psi}{ds} = u \qquad (10\text{-}41)$$

If ds coincides with the direction of flow, $\cos \alpha$ is zero and $d\psi/ds$ is zero; hence ψ is constant. Since a line of constant ψ is a streamline, it is evident that the velocity is directed along the streamlines. For an incompressible fluid, the velocity u is greatest where the streamlines on a particular graph are closest together. For axially symmetric, three-dimensional flow, the two-dimensional representation is satisfactory, as illustrated in Fig. 10-5 for flow through a nozzle.[1]

Ideal Fluids

An ideal fluid has a viscosity of zero, and the science dealing with such fluids is called *theoretical hydrodynamics.* During the latter half of the nineteenth century an elaborate mathematical theory of ideal fluids was developed, and the differential equations for flow of an ideal fluid were solved for many cases. Since air and water have low viscosities, it was reasoned that they should behave like ideal fluids except at low Reynolds numbers. Experiments showed that this was not true in many cases. For example, practical problems, like the calculation of the pressure drop for

[1] Note that the definition of ψ by Eqs. (10-39) and (10-40) automatically satisfies the equation of continuity, $\partial u_x/\partial x + \partial u_y/\partial y = 0$.

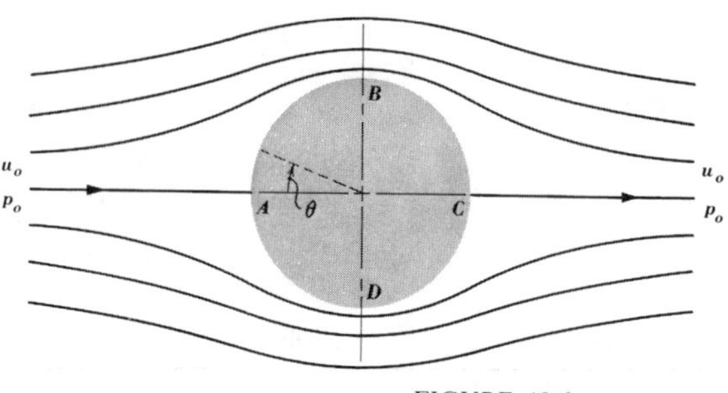

FIGURE 10-6
Ideal flow around a cylinder.

water flowing through a conduit, could not be solved by using the equations of theoretical hydrodynamics. Meanwhile, the empirical science of hydraulics, which bore very little resemblance to the theory of ideal fluids, had been developed by engineers to solve these problems.

Let us look at a few results from the theory of ideal fluids for flow past a simple shape. Figure 10-6 shows horizontal, two-dimensional flow past a cylinder of infinite length. The fluid divides and flows around the cylinder with slip at the surface, since there can be no tangential or shear stresses in an ideal fluid.

The position of the streamlines can be obtained from calculations based on a type of ideal fluid flow called irrotational, or potential, flow. Once this mathematical problem has been solved, Bernoulli's equation can be used to find the pressure distribution in the fluid. This equation becomes, for horizontal flow along a streamline in an ideal incompressible fluid,

$$\frac{u^2 - u_o{}^2}{2g_c} + \frac{p - p_o}{\rho} = 0 \qquad (10\text{-}42)$$

Because the velocity at A is zero, the pressure there is found to be

$$p = p_o + \frac{u_o{}^2\rho}{2g_c} \qquad (10\text{-}43)$$

This is the stagnation, or kinetic, pressure. At points B and D the velocity is a maximum and the pressure falls to a minimum. At point C the velocity is again zero. Since the flow is symmetrical before and behind the cylinder, the force exerted by the pressure on the upstream half of the cylinder equals that on the downstream half. As a result, there is no drag on the cylinder. Near the curved surface of the cylinder, the inner layers of fluid travel faster than the outer layers, so that there is no net rotation of fluid; hence the term irrotational flow. Furthermore, the fluid has zero viscosity, so that the sliding of these layers over each other is associated with no shearing stress. These matters will be discussed mathematically in what follows.

At a very high Reynolds number, the inertia forces outweigh the viscous forces. We set the viscosity equal to zero in the Navier-Stokes equations and obtain the equations of motion for an ideal or inviscid fluid, called the Euler equations:

$$u_x \frac{\partial u_x}{\partial x} + u_y \frac{\partial u_x}{\partial y} + u_z \frac{\partial u_x}{\partial z} + \frac{\partial u_x}{\partial \theta} = g_c X - \frac{g_c}{\rho} \frac{\partial p}{\partial x} \tag{10-44}$$

$$u_x \frac{\partial u_y}{\partial x} + u_y \frac{\partial u_y}{\partial y} + u_z \frac{\partial u_y}{\partial z} + \frac{\partial u_y}{\partial \theta} = g_c Y - \frac{g_c}{\rho} \frac{\partial p}{\partial y} \tag{10-45}$$

$$u_x \frac{\partial u_z}{\partial x} + u_y \frac{\partial u_z}{\partial y} + u_z \frac{\partial u_z}{\partial z} + \frac{\partial u_z}{\partial \theta} = g_c Z - \frac{g_c}{\rho} \frac{\partial p}{\partial z} \tag{10-46}$$

The theory of ideal fluids is useful in aerodynamics, particularly in the calculation of the lift of airfoils. It is also useful in the general study of flow past immersed bodies, for it gives the pressure distribution at the outer edge of the boundary layer. Since the theory contains the postulate of zero viscosity, it implies slip at a solid surface. For a complete solution, the effect of viscosity and shear must be considered in the boundary layer near a surface, for actually there is no slip at the surface. However, away from the surface, the assumption of an ideal fluid is often valid.

The aeronautical engineer finds the most opportunities to use the theory of ideal fluids, but it is important to all engineers in the general study of fluid dynamics. In this discussion, we shall restrict ourselves to the simple case of irrotational, or potential, flow. This name implies the absence of rotation, or vorticity, in the flow.

Potential Flow

The idea of an electric potential is familiar, and we also have dealt with the potential energy due to the field of gravity. For instance, for the z axis directed vertically upward, the body force Z is given by $-g/g_c$ in lb_f/lb. Since the change in potential energy is equal to the work done in raising a mass of one pound in the z direction, we set

$$d\Omega = Z\,dz \tag{10-47}$$

where Ω is the potential energy, $(\text{ft})(\text{lb}_f)/\text{lb}$, and

$$Z = -\frac{g}{g_c} = \frac{d\Omega}{dz} \tag{10-48}$$

Here a force is expressed as a gradient of a potential. By analogy, we define a velocity potential $\phi(x,y)$ so that the velocity is equal to the gradient of a potential. Thus

$$u_x = \frac{\partial \phi}{\partial x} \tag{10-49}$$

and

$$u_y = \frac{\partial \phi}{\partial y} \tag{10-50}$$

It is convenient to restrict this discussion to two-dimensional flow. If these expressions for the velocities are substituted into the equation of continuity,

$$\frac{\partial u_x}{\partial x} + \frac{\partial u_y}{\partial y} = 0 \tag{10-51}$$

we get

$$\frac{\partial^2 \phi}{\partial x^2} + \frac{\partial^2 \phi}{\partial y^2} = 0 \tag{10-52}$$

This is Laplace's equation. If suitable boundary conditions are known, the equation can be solved to give $\phi(x,y)$, and the velocity at any point can then be obtained from Eqs. (10-49) and (10-50). This is simple enough, but we have merely assumed so far that the potential function exists. Actually, it exists only if the flow fulfills certain requirements, which we shall now find.

Equations (10-49) and (10-50) can be differentiated to obtain

$$\frac{\partial u_x}{\partial y} = \frac{\partial^2 \phi}{\partial y\, \partial x} \tag{10-53}$$

and

$$\frac{\partial u_y}{\partial x} = \frac{\partial^2 \phi}{\partial x\, \partial y} \tag{10-54}$$

Since $\phi(x,y)$ is a single-valued function and all its second derivatives are assumed to exist, the right-hand sides of these two equations are identical, and we have

$$\frac{\partial u_y}{\partial x} - \frac{\partial u_x}{\partial y} = 0 \tag{10-55}$$

The quantity on the left-hand side of Eq. (10-55) is called the *vorticity of the fluid*; we shall see that it equals twice the angular velocity about an axis oriented in the z direction.

Consider the rotation of an element of fluid associated with the velocity gradients in Eq. (10-55). These gradients may cause shear, or the sliding of one layer

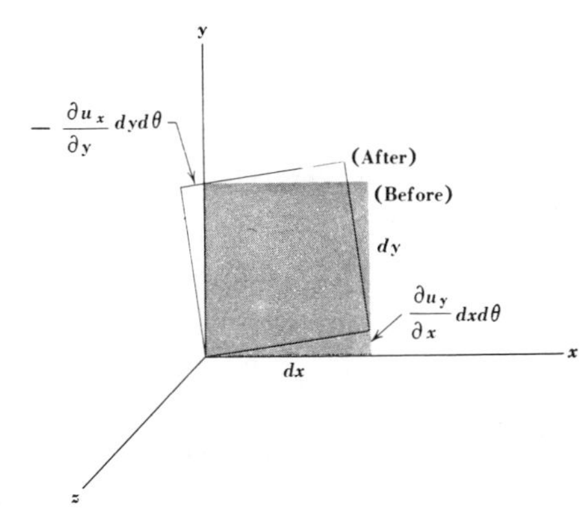

FIGURE 10-7
Fluid element in rotational flow.

over the other; Fig. 10-7 illustrates a fluid element undergoing rotation. The origin moves with the element, and the orientation of the element is shown before and after a time $d\theta$. The average rotation of a differential element is one-half the sum of the rotation of the dx segment and the dy segment. Since the angular velocity is equal to the rate of movement of the end of a radial arm divided by the radius, we have

$$\omega = \frac{1}{2}\left(\frac{\partial u_y}{\partial x} - \frac{\partial u_x}{\partial y}\right) \qquad (10\text{-}56)$$

Thus the flow illustrated by Fig. 10-7 is rotational and does not meet the requirement of Eq. (10-55) for potential flow. On the other hand, the flow in Fig. 10-8 is irrotational; the element is deformed but suffers no net rotation. For this case, the requirement

$$\frac{\partial u_y}{\partial x} - \frac{\partial u_x}{\partial y} = 0 \qquad (10\text{-}55)$$

is met; the vorticity is zero. These definitions can be extended to three dimensions.

We have now shown that the velocity distribution for potential flow can be found from the solution of Laplace's equation (10-52) and that potential flow is irrotational because of the restriction imposed by Eq. (10-55). Euler's equations can now be used to find the pressure distribution. For steady-state, two-dimensional, irrotational, incompressible flow, the substitution of (10-55) and the X equivalent of (10-48) into Eq. (10-44) gives

$$u_x \frac{\partial u_x}{\partial x} + u_y \frac{\partial u_y}{\partial x} - g_c \frac{\partial \Omega}{\partial x} + \frac{g_c}{\rho} \frac{\partial p}{\partial x} = 0 \qquad (10\text{-}57)$$

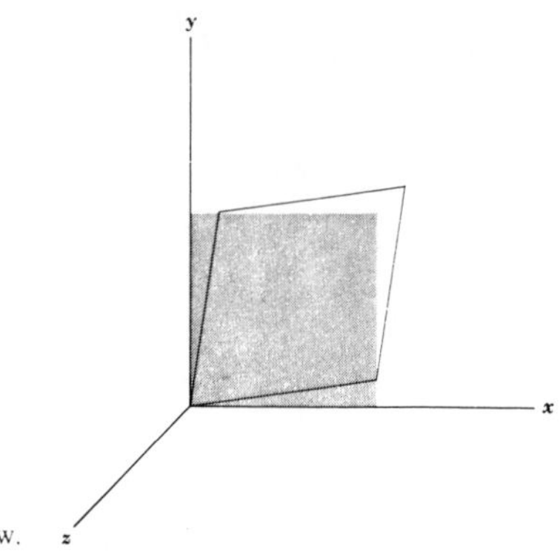

FIGURE 10-8
Fluid element in irrotational flow.

This equation can be integrated to give

$$\tfrac{1}{2}(u_x{}^2 + u_y{}^2) - g_c\Omega + \frac{g_c p}{\rho} = f_1(y) \tag{10-58}$$

Similarly, Eq. (10-45) gives

$$\tfrac{1}{2}(u_x{}^2 + u_y{}^2) - g_c\,\Omega + \frac{g_c p}{\rho} = f_2(x) \tag{10-59}$$

Since the left-hand sides of these equations are identical, $f_1(y) = f_2(x) = \text{const}$. Furthermore,

$$u_x{}^2 + u_y{}^2 = u^2 \tag{10-60}$$

and so we have

$$\frac{u^2}{2g_c} - \Omega + \frac{p}{\rho} = \text{const} \tag{10-61}$$

which is Bernoulli's equation. It is alternatively expressed, for two points in the fluid, as

$$\frac{\Delta u^2}{2g_c} - \Delta\Omega + \frac{\Delta p}{\rho} = 0 \tag{10-62}$$

Since

$$\Delta\Omega = -\frac{g}{g_c}\Delta z \tag{10-63}$$

follows from Eq. (10-48), we obtain

$$\frac{\Delta u^2}{2g_c} + \frac{g}{g_c}\Delta z + \frac{\Delta p}{\rho} = 0 \tag{10-64}$$

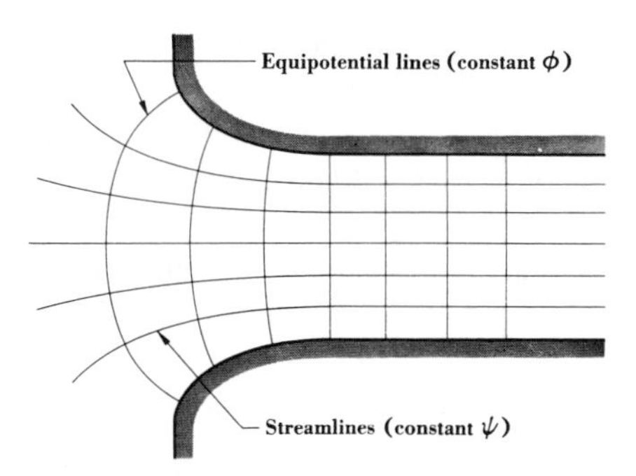

FIGURE 10-9
Flow net for a rounded entry.

which is similar to Eq. (4-30). The latter equation contains u_b rather than u, but the subscript is not really necessary, because the assumptions previously made imply that there can be no velocity variation over the planes at the inlet and outlet. For irrotational, frictionless flow in this case, u and u_b are identical.

It will be recalled that Bernoulli's equation has been used in the analysis in Chap. 6 of the pitot tube and of the flow in an orifice up to the vena contracta. For these flowmeters it is readily seen that the inertia terms do indeed outweigh the viscous terms. The fluid is accelerated by changes in the flow cross section, and there is no extended rubbing against solid surfaces. Since the flow in a jet emerging from an orifice is irrotational, the velocity across the jet is constant. Ideal-fluid theory also applies for other cases involving relatively large inertia effects, such as flow over a weir, flow from large tanks into pipe entries, and flow through the rounded entry shown in Fig. 10-9.

The equipotential lines and streamlines in Fig. 10-9 (called the *flow net*) have been obtained from the solution of Laplace's equation (10-52) for boundary conditions corresponding to the shape given. The actual flow results have been found to agree well with the theoretical results shown. For potential flow, lines of constant ϕ and lines of constant ψ are everywhere perpendicular (orthogonal).

We know that Bernoulli's equation does not apply for ordinary, steady-state, parallel flow through a pipe of constant cross section. The term lw_f, which arises from the friction, or viscous shear, must be added [Eq. (4-36)], for example, with W_s zero. For this system, there is a velocity gradient across the pipe radius but no offsetting gradient in another direction; therefore, according to Eq. (10-55), the flow is rotational. We can, of course, draw streamlines for this rotational flow; the stream function ψ exists, since it is based only on the validity of the equation of continuity. However, the potential function ϕ does not exist, for the flow is not irrotational. The rotation arising from shear, such as in laminar flow, is analogous to the rotation which can be given to a pencil held between the palms as one's hands are rubbed together.

We shall not give here the mathematical details of any solutions of Laplace's equation for ϕ. Many good discussions are available in books on fluid dynamics;[1] some knowledge of the theory of functions of a complex variable is necessary for anything more than an introduction to the subject.

PROBLEMS

10-1 The streamlines of Fig. 10-6 for flow around a cylinder can be expressed as

$$\psi = Ay\left(1 - \frac{r_o^2}{x^2 + y^2}\right)$$

where $A = 1$ m/s

r_o = radius = 0.02 m.

and the origin of the xy axes is at the center of the cylinder.

Calculate, for ideal flow, the velocity at point B on the surface of the cylinder. The flow is horizontal.

10-2 The two-dimensional flow of water at 10°C is described by a stream function which is given as a function of position by the equation

$$\psi = Axy \qquad y > 0$$

where $A = 1\ \text{s}^{-1}$ and x and y are measured in meters. This flow pattern is called *plane stagnation flow*. (*a*) What are the magnitude and direction of the velocity vector at (1,1)? (*b*) If the pressure at (1,1) is 101.3 kPa abs, what is the pressure at the origin (the stagnation pressure)?

10-3 Equation (10-5) can be applied to flow in a slit, one boundary of which moves with a constant velocity u_o. Find the equation for the velocity as a function of y, u_o, and dp/dx, and sketch the profiles obtained when dp/dx is both positive and negative for u_o positive.

This type of flow is called *Couette flow* and is of importance in the hydrodynamic theory of lubrication. If dp/dx is zero, it is called *simple* Couette flow; this case was used in connection with the first definition of viscosity, Fig. 2-1 and Eq. (2-3).

10-4 Find the equation for the velocity distribution within a film of liquid in laminar flow down a vertical, plane surface, neglecting the acceleration in the film. In this case there is no pressure gradient in the direction of flow, but the body force (gravity) is important. Find also the expression for the thickness of the film in terms of the bulk (average) velocity.

[1] V. L. Streeter, "Fluid Dynamics," McGraw-Hill Book Company, New York, 1948 (an elementary text); Lamb, op. cit. (the classical treatise on hydrodynamics); L. M. Milne-Thomson, "Theoretical Hydrodynamics," 2d ed., The Macmillan Company, New York, 1950 (emphasis on vector notation); N. Curle and H. J. Davies, "Modern Fluid Dynamics," vols. 1 & 2, D. Van Nostrand Co., Inc., Princeton, 1968.

10-5 Use Eq. (10-35) to find the position of a small sphere of density ρ_o released from rest as a function of time. The value of **Re** at the terminal (maximum) velocity is less than 1.0. What is the expression for the terminal velocity?

10-6 The potential two-dimensional flow of a fluid from a large reservoir into a rectangular channel is described by the equations

$$x = \phi + e^{\phi} \cos \psi$$

$$y = \psi + e^{\phi} \sin \psi$$

Draw the streamlines and equipotential lines (i.e., the flow net) for this flow and identify the location of the channel walls.

10-7 For the two-dimensional, potential flow past an idealized pitot tube, the flow net is defined by

$$x = -\phi + \ln \sqrt{\phi^2 + \psi^2}$$

$$y = -\psi + \tan^{-1} \frac{\psi}{\phi}$$

Sketch the streamlines for this system.

10-8 The potential function for a certain flow is given by

$$\phi = xy$$

What might be the solid boundaries causing this flow?

10-9 The potential flow around a sphere of radius r_o is given in spherical coordinates by

$$\phi = -\frac{u_o {r_o}^3}{2r^2} \cos \theta - u_o r \cos \theta$$

where θ is the angular-position variable. Find the direction and magnitude of the velocity at $r = \frac{5}{4} r_o$ and $\theta = 30°$, if $\theta = 0$ corresponds to the forward stagnation point. The velocity is the gradient of the potential, so in spherical coordinates the correct expressions must be found to relate u_r and u_θ $(u_\phi = 0)$ to ϕ. Neglecting the effect of gravity, find the expression for the pressure as a function of θ at the surface of the system.

10-10 For creeping flow around a sphere Stokes found the stream function

$$\psi = \tfrac{1}{2} u_o \left(1 - \frac{3}{2} \frac{r_o}{r} + \frac{1}{2} \frac{{r_o}^3}{r^3}\right) r^2 \sin^2 \theta \qquad (1)$$

when the fluid flows past a stationary sphere as shown in Fig. 10-6. For this axisymmetric flow the velocity is not a function of the azimuth angle ϕ of spherical coordinates, and the velocity components are related to the stream function by

$$u_r = -\frac{1}{r^2 \sin \theta} \frac{\partial \psi}{\partial \theta} \qquad (2)$$

$$u_\theta = \frac{1}{r \sin \theta} \frac{\partial \psi}{\partial r} \qquad (3)$$

$$u_\phi = 0$$

For a sphere of 0.1 mm radius in water at 20°C flowing past with a Reynolds number of 0.1, find the velocity at $r = 0.3$ mm, $\theta = 45°$, and find the maximum shear stress at the surface o' the sphere. Neglect gravity.

10-11 For potential flow around a cylinder, the velocity potential is given by

$$\phi = u_o\left(1 + \frac{r_o^2}{r^2}\right) r \cos \theta$$

If the stagnation pressure is p_o, make a plot of the pressure at the surface of the cylinder as a function of θ. Neglect gravity.

10-12 For a cylinder moving at velocity u_o through a fluid which is at rest at infinity, we find, for potential flow,

$$\psi = -\frac{u_o r_o^2}{r} \sin \theta$$

Plot a few streamlines for this flow and explain the physical significance of the diagram.

10-13 A room with a ceiling height of 4 m contains in the air a uniform dispersion of spherical dust particles each of which has a diameter of 25 μm. If the room is sealed and the air remains completely stagnant for three days, what fraction of the dust particles will still be suspended in the air at the end of that time? Assume the room remains at 15°C at all times and that each dust particle has a density of 2000 kg/m^3. The particles do not interact with each other.

10-14 A thin layer of a newtonian liquid is held between two infinite, parallel planes separated by a distance L. The lower plane is fixed but the upper plane moves in the x direction with a steady velocity u_o. If the pressure is the same at all points in the liquid a force-momentum balance will show that the relation between velocity and position for this system is

$$\frac{d^2 u_x}{dy^2} = 0$$

Using this relation derive an equation for the velocity u_x as a function of position y and use that equation to obtain a relation for u_b/u_{max}.

11

BOUNDARY-LAYER FLOW

In this chapter the flow of fluids around objects is considered in some detail. The Navier-Stokes equations have been used to find relations which describe laminar flow inside pipes, creeping flow, and ideal flow. For the last of these, it has been found that the results for flow patterns apply for flow far from surfaces or for situations where the cross section for flow is changing rapidly, so that inertial effects clearly outweigh viscous effects. However, if we investigate flow near surfaces and the drag caused by these flows, the results from potential flow are found to be inadequate.

Figure 11-1 shows a comparison of the experimental pressure distribution around a cylinder with that predicted by ideal flow. It can be seen that the theory gives results very different from the results of the experiments.

For flow past a flat plate the theory of ideal fluids gives zero drag force because there is slip at the surface. Furthermore, there is no change in velocity or pressure in the fluid. The drag actually observed for real fluids, therefore, is caused completely by viscous shear forces associated with the variation of velocity from zero at the surface of the plate to u_o in the undisturbed stream. The drag on the plate can be calculated from theory by the integration of the differential momentum balance, as will be shown in the following sections, or it can be obtained from a knowledge of the

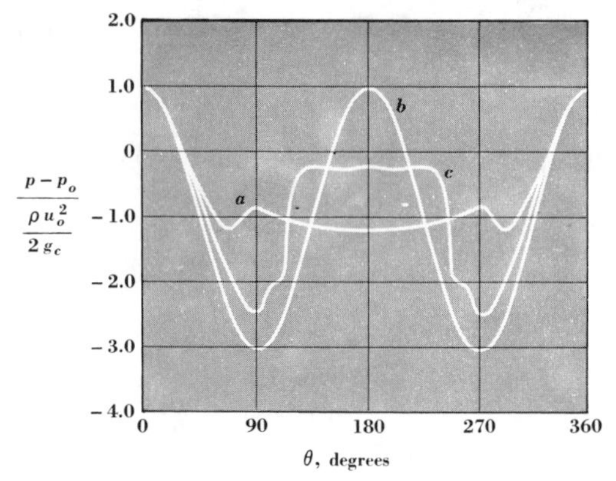

FIGURE 11-1
Pressure distribution in flow past a cylinder. (*From Hermann Schlichting, "Boundary Layer Theory," 4th ed., fig. 1.9, p. 19, McGraw-Hill Book Company, New York, 1960.*)
a Measured, **Re** $= 1.9 \times 10^5$.
b Ideal-fluid theory.
c Measured, **Re** $= 6.7 \times 10^5$.

velocity distribution near the plate and the use of the overall momentum balance, as will be shown in Chap. 12.

The Boundary Layer

The boundary layer is the region near a solid where the fluid motion is affected by the solid boundary; just where the motion ceases to be affected is, of course, subject to an arbitrary definition. In the bulk of the fluid the flow is usually governed by the theory of ideal fluids. By contrast, viscosity is important in the boundary layer, but the layer is relatively thin, so that the equations for flow in this region can be simplified and adequate solutions obtained for many cases. This division of the problem of flow past an object into these two parts, as suggested by Prandtl in 1904, has proved to be of fundamental importance in fluid dynamics.

For flow past a flat plate the thickness of the boundary layer increases from zero at the leading edge, as shown in Fig. 11-2. The Reynolds number for this case is defined as $xu_o \rho/\mu$, where x is the distance downstream measured from the leading edge. Above a certain value of $\mathbf{Re}_x$, the flow becomes turbulent, although there remains a viscous sublayer, as shown in Fig. 11-2. The transition from laminar to turbulent flow on a smooth plate occurs in the range of Reynolds numbers 2×10^5 to 3×10^6.

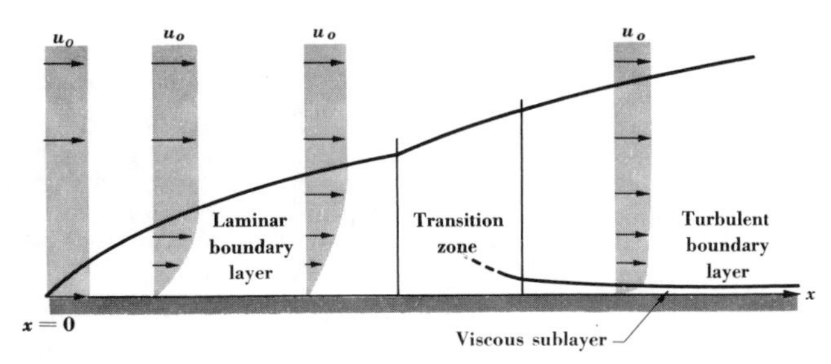

FIGURE 11-2
Sketch of the boundary layer on a flat plate.

If a fluid flowing with a uniform velocity u_o enters a conduit, a boundary layer builds up as shown in Fig. 11-3 and eventually fills the pipe. Thus, in fully developed laminar or turbulent flow, the entire radius of the pipe is in the boundary layer. Downstream from the point where the boundary layer fills the pipe, the flow pattern is independent of x, so the Reynolds number based on the distance from the inlet has no further significance. Instead, the flow is characterized by the Reynolds number based on the pipe diameter, which we have used in all previous chapters. Therefore the quantity $Du_b\,\rho/\mu$ applies only to this fully developed flow region. If the boundary layer is turbulent and fills the pipe, as it usually does except very near the entrance to the pipe, a viscous sublayer persists near the walls, just as with flow past a flat plate.

Boundary-Layer Separation

A consideration of the order of magnitude of the various terms in the equations of motion when applied to a boundary layer will lead to the conclusion that there is negligible pressure variation in the direction normal to the surface. Consequently, the pressure within the boundary layer can be approximated by the pressure given by the ideal-flow pattern outside the boundary layer.

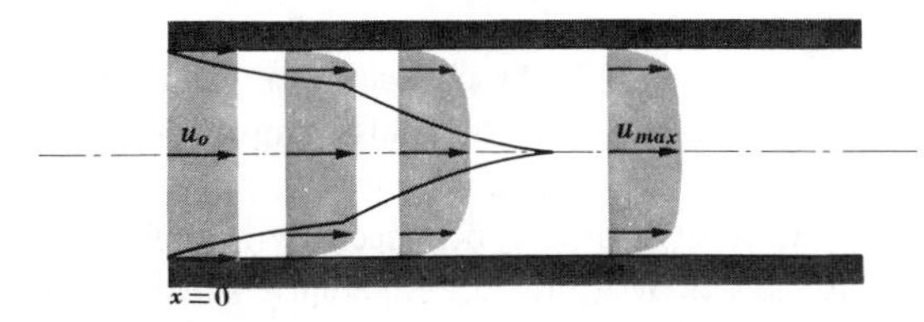

FIGURE 11-3
Boundary layer near the entrance of a conduit.

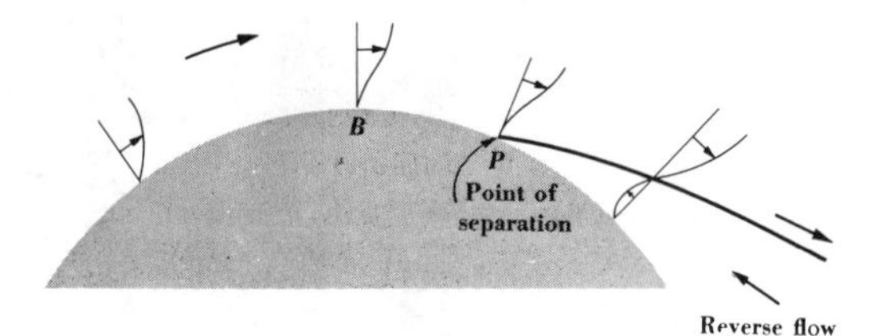

FIGURE 11-4
Separation in flow past a cylinder.

We defined the boundary layer with reference to a flat plate, but a boundary layer will also clearly exist on the surface of a cylinder, or on any other surface. In Fig. 11-4 the layers of fluid near the surface of the cylinder (of diameter D) are retarded by viscous friction, and after point B is passed, the fluid is also retarded by the unfavorable pressure gradient. These two factors are enough at all but rather low Reynolds number ($Du_o \rho/\mu$) to cause the fluid near the surface to come to rest and even to flow in the reverse direction, as shown in Fig. 11-4. The boundary layer then leaves the surface; this phenomenon is called *separation.*

If the flow rate past a cylinder is gradually increased above **Re** in the order of 1.0, separation starts at the rear stagnation point. This causes a change in the pressure and flow fields, and the separation point moves forward. The most forward position is at $\theta = 85°$, as shown in Fig. 11-5*a*. This occurs for a laminar boundary layer.

If the velocity past the cylinder is increased sufficiently to cause a transition to a turbulent boundary layer, the point of separation moves toward the rear of the cylinder, as shown in Fig. 11-5*b*. Because of the increased efficiency of momentum transfer in the turbulent flow, the velocity of the layers near the surface is increased. The higher kinetic energy of the fluid near the surface causes it to penetrate farther around the cylinder and establish a zone of higher pressure (Fig. 11-1), in accordance with the nearly ideal-flow behavior outside the boundary layer. Therefore the separation point moves to a new position past points B and D. For flow past a cylinder, most of the drag is caused by the difference in the pressure on the forward and the rear surfaces, so that the drag is actually decreased by the increase of velocity, which causes the change from a laminar to a turbulent boundary layer. Further increases in the Reynolds number, however, will cause an increase in the drag force.

The type of drag experienced by a bluff shape such as a cylinder, which is mostly caused by a pressure difference, is called form drag. The drag caused by viscous shear in the boundary layer is called skin friction; it is the only drag for flow past a flat plate. Form drag is predominant in flow past bluff, nonstreamlined

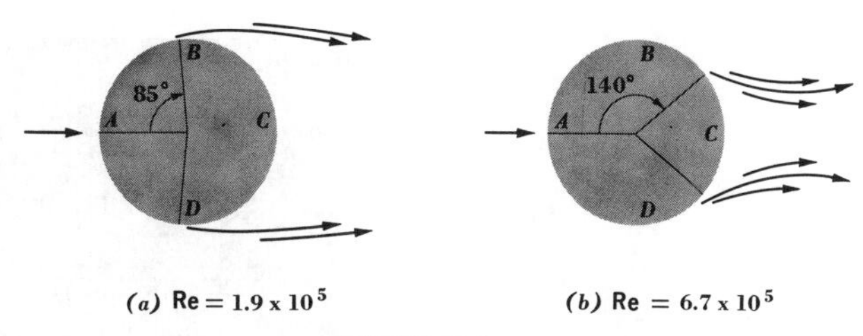

(*a*) Re = 1.9×10^5 (*b*) Re = 6.7×10^5

FIGURE 11-5
Effect of turbulence on the point of separation.

objects at all except low Reynolds numbers and is often associated with the appearance of a wake. Skin friction is predominant in flow past streamlined bodies, for which the form drag is small.

Separation of the boundary layer is associated with other cases involving flow against a pressure gradient when the cross section available to flow increases. There is therefore separation at an orifice or other sudden expansion, as described in Chaps. 5 and 6. The boundary layer occupies the entire conduit upstream from the orifice, but it separates from the walls at the sudden expansion and forms a jet.

The Drag Coefficient

The drag force per unit area for two-dimensional flow past an infinitely long cylinder would be proportional to ρu_o^2 by a momentum balance if it were assumed that the x momentum were reduced by the same proportion for all Reynolds numbers $Du_o \rho/\mu$. We could then write

$$F'_d \propto \frac{\rho u_o^2}{2g_c} \frac{A}{L} \tag{11-1}$$

where F'_d = drag force per unit length

$\dfrac{A}{L}$ = frontal area per unit length

The factor $\frac{1}{2}$ is introduced by custom. Since A/L is proportional to the diameter of the cylinder, an equivalent expression is

$$F'_d \propto \frac{\rho u_o^2 D}{2g_c} \tag{11-2}$$

However, as the Reynolds number changes, the fractional decrease in x momentum *is* affected, so we write

$$F'_d = C_D \frac{\rho u_o^2 D}{2g_c} \tag{11-3}$$

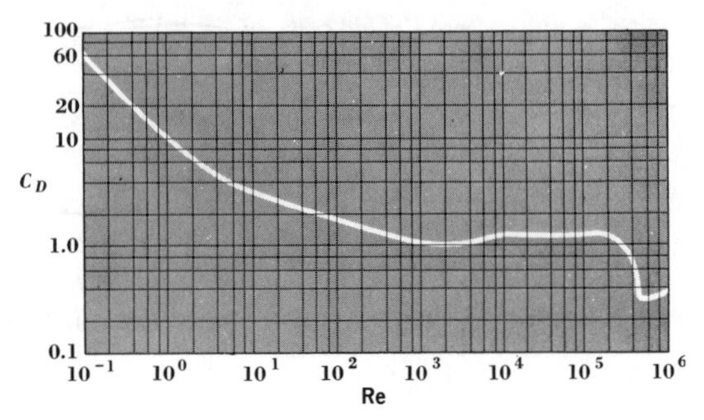

FIGURE 11-6
Drag coefficients for flow past an infinite cylinder. (*From Hermann Schlichting, "Boundary Layer Theory," 4th ed., fig. 1.4, p. 16, McGraw-Hill Book Company, New York, 1960.*)

or

$$C_D = \frac{2F'_d g_c}{\rho u_o^2 D} \tag{11-4}$$

where C_D is a function of the Reynolds number and is called the drag coefficient.

For flow past a sphere or other objects, C_D is defined slightly differently:

$$C_D = \frac{2F_d g_c}{\rho u_o^2 A} \tag{11-5}$$

F_d is the total drag on the object, and A is its frontal area.

The effects of variation of Reynolds number on the drag coefficient for flow past a cylinder are shown in Fig. 11-6. In Chap. 2 we found that the pressure drop (total drag) for laminar flow in a pipe was proportional to the bulk velocity u_b. This is approximately true for flow around a cylinder at low Reynolds numbers. Consequently, it can be seen from Eq. (11-4) that the drag coefficient C_D must be inversely proportional to u_o and the Reynolds number. For example, for creeping flow around a sphere we found that

$$F_d = \frac{6\pi\mu r_o u_o}{g_c} \tag{10-35}$$

so that the drag coefficient defined by Eq. (11-5) is

$$C_D = \frac{12\mu}{r_o u_o \rho} = \frac{24}{\mathbf{Re}} \tag{11-6}$$

Values of C_D for a wide range of **Re** are plotted in Fig. 11-6. For flow around a cylinder there is no analytical solution such as that found by Stokes for a sphere, but Fig. 11-6 shows empirically that for low Reynolds number C_D is also inversely proportional to **Re**. The portion of the curve in Fig. 11-6 with slope approximately -1 indicates the range of Reynolds numbers for which the preceding approximation is valid. As the Reynolds number is increased, viscous effects become relatively less important. The sharp drop in C_D at approximately **Re** = 500,000 corresponds to the transition from a laminar to a turbulent boundary layer, with the resulting shift of the separation point, as discussed on p. 134. In general, the relation between C_D and **Re** must be found by experiment for flow past each specific shape.

Boundary-Layer Equations

Certain terms in the Navier-Stokes equations are negligible when applied to laminar flow in a boundary layer; let us now investigate this case. These assumptions are valid only at high Reynolds numbers, for which the thickness of the boundary layer is small compared with distance from the leading edge. They are not valid at low Reynolds numbers, for which the region where viscosity is important extends relatively far from the wall, as, to mention an extreme case, in creeping flow. The theory of ideal fluids also holds at high Reynolds numbers, but since this theory requires slip of the fluid over a solid boundary, results obtained from it do not agree with physical reality for the layer of fluid near a boundary, where viscosity cannot be neglected and where the velocity is changing to its ultimate value of zero at the solid surface. We have seen that this physical unreality is particularly harmful in estimating the drag, for which ideal-fluid theory usually gives erroneous results.

We shall now see what simplifications can be made in applying the equations of motion to a two-dimensional, laminar, incompressible boundary layer at steady state. We also neglect body forces (or use the dynamic pressure); the x and y equations are, then,

$$u_x \frac{\partial u_x}{\partial x} + u_y \frac{\partial u_x}{\partial y} = -\frac{g_c}{\rho}\frac{\partial p}{\partial x} + \frac{\mu}{\rho}\left(\frac{\partial^2 u_x}{\partial x^2} + \frac{\partial^2 u_x}{\partial y^2}\right) \tag{11-7}$$

and

$$u_x \frac{\partial u_y}{\partial x} + u_y \frac{\partial u_y}{\partial y} = -\frac{g_c}{\rho}\frac{\partial p}{\partial y} + \frac{\mu}{\rho}\left(\frac{\partial^2 u_y}{\partial x^2} + \frac{\partial^2 u_y}{\partial y^2}\right) \tag{11-8}$$

Figure 11-7 is a sketch of the boundary layer. Its thickness is defined as the distance from the surface to the point where the velocity u_x has a value of $0.99u_o$; u_o is the velocity at the surface predicted by the theory of ideal fluids.

As first suggested by Prandtl,[1] since the boundary layer is presumed very thin and lies on a solid surface, u_y is very small compared with u_x, and $\partial u_x/\partial y$ is large

[1] L. Prandtl, "Über Flüssigkeitsbewegung bei sehr kleiner Reibung," *Proc. III Intern. Math. Congr.*, Heidelberg, 1904. For a good nonmathematical discussion of fluid dynamics in general, see L. Prandtl, "The Essentials of Fluid Dynamics," Hafner Publishing Company, New York, 1949.

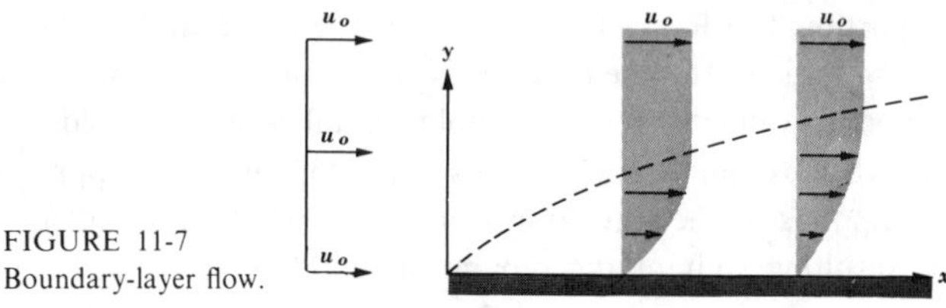

FIGURE 11-7
Boundary-layer flow.

compared with $\partial u_x/\partial x$. This means that $u_x(\partial u_x/\partial x)$ and $u_y(\partial u_x/\partial y)$ are of about the same order of magnitude, and these in turn are comparable with $\mu/\rho(\partial^2 u_x/\partial y^2)$. On the other hand, $\mu/\rho(\partial^2 u_x/\partial x^2)$ is negligible in comparison with the other terms in Eq. (11-7).

In Eq. (11-8), a similar analysis shows that all the terms containing u_y and its derivatives are small. This leads to the conclusion that $g_c/\rho(\partial p/\partial y)$ is small; in other words, the pressure varies but little from the surface to the edge of the boundary layer. This result is important, for the variation of pressure with x can be found from potential flow, so that $\partial p/\partial x$ may be considered as given and as independent of y in the boundary layer.

The problem is now reduced to that of the simultaneous solution of the equation of motion for the x direction and the equation of continuity.

$$u_x \frac{\partial u_x}{\partial x} + u_y \frac{\partial u_x}{\partial y} = -\frac{g_c}{\rho}\frac{dp}{dx} + \frac{\mu}{\rho}\frac{\partial^2 u_x}{\partial y^2} \qquad (11\text{-}9)$$

$$\frac{\partial u_x}{\partial x} + \frac{\partial u_y}{\partial y} = 0 \qquad (10\text{-}51)$$

Since dp/dx is given by the potential, frictionless-flow solution, these two equations are sufficient to solve for the two unknowns u_x and u_y. In addition it is known that at $y = 0$, $u_x = u_y = 0$, and at $y = \delta$, $u_x = u_o$. For a curved boundary, x can be measured along the surface and y normal to the surface, if the radius of curvature is large compared with δ.

Flow Past a Flat Plate

For flow past a flat plate Eq. (11-9) is further simplified in that dp/dx is zero, since u_o is constant. The solution of this problem for laminar flow, giving u_x and u_y as a function of x and y, was first obtained by Blasius[1] and later elaborated upon by Howarth.[2] Equations (11-9) and (10-51), even though much simpler than the

[1] H. Blasius, *Z. Math. u. Phys.*, **56**:1 (1908); also *NACA Tech. Mem.* 1256.
[2] L. Howarth, *Proc. Roy. Soc. London*, **A164**:547 (1938).

equations for the general case, still present considerable mathematical difficulty. The details of the solution will not be given in full, but the general procedure will be outlined.

Equations (11-9) and (10-51) can be reduced to a single partial differential equation by using the stream function, which we showed in Chap. 10 to be related to the velocities by the equations

$$u_x = \frac{\partial \psi}{\partial y} \tag{10-39}$$

$$u_y = -\frac{\partial \psi}{\partial x} \tag{10-40}$$

for an incompressible fluid. The stream function was defined so as to satisfy the equation of continuity. The substitution of Eqs. (10-39) and (10-40) into (10-51) will confirm this statement. If u_x and u_y in Eq. (11-9) are now replaced by using (10-39) and (10-40), there results

$$\frac{\partial \psi}{\partial y}\frac{\partial^2 \psi}{\partial x\,\partial y} - \frac{\partial \psi}{\partial x}\frac{\partial^2 \psi}{\partial y^2} = \nu \frac{\partial^3 \psi}{\partial y^3} \tag{11-10}$$

This third-order, nonlinear partial differential equation must be solved for $\psi(x,y)$, subject to the boundary conditions already given.

The solution of a partial differential equation such as (11-10) cannot be obtained by any set mathematical technique, but usually depends on a trial-and-error procedure involving considerable mathematical and physical intuition. In the present case, the stratagem which leads to a solution arises from the fact that the velocity profiles at various points along the plate are similar—a conclusion which we may regard as based on physical intuition. Therefore the velocity at any point along the plate should be a unique function of y/δ. The boundary-layer thickness δ is a function of x, but it increases less rapidly than the first power of x. Since there is some evidence from other solutions (not given here) of the Navier-Stokes equations that δ is proportional to $\sqrt{x}$, let us say that u_x is a function of $y/\sqrt{x}$.

$$u_x = f_1\left(\frac{y}{\sqrt{x}}\right) \tag{11-11}$$

Equation (11-10) is in terms of ψ, so we need a relation between ψ and u_x. We obtain this from Eq. (10-39). If we arbitrarily set $\psi = 0$ as the streamline at $y = 0$, $x = 0$, then we can write as an approximation, at a point x,

$$\psi = \frac{\partial \psi}{\partial y}\delta \tag{11-12}$$

The use of Eq. (10-39) and the fact that $\delta \propto \sqrt{x}$ permit us to write

$$\psi \propto u_x \sqrt{x} \qquad (11\text{-}13)$$

and, using Eq. (11-11),

$$\psi \propto \sqrt{x}\, f_1\left(\frac{y}{\sqrt{x}}\right)$$

$$\psi = (\text{const})\sqrt{x}\, f_1\left(\frac{y}{\sqrt{x}}\right) \qquad (11\text{-}14)$$

Equations (11-11) and (11-14) form a basis for the simplification of Eq. (11-10). It is found that the substitution of Eqs. (11-11) and (11-14) into (11-10) changes the latter to an *ordinary* differential equation involving $f_1(y/\sqrt{x})$ and $y/\sqrt{x}$.

In place of the simple relations (11-11) and (11-14) given above, it is customary to define the quantities used to simplify Eq. (11-10) in terms of dimensionless variables, as follows:

$$\eta = y\sqrt{\frac{u_o}{\nu x}} \qquad (11\text{-}15)$$

and

$$f(\eta) = \frac{\psi}{\sqrt{x \nu u_o}} \qquad (11\text{-}16)$$

where η can be regarded as a dimensionless position variable and $f(\eta)$ as a dimensionless stream function. Equations (11-15) and (11-16) are called similarity transformations.

Using the rules of partial differentiation[1] and the above new quantities, Eq. (11-10) is transformed by a lengthy procedure into

$$f(\eta)\frac{d^2 f(\eta)}{d\eta^2} + 2\frac{d^3 f(\eta)}{d\eta^3} = 0 \qquad (11\text{-}17)$$

or written more simply,

$$ff'' + 2f''' = 0 \qquad (11\text{-}18)$$

Equations (11-15) and (11-16) have been contrived so that x, y, ψ, ν, and u_o are all eliminated in going from (11-10) to (11-18).

The boundary conditions on Eq. (11-18) are found from the definitions

$$u_x = \frac{\partial \psi}{\partial y} = u_o f' \qquad (11\text{-}19)$$

and

$$u_y = -\frac{\partial \psi}{\partial x} = \frac{1}{2}\sqrt{\frac{\nu u_o}{x}}(\eta f' - f) \qquad (11\text{-}20)$$

[1] A succinct summary of these rules is given by H. S. Mickley, T. K. Sherwood, and C. E. Reed, "Applied Mathematics in Chemical Engineering," 2d ed., pp. 209–213, McGraw-Hill Book Company, New York, 1957.

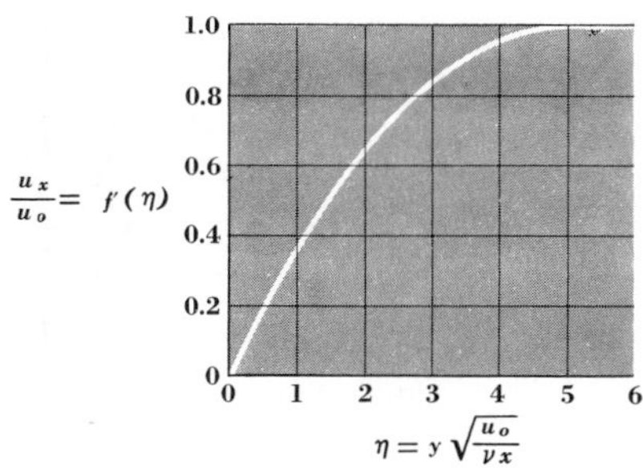

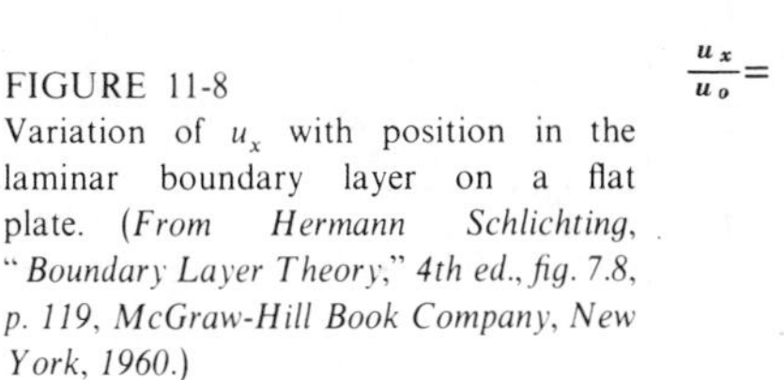
FIGURE 11-8
Variation of u_x with position in the laminar boundary layer on a flat plate. (*From Hermann Schlichting, "Boundary Layer Theory," 4th ed., fig. 7.8, p. 119, McGraw-Hill Book Company, New York, 1960.*)

Thus, at $y = 0$, where $u_x = u_y = 0$, we have $\eta = 0$ and $f = f' = 0$, and at $y = \infty$, where $u_x = u_o$, $\eta = \infty$ and $f' = 1.0$.

Equation (11-18) has not been solved to give a closed form, but the following series solution has been obtained:

$$f = 0.16603\eta^2 - 4.5943\eta^5 \times 10^{-4} + 2.4972\eta^8 \times 10^{-6} - 1.4277\eta^{11} \times 10^{-8} + \cdots \qquad (11\text{-}21)$$

The derivation of this series and the determination of the constants from the boundary values are not given here.[1] Equation (11-21) can be used with Eqs. (11-19) and (11-20) to find the values of u_x and u_y, which are represented in Figs. 11-8 and 11-9. Figure 11-8 shows that the boundary-layer thickness is approximately

$$\delta = 5.0\sqrt{\frac{\nu x}{u_o}} \qquad (11\text{-}22)$$

Drag in Flow Past a Flat Plate

The drag on a flat plate is calculated from the shear stress at the surface. At any point on the surface a distance x from the leading edge,

$$\tau_s = \frac{\mu}{g_c}\left(\frac{\partial u_x}{\partial y}\right)_{y=0} \qquad (11\text{-}23)$$

where τ_s is the shear stress at $y = 0$. Since we know u_x as a function of x and y from the solution for the laminar boundary layer, we obtain, from Eqs. (11-21) and (11-23),

$$\tau_s = 0.332\frac{\mu}{g_c}u_o\sqrt{\frac{u_o}{\nu x}} \qquad (11\text{-}24)$$

[1] See H. Schlichting and Kestin, "Boundary Layer Theory," 6th ed., chap. 7, McGraw-Hill Book Company, New York, 1968. A table of $f(\eta)$ is given on p. 129.

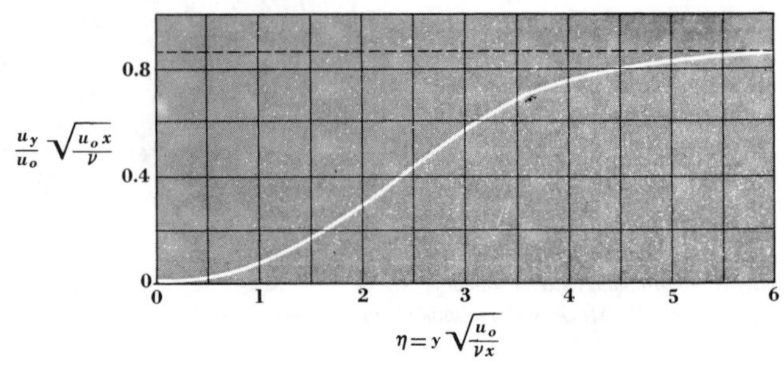

FIGURE 11-9
Variation of u_y with position in the laminar boundary layer on a flat plate. (*From Hermann Schlichting, "Boundary Layer Theory," 4th ed., fig. 7.9, p. 120, McGraw-Hill Book Company, New York, 1960.*)

where τ_s is a function of x. The total drag is given by

$$F_d = b \int_0^L \tau_s \, dx \qquad (11\text{-}25)$$

for a plate of width b and length L. This drag is caused by skin friction; there is no form drag. The substitution of Eq. (11-24) into (11-25) yields

$$F_d = 0.332 \frac{\mu b u_o}{g_c} \sqrt{\frac{u_o}{\nu}} \int_0^L \frac{dx}{\sqrt{x}}$$

$$= \frac{0.664 b u_o}{g_c} \sqrt{\mu \rho L u_o} \qquad (11\text{-}26)$$

and so

$$F_d = \frac{0.664 b}{g_c} \sqrt{\mu \rho L u_o^3} \qquad (11\text{-}27)$$

The drag coefficient related to the total drag on a plate of length L can be written for flow past one side of the plate as

$$C_D = \frac{F_d g_c}{\frac{1}{2} u_o^2 \rho A} \qquad (11\text{-}28)$$

where A equals bL. Equations (11-27) and (11-28) may be combined to give

$$C_D = 1.328 \sqrt{\frac{\nu}{L u_o}} \qquad (11\text{-}29)$$

Because the Reynolds number for this case is Lu_o/ν, the drag coefficient is given by

$$C_D = \frac{1.328}{\sqrt{\mathbf{Re}_L}} \qquad (11\text{-}30)$$

This formula, as well as all the others derived above, applies only to the laminar boundary layer, for which $\mathbf{Re}_L$ is less than about 5×10^5. The results also are valid only for positions sufficiently far from the leading edge so that L (or x) is much greater than δ, as was assumed in simplifying the Navier-Stokes equation to obtain Eq. (11-9). The results obtained above for the velocity distribution and for the drag have been verified by many experiments.

The boundary-layer flow past many other shapes has been analyzed by the general method illustrated by the flow past a flat plate. The presence of the pressure-gradient term complicates the analysis in the general case, and the subject of separation and wakes is difficult to handle analytically. The chemical engineer is interested in flow past cylinders as on the shell side of a heat exchanger and past spheres or other particles in packed and fluidized beds. At present these problems are too complicated for an analytical approach, but future progress probably will come from using the principle of boundary-layer flow. Some progress has been made in relating the solution for flow past a single sphere to the flow through a bed of spheres.

The Blasius solution for the laminar boundary layer illustrates an important point in the mathematical approach to engineering problems. The mathematician, armed with expert knowledge of analysis, is unable to solve the general Navier-Stokes equations as applied to most cases. Even the solution of the simple problem of flow past a flat plate required a knowledge of the physical situation and physical principles, plus some luck or intuition, to find out which terms could safely be dropped and what substitutions might be made to simplify the differential equation so as to arrive at its solution. This is one of the chief reasons why the engineer must acquire sufficient mathematical prowess to attack these analytical problems and not leave them solely to the mathematician.

Flow in the Entrance Section of a Conduit

The way in which the velocity profile develops has been illustrated in Fig. 11-3. For a rectangular duct of infinite width a boundary layer develops on both the top and bottom walls. This situation is similar to flow past a flat plate, and an estimate of the length of the entrance section (i.e., the length up to the point where $\delta = y_o$) might be made by using Eq. (11-22). However, this approximate procedure does not give a correct result, for the fluid in the central part of the duct does not continue to move

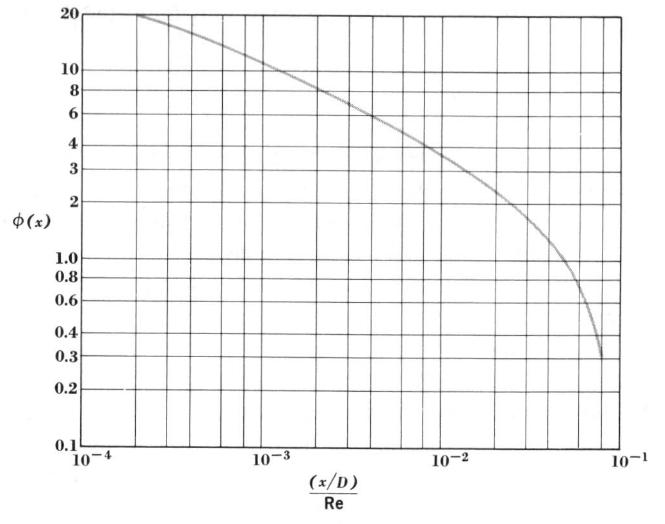

FIGURE 11-10
Function for use with Eq. (11-32). [*H. L. Langhaar, Trans. Am. Soc. Mech. Eng.*, **A64**:*55* (*1942*).]

at u_o but is accelerated in order to satisfy the equation of continuity. Similar remarks apply also to flow near the entrance to a circular tube.

Langhaar[1] has found an approximate solution of the equations of motion near the entrance to a circular pipe. The entrance length is given by

$$\frac{L_e}{D} = 0.0575\ \mathbf{Re} \qquad (11\text{-}31)$$

He also finds for the velocity profile

$$\frac{u_x}{u_o} = \frac{I_o[\phi(x)] - I_o[(r/r_i)\phi(x)]}{I_2[\phi(x)]} \qquad (11\text{-}32)$$

I_o and I_2 are modified Bessel functions of the first kind, and $\phi(x)$ is given in Fig. 11-10. Velocity distributions at several points near the entrance of a pipe are illustrated in Fig. 11-11.

[1] H. L. Langhaar, *Trans. Am. Soc. Mech. Eng.*, **A64**:55 (1942).

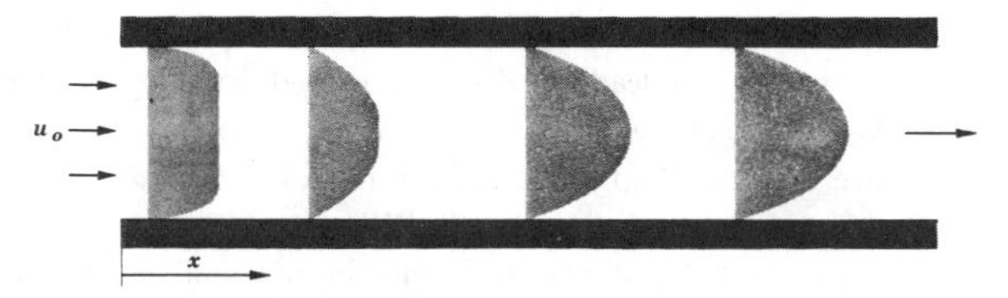

FIGURE 11-11
Velocity profiles near a pipe entrance; laminar flow.

Recapitulation

It is of interest to summarize the types of problems which can and cannot be solved by the application of the principles of fluid dynamics studied to this point.

1 The overall balances permit us to calculate an input or an output of a process when a sufficient number of the other inputs and outputs are known. The overall momentum balance, in particular, has given us a useful method for the analysis of flow through an orifice and other cases involving large inertia effects.

2 The differential momentum balance has enabled us to find solutions for the isothermal, laminar, steady-state flow of an incompressible fluid in a conduit far from its entrance. Similarly, a solution has been obtained for the laminar boundary layer on a flat plate.

3 On the other hand, we have not solved problems involving turbulent flow in a conduit or the turbulent boundary layer, or, for that matter, any boundary-layer flow past the point where separation occurs and a wake is formed. Some progress has been made in the analytical attack on these problems, but the main practical methods the engineer has for their solution depend on the idea of dimensional analysis and the use of considerable empiricism. These methods are the subject of the next few chapters of this book.

PROBLEMS

11-1 Air at 60°F and 1 atm is flowing past a flat plate at 30 ft/s. How thick is the boundary layer 1 ft from the leading edge? Calculate also the drag on the first foot of a 1-ft-wide plate and the direction and magnitude of the velocity at $x = 1$ ft and $y = \delta/2$.

11-2 Water at 68°F flows past a flat plate at 3 ft/s. Calculate the direction and magnitude of the velocity vector at a point 1 ft from the leading edge and 3×10^{-3} ft from the plate.

11-3 Water at 30°C flows past a flat plate at 1 m/s. Calculate the vorticity at a point 0.2 m from the leading edge and one-half way through the boundary layer.

11-4 What is the drag coefficient for water flowing over a flat plate at 20°C and 0.25 m/s over the length up to 0.2 m from the leading edge? What would be the point (local) drag coefficient at this point? What is τ_s (Pa)?

11-5 Air at 101.3 kPa abs and 20°C is flowing in a wind tunnel at a velocity of 10 m/s. What is the force in N/m on a 0.1-m-diameter cylinder held in this flow? If the fluid were water at 20°C, what would be the corresponding force?

11-6 Sketch on a log-log plot the drag force on a cylinder as a function of Reynolds number for a given fluid and cylinder. Be sure the slope of the curve is correctly indicated for the various regions.

11-7 Water at 20°C is flowing into a pipe at such a velocity that the Reynolds number far downstream from the entrance is 1000. Make a plot of u_x/u_o at a point where $x = \frac{1}{4}L_e$. The modified Bessel function of the first kind is given by the series

$$I_p(x) = \sum_{k=0}^{\infty} \frac{\left(\dfrac{x}{2}\right)^{2k+p}}{k!\,(k+p)!}$$

11-8 Calculate the shear stress at the wall for the conditions of Prob. 11-7 for a pipe of 0.02-m ID.

11-9 Show that Eq. (11-32) reduces to the parabolic velocity distribution for $x \gg L_e$, for which $\phi(x)$ tends toward zero.

11-10 The Blasius series solution for the velocity profile in a laminar boundary layer is often replaced by empirical polynomials such as

$$u_x/u_o = a_0 + a_1(y/\delta) + a_2(y/\delta)^2$$

Write three boundary conditions which apply to the laminar boundary layer on a flat plate, and use them to solve for the constants a_0, a_1, and a_2.

11-11 A $\frac{1}{2}$-in-diameter cylinder is placed in a wind tunnel and air flows past the cylinder at 100 ft/s, a temperature of 80°F, and a pressure of 1 atm. What is the force holding the cylinder stationary? Assume the velocity profile of the air to be flat.

12

VELOCITY DISTRIBUTION AND DRAG WITH TURBULENT FLOW

SOME FUNDAMENTAL IDEAS ABOUT TURBULENCE

Most of the practical fluid-flow problems with which an engineer deals involve turbulent, rather than laminar, flow. There are a number of laminar-flow problems for which the equations of motion have been solved exactly, and many more for which the equations can be solved by certain approximations without greatly affecting the validity of the results. Several examples illustrating these assertions are to be found in the preceding chapters. However, for turbulent flow there exists not a single exact solution. The approximate equations describing turbulent flow depend on so many assumptions that it is difficult to tell whether agreement with experiment is the result of reasonable simplifications or of fortuitous cancellation of the errors arising from the assumptions. In spite of the difficulty of obtaining a complete, theoretical solution of the problem of turbulent flow, many very useful quantitative relations have been obtained by a combination of theoretical reasoning and empiricism, and we shall proceed to develop some of the more important ones, after first discussing the basic nature of turbulence.[1]

[1] J. T. Davies, "Turbulence Phenomena," Academic Press, New York, 1972; H. Tennekes and J. L. Lumley, "A First Course in Turbulence," MIT Press, Cambridge, Mass., 1972.

The Origin of Turbulence

Turbulence usually appears in pipe flow for $\mathbf{Re} > 2100$ and for boundary-layer flow for $\mathbf{Re}_L > 5 \times 10^5$, but laminar flow has been obtained for considerably higher Reynolds numbers. The incipience of turbulence depends on the severity of disturbances in the flow, such as may occur at the pipe entrance; the more care taken to avoid disturbances, the higher is the Reynolds number at which laminar flow persists.

A theoretical analysis[1] has been developed for the calculation of the lowest possible Reynolds number for which turbulent flow may exist. The method consists in including in the analysis of laminar flow, using the equations of motion, a small sinusoidal fluctuation in the velocity; if the amplitude of the fluctuation increases with time, turbulence may develop; if the amplitude is damped, the Reynolds number is below that for which turbulence may exist. The results of this theory agree with experiment, and they are of practical importance in aerodynamics. For a streamlined body, such as an airfoil, the total drag can be appreciably reduced if laminar flow in the boundary layer can be caused to persist to a higher value of $\mathbf{Re}_L$ (i.e., farther from the leading edge). The theory also accounts for the effect of various factors on the critical $\mathbf{Re}_L$.

Above the theoretical minimum Reynolds number there exists the possibility of two solutions for the equations of motion: a laminar solution and a turbulent one. The persistence of laminar flow above the critical Reynolds number in the absence of any disturbance is an unstable case similar to the persistence of a pure vapor phase below the dew point or other phenomena of supersaturation.

Mean and Fluctuating Velocities

In this discussion of turbulent flow we shall consider only flow at steady state with respect to the mean flow; the meaning of this term will emerge from the following discussion.

In turbulent flow the velocity at a point varies chaotically with time in magnitude and direction, so that, strictly speaking, there is no such thing as steady state in turbulent flow. However, we can define a mean velocity in the x direction by

$$\bar{u}_x = \frac{1}{\theta}\int_0^\theta u_x \, d\theta \qquad (12\text{-}1)$$

where u_x is the instantaneous velocity, a function of time. Since the frequency of the velocity fluctuation is large, θ need be only a few seconds. Similar definitions can be written for the other velocity components and for the pressure. If all these mean

[1] A summary of this subject is given by H. Schlichting and Kestin, "Boundary Layer Theory," 6th ed., chaps. 16 and 17, McGraw-Hill Book Company, New York, 1968.

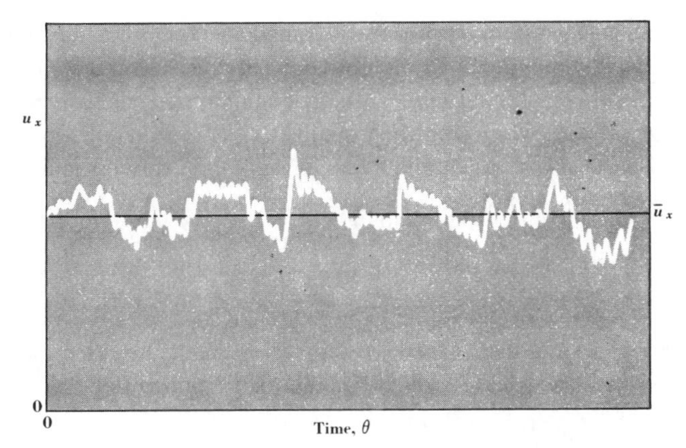

FIGURE 12-1
Velocity fluctuation in turbulent flow.

quantities are constant over successive time intervals, the turbulent flow is said to be at steady state or, strictly speaking, at steady state with respect to the mean flow.

The instantaneous variables are conveniently expressed as the sum of a mean value and a fluctuating value, as follows:

$$u_x = \bar{u}_x + u'_x \qquad u_y = \bar{u}_y + u'_y \qquad u_z = \bar{u}_z + u'_z \quad p = \bar{p} + p' \tag{12-2}$$

where u'_x, u'_y, u'_z = components of fluctuating velocity
p' = fluctuating pressure

From these definitions, it is evident that

$$\bar{u}'_x = \frac{1}{\theta}\int_0^\theta u'_x \, d\theta = 0 \tag{12-3}$$

Similarly, we find that $\bar{u}'_y$, $\bar{u}'_z$, and $\bar{p}'$ are zero. In addition, for parallel flow, we have

$$\bar{u}_y = 0 \qquad \bar{u}_z = 0 \tag{12-4}$$

so that in this case

$$u_x = \bar{u}_x + u'_x \qquad u_y = u'_y \qquad u_z = u'_z \tag{12-5}$$

A typical graph of u_x versus θ is shown in Fig. 12-1.

The Intensity of Turbulence

Turbulent flow is often described in terms of intensity, which is a measure of the importance of the fluctuating velocity relative to the mean velocity. The quantity $\bar{u}'_x$ is zero because the time average of the positive values of u'_x equals the time average of the negative ones. However, the time average of the magnitude of the fluctuating-velocity component $\overline{|u'|}$ is not zero and gives a measure of the amplitude of the oscillations of the velocity. A more common way of expressing the average fluctuating velocity is by using the quantity $\sqrt{\overline{u'^2}}$, called the *root-mean-square fluctuating velocity*. For parallel flow the intensity, or level of turbulence, is defined quantitatively by the expression

$$I = \frac{\sqrt{\frac{1}{3}(\overline{u_x'^2} + \overline{u_y'^2} + \overline{u_z'^2})}}{\bar{u}_x} \tag{12-6}$$

In the special case called *isotropic turbulence*, the three mean-square fluctuating velocities are equal, so that Eq. (12-6) is simplified to

$$I = \frac{\sqrt{\overline{u_x'^2}}}{\bar{u}_x} \tag{12-7}$$

The Scale of Turbulence

It has been found that the intensity alone does not completely characterize turbulent motion. Some method must also be available to specify the size of the eddies, usually referred to as the scale of turbulence. The velocity at one point in turbulent flow is often related to that at an adjacent point. Groups of fluid particles tend to move together and form eddies of sizes varying with the type of turbulence. Considering again the velocities at two separate points, if the second point is far enough from the first, there is no relation between the two velocities; the two points may be thought of as being in different eddies, or lumps of fluid.

Turbulence on a large scale in the atmosphere is familiar to us. Although two anemometers a mile apart in a flat region may indicate the same mean velocity, there is no correlation between the two instantaneous velocities; one varies in a random fashion with respect to the other. If the two instruments are brought closer together, a point is found where they start to indicate similar velocities at any instant, and the correlation between the two velocities increases as the distance between the points is reduced. Similar phenomena exist on a smaller scale in pipes or wind tunnels. A correlation coefficient is defined quantitatively by

$$R(y) = \frac{\overline{u_{x1}' u_{x2}'}}{\sqrt{\overline{u_{x1}'^2}}\sqrt{\overline{u_{x2}'^2}}} \tag{12-8}$$

for parallel flow in the x direction, where u_{x_1}' and u_{x_2}' are the fluctuating velocities read at the same instant at points 1 and 2, which are separated by a distance y. The correlation coefficient may vary with y as shown in Fig. 12-2.

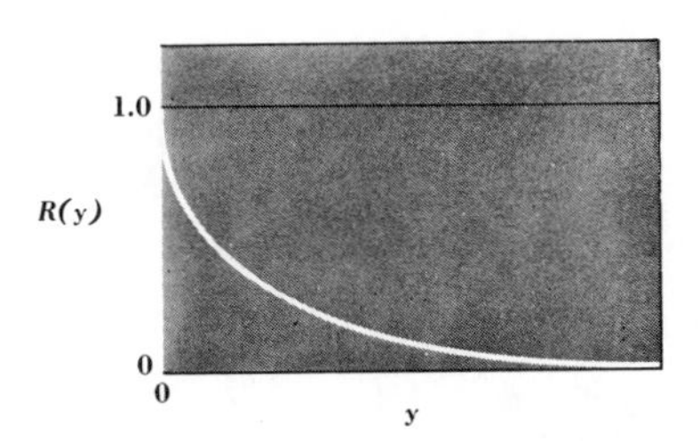

FIGURE 12-2
Variation of correlation coefficient with distance of separation.

As one might suspect, there is no definite boundary between eddies; they merge gradually into each other. A scale of turbulence is often defined in relation to $R(y)$ by the equation

$$L = \int_0^\infty R(y)\,dy \qquad (12\text{-}9)$$

An eddy in parallel flow might be considered to have a width of $2L$.

Many other correlation coefficients can be defined in terms of the various velocity components and their variation with the coordinate directions or with time. The type of analysis introduced here is greatly extended in what is called the statistical theory of turbulence.[1]

Turbulent Shear Stresses

Let us now investigate the application of the Navier-Stokes equations to the steady-state turbulent flow of an incompressible fluid. The equation of continuity (7-12) is written

$$\frac{\partial u_x}{\partial x} + \frac{\partial u_y}{\partial y} + \frac{\partial u_z}{\partial z} = 0 \qquad (12\text{-}10)$$

and then u_x is replaced by $\bar{u}_x + u'_x$, u_y by $\bar{u}_y + u'_y$, and u_z by $\bar{u}_z + u'_z$. The result is

$$\frac{\partial \bar{u}_x}{\partial x} + \frac{\partial \bar{u}_y}{\partial y} + \frac{\partial \bar{u}_z}{\partial z} + \frac{\partial u'_x}{\partial x} + \frac{\partial u'_y}{\partial y} + \frac{\partial u'_z}{\partial z} = 0 \qquad (12\text{-}11)$$

[1] Basic advanced texts on turbulence are: J. O. Hinze, "Turbulence," McGraw-Hill Book Company, New York, 1959; Y. K. Batchelor, "The Theory of Homogeneous Turbulence," Cambridge University Press, London, 1953; A. A. Townsend, "The Structure of Turbulent Shear Flow," Cambridge University Press, London, 1956. See also R. S. Brodkey, "The Phenomena of Fluid Motion," Addison-Wesley Publishing Company, Reading, Mass., 1967.

Now we take the time average of Eq. (12-11) and use the following relations:

$$\overline{\frac{\partial u_x'}{\partial x}} = \frac{\partial \bar{u}_x'}{\partial x} \qquad \overline{\frac{\partial u_y'}{\partial y}} = \frac{\partial \bar{u}_y'}{\partial y} \qquad \overline{\frac{\partial u_z'}{\partial z}} = \frac{\partial \bar{u}_z'}{\partial z}$$

$$\overline{\frac{\partial \bar{u}_x}{\partial x}} = \frac{\partial \bar{u}_x}{\partial x} \qquad \overline{\frac{\partial \bar{u}_y}{\partial y}} = \frac{\partial \bar{u}_y}{\partial y} \qquad \overline{\frac{\partial \bar{u}_z}{\partial z}} = \frac{\partial \bar{u}_z}{\partial z}$$

and also

$$\bar{u}_x' = \bar{u}_y' = \bar{u}_z' = 0$$

The results are

$$\frac{\partial \bar{u}_x}{\partial x} + \frac{\partial \bar{u}_y}{\partial y} + \frac{\partial \bar{u}_z}{\partial z} = 0 \qquad (12\text{-}12)$$

and

$$\frac{\partial u_x'}{\partial x} + \frac{\partial u_y'}{\partial y} + \frac{\partial u_z'}{\partial z} = 0 \qquad (12\text{-}13)$$

The equation of continuity is satisfied for both the mean velocities and the fluctuating velocities.

We now turn our attention to the equation of motion for the x direction (9-52).

$$u_x \frac{\partial u_x}{\partial x} + u_y \frac{\partial u_x}{\partial y} + u_z \frac{\partial u_x}{\partial z} + \frac{\partial u_x}{\partial \theta} = -\frac{g_c}{\rho}\frac{\partial p}{\partial x} + \frac{\mu}{\rho}\left(\frac{\partial^2 u_x}{\partial x^2} + \frac{\partial^2 u_x}{\partial y^2} + \frac{\partial^2 u_x}{\partial z^2}\right) \qquad (12\text{-}14)$$

The body force is omitted. It is convenient to change the left-hand side of Eq. (12-14) so that it becomes

$$\frac{\partial (u_x)^2}{\partial x} + \frac{\partial (u_x u_y)}{\partial y} + \frac{\partial (u_x u_z)}{\partial z} + \frac{\partial u_x}{\partial \theta} = \frac{-g_c}{\rho}\frac{\partial p}{\partial x} + \frac{\mu}{\rho}\left(\frac{\partial^2 u_x}{\partial x^2} + \frac{\partial^2 u_x}{\partial y^2} + \frac{\partial^2 u_x}{\partial z^2}\right) \qquad (12\text{-}15)$$

It is easy to verify that the left-hand sides of these equations are identical, provided Eq. (12-10) holds. The next step in putting the equation of motion into a more instructive form is to substitute in Eq. (12-15) for the variables in terms of their mean and fluctuating values, and then to take the time average. We shall do this term by term. For the first term we have

$$\overline{\frac{\partial (u_x)^2}{\partial x}} = \overline{\frac{\partial (\bar{u}_x + u_x')^2}{\partial x}} = \overline{\frac{\partial}{\partial x}(\bar{u}_x^{\,2} + 2\bar{u}_x u_x' + u_x'^2)} \qquad (12\text{-}16)$$

Since $\bar{u}_x$ is constant with respect to time and $\overline{u_x'}$ is zero,

$$\overline{\frac{\partial (2\bar{u}_x u_x')}{\partial x}} = 0 \qquad (12\text{-}17)$$

and for the first term in Eq. (12-15) we obtain

$$\overline{\frac{\partial u_x^{\,2}}{\partial x}} = \frac{\partial \bar{u}_x^{\,2}}{\partial x} + \frac{\partial \overline{u_x'^2}}{\partial x} \qquad (12\text{-}18)$$

In these manipulations we have used the rule that $\overline{\partial u_x/\partial x} = \partial \bar{u}_x/\partial x$, etc. From the previous discussion of the intensity of turbulence it follows that $\overline{u_x'^2}$ is not zero. even though $\bar{u}_x'^2$ is.

For the second term we get

$$\overline{\frac{\partial(u_x u_y)}{\partial y}} = \overline{\frac{\partial}{\partial y}(\bar{u}_x \bar{u}_y + u_x' \bar{u}_y + \bar{u}_x u_y' + u_x' u_y')}$$

$$= \frac{\partial(\bar{u}_x \bar{u}_y)}{\partial y} + \frac{\partial(\overline{u_x' u_y'})}{\partial y} \tag{12-19}$$

Although $\bar{u}_x'$ and $\bar{u}_y'$ are both zero, the time-average product $\overline{u_x' u_y'}$ is not zero. Except for isotropic turbulence, there is a high degree of correlation between u_x' and u_y' at a point, as will be explained later. For the remaining terms we obtain, by similar analysis,

$$\overline{\frac{\partial(u_x u_z)}{\partial z}} = \frac{\partial(\bar{u}_x \bar{u}_z)}{\partial z} + \frac{\partial(\overline{u_x' u_z'})}{\partial z} \tag{12-20}$$

$$\overline{\frac{\partial u_x}{\partial \theta}} = \frac{\partial \bar{u}_x}{\partial \theta} = 0 \tag{12-21}$$

$$\overline{\frac{\partial p}{\partial x}} = \frac{\partial \bar{p}}{\partial x} \tag{12-22}$$

and

$$\overline{\frac{\partial^2 u_x}{\partial x^2} + \frac{\partial^2 u_x}{\partial y^2} + \frac{\partial^2 u_x}{\partial z^2}} = \frac{\partial^2 \bar{u}_x}{\partial x^2} + \frac{\partial^2 \bar{u}_x}{\partial y^2} + \frac{\partial^2 \bar{u}_x}{\partial z^2} \tag{12-23}$$

Substituting Eqs. (12-18) to (12-23) into the time average of Eq. (12-15) gives

$$\frac{\partial \bar{u}_x{}^2}{\partial x} + \frac{\partial(\bar{u}_x \bar{u}_y)}{\partial y} + \frac{\partial(\bar{u}_x \bar{u}_z)}{\partial z} + \frac{\partial \overline{u_x'^2}}{\partial x} + \frac{\partial(\overline{u_x' u_y'})}{\partial y} + \frac{\partial(\overline{u_x' u_z'})}{\partial z}$$

$$= \frac{-g_c}{\rho}\frac{\partial \bar{p}}{\partial x} + \frac{\mu}{\rho}\left(\frac{\partial^2 \bar{u}_x}{\partial x^2} + \frac{\partial^2 \bar{u}_x}{\partial y^2} + \frac{\partial^2 \bar{u}_x}{\partial z^2}\right) \tag{12-24}$$

This equation has three additional terms which would not be obtained by merely replacing the variables in Eq. (12-15) by their mean values. Their meaning can be made clearer by writing the time mean of Eq. (9-11) with the body force $X = 0$ and $\partial \bar{u}_x/\partial \theta = 0$ as

$$\frac{D\bar{u}_x}{D\theta} = \frac{\partial \bar{u}_x{}^2}{\partial x} + \frac{\partial(\bar{u}_x \bar{u}_y)}{\partial y} + \frac{\partial(\bar{u}_x \bar{u}_z)}{\partial z} + \frac{\partial \overline{u_x'^2}}{\partial x} + \frac{\partial(\overline{u_x' u_y'})}{\partial y} + \frac{\partial(\overline{u_x' u_z'})}{\partial z}$$

$$= \frac{g_c}{\rho}\left(\frac{\partial \tau_{xx}}{\partial x} + \frac{\partial \tau_{yx}}{\partial y} + \frac{\partial \tau_{zx}}{\partial z}\right) \tag{12-25}$$

The time-mean stresses are the result of the averaging of the stresses arising from the action of the ordinary viscosity. Let us define each of the total mean stresses by $\bar{\tau}_{xx}{}^t$, $\bar{\tau}_{yx}{}^t$, and $\bar{\tau}_{zx}{}^t$ as the sum of the corresponding viscous stresses and turbulent stresses.

$$\bar{\tau}_{xx}{}^t = \bar{\tau}_{xx} + \bar{\tau}_{xx}{}^r \qquad (12\text{-}26)$$

$$\bar{\tau}_{yx}{}^t = \bar{\tau}_{yx} + \bar{\tau}_{yx}{}^r \qquad (12\text{-}27)$$

$$\bar{\tau}_{zx}{}^t = \bar{\tau}_{zx} + \bar{\tau}_{zx}{}^r \qquad (12\text{-}28)$$

The terms $\bar{\tau}_{xx}{}^r$, $\bar{\tau}_{yx}{}^r$, and $\bar{\tau}_{zx}{}^r$ are the turbulent-stress contributions.

We now arbitrarily write Eq. (9-11) for turbulent flow by replacing each velocity component by its time mean value and by replacing each stress component by the total mean stress.

$$\frac{\partial(\bar{u}_x{}^2)}{\partial x} + \frac{\partial(\bar{u}_x\bar{u}_y)}{\partial y} + \frac{\partial(\bar{u}_x\bar{u}_z)}{\partial z} = \frac{g_c}{\rho}\left(\frac{\partial\bar{\tau}_{xx}{}^t}{\partial x} + \frac{\partial\bar{\tau}_{yx}{}^t}{\partial y} + \frac{\partial\bar{\tau}_{zx}{}^t}{\partial z}\right) \qquad (12\text{-}29)$$

For this equation to be correct, it must be consistent with Eqs. (12-25) to (12-28). This leads to the following definitions of the turbulent-stress contributions:

$$\bar{\tau}_{xx}{}^r = -\frac{\rho\overline{u_x'^2}}{g_c} \qquad (12\text{-}30)$$

$$\bar{\tau}_{yx}{}^r = \frac{-\rho\overline{u_x'u_y'}}{g_c} \qquad (12\text{-}31)$$

$$\bar{\tau}_{zx}{}^r = \frac{-\rho\overline{u_x'u_z'}}{g_c} \qquad (12\text{-}32)$$

These stresses are also called the *Reynolds stresses*, or the *apparent stresses*. From Eqs. (9-46), (9-19), and (9-21), the total stresses for incompressible, steady, turbulent flow are

$$\bar{\tau}_{xx}{}^t = -\bar{p} + \frac{2\mu}{g_c}\frac{\partial\bar{u}_x}{\partial x} - \frac{\rho\overline{u_x'^2}}{g_c} \qquad (12\text{-}33)$$

$$\bar{\tau}_{yx}{}^t = \frac{\mu}{g_c}\left(\frac{\partial\bar{u}_x}{\partial y} + \frac{\partial\bar{u}_y}{\partial x}\right) - \frac{\rho\overline{u_x'u_y'}}{g_c} \qquad (12\text{-}34)$$

$$\bar{\tau}_{zx}{}^t = \frac{\mu}{g_c}\left(\frac{\partial\bar{u}_x}{\partial z} + \frac{\partial\bar{u}_z}{\partial x}\right) - \frac{\rho\overline{u_x'u_z'}}{g_c} \qquad (12\text{-}35)$$

Not all the terms in Eqs. (12-33) to (12-35) appear in Eq. (12-24), because some are eliminated by the equation of continuity (12-12) for the incompressible fluid.

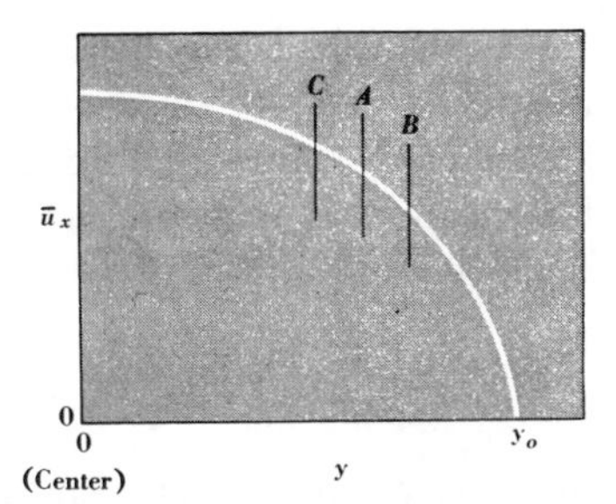

FIGURE 12-3
Turbulent velocity distribution in a slit.

Interpretation of the Reynolds Stresses

The meaning of the Reynolds stresses can be made clearer by considering the case of parallel flow, for which $\bar{u}_z$ and $\bar{u}_y$ are zero.

Consider the flow between infinite parallel plates so that $\bar{u}_x$ is a function of y only. The quantity $\bar{\tau}_{yx}{}^t$ represents the total shearing stress in the x direction acting on a plane perpendicular to the y direction. The mean velocity may be distributed as shown in Fig. 12-3. At the center line $y = 0$, $\bar{u}_x = \bar{u}_{x,\,\max}$, and at the wall $y = y_o$, $\bar{u}_x = 0$. From Eq. (12-34) the contribution of the mean flow $\bar{\tau}_{yx}$ is $\mu/g_c(d\bar{u}_x/dy)$, just as in the case of laminar flow considered in Chap. 10. The Reynolds stress is $-(\rho/g_c)\overline{u'_x u'_y}$, and we should like to show that this term is different from zero and of the correct sign.

Suppose a small packet of fluid having a mean velocity represented by A in Fig. 12-3 is transported, because of a positive u'_y, to a region B, where $\bar{u}_x$ is smaller. Since the packet will retain approximately its original velocity $\bar{u}_{xA}$, at B there is created a positive u'_x, and $u'_x u'_y$ is positive. On the other hand, if u'_y happens to be negative, the fluid is transported to a region C, where the mean velocity is greater than $\bar{u}_{xA}$; a negative u'_x is created, and $u'_x u'_y$ is again positive. It is thus easy to see that $\overline{u'_x u'_y}$ will be a positive quantity, and so, from Eq. (12-31), the mean turbulent stress $\bar{\tau}_{yx}{}^r$ is negative. This negative sign is to be expected, for $d\bar{u}_x/dy$ is negative, making the mean viscous stress $\mu/g_c(d\bar{u}_x/dy)$ also negative.

At the center of the conduit, where $d\bar{u}_x/dy$ is zero, a positive value of u'_y can no longer be associated with a positive value of u'_x, so that $\overline{u'_x u'_y}$ is zero, as it must be, for there is no shear at the center line.

Because there is no velocity variation in the z direction, all the terms of Eq. (12-35) are zero.

Regarding Eq. (12-33), there is no possible turbulent flow for which $\overline{u'^2_x}$ is zero, so that the total normal stress must be different from the mean pressure. If the intensity of turbulence decreases with x, the pressure will decrease with x less rapidly than would otherwise be the case.

Although the discussion of turbulence up to this point has given us some understanding of the behavior of turbulent flow, there is no way of calculating any fluctuating quantities and deducing a velocity distribution or a pressure drop from the equations of motion. In order to make any progress in this direction, more simplifications must be made.

Eddy Viscosity and Mixing Length

By analogy with the molecular viscosity, an eddy kinematic viscosity can be defined for parallel flow by the equation

$$\bar{\tau}_{yx}{}^{r} = \frac{\rho \nu_e}{g_c} \frac{d\bar{u}_x}{dy} \qquad (12\text{-}36)$$

so that

$$\bar{\tau}_{yx}{}^{t} = \frac{\rho}{g_c}(\nu + \nu_e) \frac{d\bar{u}_x}{dy} \qquad (12\text{-}37)$$

This quantity ν_e, introduced by Boussinesq,[1] is also called the *virtual*, or *apparent, viscosity*. Unlike the ordinary viscosity, it is not a function of state and depends strongly on position. Unfortunately, we have no way of calculating ν_e, a priori, although it can be determined experimentally from a given distribution of $\bar{u}_x$ versus y.

We next define a quantity called the *Prandtl mixing length*, which will assist us in calculating the Reynolds stresses. Still considering flow between parallel walls and referring to Fig. 12-3, we define the mixing length in the following way. Consider again a small packet of fluid which is displaced from A to B in the y direction with a velocity u'_y. In reality, the lump of fluid will gradually lose its identity, but in the definition of the mixing length, it is assumed to retain its identity until it has traveled a distance ℓ defined as the Prandtl mixing length. For the small distance involved, we say

$$\frac{d\bar{u}_x}{dy} = \frac{\bar{u}_{xB} - \bar{u}_{xA}}{\ell} \qquad (12\text{-}38)$$

As was mentioned in our previous discussion of Fig. 12-3, the packet of fluid can be assumed to retain its original velocity, so that $\bar{u}_{xB} - \bar{u}_{xA}$ is approximately $-u'_x$. Therefore we have

$$u'_x = -\ell \frac{d\bar{u}_x}{dy}$$

[1] T. V. Boussinesq, in "Mémoires présentées par divers savants à l'Académie des Sciences de l'Institut de France," vol. 23, 1877.

and, in general,

$$\overline{|u_x'|} = \ell \left|\frac{d\bar{u}_x}{dy}\right| \tag{12-39}$$

Prandtl also assumed that $\overline{|u_y'|}$ was of about the same absolute magnitude as $\overline{|u_x'|}$, so that

$$\overline{u_x' u_y'} = \ell^2 \left|\frac{d\bar{u}_x}{dy}\right|^2$$

Since the sign of $\overline{u_x' u_y'}$ depends on the sign of $d\bar{u}_x/dy$, we write this as

$$\overline{u_x' u_y'} = -\ell^2 \left|\frac{d\bar{u}_x}{dy}\right| \frac{d\bar{u}_x}{dy} \tag{12-40}$$

From Eqs. (12-31), (12-36), and (12-40), we get also

$$\bar{\tau}_{yx}{}^r = \frac{\rho \ell^2}{g_c} \left|\frac{d\bar{u}_x}{dy}\right| \frac{d\bar{u}_x}{dy}$$

and

$$\nu_e = \ell^2 \left|\frac{d\bar{u}_x}{dy}\right| \tag{12-41}$$

Although it may appear that the only result of this definition of ℓ has been to replace one empirical, noncomputable quantity with another, the mixing length is easier to estimate than ν_e. For instance, ℓ cannot be greater than the dimensions of

$\bar{u}_{max} = 3.28$ ft/sec
$y =$ distance from center line

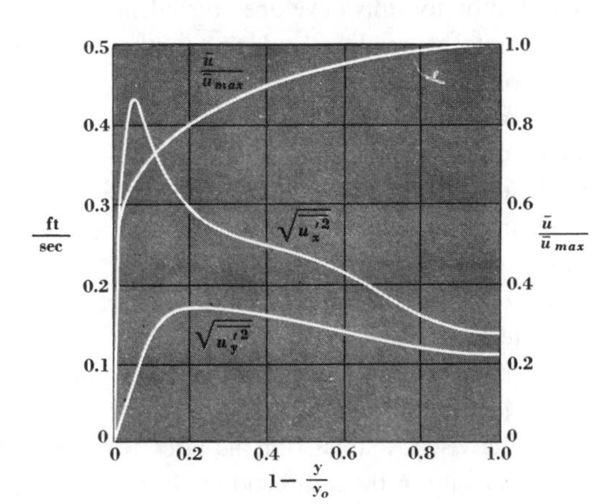

FIGURE 12-4
Intensity of turbulence and velocity profile in a rectangular duct. (*From Hermann Schlichting, "Boundary Layer Theory," 4th ed., fig. 18.3, p. 466, McGraw-Hill Book Company, New York, 1960.*)

the channel, and it should approach zero near the wall. As a matter of fact, in a later section it will be shown that some valuable results for the velocity distribution for turbulent flow in a pipe are obtained with the simple relation

$$\ell = Ky \qquad (12\text{-}42)$$

where y in this equation is the distance from the wall, and K turns out to be a universal constant, 0.4.

Some Data on Turbulence

In order to illustrate the approximate magnitude of the quantities defined in the previous paragraphs, some experimental results are shown. Figure 12-4 gives the variation of the intensity of turbulence in a rectangular duct. Figure 12-5 shows the variation of the stress components for the same duct; a simple force balance such as that leading to Eq. (2-9) proves that $\bar{\tau}_{yx}{}^t$ must be a linear function of y.

TURBULENT FLOW INSIDE A SMOOTH, CIRCULAR PIPE

The Universal Velocity Distribution from the Prandtl Mixing Length

The Prandtl mixing length will now be used to derive an equation for the velocity distribution for fully developed turbulent flow inside a circular pipe. We shall then show how this relation leads to a method of calculating the pressure drop in turbulent flow.

We start by dividing the fluid in the pipe into a central core, in which the shear stress approximately equals the Reynolds stress, and a thin, viscous sublayer near the wall, in which the influence of turbulence is negligible and the shear stress arises only from the molecular viscosity. We shall see that a fuller analysis must include a buffer zone where both stresses are important. The three zones are illustrated in Fig. 12-5. For the present only the turbulent core and the viscous sublayer are considered.

In what follows, we drop the various subscripts and superscripts on the stresses and velocities. The total stress τ is equal either to the Reynolds stress (for the core) or to the viscous stress (for the viscous sublayer); the velocity u is the time mean point velocity in the x direction. The total stress varies from the wall to the center line according to Eq. (2-9), written here as

$$\tau = \tau_s\left(1 - \frac{y}{r_i}\right)$$

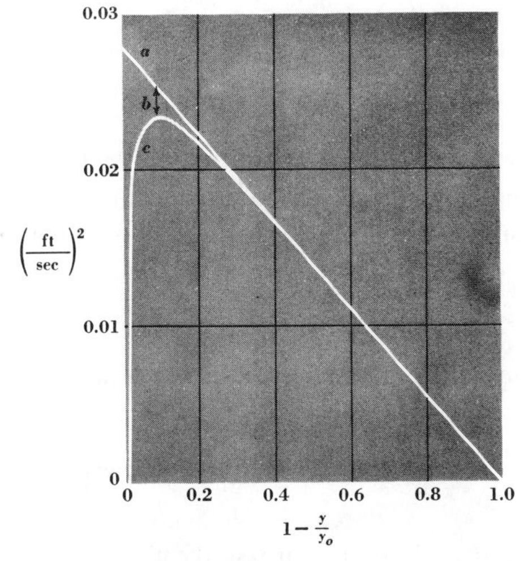

a. $g_c \bar{\tau}^t_{yx}/\rho$

b. $g_c \bar{\tau}_{yx}/\rho = \nu \dfrac{d\bar{u}_x}{dy}$

c. $g_c \bar{\tau}^r_{yx}/\rho = -\overline{u'_x u'_y}$

y is distance from center line

FIGURE 12-5
Shear stress in a rectangular duct. (*From Hermann Schlichting, "Boundary Layer Theory," 4th ed., fig. 18.4, p. 466, McGraw-Hill Book Company, New York, 1960.*)

where τ_s = stress at wall
y = distance from wall

In the viscous sublayer, which is very thin, we neglect any variation in τ and write

$$g_c \tau_s = \mu \frac{du}{dy} = \text{const} \tag{12-43}$$

This equation is integrated to give

$$g_c \tau_s y = \mu u$$

or

$$u = \frac{g_c \tau_s}{\rho} \frac{y}{\nu} \tag{12-44}$$

The velocity in the viscous sublayer is proportional to the distance from the wall; this is in contrast with the parabolic velocity profile in the entire tube when there is laminar flow. We shall find it convenient to define what is called a friction velocity by the equation

$$u^* = \sqrt{\frac{g_c \tau_s}{\rho}} \tag{12-45}$$

In terms of this quantity, Eq. (12-44) can be written as

$$\frac{u}{u^*} = \frac{yu^*}{\nu} \qquad (12\text{-}46)$$

In this equation the dimensionless velocity quotient on the left will be defined as u^+ and the dimensionless distance on the right as y^+. The latter quantity resembles a Reynolds number. It is then possible to write the velocity distribution in the viscous sublayer as

$$u^+ = y^+ \qquad (12\text{-}47)$$

Let us now turn our attention to the turbulent core, in which we neglect any viscous stresses, and write

$$\tau = \frac{\rho \ell^2}{g_c}\left(\frac{du}{dy}\right)^2 \qquad (12\text{-}48)$$

from Eqs. (12-36) and (12-41); du/dy is always positive, so the absolute-magnitude sign has been dropped.

In order to proceed, two assumptions are made, both of which are only rough approximations. Their only real justification is that they simplify the mathematics and that the final equations to which they lead agree well with experimental data. We assume that $\tau = \tau_s =$ const and that

$$\ell = Ky \qquad (12\text{-}49)$$

where K is a constant. The second assumption is reasonable, for we know that ℓ should decrease to zero at the wall, where the turbulence dies out. We then write Eq. (12-48) as

$$\frac{g_c \tau_s}{\rho} = K^2 y^2 \left(\frac{du}{dy}\right)^2 \qquad (12\text{-}50)$$

and so

$$u^* = Ky\frac{du}{dy} \qquad (12\text{-}51)$$

This equation is integrated to

$$u^* \ln y = Ku + c \qquad (12\text{-}52)$$

and c is found from the condition that u must be zero at some small value of y, say y_o:

$$u^* \ln y_o = c \qquad (12\text{-}53)$$

and

$$\frac{u}{u^*} = \frac{1}{K}\ln\frac{y}{y_o} \qquad (12\text{-}54)$$

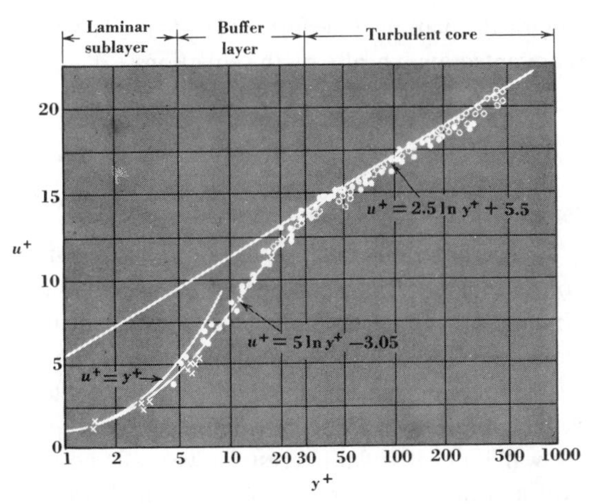

FIGURE 12-6
The universal velocity profile for flow in a smooth circular tube.

Since we are looking for a universal velocity profile, that is, one which will be of the form

$$u^+ = f(y^+) \qquad (12\text{-}55)$$

for all flow rates, i.e., values of u^* or y_o, we must modify Eq. (12-54) to meet these conditions. The variable y^+ is introduced by writing Eq. (12-54) as

$$u^+ = \frac{1}{K}\left(\ln\frac{yu^*}{\nu} - \ln\frac{y_o u^*}{\nu}\right) \qquad (12\text{-}56)$$

or

$$u^+ = \frac{1}{K}\ln y^+ + c_1 \qquad (12\text{-}57)$$

If Eq. (12-56) is to represent u^+ as a universal function of y^+, it is reasonable to suppose that $y_o u^*/\nu$, which might be written as y_o^+, is a universal constant representing the value of y^+ for a u^+ of zero. The proof of the usefulness of these assumptions depends on whether Eq. (12-57), with K and c_1 taken as universal constants, fits the data on velocity distributions over a range of y and u^*, that is, **Re**.

Figure 12-6 shows a large amount of velocity-distribution data for a range of Reynolds numbers from about 4000 to 3.2×10^6. It can be seen that Eq. (12-57) does fit the data well above $y^+ = 30$ and that the few data at low y^+ lie on the line $u^+ = y^+$ up to $y^+ = 5$. Neither equation is satisfactory in the region $5 < y^+ < 30$, which we define as the buffer region and where we represent the velocity by an empirical equation of the form of (12-57). The universal velocity distribution is thus represented analytically by the equations

$$u^+ = y^+ \qquad 0 < y^+ < 5 \qquad (12\text{-}58)$$

$$u^+ = 5.0 \ln y^+ - 3.05 \qquad 5 < y^+ < 30 \qquad (12\text{-}59)$$

$$u^+ = 2.5 \ln y^+ + 5.5 \qquad 30 < y^+ \qquad (12\text{-}60)$$

The preceding analysis has been derived from the work of L. Prandtl and T. von Kármán. The experimental velocity distributions are largely the work of J. Nikuradse. For an account of the specific contributions of those responsible for this work, refer to books by Schlichting[1] and by Knudsen and Katz.[2]

In recent years this apparently simple problem of the measurement of turbulent velocity profiles in a tube has continued to occupy researchers. The region close to the wall is of particular interest, for we shall find that it has an important influence on heat and mass transfer, especially for fluids of high Prandtl or Schmidt numbers. Techniques used include the introduction of a tracer by flash photolysis,[3] the observation of suspended particles with a microscope,[4] and the laser-Doppler technique.[5] Although Eqs. (12-58) to (12-60) are in general found to be adequate, there is apparently no layer near the wall where the flow is steady-laminar; eddies enter the region below $y^+ = 5$. In this region viscous effects are predominant, but the flow is not really laminar.

The value of the mixing length can be calculated from velocity-distribution data by means of Eqs. (12-48) and (2-9); the results are shown in Fig. 12-7. Note that Eq. (12-60) gives K a value of 0.4 and that Fig. 12-7 verifies this value for the region near the wall but that Eq. (12-49) is not correct far from the wall. The data shown are for several Reynolds numbers and for both rough and smooth pipe; the mixing length is independent of both factors for sufficiently high **Re**.

Equation (12-57) can be written for the velocity at the center of the pipe to give

$$u^+_{\text{max}} = \frac{1}{K} \ln y^+_{\text{max}} + c_1 \qquad (12\text{-}61)$$

If Eq. (12-57) is subtracted from this, there results the so-called velocity-deficiency law,

$$u_{\text{max}} - u = \frac{u^*}{K} \ln \frac{r_i}{y} \qquad (12\text{-}62)$$

[1] Schlichting, op. cit., chaps. 18–20.

[2] J. G. Knudsen and D. L. Katz, "Fluid Dynamics and Heat Transfer," McGraw-Hill Book Company, New York, 1958.

[3] A. T. Popovich and R. L. Hummel, *AIChE Journal*, **13**:854 (1967).

[4] T. K. Sherwood, K. A. Smith, and P. E. Fowles, *Chem. Eng. Sci.*, **23**:1225 (1968).

[5] J. C. Angus, D. L. Morrow, J. W. Dunning, Jr., and M. J. French, *Ind. Eng. Chem.*, **61**(2):9 (1969).

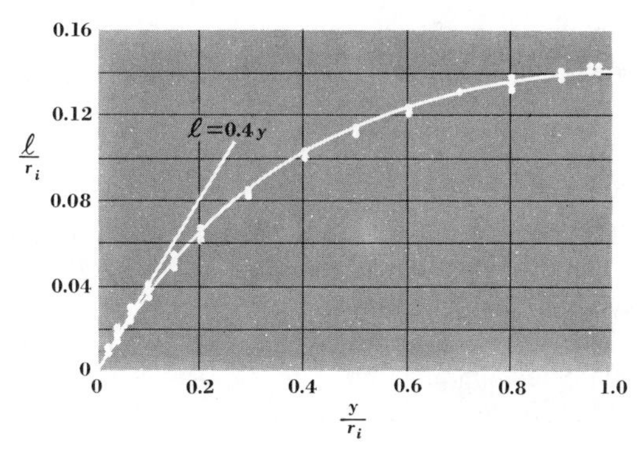

FIGURE 12-7
Mixing length as a function of radial position. (*From Hermann Schlichting, "Boundary Layer Theory," 4th ed., fig. 20.5, p. 511, McGraw-Hill Book Company, New York, 1960.*)

Since K is independent of the roughness of the pipe wall, Eq. (12-62) applies to both smooth and rough pipe. However, c_1 is a function of the roughness, as we shall see later.

The Flow Resistance from the Universal Velocity Distribution

We can easily find the bulk velocity in a pipe if we know how u varies with y. Since u_b is $(1/A) \iint u\, dA$, we write, for a circular pipe, using Eq. (12-62),

$$u_{max} - u_b = \frac{1}{\pi r_i^2} \int_0^{r_i} 2\pi(r_i - y) \frac{u^*}{K} \ln \frac{r_i}{y}\, dy \qquad (12\text{-}63)$$

This equation is integrated to give

$$u_{max} - u_b = \frac{3u^*}{2K} \qquad (12\text{-}64)$$

The substitution of Eq. (12-61) for u_{max} yields

$$\frac{1}{K} \ln \frac{r_i u^*}{\nu} + c_1 - \frac{u_b}{u^*} = \frac{3}{2K} \qquad (12\text{-}65)$$

Since $c_1 = 5.5$ and $K = 0.4$, this equation can be solved for a given u_b, r_i, and ν to give u^*. This term, the friction velocity, is related to τ_s by Eq. (12-45). The shear stress at the wall, τ_s, has been shown in Eq. (2-10) to be equal to $(-\Delta p)(r_i)/2L$, which gives the pressure drop per unit length of pipe. However, let us put Eq. (12-65) into a more convenient form by defining a coefficient of resistance, as we did a drag coefficient in Eq. (11-5). For the present case we have

$$C_D = \frac{2F_d g_c}{\rho u_b^2 A} \qquad (12\text{-}66)$$

in which A is the area of the inside wall of the pipe.

This equation is modified by using $F_d = \tau_s A$, and C_D is given the special symbol f for pipe flow:

$$f = \frac{2\tau_s g_c}{u_b^2 \rho} = \frac{2(u^*)^2}{u_b^2} \qquad (12\text{-}67)$$

This coefficient is usually called the *Fanning friction factor*. Equation (12-67) is also written as

$$u^* = u_b \sqrt{\frac{f}{2}} \qquad (12\text{-}68)$$

The substitution of Eq. (12-68) into (12-65) results in

$$\frac{1}{K} \ln \frac{D u_b}{\nu} \sqrt{\frac{f}{8}} + c_1 - \frac{3}{2K} = \sqrt{\frac{2}{f}} \qquad (12\text{-}69)$$

and the substitution of the numerical values of K and c_1, plus some rearrangement, gives the final equation[1]

$$\frac{1}{\sqrt{f}} = 4.06 \log (\mathbf{Re}\sqrt{f}) - 0.60 \qquad (12\text{-}70)$$

This equation is compared with experimental data on smooth tubes in Fig. 12-8; also shown are the empirical equations of Blasius,

$$f = 0.079\ \mathbf{Re}^{-1/4} \qquad (12\text{-}71)$$

and a relation often used in the chemical engineering literature,

$$f = 0.046\ \mathbf{Re}^{-1/5} \qquad (12\text{-}72)$$

Also shown is the relation for the laminar region,

$$f = \frac{16}{\mathbf{Re}} \qquad (12\text{-}73)$$

which is obtained from the Hagen-Poiseuille equation by a combination of Eqs. (2-10), (4-38), and (12-67).

[1] The symbol ln refers to a natural logarithm; log refers to a logarithm to the base 10.

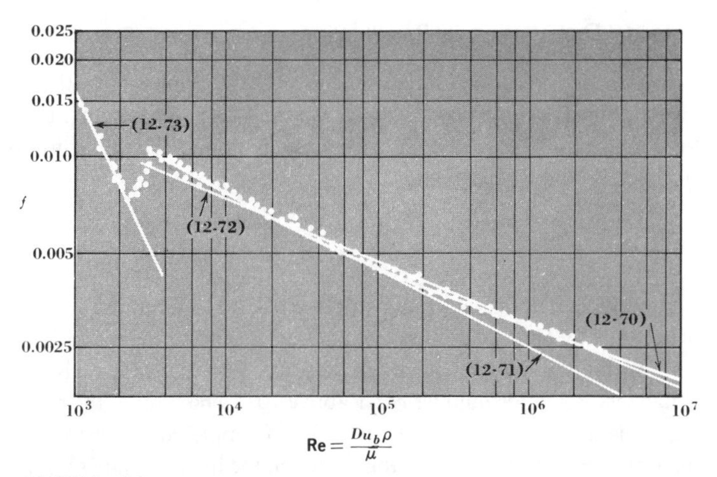

FIGURE 12-8
Friction factors for pipe flow. (*From Hermann Schlichting, "Boundary Layer Theory," 4th ed., fig. 20.1, p. 504, McGraw-Hill Book Company, New York, 1960.*)

TURBULENT FLOW INSIDE A ROUGH PIPE

In order to speak in a quantitative way about the effect of roughness on the velocity distribution or the pressure drop, we must first define a parameter which describes the roughness. We shall characterize a given roughness by the effective height of the protuberances, which we call e; the relative roughness is e/D. Although the quantity e is used to characterize a given roughness, a more exact procedure would require a description also of the spacing and orientation of the projections.

The relative roughness affects the flow in several ways. In laminar flow through commercial pipe, in which e/D is usually somewhat less than 0.01, the effect of wall roughness is negligible. The fluid fills the spaces between the protuberances, and the inner layers slide smoothly over a tube of effective diameter $D - 2e$. In turbulent flow the wall roughness also has no effect if it is smaller in height than the thickness of the viscous sublayer. In this case the pipe is said to be hydraulically smooth. However, if the irregularities enter into the main fluid stream, they increase the turbulence, change the velocity profile, and increase the flow resistance. Beyond a certain value of e, the effect of the roughness is so great that the inertia forces caused by the fluid flowing around the projections completely outweigh the viscous forces. In these circumstances the pipe is said to be completely rough. The thickness of the viscous sublayer is a function of the Reynolds number, so that the same pipe may be hydraulically smooth at one flow rate and completely rough at another.

Velocity Distribution in Rough Pipe

It has already been mentioned that Eq. (12-62) applies to both smooth and rough pipes. Since u_{max} may not be given, we shall find an equation of the form of (12-57) for rough pipe. This is done by simply assuming that y_o is proportional to e for completely rough pipe, giving

$$u^+ = \frac{1}{K} \ln \frac{y}{e} + c_2 \qquad (12\text{-}74)$$

The coefficient K has its usual value of 0.4, and Nikuradse's experiments on pipes roughened by gluing carefully sized grains of sand to the walls show that Eq. (12-74) is valid for the completely rough region. For this case c_2 is 8.5; it applies at a roughness Reynolds number eu^*/ν above 70. The limit of the hydraulically smooth region is at eu^*/ν equal to 5; there is no simple equation to express the velocity distribution in the transition range between the hydraulically smooth and completely rough condition.

Flow Resistance in Rough Pipe

In the hydraulically smooth region the roughness has no effect on f, and in the completely rough region the variation of f can be found from Eqs. (12-74) and (12-64) with y equal to r_i, combined as follows:

$$\left(\frac{1}{K} \ln \frac{r_i}{e} + c_2\right) - \frac{u_b}{u^*} = \frac{3}{2K} \qquad (12\text{-}75)$$

By replacing u_b/u^* by $\sqrt{2/f}$ [Eq. (12-68)], there results

$$\sqrt{\frac{2}{f}} = \frac{1}{K} \ln \frac{r_i}{e} + c_2 - \frac{3}{2K} \qquad (12\text{-}76)$$

Using $K = 0.4$ and $c_2 = 8.5$, Eq. (12-76) is transformed to

$$\frac{1}{\sqrt{f}} = 4.06 \log \frac{r_i}{e} + 3.36 \qquad (12\text{-}77)$$

which is valid for completely rough pipe.

In laminar flow f is inversely proportional to the velocity, since the drag is proportional to the velocity. In hydraulically smooth turbulent flow the viscous effects (skin friction) are outweighed by the inertia (form-drag) effects: f is roughly proportional to $u_b^{-1/4}$, and the drag is proportional to $u_b^{7/4}$. We have now seen that in completely rough turbulent flow f is not a function of u_b, so that the drag is proportional to u_b^2; the importance of the viscous effects has shrunk to zero.

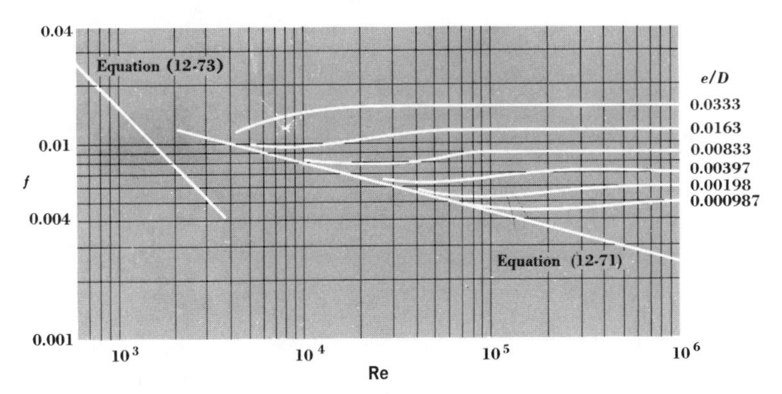

FIGURE 12-9
Friction factors for artificially roughened pipe. (*From Hermann Schlichting, "Boundary Layer Theory," 4th ed., fig. 20.18, p. 521, McGraw-Hill Book Company, New York, 1960.*)

Nikuradse's data for artificially roughened pipes are shown in Fig. 12-9. The use of the methods of this chapter for flow in commercial pipe is discussed in Chap. 14.

Results for Flow Past a Flat Plate from a Momentum Balance

In Chap. 5 we obtained the drag by the use of the overall, or integral, momentum balance; Prob. 5-3 provides an introduction to its use in boundary-layer flow. As with the overall balances, it is necessary to know the velocity distribution. Since the Navier-Stokes equations cannot be solved for the velocity distribution in the turbulent boundary layer as they were for the laminar boundary layer in Chap. 11, we use the integral momentum balance with an empirical velocity distribution in order to get the drag in turbulent flow past a flat plate. This procedure is based on the work of von Kármán and also was used by Pohlhausen for boundary-layer flow with a pressure gradient.[1]

Application of the Overall Momentum Balance to the Boundary Layer on a Flat Plate

The overall momentum balance is to be applied to a control volume as shown in Fig. 12-10. There is flow only through the faces having areas designated by A_1, A_2, and

[1] Schlichting, op. cit., chaps. 21 and 22.

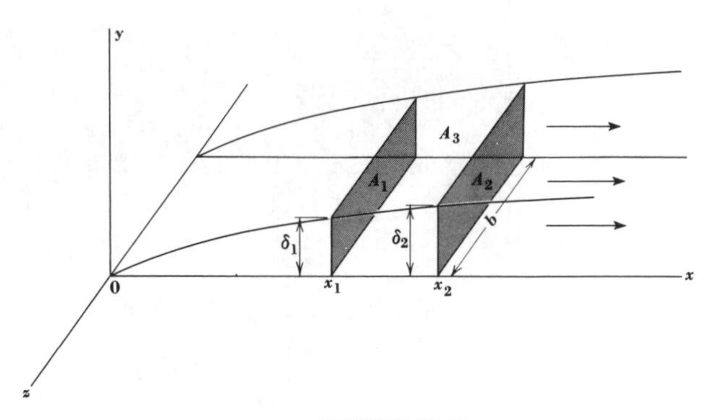

FIGURE 12-10
Control volume for boundary-layer flow.

A_3. The balance of x momentum is written from Eq. (5-3) as

$$\iint_A \frac{u_x \rho}{g_c} u \cos \alpha \, dA = \check{F}_{xd} \qquad \text{(12-78)}$$

for this steady-state flow with no pressure gradient. For A_1, $u \cos \alpha = -u_x$; for A_2, $u \cos \alpha = u_x$; and for A_3, $u_x = u_o$, the velocity outside the boundary layer. Equation (12-78) is then written

$$\iint_{A_2} u_x^2 \rho \, dA - \iint_{A_1} u_x^2 \rho \, dA + u_o \iint_{A_3} u\rho \cos \alpha \, dA = g_c \check{F}_{xd} \qquad \text{(12-79)}$$

The third integral can be replaced by using the overall mass balance. For steady state, Eq. (3-12) becomes

$$\iint_A u\rho \cos \alpha \, dA = 0 \qquad \text{(12-80)}$$

and for the present case

$$\iint_{A_2} u_x \rho \, dA - \iint_{A_1} u_x \rho \, dA + \iint_{A_3} u\rho \cos \alpha \, dA = 0 \qquad \text{(12-81)}$$

The combination of Eqs. (12-81) and (12-79) yields

$$\iint_{A_2} u_x(u_o - u_x) \, dA - \iint_{A_1} u_x(u_o - u_x) \, dA = \frac{-g_c \check{F}_{xd}}{\rho} \qquad \text{(12-82)}$$

for an incompressible fluid.

For the boundary layer on a plate of width b, $dA = b\,dy$, where y varies from zero to δ, and $-\check{F}_{xd} = \tau_{s,\,\mathrm{av}}\,b(x_2 - x_1)$. Equation (12-82) is thus modified to (where $\tau_s > 0$)

$$\int_0^{\delta_2} u_x(u_o - u_x)\,dy - \int_0^{\delta_1} u_x(u_o - u_x)\,dy = \frac{g_c \tau_{s,\,\mathrm{av}}(x_2 - x_1)}{\rho} \qquad (12\text{-}83)$$

Given a velocity distribution, the integrals can be evaluated and the shear stress $\tau_{s,\,\mathrm{av}}$ found. However, this is the average stress over the length $x_2 - x_1$, and it is desirable to know τ_s at a point, so that it can be expressed in terms of the velocity gradient at a certain x. The equation

$$g_c \tau_s = \mu\left(\frac{du}{dy}\right)_{y=0} \qquad (12\text{-}84)$$

gives τ_s at a certain x, where $(du/dy)_{y=0}$ has a certain value. Therefore we arrange Eq. (12-83) to apply at a certain x by using the definition of a derivative (notice that each definite integral is a function of x only),

$$\lim_{(x_2 - x_1)\to 0} \frac{\int_0^{\delta_2} u_x(u_o - u_x)\,dy - \int_0^{\delta_1} u_x(u_o - u_x)\,dy}{x_2 - x_1} = \frac{g_c \tau_s}{\rho} \qquad (12\text{-}85)$$

so that

$$\frac{d}{dx}\int_0^{\delta} u_x(u_o - u_x)\,dy = \frac{g_c \tau_s}{\rho} \qquad (12\text{-}86)$$

Equation (12-86) applies at any point x along the plate; δ and τ_s are functions of x.

The Momentum Balance for the Laminar Boundary Layer

Before considering the turbulent boundary layer, it is of interest to apply Eq. (12-86) to the laminar boundary layer, because the results can be compared with the exact Blasius solution (Chap. 11). We shall find that an arbitrary velocity distribution gives results very close to the Blasius solution for the variation of δ and C_D with L.

The equation for u_x as a function of y is made on the similarity assumption that u_x/u_o is everywhere the same function of y/δ.

We know that $u_x = 0$ at $y = 0$, $u_x \cong u_o$ at $y = \delta$, and $du_x/dy \cong 0$ at $y = \delta$; it is easy to verify that the following simple equation fulfills these conditions.

$$\frac{u_x}{u_o} = \frac{3}{2}\left(\frac{y}{\delta}\right) - \frac{1}{2}\left(\frac{y}{\delta}\right)^3 \qquad (12\text{-}87)$$

This equation is substituted into Eq. (12-86) to give

$$\frac{d}{dx}\left(\delta\int_0^{1.0}\left\{\left[\frac{3}{2}\left(\frac{y}{\delta}\right) - \frac{1}{2}\left(\frac{y}{\delta}\right)^3\right] - \left[\frac{3}{2}\left(\frac{y}{\delta}\right) - \frac{1}{2}\left(\frac{y}{\delta}\right)^3\right]^2\right\} d\left(\frac{y}{\delta}\right)\right)$$

$$= \frac{g_c \tau_s}{\rho u_o^2} \qquad (12\text{-}88)$$

The evaluation of the integral transforms this equation to

$$\frac{d\delta}{dx} = \frac{280}{39} \frac{\tau_s g_c}{u_o^2 \rho} \tag{12-89}$$

For the laminar boundary layer, Eq. (12-84) holds; we differentiate Eq. (12-87) to get

$$\left(\frac{du}{dy}\right)_{y=0} = \frac{3u_o}{2\delta} \tag{12-90}$$

and by Eq. (12-84) we obtain

$$g_c \tau_s = \frac{3\mu u_o}{2\delta} \tag{12-91}$$

Equations (12-91) and (12-89) give

$$\int_0^\delta \delta \, d\delta = \left(\frac{280}{39}\right)\left(\frac{3}{2}\right)\frac{\nu}{u_o}\int_0^L dx \tag{12-92}$$

and thus

$$\delta = 4.64\sqrt{\frac{\nu L}{u_o}} \tag{12-93}$$

The equation

$$C_D = 1.29\sqrt{\frac{\nu}{L u_o}} = \frac{1.29}{\sqrt{\mathbf{Re}_L}} \tag{12-94}$$

can also be derived; the method is illustrated in Chap. 11. A comparison of Eq. (12-93) with (11-22) and of (12-94) with (11-29) shows the success of the momentum method; the equations are identical except for the numerical constants. Similar results can be obtained from other arbitrarily chosen velocity distributions. Such a case is the subject of Prob. 12-8.

The Momentum Balance for the Turbulent Boundary Layer [1]

The above procedures will now be applied to the turbulent boundary layer on a flat plate. It would be desirable to adapt the logarithmic velocity distribution [Eqs. (12-58) to (12-60)] to boundary-layer flow. This can be done, but the derivation and resulting equations are long and complicated, so a simpler empirical velocity distribution will be used here.

In pipe flow the equation

$$\frac{u}{u_{\max}} = \left(\frac{y}{r_i}\right)^{1/7} \tag{12-95}$$

[1] Numerical methods are discussed by T. Cebeci and P. Bradshaw, "Momentum Transfer in Boundary Layers," Hemisphere Publishing Corp., Washington, 1977.

is valid up to **Re** $= 10^5$; it is sometimes called the *Blasius* $\frac{1}{7}$*-power law*. We shall write this equation for the boundary layer on a plate as

$$\frac{u_x}{u_o} = \left(\frac{y}{\delta}\right)^{1/7} \qquad (12\text{-}96)$$

The substitution of Eq. (12-96) into (12-86), with subsequent integration, yields

$$\frac{d\delta}{dx} = \frac{72}{7}\frac{\tau_s g_c}{\rho u_o^2} \qquad (12\text{-}97)$$

Although Eq. (12-84) holds for the fluid adjacent to the wall even in the turbulent boundary layer, Eq. (12-96) does not hold as y goes to zero.

Therefore we use the Blasius resistance law for pipe flow, which is consistent with the $\frac{1}{7}$-power velocity distribution.

$$f = 0.079\left(\frac{Du_b\rho}{\mu}\right)^{-1/4} \qquad (12\text{-}71)$$

From Eq. (12-67) we have

$$\frac{g_c\tau_s}{\rho u_b^2} = \frac{f}{2} = \frac{0.079}{2}\left(\frac{Du_b}{\nu}\right)^{-1/4} \qquad (12\text{-}98)$$

This equation, which applies for a pipe, is adapted for boundary-layer flow over a plate by the substitutions $D = 2\delta$ and $u_b = 0.817u_{\max} = 0.817u_o$, and the result is

$$\frac{g_c\tau_s}{\rho u_o^2} = 0.023\left(\frac{\delta u_o}{\nu}\right)^{-1/4} \qquad (12\text{-}99)$$

for boundary-layer flow.

Equations (12-99) and (12-97) are combined and put into integral form to give

$$\int_0^\delta \delta^{1/4}\,d\delta = \frac{72}{7}(0.023)\left(\frac{u_o}{\nu}\right)^{-1/4}\int_0^L dx \qquad (12\text{-}100)$$

and thus

$$\delta = 0.376\left(\frac{Lu_o}{\nu}\right)^{-1/5} L \qquad (12\text{-}101)$$

or

$$\delta = 0.376(\mathbf{Re}_L)^{-1/5}L \qquad (12\text{-}102)$$

An integration of the drag force from $x = 0$ to $x = L$, as has been done in several previous cases, Eqs. (11-25) *et seq*., gives

$$C_D = 0.072(\mathbf{Re}_L)^{-1/5} \qquad (12\text{-}103)$$

In the above development the turbulent boundary layer has been assumed to extend to $x = 0$, whereas there really is a certain length on the forward part of the plate in which the boundary layer is laminar.

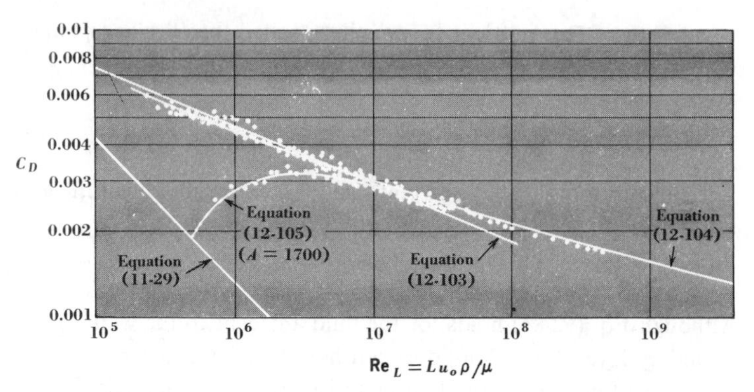

FIGURE 12-11
Drag coefficients for flow past a flat plate. (*From Hermann Schlichting, "Boundary Layer Theory," 4th ed., fig. 21.2, p. 538, McGraw-Hill Book Company, New York, 1960.*)

A comparison of the results for the laminar and for the turbulent boundary layers shows that δ increases as $L^{1/2}$ for laminar flow and as $L^{4/5}$ for turbulent flow. The drag in turbulent flow is greater than in laminar flow; F_d increases as $L^{1/2}$ in laminar flow and as $L^{4/5}$ in turbulent flow. These are quantitative arguments for attempting to maintain a laminar boundary layer on an airfoil, as discussed early in this chapter.

The Blasius formulas are not valid above $Du_b\rho/\mu = 10^5$ for pipe flow. Similarly, the use of Eqs. (12-102) and (12-103) is restricted to the range of $Lu_o\rho/\mu$ of 5×10^5 to 10^7. Just as in the case of pipe flow, if the logarithmic velocity distribution is used in the derivations, a result is obtained which is valid to very high **Re**. For the flat plate the results of such an analysis can be represented by

$$C_D = \frac{0.455}{(\log \mathbf{Re}_L)^{2.58}} \qquad (12\text{-}104)$$

Table 12-1

Transition $\mathbf{Re}_L$	A
3×10^5	1050
5×10^5	1700
1×10^6	3300
3×10^6	8700

In order to take into account the laminar boundary layer on the forward part of the plate, this equation is modified to give the Prandtl-Schlichting formula

$$C_D = \frac{0.455}{(\log \mathbf{Re}_L)^{2.58}} - \frac{A}{\mathbf{Re}_L} \qquad (12\text{-}105)$$

where A is a function of the transition $\mathbf{Re}_L$, which is governed by the intensity of the turbulence of the external flow. Values of A are given in Table 12-1.

Figure 12-11 shows some experimental data and the errors resulting from several of the preceding equations for C_D.

We shall not discuss quantitatively the flow along curved surfaces, for which there is a pressure gradient arising from the external potential flow, and the possibility of separation. This situation is discussed in a qualitative way at the beginning of Chap. 11.

PROBLEMS

12-1 At successive equal time intervals of a few milliseconds, the following velocities were measured by means of hot-wire anemometers at two points 0.05 m apart:

u_{x1}	77	78	75	75	70	73	78	83	81	77	72
u_{x2}	74	85	69	80	74	81	79	79	88	80	75

Calculate the mean velocity and the intensity of the turbulence (assume isotropic) at each point. What is the correlation coefficient for the velocities at the two points?

12-2 Find an expression for the eddy viscosity from the formula for the logarithmic velocity profile in the turbulent core and a knowledge of how τ varies with y. Sketch graphs of the eddy viscosity versus position in the pipe at constant **Re** and versus **Re** at a given position. Is the quantity v_e/u^*r_i a function of **Re**? Of position?

12-3 Proceeding as in the previous problem, find a relation for the dimensionless Prandtl mixing length ℓ/r_i. Show how this quantity varies with position and **Re**.

12-4 Point out ways in which the velocity profile predicted by Eqs. (12-58) to (12-60) is inconsistent with the profile indicated by some simple physical principles. It will be helpful to consider the behavior of du/dy.

12-5 Water is flowing at 68°F through a smooth, horizontal tube of 2-in ID. At a point $\frac{3}{4}$ in from the wall, calculate the velocity, shear stress, eddy viscosity, and mixing length for a bulk velocity of (*a*) 50 ft/s; (*b*) 5 ft/s; (*c*) 0.05 ft/s.

12-6 Water is flowing as in Prob. 12-5, but through a rough tube. Calculate the same four quantities at the same point for:

(*a*) $u_b = 50$ ft/s, $e/r_i = 0.01$
(*b*) $u_b = 5$ ft/s, $e/r_i = 0.03$
(*c*) $u_b = 5$ ft/s, $e/r_i = 0.0005$
(*d*) $u_b = 0.05$ ft/s, $e/r_i = 0.01$

12-7 Derive the Blasius resistance formula, Eq. (12-71), from the $\frac{1}{7}$-power velocity distribution, Eq. (12-95), applied to flow in a pipe.

12-8 Use the overall momentum balance to find the equations for δ and C_D analogous to Eqs. (12-93) and (12-94) for an assumed linear variation of the velocity in the laminar boundary layer on a smooth flat plate. For water at 20°C flowing by a plate at 0.6 m/s, find the thickness of the boundary layer 0.3 m from the leading edge. What is the total drag up to this point in newtons per meter of width?

12-9 Air at 20°C and 101.3 kPa abs is flowing past a smooth, flat plate at 30 m/s. If the turbulence in the main stream is such that the transition from a laminar to a turbulent boundary layer occurs at an $\mathbf{Re}_L$ of 5×10^5, how far from the leading edge is the point of transition? Calculate also the boundary-layer thickness 1 m from the leading edge and the total drag per meter of width up to this point.

12-10 The boundary layer which forms at the entrance to a circular pipe may be considered to behave like the boundary layer on a flat plate. Estimate the length needed to reach fully developed flow (i.e., for the edge of the boundary layer to reach the center line) for water at 20°C entering a pipe of 0.05-m ID at a uniform velocity of (*a*) 15 m/s; (*b*) 1.5 m/s; (*c*) 0.015 m/s.

Compare the results with the empirical rules given in Perry's Handbook. More exact treatments of this problem are to be found in texts on fluid dynamics.

12-11 D. B. Spalding [*J. Appl. Mech.*, **28**E:455 (1961)] has proposed a universal velocity distribution for turbulent flow called the *law of the wall:*

$$y^+ = u^+ + A\left[e^{\kappa u^+} - 1 - \kappa u^+ - \frac{1}{2!}(\kappa u^+)^2 - \frac{1}{3!}(\kappa u^+)^3 - \frac{1}{4!}(\kappa u^+)^4\right]$$

in which $A = 0.1108$ and $\kappa = 0.4$. (*a*) Use this relation to plot v_e/v as a function of y/r_i for fully developed turbulent flow in a tube for $\mathbf{Re} = 10^5$. (*b*) There is evidence that eddies penetrate into the viscous sublayer ($y^+ < 5$) so that v_e/v is not zero in this region but dies out as the wall is approached. If v_e/v is expanded as a MacLaurin series, show that according to the law of the wall, very near the wall, $v_e/v = \text{const}(y^+)^4$.

12-12 For the region $y^+ > 100$, L. F. Flint [*Chem. Eng. Sci.*, **22**:1127 (1967)] has suggested the equation

$$u^+ = 2.5 \ln\left[1.5\left(\frac{1 + z^2}{1 + 2z^2}\right)y^+\right] + 5.0$$

where

$$z = \frac{a^+ - y^+}{a^+}$$

and

$$a^+ = \frac{r_i u^*}{\nu}$$

Compare this velocity profile with that of Eq. (12-60) and plot v_e/v versus y/r_i for $\mathbf{Re} = 10^5$.

13

DIMENSIONAL ANALYSIS WITH APPLICATIONS IN FLUID DYNAMICS

In the preceding chapters we have treated a number of flow problems analytically by using the differential momentum balance. However, the flow problems in many engineering systems are so complicated that it is not possible to integrate the differential momentum balance. For instance, in flow past a cylinder, the curve of Fig. 11-6 for **Re** > 1.0 cannot be predicted theoretically, but must be obtained from experiment. Although the equations of motion have been solved for flow past a flat plate or in a circular pipe, they have not been solved for flow past a bank of tubes or through a bed of spheres, or for many other cases. When the analytical approach does not yield a solution, the technique of dimensional analysis often assists in the rational treatment of a problem. The advantages and limitations of the method will become apparent from the discussion and examples given later in this chapter.

Results from the Differential Momentum Balance

Even though it may not be possible to integrate the Navier-Stokes equations for a given problem, we can use them to find out how the variables must be grouped in an equation representing the solution of the problem. We begin by writing the x component of the Navier-Stokes equations for steady flow as

$$u_x \frac{\partial u_x}{\partial x} + u_y \frac{\partial u_x}{\partial y} + u_z \frac{\partial u_x}{\partial z} = g \cos \beta - \frac{g_c}{\rho} \frac{\partial p}{\partial x} + \nu \left(\frac{\partial^2 u_x}{\partial x^2} + \frac{\partial^2 u_x}{\partial y^2} + \frac{\partial^2 u_x}{\partial z^2} \right) \tag{13-1}$$

This equation is, of course, dimensionally homogeneous; each term has the dimensions $[L/\theta^2]$. Since we are interested mainly in the way in which the variables appear in an integrated form of Eq. (13-1), we write

$$\left[\frac{u^2}{L} \right] = \left[g - \frac{g_c p}{\rho L} + \frac{\nu u}{L^2} \right] \tag{13-2}$$

We have symbolized all the velocity components by a single characteristic velocity u and all the lengths by a characteristic length L. Equation (13-2) expresses a dimensional equality, but not a numerical equality. We now change the terms in Eq. (13-2) to a dimensionless form by dividing each term by u^2/L, obtaining

$$[1] = \left[\frac{gL}{u^2} - \frac{g_c p}{u^2 \rho} + \frac{\nu}{uL} \right] \tag{13-3}$$

Equation (13-3) is merely a functional relationship and can be written as

$$\frac{g_c p}{u^2 \rho} = f\left(\frac{u^2}{gL}, \frac{uL}{\nu} \right) \tag{13-4}$$

We recognize the dimensionless group uL/ν as the Reynolds number; u^2/gL is called the *Froude number*; and $g_c p/u^2 \rho$ is called the *Euler number*. Recalling that the left-hand side of (13-2) represents the effect of inertia and that the terms on the right-hand side represent, respectively, the gravity force, the pressure force, and the viscous force, it is clear that we can write

$$\mathbf{Eu} = \frac{g_c p}{u^2 \rho} \propto \frac{\text{pressure force}}{\text{inertia force}} \tag{13-5}$$

$$\mathbf{Fr} = \frac{u^2}{gL} \propto \frac{\text{inertia force}}{\text{gravity force}} \tag{13-6}$$

$$\mathbf{Re} = \frac{uL}{\nu} \propto \frac{\text{inertia force}}{\text{viscous force}} \tag{13-7}$$

Let us now see how Eq. (13-4) may be useful in the study of flow past an infinite cylinder. If we set p equal to the drag force per unit length divided by the diameter, we find

$$\mathbf{Eu} = \frac{g_c F'_d}{u^2 \rho D} = \frac{C_D}{2} \qquad (13\text{-}8)$$

The second equality is a definition of the drag coefficient from Eq. (11-4). Since gravity effects are not important in a flow problem unless there is a free liquid surface involved, Eq. (13-4) becomes

$$C_D = f_1(\mathbf{Re}) \qquad (13\text{-}9)$$

Once we have experimentally determined a curve of C_D versus **Re**, say by measuring the drag force as a function of velocity for a given fluid and a given cylinder, this curve can be used to estimate the drag force for other cylinders and other fluids. Such a curve is given in Fig. 11-6. Although the use of dimensional analysis has not given an explicit relation among the variables, the form of the equation has been limited so that only a few experimental data are needed to permit a complete solution of the problem.

The drag coefficient is also a function of the Froude number when surface waves are formed, as in the testing of a ship model or sometimes in the agitation of the liquid in an open tank.

The method of dimensional analysis just illustrated, which starts from a differential equation, is discussed more extensively by Klinkenberg and Mooy[1] and in a different way by Schlichting and Kestin.[2]

A Dimensional Basis for Similarity Transformations

The differential balances which describe certain problems can sometimes be simplified by changing them to a nondimensional form according to a procedure given by Hellums and Churchill.[3] This method not only leads to the usual dimensionless groups but also can reduce the number of independent variables. We study the method by applying it to boundary-layer flow over a flat plate. The form of the Navier-Stokes equation which applies to this boundary-layer flow contains the

[1] H. A. Klinkenberg and H. H. Mooy, Dimensionless Groups in Fluid Friction, Heat, and Material Transfer, *Chem. Eng. Progr.*, **44**:17 (1948).

[2] H. Schlichting and Kestin, "Boundary Layer Theory," 6th ed., pp. 65–67, McGraw-Hill Book Company, New York, 1968.

[3] J. D. Hellums and S. W. Churchill, *AIChE Journal*, **10**:110 (1964). Other group transformation methods are discussed by W. F. Ames, "Nonlinear Partial Differential Equations in Engineering," Academic Press, Inc., New York, 1965. See also A. G. Hansen, "Similarity Analyses of Boundary Value Problems in Engineering," Prentice-Hall, Inc., Englewood Cliffs, N.J., 1964, and S. W. Churchill, "The Interpretation and Use of Rate Data," McGraw-Hill Book Company, New York, 1974.

velocity components u_x and u_y. By defining the stream function so that the equation of continuity is satisfied, the mathematical statement of the problem is simplified so that there is only one dependent variable in one partial differential equation:

$$\frac{\partial \psi}{\partial y}\frac{\partial^2 \psi}{\partial x\, \partial y} - \frac{\partial \psi}{\partial x}\frac{\partial^2 \psi}{\partial y^2} = \nu \frac{\partial^3 \psi}{\partial y^3} \qquad (13\text{-}10)$$

The pertinent boundary conditions are

$$u_x(x, 0) = \frac{\partial \psi}{\partial y} = 0 \qquad (13\text{-}11)$$

$$u_x(x, \infty) = u_o$$

We define dimensionless variables by

$$X = \frac{x}{x_o} \qquad (13\text{-}12)$$

$$Y = \frac{y}{y_o} \qquad (13\text{-}13)$$

$$F = \frac{\psi}{\psi_o} \qquad (13\text{-}14)$$

where x_o, y_o, and ψ_o are arbitrary reference values. The substitution of these new variables into Eq. (13-10) gives

$$\frac{{\psi_o}^2}{x_o {y_o}^2}\frac{\partial F}{\partial Y}\frac{\partial^2 F}{\partial X\, \partial Y} - \frac{{\psi_o}^2}{x_o {y_o}^2}\frac{\partial F}{\partial X}\frac{\partial^2 F}{\partial Y^2} = \frac{\nu \psi_o}{{y_o}^3}\frac{\partial^3 F}{\partial Y^3} \qquad (13\text{-}15)$$

We then make each term dimensionless by dividing through by the coefficient of $\partial^3 F/\partial Y^3$:

$$\frac{y_o \psi_o}{\nu x_o}\left(\frac{\partial F}{\partial Y}\frac{\partial^2 F}{\partial X\, \partial Y} - \frac{\partial F}{\partial X}\frac{\partial^2 F}{\partial Y^2}\right) = \frac{\partial^3 F}{\partial Y^3} \qquad (13\text{-}16)$$

The boundary conditions are

$$\frac{\psi_0}{u_0 y_0}\frac{\partial F}{\partial Y}(X, 0) = 0 \qquad (13\text{-}17)$$

$$\frac{\psi_o}{u_o y_o}\frac{\partial F}{\partial Y}(X, \infty) = 1 \qquad (13\text{-}18)$$

Since the reference values can be arbitrarily chosen, we set the coefficients $y_o \psi_o / \nu x_o$ and $\psi_o / u_o y_o$ equal to unity. Note that u_o is not a reference value but a real parameter of the problem. Thus, all the reference quantities can be expressed in terms of one, say x_o, to yield

$$y_o = \sqrt{\frac{x_o \nu}{u_o}} \tag{13-19}$$

$$\psi_o = \sqrt{u_o \nu x_o} \tag{13-20}$$

The solution to our problem is now

$$F = \frac{\psi}{\sqrt{u_o \nu x_o}} = F(X, Y) = F\left(\frac{x}{x_o}, y\sqrt{\frac{u_o}{x_o \nu}}\right) \tag{13-21}$$

In this form F depends on the arbitrary value chosen for x_o; this is not physically reasonable so X and Y must be algebraically combined so that x_o is eliminated. Thus a new independent variable η is defined:

$$\eta = \frac{Y}{X^{1/2}} = y\sqrt{\frac{u_o}{\nu x}} \tag{11-15}$$

The dependent variable F must also not depend on the value of x_o, so we define a new variable f so that

$$f = \frac{F}{X^{1/2}}$$

and

$$F = f\sqrt{u_o \nu x}$$

The final dimensionless form of the problem is then

$$f = f(\eta)$$

$$u_x = \frac{\psi_o}{y_o} \frac{\partial f}{\partial Y} = u_o f'(\eta) \tag{11-19}$$

Of course, u_x must also not be a function of any reference quantity. The similarity transformation used in Chap. 11 has thus been obtained by dimensional arguments. This procedure will not always achieve such a neat simplification, but its application at the very least helps to define a problem. We shall find the method useful later for heat and mass transfer problems. It is most likely to work well for semi-infinite problems such as those related to boundary layers.

Buckingham's Pi Theorem

A dimensional analysis can also be made according to the method of Buckingham; for a rigorous mathematical derivation of the method, refer to Langhaar.[1] This method is applicable even when the pertinent differential equation is not known. We postulate that p is a function of all the variables and dimensional constants which we think may have some effect on flow:

$$p = f(L,u,\rho,\mu,g,C,\sigma,g_c) \qquad (13\text{-}22)$$

To find the form of this function by pure empiricism would involve a large number of experiments. By dimensional analysis we are able to reduce the number of experiments needed for a solution of the problem.

We have begun by assuming that the pressure (i.e., the pressure drop associated with flow) is influenced by the variables listed in Eq. (13-22). This assumption is based on some prior knowledge of the physical phenomena, but it is by no means always clear what variables should be included.[2] The velocity of sound C and the surface tension σ have been included in order to make the problem a little more general than the one treated previously.

The method is based on the premise that any equation representing the behavior of the system must be dimensionally homogeneous, so that it would be possible to write it in the form of dimensionless groups by a procedure similar to that used to get Eq. (13-4). Using π_i as a symbol for a dimensionless group, we then generalize this result and write

$$\pi_1 = f(\pi_2,\pi_3, \ldots, \pi_i) \qquad (13\text{-}23)$$

Equation (13-22) must now be written in the form of Eq. (13-23), which involves only the dimensionless groups $\pi_1, \ldots, \pi_i$.

It is first necessary to know how many groups Eq. (13-22) will yield, i.e., the value i. The value of i is given by

$$i = n - r \qquad (13\text{-}24)$$

where n equals the number of variables and r equals the maximum number of these variables which will *not* form a dimensionless group. The quantity r is often equal to the number of fundamental dimensions. For the general fluid-flow problem we are considering, Eq. (13-22) shows that $n = 9$. By trial and error we find that L, u, ρ, and g_c cannot be combined in any way so as to make a π. This can be shown by writing

$$L^a u^b \rho^c g_c^{\ d} \qquad (13\text{-}25)$$

[1] H. L. Langhaar, "Dimensional Analysis and Theory of Models," John Wiley & Sons, Inc., New York, 1951.

[2] An interesting discussion of this point, and dimensional analysis in general, is given by P. W. Bridgman, "Dimensional Analysis," chaps. 1–3, Yale University Press, New Haven, Conn., 1931.

and substituting for each variable its dimensions,

$$[L]^a \left[\frac{L}{\theta}\right]^b \left[\frac{M}{L^3}\right]^c \left[\frac{ML}{F\theta^2}\right]^d \qquad (13\text{-}26)$$

Since $[F]$ occurs only in g_c, it is evident that there is no possibility of obtaining a dimensionless group by assigning proper values to a, b, c, and d. If any one of the other variables is added to the set in (13-26), it is possible to find values of the exponents which result in a dimensionless group, as will be shown presently. Since r is 4, i must be 5; there are five possible independent dimensionless groups, $\pi_1, \ldots, \pi_5$.

There is a large number of possible sets of dimensionless groups, but from what we know about fluid dynamics, it appears desirable to have a group which depends on the presence of viscous effects (μ), one on gravity effects (g), one on high-velocity effects (C), one on surface-tension effects (σ), and one on pressure effects (p). We shall therefore choose our groups so that each of the variables μ, g, C, σ, and p appears in only one group. On the other hand, L, u, ρ, and g_c may have an effect in almost any group, so we make it possible for them to appear in any group. For π_1 we write

$$\pi_1 = L^a u^b \rho^c g_c{}^d p^e \qquad (13\text{-}27)$$

In terms of the four fundamental dimensions,

$$[1] = \left[L^a \left(\frac{L}{\theta}\right)^b \left(\frac{M}{L^3}\right)^c \left(\frac{ML}{F\theta^2}\right)^d \left(\frac{F}{L^2}\right)^e\right] \qquad (13\text{-}28)$$

The exponents on each variable must be such that the group is dimensionless, so this requires that the following equations be satisfied:

$$\text{For } [M] \qquad 0 = c + d \qquad (13\text{-}29)$$
$$\text{For } [L] \qquad 0 = a + b - 3c + d - 2e \qquad (13\text{-}30)$$
$$\text{For } [\theta] \qquad 0 = -b - 2d \qquad (13\text{-}31)$$
$$\text{For } [F] \qquad 0 = -d + e \qquad (13\text{-}32)$$

These four equations can be solved to give all the exponents in terms of one. We choose e and get $d = e$, $b = -2e$, $c = -e$, and $a = 0$. In addition, e can be arbitrarily set equal to unity, since all the π's are to appear in Eq. (13-23), which involves an unknown function of the π's. From these results we obtain

$$\pi_1 = \frac{g_c p}{u^2 \rho} = \mathbf{Eu} \qquad \text{the Euler number} \qquad (13\text{-}33)$$

For π_2 we set

$$\pi_2 = L^a u^b \rho^c g_c{}^d \mu^e \qquad (13\text{-}34)$$

which gives ($e = -1$)

$$\pi_2 = \frac{Lu\rho}{\mu} = \mathbf{Re} \qquad \text{the Reynolds number} \tag{13-35}$$

In a similar way the other π's are obtained by changing the fifth variable in Eqs. (13-27) and (13-34), successively, to g, C, and σ. The results are

$$\pi_3 = \frac{u^2}{Lg} = \mathbf{Fr} \qquad \text{the Froude number} \tag{13-36}$$

$$\pi_4 = \frac{u}{C} = \mathbf{Ma} \qquad \text{the Mach number} \tag{13-37}$$

$$\pi_5 = \frac{\rho u^2 L}{g_c \sigma} = \mathbf{We} \qquad \text{the Weber number} \tag{13-38}$$

Equation (13-23) becomes, for this problem,

$$\mathbf{Eu} = f(\mathbf{Re}, \mathbf{Fr}, \mathbf{Ma}, \mathbf{We}) \tag{13-39}$$

This set of five π's is called a complete set; every π is independent of the others. Many other π's can of course be formed from those of the complete set; for instance,

$$\mathbf{ReFr} = \frac{u^3 \rho}{\mu g} \tag{13-40}$$

This group might conceivably be useful in the analysis of a particular problem.

Fortunately, in many cases, the effects represented by many of the groups in Eq. (13-39) are negligible. The following cases, some of which have been treated analytically in previous chapters, may be mentioned:

1 Flow of an ideal fluid. This approximation can be applied to the calculation of the lift of an airfoil or the delivery of a fan, for which viscosity, gravity, surface-tension, and compressibility (Mach number) effects are negligible. Equation (13-39) becomes, for this case,

$$\mathbf{Eu} = \frac{g_c p}{u^2 \rho} = \text{const} \tag{13-41}$$

which can be recognized as a reduced form of Bernoulli's equation.

2 Flow in conduits or past totally immersed objects. In this case, only viscous friction is important.

$$\frac{g_c p}{u^2 \rho} = \phi\left(\frac{Lu\rho}{\mu}\right) \tag{13-42}$$

Recall that $g_c p/u^2 \rho$ is equivalent to C_D or f.

We have already noted that Eq. (13-4) becomes

$$C_D = f_1(\mathbf{Re}) \qquad (13\text{-}21)$$

when applied to flow past a cylinder. For flow in a smooth pipe, we have defined the friction factor f, so that

$$\frac{g_c F_d}{u^2 \rho A} = \frac{g_c \tau_s}{u^2 \rho} = \frac{f}{2} \qquad (12\text{-}66)$$

Since we know

$$\tau_s = -\frac{\Delta p D}{4L} \qquad (2\text{-}10)$$

the equation for f becomes

$$f = \frac{g_c(-\Delta p)D}{2\rho L u_b^2} = \phi(\mathbf{Re}) \qquad (13\text{-}43)$$

A plot of f versus **Re** has been given in Fig. 12-8.

3 *Flow past ship models.* Since surface waves are formed, the Froude number must be included.

$$\frac{g_c p}{u^2 \rho} = \frac{g_c F_d}{u^2 \rho A} = \phi\left(\frac{Lu\rho}{\mu}, \frac{u^2}{Lg}\right) \qquad (13\text{-}44)$$

4 *Flow at high velocity; ballistics.*

$$\frac{g_c F_d}{u^2 \rho A} = \phi\left(\frac{u}{C}\right) \qquad (13\text{-}45)$$

5 *Flow of dispersions which are breaking up.*

$$\frac{g_c p}{u^2 \rho} = \phi\left(\frac{\rho u^2 L}{g_c \sigma}\right) \qquad (13\text{-}46)$$

If two liquid phases are present, groups formed from the ratios of the properties of the two phases, such as ρ_1/ρ_2, ν_1/ν_2, etc., may also be important.

Models and Similitude

Models are useful in the design of ships and airplanes and in the scale-up of chemical process equipment. Since experiments with the full-scale object, called the *prototype*, would be difficult and expensive, it is customary to study the behavior of models. In the cases which we shall discuss, the model is geometrically similar to the prototype; the ratio of any two lengths in the model is the same as that of two corresponding lengths in the prototype. In other words, the model and the prototype have the same shape. The terms *kinematic similarity* and *dynamic similarity* are used to describe a point-to-point correspondence between the velocities and the forces, respectively, for flow past the model and the prototype. If there is complete similarity for flow past a model and its geometrically similar prototype, all the pertinent dimensionless groups will have the same values for the two flows. Equations (13-5) to (13-7) show that this means that the ratios of the pertinent forces are the same for the two flows. Flow in a given smooth circular pipe can be regarded as flow past a model. By the principle of similitude, we can then predict the effect of flow in pipes of other sizes, which can be regarded as prototypes. The shape of the pipe is fixed, so once a set of experiments has been made to determine a curve of **Eu** versus **Re**, no more experiments are necessary. Complete similarity exists at equal Reynolds numbers, for which the Euler numbers must also be equal.

The situation is different for flow past a model of a ship hull, which does not have a standard shape but is made geometrically similar to the proposed prototype. Therefore experiments with a model may be necessary every time a new shape of hull is proposed.

Let us consider the problem of estimating the drag on a 500-ft ship. Equation (13-44) can be written as

$$C_D = \phi(\mathbf{Re}, \mathbf{Fr}) \tag{13-47}$$

and the problem would be solved if **Re** and **Fr** could be varied independently so as to find the function ϕ. However, Eqs. (13-6) and (13-7) show that

$$\mathbf{Fr} = \frac{uv}{gL^2}\mathbf{Re}$$

As the velocity of a given model is varied, **Re** and **Fr** bear a fixed relation to each other if water is always the fluid, since v/gL^2 is constant. Thus the general form of ϕ cannot be found with one model using only water as the fluid.

It can also be deduced that complete similarity between the model and the prototype cannot be achieved. For a prototype of $L = 500$ ft and a model of $L = 5$ ft, it would be necessary to use a velocity for the model one-tenth that for the prototype in order to maintain **Fr** constant. However, this variation results in a value of **Re** for the model only 0.001 times the **Re** of the prototype.

Naval architects resolve this problem by an approximation which involves changing Eq. (13-47) to

$$C_D = \phi_1(\mathbf{Re}) + \phi_2(\mathbf{Fr}) \tag{13-48}$$

$\phi_1(\mathbf{Re})$ is found from experiments on flow past completely immersed bodies, for which **Fr** has no effect. Experiments are also conducted on ship models to obtain data on C_D for flow in which surface waves are important. The function $\phi_2(\mathbf{Fr})$ is found from these data and from Eq. (13-48) by difference.

Application to Mixing

The behavior of a liquid being agitated by a propeller or turbine in a tank with a free liquid surface is very complicated and cannot be predicted by an analytical solution of the equations of motion. The technique of dimensional analysis is therefore useful, and we now apply it to the problem stated as follows:

$$P = f(D,\rho,\omega,g_c,\mu,g) \qquad (13\text{-}49)$$

where P = power supplied to propeller
D = diameter of propeller
ω = angular velocity of propeller

The other symbols have their usual meanings. For this case, $n = 7$, $r = 4$, and so $i = 3$. Choosing D, ρ, ω, and g_c as possibly common to each dimensionless group, we obtain

$$\pi_1 = \frac{Pg_c}{\rho\omega^3 D^5} \qquad (13\text{-}50)$$

$$\pi_2 = \frac{D^2\omega\rho}{\mu} \qquad (13\text{-}51)$$

$$\pi_3 = \frac{D\omega^2}{g} \qquad (13\text{-}52)$$

The group π_1 is called the *power number*. The linear speed u of the turbine blades, which may be chosen as characteristic, equals $\omega D/2$; substitution for ω in π_2 makes it evident that π_2 is the Reynolds number. The group π_3 is the Froude number.

The power number is a function of both **Re** and **Fr** when there is a vortex at the liquid surface. Just as for the ship model, it is not possible to achieve complete similarity between a small mixer and its prototype, or scaled-up version, using the same liquid. However, in contrast to the case of the ship model, it is practical to conduct the experiments on the mixer model with a fluid of different ν. By a suitable choice of liquid and temperature, it is then possible to achieve complete similarity between model and prototype.

By using different liquids, it is also possible to vary **Re** and **Fr** independently. Using this method, Rushton, Costich, and Everett[1] have obtained curves for a given mixer, which are shown in Fig. 13-1.[2] Such a curve can be used to predict the

[1] J. H. Rushton, E. W. Costich, and H. J. Everett, Power Characteristics of Mixing Impellers, *Chem. Eng. Progr.*, **46**:395, 467 (1950).

[2] In Figs. 13-1 and 13-3, N (rev/s) replaces ω (rad/s).

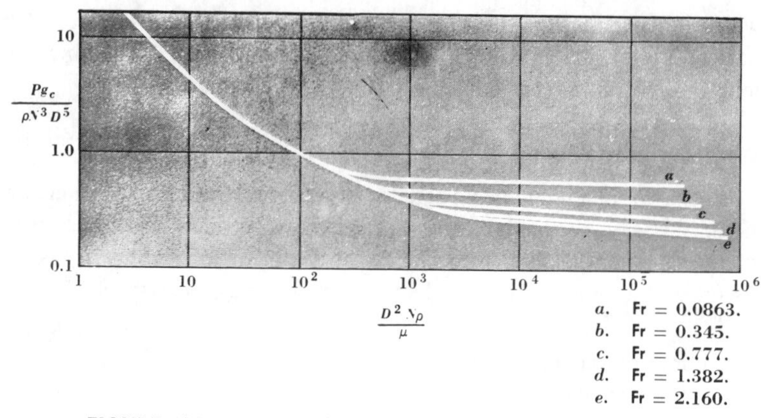

FIGURE 13-1
Power number for an unbaffled mixer. [*From J. H. Rushton, E. W. Costich, and H. J. Everett. Chem. Eng. Progr.*, **46**:*395–404 (1950)*.]

performance of only geometrically similar mixers. If the position of the impeller, depth of liquid, or ratio of impeller diameter to tank diameter is changed, the position of the curves in Fig. 13-1 may be changed.

In most industrial mixers for liquids, baffles are used to eliminate swirl, so that the Froude number is no longer important. The effect of the baffles is illustrated in Fig. 13-2. For a certain baffled mixer, the power number is a function of the Reynolds number only, as shown in Fig. 13-3. The shape of this curve resembles that for C_D as a function of **Re** for flow past a submerged object. At low Reynolds number, for which viscous effects are important, the slope of the curves in Fig. 13-1 or 13-3 approaches -1. At high Reynolds number C_D becomes constant as completely turbulent conditions are approached. A curve such as that shown in Fig. 13-3 applies only to geometrically similar mixers. As mentioned previously, if the position of the impeller, the depth of liquid, the ratio of impeller to tank diameter, or the relative size and position of the baffles is changed, a different curve may be obtained on a graph such as Fig. 13-3. Although the method discussed above permits the estimation of power consumption after scale-up, it says nothing about the effectiveness of mixing.

Rayleigh's Method

In this method Eq. (13-22) is arbitrarily written in the form

$$p = AL^a u^b \rho^c \mu^d g^e C^f \sigma^g g_c{}^h \qquad (13\text{-}53)$$

where A is a constant.

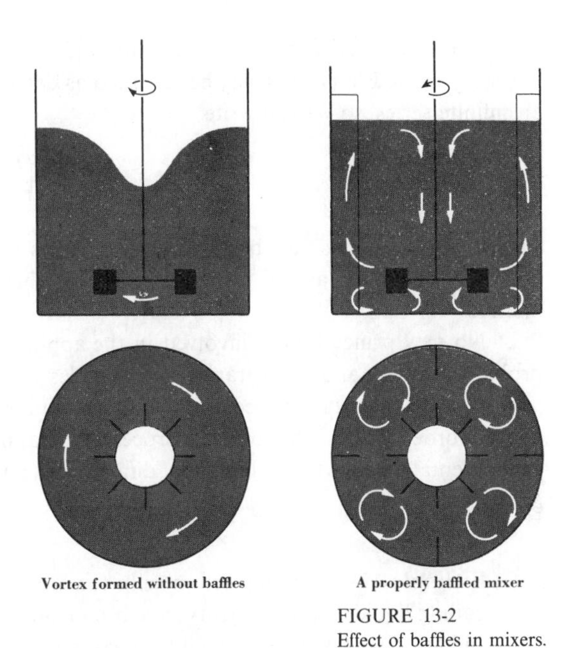

FIGURE 13-2
Effect of baffles in mixers.

Since this equation must be dimensionally homogeneous in the dimensions $[F]$, $[M]$, $[L]$, and $[\theta]$, it is possible to write four simultaneous equations and solve for four of the exponents in terms of the other four and the arbitrary exponent of 1.0 on p. This procedure gives

$$\frac{g_c p}{u^2 \rho} = A\left(\frac{Lu\rho}{\mu}\right)^{-d}\left(\frac{u^2}{Lg}\right)^{-e}\left(\frac{u}{C}\right)^{-f}\left(\frac{\rho u^2 L}{g_c \sigma}\right)^{-g} \tag{13-54}$$

where d, e, f, and g can have any values. There is no reason to suppose, in general, that $g_c p/u^2 \rho$ is the special function of other groups given by Eq. (13-54), although for

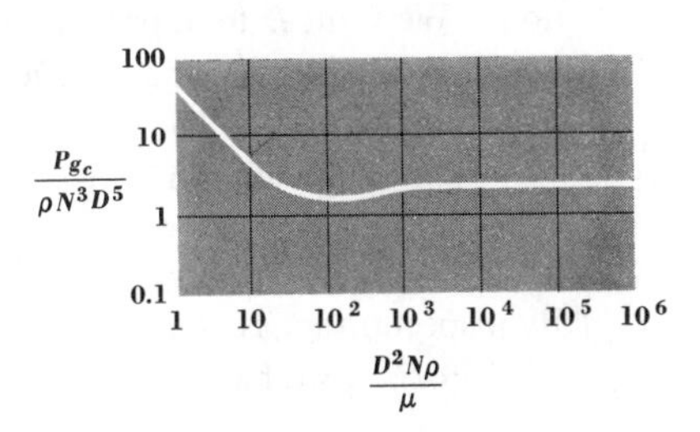

FIGURE 13-3
Power number for a baffled mixer. [*From J. H. Rushton, E. W. Costich, and H. J. Everett, Chem. Eng. Progr.*, **46**:467–476 (1950).]

many problems the function does have this form. In order to make the result more generally valid, Eq. (13-54) may be regarded as determining the form of one term in an infinite series, so we can write

$$\frac{g_c p}{u^2 \rho} = \phi\left(\frac{Lu\rho}{\mu}, \frac{u^2}{Lg}, \frac{u}{C}, \frac{\rho u^2 L}{g_c \sigma}\right) \qquad (13\text{-}55)$$

since many functions can be expanded in terms of an infinite series. Equation (13-55) is the same as the result obtained for this problem by the Buckingham method.

No new principles are involved in the application of dimensional analysis to problems in heat and mass transfer. Examples in these areas are reserved to the parts of the book dealing specifically with heat transfer and with mass transfer.

In order to show the complete process of obtaining a correlating equation from experimental data by the technique of dimensional analysis, the following example is given.

Example 13-1 In connection with fundamental studies on heat transfer to boiling liquids, measurements have been made on the rate of bubble formation and the size of bubbles formed by a gas issuing from a small orifice beneath the surface of the liquid. The method of dimensional analysis is to be used to correlate data on bubble size with the properties of the liquid.

We assume that the following equation represents the relation between bubble diameter D and the variables of the problem:

$$D = f(d,\rho,g_c,\sigma,\mu,g) \qquad (1)$$

in which d is the diameter of the orifice. It is found by trial and error that no more than four of the seven variables can be found which do not make a dimensionless group; these variables are d, ρ, g_c, and g. These variables will be chosen as the ones which may appear in any of the three $(7 - 4 = 3)$ dimensionless groups. Any dimensional constant such as g_c should always be included among these variables. The other three have been chosen by elimination. We want D to appear in only one group, and we also want to represent the effect of viscosity and the effect of surface tension so that they each appear only in one group. The variables g, g_c, d, and ρ may appear in any group.

The expression for π_1 is

$$\pi_1 = d^a \rho^b g_c{}^c g^d D^e \qquad (2)$$

By inspection, π_1 is D/d.

For π_2, we have

$$\pi_2 = d^a \rho^b g_c{}^c g^d \sigma^e \qquad (3)$$

The dimensions are substituted to yield

$$[1] = [L]^a \left[\frac{M}{L^3}\right]^b \left[\frac{ML}{F\theta^2}\right]^c \left[\frac{L}{\theta^2}\right]^d \left[\frac{F}{L}\right]^e \tag{4}$$

from which result the four equations

$$\text{For } [L] \qquad 0 = a - 3b + c + d - e \tag{5}$$

$$\text{For } [M] \qquad 0 = b + c \tag{6}$$

$$\text{For } [\theta] \qquad 0 = -2c - 2d \tag{7}$$

$$\text{For } [F] \qquad 0 = -c + e \tag{8}$$

If e is set equal to 1.0, we find $c = 1$, $b = -1$, $d = -1$, and $a = -2$, so that we get

$$\pi_2 = \frac{g_c \sigma}{\rho g d^2} \tag{9}$$

From the equation

$$\pi_3 = d^a \rho^b g_c{}^c g^d \mu^e \tag{10}$$

we obtain

$$\pi_3 = \frac{\mu^2}{\rho^2 g d^3} \tag{11}$$

in which e has been set equal to 2 in order to avoid fractional exponents. The result of the dimensional analysis is then

$$\frac{D}{d} = f\left(\frac{g_c \sigma}{\rho g d^2}, \frac{\mu^2}{\rho^2 g d^3}\right) \tag{12}$$

Let us now use Eq. (12) to correlate the data of Benzing,[1] part of which is reproduced in Table 13-1. Since these data are given in the cgs system, we get

Table 13-1

				Bubble diameter, cm	
System	Liquid viscosity, cP	Liquid density, g/cm^3	Surface tension, dyn/cm	Orifice diam., 0.293 cm	Orifice diam., 0.438 cm
Water-air	0.81	0.996	70.2	0.514	0.612
Water-H_2	0.81	0.996	70.2	0.516	0.630
7.5% ethanol sol'n-air	1.07	0.982	51.4	0.480	0.583
45% sugar sol'n-air	52.0	1.197	58.1	0.486	0.585
Wesson Oil-air	57.0	0.920	36.5	0.411	0.491

[1] R. J. Benzing, Low Frequency Bubble Formation at Horizontal Circular Orifices, *Ind. Eng. Chem.*, **47**:2087 (1955).

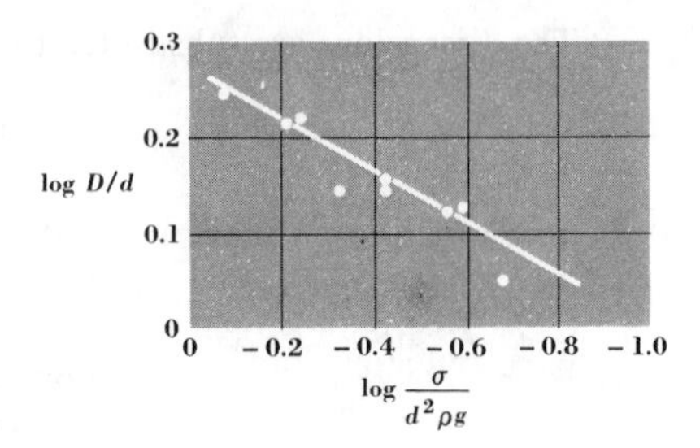

FIGURE 13-4
Correlation of data on bubble formation.

$g_c = 1.0$ in Eq. (12). We then plot in Fig. 13-4 $\log(D/d)$ versus $\log(\sigma/\rho g d^2)$. Recall that a centipoise is 0.01 g/(cm)(s) and that g is 981 cm/s^2. From the straight line which can be drawn through the points on the graph, it is possible to derive the equation

$$\frac{D}{d} = 1.88\left(\frac{\sigma}{\rho g d^2}\right)^{0.30} \tag{13}$$

If the variable $\mu^2/\rho^2 g d^3$ were important, it should be possible to draw lines through the points on Fig. 13-4 corresponding to successive values of this parameter. However, the points actually show no systematic variation with $\mu^2/\rho^2 g d^3$, so it can be concluded that D/d is independent of $\mu^2/\rho^2 g d^3$. This result is not surprising, for it is possible to show that the bubble size at a low rate of formation is governed only by a balance between the buoyant force and the force arising from surface tension.

If an equation of the form

$$\pi_1 = a\pi_2{}^b\pi_3{}^c \tag{14}$$

is assumed, the constants a, b, and c can be found by a technique known as the *method of least squares*. The exponent c is found to have a very small value (approximately 0.01), indicating that the group π_3 is of negligible importance in the equation. The values found for a and b are 1.82 and 0.25, respectively.

////

PROBLEMS

13-1 Dimensional analysis is useful in the design of centrifugal pumps. The pressure rise across a pump p (this term is proportional to the "head" developed by the pump) may be considered to be affected by the fluid density ρ, the angular velocity ω, the impeller diameter D, the volumetric rate of flow Q, and the fluid viscosity μ. Find the pertinent

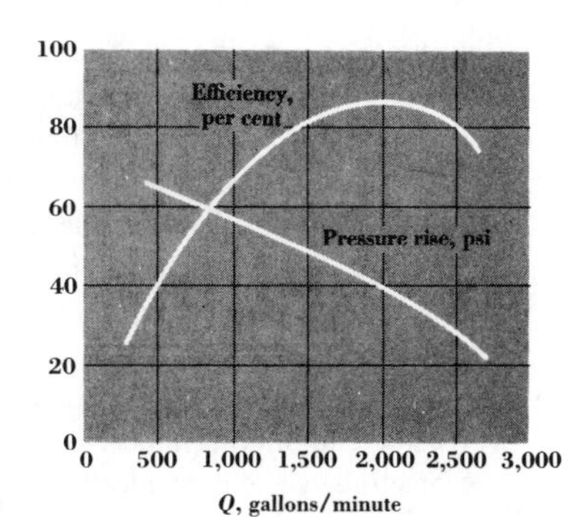

FIGURE 13-5
Performance of a centrifugal pump.

dimensionless groups, choosing them so that p, Q, and μ each appear in one group only. Find similar expressions replacing the pressure rise first by the power input to the pump, then by the efficiency of the pump.

13-2 Experiments show that the fluid viscosity actually has little effect on the performance of a centrifugal pump under usual conditions, so that one of the terms obtained in each part of Prob. 13-1 can be omitted.

Test data obtained on a centrifugal pump are shown in Fig. 13-5. Estimate the ratios of size and speed of a geometrically similar pump to the size and speed of the pump for which the data are given if it is desired to operate the second pump at maximum efficiency, 1000 gal/min, and a pressure rise of 50 psi.

13-3 A rough method of scaling up liquid-mixing tanks and impellers is to keep the power input per unit volume constant. If it is desired to increase the volume of a properly baffled liquid mixer by a ratio of 3, by what ratio must the tank diameter and impeller speed be changed? The mixers are geometrically similar, and both operate in the completely turbulent region.

13-4 By the use of the Prandtl mixing length, we showed in Chap. 12 that the velocity distribution in turbulent flow inside a pipe can be given by Eq. (12-57), which may be written as

$$\frac{u}{u^*} = \phi\left(\frac{yu^*}{\nu}\right) \qquad (A)$$

where $u^* = \sqrt{\tau_s g_c/\rho}$. Obtain Eq. ($A$) by dimensional analysis.

13-5 The time for a pendulum (consisting in this case of a bob on a string) to make a complete swing and return to any given position is called the *period* of the pendulum. Assuming that we have no knowledge of the laws of mechanics we wish, nevertheless, to derive an equation relating the period to various independent variables which might influence its value. It would seem likely that the following relation exists:

$$t = f(p, s, m, g)$$

where t = period
p = length of pendulum string
s = length of arc of pendulum
m = mass of bob on string
g = acceleration due to gravity.

Use the Rayleigh method of dimensional analysis to derive an equation which indicates the form of the functional relationship which might exist between t and the variables given above.

13-6 An unbaffled mixer is to be designed for a fluid of $\mu = 0.004$ Pa · s and $\rho = 1000$ kg/m^3. We want to study a model of this mixer in the laboratory. The model should be geometrically and dynamically similar to the prototype and on one-third scale. If we use water in the model, what water temperature and ratio of model to prototype revolutions per minute are needed?

13-7 Air bubbles in water are being made as described in Example 13-1. The orifice diameter is 0.0065 m.

(*a*) What size bubbles will be made?

(*b*) If the apparatus is moved to a planet where $g = 2.6$ m/s^2 what will be the bubble diameter?

13-8 A sphere held by a fine wire is to be coated with a viscous liquid by dipping followed by drainage to an equilibrium coating. What dimensionless groups would be useful in correlating the data on drainage time from experiments on various spheres in various liquids?

13-9 What form might the Reynolds number take for flow of a power-law nonnewtonian fluid?

13-10 Laboratory experiments indicate that a certain properly designed baffled mixer operates well at a Reynolds number of 10^4. Estimate the power needed (hp) for a geometrically and dynamically similar tank and impeller with a 2-ft-diameter impeller operating on methanol at room temperature.

14

SOME DESIGN EQUATIONS FOR THE FLOW OF INCOMPRESSIBLE FLUIDS

In the preceding chapters, an introduction to the principles of fluid dynamics has been given which should serve as a basis for the solution of many engineering problems. The overall balances permit the solution of a variety of input-output problems. The equations of motion have been solved for some simple cases, and the technique of dimensional analysis plus experiment has been introduced for the treatment of more complicated situations. As mathematical and computational techniques improve, the number of situations which can be treated analytically will continually increase, so a thorough understanding of the principles underlying these techniques is essential to the engineer.

In discussing the theory we did not want to interrupt the continuity of the reasoning by dwelling on numerous practical examples; in this chapter we present the application of the principles of fluid dynamics to the flow of incompressible fluids in conduits, past immersed objects, and through packed beds.

The Mechanical Energy Balance

The solutions of many problems are based on the mechanical energy balance. This equation, written for the steady flow of an incompressible fluid, is

$$\frac{\Delta u_b^2}{2g_c} + \frac{g}{g_c}\Delta z + \frac{\Delta p}{\rho} + lw_f + \eta_p W_s = 0 \tag{4-36}$$

for use with a pump, and is

$$\frac{\Delta u_b^2}{2g_c} + \frac{g}{g_c}\Delta z + \frac{\Delta p}{\rho} + lw_f + \frac{W_s}{\eta_t} = 0 \tag{4-33}$$

for use with a turbine. The use of these equations and their relation to the overall energy balance have been discussed in Chap. 4. The main business of this chapter is to present concisely the methods of calculating lw_f for various practical flow situations.

In the discussions of pressure loss due to friction in Chaps. 12 and 13 the Fanning friction factor was defined as

$$f = \frac{g_c(-\Delta p)D}{2u_b^2 \rho L} \tag{13-43}$$

for the flow of an incompressible fluid in a horizontal pipe of uniform diameter with no pump. This means that Δu_b^2, Δz, and W_s are zero, so there results from Eqs. (4-33) and (13-43)

$$lw_f = \frac{2fLu_b^2}{g_c D} \tag{14-1}$$

This equation relates the loss to the pipe length and the Reynolds number. It is important to realize that Eq. (14-1) can also be used when the pipe is not horizontal and when there is a work term. This statement is illustrated in the equation

$$\frac{\Delta u_b^2}{2g_c} + \frac{g}{g_c}\Delta z + \frac{\Delta p}{\rho} + \eta_p W_s + \sum \frac{2fLu_b^2}{g_c D} + \sum lw_c + \sum lw_e = 0 \tag{14-2}$$

The summation sign is inserted before the term for the loss in straight pipe because this loss may result from flow in series through several sections of pipe of various lengths and diameters. The symbol $\sum lw_c$ stands for the sum of the contraction losses, and $\sum lw_e$ refers to the sum of the expansion losses. These effects have been discussed in Chaps. 5 and 6. It will be recalled that the loss in the pump is accounted for by the efficiency η_p.

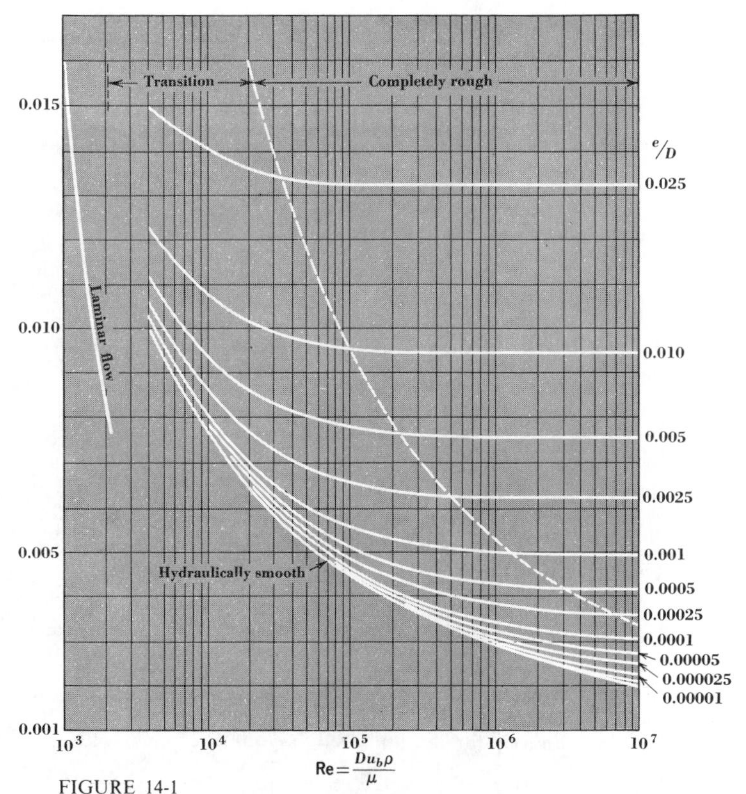

FIGURE 14-1
Friction factor vs. Reynolds number for commercial pipe. (*From J. G. Knudsen and D. L. Katz, "Fluid Dynamics and Heat Transfer," fig. 7-21, p. 176, McGraw-Hill Book Company, New York, 1950.*)

Flow in Circular Pipes

The inner surfaces of commercial pipes are not necessarily hydraulically smooth, so the friction-factor chart in Fig. 14-1 contains curves for various degrees of relative roughness. Figure 14-1 is based on the data of many investigators as correlated by Moody.[1]

[1] L. F. Moody, Friction Factors for Pipe Flow, *Trans. Am. Soc. Mech. Eng.*, **66**:671 (1944). The entire graph is also well represented by the equation of S. W. Churchill, *Chem. Eng.*, **84**(24):91 (1977):

$$f/2 = \left[\left(\frac{8}{\mathbf{Re}}\right)^{12} + \frac{1}{(A+B)^{3/2}}\right]^{1/12}$$

where

$$A = \left[2.457 \ln\left(\frac{1}{(7/\mathbf{Re})^{0.9} + 0.27e/D}\right)\right]^{16}$$

$$B = \left(\frac{37530}{\mathbf{Re}}\right)^{16}$$

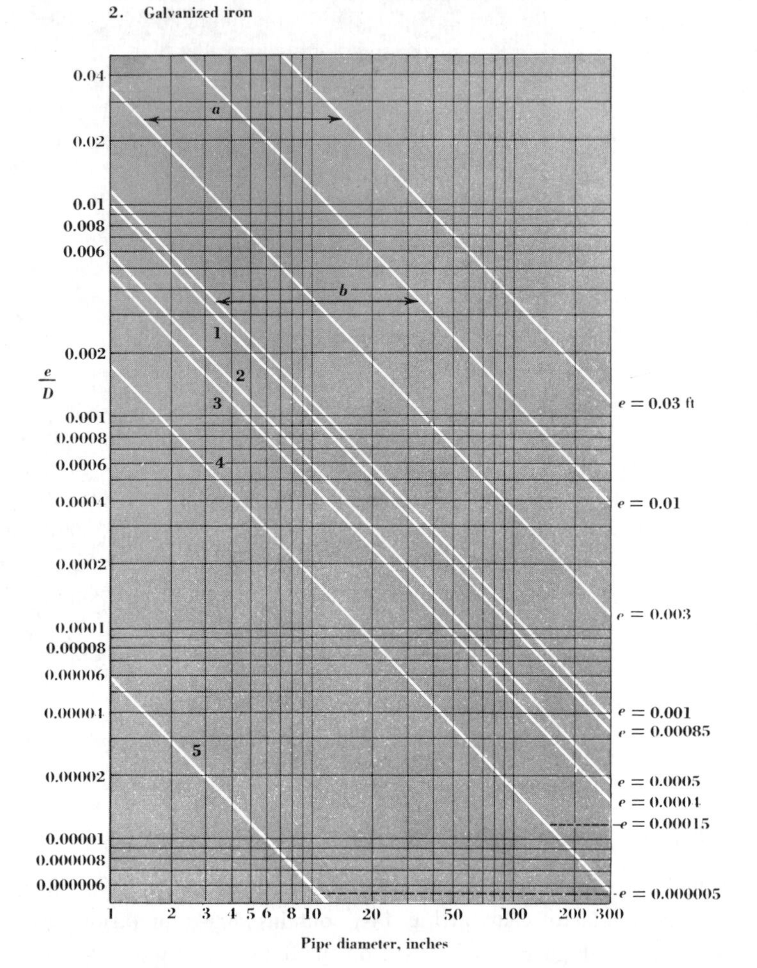

FIGURE 14-2
Relative roughness of commercial pipe. [*From L. F. Moody, Trans. Am. Soc. Mech. Eng.*, **66**:*671* (*1944*).]

The curves for rough pipe have different shapes from those obtained by Nikuradse for pipes artificially roughened with sand (Fig. 12-9). The curves for commercial pipe become horizontal in the completely rough region, but they differ from those of Nikuradse in the transition region between rough and hydraulically smooth. The relative roughness parameters shown in Fig. 14-1 are the same as those for the lines in Fig. 12-9, which were defined to coincide with them in the completely rough zone. They were not found by actually measuring the heights of protuberances in commercial pipe. Figure 14-2 enables one to find the relative roughness of various kinds of pipe.

The line in Fig. 14-1 for laminar flow is identical with Eq. (12-73); the line for turbulent flow for the hydraulically smooth regime is identical with Eq. (12-70); and the lines for the completely rough regime are identical with Eq. (12-77).

If expansion and contraction losses make up a large fraction of the overall loss, as in a short heat exchanger, these losses should be calculated as outlined in Chap. 5 and as given in detail by Kays.[1] In many cases, however, the losses at expansions, contractions, and fittings of various kinds are a small portion of the overall loss, and Fig. 14-3 can be used. This nomograph gives the approximate equivalent length of straight pipe which would have the same effect as a given fitting. These lengths are added to the lengths of straight pipe between fittings, and the sums are used for the pertinent values of L in Eq. (14-2). In this procedure the effects of $\sum lw_c$ and $\sum lw_e$ are included in the equivalent length, so these terms may be omitted. The following example will explain more fully the use of these methods.

Example 14-1 Water at 60°F is pumped at 100 gal/min from a reservoir through a system of piping into an open tank, the level of which is maintained constant at 17 ft above the level of the reservoir. From the reservoir to the pump 3-in, schedule-40 steel pipe is used, and from the pump to the elevated tank 2-in, schedule-40 pipe is used. The pipe lengths and fittings are shown in Fig. 14-4. Calculate the power consumption of the pump if its efficiency is 70 percent.

The solution of this problem is begun by writing the mechanical energy balance between the two planes at the liquid surfaces indicated by 1 and 2 in the figure. Thus Δp is zero, Δu_b is zero, and Δz is 17 ft. Equation (4-36) then yields

$$17 + lw_f + 0.70W_s = 0 \qquad (1)$$

Once we have determined lw_f, this equation will permit the computation of W_s, and thus the power input. The parts of this loss term are represented by the last three terms on the left side of Eq. (14-2). Using Fig. 14-3, we find that for

[1] W. M. Kays, *Trans. Am. Soc. Mech. Eng.*, **72**:1067 (1950).

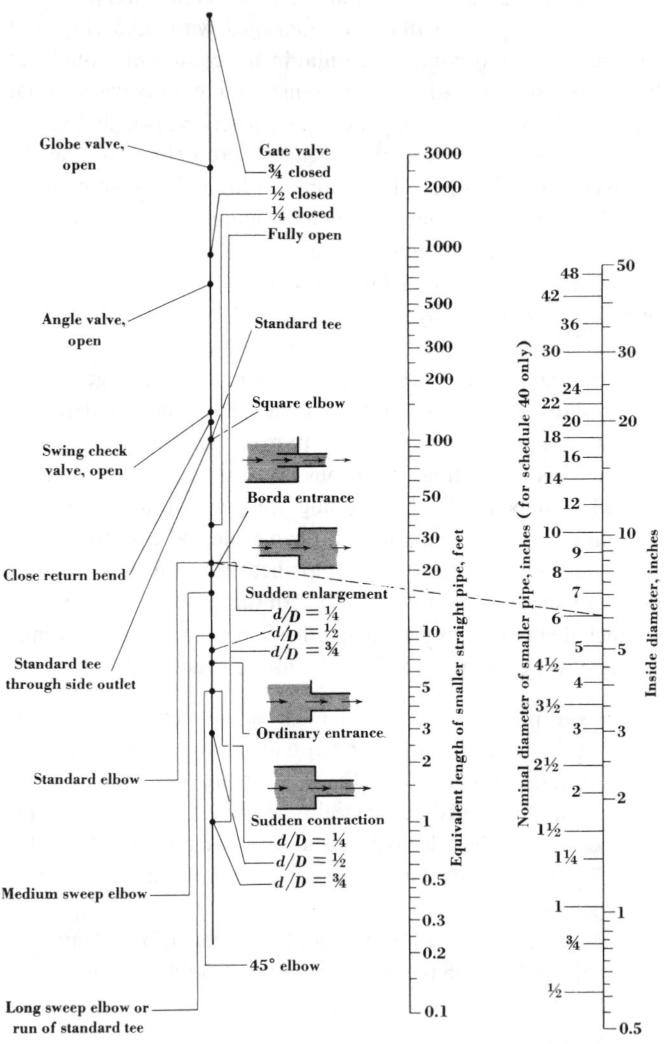

FIGURE 14-3
Equivalent lengths for friction losses. (Crane Co.)

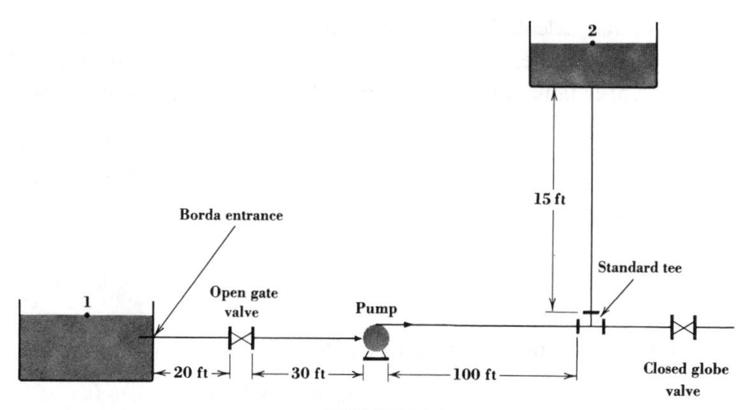

FIGURE 14-4
Piping systems for Example 14-1. (Not to scale.)

the 3-in pipe, the equivalent length of the Borda entrance is 8 ft, and that of the open gate valve is 1.7 ft. The equivalent length of 3-in pipe is, then, 8 + 20 + 1.7 + 30, or 59.7 ft. We also calculate the following quantities for the 3-in pipe:[1]

$$w = \frac{(100)(62.6)}{(7.48)(60)} = 13.97 \text{ lb/s}$$

$$u_b = \frac{13.97}{(7.393/144)(62.6)} = 4.35 \text{ ft/s}$$

$$\mathbf{Re} = \frac{Du_b\rho}{\mu} = \frac{(3.068/12)(4.35)(62.4)}{(1.05)(6.72 \times 10^{-4})} = 9.87 \times 10^4$$

From Fig. 14-2, we find

$$\frac{e}{D} = 0.0006$$

From Fig. 14-1, the friction factor is

$$f = 0.0052$$

and so, for the 3-in pipe, the loss is

$$lw_{f3} = \frac{2fLu_b^2}{g_c D} = \frac{(2)(0.0052)(59.7)(4.35)^2}{(32.2)(3.068/12)}$$

$$= 1.43(\text{ft})(\text{lb}_f)/\text{lb}$$

[1] Pipe diameters and cross-sectional areas, viscosities, and densities are from Perry's Handbook or the tables in the appendix.

A similar calculation is made to find the loss in the 2-in pipe. For the tee, $L = 12$ ft. There is no suitable point on Fig. 14-3 for the expansion loss into the upper tank, for which $d/D = 0$, but from Eq. (5) of Example 5-2, we find

$$lw_e = \frac{u_{b1}^2}{2g_c}\left(1 - \frac{A_1}{A_2}\right)^2$$

Since $A_1/A_2 = 0$, we have

$$lw_e = \frac{u_{b1}^2}{2g_c}$$

For the 2-in pipe, we now calculate

$$u_b = 4.35(3.068/2.067)^2 = 9.56 \text{ ft/s}$$

$$\mathbf{Re} = \frac{(2.067/12)(9.56)(62.6)}{(1.05)(6.72 \times 10^{-4})} = 1.46 \times 10^5$$

Figure 14-2 yields

$$\frac{e}{D} = 0.0009$$

and so we have, from Fig. 14-1,

$$f = 0.0053$$

For the 2-in pipe the loss is

$$lw_{f2} = \frac{(2)(0.0053)(100 + 12 + 15)(9.56)^2}{(32.2)(2.067/12)} + \frac{(9.56)^2}{2(32.2)}$$

$$= 22.4 + 1.43 = 23.8(\text{ft})(\text{lb}_f)/\text{lb}$$

The total loss term is then

$$lw_f = lw_{f3} + lw_{f2} = 1.4 + 23.8 = 25.2(\text{ft})(\text{lb}_f)/\text{lb}$$

This result is substituted into Eq. (1) to give

$$17 + 25.2 + 0.70W_s = 0$$

$$W_s = -60.3(\text{ft})(\text{lb}_f)/\text{lb}$$

The power requirement is then given by

$$\frac{(60.3)(13.97)}{550} = 1.53 \text{ hp} \qquad ////$$

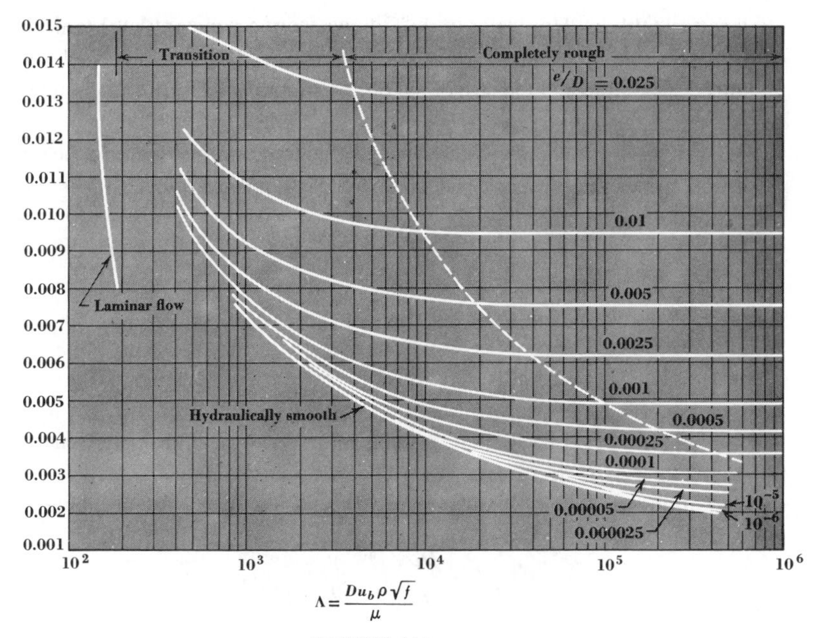

FIGURE 14-5
Friction factor versus Kármán number for commercial pipe.

In most problems the properties of the fluid are known; in the particular case just illustrated it was a straightforward procedure to calculate **Re** and obtain f from Fig. 14-1 and then find lw_f from Eq. (14-1). On the other hand, if the horsepower of the pump had been given instead of the flow rate, it would have been necessary to use a trial-and-error procedure to obtain the velocity (i.e., the flow rate) since u_b must be known to calculate both the friction factor and the Reynolds number.

In order to solve a problem in which the velocity is the unknown (that is, lw_f is given), the data of Fig. 14-1 are plotted so that the velocity need not be known to calculate the abscissa. This is done by defining a variable Λ as follows:

$$\Lambda = \mathbf{Re}\sqrt{f} = \frac{D\rho}{\mu}\sqrt{\frac{lw_f g_c D}{2L}} \qquad (14\text{-}3)$$

From the data in Fig. 14-1 it is easy to calculate values of Λ and find the corresponding values of f. Figure 14-5 is a plot of f versus Λ; its use is illustrated in the following example.

Example 14-2 Hot water at 43°C flows from a constant-level tank through 2-in, schedule-40 steel pipe, from which it emerges 12.2 m below the level in the tank. The equivalent length of the piping system is 45.1 m. Calculate the rate of flow in m^3/s.

If the input plane is at the level of the liquid in the tank and the output is at the end of the pipe, the mechanical energy balance reduces to

$$\Delta z \frac{g}{g_c} + \frac{\Delta u_b^{\,2}}{2g_c} + lw_f = 0 \tag{1}$$

or, since u_{b1} is zero and Δz is -12.2 m,

$$(-12.2)\left(\frac{9.805}{1}\right) + \frac{u_b^{\,2}}{2g_c}\left(1 + \frac{4fL}{D}\right) = 0 \tag{2}$$

Equation (2) shows that the relative importance of the kinetic energy term increases as L/D decreases.

In order to find f, and thus u_b, Fig. 14-5 is used. We calculate

$$\mathbf{Re}\sqrt{f} = \frac{D\rho}{\mu}\sqrt{\frac{lw_f g_c D}{2L}} \tag{3}$$

$$\mathbf{Re}\sqrt{f} = \frac{(0.0525)(987)}{0.621 \times 10^{-3}}\sqrt{\frac{(120)(1)(0.0525)}{(2)(45.11)}} \tag{4}$$

$$\mathbf{Re}\sqrt{f} = 2.20 \times 10^4$$

From Fig. 14-5, with $e/D = 0.0009$, $f = 0.0049$.

In the above calculation lw_f has been taken as 120, found from Eq. (1), neglecting the kinetic energy term. If the latter is important, a correction can be made for it later. Equation (2) now becomes

$$-120 + \frac{u_b^{\,2}}{2}\left[1 + \frac{(4)(0.0049)(45.11)}{0.0525}\right] = 0$$

and

$$-120 + \frac{u_b^{\,2}}{2}(1 + 16.84) = 0$$

and finally

$$u_b = \sqrt{\frac{(120)(2)}{17.84}} = 3.67 \text{ m/s}$$

From this value of u_b a new value of lw_f can be calculated from Eq. (1) and a new value of $\mathbf{Re}\sqrt{f}$ obtained. However, the latter is changed only by a few percent, so f will not be changed significantly from its old value. Therefore the correct answer is $u_b = 3.67$ m/s, and the flow rate is

$$(3.67)(0.00216) = 0.00794 \text{ m}^3/\text{s} \qquad ////$$

The group $\mathbf{Re}\sqrt{f}$ is sometimes called the Kármán number. If $1/\sqrt{f}$ is plotted versus log $(\mathbf{Re}\sqrt{f})$, Eq. (12-70) shows that a straight line is obtained in the hydraulically smooth zone. Such a curve is given by Rouse,[1] for example.

It is left as an exercise for the reader to show that a plot of f versus $\mathbf{Re}\ f^{1/5}$ might be convenient for a case in which the total frictional loss, length, and volumetric flow rate are fixed but the pipe diameter is unknown.

Flow in Parallel Conduits

Flow in a network of pipe is often a complicated situation and may require the solution of many simultaneous, nonlinear equations. High-speed computing machines are useful in the calculation of the flows in the various parts of a large distribution system. The following simple example illustrates some of the principles involved.[2]

Example 14-3 Water at 68°F flows from a constant-head tank through an equivalent length of 50 ft of 2-in, schedule-40 pipe to a point 15 ft below the liquid surface. At this point the line is divided; part of the water flows through an equivalent length of 30 ft of $1\frac{1}{2}$-in pipe to discharge at a point 30 ft below the level of the liquid in the tank, and the other part flows in parallel through an equivalent length of 40 ft of 1-in pipe to discharge at a point 20 ft below the level in the tank. Both lines discharge at atmospheric pressure. Find the bulk velocities in the three sections.

Referring to the 2-in pipe as section 1, the $1\frac{1}{2}$-in pipe as section 2, and the 1-in pipe as section 3, we write the mechanical energy balance for each section:

$$-15 + \frac{\Delta p_1}{\rho} + lw_{f1} = 0 \qquad (1)$$

$$-15 + \frac{\Delta p_2}{\rho} + lw_{f2} = 0 \qquad (2)$$

$$-5 + \frac{\Delta p_3}{\rho} + lw_{f3} = 0 \qquad (3)$$

[1] H. Rouse and J. W. Howe, "Basic Mechanics of Fluids," p. 145, John Wiley & Sons, Inc., New York, 1953.

[2] An efficient computer technique is described by B. Gay and P. Middleton, *Chem. Eng. Sci.*, **26**:109 (1971). The classical method is that of Hardy Cross, Analysis of Flow in Networks of Conduits or Conductors, *Univ. Illinois Eng. Exp. Stn. Bull.* 286, 1936.

In order not to complicate the problem, kinetic energy effects and losses at fittings, contractions, and expansions are omitted.

Since the overall Δp from the surface of the liquid in the tank to the outlets of the two pipes is zero and Δp_2 must equal Δp_3, we have

$$\Delta p_1 = -\Delta p_2 = -\Delta p_3 \tag{4}$$

In addition, the overall mass balance gives

$$w_1 = w_2 + w_3$$

and

$$u_{b1} D_1^2 = u_{b2} D_2^2 + u_{b3} D_3^2 \tag{5}$$

We must now proceed by trial and error. First Δp_1 is assumed, so that lw_{f1}, lw_{f2}, and lw_{f3} can be calculated from Eqs. (1) to (3). By the method illustrated in Example 14-2, u_{b1}, u_{b2}, and u_{b3} are then calculated.

If Eq. (5) is not satisfied using these values, a new Δp_1 is chosen until one is found for which Eq. (5) is satisfied.

Let us begin by choosing $\Delta p_1 = 0$, a reasonable choice which has the added virtue of simplifying the calculations. From Eqs. (1) to (3) we get $lw_{f1} = 15$, $lw_{f2} = 15$, and $lw_{f3} = 5$. We then find, for section 1,

$$\mathbf{Re}\sqrt{f} = 1.46 \times 10^4 \qquad f = 0.0050$$

$$u_{b1} = 12.9 \text{ ft/s}$$

For section 2,

$$\mathbf{Re}\sqrt{f} = 1.29 \times 10^4 \qquad f = 0.0053$$

$$u_{b2} = 14.3 \text{ ft/s}$$

For section 3,

$$\mathbf{Re}\sqrt{f} = 3.41 \times 10^3 \qquad f = 0.0065$$

$$u_{b3} = 5.21 \text{ ft/s}$$

Substituting the appropriate diameters, Eq. (5) becomes

$$u_{b1} = 0.607 u_{b2} + 0.257 u_{b3} \tag{6}$$

If the three velocities found above are substituted in Eq. (6), the right-hand side is 10.0 and the left-hand side is 12.9.

This result shows that the velocity in the 2-in pipe is too large, so that Δp_1 should be chosen greater than zero. This change will *reduce* the friction loss in section 1. If $\Delta p_1/\rho$ is chosen as 3.5 (ft)(lb_f)/lb, corresponding to $\Delta p_1 = +1.5$ psi, the resulting velocities will satisfy Eq. (6). The velocities are $u_{b1} = 11.3$ ft/s, $u_{b2} = 15.9$ ft/s, and $u_{b3} = 6.79$ ft/s.

In obtaining this result, the trial-and-error process has been simplified by assuming that the f's remain constant as the u_b's are changed by the relatively small factors involved. ////

Flow in Conduits Other Than Circular

The following remarks apply only to turbulent flow; laminar flow for several cases has been discussed in Chap. 10.

Figures 14-1 and 14-5 can be used for turbulent flow in many noncircular conduits by replacing the diameter D by the equivalent diameter D_{eq} defined by the relation

$$D_{eq} = 4r_H \tag{14-4}$$

In this equation r_H is the hydraulic radius, in turn defined by the equation

$$r_H = \frac{A}{l_p} \tag{14-5}$$

where l_p = wetted perimeter of conduit
A = cross-sectional area of conduit

For a circular pipe it may be verified that this definition leads to the consistent result that $D = D_{eq}$.

As an example, for flow in an annulus, D_{eq} equals $D_2 - D_1$, and thus we obtain

$$lw_f = \frac{2fLu_b^2}{g_c(D_2 - D_1)} \tag{14-6}$$

and

$$\mathbf{Re} = \frac{(D_2 - D_1)u_b\rho}{\mu} \tag{14-7}$$

The hydraulic radius can also be used with the friction-factor charts for flow in open channels or partially filled pipes.

Flow past Immersed Bodies

In Chap. 11 we discussed at some length flow past a cylinder; the results are expressed quantitatively in Fig. 11-6. In Chap. 13 we showed that the form of the coordinates of Fig. 11-6 could be obtained by dimensional analysis.

It is of interest also to know the drag in flow past spheres and disks; Fig. 14-6 presents the relation between C_D and **Re**, where

$$C_D = \frac{2F_d g_c}{\rho u_o^2 A} \tag{11-5}$$

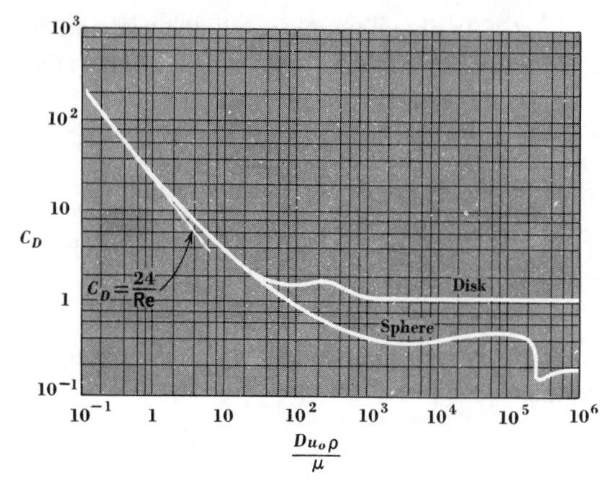

FIGURE 14-6
Drag coefficients for spheres and disks. (*From H. Rouse and J. W. Howe, "Basic Mechanics of Fluids," fig. 107, p. 181, John Wiley & Sons, Inc., New York, 1953.*)

where $A = \pi D^2/4$, and where

$$\mathbf{Re} = \frac{Du_o\rho}{\mu}$$

In these equations, D is the diameter of the sphere or disk.

The curve for the sphere is similar to that for the cylinder; there is a sudden drop in C_D at $\mathbf{Re} \cong 3 \times 10^5$, where the point of separation moves downstream as the boundary layer becomes turbulent. The curve for the disk does not have this sudden drop, for the point of separation can occur only at the sharp edge of the disk.

For $\mathbf{Re} < 1.0$, Stokes' equation, discussed in Chap. 10, applies for both bodies.

Flow Normal to Banks of Tubes

Flow normal to a bank of tubes is encountered in air heaters and other heat exchangers. Figure 14-7*a* and *b* shows one way in which the fluid on the shell side of a heat exchanger may be caused to flow normal to the tubes by using baffles. There is flow across N_1 rows of tubes between baffles and flow parallel to the tubes in the baffle windows. There is also some leakage between the baffles and the tubes.

The Reynolds number and friction factor for flow past tube banks are defined in the literature in several ways. Some of the possible ways of defining the diameter in **Re** and f are as follows:

1 The flow is considered to be the sum of the effects obtained in flow past a single cylinder; the important dimension is thus the tube diameter.

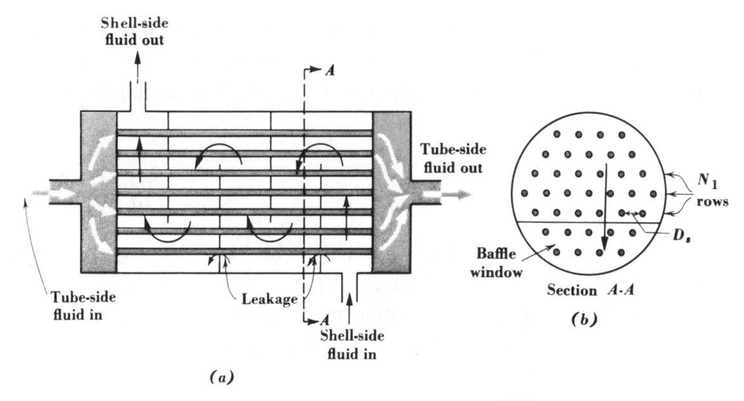

FIGURE 14-7
Heat exchanger with baffles.

2 The flow is considered to be the sum of the effects obtained in flow through rectangular orifices formed by the spaces between the tubes; the important dimension is thus the tube spacing D_s (Fig. 14-7*b*).

3 The situation is considered as flow through a noncircular conduit, and the hydraulic radius is defined as the free volume in the tube bank divided by the exposed surface area of the tubes. The important dimension is thus the equivalent diameter, $D_{eq} = 4r_H$.

No attempt is made here to give a comprehensive survey of this problem, for which many correlations have been proposed. As an example, the following equation given by Donohue[1] is presented for turbulent flow across a bank of N_1 rows of staggered tubes:

$$f' = 0.99\left(\frac{Du_{b,\max}\rho}{\mu}\right)^{-0.2} \tag{14-8}$$

for **Re** > 500. For this type of system laminar flow exists for **Re** < 100. The term $u_{b,\max}$ is the velocity based on the minimum cross-sectional area normal to the flow. The friction factor is defined by

$$f' = \frac{g_c(lw_f)}{2N_1 u_{b,\max}^2}\left(\frac{\mu}{\mu_s}\right)^{0.14} \tag{14-9}$$

where μ_s is the fluid viscosity at the average surface temperature of the tubes. The average bulk fluid temperature is used to find μ and ρ.

[1] D. A. Donohue, Heat Transfer and Pressure Drop in Heat Exchangers, *Ind. Eng. Chem.*, **41**:2499 (1949).

The factor $(\mu/\mu_s)^{0.14}$ can be used with turbulent flow in general to correct the friction factor for the effect of temperature variation in the fluid in the direction normal to the flow. For instance, for flow inside a heat-exchanger tube, the ordinate of Fig. 14-1 can be interpreted as

$$\frac{g_c l w_f D}{2u_b{}^2 L}\left(\frac{\mu}{\mu_s}\right)^{0.14}$$

This empirical method is based on the fact that the viscosity in the region of high shear near the wall has a large effect on f.

A survey of some of the numerous equations which have been proposed for flow across tube banks is given by Knudsen and Katz.[1]

Flow through Packed Beds

Flow of a fluid through a packed bed of granular particles occurs frequently in chemical processes. Some examples are flow through a fixed-bed catalytic reactor, flow through an adsorption or absorption column, and flow through a filter cake. An understanding of flow through packed beds is important in the study of fluidization and sedimentation. Two fluids may flow simultaneously, as in the absorption of a component in a gas by a liquid in a packed tower. In the present chapter, we treat only the flow of a single homogeneous fluid.

As for flow past tube banks, there exist many correlations for calculating the pressure drop in flow through a packed bed. This problem can be considered in the three ways mentioned in connection with flow past tube banks; we choose here a method based on the use of the hydraulic radius. This method is applicable to incompressible beds composed of nearly spherical particles; the bed porosities (fractional-void space) may range from about 0.3 to 0.6. More comprehensive discussions and reviews of the extensive literature are found in specialized treatises.[2] A successful method for the prediction of the behavior of flow through a bed from flow past a single sphere is given by Ranz.[3]

For a packed bed of N_p particles, the hydraulic radius is defined as

$$r_H = \frac{\text{void volume of bed}}{\text{surface area of packing}} \tag{14-10}$$

[1] J. G. Knudsen and D. L. Katz, "Fluid Dynamics and Heat Transfer," chap. 11, McGraw-Hill Book Company, New York, 1958.

[2] M. Leva, M. Weintraub, M. Grummer, M. Pollchik, and H. H. Storch, Fluid Flow through Packed and Fluidized Systems, *U.S. Bur. Mines Bull.* 504, 1951; P. C. Carman, "Flow of Gases through Porous Media," Academic Press, Inc., New York, 1956; A. E. Scheidegger, "The Physics of Flow through Porous Media," The Macmillan Company, New York, 1957.

[3] W. E. Ranz, Friction and Transfer Coefficients for Single Particles and Packed Beds, *Chem. Eng. Progr.*, **48**:247 (1952).

If the volume of a particle is v_p and the surface area of a particle is S_p, the specific surface of a particle is defined by

$$S_v = \frac{S_p}{v_p} \qquad (14\text{-}11)$$

For a spherical particle

$$S_v = \frac{6}{D} \qquad (14\text{-}12)$$

In considering packed beds of nonspherical particles, we shall use D as the effective particle diameter defined by the equation

$$D = \frac{6}{S_v} \qquad (14\text{-}13)$$

We now substitute in Eq. (14-10) to find the hydraulic radius.

$$r_H = \frac{\varepsilon v_p N_p/(1-\varepsilon)}{S_p N_p} = \frac{\varepsilon}{1-\varepsilon}\frac{v_p}{S_p} \qquad (14\text{-}14)$$

Here ε is the void fraction. This equation can be written as

$$r_H = \frac{\varepsilon}{(1-\varepsilon)S_v} = \frac{1}{6}\frac{\varepsilon}{1-\varepsilon}D \qquad (14\text{-}15)$$

Since we know by definition that

$$\mathbf{Re} = \frac{4r_H u_b \rho}{\mu}$$

we obtain

$$\mathbf{Re} = \frac{4\varepsilon}{6(1-\varepsilon)}\frac{Du_b\rho}{\mu} \qquad (14\text{-}16)$$

In this equation, u_b is the average interstitial velocity at any cross section in the bed; it is more convenient to define a superficial velocity, u_{bs}, based on the cross section of the empty container,

$$u_{bs} = \varepsilon u_b \qquad (14\text{-}17)$$

and the Reynolds number is then

$$\mathbf{Re} = \frac{4}{6(1-\varepsilon)}\frac{Du_{bs}\rho}{\mu} \qquad (14\text{-}18)$$

By a similar process, the equation for the friction factor becomes

$$f = \frac{4g_c r_H(lw_f)}{2Lu_b^2} = \frac{g_c D(lw_f)}{3Lu_{bs}^2}\frac{\varepsilon^3}{1-\varepsilon} \tag{14-19}$$

Ergun[1] has defined a Reynolds number and a friction factor similar to those given above but without the numerical constants.

$$\mathbf{Re}_p = \frac{Du_{bs}\rho}{\mu(1-\varepsilon)} \tag{14-20}$$

and

$$f_p = \frac{g_c D(lw_f)\varepsilon^3}{Lu_{bs}^2(1-\varepsilon)} \tag{14-21}$$

For laminar flow $\mathbf{Re}_p < 1.0$; in this region, by analogy to flow in many other systems, we set f_p equal to a constant divided by $\mathbf{Re}_p$. An analysis of many experimental data indicates that the constant is 150, so for laminar flow we have

$$f_p = \frac{150}{\mathbf{Re}_p} \qquad \text{or} \qquad \frac{g_c D^2 \rho(lw_f)\varepsilon^3}{\mu L u_{bs}(1-\varepsilon)^2} = 150 \tag{14-22}$$

This is referred to as the *Kozeny-Carman equation*. For a given bed and fluid, it predicts that the flow rate is proportional to the pressure drop, which is Darcy's law.

For completely turbulent flow, it is reasoned that f_p should approach a constant value and that all packed beds have the same relative roughness. The constant is found by experiment to be 1.75, so we have

$$f_p = 1.75 = \frac{g_c D(lw_f)\varepsilon^3}{Lu_{bs}^2(1-\varepsilon)} \tag{14-23}$$

This is called the *Burke-Plummer equation*.

A consideration of flow at intermediate Reynolds numbers led Ergun to propose as a general equation

$$f_p = \frac{150}{\mathbf{Re}_p} + 1.75 \tag{14-24}$$

based on a relation suggested by Reynolds.[2]

There is no abrupt transition from laminar to turbulent flow in a packed bed. Equation (14-24) reduces to (14-23) for $\mathbf{Re}_p > 10^4$ and to Eq. (14-22) for $\mathbf{Re}_p < 1.0$.

If all the particles are not the same size, we define the mean specific surface as

$$S_{vm} = \sum x_i S_{vi} \tag{14-25}$$

[1] S. Ergun, Fluid Flow through Packed Columns, *Chem. Eng. Progr.*, **48**:89 (1952).

[2] O. Reynolds, "Papers on Mechanical and Physical Subjects," p. 83, Cambridge University Press, London, 1900.

in which x_i is the volume fraction. In terms of the mean effective diameter, we have

$$D_m = \frac{6}{S_{vm}} = \frac{6}{\sum x_i(6/D_i)} = \frac{1}{\sum (x_i/D_i)} \tag{14-26}$$

For example, the mean diameter for a mixture of 1- and 2-cm spheres containing equal volumes of each (i.e., equal masses, if all the spheres have same density) is 1.33 cm, rather than 1.50 cm as given by a simple arithmetic average.

Example 14-4 Hydrocarbon vapors are cracked in a catalytic reactor which consists of a randomly packed bed of pellets shaped as cubes 0.5 cm long. The material of which the pellets are made has a density of 1600 kg/m^3 and the bulk density of the packed bed is 960 kg/m^3. The container has a cross-sectional area of 0.1 m^2, and the bed is 2 m deep. Find the pressure drop through the bed if the superficial velocity is 1 m/s.

The density of the vapor is 0.642 kg/m^3, and its viscosity is 1.5×10^{-5} Pa · s.

We begin by calculating the void fraction.

$$\varepsilon = \frac{\text{volume void}}{\text{volume bed}} = \frac{(\text{mass bed}/\rho_{\text{bed}}) - (\text{mass bed}/\rho_{\text{catalyst}})}{\text{mass bed}/\rho_{\text{bed}}}$$

$$= \frac{\rho_c - \rho_B}{\rho_c} = \frac{1600 - 960}{1600} = 0.40$$

Next we find the specific surface and thus D.

$$S_v = \frac{(6)(0.5)^2}{(0.5)^3} = \frac{6}{0.5} = 12 \text{ cm}^{-1} = 1200 \text{ m}^{-1}$$

$$D = \frac{6}{S_v} = 0.5 \text{ cm} = 0.005 \text{ m}$$

Now we calculate $\mathbf{Re}_p, f_p$, and then $-\Delta p$.

$$\mathbf{Re}_p = \frac{Du_{bs}\rho}{\mu(1-\varepsilon)} = \frac{(0.005)(1)(0.642)}{(1.5 \times 10^{-5})(0.6)} = 357$$

$$f_p = \frac{150}{\mathbf{Re}_p} + 1.75 = 0.42 + 1.75 = 2.17$$

$$-\Delta p = \frac{f_p L u_{bs}^2 (1-\varepsilon)\rho}{D\varepsilon^3 g_c} = \frac{(2.17)(2)(1)^2(0.6)(0.642)}{(0.005)(0.4)^3(1)}$$

$$= 5224.3 \text{ Pa}$$

////

Note on the Application of the Equations to Gases

Equation (13-43) can easily be adapted for use with the steady flow of a gas in a pipe of uniform cross section. As the pressure falls in the downstream direction, the density of the gas decreases and, according to the overall mass balance, its velocity increases. It is therefore convenient to replace the velocity by the mass velocity, which is not affected by pressure or temperature. For a gas, we then write Eq. (13-43) for a differential length of uniform pipe as

$$-\rho\, dp = \frac{2fG^2\, dL}{g_c D}$$

The density is a function of the pressure and may be replaced in most circumstances by using the ideal-gas equation, so that we obtain

$$-\frac{M_m}{RT}\int_{p_1}^{p_2} p\, dp = \frac{2fG^2}{g_c D}\int_0^L dL \tag{14-27}$$

for flow at constant temperature through a length of pipe, L.

Upon integration we obtain

$$-\frac{M_m}{2RT}(p_2{}^2 - p_1{}^2) = \frac{2fLG^2}{g_c D} \tag{14-28}$$

and

$$-\frac{M_m}{2RT}(p_2 + p_1)(p_2 - p_1) = \frac{2fLG^2}{g_c D}$$

and finally

$$-\Delta p = \frac{2fLG^2}{g_c D \rho_{av}} \tag{14-29}$$

where ρ_{av} is the density at the arithmetic mean pressure. If the temperature changes moderately, the arithmetic mean temperature can be used in figuring ρ_{av}.

In the above derivation, it has been assumed that the Reynolds number DG/μ does not change with pressure, and thus f remains constant. Recall that μ is not a function of pressure for an ideal gas.

It is extremely important to realize that Eq. (14-29) applies only to cases for which the relative pressure change $(-\Delta p)/p$ is small enough so that it does not cause a large change in velocity. If the downstream velocity becomes very large, the kinetic energy term, which has been omitted in the above derivation, becomes important, and Eq. (14-29) no longer applies. As the relative pressure drop rises above 20 or 30 percent, so-called compressibility effects become important; the velocity in adiabatic flow in a uniform pipe cannot exceed the velocity of sound.

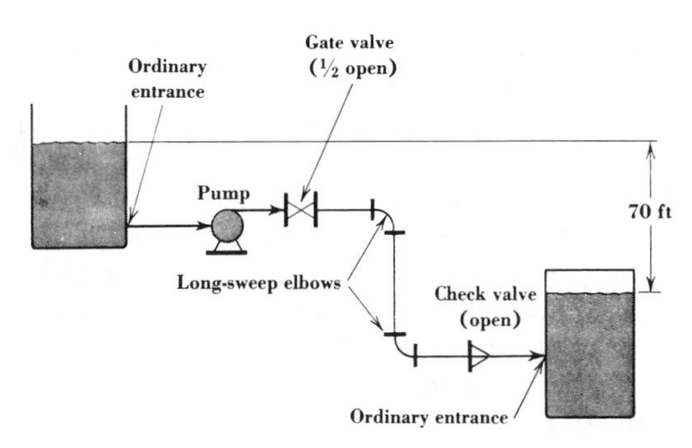

FIGURE 14-8
Sketch of piping system for Prob. 14-1. (Not to scale.)

In summary, it can be said that the equations in this book derived for the flow of incompressible fluids in a uniform conduit can be used for gases if $(-\Delta p)/p$ is not much more than 0.1 and if the density figured at the arithmetic average pressure is used. It is convenient to replace u_b by G/ρ wherever possible.

PROBLEMS

14-1 A 35°API distillate is being transferred from a storage tank at 1 atm abs pressure to a pressure vessel at 50 psig by means of the piping arrangement shown in Fig. 14-8.

The liquid flows at the rate of 23,100 lb/h through 3-in, schedule-40 steel pipe; the length of straight pipe is 450 ft. Calculate the minimum horsepower input to a pump having an efficiency of 60 percent. The properties of the distillate at 80°F are viscosity, 3.4 cP; density, 52 lb/ft^3.

14-2 Water is to be withdrawn from a large main, in which the pressure is 35 psig, and carried through 175 ft of pipe to discharge to the atmosphere at a point 22 ft above the main. What is the minimum diameter of schedule-40 steel pipe required to assure a flow of 275 gal/min?

14-3 Water is being heated from 80 to 150°F as it flows in the annulus of a horizontal-concentric-pipe heat exchanger. The outer pipe is 2-in, schedule-40, and the inner pipe is 1-in, schedule-40, and the equivalent length of the exchanger is 200 ft. What pressure drop is necessary to cause the water to flow at 45 gal/min?

14-4 Water is drawn from a reservoir and pumped an equivalent length of 2 mi through a horizontal, circular concrete duct of 10-in ID ($e = 0.01$ ft). At the end of this duct, the flow is divided into a 4- and a 3-in, schedule-40 steel pipe. The junction of the duct and the two pipes is at the level of the surface of the reservoir.

The 4-in line has an equivalent length of 200 ft and rises to a point 50 ft above the surface of the water in the reservoir, where the flow discharges to the atmosphere. This flow must be maintained at a rate of 1000 gal/min.

The 3-in line discharges to the atmosphere at a point 700 ft from the junction at the level of the surface of the water in the reservoir.

Calculate the horsepower input to the pump, which has an efficiency of 70 percent. Kinetic energy effects can be neglected.

14-5 A shell-and-tube heat exchanger is made from 70 tubes, each 4.25 m long and made of 1-in, 16-gauge standard condenser tubing. These tubes are mounted inside a 0.40-m (ID) shell, and fluid leaves and enters the header spaces at the ends of the exchangers by 3-in pipes attached at the shell center line. If water flows through the tube side of the exchanger at a rate of 0.02 m^3/s and an average temperature of 38°C, estimate the overall pressure drop.

14-6 Air is heated from 25 to 50°C by blowing it over a bank of 40 rows of 1-in, 10-gauge condenser tubing. The tubes are 0.6 m long and are placed vertically. There are 10 tubes per row set 0.01 m apart; there is 0.01 m between the outside tubes and the walls of the duct which contains the tube bank. The circulating fan supplies the air at a pressure of 25.7 mmHg. At what rate is heat being added to the air?

14-7 The tubes in the heater of Prob. 14-6 are heated by hot water which enters the tubes from the bottom. If the flow rate of the water is 0.03 m^3/s and it enters at 93°C, estimate the pressure drop through the tubes.

14-8 Turpentine flows from a constant-head tank through 24 m of $1\frac{1}{2}$-in, schedule-40 steel pipe to a second constant-head tank, where the level is 15 m below that in the first. There is an ordinary entrance at each tank, and there are two standard elbows and one open gate valve in the line.

(*a*) Find the flow rate in the line.

(*b*) Find the flow rate if a 0.01-m orifice is inserted in a horizontal section of the pipe. If a mercury manometer is used with corner taps, what is the reading on this manometer?

14-9 A discussion of the economic pipe diameter is given in Perry's Handbook,[1] pp. 384 to 385. Use the method outlined there and Eq. (12-72) for f and show the details of the derivation of a dimensionally homogeneous equation for the economic pipe diameter. Check your equation against the nomograph in Perry, p. 386.

14-10 The permeability K of a packed bed is defined for laminar flow by the equation

$$u_{bs} = K \frac{-\Delta p}{L\mu}$$

for horizontal flow. Make a suitable graph which shows the way in which the permeability varies with porosity for various particle diameters, assuming that the Kozeny-Carman equation is valid.

14-11 Calculate the pressure drop for air at 38°C and 101.3 kPa flowing at 0.0655 kg/s through a bed of 0.0127-m spheres. The bed is 0.10 m in diameter and 0.20 m high and has a porosity of 0.38.

14-12 A gravity filter is made from a bed of granular particles. Fifty percent by weight of the particles have a specific surface of 800 m^{-1}, and the rest have a specific surface of

[1] References in this problem are to Perry, 3rd ed. See also discussion in Perry, 5th ed., p. 5-31.

1200 m^{-1}. The bed porosity is 0.43. If the bed is 0.3 m in diameter and 1.5 m deep, at what rate will water at 25°C flow through the bed? There is a head of 0.25 m of water above the bed.

14-13 Water flows through a bed of spheres under the influence of gravity. The bed, contained in a vertical cylindrical vessel, consists of $\frac{3}{4}$-in-diameter spheres packed in such a way that the porosity is 0.40. The water enters and leaves the bed at atmospheric pressure and, while occupying all the pores of the bed, does not extend in depth above the bed. Assuming the motion of the water to be turbulent, determine the superficial velocity attained.

14-14 A water circulation system pumps water at 70°F at the rate of 9.45 gal/min around a closed loop consisting of 200 ft of $\frac{1}{2}$-in, schedule-40 steel pipe. What is the size of the pump in hp if the pump efficiency is 0.70?

14-15 A horizontal pipeline 20 mi long delivers 5000 bbl/d of gasoline. The pressure drop over the line is 500 psi. Find the new capacity of the pipeline if a parallel, identical line is laid along the last 12 mi of the line. The pressure drop is to remain 500 psi and the flow is turbulent and in the fully rough region in all cases.

14-16 Water at 70°F flows into a rectangular tank from an overhead outlet at the rate of $5 \text{ ft}^3/\text{min}$. The tank is 3 ft wide, 2 ft deep, and 4 ft long. At the bottom is an open drain connected to an equivalent length of 75 ft of 1-in (ID) plastic tubing. The drain discharges into a sewer at atmospheric pressure at a point 30 ft lower in elevation than the bottom of the tank. Will the tank overflow if allowed to fill at the specified rate?

14-17 A double-pipe heat exchanger consists of a 1-in, schedule-40 steel pipe inside a 2-in, schedule-40 steel pipe. Both pipes are set on the same axis and water flows in both in the same direction with the same bulk velocity of 10 ft/s and the same inlet pressure. If a leak should develop anywhere in the system, which stream will flow into the other?

14-18 A letter to the editor of the journal *Chemical Engineering*[1] asks the question, "If two vertical drains of different length are connected at the same level to the same tank of liquid, which drain will carry the greater discharge rate?" Determine the answer to this problem assuming that both drains have the same diameter, entrance and exit losses from the drains are negligible, and flow in the drains is in the turbulent, fully rough region at all times. Assume also that the depth of liquid in the tank is constant and that both drains discharge at atmospheric pressure. Do not neglect kinetic energy effects.

14-19 Repeat Prob. 14-18 for the case of a liquid likely to be in laminar flow. As is often the case in laminar flow systems, kinetic energy changes can be neglected.

14-20 Water at 60°F flows in laminar motion through 600 ft of horizontal, $\frac{1}{2}$-in, schedule-40 steel pipe and then discharges through a stationary nozzle into the atmosphere. On leaving the nozzle the water rises 22 ft straight up in the air before falling back to the ground. If the pressure at the upstream end of the pipe is 10 psig, what is the flow rate in gal/min from the nozzle?

14-21 A pump with a constant discharge pressure of 4930 kPa abs is pumping oil into a line of

[1] J. Heitala, Chemical Engineering **69**(23):7 (1962).

uniform diameter at the rate of 3.7×10^{-3} m^3/s. The line is 160 km long, horizontal, and discharges into an open storage tank at atmospheric pressure. One day it is noticed that the discharge rate into the tank has fallen to 2.8×10^{-3} m^3/s. A telephone call to the operator of the pump station at the other end of the line establishes that the pump is still discharging oil into the line at 4930 kPa abs, but the discharge rate from the pump has risen to 5.2×10^{-3} m^3/s. Obviously there is a leak somewhere in the line. Find the location of this leak assuming flow is viscous under all conditions.

14-22 A pneumatic system constructed of capillary tubing is arranged as shown schematically in the diagram below. The lengths and diameters of the lines are given in the accompanying table except for the inside diameter of line 4. Assuming laminar flow in all parts of the system and negligible kinetic and potential energy changes, find the ID of line 4 which would give zero velocity in line 5.

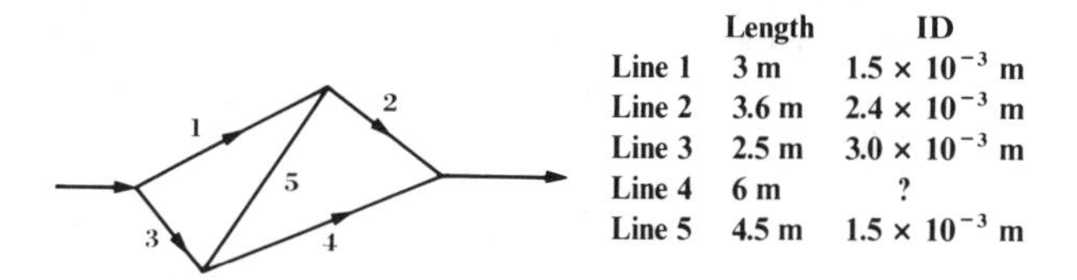

	Length	ID
Line 1	3 m	1.5×10^{-3} m
Line 2	3.6 m	2.4×10^{-3} m
Line 3	2.5 m	3.0×10^{-3} m
Line 4	6 m	?
Line 5	4.5 m	1.5×10^{-3} m

14-23 A hollow metal sphere 0.6 m in diameter is released at a considerable depth in a lake. What is the maximum upward velocity attainable by the sphere, assuming that it ascends without rotation? The sphere and its contents weight 110 kg.

14-24 Find the terminal velocity for an 80 kg person descending by parachute from an airplane. Assume the parachute has a weight of 10 kg, a diameter when fully open of 10 m, and the drag characteristics of a circular disk. Mean air density may be taken as 1.28 kg/m^3 and air viscosity as 1.734×10^{-5} Pa·s.

14-25 A 1.5-m-diameter spherical balloon, fastened to the ground by a cable, exerts an upward pull of 9 N on a calm day. On a particular day, however, when a wind is blowing, the cable holding the balloon makes an angle of 60° with the ground. Find the velocity of the wind, neglecting drag on the cable. Air density is 1.19 kg/m^3; air viscosity is 1.85×10^{-5} Pa·s.

14-26 Derive an equation for the drag coefficient for steady laminar flow of a newtonian fluid down a vertical plane. The velocity profile for this system is given in Prob. 3-4. Using the nomenclature of Prob. 3-4 put the final answer in the form $C_D = a\,\mathbf{Re}^b$ where a and b are constants and $\mathbf{Re} = L\rho u_b/\mu$.

14-27 Water at 68°F flows from an open reservoir through an ordinary entrance into a straight, 1-in (actual ID) horizontal, hydraulically smooth pipe. The water discharges to the atmosphere, and the height of water above the pipe entrance is 40 ft. There is an orifice in the pipeline, and its manometer, attached as shown in Fig. 6-4, reads 26.6 inHg; $\beta = 0.5$. Calculate the actual length of the pipe from reservoir to discharge point.

14-28 A viscous oil ($\rho = 55$ lb/ft^3, $\mu = 10$ cP) runs out of a constant-pressure tank through 200 ft of horizontal 1-in (actual ID) pipe at 1.0 gal/min. To increase the flow a second 1-in pipe is connected to the tank and runs in parallel with the first pipe for 100 ft, where it joins the first pipe. Neglect the effect of fittings, entrance, etc. What is the flow rate with the added length of pipe?

14-29 A pump of 70 percent efficiency is used to send water through a 3-in galvanized iron pipe of 100 ft equivalent length. The suction pressure is 1 atm abs, and the pipe discharges to the atmosphere at the level of the pump inlet.

(*a*) Calculate the horsepower required to pump water at 68°F at 132 lb/s.

(*b*) Estimate how much more one could afford to spend on drawn tubing instead of galvanized iron pipe. There are 0.745 kW in 1 hp; assume electricity costs 1 ¢/kWh.

14-30 Water at 68°F is being pumped through an equivalent length of 120 ft of 2-in, schedule-40 steel pipe from an open tank to a second open tank on the second floor of a building. The level of the liquid in the second tank is 30.0 ft above that in the first tank, and the flow rate is 200 gal/min. If the rate at which energy is put into the shaft of the pump is measured as 7.50 hp, what is the efficiency of the pump? It is of course understood that the equivalent length of 120 ft does not include the effect of the energy lost in the pump.

14-31 Oil having a density of 50 lb/ft^3 and a viscosity of 8.0 cP runs from a main to a shower head, located 18 ft above it, through 100 ft (equivalent length) of 1-in, schedule-40 steel pipe. The pressure drop across the shower head to the atmosphere is 12 psi. What must the pressure in the main be to produce a flow of 15 gal/min? Neglect contraction and enlargement losses.

14-32 Water is pumped from a reservoir into the bottom of an elevated open-top tank at a rate of 100 gal/min. The liquid level in the tank remains substantially constant at a height of 8 ft. The pump in use is rated at 8 hp and operates at 70 percent efficiency. The system is composed of 200 ft of 3-in standard piping and contains one long-sweep elbow, three gate valves, one globe valve, one tee, and one medium-sweep elbow. How high is the tank above the level of the pump? Water temperature is 60°F.

14-33 An SAE 10 lubricating oil is being cooled from 300 to 100°F in an annular heat exchanger. The heat exchanger is made of 1-in nominal-diameter standard pipe inside 2-in-nominal-diameter standard pipe (schedule-40), and is 200 ft long. The oil is flowing in the annular space at the rate of 45 gal/min. What is the pressure drop through the heat exchanger, in the annular space, in pounds per square inch?

Data:

1 Average density: 55 lb/ft^3

2 Average viscosity: 6 cP

14-34 **Air flows through a horizontal capillary tube at 25°C; its inlet pressure is 110 kPa and** its outlet pressure is 100 kPa. The tube is smooth and has an ID of 150 μm and a length of 50 cm. Find the velocity of the air in meters per second.

14-35 Water at 20°C is maintained at constant level in an open cylindrical tank of 50 cm inside diameter to a depth of 1 m. A 1-mm ID capillary 1 m long extends vertically downward from the center of the bottom of the tank. How fast will the water flow out through the capillary in milliliters per minute? Neglect kinetic energy effects; all the lost work (friction) occurs in the capillary.

14-36 A centrifugal pump has the performance curve given by Fig. 13-5. Find the maximum flow of water at 68°F that can be obtained from the pump if an orifice ($D_o = 1.25$ in) is placed in the outlet line, which is made of 2-in, schedule-40 steel pipe. The pump suction is at atmospheric pressure. The orifice coefficient is 0.61.

14-37 Water is flowing out of an open constant-head tank through a straight piece of horizontal pipe which discharges at atmospheric pressure. We want to increase the flow rate, and someone has suggested that a pipe of the same diameter be hooked up in parallel with the last half of the original pipe. For the original single pipe, **Re** was 10^6 and e/D was 0.010. What is the ratio of the new flow rate from the tank to the old one? Neglect all losses from fittings, entrances, etc., and the velocity heads.

14-38 A pump having the performance given by Fig. 13-5 is pumping water at 1000 gal/min at 68°F from a constant-level tank through a length of 3-in, schedule-40 S pipe which discharges at the same level as the pond. What is the equivalent length of the pipe and what horsepower must be delivered to the pump shaft?

14-39 The performance of a centrifugal pump is given by Fig. 13-5. This pump takes in water ($\mu = 1.0$ cP) through a short length of 5-in, schedule-40 S pipe from an inlet pressure of 10 psig and delivers it to a system made entirely of 5-in, schedule-40 S steel pipe. From the pump there is a pipe of equivalent length 209 ft; at this point the line branches: one part flows through a pipe of equivalent length 68 ft, and the other through one of equivalent length 100 ft. Discharge from both pipes is to the atmosphere, and the whole system is horizontal. Find the flow (gal/min) through the pump. Assume the flow is everywhere turbulent and that f is always equal to 0.004.

14-40 A pump supplied with energy at the rate of 2 hp operates with an efficiency of 60 percent in pumping water around a closed loop at the rate of 30 gal/min. What is the pressure drop over the pump?

14-41 (*a*) What is the velocity of water flowing downward at 20°C in laminar flow through a vertical 1.5-mm tube when it is flowing only under the influence of gravity?

(*b*) What is the maximum diameter a tube can have in part (*a*) and still maintain laminar flow?

14-42 A viscous liquid stored in a tank at atmospheric pressure is to be drained by laminar flow through two vertical pipes which are connected to the bottom of the tank. Both pipes are identical except that one is twice as long as the other, and both discharge into the atmosphere. It is thought that the short pipe should offer less resistance to flow but that the longer one should provide a greater potential head, so it is uncertain which would have the greater flow rate. Derive an expression for the bulk velocity in each pipe and indicate which velocity is greater. Assume that there are no frictional losses due to valves or to entrance or exit effects. Since the flow is laminar, the kinetic energy of the liquid is negligible at all times.

14-43 Water is being pumped around a loop at the rate of 26.9 gal/min. In addition to the pump, the system consists of a total of 200 ft of 1-in, schedule-40 steel pipe, four elbows, and a partly open gate valve. The energy supplied to the pump is 2 hp and it operates with an efficiency of 40 percent. Is the gate valve wide open? Water density is 62.4 lb/ft^3 and water viscosity is 1 cP. Pump efficiency includes inlet and outlet losses at pump.

14-44 A pump is pumping water at 20°C through a horizontal 2-in, schedule-40 steel pipe at the rate of 7 L/s. If an identical second pump is put into the line in series with the first, what will be the flow rate? Assume that the pressure drop over each pump is the same and that the pressure drop over the first pump is unchanged by the presence of the second pump. Also assume that in all cases flow is in the "completely rough" region of the friction factor plot.

15

FILTRATION

In this chapter we turn to the study of a unit operation which illustrates important applications of the principles of fluid dynamics. *Filtration* is characterized by the relative flow of a fluid and a bed of particles.

Filtration is one of a class of mechanical separations which involve the physical removal of a component as a separate phase, such as the separation of a solid from a liquid. Additional mechanical separations are centrifugation, sedimentation, screening, and flotation. The other category of separation depends on the tendency of a soluble component to concentrate in one phase or another. Examples of this class of unit operations are distillation, gas absorption, and liquid-liquid extraction.

Filtration on an industrial scale is similar to a simple laboratory filtration. A slurry is forced against a filter medium, which is a thin barrier or cloth made of natural, synthetic, or metallic fibers. The pores of the medium, or septum, are small enough to prevent the passage of some of the solid particles; others impinge on the fibers. Consequently, a cake builds up on the filter, and after the initial deposition, the cake itself serves as the barrier. The capacity of the device is governed by the rate of flow of the fluid filtrate through the ever-thickening bed formed by the solid particles.

Fluidization is another important unit operation and is used to achieve an intimate, uniform contact between a fluid and solid particles. As the velocity of a fluid passing up through a bed of particles is increased, a point is reached at which the upward force is sufficient to lift the particles and expand the bed. The individual particles are then no longer in continual contact, but are more or less free to move about through the bed. If the fluid velocity is increased further, the void fraction increases toward unity; the particles eventually are sufficiently separated so that they behave as single particles. If the upward force on a particle becomes significantly greater than its weight, it is swept completely out of the bed. Thus fluidization resembles flow through a packed bed at the minimum fluidizing velocity (a dense bed) and flow past a single particle at high velocity (an expanded bed).

Sedimentation, like filtration, is a means of separating a fluid from particles suspended in it. The suspension is placed in a tank, and the particles allowed to settle out; the fluid can then be removed from above the solid bed. It is interesting to notice that the sequence of events in a batch sedimentation is the reverse of that described above for fluidization. At first the particles in a suspension behave independently, but as they congregate in the bottom of the vessel, the effects of neighboring particles become important, as in fluidization. In sedimentation this situation is called *hindered settling*. The final state is that of a packed bed resembling a filter cake if the process is allowed to continue long enough.

In this chapter we shall analyze the unit operation of filtration by combining the principles covered so far with some additional empirical information. Although typical equipment will be described, the reader is referred to Perry's Handbook for equipment details, as well as for discussions of the other unit operations involving mechanical separations.[1]

General Remarks

In this section we consider the filtration of a liquid containing an appreciable amount (say 1 percent or more by volume) of suspended solid particles. This filtration is characterized by the formation of a cake of solid particles, which is removed from the apparatus intermittently or continuously. The other principal type of filtration is used to clarify or clean fluids containing relatively little solid, say up to 100 ppm by volume. In this operation, the particles are retained on or within the filter medium instead of by a cake of previously deposited solid. The filter can operate for relatively long periods before the pores become filled. Examples of this kind of filtration are air cleaners in a heating system and the filters used to remove dirt from a liquid ahead of a pump.

[1] For a good introduction to fluidization, see D. Kunii and O. Levenspiel, "Fluidization Engineering," John Wiley & Sons, Inc., New York, 1969.

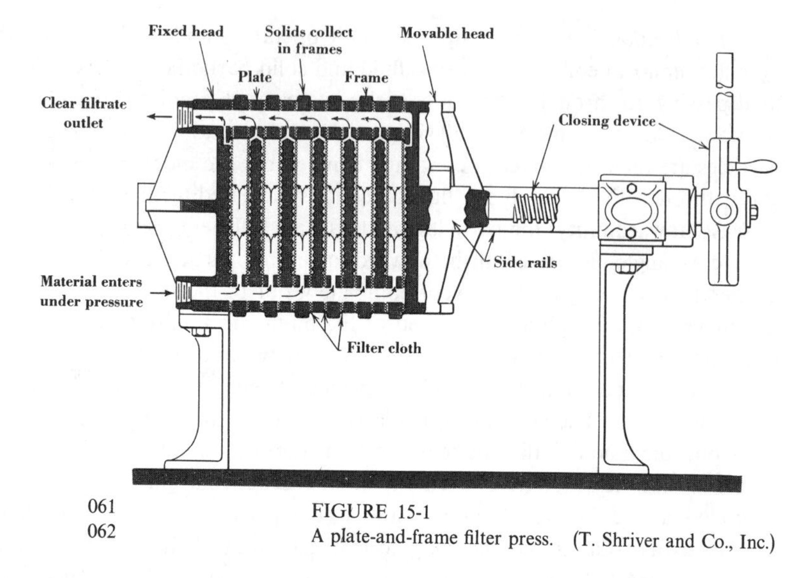

061
062

FIGURE 15-1
A plate-and-frame filter press. (T. Shriver and Co., Inc.)

Three of the more important kinds of liquid-cake-filtration equipment are shown in Figs. 15-1 to 15-3. A plate-and-frame filter press is illustrated schematically in Fig. 15-1. The plates, frames, and filter cloths are pressed together as shown, and the entering slurry passes through lines leading to the frames. The cake is deposited against the filter cloths, which form the faces of the frames. The filtrate flows through the grooved channels on the faces of the plates and out through the openings at the tops of the plates. When the frames have been filled with cake, washing can be done by introducing the washing liquid through washing plates. It flows through the entire cake thickness. The press is then opened, the solid removed, and another cycle of filtration started. Pressure up to about 150 psi can be used with this type of filter.

A leaf-type filter is shown in Fig. 15-2. The slurry, under pressure which can be considerably higher than 150 psi, is forced into the shell surrounding the filter leaves. The filtrate is removed as shown from inside the leaves, and the cake forms on the outside. In some models the leaves are rotated to agitate the slurry and to improve the uniformity of the cake. The wash liquid follows the same path as the filtrate. When sufficient cake has been deposited, the shell is opened and the cake removed; sometimes air is blown into the leaves to aid in dislodging the cake.

The two filters already shown suffer from the disadvantages common to batch operations. The advantages of continuous operation are achieved by the use of a rotary vacuum filter[1] of the type shown in Fig. 15-3. A vacuum is maintained on the

[1] Drum filters are analysed by D. A. White, *Chem. Eng. Sci.*, **31**:419 (1976).

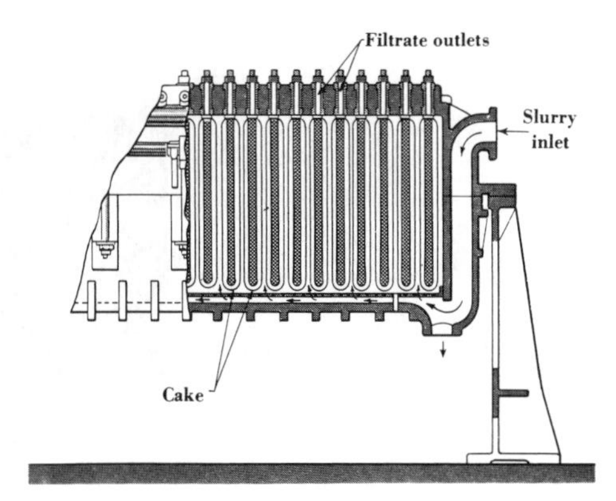

FIGURE 15-2
A leaf filter. (*From W. L. Badger and J. T. Banchero, "Introduction to Chemical Engineering," p. 565, McGraw-Hill Book Company, New York, 1955.*)

interior of the rotating drum, which is covered on its curved surface by the filter medium. The drum dips into a tank of slurry, and the filtrate passes through the medium and is removed through the axle of the filter. The cake is continuously washed and removed as shown. Offsetting the advantage of continuous operation is the fact that the filter illustrated has a maximum pressure differential of only 1 atm. It is not suitable for highly viscous filtrates or for liquids which cannot be exposed to the atmosphere. These disadvantages can be overcome by enclosing the filter in a shell which is maintained above atmospheric pressure.

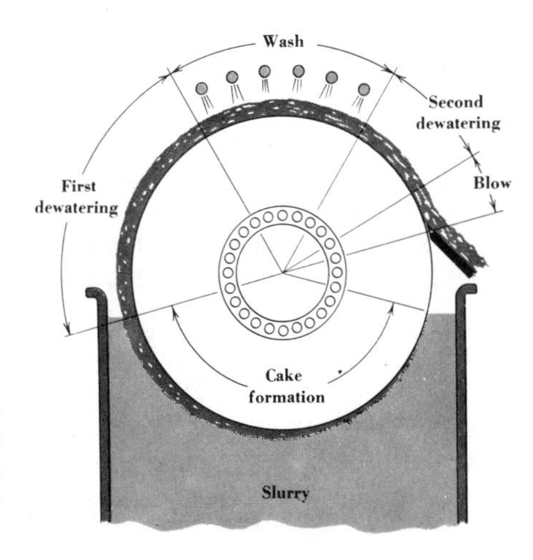

FIGURE 15-3
A continuous rotary filter. (*From W. L. Badger and J. T. Banchero, "Introduction to Chemical Engineering," p. 569, McGraw-Hill Book Company, New York, 1955.*)

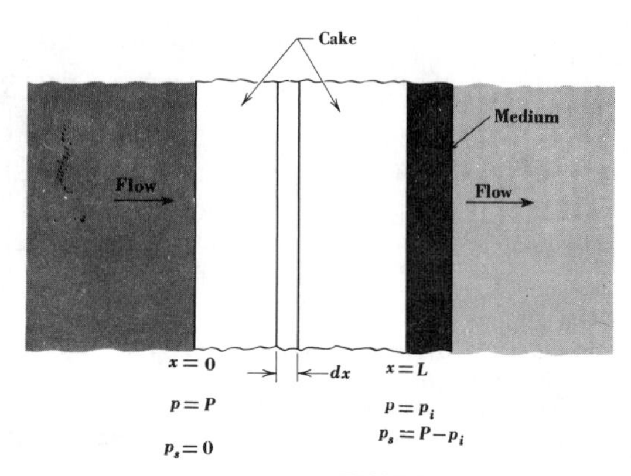

FIGURE 15-4
Section through a filter cake.

Basic Equations[1]

Figure 15-4 is a sketch of a section through a filter cake. In order to simplify the notation, the pressure on the downstream side of the medium is taken as zero; p_i is the pressure at the interface between the medium and the cake, and P is the upstream pressure.

As is often the case, the analysis of filtration is begun by considering a mass balance. It will be useful to know the relations among the mass of cake, its thickness, and the volume of filtrate which corresponds to this cake. A differential element of cake is considered containing a mass of solid dM_c. This mass of solid is the solid carried by a volume of filtrate dV which has passed through the element of cake, plus the mass of solid carried by the liquid which adheres to the cake and remains in the element. This equality is written as

$$dM_c = \frac{s}{1 - s} \rho \, dV + \frac{s}{1 - s} \frac{1 - s'}{s'} dM_c \tag{15-1}$$

where ρ = density of filtrate
s = mass fraction of solid in slurry
s' = mass fraction of solid in wet cake

Equation (15-1) can be rearranged to

$$dM_c = \frac{ss'}{s' - s} \rho \, dV \qquad (s' > s) \tag{15-2}$$

[1] J. B. Poole and D. Doyle, "Solid-Liquid Separation," Her Majesty's Stationery Office, London (1966) is a good reference for more details on the theory of filtration. See also F. M. Tiller and J. R. Crump, *Chem. Eng. Progr.*, **73**(10):65 (1977).

We shall find it useful to express the mass of cake in terms of distance through the cake, x, by using the void fraction ε to obtain

$$dM_c = (1 - \varepsilon)\rho_s A\, dx \qquad (15\text{-}3)$$

where ρ_s = density of dry solid
A = total area normal to flow

It will be convenient to have a relation between the cake porosity and the mass fraction of solid in the cake,

$$s' = \frac{(1 - \varepsilon)\rho_s}{(1 - \varepsilon)\rho_s + \varepsilon\rho} \qquad (15\text{-}4)$$

We shall also use the quantity $m = 1/s'$, which is the mass ratio of cake when wet to cake when dry. We are now ready to consider flow through the cake. Since the particles making up a filter cake are relatively small, the flow is usually laminar and follows the Kozeny-Carman equation, written for a differential thickness as

$$\frac{d(lw_f)}{dx} = \frac{150\mu u_{bs}(1 - \varepsilon)^2}{g_c D^2 \rho \varepsilon^3} \qquad (14\text{-}22)$$

Using the mechanical energy balance for the usual case in which only pressure head is important and substituting $6/S_v$ for D, we get the following equation for the pressure drop over the cake:

$$-\frac{dp}{dx} = \frac{4.17 S_v{}^2 \mu u_{bs}(1 - \varepsilon)^2}{g_c \varepsilon^3} \qquad (15\text{-}5)$$

in which S_v is the specific surface defined in Chap. 14.

For many cakes the porosity and specific surface vary with position inside the cake. This effect is caused by the compressive pressure on the solid particles, which varies with position. This pressure, written as p_s, is the mechanical stress tending to compress the cake in the x direction, and it is defined by

$$p_s = P - p \qquad (15\text{-}6)$$

The "solid pressure" p_s increases from zero at $x = 0$ to a maximum value of $P - p_i$ at the medium. We write Eq. (15-5) in terms of p_s:

$$\frac{dp_s}{dx} = \frac{k S_v{}^2 \mu u_{bs}(1 - \varepsilon)^2}{g_c \varepsilon^3} \qquad (15\text{-}7)$$

in which the constant 4.17 has been replaced by k. In most discussions of filtration, k is given a value of 5.0 for isotropic cakes of $0.3 < \varepsilon < 0.6$. Higher values of k are found for fibrous materials ($\varepsilon > 0.6$), and ideally k should be determined for a given material.[1]

It is conventional to characterize filter cakes by a term α, known as the specific cake resistance. To show the basis for the definition of this quantity, we express the distance x in Eq. (15-7) in terms of the mass of dry cake M_c, which is contained between $x = 0$ and $x = x$. Equation (15-3) is combined with Eq. (15-7) to obtain

$$\frac{dp_s}{dM_c} = \frac{kS_v{}^2\mu u_{bs}(1 - \varepsilon)}{g_c A\rho_s\varepsilon^3} \tag{15-8}$$

The specific cake resistance, which has units of feet per pound, is defined by

$$\alpha = \frac{kS_v{}^2(1 - \varepsilon)}{\rho_s\varepsilon^3} \tag{15-9}$$

Equation (15-8) can then be written as

$$\frac{dp_s}{dM_c} = \frac{\alpha\mu u_{bs}}{g_c A} \tag{15-10}$$

and α is seen to include all properties dependent on the cake. It is also a function of p_s for certain substances.

Filter performance is usually specified in terms of filtrate volume, which is related to the mass of the cake by Eq. (15-2); Eq. (15-10) becomes

$$\frac{dp_s}{dV} = \frac{\alpha\mu qs\rho}{g_c A^2(1 - ms)} \tag{15-11}$$

in which u_{bs} has been replaced by q/A, q being the volumetric rate of flow. To apply this equation to a cake of finite thickness we integrate Eq. (15-11) between 0 and V, where V is the volume of filtrate equivalent to a cake of thickness L. It should be emphasized that this integration is performed across the cake at any instant for which P and q are given. We obtain from Eq. (15-11)

$$\int_0^{P-p_i} \frac{dp_s}{\alpha} = \frac{\mu qs\rho}{A^2 g_c(1 - ms)} \int_0^V dV \tag{15-12}$$

Thus we have

$$P - p_i = \frac{\alpha_{av}\mu qs\rho V}{A^2 g_c(1 - ms)} \tag{15-13}$$

where

$$\alpha_{av} = \frac{P - p_i}{\int_0^{P-p_i}(dp_s/\alpha)} \tag{15-14}$$

[1] H. P. Grace, *Chem. Eng. Progr.*, **49**:303 (1953).

The quantity p_i is related to the pressure drop over the filter medium. For a given value of q, it is the upstream pressure at the start of filtration. However, since q may vary during the course of a filtration, p_i may not be a constant; it is desirable to relate this quantity to q and to the properties of the medium. It is customary to write the following equation, for which the fluid pressure on the downstream side of the medium is zero as usual:

$$p_i = \frac{\mu R_m q}{g_c A} \qquad (15\text{-}15)$$

The quantity R_m defined by this equation is called the *resistance of the medium* and is a constant for a given medium and slurry. A comparison between Eqs. (15-15) and (15-13) shows that the resistance of the medium is

$$R_m = \frac{\alpha_{av} s \rho V_m}{(1 - ms)A} \qquad (15\text{-}16)$$

This equation shows that the resistance of the medium may be expressed in terms of the resistance of a hypothetical layer of cake, which corresponds to the collection of a hypothetical volume of filtrate V_m. The resistance of a typical filter medium is equivalent to a few tenths of an inch of cake.

We use the definition of R_m to replace p_i in Eq. (15-13) and obtain

$$P = \left[\frac{s\rho V \alpha_{av}}{(1 - ms)A} + R_m\right] \frac{\mu q}{g_c A} \qquad (15\text{-}17)$$

If α is constant throughout the cake (in other words, α is independent of the mechanical compressive stress p_s) and equal to α_{av}, the cake is said to be *incompressible*. Such cakes are formed from slurries of regularly shaped, crystalline particles. Unfortunately, most slurries form *compressible* filter cakes, for which α is a function of p_s, and α_{av} must be evaluated for the conditions of interest by the use of Eq. (15-14). This procedure requires a knowledge of α as a function of p_s and will be discussed later in this chapter.

In interpreting the results of filtration experiments on an incompressible slurry with a given filtering apparatus, it is simpler to write Eq. (15-17) as

$$P = (K_1 V + K_2)q \qquad (15\text{-}18)$$

where

$$K_1 = \frac{s\rho\mu\alpha_{av}}{(1 - ms)A^2 g_c} \qquad (15\text{-}19)$$

and

$$K_2 = \frac{R_m \mu}{A g_c} \qquad (15\text{-}20)$$

For a given incompressible slurry K_1 and K_2 are constant. Computations using Eq. (15-18) can be relatively simple, for K_1 and K_2 can be expressed using arbitrary units such as pounds per square inch and gallons. However, if it is desired to use Eqs. (15-19) and (15-20) to calculate, for example, α_{av} and R_m from experimental values of K_1 and K_2, consistent units must be used. These equations show the effect of the several variables on K_1 and K_2; for instance, if the temperature is changed so that the viscosity doubles, K_1 and K_2 will double.

The result of the analysis thus far is an expression [Eq. (15-17) or (15-18)] giving the instantaneous pressure drop P across the filter as a function of the rate of flow q and the cake thickness corresponding to the filtrate volume V. During an actual filtration cycle the cake thickness increases from 0 to L_f, corresponding to an increase in filtrate collected from 0 to V_f. The pressure and flow rate may change during the deposition of the cake. The time θ_f required for the deposition of the cake and the collection of a volume V_f is obtained by writing the rate as

$$q = \frac{dV}{d\theta} \tag{15-21}$$

and integrating this equation to give

$$\int_0^{\theta_f} d\theta = \int_0^{V_f} \frac{dV}{q} \tag{15-22}$$

The flow rate q may be replaced using Eq. (15-18).

$$\theta_f = \int_0^{V_f} \frac{(K_1 V + K_2)\, dV}{P} \tag{15-23}$$

We now have the general equations needed to analyze most filtration problems. We consider next their application to the various types of filter operation.

Incompressible Filter Cakes

Constant-pressure operation If slurry is fed to a filter from a constant-head tank, the pressure at the upstream face of the cake is constant. Since the pressure at the downstream face of the medium is usually 1 atm, the quantity P in the equations given in the preceding section is known and constant. The effect of the pressure drops in the channels leading to and from the filter cake should also be considered, but it is usually sufficiently accurate to include these effects in R_m (or K_2).

In most industrial filtrations the pressure is furnished by a centrifugal pump rather than a constant-head tank. However, constant-pressure filtration is used in much laboratory and research work and, of necessity, with continuous rotary filters.

For an incompressible cake Eq. (15-23) can be integrated analytically, since P, K_1, and K_2 are constant. The result is

$$\theta_f = \frac{K_1}{2P} V_f^{\,2} + \frac{K_2}{P} V_f \tag{15-24}$$

The values of K_1 and K_2 for a given slurry can be found from data on V_f versus θ_f collected from experiments at constant pressure with a laboratory filtration apparatus. To calculate K_1 and K_2, Eq. (15-24) is written as

$$\frac{\theta_f}{V_f} = \frac{K_1}{2P} V_f + \frac{K_2}{P} \qquad (15\text{-}25)$$

and the experimental data are used to make a graph of θ_f/V_f versus V_f. The slope and intercept of the best straight line through the points can be used to calculate K_1 and K_2.

The values of K_1 and K_2 obtained apply only to the slurry and filter used. If only the size of the filter, that is, A, is changed, it is simple to calculate new values of K_1 and K_2 based on Eqs. (15-19) and (15-20). For instance, if the filter area is increased by a factor of 10, K_1 is reduced by a factor of 100 and K_2 by a factor of 10. Changes in μ, s, and m can be similarly accounted for. For an incompressible cake α_{av} is a constant for a given suspended solid, and R_m is constant for a given medium. If a reliable value of α_{av} is known from other experiments, it can be used in Eq. (15-19) to calculate K_1 directly.

Constant-rate operation If slurry is fed to a filter by means of a positive displacement pump, the rate, which we shall call q_o, is approximately constant. The integration of Eq. (15-22) for this type of operation gives

$$V_f = q_o \theta_f \qquad (15\text{-}26)$$

or at any moment during the filtration,

$$V = q_o \theta \qquad (15\text{-}27)$$

Equation (15-18) shows that the filtration pressure P increases from p_i linearly with volume of filtrate collected at a constant rate, $q = q_o$. The pressure may be related to time by combining Eqs. (15-27) and (15-18) to give

$$P = K_1 q_o^2 \theta + K_2 q_o \qquad (15\text{-}28)$$

A plot of P versus θ for a known q_o obtained from data on a constant-rate filtration permits the calculation of K_1 and K_2. These constants can also be calculated from values of α_{av} and R_m, if they are available. Note that p_i is constant and equal to $K_2 q_o$; it does not decrease with time as in constant-pressure filtration.

Variable-rate – variable-pressure operation If slurry is fed by a centrifugal pump, the pressure and rate are related by a curve such as that shown in Fig. 15-5.

Values of K_1 and K_2 can be found from a filtration experiment measuring P at various values of V. The curve of P as a function of q is determined by the characteristics of the pump used. Equation (15-18) shows that values of P/q can be plotted

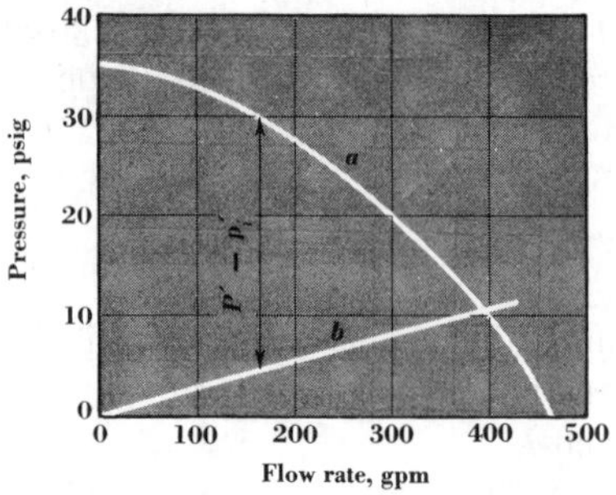

FIGURE 15-5
Performance curve for a centrifugal pump.
a Filtration pressure, P'.
b Pressure at medium, p_i'.

against V, permitting the computation of K_1 and K_2 from the slope and intercept of the best straight line through the points.

The time θ_f required to collect a volume of filtrate V_f is given by the graphical or numerical integration of Eq. (15-22). The variation of q with V is given by the simultaneous solution of Eq. (15-18) and the relation given by the line representing filtration pressure as a function of flow rate in Fig. 15-5.

The filtration cycle After the desired quantity of cake has been collected, the mother liquor is usually removed by washing with a suitable solvent. Washing occurs at constant rate and pressure, and Eq. (15-18) is easily adapted to give the relation between the wash rate q_w and the pressure P_w:

$$P_w = (K_1 V_f + K_2)q_w \qquad (15\text{-}29)$$

Recall that V_f is a constant, representing the total filtrate collected during the deposition of the cake. The time of washing is given by

$$\theta_w = \frac{V_w}{q_w} \qquad (15\text{-}30)$$

After washing has been completed, additional time is needed for draining and dumping the cake and cleaning and refilling the filter. If the sum of these times is θ_d, the total cycle time θ_c is given by

$$\theta_c = \theta_f + \theta_w + \theta_d \qquad (15\text{-}31)$$

The capacity C of the filter is defined in terms of filtrate volume as

$$C = \frac{V_f}{\theta_c} \qquad (15\text{-}32)$$

We shall consider in Example 15-3 the problem of finding the value of V_f which gives the maximum capacity C.

The same parts of the filtration cycle also occur with a continuous rotary filter. In this apparatus the times θ_f, θ_w, and θ_d are determined by the arrangement of the various compartments in the rotating drum, by the radius of the drum, and by its speed of rotation.

Example 15-1 A small leaf filter is used to find the filtering characteristics of a slurry which forms an incompressible cake. A test made on a suspension was run at a constant rate of 0.3 gal/min of filtrate and gave the following results:

Time, min	1	2	3	5	10	20
Pressure drop across filter, psi	7.0	11.1	15.0	23.2	43.0	81.7

Using the same suspension and filter, the filtration cycle will be run in the following way. The filtrate is forced through at a constant rate of 0.5 gal/min until the pressure differential reaches 50 psi. The filtration is then continued at constant pressure (50 psi) until the total filtrate collected is 15 gal. The cake is then washed with 3.0 gal of water; 10 min is needed to dump and clean the filter. Find the capacity of this filtration cycle in gal/h.

Since the cake is incompressible, K_1 and K_2 in Eq. (15-28) are constant, and a graph of the test data given yields a straight line when P is plotted versus θ. The intercept is at $P = 2.92$ psi, so that we have

$$K_2 q_o = 2.92$$

$$K_2 = 2.92/0.3 = 9.73$$

The slope is 4.01, giving

$$K_1 q_o{}^2 = 4.01$$

$$K_1 = 4.01/0.09 = 44.6$$

For the constant-rate portion of the proposed filtration cycle, with $q_o = 0.5$ gal/min, Eq. (15-28) becomes

$$P = (44.6)(0.5)^2\theta + 9.73(0.5)$$

$$= 11.15\theta + 4.86$$

For $P = 50$ psi, this equation gives $\theta = 4.04$ min, and so $V_{f_1} = 2.02$ gal. We next consider the constant-pressure part of the filtration. Since Eq. (15-24) is based on constant-pressure filtration right from the start, it does not apply; we must use Eq. (15-23), written as

$$\theta_f - \theta_{f_1} = \int_{V_{f1}}^{V_f} \frac{(K_1 V + K_2)\, dV}{P}$$

In θ_{f_1} min (4.04) we have collected V_{f_1} gal (2.02) of filtrate. The above equation permits us to find the total time of filtration θ_f needed to obtain V_f gal (15.0) of filtrate.

$$\theta_f = \frac{K_1}{2P}(V_f{}^2 - V_{f_1}{}^2) + \frac{K_2}{P}(V_f - V_{f_1}) + \theta_{f_1}$$

$$= \frac{(44.6)}{2(50)}(225 - 4.07) + \frac{9.73}{50}(15 - 2.02) + 4.04$$

$$= 98.7 + 2.53 + 4.04 = 105.3 \text{ min}$$

Readers should be sure they understand why it would have been wrong to use Eq. (15-24) with $V_f = 12.98$ gal.

The time θ_f in Eq. (15-31) has been calculated; since θ_d is given as 10 min, it remains to calculate θ_w. Equations (15-29) and (15-30) are combined to give

$$\theta_w = \frac{V_w(K_1 V_f + K_2)}{P}$$

and so we have

$$\theta_w = \frac{(3.0)[(44.6)(15) + 9.73]}{50}$$

$$= 40.7 \text{ min}$$

The total cycle time is

$$\theta_c = 105.3 + 40.7 + 10$$

$$= 156.0 \text{ min}$$

and the capacity is

$$C = 15/156 = 0.0962 \text{ gal/min, or } 5.77 \text{ gal/h}$$ ////

Example 15-2 A suspension containing 225 g of carbonyl iron powder, Grade E, per liter of a solution of 0.01 N NaOH is to be filtered, using a leaf filter. Estimate the size (area) of the filter needed to obtain 50 kg of dry cake in 1 h of filtration at a constant pressure drop of 138 kPa.

Grace[1] has obtained experimental data for this slurry and found that the cake is incompressible, that α_{av} is 6.7×10^{10} m/kg, and that the void fraction ε is 0.40. The resistance of the medium is taken as that offered by 2.5 mm of cake.

[1] Ibid.

Since the cake is incompressible, Eq. (15-24) will apply, and Eqs. (15-19) and (15-29) can be used to find K_1 and K_2. We first calculate K_1 by

$$K_1 = \frac{s\rho\mu\alpha_{av}}{(1 - ms)A^2 g_c}$$

Assuming the filtrate to have the properties of pure water, we find

$$\rho = 1000 \text{ kg/m}^3$$
$$\mu = 0.001 \text{ Pa}\cdot\text{s}$$
$$s = 0.225/1.225 = 0.184$$

Eq. (15-4)($s' = 1/m$) gives:

$$m = \frac{(1 - \varepsilon)\rho_s + \varepsilon\rho}{(1 - \varepsilon)\rho_s} = \frac{(0.6)(7) + (0.4)(1)}{(0.6)(7)}$$
$$= 1.093$$

(The specific gravity of the powder has been taken as 7.)
The value for K_1 is then

$$K_1 = \frac{(0.184)(1000)(0.001)(6.7 \times 10^{10})}{[1 - (1.093)(0.184)](A^2)(1)}$$
$$= \frac{1.543 \times 10^{10}}{A^2}$$

The value of K_2 is found from

$$K_2 = \frac{R_m \mu}{A g_c}$$

and

$$R_m = \frac{\alpha_{av} s\rho V_m}{(1 - ms)A}$$

In the latter equation, we must find V_m. A material balance gives, as in Eqs. (15-2) and (15-3),

$$\frac{V_m}{A} = \frac{(1 - \varepsilon)\rho_s L_m(1 - ms)}{s\rho}$$
$$= \frac{(0.6)(7)(0.0025)(0.799)}{(0.184)(1)} = 0.0456$$

Thus

$$R_m = \frac{(6.7 \times 10^{10})(0.184)(1000)(0.0456)}{(0.799)} = 7.04 \times 10^{11}$$

We get, finally,

$$K_2 = \frac{(7.04 \times 10^{11})(0.001)}{(A)(1)} = \frac{7.04 \times 10^8}{A}$$

In order to use Eq. (15-24),

$$\theta_f = \frac{K_1}{2P} V_f^{\,2} + \frac{K_2}{P} V_f$$

we must calculate the value of V_f corresponding to $M_{cf} = 50$ kg.

$$V_f = \frac{1 - ms}{s\rho} M_{cf}$$

$$= \frac{(0.799)}{(0.184)(1000)}(50) = 0.217 \text{ m}^3$$

Equation (15-24) can finally be used:

$$3600 = \frac{(1.543 \times 10^{10})}{A^2(2)(13.8 \times 10^4)}(0.217)^2 + \frac{7.04 \times 10^8}{A(13.8 \times 10^4)}(0.217)$$

and this becomes

$$A^2 - 0.308A - 0.731 = 0$$

$$A = 1.02 \text{ m}^2$$

A leaf filter having about this surface might consist of 10 leaves, each roughly 0.26 m in diameter. ////

Example 15-3 The filter described in the preceding example is to be operated at maximum capacity. Fifteen minutes is needed for dumping and cleaning, and the volume of wash water used in any one cycle is one-tenth the volume of filtrate collected. All other conditions are as given in Example 15-2. From the value of the area $A = 1.02$ m^2, we find

$$K_1 = (1.543 \times 10^{10})/(1.02)^2 = 1.48 \times 10^{10}$$

$$K_2 = (7.04 \times 10^8)/(1.02) = 6.90 \times 10^8$$

Equation (15-24), with $P = 13.8 \times 10^4$ Pa, gives

$$\theta_f = 5.36 \times 10^4 V_f^{\,2} + 0.5 \times 10^4 V_f$$

From Eqs. (15-29) and (15-30) we obtain

$$\theta_w = \frac{0.1V_f(K_1 V_f + K_2)}{P} = 1.07 \times 10^4 V_f^2 + 0.05 \times 10^4 V_f$$

Using $\theta_d = 900$, the time for a cycle is

$$\theta_c = 6.43 \times 10^4 V_f^2 + 0.55 \times 10^4 V_f + 900$$

and the capacity is

$$C = \frac{V_f}{\theta_c} = \frac{V_f}{6.43 \times 10^4 V_f^2 + 0.55 \times 10^4 V_f + 900}$$

This equation shows that the capacity is zero for $V_f = 0$, rises to a maximum value, and approaches zero as V_f becomes very large. For small values of V_f, an excessive portion of the time is spent in cleaning and dumping, and for large values of V_f, the cake is thick and q becomes very small. The maximum capacity is found by the condition $dC/dV_f = 0$; the differentiation of the equation for C and the substitution of $dC/dV_f = 0$ gives $V_f = 0.12$ m^3, corresponding to 0.007 m of cake. The maximum capacity is 4.4×10^{-5} m^3/s; each cycle lasts 0.681 h. ////

Compressible Filter Cakes

If α is not constant but is a function of p_s, the filter cake is said to be compressible. With incompressible cake it is possible to perform small-scale filtration experiments to find values of K_1 and K_2, and then use these values in the design of another filter operating with a changed A, P, or q. With a compressible cake the results of such an experiment can be extrapolated only to a different value of A (i.e., filter size). The relation between P and q must be the same in the two filters. This is so because α_{av}, and thus K_1 and K_2, are functions of both P and q. In order to define the behavior of a filter for a given compressible substance, many filtration experiments must be made. Therefore it is better to use the more basic equations (15-14) and (15-17) and find the relation between α and p_s by means of the experimental procedure discussed below.

The relation between α and p_s can be found using the apparatus sketched in Fig. 15-6; this procedure is called the *compression-permeability technique* and was applied to the study of filter cakes by Ruth.[1] A cake of the material to be studied is

[1] B. F. Ruth, *Ind. Eng. Chem.*, **38**:564 (1946).

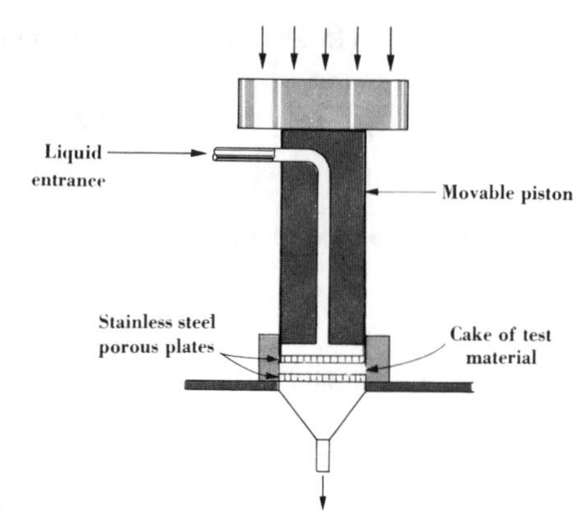

FIGURE 15-6
Compression-permeability apparatus.

contained in a cylinder and held between a fixed porous stainless steel plate and a movable one. The latter is attached to a piston, and the solid pressure p_s can be varied by applying a known force to the piston by means of a hydraulic press. For a given value of p_s the specific cake resistance is determined by measuring the pressure drop experienced by a liquid flowing across the cake at a known rate. The cake porosity ε is also measured. The pressure drop across the cake is kept low compared with p_s by using a thin cake and a low rate of flow.

Compression-permeability measurements have been made by Grace[1] for a number of substances. Figures 15-7 and 15-8 show some of his results.

Let us now consider how the data of these figures can be used to calculate the time θ_f required to obtain a quantity of filtrate V_f. The method of feeding to the proposed filter will determine the relation between P and q. Some experiments must be run with the filter medium to find the value of R_m; the pressure p_i can then be calculated from Eq. (15-15) for a given q. With the relations between p_i and q, and α and p_s, known, Eq. (15-14) is next used to calculate α_{av} for the series of P's and q's which are to be used during the cycle. Equation (15-17) is then used to find the V's which correspond to the various values of q. Finally, the filtration time θ_f is found from Eq. (15-22) and the relation between q and V. This procedure is illustrated in Example 15-4. If the curve for α versus p_s happens to be a straight line on a log-log plot, the integral in Eq. (15-14) can be evaluated analytically by substituting for α an expression of the form

$$\alpha = \beta p_s{}^{\gamma} \qquad (15\text{-}33)$$

in which β and γ are empirical constants obtained from the best-fitting straight line.

[1] Loc. cit.

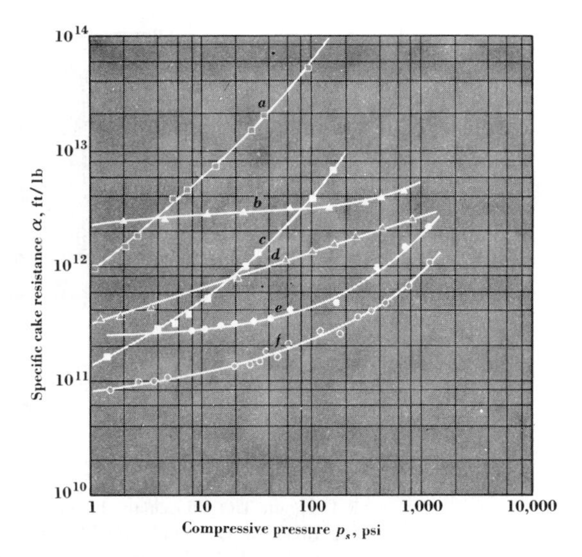

FIGURE 15-7
Specific resistance of compressible filter cakes as a function of solid compressive pressure. [*From H. P. Grace, Chem. Eng. Progr.*, **46**:*467–476* (*1950*).]

a ZnS type *B*, pH = 9.07.
b R-110 grade TiO_2, 50 g/L of 0.01 *N* HCl, pH = 3.45.
c ZnS type *A*, pH = 9.10.
d R-110 grade TiO_2 (flocculated), 50 g/L of distilled water, pH = 7.8.
e Superlight $CaCO_3$, 50 g/L of 0.01 *M* $Na_4P_2O_7$, pH = 10.3.
f Superlight $CaCO_3$ (flocculated), 50 g/L of distilled water, pH = 9.8.

The data obtained from filtration experiments do not have a high degree of reproducibility. For instance, the state of agglomeration of the particles in the suspension is difficult to control, and it apparently has an important effect on the porosity and specific resistance of the cake formed. It is therefore advisable to use a liberal safety factor in the design of filtration equipment.

Example 15-4 A battery of leaf filters having a total area of 2500 ft^2 is to be used to filter a suspension of 50 g of R-110 grade TiO_2 (flocculated) per liter of water. The properties of this suspension are given in Figs. 15-7 and 15-8. The filter will be fed by a pump having the characteristic curve shown in Fig. 15-5. Find the time required to obtain 20,000 lb of dry cake. For R_m a value of 2×10^{11} ft^{-1} may be used; m is 1.64.

During the course of the filtration the pressure and rate will follow the curve given in Fig. 15-5. Values of α_{av} as a function of P must be calculated

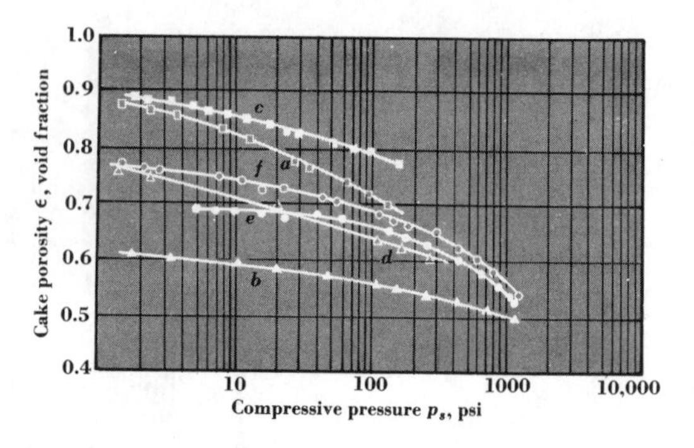

FIGURE 15-8
Void fraction of compressible filter cakes as a function of solid compressive pressure. [*From H. P. Grace, Chem. Eng. Progr.*, **49**:*303 (1953)*.]
a ZnS type *B*, pH = 9.07.
b R-110 grade TiO_2, 50 g/L of 0.01 *N* HCl, pH = 3.45.
c ZnS type *A*, pH = 9.10.
d R-110 grade TiO_2 (flocculated), 50 g/L of distilled water, pH = 7.8.
e Superlight $CaCO_3$, 50 g/L of 0.01 *M* $Na_4P_2O_7$, pH = 10.3.
f Superlight $CaCO_3$ (flocculated), 50 g/L of distilled water, pH = 9.8.

from Eq. (15-14), and so p_i must be known for every P and its corresponding q. Equation (15-15) is used to calculate p_i.

$$p_i = \frac{\mu R_m q}{g_c A}$$

$$= \frac{(1)(2.42)(2 \times 10^{11})q}{(4.16 \times 10^8)(2500)} = 0.466q$$

It will be convenient to use in the calculations the pressure in pounds per square inch and the rate in gallons per minute, defined as p_i' and q' (the primes on any pressure symbol will indicate pounds per square inch).

$$p_i' = \frac{(0.466)(60)q'}{(144)(7.48)} = 0.0260q'$$

This equation is plotted as a straight line in Fig. 15-5, and it is thus possible to obtain $P' - p_i'$ as a function of q'. Filtrate (water) will flow through the empty filter at 400 gal/min with a pressure drop of 10.4 psi.

Since the curve of α versus p_s in Fig. 15-7 is a straight line, Eq. (15-33) can be used to represent the variation of α, giving

$$\alpha = 3.52 \times 10^{11}(p_s')^{0.266}$$

The substitution of this expression into Eq. (15-14) gives

$$\alpha_{av} = \frac{3.52 \times 10^{11}(P' - p_i')}{\int_0^{P'-p_i'} \frac{dp_s'}{(p_s')^{0.266}}}$$

The value of α_{av} is not affected by the units used for the pressures. Integration gives

$$\alpha_{av} = 2.58 \times 10^{11}(P' - p_i')^{0.266}$$

and the following table can be constructed:

q'	P'	$P' - p_i'$	$\alpha_{av} \times 10^{-11}$
400	10.4	0	
300	20.0	12.4	5.04
200	27.6	22.6	5.92
100	33.0	30.5	6.40
50	34.5	33.0	6.54

It is now possible to relate the filtrate volume collected to the rate of flow by Eq. (15-17).

$$P = \left[\frac{s\rho V \alpha_{av}}{(1 - ms)A} + R_m\right]\frac{\mu q}{g_c A}$$

For this problem, $s = 0.0476$, and $1 - ms = 0.922$. Substituting the appropriate values and rearranging, we obtain

$$V = \frac{0.598 \times 10^{16}}{\alpha_{av}}\left(\frac{P'}{q'} - 0.0260\right)$$

We now have sufficient information to find V as a function of q'.

q'	V
400	0
300	482
200	1132
100	2850
50	6070

The time required is found from

$$\theta_f = \frac{7.48}{60}\int_0^{V_f} \frac{dV}{q'}$$

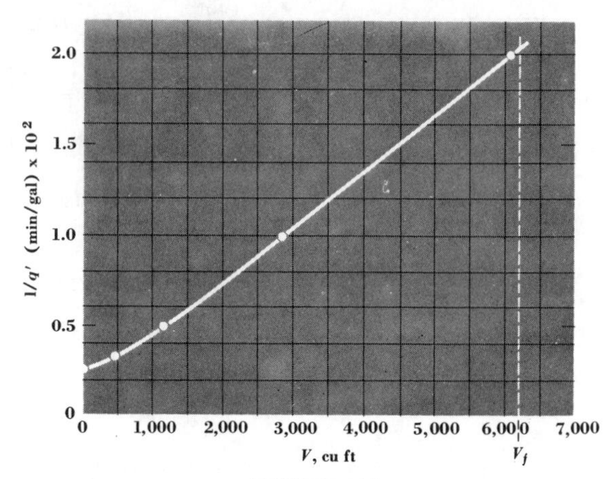

FIGURE 15-9
Graphical integration for Example 15-4.

by a graphical integration. The volume of filtrate is given by

$$V_f = \frac{(1 - ms)M_{cf}}{s\rho}$$

$$= \frac{(0.922)(20{,}000)}{(62.4)(0.0476)} = 6220 \text{ ft}^3$$

The integration, which is illustrated by Fig. 15-9, yields $\theta_f = 8.53$ h. This corresponds to about $1\frac{1}{2}$ in of cake. A cycle for maximum capacity would require a much shorter filtration time. ////

PROBLEMS

15-1 A test of a certain leaf filter shows that if the pressure is maintained constant at 50 psig, the initial rate of collection is 10 gal/min, and 1 h is needed to collect 100 gal of filtrate. What is the maximum hourly capacity of this filter if 15 gal of wash water is used and the time for dumping and cleaning is 20 min?

15-2 A small leaf filter is run at constant rate. It is found that the initial pressure is 34.5 kPa gage and the pressure after 20 min of operation, during which 0.14 m^3 of filtrate is collected, is 345 kPa gage. If this filter is used with the same slurry in a constant-pressure filtration at 345 kPa gage, how much filtrate is collected in 20 min?

15-3 Find the quantity of cake which can be obtained in 1 h of operation of a leaf filter with an area of 1000 ft^2 working at 150 psig on a slurry of 100 g/L of ZnS, type B. Use the data in Figs. 15-7 and 15-8. The resistance of the medium is equivalent to 0.05 in of cake.

15-4 (*a*) A homogeneous sludge, 15 percent solid by weight, forming a uniform incompressible cake, is filtered through a batch leaf filter at a constant difference in pressure of 60 psi. One inch of cake is deposited in 1 h with a filtrate volume of 1800 gal. Three minutes is required to drain the liquor from the filter. Two minutes is required to fill the filter with water. Washing proceeds exactly as filtration, using 400 gal. Opening, dumping, and closing take 6 min. Assume the filtrate to have the same properties as the wash water, and neglect the resistances of the filter cloth and flow line. Determine the maximum daily capacity of this filter operating on this slurry. Assume the ratio of wash water to filtrate is constant at a value of 400/1800.

(*b*) The sludge feed of part (*a*) is available in unlimited quantities. It is a waste stream from a plant and is ordinarily discarded. A neighboring plant is willing to buy the solid at \$4.65/1000 lb and will buy whatever quantity is available. A laborer, who is paid \$2 per hour, must be on hand for all parts of the cycle except during filtering and washing, when his time will be charged to other parts of the plant. Calculate the optimum batch size, the daily production of solid, and the daily profit. The filter of part (*a*) will be used. Take into account only the income from the sale and the cost of the laborer's time.

15-5 A continuous rotary filter is being used with a slurry which forms an incompressible cake. The resistance of the medium is negligible. What is the effect on the capacity of the filter of doubling the speed of rotation?

15-6 In the operation of a leaf filter producing an incompressible cake at constant pressure, determine separately the effects of the following changes on the amount of filtrate that can be collected in a given time. Assume that the resistance of the filtration medium is zero and show the equations used to justify your conclusions.

(*a*) Double the filtrate viscosity.
(*b*) Double the area of the filter.
(*c*) Double the filtration time.

15-7 A slurry with a density of 1280 kg/m^3 flows by gravity from a tank in which a constant depth of liquid is maintained to a leaf filter located 20 m below the liquid level in the tank. The filter, operating continuously after starting up, produces 0.040 m^3 of filtrate during the first minute of operation and 0.016 m^3 during the second minute. Determine the volume of filtrate produced during a total of 10 min of continuous operation. Assume the filter cake to be incompressible and assume that both frictional losses and kinetic energy changes in the line carrying the slurry to the filter are negligible. The pressure at the liquid surface in the tank is atmospheric as is the pressure at the filter discharge.

15-8 Ten liters of dilute slurry are to be filtered using a laboratory Büchner funnel. To shut off the filter, change the filter paper, and start up again takes 1 min. In the first run it takes 12.5 min to filter 0.5 L of slurry. Assuming constant pressure drop over the Büchner funnel and negligible resistance of the filter paper to flow, what is the shortest length of time theoretically required to do the job? The cake is not to be washed.

15-9 A rotating vacuum disk filter operates at constant pressure with the disks continually half-submerged. The filter cloth is attached to both the sides of the disks but there is no flow through the outer rim or disk periphery. Derive an expression for the rate of filtrate production per disk as a function of the diameter of the disks, the speed of rotation, and the appropriate filtration constants. Neglect the area of the shaft supporting the disks, the effect of varying hydrostatic head on pressure drop, and the resistance to flow offered by the filter cloth.

15-10 A leaf filter is operated at constant rate, and 100 gal of filtrate are collected after 20 min. The resistance of the medium can be neglected. The final pressure (at 20 min) is 50 psig.

(*a*) If the filter is run on the same slurry, what constant pressure should be used to collect 100 gal of filtrate in 20 min?

(*b*) What is the washing rate at the pressure of part (*a*) after 20 min of filtration?

PART TWO

Heat Transfer

16

INTRODUCTION TO HEAT TRANSFER

Scientists and philosophers have speculated on the nature of heat for many centuries. Galileo is said to have constructed one of the first thermometers, a device in which the expansion of air was used as a measure of temperature. Numerous other temperature measuring instruments were proposed and used before Fahrenheit described the first mercury thermometer in 1724.

Among the early investigators was Franklin, who in 1757 reported a series of observations on the thermal conductivity of metallic, ceramic, and wooden materials, with particular reference to teapot handles. Newton proposed in 1701 that the rate of cooling of a body was proportional to the temperature difference between the body and its surroundings, a relation which was true within the limits of experimental accuracy attainable in the eighteenth century. Scheele (1724–1786) observed that heat from a stove could move through air without warming it but could be absorbed by glass. He proposed that heat transferred in this fashion be called *radiant* heat.

The examples given above are typical of the numerous studies of heat transfer which were begun as a result of observations of a nonindustrial world. Although engineers are primarily concerned with applying the principles of heat transfer to industrial problems, they should not overlook the fact that the natural world furnishes many examples of heat-transfer phenomena which can increase their understanding of principles that apply elsewhere.

Mechanisms

Heat transfer is the transfer of energy occurring as a result of a driving force which we call temperature difference. There are three mechanisms by which heat transfer can occur:

Conduction Heat can be conducted through solids, liquids, and gases. Most illustrations of heat conduction are chosen with an opaque solid as the conducting substance, because in such materials conduction is the only method by which heat can be transferred. Heat is conducted by the transfer of energy of motion between adjacent molecules. The kinds of motion depend on the molecular state of the system and range from the vibration of atoms in a crystal lattice to the random motion of molecules in a gas. It is known that there are supplementary mechanisms in certain cases, as, for example, the transfer of energy by free electrons in metallic solids.

Convection This term implies transfer of heat due to bulk transport and mixing of macroscopic elements of liquid or gas. Because motion of a fluid is involved, heat transfer by convection is partially governed by the laws of fluid mechanics. If convection is induced by density differences resulting from temperature differences within the fluid, it is said to be *natural convection.* However, if the motion of the fluid is the result of an outside force, as might be exerted by a pump impeller, then it is called *forced convection.*

Radiation The transfer of heat by radiation is energy transfer by electromagnetic radiation, or photons, having a certain range of wavelengths. It follows that the same laws that govern the special range of radiation which we call visible light also govern the radiation of the energy we call heat. Although energy can be transferred by radiation through gases, liquids, and solids, these media absorb some or all of the energy, so that energy is radiated most efficiently through empty space.

Illustrations

Both our natural and humanly constructed surroundings contain numerous illustrations of the separate and combined mechanisms of heat transfer. For example, the temperature in the earth's crust has been observed experimentally to increase with depth at a rate which is usually between 0.01 and 0.02°F ft. This indicates that heat is being conducted continuously to the earth's surface from its interior. This heat transfer has little effect on surface conditions because the conduction rate [roughly 0.01 Btu (h)(ft^2)] is far less than the rate at which heat is transferred to and from the surface by other mechanisms.[1] Examples of industrial heat-transfer processes in which conduction is the predominant mechanism are thermal curing of rubber, heat treatment of steel forgings, and heat flow through the walls of heat exchangers.

An example illustrating a process which is primarily convective is the operation of the steam "radiator." In this apparatus condensing steam in the radiator gives up heat which, after flowing through the radiator wall by conduction, is discharged into the atmosphere. Some heat is lost from the surface by radiation, but contrary to the name of the device, this amount is small compared with the total amount of heat given up. In fact, the common practice of painting radiators with aluminum or other light-colored paint does much to lessen radiation, as we shall see later. Heat is transferred by conduction into the air surrounding the radiator, and this step is followed by natural convection caused by the resulting density differences. If the air were moved by the action of a fan, the transfer would be achieved by combined forced and natural convection.

The most obvious example of the radiation mechanism is the transport of heat to the earth from the sun. The difficulty of transporting heat a distance of 93 million miles through a near vacuum rules out serious thought of contributions by conduction and convection. It is of interest to note that measurements of solar radiation striking the earth, which vary with time and location, indicate a heat-transfer rate normal to the sun's rays of approximately 440 Btu (h)(ft^2).

An industrial example of a heat-transfer process in which radiation is predominant is the operation of a furnace for heating petroleum. The tube still, as it is called, is essentially a room constructed of fire-resistant materials into which a combustible mixture of gases is injected and burned. Petroleum flows through a network of tubes suspended near the walls and ceiling. The combustion gases, which may have a temperature as high as 2000°F, transfer heat to the outer tube surfaces primarily by the mechanism of radiation and to a lesser extent by conduction and convection.

Heat is seldom transferred by a single mechanism, but usually by a combination of mechanisms operating in series or in parallel, as in the examples just discussed. The analysis of many heat-transfer problems, however, does not require that all parts be solved with equal care. With experience, the engineer will usually be able to recognize the predominant mechanisms and can base his calculations on these while neglecting many minor effects.

[1] B. Gutenberg (ed.), "Internal Constitution of the Earth," Dover Publications, Inc., New York, 1951.

17

CONDUCTION AND THERMAL CONDUCTIVITY

The Fourier Equation for Heat Conduction

The basic equation for steady-state heat conduction is known as *Fourier's equation.*[1] Written for conduction in one direction, it is

$$q = -kA\frac{dt}{dx} \tag{17-1}$$

where q = heat-conduction rate in x direction
A = cross-sectional area normal to heat flow
dt/dx = temperature gradient in x direction
k = a proportionality constant known as the thermal conductivity of the conducting medium.

Steady-state conduction means that the heat-transfer rate q through a given cross section in the system does not vary with time.[2] The thermal conductivity k is a function of the molecular state of the medium; for a single-phase system it is usually

[1] J. B. J. Fourier, "Théorie analytique de la chaleur," Gauthier-Villars, Paris, 1822.
[2] Electrical engineers and physicists use the term steady state to designate, in addition, systems in which there is a steady periodic variation.

considered to be dependent on temperature and pressure. If the temperature varies greatly across the system, this may cause a significant variation of k with position. The cross-sectional area normal to heat flow may also be dependent on position. The minus sign in the equation is present because it is usually convenient to make q positive in the positive x direction.

In applying Eq. (17-1) to solids, we are usually able to assume that the thermal conductivity at any given point depends only on the temperature, pressure, and composition at that point; the orientation of the substance relative to the direction of heat flow is of no consequence. Materials having thermal conductivities which are independent of the direction of flow of heat are said to be *isotropic*. For three-dimensional heat flow in isotropic solids we can imagine the existence inside the solid of hypothetical, isothermal surfaces between which heat is always conducted in directions normal to these surfaces. This is not the case, however, for certain single crystals which have different thermal conductivities in the different directions of the crystal axes,[1] nor is it true for wood or for objects made of laminated materials. These substances are said to be nonisotropic. Eckert and Drake[2] show that for these substances the heat flux vector is generally not normal to the isothermal surfaces.

Thermal Conductivity

In this chapter we shall look only briefly at the subject of thermal conductivity. For the purpose of this book it is enough that a student have some knowledge of the conduction mechanisms and the magnitudes of the thermal conductivity in various systems. For information beyond this, the student is referred to textbooks on modern physics and physical chemistry in which detailed discussions of the properties of matter rightly belong.

Solids The thermal conductivity of homogeneous solids varies widely, as may be seen in the few examples[3] shown in Table 17-1. It is typical that metallic solids have higher thermal conductivities than nonmetallic solids.

Among the metallic solids, inspection of thermal-conductivity data reveals that the solids with high thermal conductivities are also the ones which are known to have high electrical conductivities. This was first reported by Wiedemann and Franz in 1853 when they noted that the ratios of thermal conductivity to electrical conductivity for metals at a specified temperature were approximately the same. The fact

[1] The thermal conductivity of a quartz crystal parallel to the main crystallographic axis is 11 Btu/(h)(ft)(°F) at 0°C. Normal to the axis the value is 6, while the value for quartz glass is 1.1. From M. Jakob, "Heat Transfer," vol. 1, p. 96, John Wiley & Sons, Inc., New York, 1949.

[2] E. R. G. Eckert and R. M. Drake, Jr., "Analysis of Heat and Mass Transfer," chap. 1, McGraw-Hill Book Company, New York, 1972.

[3] More extensive data on thermal conductivity of solids, liquids, and gases are given in Appendix 3, Tables A-11 to A-15.

that this relation does not apply to nonmetallic solids has led to the supposition that heat is conducted through solids by more than one mechanism. Current theories hold that the two principal carriers of heat in solids are electrons and lattice waves (phonons). However, heat can also be transmitted in solids by magnetic excitations and by electromagnetic radiation. The total thermal conductivity is, thus, the sum of the contributions from the four modes of transport. The electron transport mechanism, which applies only to electrical conductors, assumes that heat, like electricity, is conducted by free electrons which move through the lattice of a metal in a manner analogous to molecules moving in a gas. This component of the total thermal conductivity, often referred to as the *electronic thermal conductivity*, is proportional in magnitude to the electron mean free path. In a perfect crystal lattice this would be infinite, but in actual materials it is limited by the electron scattering produced by chemical impurities, physical defects, and thermal vibrations.

This theory, which is the basis of the Wiedemann-Franz relation, is supported by the fact that carefully grown crystals of very pure metallic elements often have very high thermal conductivities. For example, values for the thermal conductivity of copper at very low temperatures have been measured in the range of 5000 to 7000 Btu/(h)(ft)(°F). It is assumed that the lattice dislocations more prevalent in commercially prepared copper cause the decrease from these values to the commonly used figure of 220 Btu/(h)(ft)(°F).

The second mechanism by which heat is conducted in solids is the transmission of energy of vibration between adjacent atoms or molecules in the direction of decreasing temperature. This transmission of energy of vibration can be treated as occurring from a superposition of lattice (or displacement) waves of a wide range of

Table 17-1 THERMAL CONDUCTIVITIES OF SOLIDS [k, Btu/(h)(ft)(°F)]†

Material	t, °F	k	t, °F	k
Metals:				
Aluminum	32	117	212	119
Copper	32	224	212	218
Iron (pure)	64	39.0	212	36.6
Steel (1% C)	64	26.2	212	25.9
Silver	32	242	212	238
Nonmetals:				
Asbestos (density 36.0 lb/ft^3)	32	0.087	212	0.111
Brick (fused alumina)	800	1.8		
Corkboard (density 10 lb/ft^3)	86	0.025		

SOURCE: R. H. Perry, "Chemical Engineers' Handbook," 5th ed., pp. 3-219 to 3-221, McGraw-Hill Book Company, New York, 1973.

† SI units: Multiply values of thermal conductivity given in table by 1.731 to convert to W/m·K.

frequencies. Each wave obeys the dynamical equation of a harmonic oscillator whose energy is not continuously variable but is, rather, regarded as made up of an integral number of quanta, known as *phonons*. Thus, the energy of vibration can be considered as the energy possessed by a gas of phonons which diffuses in the manner of free electrons and has a mean free path of its own. Factors affecting phonon diffusion such as lattice defects, therefore, limit the contribution of lattice thermal conductivity. The lattice mechanism is, of course, present, but not usually significant, when heat is being conducted through metallic solids. However, it is the controlling mechanism for nonmetallic substances. Although heat-transfer rates for these substances are small compared with the rates obtained through metals, the thermal conductivities of nonmetallic substances are not insignificant, as are the electrical conductivities.

The coupling of magnetic dipoles of adjacent atoms serves as a third mechanism of heat conduction. Cooperative effects between magnetic moments in a lattice lead to the concept of *magnons* as carriers of heat.

Finally, as might be expected, the transmission of electromagnetic radiation in translucent materials in the form of photons adds to the total effective thermal conductivity of solids. This energy is principally in the infrared portion of the electromagnetic spectrum. If the solid either has no absorptive capacity for the radiation, or if it is completely opaque, there will be no augmentation of thermal conductivity. However, for substances between these two extremes, each volume element of the solid will emit, absorb, and re-radiate photons. This is considered to be the reason why the thermal conductivities of glasses tend to increase rapidly at high temperatures.

Liquids The thermal conductivities of most liquids are rather small, the exception being liquids regarded as metallic. A few values for common substances are given in Table 17-2.

Table 17-2 THERMAL CONDUCTIVITIES OF LIQUIDS [k, Btu/(h)(ft)(°F)]

Material	t, °F	k	t, °F	k
Benzene	86	0.092	140	0.087
n-Hexane	86	0.080	140	0.078
Mercury	82	4.8	140	5.6
n-Nonane	86	0.084	140	0.082
Sodium	212	49	410	46
Water	32	0.343	200	0.393

SOURCE: R. H. Perry, "Chemical Engineers' Handbook," 5th ed., p. 3-214, McGraw-Hill Book Company, New York, 1973.

• sound [sáund] n) a long, narrow passage of water joining two larger bodies of water

Although the theories of heat conduction in liquids are under intensive study, most equations presently available for predicting thermal conductivities of liquids are semiempirical correlations based on experimental results for specific classes of liquids. One of the earliest theories was that of Bridgman[1] who proposed a mechanism whereby energy is handed down rows of molecules with the velocity of sound. The overall energy-transfer rate is considered to be the rate of transfer per row multiplied by the number of rows per unit of cross-sectional area. This product divided by dt/dx provides an expression for the thermal conductivity k of the liquid. Although the model proposed is simple, the equation can be used to predict the thermal conductivities for many liquids, such as water and acetone, within 10 percent of the measured values.

A more sophisticated model is that of Horrocks and McLaughlin[2] who assumed the liquid to have a lattice structure through which excess energy due to a temperature gradient is transferred both by motion of molecules from cell to cell and by a vibrational mechanism in which the molecule vibrates within its cell. The former term is shown to be negligibly small. The intracellular contribution, calculated from intermolecular forces, gives good results for simple liquids.

Gases The conduction of heat in a gas is basically the mechanism of a random walk (diffusion and collision). The molecules of high-temperature gas diffuse among the molecules of low-temperature gas, collide with them, and give up kinetic energy. In a system behaving according to the simple kinetic theory of gases, the number of molecules per unit of volume is directly proportional to the pressure on the system. As the mean free path of a molecule during a walk in such a system is inversely proportional to the pressure, it follows that the thermal conductivity should be independent of pressure. This has been found to be approximately true of most gases near atmospheric pressure. It is not true, however, at very low pressures where the thermal conductivity of a gas obviously must approach zero, nor is it true at even moderately low pressures when the dimensions of the container are less than the mean free path of the gas molecules (a so-called *Knudsen gas*). At high pressures, where the simple kinetic theory of gases does not apply, it is to be expected that there would be a dependence upon pressure.

It is worth noting that there is a relationship between the viscosity and thermal conductivity of a gas. Gas viscosity is a measure of the drag exerted by diffusion of gas molecules from a zone moving at one mean velocity into gas molecules in a zone moving at a different mean velocity. This is similar to heat conduction, except that in the case of conduction, the quantity being transported is kinetic energy associated with random motion of the molecules rather than directed momentum.

[1] P. W. Bridgman, *Proc. Natl. Acad. Sci. U.S.*, **9**:341 (1923).
[2] J. K. Horrocks and E. McLaughlin, *Trans. Faraday Soc.*, **56**:206 (1960).

Table 17-3 gives the thermal conductivities of some common gases. Since the mechanism of heat conduction in a gas is a function of its tendency to diffuse, it is to be expected that light gases such as hydrogen would have relatively high thermal conductivities. This is what actually happens. The data given in Table 17-3 for the paraffin hydrocarbons show a typical decrease of thermal conductivity with increase of molecular weight in a homologous series.

The early methods for theoretical evaluation of thermal conductivities of gases were based on simple kinetic mean-free-path assumptions. More sophisticated and rigorous methods based on statistical theory have been developed, such as the Chapman-Enskog theory which represents thermal conductivity results as an infinite series requiring a knowledge of potential functions and collision integrals for evaluation.

General Remarks on Thermal Conductivity

As was mentioned earlier, thermal conductivity is, in general, a function of temperature and pressure. The effect of pressure on the conductivities of solids and liquids has received very little attention from engineers, probably because of their primary concern with applications at atmospheric pressure and also partly because of the masking effects of other variables such as the presence of impurities. As mentioned earlier, the thermal conductivity of an ideal gas is independent of pressure. The effects of high pressure on the thermal conductivities of gases have been the subject of recent investigations; in general, conductivities have been found to increase with pressure.

Thermal conductivities of gases, liquids, and solids are moderately dependent on temperature, as can be seen by inspection of Tables 17-1 to 17-3. In general, an increase in temperature causes the conductivity of a gas to increase and the conductivity of a solid or liquid to decrease. However, there are many exceptions to

Table 17-3 **THERMAL CONDUCTIVITIES OF GASES [k, Btu/(h)(ft)(°F), at approximately atmospheric pressure]**

Material	t, °F	k	t, °F	k
Hydrogen	32	0.100	212	0.129
Methane	32	0.0175	122	0.0215
n-Butane	32	0.0078	212	0.0135
n-Hexane	32	0.0072	68	0.0080
Air	32	0.0140	212	0.0183

SOURCE: J. H. Perry, "Chemical Engineers' Handbook," 4th ed., p. 3-206, McGraw-Hill Book Company, New York, 1963.

these generalizations; in fact, there are some substances for which conductivities pass through maxima or minima with change in temperature.

In considering thermal conductivities it is apparent that the magnitudes of the values decrease markedly as we consider them in the order of decreasing density. Most of the common substances encountered in engineering work have conductivities to be found in the following ranges:

	[k, Btu/(h)(ft)(°F)]
Gases	0.001–0.1
Liquids	0.01–1.0
Solids	1.0–100

One result of this difference in magnitude is that within a given two-phase material the effective thermal conductivity may be quite difficult to predict. A conductivity computed simply on the basis of weight or volume fraction may not be even close to the correct value. Rather elaborate geometrical models have been postulated and analyzed mathematically to determine effective thermal conductivity, but even these have not proved very successful. The principal reason for this, aside from oversimplification of the model, lies in the fact that in a two-phase system, such as porous insulation, some heat is transferred by convection and radiation as well as by conduction. A good discussion of this subject is given by Jakob.[1]

An interesting example of conduction in a two-phase system is found in studies of conduction through fine powders. At sufficiently high pressures the mean free path of the gas molecules is much smaller than the dimensions of the cavities in the powder and the conductivity of the system lies between the values for the solid and the gas. Below a certain pressure, which is a function of the pore size, the gas in the pores behaves as a Knudsen gas, with the result that its thermal conductivity is lower than the conductivity of the gas at the same pressure in a large free volume. As a result, the conductivity of the porous system may be lower than either the conductivity of the solid or the gas as measured under ordinary conditions.

The conductivities of homogeneous mixtures of solids, liquids, or gases are often as hard to predict as the conductivities of two-phase systems. Some experimental measurements have been made employing mixtures, but in general, few experimental data or theoretical methods are available for liquids and solids. Theoretical methods for mixtures of gases are somewhat more reliable.

Tabulations of thermal conductivities and summaries of theories and experimental techniques for determining thermal conductivity have recently become more extensive. Reid and Sherwood[2] provide a discussion of various theories of conduc-

[1] M. Jakob, "Heat Transfer," vol. I, chap. 6, John Wiley & Sons, Inc., New York, 1949.

[2] R. C. Read, J. M. Prausnitz, and T. K. Sherwood, "The Properties of Gases and Liquids," 3rd ed., McGraw-Hill Book Company, New York, 1977.

tivity and a summary of correlation equations. The most extensive collection of measurements of thermal conductivity, however, is to be found in the TPRC Data Series (vols. I, II, III).[1] This set of publications, the result of a literature search of several thousand journals, contains thermal conductivity results for metallic elements and alloys (Vol. I), nonmetallic solids (Vol. II), and nonmetallic liquids and gases (Vol. III). Each volume contains, in addition, summaries of theoretical methods and experimental techniques for measurement of thermal conductivity. Results for materials not listed in the Data Series may frequently be found by consulting a companion volume,[2] which supplies references to 33,700 papers on the subject.

[1] "Thermophysical Properties of Matter," The Thermophysical Properties Research Center Data Series, vols. I–III, Plenum Press, New York, 1970.

[2] Y. S. Touloukian, J. K. Gerritsen, and N. Y. Moore, "Thermophysical Properties Research Literature Retrieval Guide," 6 vols., Plenum Press, New York, 1979.

18

STEADY-STATE HEAT CONDUCTION

In this chapter the Fourier conduction equation will be used in the analysis of one-dimensional, steady-state heat flow in solid systems that are geometrically simple. The flat wall and the hollow cylinder, in addition to being simple, are the commonest shapes with which engineers must deal.

Flat Wall

In the application of the Fourier conduction equation,

$$q = -kA\frac{dt}{dx} \qquad (18\text{-}1)$$

to a flat wall, it is apparent that if heat is flowing normal to the principal surfaces, the area term A is constant. Furthermore, if the conductivity is assumed to be constant, then q at any cross section is proportional to the temperature gradient dt/dx. If energy is neither generated nor accumulated in the wall, q is identical at all cross sections and so is dt/dx. A sketch of the system and a graphical representation of its temperature profile are given in Fig. 18-1.

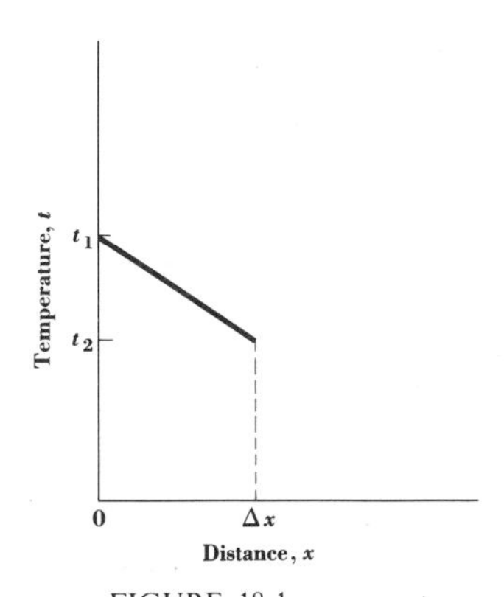

FIGURE 18-1
Heat conduction in a flat wall.

If the thermal conductivity varies with temperature, dt/dx is not constant. If heat flow and area remain constant at all cross sections and k increases with decreasing temperature, the temperature gradient must diminish in the direction of decreasing temperature. Thus the curve representing temperature in steady-state flow for this system is concave upward.

The integration of Eq. (18-1) is readily performed when q, k, and A are constant, and this gives

$$q = \frac{kA(t_1 - t_2)}{\Delta x} \tag{18-2}$$

The equation can also be integrated and solved for the temperature at a point x.

$$t = -\frac{q}{kA}x + t_1 \tag{18-3}$$

This equation indicates the linearity of temperature with distance when the conditions are as stated.

Multilayer Flat Wall

When heat is conducted through a flat wall consisting of layers of different substances, a situation is produced comparable with an electrical system in which a number of resistances are connected in series. The thermal system is shown in Fig. 18-2.

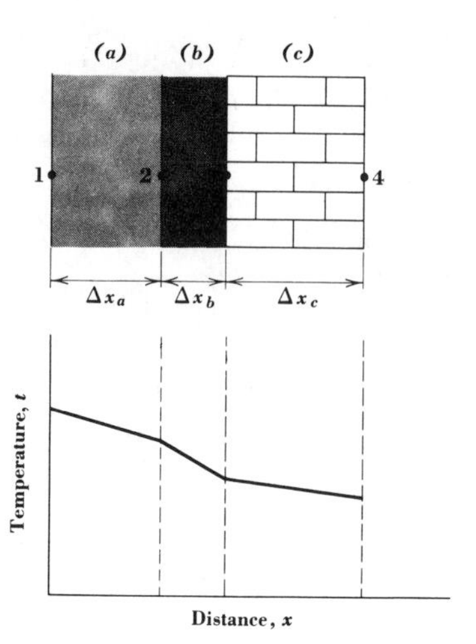

FIGURE 18-2
Conduction in a multilayer flat wall.

The Fourier equation can be written for each of these layers, and for steady state the heat flow q is the same for all three walls. By analogy with Eq. (18-2) we can write

$$q = k_a A \frac{t_1 - t_2}{\Delta x_a} = k_b A \frac{t_2 - t_3}{\Delta x_b} = k_c A \frac{t_3 - t_4}{\Delta x_c} \tag{18-4}$$

The overall temperature difference is usually specified in engineering systems, so an equation for heat flow as a function of $t_1 - t_4$ will be derived. Rewriting the equalities given in (18-4) we get

$$t_1 - t_2 = q \frac{\Delta x_a}{k_a A}$$

$$t_2 - t_3 = q \frac{\Delta x_b}{k_b A}$$

$$t_3 - t_4 = q \frac{\Delta x_c}{k_c A}$$

Adding these equations gives

$$t_1 - t_4 = q\left(\frac{\Delta x_a}{k_a A} + \frac{\Delta x_b}{k_b A} + \frac{\Delta x_c}{k_c A}\right)$$

or

$$q = \frac{t_1 - t_4}{\Delta x_a/k_a A + \Delta x_b/k_b A + \Delta x_c/k_c A} \tag{18-5}$$

If the integrated form of the Fourier equation (18-2) is considered analogous to Ohm's law for electric conduction, then the quantity $\Delta x/kA$ is a measure of the resistance to heat flow. The denominator of Eq. (18-5) is the overall resistance, which is the sum of the individual resistances.

Figure 18-2 shows temperature gradients for the different layers of solids. Because q/A is the same for all layers, it follows that $k(\Delta t/\Delta x)$ is the same for all layers; thus $\Delta t/\Delta x$ is inversely proportional to the thermal conductivity.

An important application of heat conduction through multilayer walls occurs in the use of fins which are attached by crimping, soldering, or welding to the outer surface of a tube to increase the heat-transfer rate. Vibration or exposure to cycles of increasing and decreasing temperature (thermal cycles) can cause these fins to become loosened from the tube surface, with a serious decrease in the heat-transfer rate.

Example 18-1 A cold-storage room has walls constructed of a 4-in layer of corkboard contained between double wooden walls, each $\frac{1}{2}$ in thick. Find the rate of heat loss in Btu/(h)(ft^2) if the wall surface temperature is 10°F inside the room and 70°F outside the room. In addition, find the temperature at the interface between the outer wall and the corkboard.

Although thermal conductivity is a function of temperature, it is often assumed to be constant at the arithmetic average temperature of the layer involved. The conductivities of many materials have not been measured over a temperature range; in addition, other factors such as the density (in the case of corkboard) and the presence of impurities (e.g., moisture) can have an effect on the conductivity. Data limitations such as these often have a direct effect on the accuracy of the solution and guide the engineer in determining the degree of simplification he can use in his calculations.

In this problem the thermal conductivity of corkboard is given "at approximately room temperature" in Perry (p. 3-220) as 0.0225 and 0.025 Btu/(h)(ft)(°F), corresponding to densities of 7.0 and 10.6 lb/ft^3. A value of 0.024 will be chosen for the solution. The thermal conductivity of several varieties of wood is given in Perry, p. 3-219. Fir is one of the cheaper woods and is commonly used in wall construction; its thermal conductivity of 0.062 Btu/(h)(ft)(°F) (given only at 60°C) will be used here.

$$\text{Thermal resistance of each wood layer} = \frac{\Delta x}{kA} = \frac{(\frac{1}{2})(\frac{1}{12})}{(0.062)(1)}$$

$$= 0.67(\text{h})(°\text{F})/\text{Btu}$$

$$\text{Thermal resistance of corkboard} = \frac{\frac{4}{12}}{(0.024)(1)}$$

$$= 13.9(\text{h})(°\text{F})/\text{Btu}$$

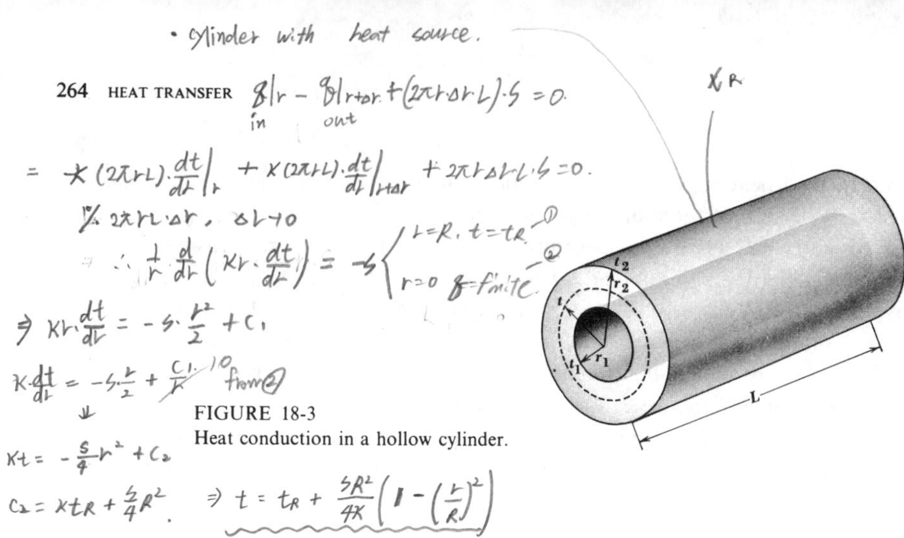

FIGURE 18-3
Heat conduction in a hollow cylinder.

The heat loss, using Eq. (18-5), is

$$q = \frac{70 - 10}{0.67 + 13.9 + 0.67} = 3.9 \text{ Btu/h}$$

The temperature at the interface between the outer wooden wall and the corkboard can be determined by rearranging the equations for the individual temperature drops which led to Eq. (18-5). This gives the temperature drop over the outer layer of wood,

$$t_1 - t_2 = \frac{(\Delta x/kA)_{\text{wood}}(t_1 - t_4)}{\sum \Delta x/kA}$$

$$= (0.67/15.3)(60)$$

$$= 2.6°\text{F}$$

Therefore the interface temperature t_2 is 67.4°F. ////

Hollow Cylinder

The conduction of heat in the walls of a hollow cylinder can be described mathematically by the Fourier equation written in rectangular coordinates [Eq. (18-1)], but the equation is usually written in cylindrical coordinates for convenience. For the example shown in Fig. 18-3, the equation is

$$q = -kA\frac{dt}{dr} \tag{18-6}$$

The area normal to heat flow is

$$A = 2\pi rL$$

The equation can be integrated and solved for the heat-transfer rate q in terms of the boundary conditions:

$$q\int_{r_1}^{r_2}\frac{dr}{r} = -2\pi Lk\int_{t_1}^{t_2} dt \qquad (18\text{-}7)$$

$$q = \frac{2\pi Lk(t_1 - t_2)}{\ln (r_2/r_1)} \qquad (18\text{-}8)$$

This equation can be multiplied by $r_2 - r_1$ in both numerator and denominator to produce

$$q = \frac{2\pi Lk(r_2 - r_1)}{\ln (r_2/r_1)}\frac{t_1 - t_2}{r_2 - r_1}$$

$$= k\frac{A_2 - A_1}{\ln (A_2/A_1)}\frac{t_1 - t_2}{r_2 - r_1} \qquad (18\text{-}9)$$

$$q = kA_{lm}\frac{\Delta t}{\Delta r} \qquad (18\text{-}10)$$

The area term in Eq. (18-10), A_{lm}, is the log mean area $(A_2 - A_1)/[\ln (A_2/A_1)]$. The temperature difference Δt is $t_1 - t_2$ instead of $t_2 - t_1$. This usage is contrary to mathematical practice but is conventional in writing integrated heat-transfer equations for which the direction of heat flow is usually clearly understood.

Equation (18-10) is similar in form to Eq. (18-2) for conduction in a flat wall, the principal difference being in the area terms. In most engineering applications (e.g., pipes), $r_2/r_1 \ll 2$. In these circumstances the arithmetic mean area may be used in Eq. (18-10), with a consequent error in q less than 4 percent.

Integration of Eq. (18-6) to give a relation for the temperature t at any radial position r shows that t is a linear function of $\ln r$, rather than of r as in the case of a flat wall.

Multilayer Cylinder

The analysis of multilayer cylinders is done by combining the modes of analysis for multilayer flat walls and single hollow cylinders.

Consider the case of three concentric hollow cylinders, e.g., a pipe with two layers of insulation around it. The thicknesses of the three layers will be designated Δr_a, Δr_b, and Δr_c, and the temperature drops over the individual layers Δt_a, Δt_b, and

Δt_c. The total heat-transfer rate, which will be the same for all the cylinders, can be written by analogy with Eq. (18-10).

$$q = \left(kA_{lm}\frac{\Delta t}{\Delta r}\right)_a = \left(kA_{lm}\frac{\Delta t}{\Delta r}\right)_b = \left(kA_{lm}\frac{\Delta t}{\Delta r}\right)_c \tag{18-11}$$

The individual temperature drops may be found by rearranging (18-11).

$$\Delta t_a = q\left(\frac{\Delta r}{kA_{lm}}\right)_a$$

$$\Delta t_b = q\left(\frac{\Delta r}{kA_{lm}}\right)_b$$

$$\Delta t_c = q\left(\frac{\Delta r}{kA_{lm}}\right)_c$$

Adding and rearranging these three equations gives

$$q = \frac{\Delta t_{\text{overall}}}{(\Delta r/kA_{lm})_a + (\Delta r/kA_{lm})_b + (\Delta r/kA_{lm})_c} \tag{18-12}$$

If the resistance concept is applied to this problem, the individual resistances are the terms written as $\Delta r/kA_{lm}$, and as in the case of the multilayer flat wall, the overall resistance is simply the sum of the individual resistances in series.

Hollow Sphere

Steady-state conduction normal to the walls of a hollow sphere can be readily analyzed and expressed in an equation analogous to Eq. (18-2) for a flat wall and Eq. (18-10) for a hollow cylinder. Equation (18-6) is used as the starting point in the analysis; area normal to heat flow is $4\pi r^2$. For the case of a hollow sphere with temperature t_1 at inner radius r_1, and temperature t_2 at outer radius r_2, we write

$$q\int_{r_1}^{r_2}\frac{dr}{r^2} = -4\pi k\int_{t_1}^{t_2} dt \tag{18-13}$$

$$q\left(\frac{1}{r_1} - \frac{1}{r_2}\right) = 4\pi k(t_1 - t_2) \tag{18-14}$$

$$q = 4\pi k(t_1 - t_2)\frac{r_1 r_2}{r_2 - r_1} \tag{18-15}$$

$$= kA_{gm}\frac{\Delta t}{\Delta r} \tag{18-16}$$

The area term in Eq. (18-16), A_{gm}, is the geometric mean area, $\sqrt{A_1 A_2}$. Integration of Eq. (18-13) to give the temperature t at any radial position r inside the wall of the sphere shows that the temperature is a linear function of $1/r$.

The analysis of a multilayer, spherical system is made in the same manner as for cylinders and flat surfaces. For a system of three concentric spherical layers of materials a, b, and c, we obtain

$$q = \frac{\Delta t_{\text{overall}}}{(\Delta r/kA_{gm})_a + (\Delta r/kA_{gm})_b + (\Delta r/kA_{gm})_c} \tag{18-17}$$

Thermal Contact Resistance

In the analysis of multilayer systems in this chapter it has been assumed that perfect contact is made between adjacent layers; i.e., adjacent surfaces are considered to be at the same temperature. This is an assumption which, while never completely correct, is adequate for the majority of systems in industry. Nevertheless, in systems with very high heat fluxes, such as are found in nuclear power generation, the temperature drop between contacting layers may be large enough to be significant. The effects of contact resistance, as it is called, may also be significant in systems for the experimental measurement of thermal conductivity.[1]

Any two solid surfaces brought together will touch only at certain points where contact is made by protuberances on the surfaces. Heat flowing between the two solids thus flows by conduction through the protuberances at the contact points and by conduction through the fluid filling the spaces where solid contact is not made. If these fluid spaces are large enough, convection cells could conceivably be set up in these parts of the interfacial region. Furthermore, if the system were operating at high temperatures, radiation across the gap could add significantly to the heat-transfer rate. Both of these mechanisms would diminish the effective resistance to heat transfer at the junction.

Although thermal contact resistance has been studied considerably in recent years, no equations are available for predicting the effects in general cases. This is mainly because the contacting surfaces vary in the shape and number of their protuberances. Additional complicating variables include the pressure with which the surfaces are brought together (increasing pressure decreases heat-transfer resistance) and the presence of adsorbed gases on the surfaces (which increases heat-transfer resistance).

Although correlations for accurate prediction of thermal contact resistance are not available, the maximum resistance can be determined if the maximum gap width

[1] *Proceedings of the Eighth Conference on Thermal Conductivity*, Purdue University, Plenum Press, New York, 1968.

is known or can be estimated. In this case the gap is merely treated as an additional series conduction resistance in Eqs. (18-5), (18-12), and (18-17). The fluid in the gap is assumed to be stagnant and continuous. Any actual departure from this condition would enhance the heat-transfer rate and cut down the resistance and temperature drop across the gap.

Example 18-2 A 6-in, schedule-80 steel pipe is covered with a 0.1-m layer of 85 percent magnesia insulation. The temperature of the inner surface of the pipe is 250°C, and the temperature of the outer surface of the insulation is 40°C. Calculate the rate of heat loss per meter of pipe and the temperature at the interface between the pipe and the insulation.

The thermal conductivity of the steel pipe can be taken as 44.8 W/m·K (Perry, p. 3-220), and that of the 85 percent magnesia as 0.066 (Perry, p. 3-221). The OD of the pipe is 0.1683 m, and the ID is 0.1463 m.

For the pipe
$$A_{lm} = \pi \frac{0.1683 - 0.1463}{\ln (0.1683/0.1463)} = 0.49 \text{ m}^2$$

For the insulation
$$A_{lm} = \pi \frac{0.3683 - 0.1683}{\ln (0.3683/0.1683)} = 0.80 \text{ m}^2$$

Resistance of pipe
$$= \frac{\Delta r}{kA_{lm}} = \frac{0.011}{(44.80)(0.49)} = 0.00050 \text{ K/W}$$

Resistance of insulation
$$= \frac{0.1}{(0.066)(0.80)} = 1.89 \text{ K/W}$$

Rate of heat loss calculated from Eq. (18-12)
$$= q = \frac{210}{0.0005 + 1.89} = 111 \text{ W}$$

The thermal resistance of the insulation is so large compared with the resistance of the steel that it is of little importance in the problem to evaluate accurately such terms as the log mean area of the steel. Although the arithmetic mean area would suffice for calculation of the steel resistance,

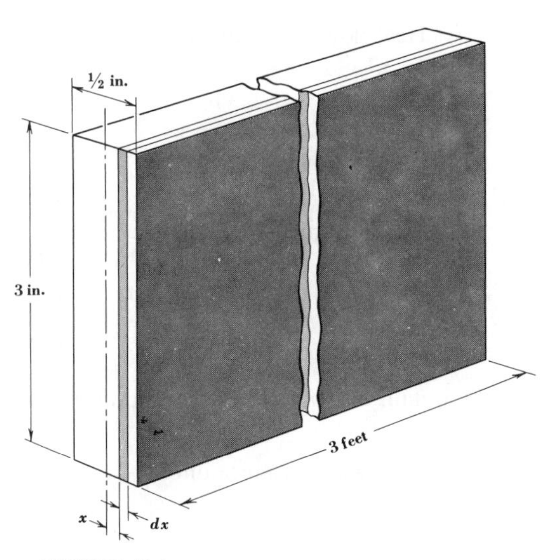

$R = \rho_\Omega \cdot \frac{l}{A}$

FIGURE 18-4
Heat conduction in a bar with generation (Example 18-3).

the use of the arithmetic mean area for the insulation would have introduced an error of 10 percent.

It is apparent that the temperature drop over the steel is very small. Therefore the temperature at the steel-insulation interface is nearly the same as the temperature on the outside surface of the steel.

$$\text{Interface temperature} = 250 - \left(\frac{0.00050}{1.89 + 0.00050}\right)(210)$$

$$= 249.94°\text{C} \qquad ////$$

Example 18-3 A heating element is constructed from carbon in the shape of a bar 3 in wide, $\frac{1}{2}$ in thick, and 3 ft long. When a potential of 12 V is applied to the ends of the bar, its surface reaches a uniform temperature of 1400°F, as indicated by an optical pyrometer. What is the temperature at the center of the bar? The electrical resistivity of the bar is $1.30 \times 10^{-4}(\Omega)(\text{ft})$, and its thermal conductivity is 2.9 Btu/(h)(ft)(°F).

Only heat conduction normal to the largest faces of the bar will be considered, since heat leaves the bar principally through these faces. A differential equation is obtained by writing an energy balance on a differential segment dx of the bar, as shown in Fig. 18-4.

Heat flow into element

$$= -kA\frac{dt}{dx} \qquad \text{Btu/h} \qquad (1)$$

where $A = (\frac{3}{12})(3) = 0.75\ \text{ft}^2$.

$$\text{Heat flow out of element} = -kA\left(\frac{dt}{dx} + d\frac{dt}{dx}\right)$$

$$= -kA\left(\frac{dt}{dx} + \frac{d^2t}{dx^2}dx\right) \qquad \text{Btu/h} \qquad (2)$$

To obtain the rate of heat generation in the element:

$$\text{Electrical resistance of bar} = \frac{(1.30 \times 10^{-4})(3)}{(\frac{1}{24})(\frac{3}{12})}$$

$$= 0.0375\ \Omega$$

Rate of heat generation in entire bar

$$= \frac{(12)^2}{0.0375}$$

$$= 3830\ \text{W}\ [\text{or } 419{,}000\ \text{Btu/(h)(ft}^3)]$$

Rate of heat generation in element

$$= 419{,}000A\ dx \qquad \text{Btu/h} \qquad (3)$$

A heat balance on the element gives the following equation:

Heat flow in + rate of heat generation = heat flow out

The terms in this expression are given as Eqs. (1), (2), and (3). These are combined to give

$$-kA\frac{dt}{dx} + 419{,}000A\ dx = -kA\left(\frac{dt}{dx} + \frac{d^2t}{dx^2}dx\right)$$

which reduces to

$$\frac{d^2t}{dx^2} = -\frac{419{,}000}{k} \qquad (4)$$

This equation could have been obtained by simplifying the general form of the differential energy balance (8-14).

Equation (4) becomes

$$\frac{d^2t}{dx^2} = -\frac{419{,}000}{2.9}$$

$$= -145{,}000°\text{F/ft}^2$$

This is integrated to give

$$\frac{dt}{dx} = -145{,}000x + C_1$$

If the bar is symmetrical, $dt/dx = 0$ at $x = 0$, so $C_1 = 0$. A second integration is performed to get the temperature

$$t = \frac{-145{,}000x^2}{2} + C_2$$

As a second boundary condition we use $x = 1/48$ ft, $t = 1400°$F, and obtain

$$C_2 = 1400 + (145{,}000/2)(1/48)^2$$

$$= 1431$$

The equation for the temperature distribution in the bar is

$$t = -72{,}500x^2 + 1431$$

At the center of the bar the temperature is 1431°F. ////

PROBLEMS

18-1 The outer wall of the cold-storage room described in Example 18-1 is exposed to air which is at a temperature of 70°F and has a dew point of 60°F. If the air is permitted to diffuse freely through the walls of the wood and cork, it will start to condense at the point where the wall temperature is 60°F and it will freeze at 32°F. Find the zones of moisture and frost in the wall.

18-2 To prevent free diffusion of air and moisture within cold-storage insulation, its surfaces are usually sealed with some material which is impervious to diffusion. Assuming that the corkboard of Example 18-1 and Prob. 18-1 is treated in this way, though the wood is still pervious, find the minimum thickness of cork which will prevent condensation of moisture at the outer cork-wood interface. All the other conditions of Example 18-1 and Prob. 18-1 apply.

18-3 A furnace is to be built with a layer of firebrick ($k = 1.50$ W/m · K) on the inside. This is covered with a 0.20-m layer of insulating brick ($k = 0.15$ W/m · K) and then by 0.15 m of building brick ($k = 1.00$ W/m · K) on the exterior. The interior of the furnace is at 1200°C, and the exterior is at 70°C. Determine the thickness of firebrick necessary to keep the temperature of the insulating brick below 900°C. Calculate the inside temperature of the building brick.

18-4 The term *thermopane* is given to a window which in a certain case consists of two layers of glass, each $\frac{1}{4}$ in thick, separated by a layer of dry, stagnant air, also $\frac{1}{4}$ in thick. If the temperature drop over the composite system is 30°F, find the heat loss through such a window, 10 ft long and 4 ft wide. If the window replaces a single glass pane, $\frac{3}{8}$ in thick, at a difference in cost of $120, find the number of days of operation under the specified temperature conditions to pay for the window. Coal with a heating value of 13,200 Btu/lb costs $20 per ton. The furnace efficiency can be assumed to be 50 percent. The thermal conductivity of the window glass of 0.5 Btu/(h)(ft)(°F).

18-5 A brewery fermentation tank 12 m in diameter is situated in a room which has a temperature of 15°C. The tank is constructed of 0.010-m welded steel plate with a 0.012-m glass lining. The temperature at the interface between the glass and the contents of the tank is known to be 50°C. Assuming that the resistance offered by the air film on the outside of the tank equals the combined resistances of the glass and steel, calculate the temperature at the glass-steel interface and the temperature at the steel-air interface. The thermal conductivity of the glass is 0.80 W/m · K.

18-6 A cylindrical exhaust duct has a constant inside temperature of 600°F. It is insulated on the exterior with a 4-in layer of rock wool which has an outer surface temperature of 100°F. The duct has an ID of 3.5 in and $\frac{1}{4}$-in-thick walls of a ceramic material with a thermal conductivity of 0.88 Btu/(h)(ft)(°F). Find the rate of heat loss from the duct per lineal foot. The thermal conductivity of rock wool is represented by the equation $k = 0.025 + 0.00005t$, where t is in °F and k is in Btu/(h)(ft)(°F).

18-7 A 60-ft length of Nichrome IV ribbon, 0.051 in thick and 0.50 in wide, is used as a heating element. The surface temperature of the ribbon is 1400°F when the voltage drop is 110 V. Find the maximum temperature in the ribbon under these conditions. The product of electrical resistivity and thermal conductivity of Nichrome IV is 25.7×10^{-6}(Btu)(Ω)/(h)(°F).

18-8 A fuel element proposed for use in a nuclear power reactor consists of a 0.05-m-diameter cylindrical core of fuel supported by a surrounding layer of 0.006-m-thick aluminum cladding. The outside of the aluminum will be in contact with a heat transfer fluid which will keep the outer surface temperature of the cladding at 100°C. The fuel in the core gives off heat at the rate of 6.0×10^8 W/m³. Because of the adverse effect of temperature on the strength of the aluminum, the cladding temperature must not exceed 400°C at any point. Is the proposed design satisfactory? For aluminum the thermal conductivity may be expressed as $k = K(1 + \alpha t)$ where $K = 200$ W/m · K and $\alpha = -8.0 \times 10^{-4}(°\text{C})^{-1}$ in the temperature range 100 to 400°C.

18-9 A 0.05-m-diameter sphere of radioactive material generates heat at 30 W. If the sphere is enclosed in a spherical shell of insulation ($k = 0.2$ W/m · K), what will be its temperature if the OD of the insulation is 0.6 m? The outer surface of the insulation is at 40°C. Assume the radioactive sphere to be at a uniform temperature.

18-10 A cube, 1-in long on each side, is constructed of equal-sized, alternating, parallel layers of copper [$k = 220$ Btu/(h)(ft)(°F)] and glass [$k = 0.5$ Btu/(h)(ft)(°F)]. There are in the cube five layers of copper and five layers of glass. A question has arisen whether this cube would be a better conductor of heat (for a fixed overall Δt) if it were oriented so that heat was conducted parallel to the alternating layers or normal to them. Find the ratio of the two conduction rates.

18-11 Heat is generated in a fluid contained in a spherical vessel at the rate of 6×10^7 Btu/(h)(ft^3). The vessel has an inside diameter of 3 in. The temperature of the outer surface of the vessel is kept at a uniform temperature of 80°F by spraying it with water. How thick must the wall of the vessel be if the interior surface is to be at 600°F? The thermal conductivity of the metal in the wall is given by $k = 120 (1 - 5 \times 10^{-4}\ t)$ where t is in °F and k is in Btu/(h)(ft)(°F).

19

UNSTEADY-STATE HEAT CONDUCTION

Unsteady-state heat conduction is of importance to engineers in many circumstances. It may control the rate at which process equipment is brought to stable operating conditions, and it is also important in determining the processing time of many solid articles. For example, the curing time of objects made of molded plastic or rubber is often dependent on the time required to bring the center to some specified temperature without causing thermal damage to the material at the surface. There are also many applications of unsteady-state-conduction theory in the heat treating and casting of metals.

A slightly different class of problems is characterized by a periodic variation of temperature. Internal-combustion engines, compressors, and automatic-firing weapons generate heat periodically; the dissipation of this heat causes regular periodic temperature fluctuations in the surroundings. Another example is the effect of diurnal, atmospheric temperature variations on large structures such as bridges and on small structures such as growing plants.

Fundamental Equations

The basic differential equation for unsteady-state conduction in a solid is a special case of the differential energy balance which was derived in Chap. 8. The same equation applies to a stagnant, incompressible fluid. For constant thermal conductivity and no heat generation the differential energy balance was written

$$\frac{\partial t}{\partial \theta} = \frac{k}{\rho C_p}\left(\frac{\partial^2 t}{\partial x^2} + \frac{\partial^2 t}{\partial y^2} + \frac{\partial^2 t}{\partial z^2}\right) \tag{8-12}$$

Special cases which are of interest include the problem of unsteady-state conduction in only one direction, for which Eq. (8-12) reduces to

$$\frac{\partial t}{\partial \theta} = \frac{k}{\rho C_p}\frac{\partial^2 t}{\partial x^2} \tag{19-1}$$

and the case of steady-state conduction in three directions, where $\partial t/\partial \theta = 0$, and Eq. (8-12) becomes

$$\frac{\partial^2 t}{\partial x^2} + \frac{\partial^2 t}{\partial y^2} + \frac{\partial^2 t}{\partial z^2} = 0 \tag{19-2}$$

When Eq. (8-12) is written for steady-state conduction in only one direction, we obtain

$$\frac{d^2 t}{dx^2} = 0$$

which is readily integrated to give a linear relation between temperature and distance. This was shown earlier in Eq. (18-3), obtained by integrating the Fourier conduction equation.

The differential energy balance can also be obtained in cylindrical coordinates by writing an energy balance on a control volume which has the shape of a hollow cylinder. The equation obtained for no flow and no heat generation is[1]

$$\frac{\partial t}{\partial \theta'} = \frac{k}{\rho C_p}\left(\frac{\partial^2 t}{\partial r^2} + \frac{1}{r}\frac{\partial t}{\partial r} + \frac{1}{r^2}\frac{\partial^2 t}{\partial \theta^2} + \frac{\partial^2 t}{\partial z^2}\right) \tag{19-3}$$

If there is negligible axial conduction, $\partial^2 t/\partial z^2 = 0$, and if there is no variation of temperature with angular position, $\partial^2 t/\partial \theta^2 = 0$. The derivation of Eq. (19-3) is left as an exercise for the reader. Many engineering systems involve cylindrical shapes, and Eq. (19-3) and its simplifications are more useful in describing these systems than Eq. (8-12), which is expressed in rectangular coordinates.

[1] See Appendix 2, Table A-3.

Example 19-1 Problems of heating or cooling spheres are best solved using the differential energy balance written in spherical coordinates. Derive this equation for the case where there is no variation of temperature with angular position.

The center of the sphere is chosen as the origin of the coordinate system. At a radial distance r from the center, we have a spherical surface at a uniform temperature t. At distance $r + dr$ is another surface at temperature $t + dt$. These two surfaces can be considered as the boundaries of the control volume.

$$\text{Rate of heat flow into control volume} = -k(4\pi r^2)\frac{\partial t}{\partial r} \tag{1}$$

Rate of heat flow out of control volume

$$= -k[4\pi(r + dr)^2]\left[\frac{\partial t}{\partial r} + d\left(\frac{\partial t}{\partial r}\right)\right]$$

$$= -4\pi k(r^2 + 2r\,dr + dr^2)\left(\frac{\partial t}{\partial r} + \frac{\partial^2 t}{\partial r^2}\,dr\right) \tag{2}$$

The net heat flow in can be found by subtracting (2) from (1). After this is done, all second- and third-order differentials can be dropped (because they are negligible), giving

$$\text{Net heat flow} = 4\pi k\left(r^2\frac{\partial^2 t}{\partial r^2}\,dr + 2r\frac{\partial t}{\partial r}\,dr\right) \tag{3}$$

The rate of accumulation of energy in the control volume is

$$(4\pi r^2\,dr)\rho C_p\frac{\partial t}{\partial \theta} \tag{4}$$

which is equal to the net heat flow in.

Therefore we have

$$4\pi k\left(r^2\frac{\partial^2 t}{\partial r^2}\,dr + 2r\frac{\partial t}{\partial r}\,dr\right) = (4\pi r^2\,dr)\rho C_p\frac{\partial t}{\partial \theta} \tag{5}$$

from which we obtain the equation for unsteady conduction,[1]

$$\frac{\partial t}{\partial \theta} = \frac{k}{\rho C_p}\left(\frac{\partial^2 t}{\partial r^2} + \frac{2}{r}\frac{\partial t}{\partial r}\right) \qquad //// \tag{6}$$

[1] Compare Appendix 2, Table A-3.

Solutions of the Fundamental Equations[1]

The equations for unsteady conduction which have been presented are all partial differential equations. There are several methods of solving these equations. Some are susceptible to solution after an adroit transformation of the original variables into a new set of variables reduces the original partial differential equation into an ordinary differential equation. This technique has been illustrated earlier, in Chap. 11, in the analysis of laminar boundary-layer flow. Other common techniques are the methods of separation of variables and of integral transforms, usually Laplace transforms. The technique of transformation of variables is illustrated below in the case of heat conduction into a slab of infinite thickness. Solution by separation of variables is shown in Example 19-2 for conduction into a slab of finite thickness.

Conduction into a Plate of Infinite Thickness

The system to be considered is a solid, initially at a uniform temperature t_0 throughout, extending from $x = 0$ to $x = \infty$. The surface temperature at $x = 0$ is suddenly raised to and held at a higher temperature t_s. It is assumed that either because of insulation on the faces normal to the y and z coordinates or because the system is infinite in these directions, we need consider conduction in only the x direction. Thus, the conduction equation to be solved is Eq. (19-1).

To reduce it to a simpler form, we arbitrarily define a new variable n, which combines the independent variables θ and x in the expression

$$n = \frac{x}{\sqrt{4\alpha\theta}} \tag{19-4}$$

in which α is the thermal diffusivity $k/\rho C_p$. The derivatives in Eq. (19-1), when expressed in terms of the new variable n, become

$$\frac{\partial t}{\partial \theta} = -\frac{n}{2\theta}\frac{\partial t}{\partial n} \tag{19-5}$$

$$\frac{\partial^2 t}{\partial x^2} = \frac{1}{4\alpha\theta}\frac{\partial^2 t}{\partial n^2} \tag{19-6}$$

[1] Methods of solving the heat-conduction equations are presented in most textbooks on applications of mathematics in engineering. Three books that might be recommended are H. S. Mickley, T. K. Sherwood, and C. E. Reed, "Applied Mathematics in Chemical Engineering," 2d ed., McGraw-Hill Book Company, New York, 1957; C. R. Wylie, "Advanced Engineering Mathematics," 3d ed., McGraw-Hill Book Company, New York, 1966; and H. S. Carslaw and J. C. Jaeger, "Operational Methods in Applied Mathematics," 2d ed., Oxford University Press, London, 1947. Also recommended is V. Arpaci, "Conduction Heat Transfer," Addison-Wesley Publishing Company, Reading, Mass., 1966.

These are substituted in Eq. (19-1) to give the ordinary differential equation

$$\frac{d^2t}{dn^2} + 2n\frac{dt}{dn} = 0 \qquad (19\text{-}7)$$

This is further simplified by substituting a new variable $p = dt/dn$ to give

$$\frac{dp}{dn} + 2np = 0 \qquad (19\text{-}8)$$

which is readily integrated to yield

$$p = C_1 e^{-n^2} \qquad (19\text{-}9)$$

A second integration gives

$$t = C_1 \int_0^n e^{-n^2}\,dn + C_2 \qquad (19\text{-}10)$$

Two boundary conditions are available for evaluation of C_1 and C_2. The first to be used ($t = t_s$ at $x = 0$, $n = 0$) gives $C_2 = t_s$. The second ($t = t_0$, $\theta = 0$, $n = \infty$) results in

$$t_0 = C_1 \int_0^\infty e^{-n^2}\,dn + t_s \qquad (19\text{-}11)$$

$$= C_1 \frac{\sqrt{\pi}}{2} + t_s \qquad (19\text{-}12)$$

from which $C_1 = (-2/\sqrt{\pi})(t_s - t_0)$. Thus, the equation for the temperature distribution in the slab is

$$\frac{t_s - t}{t_s - t_0} = \frac{2}{\sqrt{\pi}} \int_0^{x/\sqrt{4\alpha\theta}} e^{-n^2}\,dn \qquad (19\text{-}13)$$

The right-hand side of Eq. (19-13) is known as the *Gauss error integral* or *probability function.* Equation (19-13) is often written as

$$\frac{t_s - t}{t_s - t_0} = \operatorname{erf}\frac{x}{\sqrt{4\alpha\theta}}$$

Example 19-2 A rubber sheet $\frac{1}{2}$-in thick is to be cured at 292°F for 50 min. If the sheet is initially at 70°F and heat is applied from both surfaces, find the time required for the temperature at the center of the sheet to reach 290°F. It can be assumed that the surfaces are brought to 292°F as soon as curing is begun and held at that temperature throughout the process. The thermal diffusivity $k/\rho C_p$ of rubber can be taken as 0.0028 ft²/h.

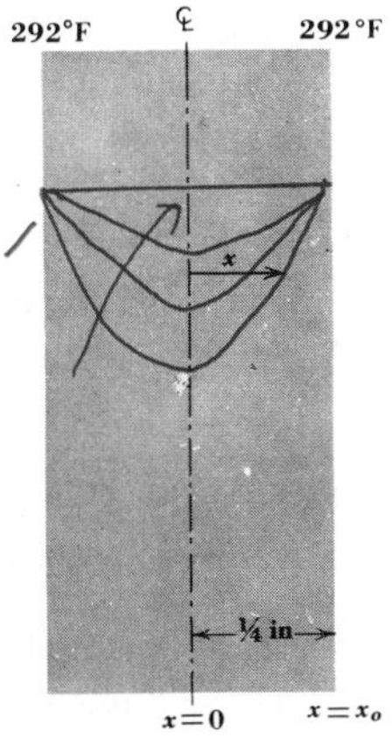

$$G.E \Rightarrow \frac{\partial t}{\partial \theta} = \alpha \frac{\partial^2 t}{\partial x^2} \quad \begin{cases} \text{I.C. } \theta = 0, \ t = t_0 \\ \text{B.C } x = \pm l \ \ t = t_s \\ \quad x = 0 \ \ \frac{dt}{dx} = 0 \end{cases}$$

- dimensionless variable

$$Y = \frac{t_s - t}{t_s - t_0} \qquad \tau = \frac{\alpha \theta}{x_0^2} \qquad \eta = \frac{x}{x_0}$$

$$\frac{\partial t}{\partial \theta} = \frac{\partial t}{\partial Y} \cdot \frac{\partial Y}{\partial \tau} \cdot \frac{\partial \tau}{\partial \theta} = (t_0 - t_s) \cdot \frac{\alpha}{x_0^2} \cdot \frac{\partial Y}{\partial \tau}$$

$$\frac{\partial t}{\partial x} = \frac{\partial t}{\partial Y} \cdot \frac{\partial Y}{\partial \eta} \frac{\partial \eta}{\partial x} = (t_0 - t_s) \cdot \frac{1}{x_0} \cdot \frac{\partial Y}{\partial \eta}$$

$$\frac{\partial}{\partial x}\left(\frac{\partial t}{\partial x}\right) = \frac{\partial}{\partial x}\left((t_0 - t_s)\frac{1}{x_0}\frac{\partial Y}{\partial \eta}\right) = \frac{t_0 - t_s}{x_0^2} \frac{\partial^2 Y}{\partial \eta^2}$$

⇒ G.E에 대입 ⇓ t=ts (2)

292°F ℄ 292°F

x

¼ in

x=0 x=xo

FIGURE 19-1
Unsteady-state heat conduction in a rubber sheet (Example 19-2).

The differential equation for unsteady heat conduction in one direction is

$$\frac{\partial t}{\partial \theta} = \frac{k}{C_p \rho} \frac{\partial^2 t}{\partial x^2} \tag{19-1}$$

The origin is chosen at the center of the sheet, so that x is the distance from the center, as shown in Fig. 19-1, and x_0 is the half thickness.

In solving partial differential equations it is usually convenient to define the variables so that they range between zero and 1 or between zero and infinity and are dimensionless. Therefore Eq. (19-1) will be expressed in terms of the following redefined variables:

$$Y = \frac{292 - t}{292 - 70} \qquad n = \frac{x}{x_o} \qquad \tau = \frac{k\theta}{C_p \rho x_o{}^2} \tag{1}$$

Equation (19-1) becomes

$$Y = Y(\tau, \eta)$$

$$Y = T(\tau) \cdot N(\eta) \rightarrow$$

$$N \cdot \frac{\partial T}{\partial \tau} = T \cdot \frac{d^2 N}{d\eta^2}$$

$$\frac{\partial Y}{\partial \tau} = \frac{\partial^2 Y}{\partial n^2} \tag{2}$$

as can be proved by substituting the new variables (1) in Eq. (2) and performing the indicated partial differentiations.

A solution to Eq. (2) is obtained by postulating that such a solution can be written as the product of two terms,

$$Y = \tau N \tag{3}$$

where τ is a function of only the quantity τ, in which time is the variable, and N

$$\frac{1}{T}\frac{dT}{d\tau} = \frac{1}{N} \cdot \frac{d^2 N}{d\eta^2} = -a^2 \rightarrow 6.9$$

is a function of only n, in which distance is the variable. If $Y = \tau N$ is a solution, it can be differentiated as stated in (2) and yields

$$N \frac{\partial \tau}{\partial \tau} = \tau \frac{\partial^2 N}{\partial n^2} \tag{4}$$

This equation can be rearranged to give only terms which depend on τ on the left side of the equation and only terms which are a function of n on the right side.

$$\frac{1}{\tau}\frac{\partial \tau}{\partial \tau} = \frac{1}{N}\frac{\partial^2 N}{\partial n^2} \tag{5}$$

Since time and distance are independent of each other in this problem, the left side of this equation is independent of distance and the right side is independent of time. Hence each side must be constant. This constant we shall call, for convenience, $-a^2$. Thus we can write two ordinary differential equations,

$$\frac{d\tau}{d\tau} + a^2\tau = 0 \tag{6}$$

and

$$\frac{d^2N}{dn^2} + a^2 N = 0 \tag{7}$$

The solution of Eq. (6) is simply

$$\tau = C_1 e^{-a^2\tau} \tag{8}$$

The student of differential equations will recognize Eq. (7) as a second-order, ordinary differential equation with constant coefficients. The solution is written as

$$N = C_2 \sin an + C_3 \cos an \tag{9}$$

Having obtained expressions for τ and N, we can now write the solution for the partial differential equation (2), which we originally postulated as Eq. (3).

$$\begin{aligned} Y &= \tau N \\ &= C_1 e^{-a^2\tau}(C_2 \sin an + C_3 \cos an) \end{aligned} \tag{10}$$

This equation contains four constants, C_1, C_2, C_3, and a; however, C_1 can be combined with C_2 and C_3, so that only three boundary conditions are needed to complete the solution. These are:

1 The system is symmetrical, so no heat is conducted across the center line ($x = 0$). Therefore, at $n = 0$, $\partial Y/\partial n = 0$.

2 At the surface ($x = x_o = \frac{1}{48}$), the temperature is constant at 292°F at all times. Thus, when $n = 1$, $Y = 0$.
3 Initially ($\theta = 0$), the entire sheet is at 70°F. Therefore, when $\tau = 0$, $Y = 1$.
Boundary condition 1 is used first.

$$\frac{\partial Y}{\partial n} = C_1 e^{-a^2\tau}(C_2 a \cos an - C_3 a \sin an)$$

When $n = 0$, $\sin an = 0$; however, $\cos an = 1$. Therefore

$$\frac{\partial Y}{\partial n} = aC_1 C_2 e^{-a^2\tau}$$

For this to be equal to zero and still preserve a general solution, the term C_2 must equal zero. After this simplification the general solution can be written

$$Y = Ae^{-a^2\tau} \cos an \qquad (11)$$

where A includes both C_1 and C_3.

Boundary condition 2 is next employed.

$$0 = Ae^{-a^2\tau} \cos a$$

This condition is realized for many values of a such as $\pi/2$, $3\pi/2$, $5\pi/2$, etc.

The substitution of one of these values for a in Eq. (11) would meet the requirement of the second boundary condition, but it is not possible to represent an arbitrary temperature distribution in the slab by a cosine curve. However, it is possible to represent an arbitrary function by an infinite series of cosine terms, provided each term is multiplied by a suitable coefficient. Therefore we can write the general solution as

$$Y = A_1 e^{-(\pi/2)^2\tau} \cos\left(\frac{\pi}{2}\right)n + A_2 e^{-(3\pi/2)^2\tau} \cos\left(\frac{3\pi}{2}\right)n + A_3 e^{-(5\pi/2)^2\tau} \cos\left(\frac{5\pi}{2}\right)n + \cdots + A_i e^{-[(2i-1)\pi/2]^2\tau} \cos\left[\frac{(2i-1)\pi}{2}\right]n + \cdots \qquad (12)$$

where i is an integer. It can be verified by differentiation and substitution that this sum still satisfies the partial differential equation and the first two boundary conditions.

The final step is to evaluate the constant which serves as a coefficient of each term in the series. This is done using the third boundary condition.

Equation (12) becomes

$$1 = A_1 \cos\left(\frac{\pi}{2}\right)n + A_2 \cos\left(\frac{3\pi}{2}\right)n + A_3 \cos\left(\frac{5\pi}{2}\right)n + \cdots$$

$$+ A_i \cos\left(\frac{2i-1}{2}\right)\pi n + \cdots \tag{13}$$

Both sides of this equation are multiplied by $\cos\,[(2i-1)/2]\pi n\, dn$ and integrated over the range of 0 to 1. The left-hand side becomes

$$\int_0^1 \cos\left(\frac{2i-1}{2}\right)\pi n\, dn = \frac{2}{2i-1}\frac{1}{\pi}\left[\sin\left(\frac{2i-1}{2}\right)\pi n\right]_0^1$$

$$= -\frac{2}{2i-1}\frac{1}{\pi}(-1)^i$$

The first term on the right-hand side of the equation can be integrated with the aid of a table of integrals to give

$$\int_0^1 A_1 \cos\left(\frac{\pi}{2}\right)n \cos\left(\frac{2i-1}{2}\right)\pi n\, dn$$

$$= A_1\left[\frac{\sin\,(i-1)\pi n}{(2i-2)\pi} + \frac{\sin\, i\pi n}{2i\pi}\right]_0^1$$

$$= A_1\left[\frac{\sin\,(i-1)\pi}{(2i-2)\pi} + \frac{\sin\, i\pi}{2i\pi}\right]$$

$$= 0$$

The integral is zero for all values of the integer i other than 1. The second term on the right-hand side vanishes in the same manner as the first, and so do all succeeding terms except the ith term. This becomes

$$A_i \int_0^1 \cos^2\left(\frac{2i-1}{2}\right)\pi n\, dn$$

$$= A_i \frac{2}{2i-1}\frac{1}{\pi}\left[\frac{\pi}{2}\frac{2i-1}{2}n + \tfrac{1}{4}\sin\,(2\pi)\frac{2i-1}{2}n\right]_0^1$$

$$= A_i \frac{2}{2i-1}\frac{1}{\pi}\left(\frac{\pi}{2}\frac{2i-1}{2} + 0 - 0 - 0\right)$$

$$= \frac{A_i}{2}$$

Thus Eq. (13) reduces to

$$-\frac{2}{2i-1}\frac{1}{\pi}(-1)^i = \frac{A_i}{2}$$

from which

$$A_i = \frac{-4(-1)^i}{(2i-1)\pi}$$

$$\left(\text{That is, } A_1 = \frac{4}{\pi},\ A_2 = -\frac{4}{3\pi},\ A_3 = \frac{4}{5\pi},\ \text{etc.}\right)$$

The general solution is therefore

$$Y = \frac{4}{\pi}e^{-(\pi/2)^2\tau}\cos\left(\frac{\pi n}{2}\right) - \frac{4e^{-(3\pi/2)^2\tau}}{3\pi}\cos\left(\frac{3\pi n}{2}\right) + \frac{4e^{-(5\pi/2)^2\tau}}{5\pi}\cos\left(\frac{5\pi n}{2}\right) - \cdots \tag{14}$$

Equation (14) written in compact form is

$$Y = \sum_{i=1}^{i=\infty} \frac{-2(-1)^i}{[(2i-1)/2]\pi} e^{-[(2i-1)\pi/2]^2\tau} \cos\left(\frac{2i-1}{2}\right)\pi n \tag{15}$$

Equation (15) is now applied to the specific example.

$$Y = \frac{292-290}{292-70} = 0.0090$$

$$n = 0$$

Therefore $\quad 0.0090 = \frac{4}{\pi}e^{-(\pi/2)^2\tau} - \frac{4}{3\pi}e^{-(3\pi/2)^2\tau} + \frac{4}{5\pi}e^{-(5\pi/2)^2\tau} - \cdots$

The solution for τ (which contains the time variable θ) must be obtained by trial and error. As a first approximation, only the first term on the right-hand side will be considered. This gives

$$\tau = -\left(\frac{2}{\pi}\right)^2 \ln\frac{(\pi)(0.0090)}{4}$$

$$= 2.01$$

$$\theta = \frac{C_p\rho}{k}x_o{}^2\tau$$

$$= \frac{2.01}{(0.0028)(48)^2}$$

$$= 0.313 \text{ h } (18.7 \text{ min})$$

It is necessary to check the relative magnitude of the terms in the series solution to see if all terms other than the first one are negligible.

When $\tau = 2.01$, the series becomes

$$Y = \frac{4}{\pi} e^{-4.97} - \frac{4}{3\pi} e^{-44.8} + \frac{4}{5\pi} e^{-124} - \cdots$$

$$= (1.27)(0.00694) - (0.424)(3.50 \times 10^{-20}) + \cdots$$

The validity of the approximation which employed only the first term of the series is apparent. Although not all series solutions converge so rapidly as in the example shown here, they frequently do so. This can often be established by a visual inspection of the exponents in the series (as in this case).

The temperature at any point in the rubber sheet at any instant can be determined by substitution of the appropriate values for n and τ in Eq. (15). Table 19-1 was prepared to show the development of the temperature profiles as heating proceeds. It is apparent that the results are contradictory at the surface at zero time. This is a result of the boundary conditions imposed on the mathematical solution.

The results are illustrated graphically in Fig. 19-2. At the center the temperature gradient is zero since the slab is being heated uniformly from both sides. The same condition would result if a sheet of half the thickness of the sheet in the example were heated from one side only, while the other side was covered with a layer of perfect insulation. ////

Although the preceding example is rather lengthy, it is a typical illustration of the application of the technique of separation of variables to a problem in unsteady-

Table 19-1 TEMPERATURE DISTRIBUTION IN RUBBER SHEET (Example 19-2)

Time elapsed, min	$\frac{4}{\pi} e^{-(\pi/2)^2\tau}$	Temperature in sheet, °F				
		$n = 0$ $\cos \frac{n\pi}{2} = 1$ (center line)	$n = \frac{1}{4}$ $\cos \frac{n\pi}{2} = 0.924$	$n = \frac{1}{2}$ $\cos \frac{n\pi}{2} = 0.707$	$n = \frac{3}{4}$ $\cos \frac{n\pi}{2} = 0.383$	$n = 1$ $\cos \frac{n\pi}{2} = 0$ (surface)
0	1	70	70	70	70	
1	9.76×10^{-1}	75	92	139	209	292
5	3.37×10^{-1}	217	223	239	263	292
10	9.02×10^{-2}	272	273	278	284.3	292
20	6.37×10^{-3}	290.6	290.7	291	291.5	292
30	4.52×10^{-4}	291.9	291.9	291.9	292	292
40	3.19×10^{-5}	292	292	292	292	292

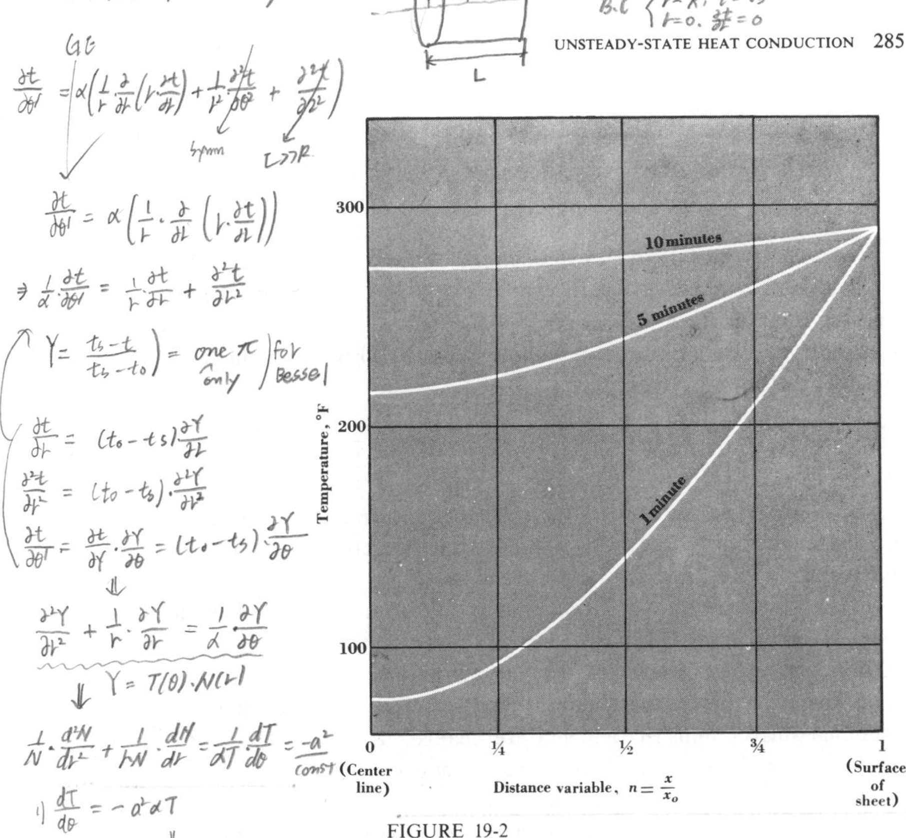

FIGURE 19-2
Illustration of temperature profiles calculated in Example 19-2.

state heat transfer. The mathematical manipulations were shown to be rather elementary. Actually, the student who has studied the use of Fourier series will recognize that the constant A_i, which was determined rather laboriously in the example, is a Fourier constant and can be found with much less effort when the appropriate mathematical techniques are known.

Solutions to unsteady-state conduction problems involving flat, parallel surfaces usually contain series of sine or cosine terms. On the other hand, systems with cylindrical boundaries generally give rise to expressions which are known as *Bessel functions* (called in German *Zylinderfunktionen*). These functions are available in tabulated form.[1]

If a solid cylinder of length sufficient to eliminate end effects receives heat uniformly from its surface, which is held at constant temperature, the differential

[1] E. Jahnke and F. Emde, "Tables of Functions," Dover Publications, New York, 1945; W. Flügge, "Four-Place Tables of Transcendental Functions," McGraw-Hill Book Company, New York, 1954.

equation relating temperature, time, and radial position derived from Eq. (19-3) is

$$\frac{\partial t}{\partial \theta} = \frac{k}{\rho C_p}\left(\frac{\partial^2 t}{\partial r^2} + \frac{1}{r}\frac{\partial t}{\partial r}\right) \tag{19-14}$$

This equation, when treated in the manner outlined in Example 19-2, gives rise to two solutions, one containing an exponential time variable and the other a Bessel function in which the argument contains the radius variable. After these solutions have been combined, the general solution takes the form of an infinite series in order to satisfy the boundary conditions.

$$\frac{t_s - t}{t_s - t_o} = \frac{2}{R}\sum_{i=1}^{i=\infty} \frac{e^{-a_i^2 k\theta/\rho C_p}}{a_i J_1(a_i R)} J_0(a_i r) \tag{19-15}$$

J_0 and J_1 are Bessel functions of the first kind (zero and first order, respectively), and a_i is the ith root of $J_0(a_i R) = 0$. The initial temperature is t_o, the surface temperature (constant) is t_s, and the radius of the rod is R.

The solutions to many common problems in heat conduction have been presented graphically. Typical plots are shown in Figs. 19-3, 19-4, and 19-5. Examination of the graphs of the various authors reveals the presence of the same groups used in the solutions given above. The ordinate is usually the fractional unaccomplished temperature change, $(t_s - t)/(t_s - t_0)$; the abscissa is the group $\alpha\theta/x_0^2$ (or $\alpha\theta/r_0^2$ for cylinders and spheres). This particular dimensionless group is known as the *Fourier number*. The distance variable n $(x/x_0$ or $r/r_0)$ is a parameter within each family of curves. Each family of curves is determined by the boundary conditions, i.e., whether or not a convective resistance (represented by a heat transfer coefficient h) exists at the surface. The quantity m characterizing each family of curves is the dimensionless group k/hx_0 (or k/hr_0 for cylinders and spheres); this group is the inverse of the Biot number hx_0/k.

Example 19-3 Find the solution to the problem given in Example 19-2 using Fig. 19-3.

$$\frac{\alpha\theta}{x_0^2} = \frac{0.0028\theta}{(1/48)^2} = 6.44\theta$$

$$\frac{t_s - t}{t_s - t_0} = \frac{292 - 290}{292 - 70} = 0.0090$$

$$n = \frac{x}{x_0} = 0$$

$$m = \frac{k}{hx_0} = 0$$

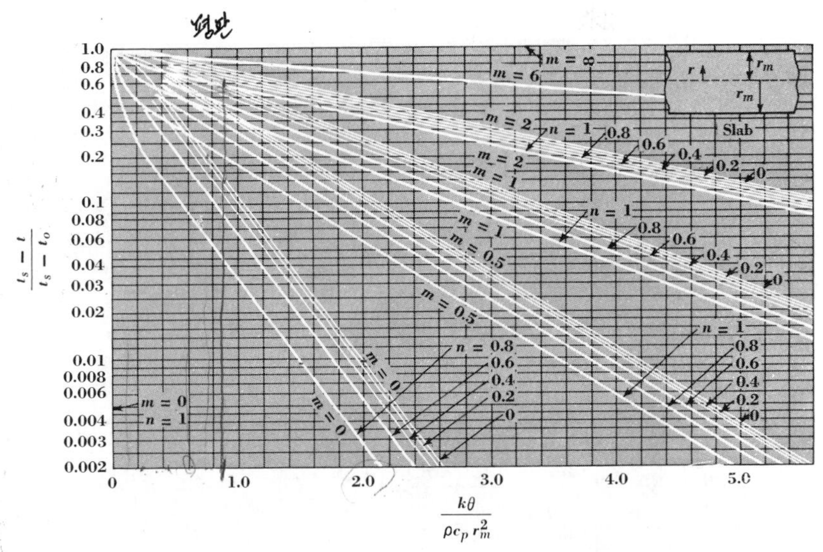

FIGURE 19-3
Gurney-Lurie chart[1] for large slab.

The modulus m is zero for this case because the assumption of constant surface temperature implies that surface resistance to heat transfer is negligible ($h = \infty$). From Fig. 19-3 we obtain

$$\frac{\alpha\theta}{x_0^2} = 2.0$$

Therefore

$$\theta = \frac{2.0}{6.44}$$

$$= 0.31 \text{ h } (19 \text{ min}) \qquad ////$$

Newman's Rule

The solutions discussed to this point have been solutions to differential equations containing three variables. Obviously, not all cylinders and plates encountered in industry have sufficient length-thickness ratios so that they can be treated as infinitely wide or long. Fortunately, the technique for solving many problems of systems with finite dimensions in all directions is rather simple. This technique is

[1] *Ind. Eng. Chem.*, **15**:1170 (1923).

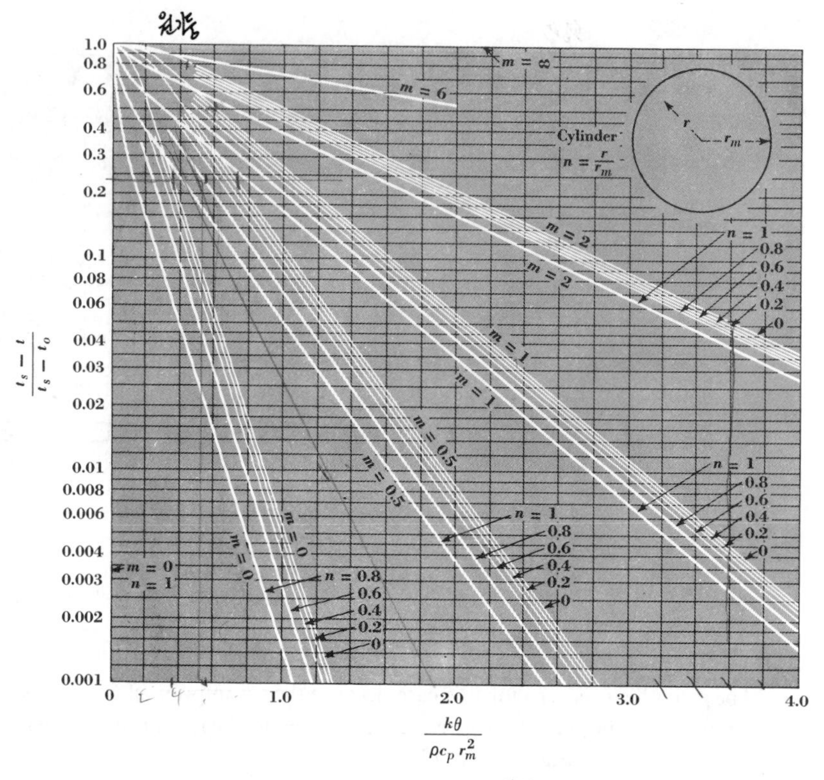

FIGURE 19-4
Gurney-Lurie chart for long cylinder.

often called *Newman's rule.* Newman has shown[1] that if a brick-shaped object is heated or cooled, the general solution describing temperature as a function of time θ and three distance variables x, y, and z can be written as

$$Y = Y_x Y_y Y_z$$

where $Y_x = f_1(x,\theta)$ is the equation already derived for one-dimensional unsteady-state conduction in Example 19-2. $Y_y = f_2(y,\theta)$ and $Y_z = f_3(z,\theta)$ are equations for conduction in the y and z directions, respectively, and are identical with Y_x in all respects, except that the term representing one-half the thickness in the x direction is replaced by the half-slab thickness in the y and z directions.

[1] A. B. Newman, *Trans. Am. Inst. Chem. Eng.*, **27**:310 (1931); H. S. Carslaw and J. C. Jaeger, "Conduction of Heat in Solids," 2d ed., Oxford University Press, London, 1959.

Example 19-4 Find the time for the temperature at the center of a cube of rubber ($\frac{1}{2}$-in side) to reach 290°F under the same conditions as Examples 19-2 and 19-3.

Since the dimensions are the same in all directions, $Y_x = Y_y = Y_z$. Therefore $Y = Y_x^{\,3}$.

Since

$$Y = 0.0090$$

$$Y_x = 0.208$$

As before, $x/x_0 = 0$ and $k/hx_0 = 0$. From Fig. 19-3

$$\frac{\alpha\theta}{x_0^{\,2}} = 0.73$$

$$= 6.44\theta$$

$$\theta = \frac{0.73}{6.44}$$

$$= 0.113 \text{ h (7 min)}$$

The basis for the validity of Newman's Rule can be demonstrated by a brief extension of the analysis given in Example 19-2. Assume that, instead of one-dimensional conduction in a thin sheet of thickness $2x_0$, the problem in Example 19-2 involved three-dimensional conduction in a brick of dimensions $2x_0$, $2y_0$, and $2z_0$. The brick is initially at some uniform temperature until the faces normal to the x direction are suddenly raised to some new, higher temperature and held at that condition. At the same instant, the faces normal to the y direction are raised to and then held at a second temperature and the faces normal to the z direction are raised to and held at a third temperature.

The differential equation for unsteady heat in three directions is

$$\frac{\partial t}{\partial \theta} = \frac{k}{C_p \rho}\left(\frac{\partial^2 t}{\partial x^2} + \frac{\partial^2 t}{\partial y^2} + \frac{\partial^2 t}{\partial z^2}\right) \tag{8-12}$$

This can be rewritten in the manner of Eq. (2) in Example 19-2.

$$\frac{\partial Y}{\partial \tau} = \frac{\partial^2 Y}{\partial x^2} + \frac{\partial^2 Y}{\partial y^2} + \frac{\partial^2 Y}{\partial z^2} \tag{1}$$

A solution to the problem given above is postulated as follows

$$Y = \bar{\tau} Y_x Y_y Y_z \tag{2}$$

in which $\bar{\tau}$ is a function only of τ and Y_x, Y_y, and Y_z are functions, respectively, of x, y, and z. Differentiation of Eq. (2) as indicated in Eq. (1) gives

$$Y_x Y_y Y_z \frac{\partial \bar{\tau}}{\partial \tau} = \bar{\tau} Y_y Y_z \frac{\partial^2 Y_x}{\partial x^2} + \bar{\tau} Y_x Y_z \frac{\partial^2 Y_y}{\partial y^2} + \bar{\tau} Y_x Y_y \frac{\partial^2 Y_z}{\partial z^2}$$

which is rearranged to give

$$\frac{1}{\bar{\tau}} \frac{\partial \bar{\tau}}{\partial \tau} = \frac{1}{Y_x} \frac{\partial^2 Y_x}{\partial x^2} + \frac{1}{Y_y} \frac{\partial^2 Y_y}{\partial y^2} + \frac{1}{Y_z} \frac{\partial^2 Y_z}{\partial z^2} \tag{3}$$

Each part of Eq. (3) is a function of one of the four independent variables and hence, it is reasoned, each of the four terms is a constant, chosen for convenience as follows:

$$\frac{1}{Y_x} \frac{\partial^2 Y_x}{\partial x^2} = -a_1^2$$

$$\frac{1}{Y_y} \frac{\partial^2 Y_y}{\partial y^2} = -a_2^2$$

$$\frac{1}{Y_z} \frac{\partial^2 Y_z}{\partial z^2} = -a_3^2$$

and

$$\frac{1}{\bar{\tau}} \frac{\partial \bar{\tau}}{\partial \tau} = -(a_1^2 + a_2^2 + a_3^2)$$

These four expressions can be written as ordinary differential equations for which the solutions are

$$\bar{\tau} = C_1 e^{-(a_1^2 + a_2^2 + a_3^2)\tau} \tag{4}$$

$$Y_x = C_2 \sin a_1 x + C_3 \cos a_1 x \tag{5}$$

$$Y_y = C_4 \sin a_2 y + C_5 \cos a_2 y \tag{6}$$

$$Y_z = C_6 \sin a_3 z + C_7 \cos a_3 z \tag{7}$$

These individual solutions are combined as indicated in Eq. (19-8) to give the general solution

$$Y = C_1 e^{-(a_1^2 + a_2^2 + a_3^2)\tau} (C_2 \sin a_1 x + C_3 \cos a_1 x)$$
$$(C_4 \sin a_2 y + C_5 \cos a_2 y)(C_6 \sin a_3 z + C_7 \cos a_3 z) \tag{8}$$

////

This general solution for three-dimensional conduction can thus be seen to consist of the products of each of the one-dimensional solutions. The general solution will not

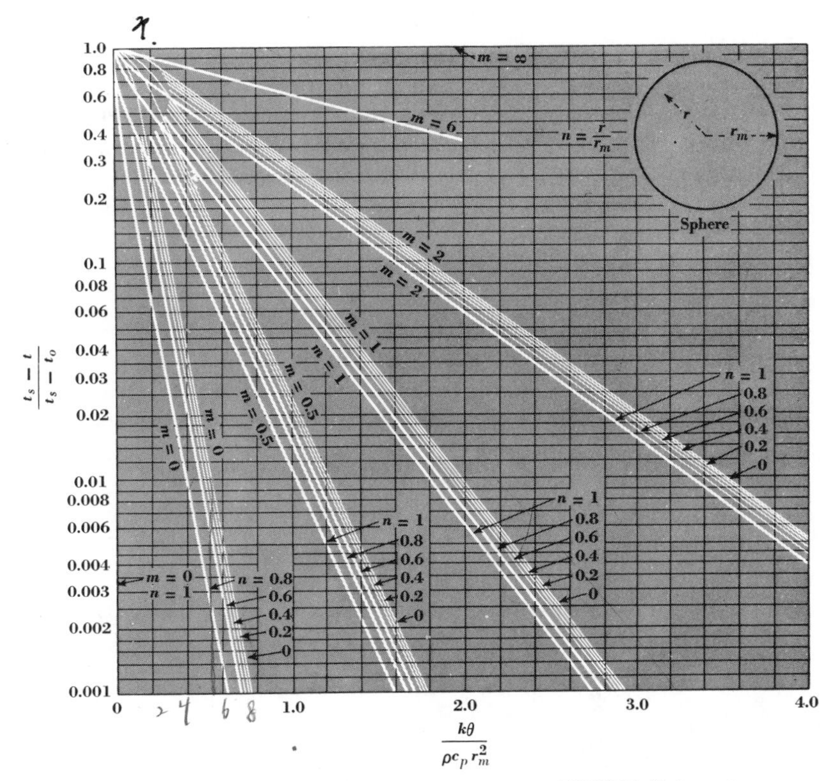

FIGURE 19-5
Gurney-Lurie chart for spheres.

be pursued further here other than to observe that if each distance variable has boundary conditions of the same form, but is otherwise independent of the other distance variables, the forms of each of the three solutions will be the same.

PROBLEMS

19-1 A solid, steel cylinder 0.05 m in diameter is at an initial temperature of 700°C throughout when it is suddenly quenched in water at 50°C. If the surface of the steel is assumed to fall immediately to 50°C, find the time (in seconds) required for the temperature at the center of the cylinder to reach 200°C. The cylinder can be assumed to have a large ratio of length to diameter.

19-2 A cylinder of the same dimensions as that in Prob. 19-1 is quenched under the same conditions. However, the resistance to heat transfer at the surface is to be accounted for by using an average surface heat-transfer coefficient of 7100 $W/m^2 \cdot K$. Find the time for the center of the cylinder to reach 200°C.

19-3 A concrete beam (8 by 12 in) is exposed to a temperature of 1500°F in a burning building. Determine the exposure time required for the temperature at the center of the beam to reach 600°F if the beam is at an initial temperature of 70°F. The thermal diffusivity of concrete can be taken as 0.022 ft^2/h.

19.1

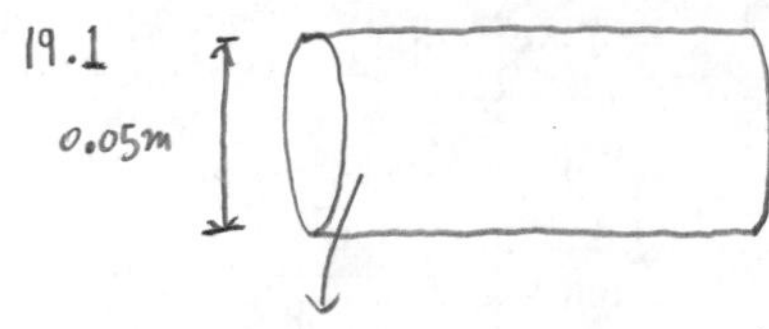

$t_o = 700°C$
$t_s = 50°C$
중심온도 200°C　L >> R.
⇓
열이 r방향으로만 전달
⟨t_s는 θ>0 -200°C⟩

$K_{steel} = 44.8 W/m \cdot k$ (from 268 (table A-11))

$C_p = 502 J/kg \cdot k$ (P807). $\rho = 7800 kg/m^3$.

$n = \frac{r}{r_m} = 0$

$m = \frac{K}{h \cdot r_m} = 0$ (∵ h → ∞ : 곧바로 표면온도가 변하고, 그온도 유지)

$Y = \frac{t_s - t}{t_s - t_o} = \frac{50-200}{50-700} = 0.23 (Y)$　$x = 0.35 = \frac{k\theta}{\rho \cdot C_p \cdot r_m^2}$

→ $\theta = \frac{(0.35)\cdot(7800)\cdot(502)\cdot(0.025)^2}{44.8}$

$= 19.1 sec.$

19.2　$h = 7100 w/m \cdot k$

$m = \frac{K}{h \cdot r_m} = \frac{44.8}{(7100)\cdot(0.025)} = 0.25$. - using interpolation:

$x = \frac{K\theta}{\rho \cdot C_p \cdot r_m^2} = 0.55$　$\theta = 30.0 (sec)$

x　0.73　0.35　0.55　0　0.25　0.5　m

20

NUMERICAL, GRAPHICAL, AND ANALOG METHODS IN THE ANALYSIS OF HEAT CONDUCTION

Many problems in steady- and unsteady-state conduction are difficult to solve analytically. The differential equations, based on the differential energy balance, can be written readily enough. However, the mathematical boundary conditions are difficult to apply when the physical boundaries of the object are complex and when the temperatures at these boundaries are not uniform. In general, few analytical solutions exist for heat-transfer systems which do not possess some elements of symmetry in both shape and temperature.

When the analytical solutions cannot be obtained, the engineer may resort to an alternative method such as numerical or graphical analysis, or he may construct an electrical or hydraulic analog of the system and make direct measurements of its performance. In this chapter, we shall examine several numerical and graphical methods and, at the end of the chapter, we shall discuss briefly the use of analogs.

It should not be assumed that the numerical and graphical solutions are necessarily less accurate than the analytical solutions. In fact, many (though not all) of them can be made as accurate as desired by mere repetition of routine steps which, if performed an infinite number of times, would give an exact solution. This characteristic is often an advantage, because the accuracy required of a solution is usually known beforehand, and the solution can be stopped when this is reached.

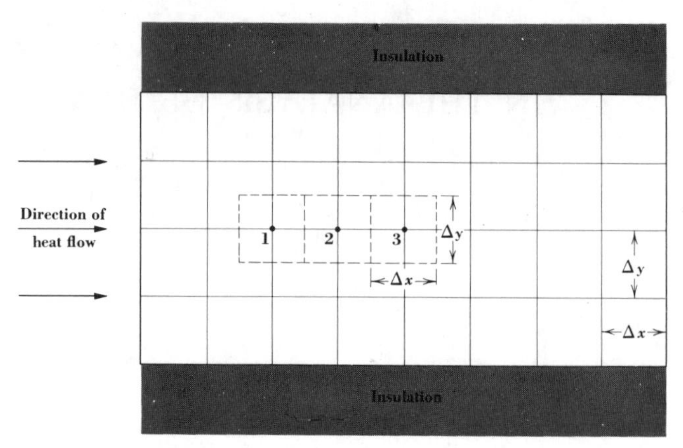

FIGURE 20-1
Relaxation technique applied to one-dimensional conduction.

An analytical solution, on the other hand, must usually be carried out to the end before any answer whatsoever is obtained.

The principal disadvantage of numerical methods is that they are tedious to perform, but with the use of a high-speed digital computer, this disadvantage is overcome. Sometimes numerical procedures, however, do not readily converge to the correct result, no matter how many iterations are performed.

Steady-State Conduction

A number of methods of solving steady-state conduction problems are described by Jakob.[1] We shall consider only the relaxation technique, which is a simple but powerful tool. There are applications of this method in many branches of engineering and science,[2] but we shall restrict our discussion to its use in solving heat-conduction problems.

We shall consider first the application of the relaxation technique in one-dimensional heat flow. The system shown in Fig. 20-1 is a solid of uniform thermal conductivity, insulated on four sides so that heat is conducted in only the x direction. A square grid is drawn on the figure which represents the solid. The sides of the squares, Δy and Δx, are equal; a thickness Δz in the z direction will be

[1] M. Jakob, "Heat Transfer," vol. 1, John Wiley & Sons, Inc., New York, 1949.

[2] R. V. Southwell, "Relaxation Methods in Engineering Science," Oxford University Press, London, 1940. See also V. Arpaci, "Conduction Heat Transfer," Chap. 9, Addison-Wesley Publishing Company, Reading, Mass., 1966.

assumed. As the heat in the element surrounding point 1 flows into the block surrounding point 2, the rate of heat flow will be

$$q_{12} = k\,\Delta z\,\Delta y\,\frac{t_1 - t_2}{\Delta x} \tag{20-1}$$

The rate of heat flow from the block of solid around point 2 to the block around point 3 will be

$$q_{23} = k\,\Delta z\,\Delta y\,\frac{t_2 - t_3}{\Delta x} \tag{20-2}$$

The rate at which heat accumulates at point 2 is

$$q_{12} - q_{23} = k\,\Delta z(t_1 + t_3 - 2t_2) \tag{20-3}$$

since $\Delta y = \Delta x$. Equation (20-3) can also be written as

$$\frac{q_{12} - q_{23}}{k\,\Delta z} = t_1 + t_3 - 2t_2 \tag{20-4}$$

If the system is at steady state, q_{12} equals q_{23}, and Eq. (20-4) can be rearranged to give

$$t_2 = \frac{t_1 + t_3}{2} \tag{20-5}$$

If an arbitrary temperature distribution is chosen for all the points in the system, it can be tested for correctness using Eq. (20-5). If the temperature t_2 is in error, it may be revised by adding or subtracting $\frac{1}{2}[(q_{12} - q_{23})/k\,\Delta z]$ [as found from Eq. (20-4) with the initial values of t_1, t_2, and t_3] to the temperature t_2. The quantity $(q_{12} - q_{23})/(k\,\Delta z)$ has the dimension of temperature, and the addition of the amount indicated will, of necessity, make Eq. (20-5) valid at that point. If all but one of the temperatures at the nodes of the grid are correct, then alterations at this point will, of course, destroy the consistency among the temperatures at the remaining points. However, the magnitude of the correction diminishes by halves with succeeding points and, if not too great initially, soon vanishes as a significant quantity. The relaxation technique is illustrated in Example 20-1.

Example 20-1 A solid rod conducts heat from a heat source at 800°F to a heat sink at 200°F. The sides of the rod are insulated as shown in Fig. 20-2. Find the temperature distribution in the rod.

We know by the use of Fourier's heat-conduction equation that the temperature distribution in the rod will be a linear function of position. However, we shall assume for the purposes of illustrating the technique that the tempera-

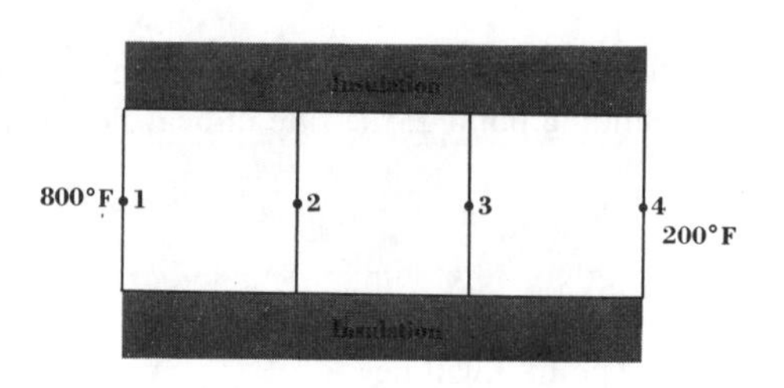

FIGURE 20-2
Relaxation model (Example 20-1).

ture at point 2 is 700°F and at point 3 is 300°F. The calculation can be made as shown in Table 20-1. The quantity $(q_{12} - q_{23})/(k\ \Delta z)$ will be designated q'_2 and, similarly, $(q_{23} - q_{34})/(k\ \Delta z)$ will be called q'_3. These quantities are sometimes referred to as *residuals*.

Using the initial temperature distribution (step 0) the quantity q'_2 is calculated using Eq. (20-4) and q'_3 is found from a similar expression. Half of

Table 20-1 ONE-DIMENSIONAL RELAXATION

Step	Point 1	Point 2		Point 3		Point 4
	t_1	t_2	q'_2	t_3	q'_3	t_4
0	800	700	−300	300	300	200
1	800	550	150	450	−150	200
2	800	625	−75	375	75	200
3	800	587	39	413	−39	200
4	800	607	−21	393	21	200
5	800	596	12	404	−12	200
6	800	602	−6	398	6	200
7	800	599	3	401	−3	200
8	800	600.5	...	399.5	...	200

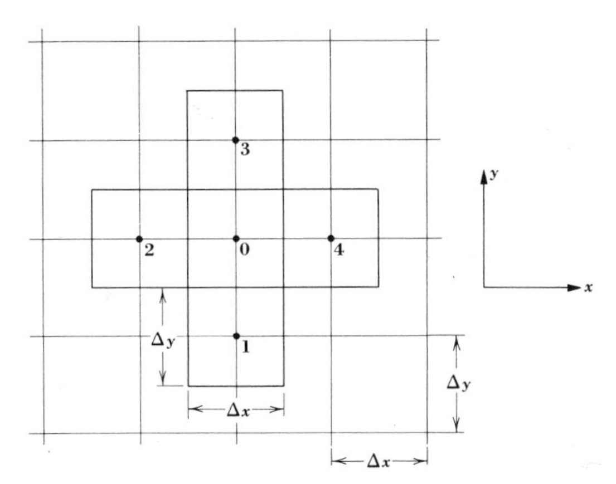

FIGURE 20-3
Relaxation techniques applied to two-dimensional conduction.

each of these quantities is then added or subtracted, as the sign indicates, to the respective temperatures t_2 and t_3 (step 1). Using the new temperature distribution, new values of q'_2 and q'_3 are calculated and new values of t_2 and t_3 once more computed (step 2). The repetition of this simple process gives figures nearly equal to the correct temperatures after eight steps. ////

The procedure shown in Example 20-1, although simple to describe, can be replaced by more sophisticated procedures[1] which can shorten the number of steps. However, the principles are the same as illustrated in Table 20-1.

The use of relaxation in two-dimensional heat flow is illustrated with reference to Fig. 20-3.

We consider the solid to be conducting heat in the x and y directions but not in the z direction. If the solid has a thickness Δz in the z direction, we can write

$$q_{10} = k \, \Delta z \, \Delta x \frac{t_1 - t_o}{\Delta y} \tag{20-6}$$

$$q_{20} = k \, \Delta z \, \Delta y \frac{t_2 - t_o}{\Delta x} \tag{20-7}$$

$$q_{03} = k \, \Delta z \, \Delta x \frac{t_o - t_3}{\Delta y} \tag{20-8}$$

[1] F. Kreith, "Principles of Heat Transfer," 2d ed., International Textbook Company, Scranton, Pa., 1966.

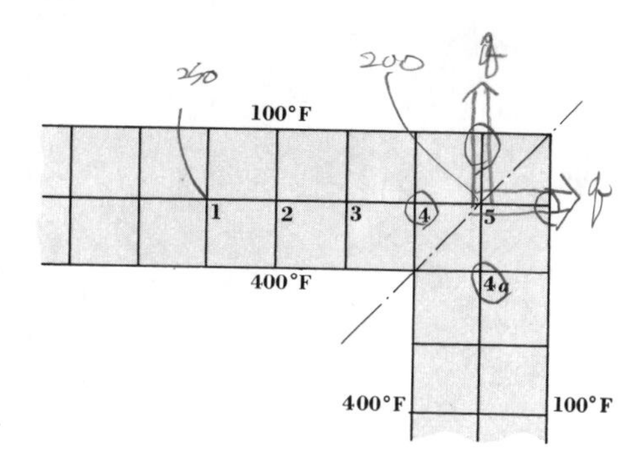

FIGURE 20-4
Relaxation model (Example 20-2).

$$q_{04} = k \, \Delta z \, \Delta y \frac{t_o - t_4}{\Delta x} \qquad (20\text{-}9)$$

If we subtract Eqs. (20-8) and (20-9) from Eqs. (20-6) and (20-7), we obtain an equation for the net heat flux into the element.

$$q_{10} + q_{20} - q_{03} - q_{04} = k \, \Delta z(t_1 + t_2 + t_3 + t_4 - 4t_o) \qquad (20\text{-}10)$$

As before, Δx equals Δy, so these quantities have been canceled. We rearrange Eq. (20-10) and define a term q'_o as

$$q'_o = \frac{q_{10} + q_{20} - q_{03} - q_{04}}{k \, \Delta z} = t_1 + t_2 + t_3 + t_4 - 4t_o \qquad (20\text{-}11)$$

If the heat is being transferred at steady state, q'_o is zero and t_o is the arithmetic average of the temperatures at the four surrounding points.

An alternate approach to this problem, more suitable to implementation on the computer, is to set the residual q'_o quantities at all nodal points equal to zero. This generates a system of linear equations for the temperatures at the nodal points which can be solved using either direct or iterative methods.

Example 20-2 The wall of a furnace has an inside temperature of 400°F and an outside temperature of 100°F. Find the centerline temperature of the wall.

The wall is shown in Fig. 20-4 with a square grid with sides equal to half the wall thickness. Because of the plane of symmetry through the corner, the temperature at point 4 will be the same as at point 4*a*. At a considerable distance from the corner the mid-wall temperature will obviously be 250°F. If we choose all the centerline temperatures as 250°F, we shall introduce some error, because the temperature at point 5 would seem to be lower. Therefore we choose $t_5 = 200$°F and $t_1 = t_2 = t_3 = t_4 = 250$°F. The solution to the problem is shown in Table 20-2.

The procedure followed is to use the initial temperature distribution to calculate the q' values. For example, we find in step 0 that

$$q'_5 = 250 + 100 + 100 + 250 - (4)(200)$$
$$= -100$$

and

$$q'_4 = 250 + 100 + 200 + 400 - (4)(250)$$
$$= -50$$

and so on.

We then add or subtract one-fourth the value of q' to or from each corresponding temperature to obtain the new temperature profile, with which we repeat the calculations until the values of q' are negligible. ////

The accuracy of the temperatures found in Example 20-2 can be increased by use of a smaller grid. If the boundary of the system had been curved, a smaller grid would, in fact, have been necessary. However, the temperatures found with the large grid are a good basis for guessing the initial temperature distribution in the smaller grid. In Example 20-2 the calculations are simplified by the fact that although there is heat flow in two directions, the unknown temperatures extend only in one direction. With a smaller grid for this system, the temperatures would have to be "relaxed" in both directions. In this case the tabular form for calculating the temperatures, shown in the example, would be rather cumbersome. A more

Table 20-2 TWO-DIMENSIONAL RELAXATION

Step	Point									
	1		2		3		4		5	
	t_1	q'_1	t_2	q'_2	t_3	q'_3	t_4	q'_4	t_5	q'_5
0	250	0	250	0	250	0	250	-50	200	-100
1	250	0	250	0	250	-13	237	-23	175	-26
2	250	0	250	-3	247	-7	231	-9	168	-10
3	250	-1	249	-1	245	-2	229	-6	165	-2
4	250	-1	249	-2	244	0	227	0	164	-2

convenient method would be to sketch the shape of the solid section on a large sheet of paper and to construct beside each point a table for calculations of t and q'. The physical locations of the points thus are much more obvious in the calculations, and there is less chance for error. If any arithmetic error is discovered, the calculations need not be repeated but merely continued. The error will eventually be eliminated, and the only penalty will be in the increased number of routine calculations that must be made.

Other factors affecting heat transfer can be treated by relaxation calculations. For example, if heat is being generated in the solid at the rate of q_r Btu/(h)(ft^3), the heat-balance equation, comparable with Eq. (20-11) for two-dimensional conduction, becomes

$$q'_o = t_1 + t_2 + t_3 + t_4 - 4t_o + \frac{q_r(\Delta x)^2}{k} \qquad (20\text{-}12)$$

The initial temperature distribution is guessed and then relaxed by revising t_o values so that q'_o in Eq. (20-12) is zero.

Three-dimensional conduction can also be analyzed by relaxation techniques. A three-dimensional grid gives six points surrounding each interior nodal point. The relaxation equation (without generation) is

$$q'_o = t_1 + t_2 + t_3 + t_4 + t_5 + t_6 - 6t_o \qquad (20\text{-}13)$$

In the cases we have chosen, the temperatures at the boundaries of the system have been known. If, however, the solid surfaces are in contact with a fluid, we are likely to know instead the temperature of the fluid. This type of problem can be solved by relaxation methods but requires the use of a coefficient of heat transfer, which we have not yet discussed. The procedure for doing this is rather simple, and it will be left to the student to derive the appropriate relaxation equation after he has learned something about convective coefficients.

Although a knowledge of the temperature distribution in a solid may be desired for some other purpose, it is frequently used to calculate the heat flux. This can be found from the temperature distribution by simply multiplying the temperature gradient normal to any isothermal plane in the solid by thermal conductivity, as indicated by the Fourier conduction equation.

Unsteady-State Conduction

Numerical methods have been developed for solving problems of unsteady-state heat conduction, and graphical techniques have been suggested which are equivalent to the numerical solutions. We shall consider only sufficient cases to make clear the

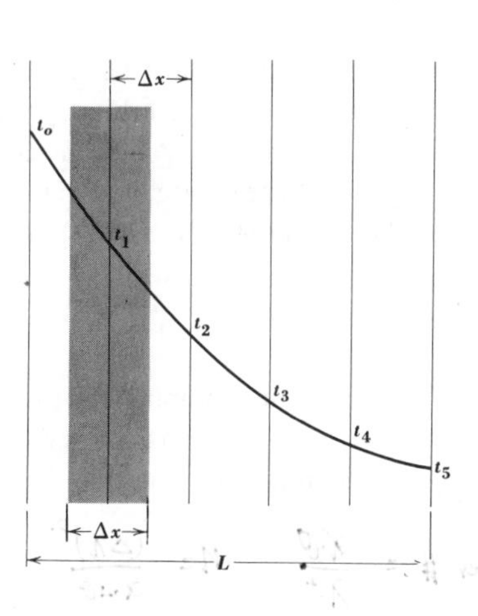

FIGURE 20-5
Unsteady-state heat conduction.

methods employed; for the numerous specialized applications of the method, we refer the student to literature in this field.[1]

The numerical method suggested by Dusinberre will be applied to a slab having a uniform cross-sectional area A and a thickness L. The initial temperature profile for the system is illustrated in Fig. 20-5. The total thickness L is divided into a number of sections of equal thickness Δx.

At an instant of time the temperature profile is as shown by the line through temperatures t_o, t_1, t_2, etc. A heat balance is written for the shaded element of thickness Δx, which has plane 1 as its center. The temperature gradient at the left face of the element is taken as $(t_o - t_1)/\Delta x$; thus the rate of heat flow is $kA[(t_o - t_1)/\Delta x]$. The temperature gradient at the right side of the element is taken as $(t_1 - t_2)/\Delta x$, so that the rate of heat flow out is $kA[(t_1 - t_2)/\Delta x]$. The rate of accumulation of energy in the element, expressed in finite differences, is $[A\,\Delta x\,\rho C_p(t_1' - t_1)]/\Delta\theta$, in which $t_1' - t_1$ is the increase in the temperature of the element during time, $\Delta\theta$. These three terms can be written as an equation for the conservation of energy,

$$\frac{kA(t_o - t_1)}{\Delta x} - \frac{kA(t_1 - t_2)}{\Delta x} = \frac{A\,\Delta x\,\rho C_p(t_1' - t_1)}{\Delta\theta}$$

which simplifies to

$$t_1' - t_1 = \frac{k}{\rho C_p}\frac{\Delta\theta}{(\Delta x)^2}(t_o + t_2 - 2t_1) \qquad (20\text{-}14)$$

[1] G. M. Dusinberre, "Heat Transfer Calculations by Finite Differences," 2d ed., International Textbook Company, Scranton, Pa., 1961; G. A. Hawkins and J. T. Agnew, The Solution of Transient Heat Conduction Problems by Finite Differences, *Purdue Univ. Eng. Bull., Res. Ser.*, no. 98, 1946.

Numerical values must be chosen for both Δx and $\Delta\theta$. Although it might appear that these choices are arbitrary, there are limitations for reasons given below. The distance increment Δx is usually chosen to give some whole number of increments. The choice of $\Delta\theta$ remains. If the quantity $(\rho C_p/k)(\Delta x)^2/\Delta\theta$ is expressed by the symbol M, Eq. (20-14) becomes

$$t_1' = \frac{t_o + t_1(M-2) + t_2}{M} \tag{20-15}$$

The choice of a number for the modulus M therefore fixes the increment of time $\Delta\theta$. If M is chosen as 2, Eq. (20-15) reduces to $t_1' = (t_o + t_2)/2$ and the calculation of temperatures is reduced to the taking of arithmetic averages. In this case we have

$$\Delta\theta = \frac{\rho C_p}{k}\frac{(\Delta x)^2}{2} \tag{20-16}$$

If more accuracy is desired, a smaller value of Δx, giving a finer space mesh with more nodal points, must be used. For a given value of M this will produce shorter time steps.

The reciprocal of the modulus M is seen to be the Fourier number, previously encountered in the analytical solution of unsteady-state conduction problems. As implied in the preceding discussion, choice of values of M is restricted. It is pointed out by Arpaci[1] that values of $M < 2$, when used with Eq. (20-15) for the evaluation of temperature distribution in a solid, result in the accumulation and amplification of the numerical errors as the solution progresses. Such a solution is said to be *unstable*, whereas a choice of $M \geq 2$ leads to a solution in which the errors diminish and the solution is said to be *stable*.

The numerical limitations on the choice of values for M can be viewed in a different manner; they can be made more evident by substituting some numerical values for the quantities in Eq. (20-15). Let $t_0 = 60$, $t_1 = 39$, $t_2 = 30$ represent the temperature distribution in a slab shown in Fig. 20-5 at some instant in time. If we choose $M = 2$, then the value for t_1' given by Eq. (20-15) would be 45 which would represent the highest value, i.e., steady-state, that could be achieved at point 1, if points 0 and 2 were held at 60 and 30 indefinitely. If we choose $M = 3$, we obtain a midpoint temperature $t_1' = 43$, while the choice $M = 1$ gives $t_1' = 51$. Obviously, the value of the temperature calculated when $M = 1$ is thermodynamically unobtainable (given $t_0 = 60$ and $t_2 = 30$), as are all midpoint temperatures obtained when $M < 2$. On the other hand, temperatures calculated at the midpoint with values of $M > 2$ are all perfectly feasible; the limiting case is $M = 2$.

The graphical solution of a problem using $M = 2$ is known as the *Schmidt* method. A plot is made of initial temperature distribution, and the points two

[1] Arpaci, op. cit., p. 503.

planes apart are joined in alternate steps. Both the numerical and graphical procedures are illustrated in Example 20-3.

Example 20-3 A system similar to that illustrated in Example 19-2 will be used to illustrate both the numerical and graphical procedures. A rubber sheet $\frac{1}{2}$-in thick, initially at 70°F, is heated by bringing both surfaces to a temperature of 292°F at the same instant and holding them at that temperature. The thermal diffusivity of rubber, $k/\rho C_p$, is 0.0028 ft²/h.

If $M = 2$ is chosen and the sheet is considered as six slabs, each of thickness 0.0833 in, the time increment is

$$\Delta\theta = \frac{(\Delta x)^2}{2} \frac{\rho C_p}{k}$$

$$= \frac{(0.0833/12)^2}{(2)(0.0028)}$$

$$= 0.00858 \text{ h } (0.515 \text{ min})$$

If we carried the solution to the point of finding how long it took for the centerline temperature to reach 290°F (as in Example 19-2), the number of increments would be 18.7/0.515 = 36.3 (the required time, as found in Example 19-2, was 18.7 min). This is such a large number of steps that the solution will not be carried to that extent. However, in the same example we see that the centerline temperature was 217°F after 5 min (9.7 increments), so the checking of this figure will be our present goal. The results are shown in Table 20-3.

The results in Table 20-3 indicate that after $9(\Delta\theta)$ and $10(\Delta\theta)$ time increments the centerline temperature was 223°F. This checks approximately

Table 20-3 NUMERICAL ANALYSIS OF UNSTEADY-STATE CONDUCTION IN A RUBBER SHEET

$\Delta\theta$	t_o	t_1	t_2	t_3	t_4	t_5	t_6
0	292	70	70	70	70	70	292
1	292	181	70	70	70	181	292
2	292	181	125	70	125	181	292
3	292	209	125	125	125	209	292
4	292	209	167	125	167	209	292
5	292	230	167	167	167	230	292
6	292	230	199	167	199	230	292
7	292	246	199	199	199	246	292
8	292	246	223	199	223	246	292
9	292	258	223	223	223	258	292
10	292	258	240	223	240	258	292

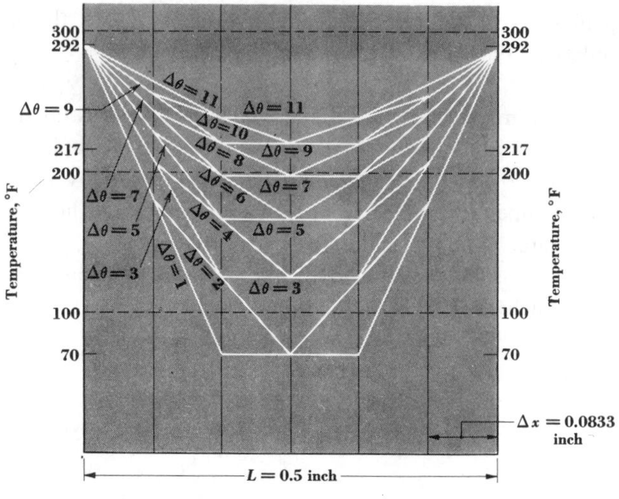

FIGURE 20-6
Graphical solution of heat-conduction problem (Example 20-3).

the analytical results of Example 19-2, in which the centerline temperature was found to be 217°F after 5 min (9.70 $\Delta\theta$). If the results of Table 20-3 are plotted as centerline temperature versus time, a smooth curve drawn through the points will eliminate the stepwise character of the results and permit interpolation.

The graphical solution is shown in Fig. 20-6, and it coincides with the results of the numerical and analytical solutions. ////

Numerical and graphical solutions are available for systems in which there is unsteady-state heat transfer by two-dimensional conduction, conduction with convection at a surface, and conduction with heat generation in the solid. Solutions are also available for transient conduction in cylinders, composite slabs, and composite cylinders and spheres.

An awkward feature of the numerical solution shown in Table 20-3 is that each temperature value found for the interior of the slab appears always to represent the temperature at two adjacent values of $\Delta\theta$. This can be avoided by the simple expedient of arbitrarily designating the surface temperatures (t_0 and t_6) at zero time as the arithmetic average of 292°F [t_0 and t_6 after $1(\Delta\theta)$] and 70°F (t_1 and t_5 at $\theta = 0$). Use of this value (181°F) for t_0 and t_6 at $\theta = 0$ gives a value for t_1 and t_5 of 125°F at $1(\Delta\theta)$, rather than 181°F as shown in Table 20-3. The value 181°F is then found to be the temperature for t_1 and t_5 at $2(\Delta\theta)$. This technique can also be employed to modify the graphical solution shown in Fig. 20-6.

Numerical solutions to problems of significance are almost always obtained using a digital computer. The main purpose of the preceding discussion has been to focus attention on the basic equations and techniques which are to be programmed on the computer. A program written for the rubber sheet problem gave the results presented in Table 20-4. These confirm, as predicted at the beginning of Example 20-3, that the center of the sheet reaches 290°F in approximately 18.7 min ($35\ \Delta\theta = 18.0$ min; $36\ \Delta\theta = 18.5$ min). If the program is run with a value for the modulus $M = 6$, the time increment becomes 0.172 min. The computer solution[1] for this case gives a centerline temperature of 290.1°F after 109 time increments (18.7 min).

Analog Methods

Heat-conduction problems can often be solved by using systems which are analogs of the heat-transfer system. The criterion for two systems to be mathematical analogs is that they must be described by the same mathematical equations. For example, the two-dimensional Laplace equation, $\partial^2\phi/\partial x^2 + \partial^2\phi/\partial y^2 = 0$, can be written for steady-state heat conduction (ϕ is temperature), or for the voltage distribution in an electric conductor (ϕ is voltage), or for the flow field of an incompressible ideal fluid (ϕ is velocity potential).

The flow lines in a heat-conduction system can be obtained by constructing

Table 20-4 COMPUTER RESULTS OF ANALYSIS OF UNSTEADY-STATE CONDUCTION IN A RUBBER SHEET

$\Delta\theta$	t_0	t_1	t_2	t_3	t_4	t_5	t_6
0	292	70	70	70	70	70	292
1	292	181	70	70	70	181	292
2	292	181	125.5	70	125.5	181	292
⋮							
30	292	289.9	288.9	287.9	288.9	289.9	292
31	292	290.4	288.9	288.9	288.9	290.4	292
32	292	290.4	289.6	288.9	289.6	290.4	292
33	292	290.8	289.6	289.6	289.6	290.8	292
34	292	290.8	290.2	289.6	290.2	290.8	292
35	292	291.7	290.2	290.2	290.2	291.7	292
36	292	291.7	291.1	290.2	291.1	291.7	292

[1] For an extensive discussion of computer solutions to heat transfer problems, see J. A. Adams and D. F. Rogers, "Computer Aided Analysis in Heat Transfer," McGraw-Hill Book Company, New York, 1973.

either a liquid-flow analog or an electric analog. The liquid flow analog system is obtained by constructing a plate of shape similar to a plane in the heat-transfer system. A sheet of glass is placed a very short distance (say $\frac{1}{16}$ in) above and parallel to the plate. Liquid flows into the space between the plate and the glass at boundaries representing the heat source and out at other boundaries representing the heat sink. Insulated surfaces are represented by walls over which the fluid cannot flow. Before the glass is installed, soluble colored crystals (e.g., potassium permanganate) are sprinkled over the plate. When the flow (known as *Hele-Shaw flow*) begins, these give off streaks of colored fluid which indicate the flow lines in the system. Flow must, of course, be laminar and must fill the space between the plates. When the flow lines are fully established, the plate is photographed. The isotherms can then be drawn at right angles to the flow lines, and by using a technique known as the *method of curvilinear squares*, the heat flux can be calculated.[1]

The flow lines can also be found by use of an electric-analog field plotter. In this method a thin sheet of electrically conducting paper is cut to the same shape as the heat-conduction system. Electrodes are attached to highly conductive wires at edges representing isothermal boundaries of the heat-transfer system. When the voltage is applied to these electrodes, the locations of constant-voltage lines on the paper are detected and plotted by means of a stylus attached to a voltage-detecting instrument. Perforations are made on the paper while tracing equipotential lines which are, of course, analogous to the isotherms in the heat-transfer system. The flow lines are drawn perpendicular to the isotherms. Alternatively, the flow lines can be located by attaching the electrodes to the sides of the paper representing the insulated surfaces and thus locating a new set of constant-voltage lines; these will be at right angles to the first set and will satisfactorily represent lines of heat flow.

A three-dimensional steady-state conduction system can be represented by an electrolytic-bath analog. The heat-transfer system is reproduced in the shape of a tank filled with electrolyte. The walls of the tank are constructed of conducting or insulating materials representing the boundary conditions of the heat-transfer system. The location of equipotential surfaces in the electrolytic model corresponds with the location of isothermal surfaces in the heat-transfer system. Heat-transfer rates can be obtained by either of two methods: by calculating the heat flow between the isothermal surfaces; or, more directly, by putting a second electrolytic tank in series with the original electrolytic tank. The second tank is modelled after a simple thermal system that can be readily analyzed, such as a cubic tank with conduction between two parallel planes. The ratio of the voltage drops measured for the two electrolytic tanks is a shape factor which applies also to the heat-transfer rates for the

[1] An interesting discussion of liquid-flow analogs is given by A. D. Moore, *J. Appl. Phys.*, **20**:790 (1949).

cubic model and the system for which the heat flow rate is being sought. A more extensive discussion of the method is provided by Jakob.[1]

The methods described above are known as *geometric* analogs. In this class of analogs the systems are essentially homogeneous and the physical phenomena under observation, whether in the original or the analog, are "distributed" throughout the system. In this sense the geometric analog shares with the analytical solution of the differential equations the characteristic that the solutions can be evaluated at all points in the system. A different kind of analog method is known as the *network* analog. In this type the system is represented by a network of lumps or nodes, each of which offers resistance to and capacitance for some kind of flow in such manner that the network as a whole is the equivalent of the thermal system being studied. The representation of the system can be done physically or numerically. Physical representation is usually done by an electrical network with resistors and capacitors or by a fluid network with lengths of tubing and reservoirs. Mathematical representation is done as already demonstrated in this chapter in the discussion of relaxation calculations and unsteady-state heat conduction. Representation of a portion of the thermal system by an equivalent numerical, electrical, or fluid element is called *lumping*.

Construction of a network of resistors to represent two- or three-dimensional steady-state heat conduction parallels closely the building of a numerical grid, as described in the discussion on relaxation techniques. Convective resistance to heat transfer at the boundaries of the system can be represented by additional resistors for which the resistance may vary with location, if required to represent a varying convective resistance.

Unsteady-state conduction problems can also be solved by electrical[2] and fluid-flow analogs. In both cases networks are set up which simulate the heat-conduction system. We have seen that the equation for unsteady-state conduction in one direction is

$$\frac{\partial t}{\partial \theta} = \alpha \frac{\partial^2 t}{\partial x^2} \tag{19-1}$$

The electric potential E in a small wire in which a current is flowing can be represented by

$$\frac{\partial E}{\partial \theta} = \frac{1}{RC} \frac{\partial^2 E}{\partial x^2} \tag{20-17}$$

where R = resistance per unit length

C = capacitance per unit length

[1] Loc. cit.

[2] M. Jakob and G. A. Hawkins, "Elements of Heat Transfer," 3d ed., chap. 4, John Wiley & Sons, Inc., New York, 1957.

Self-inductance and leakage conductance are assumed to be negligible. In the electrical analog of the thermal system, voltage corresponds to temperature difference, electric current corresponds to heat flow, electric resistance corresponds to thermal resistance, and electric capacitance corresponds to thermal capacitance. Electric circuits used as analogs for heat-transfer systems are said to be "passive" circuits if they contain only resistors and capacitors. If, however, they involve elements such as dc amplifiers, function generators, and function multiplier-dividers, they are referred to as "active" circuits. All passive circuits suffer from loading errors. Hence, they are coupled with electronic amplifiers with high-input impedances to minimize these errors. Application of analog computers to the solution of heat-transfer problems is discussed by Arpaci,[1] Gebhart,[2] and Moyle.[3] Recent trends indicate growing use of combinations of analog circuits with small digital computers for solution of complex problems. Such systems require fewer analog components; in addition, the programming of the digital computer is likely to require less time than the assembly of a complete analog network.

The use of liquid-flow networks as analogs of heat-transfer systems has been less common than the use of the electrical networks, probably because of the greater ease of making electrical measurements. The hydraulic-analog method has been described by Juhasz.[4]

PROBLEMS

20-1 Using the conditions shown in Example 20-2, find the temperature on a grid of side length equal to one-quarter the furnace-wall thickness.

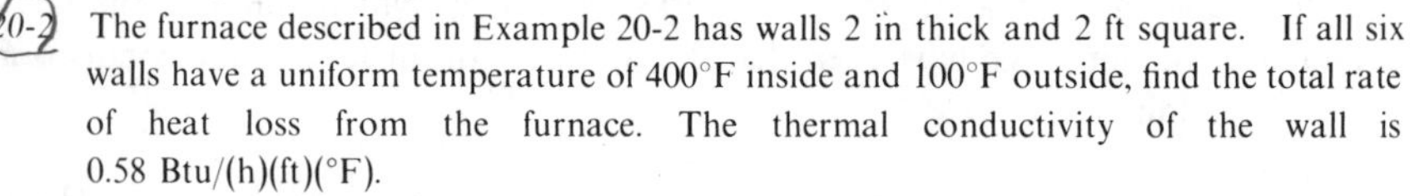

20-2 The furnace described in Example 20-2 has walls 2 in thick and 2 ft square. If all six walls have a uniform temperature of 400°F inside and 100°F outside, find the total rate of heat loss from the furnace. The thermal conductivity of the wall is 0.58 Btu/(h)(ft)(°F).

20-3 Devise a method of applying relaxation methods to steady-state, one-dimensional heat flow through a slab consisting of two layers of different thermal conductivity.

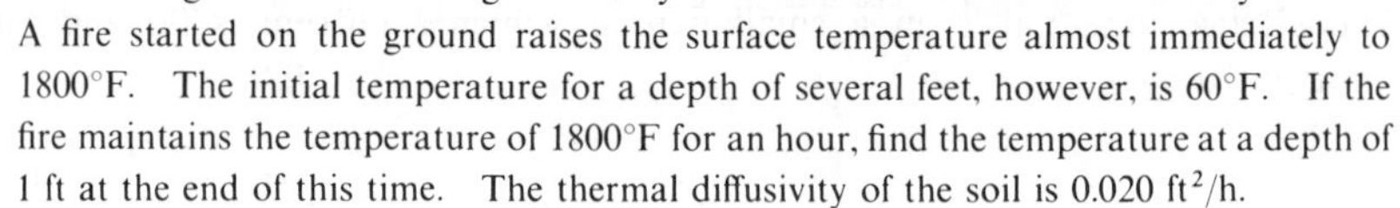

20-4 A fire started on the ground raises the surface temperature almost immediately to 1800°F. The initial temperature for a depth of several feet, however, is 60°F. If the fire maintains the temperature of 1800°F for an hour, find the temperature at a depth of 1 ft at the end of this time. The thermal diffusivity of the soil is 0.020 ft^2/h.

[1] Loc. cit.

[2] B. Gebhart, "Heat Transfer," 2d ed., McGraw-Hill Book Company, New York, 1971.

[3] M. P. Moyle, "Introduction to Computers for Engineers," John Wiley & Sons, Inc., New York, 1967.

[4] I. S. Juhasz, "Hydraulic Analogy for Transient Cross-Flow Heat-Exchangers," Paper 57-A-125, Meeting of the American Society of Mechanical Engineers, New York, December 1957.

20-5 A brick wall 1 ft thick is initially at 70°F. Suddenly the temperature at one side is raised to 400°F, and the temperature at the other side is raised at the same instant to 300°F. How long will it take the centerline temperature to read 100°F if the surface temperatures specified above are constant? The thermal diffusivity of the brick is 0.016 ft^2/h.

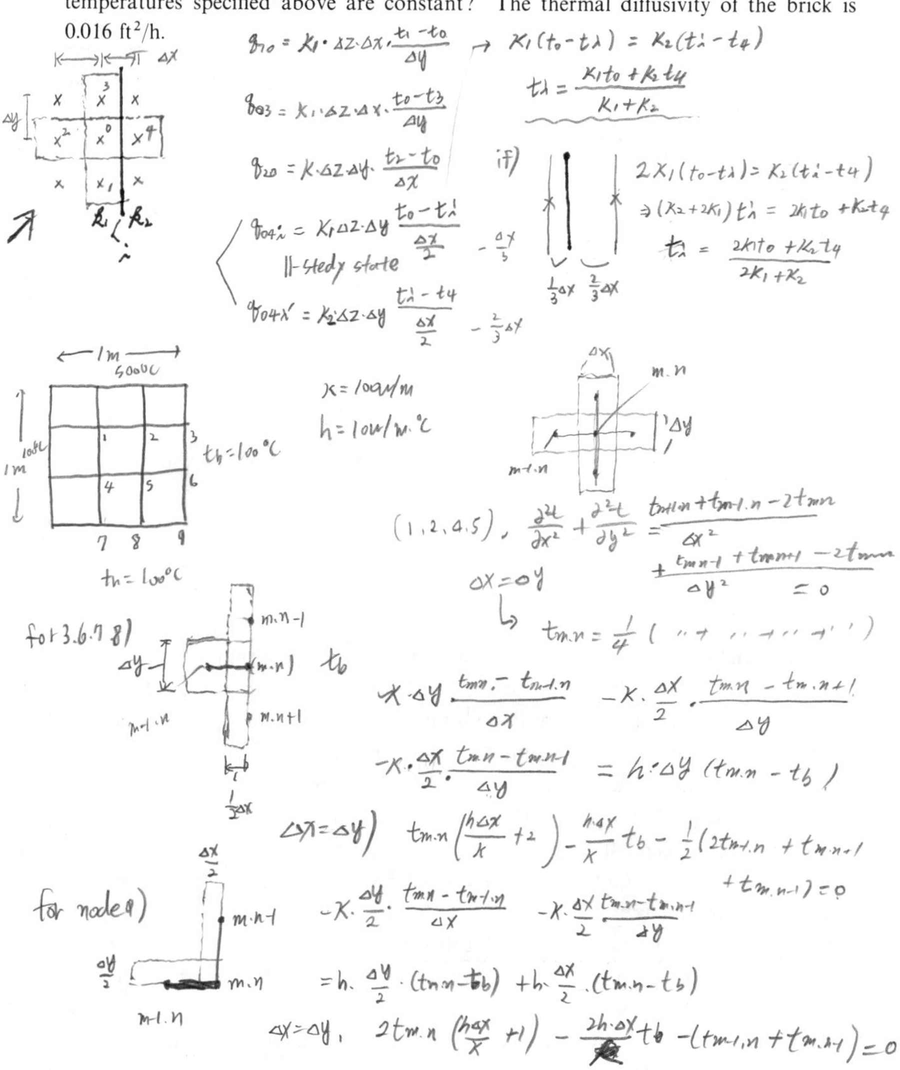

21

CONVECTIVE HEAT-TRANSFER COEFFICIENTS

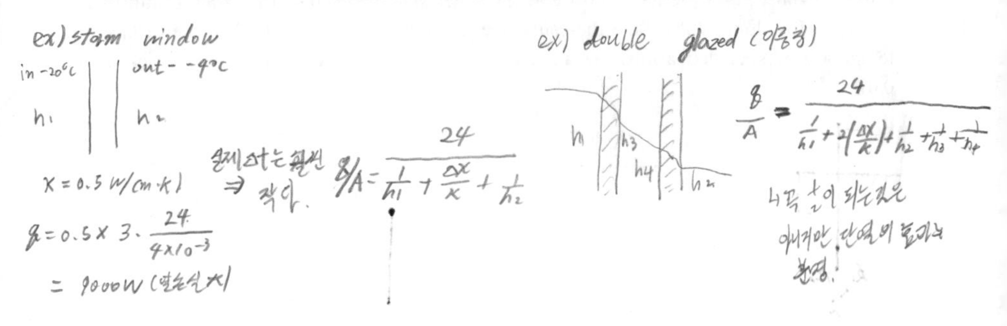

In most transport processes heat transfer in fluids is accompanied by some form of fluid motion so that the heat transfer does not occur by conduction alone. The resulting transfer of heat by simultaneous conduction and fluid flow has been defined as forced convection when the motion arises principally from a pressure gradient caused by a pump or blower, and as natural convection when the motion arises only from the density differences associated with the temperature field.

Whether the heat-transfer mechanism is natural or forced convection, the fluid motion can be described by the equations of fluid mechanics. At low velocities the flow is laminar throughout the system; at high velocities it is generally thought to be laminar near the heating surface and turbulent some distance away. Although fluid velocities in natural convection are usually lower than in forced convection, it is incorrect to think of natural convection as causing only laminar flow; turbulence does occur when the critical Reynolds number for the system is exceeded.

All problems of convective heat transfer can be expressed in terms of the differential mass, energy, and momentum balances. However, the mathematical difficulties connected with the integration of these simultaneous nonlinear partial differential equations are such that analytical solutions exist only for simplified

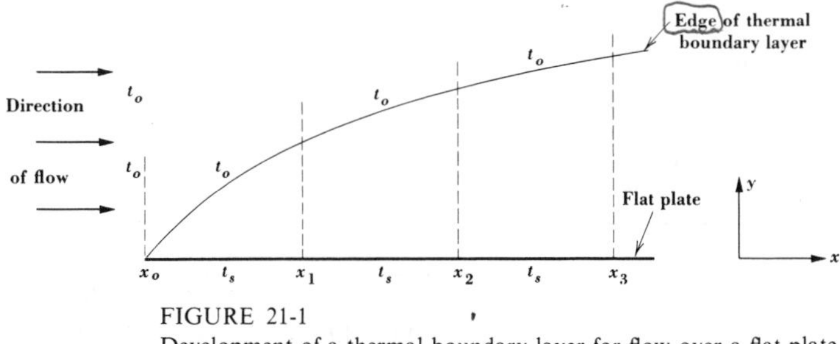

FIGURE 21-1
Development of a thermal boundary layer for flow over a flat plate.

cases. High-speed electronic computers can often be used for the numerical solution of the equations, but even these powerful tools have so far permitted the solution of only relatively simple problems.

One of the problems simplest to deal with analytically is that of heat transfer between a fluid and a flat plate when the fluid is flowing parallel to the plate. Such a system is illustrated in Fig. 21-1. The fluid approaching the plate has a uniform temperature t_o, and the plate has a uniform temperature t_s; for this discussion we shall take t_o as being greater than t_s. The fluid adjacent to the plate cools, and the zone of fluid in which the temperature lies between t_o and t_s is known as the thermal boundary layer. Under special circumstances the boundaries of this region coincide with those of the velocity boundary layer; this matter is discussed at length in Chap. 22.

As the fluid moves by the plate and the distance from the leading edge of the plate increases, the thickness of the thermal boundary layer increases. The temperature profiles at various cross sections normal to the plate also change. A number of profiles are shown in Fig. 21-2. These profiles depend on flow conditions as well as on the thermal properties of the fluid. Regardless of these factors, however, the general shape of a developing thermal boundary on a flat plate is as shown in Fig. 21-1.

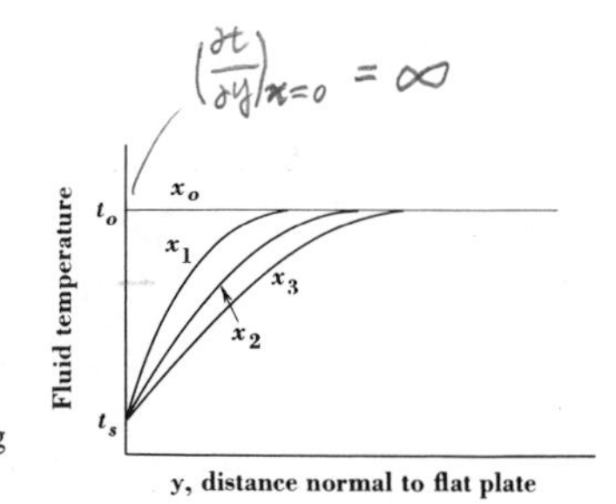

FIGURE 21-2
Temperature profiles in a developing thermal boundary layer on a flat plate.

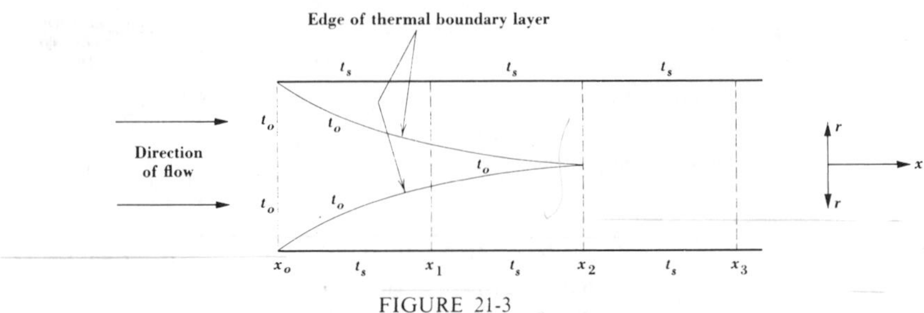

FIGURE 21-3
Development of a thermal boundary layer for flow in a pipe.

Another important heat-transfer system is one in which heat is transferred between a fluid and the wall of a pipe in which it is flowing. If the fluid enters at a uniform temperature t_o and the pipe wall is at some lower temperature t_s, the development of a thermal boundary layer is as depicted in Fig. 21-3. The thermal boundary layer thickens as the distance from the pipe inlet increases, and the layer finally comes together at the center of the pipe. The axial distance from the pipe inlet to this point x_2 is known as the *thermal-entrance* length. Beyond this point the temperature profile becomes flatter; if the pipe is long enough, the profile becomes completely flat at a uniform temperature t_s. The temperature profiles near the entrance are depicted in Fig. 21-4.

As mentioned above, the temperature profile at a point in a flow system is influenced by the velocity profile. This influence is indicated by the presence of the velocity terms in the differential energy balance, Eq. (8-11). It is not so apparent, however, that the velocity profile is influenced by the temperature profile. The Navier-Stokes equations do not in fact contain temperature explicitly, but they do contain temperature-dependent terms, notably those including viscosity. Consequently, the velocity profile of an isothermal system may differ substantially from

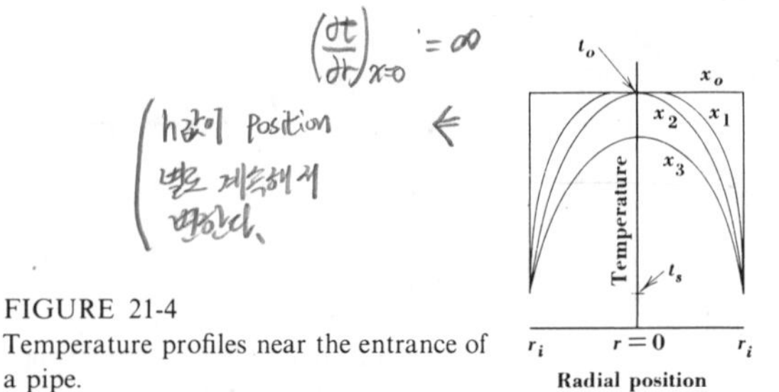

FIGURE 21-4
Temperature profiles near the entrance of a pipe.

the velocity profile of a system in which heat is being transferred. It is this nonisothermal velocity profile, of course, that provides the velocity terms to be used in solving the differential momentum and energy balances for a nonisothermal system. Needless to say, this additional complication must frequently be overlooked to permit a solution to be obtained; this simplification, however, may introduce a serious error when the viscosity of the fluid is strongly dependent on temperature. Frequently the temperature gradient is greatest near the wall, and it is in this region that the velocity gradient is also greatest. The effect of temperature on the viscosity of the fluid at the wall may therefore have a pronounced effect on both the velocity and temperature profiles of the system.

Individual Heat-transfer Coefficients

Although the analysis is complicated for steady-state laminar-flow systems, even greater difficulties are encountered in systems with turbulent flow wherein the velocity and temperature at any point are functions of time. In the study of the isothermal flow of fluids, the difficulties caused by the time-dependent nature of turbulent systems led to the use of dimensional analysis and the concept of the mixing length. In order to get around the difficulties encountered in the solution of heat-transfer problems in turbulent systems, it is common practice in engineering to write the rate of heat transfer in terms of a heat-transfer coefficient h as

$$q = hA(t_s - t_m) \qquad (21\text{-}1)$$

The surface temperature is t_s, and t_m represents the temperature of the fluid medium at some distance from the surface. For this case t_s is greater than t_m; if t_m is greater than t_s, the positions of the two terms may be reversed to keep q positive. The definitions of t_m for specific cases are set by convention and will be discussed presently.

Since h and t_s often vary from one point to another on the heat-transfer surface, Eq. (21-1) should be thought of as applying at a point on a surface and put in the form

$$dq = h(t_s - t_m)\,dA \qquad (21\text{-}2)$$

Because the quantities h, $t_s - t_m$, and dq can often be expressed as functions of temperature, the integral form of Eq. (21-2) is written as

$$\int_0^q \frac{dq}{h(t_s - t_m)} = \int_0^A dA \qquad (21\text{-}3)$$

At this moment we shall not be concerned with the integration of Eq. (21-2); this subject will be considered in later discussions on applications, such as heat exchangers.

Although Eq. (21-1) will frequently be used in what follows in this chapter, the possible restriction of its application to a differential area if $h(t_s - t_m)$ varies should be kept in mind.

The concept of a heat-transfer coefficient is useful, but it does not really allow us to avoid the basically complicated nature of the problem. The quantity h is a function of the properties of the fluid, the geometry and roughness of the surface, and the flow pattern of the fluid. Several methods are available for evaluating h: analytical methods are used for laminar-flow systems; for turbulent systems we shall use integral methods, mixing-length theory, and dimensional analysis.

Another way of looking at convective heat transfer is to write the conduction equation for the fluid at the solid surface. For convenience we shall use the system shown in Fig. 21-1, for which we write

$$q = kA\left(\frac{dt}{dy}\right)_{y=0} \tag{21-4}$$

Although we know k for the fluid, $(dt/dy)_{y=0}$ is unknown, so this equation is not directly useful for calculating q. In considering conduction in a solid, we were able to integrate the Fourier equation (18-1), but in the flowing fluid at values of $y > 0$, energy is also being carried by convective transport, so the energy balance on an element of fluid leads to the relatively complicated differential energy balance, Eq. (8-11), rather than the simple conduction equation (18-1).

It is of interest to combine Eq. (21-4) with a form of Eq. (21-1) applicable to the flat-plate problem of Fig. 21-1. This is written as

$$dq = k\,dA\left(\frac{dt}{dy}\right)_{y=0} = h(t_o - t_s)\,dA$$

which can be rearranged to give

$$h = \frac{k}{t_o - t_s}\left(\frac{dt}{dy}\right)_{y=0} \tag{21-5}$$

or in an equivalent form,

$$h = k\left\{\frac{d[(t - t_s)/(t_o - t_s)]}{dy}\right\}_{y=0} \tag{21-6}$$

Equation (21-6) is written in terms of the dimensionless unaccomplished temperature change from the surface to the edge of the thermal boundary layer. It is apparent from these relations that any factor which would cause an increase in the temperature gradient at the wall, such as an increase in the velocity of flow past the heated surface, would also cause an increase in the value of the heat-transfer coefficient. Equation (21-6) also shows how the heat-transfer coefficient can vary with position. For both the flat plate and the heated pipe, if the wall temperature is constant at all points, the quantity $\{d[(t - t_s)/(t_o - t_s)]/dy\}_{y=0}$ is infinite at the leading edge so that the coefficient h is infinite at this point. As the thermal boundary layer thickens, both the gradient at the wall and the heat-transfer coefficient diminish. This behavior is illustrated by the curves of Figs. 21-2 and 21-4. In pipes the local, or point, heat-transfer coefficient approaches some constant value as the distance from the pipe inlet increases. For long pipes the contribution of the entrance effect is often small and is frequently neglected in engineering calculations.

Let us now complete the definition of h for some specific cases by defining t_m for some other situations. For flow through a heated conduit with a wall temperature t_s, we write

$$q = hA(t_s - t_b) \tag{21-7}$$

where t_b is the bulk (mixing-cup) temperature.

For natural convection adjacent to a hot surface at t_s we set

$$q = hA(t_s - t_\infty) \tag{21-8}$$

where t_∞ is the temperature of the fluid far from the surface, i.e., at the edge of the thermal boundary layer.

The concept of the convective coefficient is also applied to heat transfer by boiling and condensation. Experimental evidence shows that when condensation of a pure vapor occurs on the exterior surface of a cold tube, almost the entire resistance to heat transfer outside the tube occurs in the film of condensate. This is the basis of an analytical model, which gives reasonably accurate results in many systems. The condensing coefficient is defined by the equation

$$q = hA(t_{sv} - t_s) \tag{21-9}$$

where t_{sv} = temperature of saturated vapor

t_s = temperature of tube surface

In this case, as in that illustrated by Fig. 21-1, the temperature difference in the defining equation for h is arbitrarily reversed to keep q positive.

For heat transfer to a boiling fluid the defining equation is

$$q = hA(t_s - t_{sl}) \tag{21-10}$$

where t_{sl} is the saturation liquid temperature. The mechanism of boiling is believed to be by heat transfer from the solid to the liquid, then from the liquid to the interface of each growing bubble. The application of Eq. (21-10) is complicated by the very strong dependence (sometimes to the third power) of h on the temperature difference $t_s - t_{sl}$.

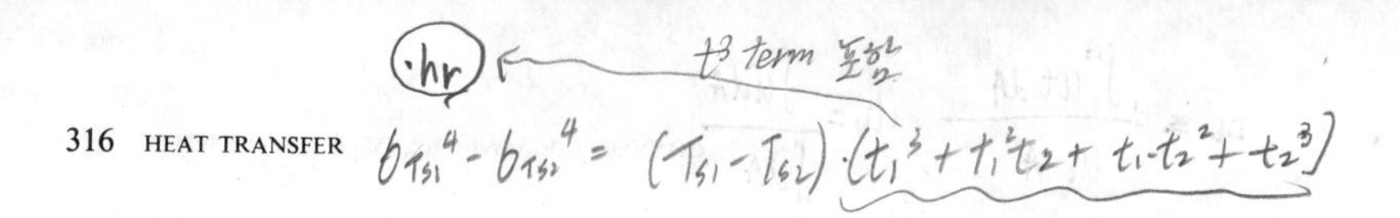

Finally, the concept of the heat-transfer coefficient has been extended to the design of systems where radiant-heat transfer is a contributing mechanism. The equation for radiant-heat transfer between two surfaces at t_{s1} and t_{s2} can be written as a pseudoconvection process by setting

$$q = h_r A(t_{s1} - t_{s2}) \qquad (21\text{-}11)$$

where the subscript r is added to the coefficient to make its meaning clear. It will be shown later in the discussion of radiant-heat transfer that h_r can be calculated from fundamental considerations. Although it has a third-power dependence on temperature and is therefore variable with the temperature difference (as the boiling coefficient was), in many cases h_r can be assumed to be approximately constant at some average value throughout the system. This procedure often has the distinct advantage of preserving the linearity of a system of heat-transfer equations. As a result it is possible to obtain analytical solutions to problems which otherwise would have to be analyzed numerically.

The coefficient h is often called in the engineering literature the *individual film coefficient*. This expression is best understood with reference to heat transfer to a turbulent fluid flowing in a conduit. If the resistance to heat flow is thought of as existing only in a laminar film, the coefficient h is equivalent to $k/\Delta x_e$, where Δx_e is the equivalent thickness of a stationary film just thick enough to offer the resistance corresponding to the observed value of h. Since there is often an appreciable resistance in the turbulent core and since the stationary film has not even an approximate physical counterpart in laminar flow, boiling, and radiation, we shall refer in this book to h as an *individual*, or a *convective, heat-transfer coefficient*. However, it is not uncommon in engineering practice to use the term *film coefficient*, and the fictitious nature of the film should be kept in mind.

Range of Individual Coefficients

It is of considerable value to acquire some notion of the magnitude of the various coefficients. Heat transfer is generally a process in which several mechanisms are combined in both series and parallel. Frequently, the magnitude of one or two resistances[1] is so much greater than those of the remainder of the process that an adequate analysis is obtained by neglecting all minor resistances. Table 21-1 is reproduced from McAdams[2] to indicate the relative magnitude of various types of convective coefficients.

[1] The thermal resistance in convection is $1/hA$, in accord with the discussion in Chap. 18.

[2] W. H. McAdams, "Heat Transmission," 3d ed., McGraw-Hill Book Company, New York, 1954.

As an example of the idea of a controlling resistance, if air were heated in a device in which condensing steam gave up heat on one side of a metal surface and the air received the heat on the other side, it is apparent that the resistance due to the condensing steam would be negligible compared with the resistance of the air.

Example 21-1 The analysis of unsteady-state heating or cooling of solid objects can often be simplified by assuming that the resistance to heat conduction inside the solid is negligible compared to the convective heat-transfer resistance in the surrounding fluid. The circumstances necessary to justify this assumption are a high thermal conductivity for the solid and a low convective coefficient in the adjacent fluid. A metallic object being heated or cooled in air often constitutes such a system.

To illustrate, we shall calculate the time required for a mercury thermometer initially at 70°F, placed in an oven at 400°F, to reach 279°F. The thermometer bulb has a diameter of 0.24 in and will be assumed to be adequately represented as an infinitely long cylinder with negligible resistance to heat transfer either in the glass which is very thin or in the mercury, which has an adequately high thermal conductivity [$k = 5.6$ Btu/(h)(ft)(°F) at 140°F]. The mean convective heat-transfer coefficient between the outside of the thermometer and the air in the oven will be taken as $h = 2$ Btu/(h)(ft^2)(°F).

A heat balance on the thermometer bulb can readily be written

$$hA(400 - t) = \rho C_p V \frac{dt}{d\theta} \tag{1}$$

$$\frac{dt}{400 - t} = \frac{hA}{\rho C_p V}\, d\theta \tag{2}$$

Table 21-1 APPROXIMATE RANGE OF SOME INDIVIDUAL COEFFICIENTS

	h, Btu/(h)(ft^2)(°F)†
Steam, dropwise condensation	5000–20,000
Steam, film-type condensation	1000–3000
Water, boiling	300–9000
Organic vapors, condensing	200–400
Water, heating	50–3000
Oils, heating or cooling	10–300
Steam, superheating	5–20
Air, forced convection	2–15
Air, natural convection	0.5–2

† SI units: Multiply values of coefficients given in table by 5.678 to get W/m^2·K.

If the heat-transfer coefficient and the physical properties can be considered constant with time at some average values, we can integrate Eq. (2) to give

$$\ln \frac{400 - 279}{400 - 70} = -\frac{hA}{\rho C_p V}\theta \qquad (3)$$

For a cylinder, $A/V = (\pi D)/(\pi D^2/4) = 4/D$. The other physical properties are $\rho = 849$ lb/ft^3 and $C_p = 0.033$ Btu/(lb)(°F). These values are substituted in Eq. (3) as follows:

$$\theta = \frac{(849)(0.033)}{(2)(4/0.020)} \ln \frac{330}{121}$$

$$= 0.0707 \ln 2.72$$

$$= 0.0707 \text{ h } (4.24 \text{ min})$$

Equation (3) could have been written as

$$\frac{400 - 279}{400 - 70} = \exp\left[-\frac{hA\theta}{\rho C_p V}\right] \qquad (4)$$

$$= \exp\left[-\left(\frac{hL}{k}\right)\left(\frac{k\theta}{\rho C_p L^2}\right)\right] \qquad (5)$$

in which hL/k = Biot number, $(k/\rho C_p)\theta/L^2$ = Fourier number. The quantity L in these two dimensionless groups is any length characteristic of the geometry of the system. For an infinite cylinder, $L = V/A = D/4$.

The group of terms $\rho C_p V/hA$ has the dimension of time and is known as the *time constant* for the system. As can be seen by examining Eq. (4), the time constant represents the time required for the temperature difference driving force to fall to e^{-1}, or 0.368 of its initial value. Put another way, the time constant represents the time required for the change in the temperature of the thermometer to become 63.2 percent complete. In the example chosen, this time is 4.2 min. ////

Overall Heat-Transfer Coefficients

The problem of conduction through a series of resistances has already been considered. It is now appropriate to study systems in which heat is transferred by a series of conduction and convection mechanisms.

Figure 21-5 illustrates a system in which heat flows from a fluid at a bulk temperature t_1, through the pipe wall, through a layer of insulation, and finally into a fluid at bulk temperature t_5. The interfacial temperatures t_2, t_3, and t_4 are as shown.

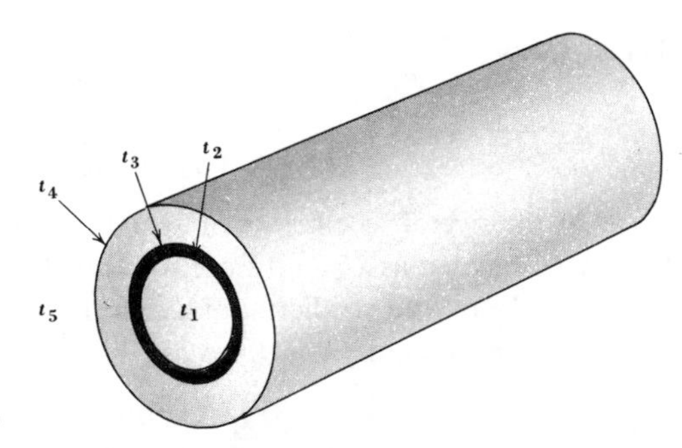

FIGURE 21-5
Heat transfer through an insulated pipe.

The heat-transfer coefficients h_i and h_o refer to the inside and outside coefficients, respectively. The steady-state heat flow rate is

$$q = h_i A_i(t_1 - t_2) = k_b A_{b,\,lm} \frac{t_2 - t_3}{\Delta r_b} = k_c A_{c,\,lm} \frac{t_3 - t_4}{\Delta r_c} = h_o A_o(t_4 - t_5) \qquad (21\text{-}12)$$

The individual temperature drops can be found by rearranging (21-12).

$$t_1 - t_2 = q \frac{1}{h_i A_i}$$

$$t_2 - t_3 = q \frac{\Delta r_b}{k_b A_{b,\,lm}}$$

$$t_3 - t_4 = q \frac{\Delta r_c}{k_c A_{c,\,lm}}$$

$$t_4 - t_5 = q \frac{1}{h_o A_o}$$

These equations are added and rearranged to give

$$q = \frac{t_1 - t_5}{1/h_i A_i + \Delta r_b/k_b A_{b,\,lm} + \Delta r_c/k_c A_{c,\,lm} + 1/h_o A_o} \qquad (21\text{-}13)$$

Each term in the denominator can be considered as a resistance, so that Eq. (21-13) can be rewritten as

$$q = \frac{\Delta t_{\text{overall}}}{\sum R} \qquad (21\text{-}14)$$

However, a more common procedure is to multiply the right-hand side of Eq. (21-13) by either A_i/A_i or A_o/A_o. Using the latter term, we obtain

$$q = \frac{A_o\, \Delta t_{\text{overall}}}{A_o/h_i A_i + \Delta r_b A_o/k_b A_{b,\,lm} + \Delta r_c A_o/k_c A_{c,\,lm} + 1/h_o} \qquad (21\text{-}15)$$

The quantity

$$\frac{1}{A_o/h_i A_i + \Delta r_b A_o/k_b A_{b,\,lm} + \Delta r_c A_o/k_c A_{c,\,lm} + 1/h_o}$$

is called the overall coefficient of heat transfer based on the outside area and is usually designated as U_o. If Eq. (21-13) had been multiplied by A_i/A_i, a quantity called U_i, the overall coefficient of heat transfer based on the inside area, would have resulted. This quantity is defined as

$$U_i = \frac{1}{1/h_i + \Delta r_b A_i/k_b A_{b,\,lm} + \Delta r_c A_i/k_c A_{c,\,lm} + A_i/h_o A_o} \tag{21-16}$$

The definitions of these quantities, U_o and U_i, make it possible to write

$$q = U_o A_o \, \Delta t_{\text{overall}} \tag{21-17}$$

$$q = U_i A_i \, \Delta t_{\text{overall}} \tag{21-18}$$

The ratios $A_i/A_{b,\,lm}$, A_o/A_i, A_i/A_o, etc., can be replaced by the ratios of radii $r_i/r_{b,\,lm}$, r_o/r_i, r_i/r_o. In many cases these ratios are nearly unity and can be taken as such. This is about the only basis for deciding whether to calculate an outside or an inside coefficient. For example, if the inside resistance is much greater than the other three resistances (as shown by h_i being much smaller than h_o, $k_b/\Delta r_b$, and $k_c/\Delta r_c$), one can write

$$U_i = \frac{1}{1/h_i + \Delta r_b/k_b + \Delta r_c/k_c + 1/h_o} \tag{21-19}$$

as a satisfactory approximation. The basis for doing this is that the last three terms of the denominator are small in comparison with $1/h_i$, so that the ratios of radii would represent negligible corrections. In certain problems it is reasonable to assume that all the terms but one can be ignored, and we write $U_i = h_i$ or $U_o = h_o$, depending on which term remains.

The illustration chosen here contains four resistances in series. However, the overall coefficient can be used to represent as many resistances in series as desired. Such cases can be analyzed using the procedures followed here, but the nature of additional terms in the overall coefficient is rather obvious, so that these terms can be written by analogy with those given in Eqs. (21-15) and (21-16).

If the system is one of flat, parallel walls, the areas cancel, so that the expression

$$U_i = U_o = \frac{1}{1/h_i + \Delta x_b/k_b + \Delta x_c/k_c + 1/h_o} \tag{21-20}$$

is rigorous.

The overall coefficient could be defined in terms of one of the log mean areas, but this is seldom done. In fact, when tabulations of overall coefficients are encountered, they frequently do not state whether the values given are for U_o or U_i because the uncertainty in the individual coefficients is somewhat greater than the effect of the area ratio. In such a case the overall coefficient is usually assumed to be U_o.

Tabulations of overall coefficients are given in Perry[1] and in Nelson.[2] The sample in Table 21-2 is taken from Perry.[3] Values such as those quoted in Table 21-2 are recommended only for preliminary design estimates, although it is not uncommon for information of this sort to be the only kind obtainable.

Fouling Coefficients

The tendency of certain fluids to form deposits on heat-transfer surfaces is a serious problem in the design of heat-exchange equipment. It can happen that the thermal resistance due to the deposit is greater than the sum of all the other resistances.

In some cases a hard scale is deposited on heat-transfer surfaces in a boiler or evaporator. Coke is often deposited inside the tubes of oil heaters in a refinery. The resistance of such layers can be represented using a heat-transfer coefficient equal to the thermal conductivity of the scale divided by its thickness. This type of scale can be removed by sandblasting, by the use of pneumatic cleaning tools, and occasionally by pumping a chemical cleaning solution through the equipment.

Another type of fouling is the porous deposit formed from mud, soot, and even vegetable matter. The thermal conductivities of these materials may be high, but the fluid contained in the pores frequently has a much lower thermal conductivity. The effective thermal conductivity, therefore, may be almost as low as that of the fluid. These deposits can sometimes be removed by blowing with steam, air, or hot water. The growth of vegetable material in condensers is inhibited by chlorinating the water.

Table 21-2 OVERALL COEFFICIENTS OF HEAT TRANSFER

	U, Btu/(h)(ft^2)(°F)
Stabilizer reflux condenser	94
Oil preheater	108
Reboiler (condensing steam to boiling water)	300–800
Air heater (molten salt to air)	6
Steam-jacketed vessel evaporating milk	500

[1] Pages 10-39 to 10-42 (5th ed.).
[2] W. L. Nelson, "Petroleum Refinery Engineering," 4th ed., p. 522, McGraw-Hill Book Company, New York, 1949.
[3] Pages 480–482 (3d ed.).

The presence of fouling deposits is anticipated in the design of heat exchangers by the use of a fouling coefficient which has the form of a heat-transfer coefficient. An equation for the heat flux through the scale is written

$$q = h_d A \, \Delta t_{\text{scale}} \qquad (21\text{-}21)$$

If the fouling resistance is incorporated into the analysis of a series of resistances, the final equation for heat transfer is the same as Eqs. (21-17) and (21-18). However, the overall coefficient contains an extra term and is written

$$U_o = \frac{1}{\dfrac{A_o}{h_i A_i} + \dfrac{A_o}{h_{di} A_i} + \dfrac{\Delta r_b}{k_b}\dfrac{A_o}{A_{b,\,lm}} + \dfrac{\Delta r_c}{k_c}\dfrac{A_o}{A_{c,\,lm}} + \dfrac{1}{h_o}} \qquad (21\text{-}22)$$

The fouling coefficient for the inside surface is h_{di}. If fouling occurred simultaneously on the outer surface, the term $1/h_{do}$ would need to be added to the denominator.

Tabulations of fouling coefficients are given in many references, including Nelson,[1] The Standards of the Tubular Exchanger Manufacturers Association,[2] and Perry.[3] Nelson gives an interesting discussion of the problem. It should be noted, however, that he refers to the fouling coefficient in terms of the "fouling factor," which is $1000/h_d$. Both terms are in use in industry. Nelson shows that fouling is often as much dependent on the velocity of the fluid past the surface as it is on the nature of the fluid. A sample tabulation of fouling coefficients is given in Table 21-3.

It is apparent that the fouling resistance will usually increase with time until cleaning is necessary; however, there are some cases for which the resistance ceases to increase after some time because the rate of deposition is balanced by the rate of

Table 21-3 FOULING COEFFICIENTS

	h_d, Btu/(h)(ft²)(°F)
Overhead vapors from crude-oil distillation	1000
Dry crude oil (300–100°F):	
Velocity under 2 ft/sec	250
Velocity 2–4 ft/sec	330
Velocity over 4 ft/sec	500
Air	500
Steam (non-oil-bearing)	200
Water, Great Lakes, over 125°F	500

[1] *Ibid.*, pp. 489–492.
[2] Tubular Exchanger Manufacturers Association, New York, 1949.
[3] Page 10-38.

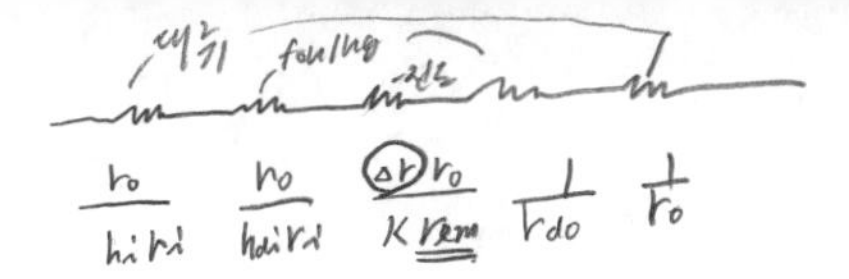

removal by scouring. No indication of the effect of time is given in Table 21-3: this omission is customary in published data because of the approximate nature of the figures available.

Example 21-2 A reflux condenser contains $\frac{3}{4}$-in. 16-gauge copper tubes in which cooling water circulates. Hydrocarbon vapors condense on the exterior surfaces of the tubes. Find the overall heat-transfer coefficient U_o. The inside convective coefficient can be taken as 4500 W/m$^2 \cdot$ K, and the outside coefficient as 1500 W/m$^2 \cdot$ K.

Appropriate fouling coefficients from Table 21-3 are $h_{do} = 5700$ and $h_{di} = 2840$. The following quantities can be calculated for use in Eq. (21-22):

$$\frac{1}{h_o} = \frac{1}{1500} = 0.00067 \qquad \frac{1}{h_{do}} = \frac{1}{5700} = 0.00018$$

$$\frac{\Delta r}{k}\frac{r_o}{r_{lm}} = \left(\frac{0.00165}{380}\right)\left(\frac{0.0191}{0.0175}\right) = 0.0000047$$

$$\frac{r_o}{h_i r_i} = \frac{0.0191}{(4500)(0.0157)} = 0.00027$$

$$\frac{r_o}{h_{di} r_i} = \frac{0.0191}{(2840)(0.0157)} = 0.00043$$

These quantities are substituted in Eq. (21-22) to give

$$U_o = \frac{1}{0.00043 + 0.00027 + 0.0000047 + 0.00067 + 0.00018}$$

$$= \frac{1}{0.00155} = 645 \text{ W/m}^2 \cdot \text{K}$$

The fractional resistance due to fouling is $0.00061/0.00155 = 0.39$; the resistance of the metal wall is negligible. U_i is easily calculated since $U_i A_i = U_o A_o$, giving $U_i = (645)(0.0191/0.0157) = 785$ W/m$^2 \cdot$ K.

Heat Exchangers

The design of heat-exchange apparatus is discussed in Chap. 27. However, a brief look at the design of one of the simpler pieces of equipment, the double-pipe heat exchanger, is appropriate at this point.

The determination of the required heat-transfer area is one of the principal objectives in the design of heat exchangers. The basic equation for determining area

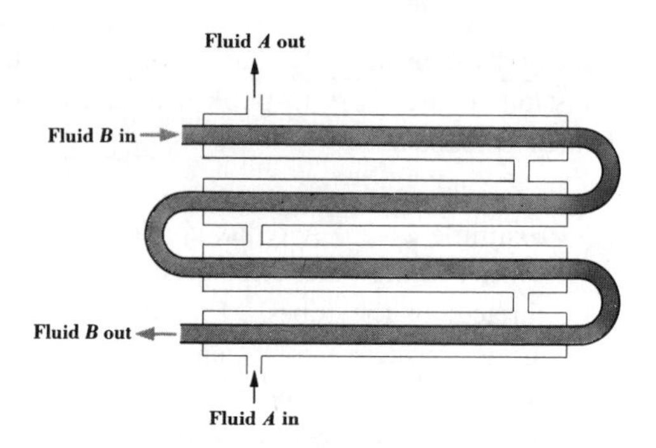

FIGURE 21-6
Double-pipe heat exchanger.

is Eq. (21-3). It is, however, more common to write Eq. (21-17) for a differential segment of the exchanger for which the outside tube area is dA_0 and integrate as follows:

$$\int_0^q \frac{dq}{U_o \, \Delta t_{\text{overall}}} = \int_0^{A_o} dA_o \qquad (21\text{-}23)$$

This equation contains the overall coefficient of heat transfer U_o, based on the outside tube area A_o, and the difference between the mixing-cup temperatures of the hot and cold fluids, Δt.† The reason for not using Eq. (21-17) directly is that both Δt and U_o may vary from one end of the exchanger to the other. Equation (21-18), which defined U_i in terms of A_i, can also be expressed in differential form and integrated in the manner shown above. We shall limit our subsequent analysis to equations based on the outside area; however, it should be remembered that a similar treatment exists in terms of A_i and U_i.

Under steady conditions the mixing-cup temperatures of the hot and cold fluids in a heat exchanger are assumed to be fixed at any cross section normal to the flow. If we designate these temperatures as t_h and t_c, the overall temperature difference in Eq. (21-23) is $t_h - t_c$. For convenience, this quantity will be written as

$$\Delta t = t_h - t_c \qquad (21\text{-}24)$$

The double-pipe heat exchanger consists of two concentric tubes through which the hot and cold fluids flow. The fluids may flow in the same direction, i.e., concurrent, or they may flow in opposite directions, i.e., countercurrent. The double-pipe exchanger can be made from a concentric pair of single lengths of pipe or from a number of lengths arranged in a vertical row with the sections connected at alternate ends, as shown in Fig. 21-6.

† To simplify the notation, Δt, as defined by Eq. (21-24), is used in this chapter to designate $\Delta t_{\text{overall}}$.

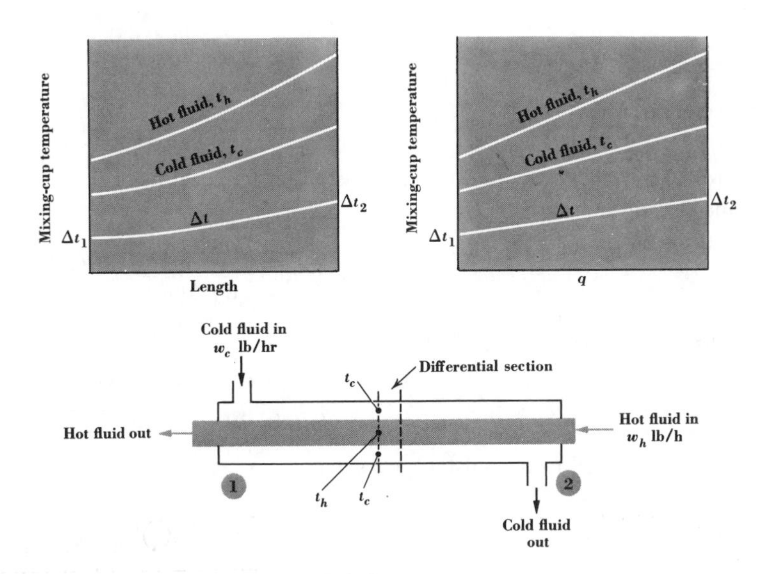

FIGURE 21-7
Temperature distribution in a double-pipe heat exchanger.

The exchanger shown in Fig. 21-6 can be regarded mathematically as one long run of two concentric pipes. The analysis below will be based on this simplification. Two fluids with constant but different inlet temperatures enter opposite ends of the exchanger shown in Fig. 21-7. In this system, the hot fluid enters the inner pipe at a constant rate of w_h lb/h and the cold fluid enters the annulus at a constant rate of w_c lb/h. The choice of path for each fluid depends on considerations such as corrosion, fluid pressure, and permissible pressure drop. The temperatures of the fluids as a function of length are shown in Fig. 21-7. The fluid temperatures usually do not vary linearly with distance from the inlet.

A heat balance can be written for the differential section of the exchanger shown in Fig. 21-7.

$$dq = w_c C_{pc}\, dt_c = w_h C_{ph}\, dt_h = U_o(t_h - t_c)\, dA_o \tag{21-25}$$

Equation (21-25) shows that if the specific heats are constant, the individual mixing-cup temperatures are linear with respect to q, as illustrated in Fig. 21-7. The difference between the mixing-cup temperatures, Δt, is also linear with respect to q. Consequently, we can express the derivative of Δt with respect to q in terms of the overall change in Δt and the total heat-transfer rate q in the exchanger. The derivative can be written as

$$\frac{d(\Delta t)}{dq} = \frac{\Delta t_2 - \Delta t_1}{q} \tag{21-26}$$

From Eq. (21-25) we obtain an expression for dq in terms of the overall coefficient and substitute it in Eq. (21-26) to give

$$\frac{d(\Delta t)}{U_o \, \Delta t \, dA_o} = \frac{\Delta t_2 - \Delta t_1}{q} \tag{21-27}$$

This equation is rearranged to give

$$\int_{\Delta t_1}^{\Delta t_2} \frac{d(\Delta t)}{U_o \, \Delta t} = \frac{\Delta t_2 - \Delta t_1}{q} \int_0^{A_o} dA_o \tag{21-28}$$

If U_o is assumed to be constant, we get

$$q = U_o A_o \left[\frac{\Delta t_2 - \Delta t_1}{\ln (\Delta t_2/\Delta t_1)} \right] \tag{21-29}$$

The quantity in brackets is the logarithmic mean temperature difference.

The same equation would have been obtained if the flow of the two streams in Fig. 21-7 had been concurrent. If $w_c C_{pc}$ and $w_h C_{ph}$ are equal, Eq. (21-25) shows that any change in temperature of the hot fluid along the exchanger is accompanied by equal change in the temperature of the cold fluid. In such a case, with countercurrent flow, Δt is the same at all cross sections in the exchanger, so that Δt_2 and Δt_1 are equal, making the right-hand side of Eq. (21-29) indeterminate. This presents no problem, however, because if Δt and U_o are both constant, Eq. (21-25) can be integrated directly and gives for the whole exchanger

$$q = U_o A_o \, \Delta t \tag{21-17}$$

Example 21-3 Crude oil flows at the rate of 2000 lb/h through the inside pipe of a double-pipe heat exchanger and is heated from 90 to 200°F. The heat is supplied by kerosene, initially at 450°F, flowing through the annular space. If the temperature of approach (minimum temperature difference between fluids) is 20°F, determine the heat-transfer area and the required kerosene flow rate for (*a*) concurrent flow and (*b*) countercurrent flow.

Data: Overall coefficient $U_o = 80$ Btu/(h)(ft^2)(°F)
Specific heat of crude oil = 0.56 Btu/(lb)(°F)
Specific heat of kerosene = 0.60 Btu/(lb)(°F)

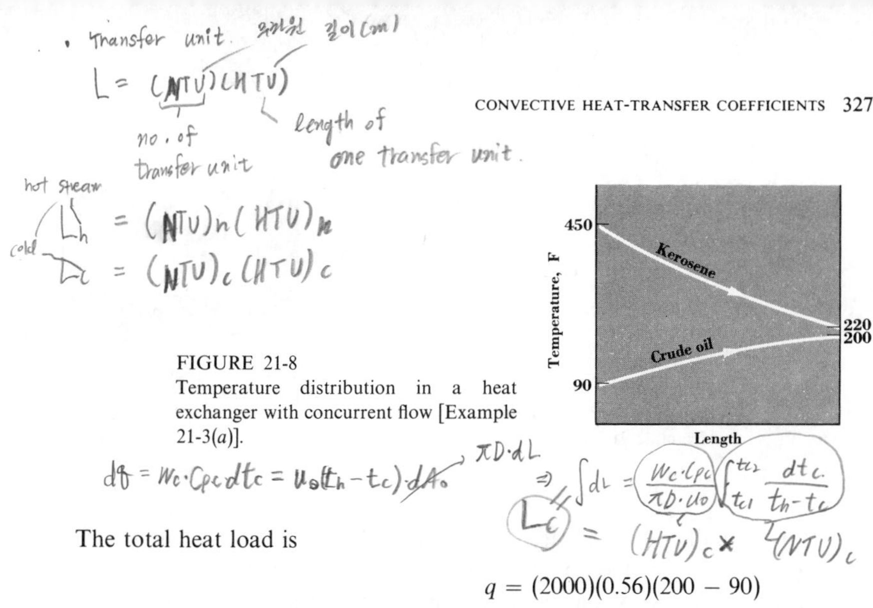

FIGURE 21-8
Temperature distribution in a heat exchanger with concurrent flow [Example 21-3(*a*)].

The total heat load is

$$q = (2000)(0.56)(200 - 90)$$
$$= 123{,}000 \text{ Btu/h}$$

(*a*) *Concurrent flow.* The temperature distribution is represented in Fig. 21-8. The exit temperature of kerosene is 220°F. The kerosene flow rate is

$$w_h = \frac{123{,}000}{(0.60)(450 - 220)}$$
$$= 891 \text{ lb/h}$$

The terminal temperature differences are $\Delta t_1 = 20°$ and $\Delta t_2 = 360°$ The tube area is found from Eq. (21-29).

$$A_o = \frac{123{,}000}{80[(360 - 20)/\ln(\frac{360}{20})]}$$
$$= 13.1 \text{ ft}^2$$

(*b*) *Countercurrent flow.* The temperature distribution is shown in Fig. 21-9. The exit temperature of kerosene in this case is 110°F, and the kerosene flow rate is

$$w_h = \frac{123{,}000}{(0.60)(450 - 110)}$$
$$= 603 \text{ lb/h}$$

The terminal temperature differences are $\Delta t_1 = 20°$ and $\Delta t_2 = 250°$. The tube area is

$$A_o = \frac{123{,}000}{80[(250 - 20)/\ln(\frac{250}{20})]}$$
$$= 16.9 \text{ ft}^2$$

////

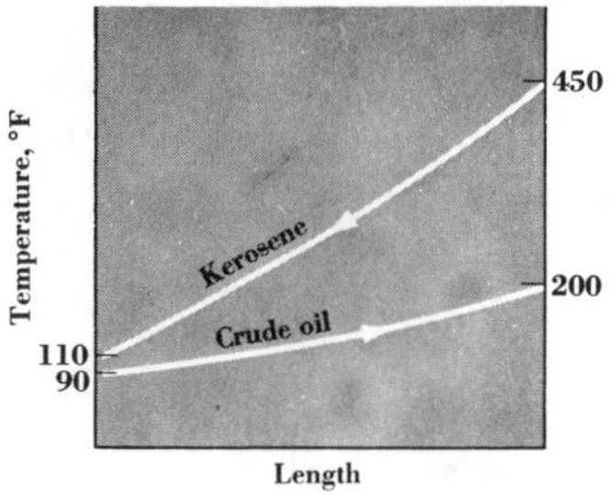

FIGURE 21-9
Temperature distribution in a heat exchanger with countercurrent flow [Example 21-3(*b*)].

If the overall coefficient varies within the exchanger, Eq. (21-29) is not rigorous, although it is often used as an approximation. In some circumstances the convective coefficient of one of the fluids can be expressed as a linear function of the mixing-cup temperature of the fluid. If this fluid offers the major resistance to heat transfer, the overall coefficient is approximately equal to the convective coefficient (as was shown earlier in this chapter), so that U_o is a linear function of the mixing-cup temperature of that fluid. Since the temperature of each fluid and the temperature difference Δt are linear functions of q, it can be shown algebraically that Δt is a linear function of each temperature. Under these circumstances the overall coefficient is a linear function of Δt. If we substitute such an expression,

$$U_o = a + b\,\Delta t$$

in Eq. (21-28), we obtain

$$\int_{\Delta t_1}^{\Delta t_2} \frac{d(\Delta t)}{\Delta t(a + b\,\Delta t)} = \frac{\Delta t_2 - \Delta t_1}{q} \int_0^{A_o} dA_o \qquad (21\text{-}30)$$

The integration of the left-hand side yields

$$\left[\frac{1}{a}\ln\frac{\Delta t}{a + b\,\Delta t}\right]_{\Delta t_1}^{\Delta t_2} = \left[\frac{1}{a}\ln\frac{\Delta t}{U_o}\right]_{(\Delta t/U_o)_1}^{(\Delta t/U_o)_2} = \frac{1}{a}\ln\frac{U_{o1}\,\Delta t_2}{U_{o2}\,\Delta t_1} \qquad (21\text{-}31)$$

The constant a may be found by algebraic manipulation of the definition of U_o as a linear function to give

$$a = \frac{U_{o1}\,\Delta t_2 - U_{o2}\,\Delta t_1}{\Delta t_2 - \Delta t_1} \qquad (21\text{-}32)$$

We substitute this expression for the constant a in Eq. (21-31) and write the integrated form of Eq. (21-30),

$$\frac{\Delta t_2 - \Delta t_1}{U_{o1}\,\Delta t_2 - U_{o2}\,\Delta t_1}\ln\frac{U_{o1}\,\Delta t_2}{U_{o2}\,\Delta t_1} = \frac{\Delta t_2 - \Delta t_1}{q} A_o \qquad (21\text{-}33)$$

which is rearranged to give

$$q = \frac{U_{o1}\,\Delta t_2 - U_{o2}\,\Delta t_1}{\ln\,(U_{o1}\,\Delta t_2/U_{o2}\,\Delta t_1)} A_o \qquad (21\text{-}34)$$

Thus it is not necessary to know the constants of the linear relation between the quantities U_o and Δt, but merely the terminal values of these quantities. If U_o is constant, Eq. (21-34) reduces to Eq. (21-29).

If the variation of U_o is so great that it cannot be represented adequately by a linear function of Δt, it is necessary to integrate Eq. (21-23) graphically or numerically. The quantity dq is replaced by either $w_h C_{ph}\,dt_h$ or $w_c C_{pc}\,dt_c$, and the equation becomes

$$w_h C_{ph} \int_{t_{h1}}^{t_{h2}} \frac{dt_h}{U_o(t_h - t_c)} = \int_0^{A_o} dA_o \qquad (21\text{-}35)$$

A relation between t_h and t_c at any cross section of the exchanger can be found by writing an enthalpy balance to include both fluids between that cross section and either end of the exchanger. This, in effect, is an integration of the middle terms of Eq. (21-25). The quantity U_o is calculated at that cross section from the individual coefficients expressed as resistances in the manner shown earlier in this chapter.

PROBLEMS

21-1 A double-pipe heat exchanger is to be designed to cool 5 gal/min of hot oil from 250°F to 120°F using 10 gal/min of cooling water available at 70°F. The exchanger is to consist of 12-ft-long sections of $\frac{3}{4}$-in, 16-gauge copper tubing inside $1\frac{1}{2}$-in, 16-gauge tubing, with the water flowing in the annular space. How many sections are required if the flow of the two streams is (*a*) concurrent, (*b*) countercurrent? The overall heat-transfer coefficient $U_o = 105$ Btu/(h)(ft²)(°F), the specific heat of the oil is 0.55 Btu/(lb)(°F), and the density of the oil is 52 lb/ft³. Assume that there is no heat transfer to or from the surroundings.

21-2 A jacketed kettle is used for removing 0.20 kg/s of water from an aqueous salt solution. Heat is supplied by steam condensing in the jacket at 120°C. The solution in the kettle boils at 104°C and has a latent heat of 2250 kJ/kg. The heat-transfer surface has an area of 50 m², is constructed of steel, and is 0.0060 m thick. Calculate the increase in capacity that might be expected if the heat-transfer surface is replaced by one of copper and all other conditions are unchanged.

21-3 An experimental study of the effect of scale on heat flux was conducted using a 1-in, 18-gauge copper tube. Water flowed inside the tube, and steam condensed on the outside. The overall coefficient U_o was determined over a wide range of water velocities for both clean and fouled tubes and was found to follow the equations

$$\frac{1}{U_o} = 0.00030 + \frac{1}{268 u_b^{0.8}} \qquad \text{clean tube}$$

$$= 0.00080 + \frac{1}{268 u_b^{0.8}} \qquad \text{fouled tube}$$

Find the fouling coefficient, the steam-side coefficient (assumed constant), and the water-side coefficient for a velocity $u_b = 2$ ft/s.

21-4 Tallow is stored in a vertical tank 3 m in diameter and 9 m high. The tank is insulated with a 0.05 m layer of 85 percent magnesia. To prevent freezing, the tallow is maintained at a temperature of 50°C by a heating coil consisting of a $\frac{3}{4}$-in, 20-gauge copper tube containing condensing steam at 135.8 kPa abs. Assuming that the minimum outdoor temperature is −20°C and that the tallow is always at a uniform temperature, compute the length of copper tubing needed to maintain the tank at 50°C in coldest weather. Neglect heat losses from the top and bottom of the tank.

Heat-transfer coefficient	h, W/m$^2 \cdot$K
Steam condensing inside coil	4540
Tubing wall to molten tallow	230
Molten tallow to tank wall	230
Outer surface of insulation to surroundings	10

21-5 A single-effect evaporator is removing 5000 gal water/week from a colloidal suspension which deposits scale on the steam-heating chest. It operates continuously except for 6 h, once a week, to clean the scale from the chest. Just before cleaning, the overall heat-transfer coefficient was found to be 50, and 1 h after resuming operation it was 230 Btu/(h)(ft^2)(°F).

Assuming that none of the scale formed on the chest falls off during the run and that scale deposition is proportional to the total amount of heat transferred, calculate the length of a run for maximum capacity. What would this capacity be, expressed as gallons of water evaporated per week?

21-6 A rectangular metal fin 0.03 m high and 0.003 m thick has a thermal conductivity of 45 W/m·K and a uniform base temperature of 120°C. It is exposed to an air stream at 20°C with a velocity past the fin such that the convective coefficient of heat transfer h is 85 W/m$^2 \cdot$K. Assuming that the temperature gradient parallel to the base of the fin is negligible, compute the temperature at the tip of the fin, the total heat flow from the fin, and the fin efficiency. The fin efficiency is defined as the actual heat flow from the fin divided by the heat flow if the fin were everywhere at the same temperature as at its base.

21-7 A 1-in-diameter rod 2 ft long is supported by two brass plates maintained at 600 and 500°F. The rod is in an atmosphere at 100°F and loses heat under conditions such that the combined heat-transfer coefficient covering both convection and radiation from its surface is 4 Btu/(h)(ft^2)(°F). Find the rate at which heat flows from the rod to the air in Btu/hr, making the following assumptions:

1 The thermal conductivity of rod, k, is constant at 100 Btu/(h)(ft)(°F).
2 The radial-temperature variation in the rod is negligible.

21-8 One end of a 0.01-m-diameter aluminum rod is being heated in a furnace which is at a uniform temperature of 320°C. The combined heat-transfer coefficient covering both convection and radiation from the furnace to the rod is 36 W/m$^2 \cdot$K. The rod extends into the furnace to a depth of 0.20 m. Outside the furnace the rod is exposed to air at a constant temperature of 25°C, and the combined heat-transfer coefficient here is 12 W/m^2·K. The rod is quite long, so its cold end may be assumed to be at room temperature. Find the temperature at the hot end of the rod assuming:

1 Heat flow and temperature distribution are independent of time.
2 The furnace wall is of zero thickness.
3 Each heat-transfer coefficient given above is uniform over the entire surface of the section of the rod to which it applies.
4 The radial-temperature gradient in the rod is zero.
5 The thermal conductivity of aluminum is constant at $k = 200$ W/m^2·K.

21-9 If an automobile radiator were made of aluminum instead of steel, what would be the approximate increase in cooling capacity?

21-10 A small metallic object at 25°C is dropped into a bath containing a large amount of hot oil at 200°C. The temperature of the metallic object, after 4 s of immersion, has risen to 140°C. Find the length of time required for the object to change temperature from 25 to 198°C assuming its temperature is uniform throughout at all times and the convection coefficient in the oil is constant.

21-11 Water is flowing in a thin-walled copper tube under conditions such that the coefficient of convective heat transfer from the water to the tube wall is represented by the empirical equation $h = 25u_b^{1/3}$ where u_b is given in ft/s and h is in Btu/(h)(ft^2)(°F). Heat from the water flows through the tube wall to the surrounding atmosphere at a rate such that the convection coefficient from the outer surface of the tube to the atmosphere is given by $h = 0.5(\Delta t)^{1/4}$ where h is given in Btu/(h)(ft^2)(°F) and Δt, the temperature drop between the tube wall and the atmosphere, is expressed in °F. Assuming the resistance to heat transfer by the tubing wall is negligible, find the minimum water velocity necessary to keep ice from forming in the pipe when the bulk water temperature is 35°F and the atmospheric temperature is −25°F.

21-12 Lead shot are formed by spraying molten lead from a shot tower and letting it cool and partially solidify as it falls through air. The molten droplets of lead are spherical and have a diameter of 0.01 ft. They enter the tower at a temperature of 650°F and fall for 1 s through air at 70°F before hitting the bottom of the tower. Find the weight fraction of solid in each pellet when it reaches the bottom. To simplify the problem assume the heat conduction rate within each lead pellet is so large compared to the heat transfer rate in the air around the pellet that the lead (either molten or solid, or both) may be considered to be at a uniform temperature throughout at any instant. The convective coefficient $h = 65$ Btu/(h)(ft^2)(°F). For lead, density = 710 lb/ft^3, specific heat = 0.034 Btu/(lb)(°F), melting point = 622°F, heat of fusion = 10.7 Btu/lb.

21-13 A sheet of glass coated on one side with an exceedingly thin, but transparent, electrically conducting layer is mounted vertically in a room in which the air is at 20°C. An electric current is passed through the conducting layer to generate heat at the rate of 2500 W/m^2 of conducting material. If the mean convection heat-transfer coefficients on both sides of the sheet are taken as $h = 11$ W/m^2·K find the temperature on both sides of the glass. The glass has a thickness of 0.006 m, but the electrically conducting layer may be considered to have zero thickness. The thermal conductivity of the glass is 0.7 W/m · K.

21-14 A form of treatment for bruised muscles is immersion in a stirred bath of hot liquid paraffin wax. In a particular case, the injured arm or leg immersed in the hot wax can be represented mathematically as an infinitely long, hollow cylinder with an OD of 4 in, an ID of 3.5 in and a thermal conductivity $k = 0.40$ Btu/(h)(ft)(°F). The inner surface of the cylinder is maintained at 98.6°F by a constant flow of cooling fluid. The liquified wax in the tank has a constant temperature of 135°F and will solidify if cooled to 120°F. Assuming that the convective heat-transfer coefficient of the liquid wax is always $h = 10$ Btu/(h)(ft^2)(°F), find out how much wax, if any, solidifies on the surface of the hollow cylinder. The thermal conductivity of solid paraffin wax is 0.14 Btu/(h)(ft)(°F).

21-15 The circulation of the blood in the finger maintains a temperature of 37°C at a short distance, say 0.0030 m, below the surface of the skin. If the nerve endings, which are temperature indicators, are 0.0015 m below the surface of the skin, and these attain a temperature of 43°C, your distress might be considered to be a limiting condition. Using these criteria, find the maximum temperature of water in which you could hold your finger assuming the system is adequately modelled as flat, parallel planes.

Data

1 Thermal conductivity of flesh and blood = 0.60 W/m · K.

2 Convective coefficient for finger dipped in water, $h = 570$ W/m^2 · K.

21-16 (*a*) A single pane of glass [$k = 0.50$ Btu/(h)(ft)(°F)], $\frac{1}{4}$ in thick, separates the air in a room, which is at a bulk temperature of 70°F, from the air outdoors, where the bulk temperature is 0°F. What temperature would a thermocouple attached to the surface of the glass inside the room read? Assume the heat-transfer coefficients, h_i and h_o are both uniform and constant at 1 Btu/(h)(ft^2)(°F).

(*b*) It has been suggested that the outside bulk atmospheric temperature could be determined approximately by fastening a thermocouple to the inner surface of the window as in part (*a*) and covering it with a thick layer of insulation [$k = 0.06$ Btu/(h)(ft)(°F)]. What thickness of insulation would be required to give a reading of the thermocouple within 5° of the bulk outdoors temperature? Assume the convection coefficients are the same as in part (*a*).

(*c*) What, in fact, would happen to each of the convection coefficients given in part (*a*) if the insulation were added as suggested in part (*b*)?

21-17 100 kg of water in a closed insulated autoclave is heated from 20 to 90°C by a heating coil made of 12 m of 1-in, 15-BWG-gauge copper tubing (ID = 2.174 cm). When steam at 100°C is used in the coil, it is found that it takes 40 min to heat the water. The steam-side coefficient h_i is known to be 12,000 W/m^2 · K.

(*a*) If a propeller agitator stirs the water, the water-side coefficient h_o doubles. How long would the heating now require?

(*b*) If high-pressure steam at 150°C is used (with no agitation), how long would the heating take?

21-18 A heat pipe consists of a closed copper tube 18 in long containing a pure fluid at a pressure such that it boils and condenses at 120°F. All the heat enters at one end and leaves at the opposite end. The rest of the tube is insulated. The hot end of the tube is at 200°F and the cold end is at 70°F. If the boiling coefficient is $h = 500$ Btu/(h)(ft^2)(°F), what is the condensing coefficient? If the heat pipe had to be replaced by a solid cylindrical rod with the same dimensions and boundary conditions, what would the thermal conductivity of the rod need to be to remove heat at the same rate as the heat pipe?

21-19 The wall of a house has an average thermal conductivity of 0.10 Btu/(h)(ft)(°F). The bulk outside air temperature is 0°F and the bulk inside air temperature is 70°F. If the inside wall surface is at a temperature of 50°F, what is the wall thickness? Assume that the heat transfer coefficient on each surface can be represented by the empirical equation $h = 0.18\ (\Delta t)^{1/3}$, where h is in Btu/(h)(ft^2)(°F) and Δt is in °F.

21-20 A refrigerant is flowing through a $\frac{3}{4}$-in copper tube under conditions such that its mixing-cup temperature is 15°F. Condensation of moisture from the atmosphere on the outside of the tube is considered a nuisance, so it is decided to insulate the tube with glass wool [$k = 0.025$ Btu/(h)(ft)(°F)]. Find the minimum thickness of glass wool necessary to prevent condensation on the outer surface of the insulation. Air in the room is at 70°F but it has a dew-point temperature of 45°F. The natural convection heat-transfer coefficient between the insulation surface and the air in the room is represented by $h = 0.5\ (\Delta t)^{1/4}$, in which Δt is the temperature drop in °F between the insulation surface and the air in the room, and h is in Btu/(h)(ft^2)(°F). The convective coefficient between the inner wall of the tube and the refrigerant is sufficiently large that the resistance to heat transfer in the refrigerant is negligible. The resistance of the copper tube to heat transfer is also negligible.

21-21 Fissionable material generating heat at a steady rate of 5.8×10^7 Btu/(h)(ft^3) is to be formed into thin slabs which will be held in a frame with cooling fluid circulating past each of the two principal faces. If the maximum allowable temperature of the fissionable material is 1000°F, find the maximum slab thickness for the following conditions:

1 Bulk temperature of cooling fluid is 80°F.
2 Convective heat transfer coefficient on each face: $h = 1000$ Btu/(h)(ft^2)(°F).
3 Thermal conductivity of fissionable material: $k = 20$ Btu/(h)(ft)(°F).
4 Fissionable material and cooling fluid are in direct contact.

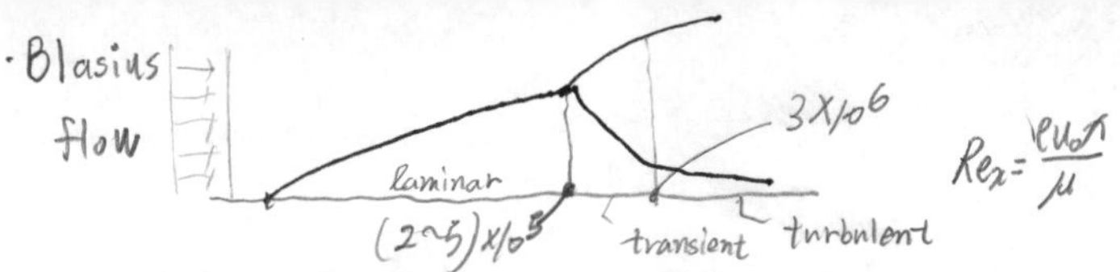

22

HEAT TRANSFER WITH LAMINAR FLOW

In this chapter we shall study some problems of heat transfer in a fluid in laminar flow. In Chaps. 10 and 11 we saw that the equation of continuity could be combined with the Navier-Stokes equations to give solutions for problems of isothermal laminar flow in some important situations. To solve heat-transfer problems, the differential energy balance must be considered simultaneously. As before, we shall see that the problems involving laminar flow are easier to solve than problems with turbulent flow.

The heat-transfer problems in laminar flow of interest to engineers will be considered in two categories dependent on flow pattern:

1 The flow pattern varies with the distance from the leading edge of the system. This occurs with all external flow systems and also near the inlet of all internal flow systems.

2 The flow pattern is the same at all cross sections normal to flow. This is called "developed" flow and occurs only in internal flow systems far from the entrance.

All flow systems must have some entrance effect. Furthermore, the physical properties of all fluids are dependent on temperature and differ at all points in a nonisothermal system. Therefore, in a system transferring heat, a uniform flow pattern is never developed even at great distances from the entrance. From this we conclude that the concept of "developed" laminar flow with heat transfer is only an idealization. However, the solution of the differential balances when the physical properties vary throughout the system is quite difficult, so in our discussions we shall assume that these properties are constant. For these idealized systems developed flow is an attainable condition.

Most of our attention will be given in this chapter to one example of each class of problem cited above. First, we shall consider heat transfer between a fluid and a flat plate. From the differential energy balance we shall obtain a solution in which the temperature of the fluid is a function of such independent variables as position and free-stream velocity. As a second example, we shall apply the differential energy balance (in cylindrical coordinates) to a fluid being heated in a pipe. The solution for this system also gives the temperature as a function of position and velocity.

Although these solutions of the differential energy balance provide enough information to solve most practical problems, engineers prefer to work with heat-transfer coefficients. As a result, it is customary to express heat-transfer performance, even for laminar flow, in terms of these coefficients. The procedure for converting an equation expressing temperature as a function of position into an equation for the heat-transfer coefficient will be illustrated later in this chapter. The heat-transfer coefficients for laminar flow have a strong dependence on position. This is not usually the case for heat transfer with turbulent flow.

Heat Transfer with a Developing Velocity Distribution

Laminar flow parallel to a flat plate The problem of heat transfer to a fluid flowing in laminar motion parallel to a flat plate has been solved by Pohlhausen.[1] His solution is based upon the differential balances of mass, energy, and momentum which were derived earlier in Chaps. 7 to 9. The application of the balances of mass and momentum to the problem of determining the velocity field for isothermal, laminar flow past a flat plate was described in Chap. 11. To extend the discussion to include heat transfer from the flat plate to the fluid, we start by writing the differential energy balance for incompressible flow without heat generation.

$$u_x \frac{\partial t}{\partial x} + u_y \frac{\partial t}{\partial y} + u_z \frac{\partial t}{\partial z} + \frac{\partial t}{\partial \theta} = \frac{k}{\rho C_p}\left(\frac{\partial^2 t}{\partial x^2} + \frac{\partial^2 t}{\partial y^2} + \frac{\partial^2 t}{\partial z^2}\right) \qquad (8\text{-}11)$$

[1] E. Pohlhausen, *Z. angew. Math.u. Mech.*, **1**:115 (1921). See also J. G. Knudsen and D. L. Katz, "Fluid Dynamics and Heat Transfer," McGraw-Hill Book Company, New York, 1958.

If the flow is two-dimensional, $u_z = 0$. At steady state $\partial t/\partial \theta = 0$. In addition, conduction is neglected in all directions except normal to the plate, so $\partial^2 t/\partial x^2 = \partial^2 t/\partial z^2 = 0$. The resulting equation is

$$u_x \frac{\partial t}{\partial x} + u_y \frac{\partial t}{\partial y} = \frac{k}{\rho C_p} \frac{\partial^2 t}{\partial y^2} \tag{22-1}$$

This equation is similar to the simplified momentum equation derived earlier,

$$u_x \frac{\partial u_x}{\partial x} + u_y \frac{\partial u_x}{\partial y} = \frac{\mu}{\rho} \frac{\partial^2 u_x}{\partial y^2} \tag{22-2}$$

When Eq. (22-2) was combined with the continuity equation (10-44), a solution was obtained for the velocity distribution in the laminar boundary layer. This is shown in Fig. 11-8, which presents the dimensionless velocity u_x/u_o as a function of $y\sqrt{u_o/\nu x}$ (designated as η).

If the fluid has a Prandtl number $C_p\mu/k$ of 1, then it can be seen that the thermal diffusivity $k/\rho C_p$ and the kinematic viscosity μ/ρ are equal. Furthermore, if the temperature t in Eq. (22-1) is replaced by a dimensionless variable $(t_s - t)/(t_s - t_o)$, it follows that Eqs. (22-1) and (22-2) have the same boundary conditions. We consider here only the case for which the plate is at a uniform temperature t_s; for convenience we choose t_s as greater than the fluid temperature t_o at the edge of the thermal boundary layer.

$$\text{At } y = 0 \qquad \frac{t_s - t}{t_s - t_o} = 0, \quad \frac{u_x}{u_o} = 0$$

$$\text{At } y = \infty \qquad \frac{t_s - t}{t_s - t_o} = 1, \quad \frac{u_x}{u_o} = 1$$

$$\text{At } x = 0 \qquad \frac{t_s - t}{t_s - t_o} = 1, \quad \frac{u_x}{u_o} = 1$$

We see that the differential equations for energy and momentum, (22-1) and (22-2), are identical when applied to fluids with a Prandtl number of 1. Therefore they have identical solutions; i.e., for any point (x,y) in the flow system, the dimensionless temperature and velocity variables $(t_s - t)/(t_s - t_o)$ and u_x/u_o are equal. Thus the velocity-profile solution shown in Fig. 11-8 is equally valid for the determination of temperature profiles. This implies that in this case the transfers of heat and momentum are directly analogous, and the thermal and hydrodynamic boundary layers are of equal thickness. These results are significant because many gases and some liquids (such as water at 350°F) have a Prandtl number of approximately 1. In general, however, liquids have Prandtl numbers in the range of 0.001 to 1000 and sometimes beyond. Pohlhausen extended the solution to cover fluids with a Prandtl number not equal to 1 by the procedure outlined below.

We use the variables defined in Chap. 11:

$$\eta = y\sqrt{\frac{u_o}{\nu x}} \tag{11-15}$$

$$f(\eta) = \frac{\psi}{\sqrt{x\nu u_o}} \tag{11-16}$$

$$u_x = \frac{\partial \psi}{\partial y} = u_o f'(\eta) \tag{11-19}$$

$$u_y = -\frac{\partial \psi}{\partial x} = \frac{1}{2}\sqrt{\frac{\nu u_o}{x}}[\eta f'(\eta) - f(\eta)] \tag{11-20}$$

and the dimensionless temperature $(t_s - t)/(t_s - t_o)$. The differential energy balance, Eq. (22-1), then reduces to the following equation:

$$\frac{d^2\left(\dfrac{t_s - t}{t_s - t_o}\right)}{d\eta^2} + \frac{\mathbf{Pr}[f(\eta)]}{2}\frac{d\left(\dfrac{t_s - t}{t_s - t_o}\right)}{d\eta} = 0 \tag{22-3}$$

This ordinary differential equation can be integrated by standard procedures. For simplicity, we shall express the unaccomplished temperature change $(t_s - t)/(t_s - t_o)$ by the symbol Y. Equation (22-3) and the boundary conditions expressed in terms of this variable are

$$\frac{d^2 Y}{d\eta^2} + \frac{\mathbf{Pr}[f(\eta)]}{2}\frac{dY}{d\eta} = 0 \tag{22-4}$$

$$y = 0, \qquad \eta = 0, \qquad Y = 0$$

$$y = \infty, \qquad \eta = \infty, \qquad Y = 1$$

We let $dY/d\eta = p$ and rewrite Eq. (22-4) as

$$\frac{dp}{d\eta} + \frac{\mathbf{Pr}[f(\eta)]}{2}p = 0 \tag{22-5}$$

$$\frac{dp}{p} = -\frac{\mathbf{Pr}[f(\eta)]}{2}d\eta \tag{22-6}$$

This is integrated to give

$$p = C_1 \exp\left(-\mathbf{Pr}/2\int_0^\eta f(\eta)\,d\eta\right)$$

$$= \frac{dY}{d\eta} \tag{22-7}$$

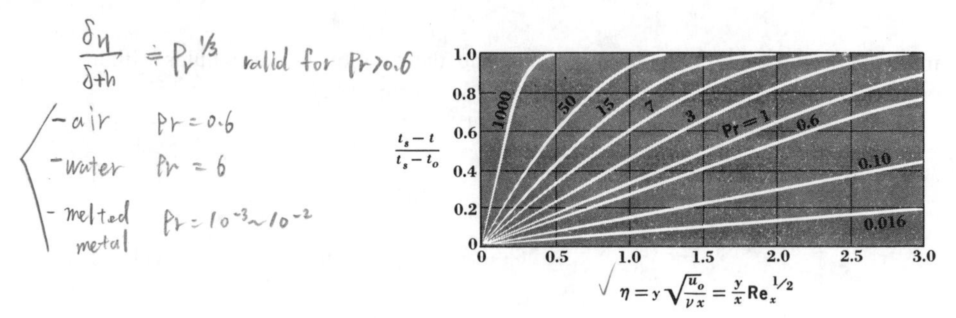

FIGURE 22-1
Temperatures for laminar flow past a flat plate at uniform temperature; Prandtl number range, 0.016 to 1000.

Integrating a second time gives

$$Y = C_1 \int_0^\eta \exp\left(-\frac{\mathbf{Pr}}{2}\int_0^\eta f(\eta)\,d\eta\right)d\eta + C_2 \tag{22-8}$$

Substitution of the first boundary condition ($\eta = 0$, $Y = 0$) shows that $C_2 = 0$. Substitution of the second boundary condition ($\eta = \infty$, $Y = 1$) provides an expression for C_1.

$$C_1 = \frac{1}{\int_0^\infty \{\exp[-(\mathbf{Pr}/2)\int_0^\eta f(\eta)\,d\eta]\}\,d\eta} \tag{22-9}$$

Thus, the complete solution for the temperature profile in the boundary layer is

$$Y = \frac{t_s - t}{t_s - t_o} = \frac{\int_0^\eta \{\exp[-(\mathbf{Pr}/2)\int_0^\eta f(\eta)\,d\eta]\}\,d\eta}{\int_0^\infty \{\exp[-(\mathbf{Pr}/2)\int_0^\eta f(\eta)\,d\eta]\}\,d\eta} \tag{22-10}$$

The values of $f(\eta)$ which were determined for the isothermal-flow problem are represented by a series given earlier in this book [Eq. (11-21)]. They were used by Pohlhausen to obtain the temperature profiles for fluids with a wide range of Prandtl numbers. A number of such curves are shown in Fig. 22-1. The quantity η used in Fig. 22-1 can also be written as $(y/x)\sqrt{u_o x/\nu}$, in which two dimensionless groups, y/x and $u_o x/\nu$, can be discerned. The latter group is a local Reynolds number and is usually written $\mathbf{Re}_x$, indicating that it applies at a distance x from the leading edge of the plate.

We shall now use the relation shown in Eq. (22-10) to find the heat-transfer coefficient. This quantity was defined in the previous chapter by the equation

$$h = k\left|\frac{d[(t - t_s)/(t_o - t_s)]}{dy}\right|_{y=0} \tag{21-6}$$

In view of the boundary conditions of this system, we rewrite Eq. (21-6) as

$$h = k\left\{\frac{d[(t_s - t)/(t_s - t_o)]}{dy}\right\}_{y=0} \tag{22-11}$$

This is rewritten substituting $\eta\sqrt{\nu x/u_0} = y$ to give

$$h = k\sqrt{\frac{u_0}{\nu x}}\left[\frac{dY}{d\eta}\right]_{\eta=0} \tag{22-12}$$

The derivative in this equation has already been given in Eq. (22-7). Applying the condition $\eta = 0$ to this expression we find the exponential term to be equal to unity and the derivative to be $dY/d\eta = C_1$. Therefore, we can write

$$h = k\sqrt{\frac{u_0}{\nu x}}\, C_1 \tag{22-13}$$

The value of h given by this equation can be seen to be a function of the distance x from the leading edge of the plate. Thus, h is called the *local heat-transfer coefficient* and its dependence on x emphasized by writing it as h_x. The quantity $h_x x/k$ is an important dimensionless group in heat transfer, known as the *Nusselt number*. The quantity x is a characteristic length. When the Nusselt group contains a heat-transfer coefficient dependent on position h_x, then the Nusselt number is likewise dependent on position and is written $\mathbf{Nu}_x$. It is apparent, therefore, that Eq. (22-13) can be written entirely in terms of dimensionless groups.

$$\frac{h_x x}{k} = \sqrt{\frac{x u_0}{\nu}}\, C_1 \tag{22-14}$$

$$\mathbf{Nu}_x = (\mathbf{Re}_x)^{1/2}\left(\int_0^\infty \left\{\exp\left[-\frac{\mathbf{Pr}}{2}\int_0^\eta f(\eta)\, d\eta\right]\right\} d\eta\right)^{-1} \tag{22-15}$$

This expression is valid for all Prandtl numbers, but for Reynolds numbers only in the laminar boundary-layer range, i.e., below 5×10^5.

Equation (22-15), although of wide validity, is tedious to evaluate. For fluids with the more limited range, $\mathbf{Pr} > 0.6$, the curves on Fig. 22-1 can be represented empirically by a single line relating the temperature variable $(t_s - t)/(t_s - t_o)$ to the quantity $(y/x)(\mathbf{Re}_x)^{1/2}\mathbf{Pr}^{1/3}$, as shown in Fig. 22-2. This procedure is based on the equation found by Pohlhausen for the ratio of the thickness of the hydrodynamic boundary layer δ to the thickness of the thermal boundary layer δ_{th}:

$$\frac{\delta}{\delta_{th}} = \mathbf{Pr}^{1/3} \tag{22-16}$$

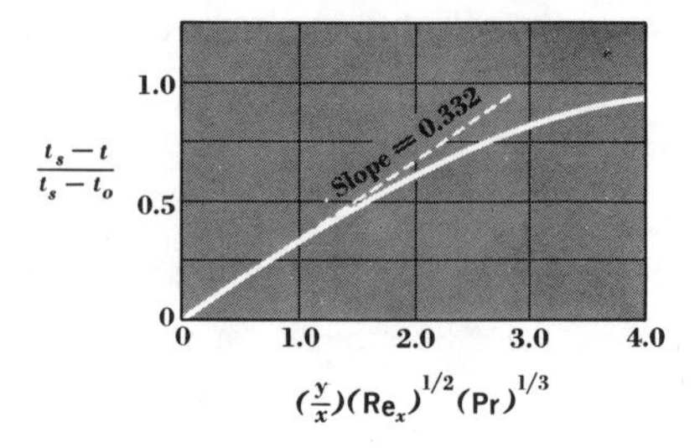

FIGURE 22-2
Temperatures for laminar flow past a flat plate at uniform temperature; $\mathbf{Pr} > 0.6$.

This approximation is valid for $\mathbf{Pr} > 0.6$; the principal substances thus excluded are liquid metals.[1]

The quantity $\left\{\frac{\partial[(t_s - t)/(t_s - t_o)]}{\partial y}\right\}_{y=0}$ can be obtained for fluids with $\mathbf{Pr} > 0.6$ by taking the slope of the curve in Fig. 22-2 at $(y/x)(\mathbf{Re}_x)^{1/2}\mathbf{Pr}^{1/3} = 0$. This slope is 0.332.

$$\left\{\frac{\partial[(t_s - t)/(t_s - t_o)]}{\partial y}\right\}_{y=0} = \frac{0.332}{x}(\mathbf{Re}_x)^{1/2}\mathbf{Pr}^{1/3} \tag{22-17}$$

Consequently, the local heat-transfer coefficient h_x can be represented by the equation

$$h_x = \frac{0.332k}{x}(\mathbf{Re}_x)^{1/2}\mathbf{Pr}^{1/3} \tag{22-18}$$

and the local Nusselt number by

$$\mathbf{Nu}_x = 0.332(\mathbf{Re}_x)^{1/2}\mathbf{Pr}^{1/3} \tag{22-19}$$

Equation (22-18) gives the local heat-transfer coefficient h_x at a distance x from the leading edge of the plate. If the effective mean coefficient h_m for the entire section of the plate between the leading edge and point x is desired, this can be obtained as follows:

$$q = h_m A(t_s - t_o) = \int_0^L h_x(t_s - t_o)b\,dx \tag{22-20}$$

where b is the width of the plate in the z direction. Since the plate is at a uniform temperature, $t_s - t_o$ is constant. We get, then,

[1] S. W. Churchill and H. Ozoe, *AIChE Journal*, **19**:177 (1973), confirm Eq. (22-16) for $\mathbf{Pr} \to \infty$ and indicate that $\delta/\delta_{th} = \mathbf{Pr}^{1/2}$ for $\mathbf{Pr} \to 0$.

$$h_m = \frac{b}{A} \int_0^L h_x \, dx$$

$$= \frac{0.332k}{L} \left(\frac{u_o}{\nu} \right)^{1/2} \left(\frac{C_p \mu}{k} \right)^{1/3} \int_0^L \frac{dx}{x^{1/2}} \qquad (22\text{-}21)$$

Therefore the equation for the mean coefficient h_m is

$$h_m = \frac{0.332k}{L} \left(\frac{u_o}{\nu} \right)^{1/2} \left(\frac{C_p \mu}{k} \right)^{1/3} (2L^{1/2})$$

$$= \frac{0.664k}{L} \left(\frac{u_o L}{\nu} \right)^{1/2} \left(\frac{C_p \mu}{k} \right)^{1/3} \qquad (22\text{-}22)$$

This may be written in terms of dimensionless numbers as

$$\mathbf{Nu}_m = 0.664(\mathbf{Re}_L)^{1/2}\mathbf{Pr}^{1/3} \qquad (22\text{-}23)$$

The symbol $\mathbf{Re}_x$ is used in equations involving point coefficients. The symbol $\mathbf{Re}_L$ is used in equations involving average coefficients which apply over a surface of length L.

It was pointed out in Chap. 11 that the laminar boundary layer persists on smooth plates to a Reynolds number of approximately 5×10^5. Obviously, the analysis of heat transfer given in this chapter is governed by the same restrictions. Furthermore, it should not be forgotten that Eqs. (22-16) to (22-23) apply only for $\mathbf{Pr} > 0.6$.

Example 22-1 Air at 70°F with a uniform free-stream velocity of 50 ft/s is moving parallel to a smooth flat plate heated to a uniform surface temperature of 212°F. Find the length of the laminar boundary layer on the plate, the thickness of the hydrodynamic and thermal boundary layers at the critical length, the local heat-transfer coefficient at the critical length, and the mean heat-transfer coefficient over the portion of the plate covered by the laminar boundary layer.

Air at 150°F has the following physical properties:[1]

$$\nu = 0.21 \times 10^{-3} \text{ ft}^2/\text{s}$$

$$k = 0.0164 \text{ Btu/(h)(ft)(°F)}$$

$$\mathbf{Pr} = 0.72$$

[1] F. Kreith, "Principles of Heat Transfer," International Textbook Company, Scranton, Pa., 1958.

The length of laminar boundary layer is found by taking the critical Reynolds number $\mathbf{Re}_x$ as 5×10^5. Solving for x, we get

$$x = (5 \times 10^5)\frac{\nu}{u_o} = \frac{(5 \times 10^5)(0.21 \times 10^{-3})}{50}$$

$$= 2.1 \text{ ft}$$

The thickness of the hydrodynamic boundary layer can be found from Eq. (11-22).

$$\delta = 5.0\sqrt{\frac{\nu x}{u_o}}$$

At $x = 2.1$ ft, we find

$$\delta = 5\sqrt{\frac{(0.21 \times 10^{-3})(2.1)}{50}}$$

$$= 0.0148 \text{ ft}$$

The thermal-boundary-layer thickness at $x = 2.1$ ft is found from Eq. (22-16).

$$\frac{\delta}{\delta_{th}} = \mathbf{Pr}^{1/3}$$

$$\delta_{th} = \frac{0.0148}{(0.72)^{1/3}}$$

$$= \frac{0.0148}{0.895}$$

$$= 0.0165 \text{ ft}$$

The local heat-transfer coefficient at $x = 2.1$ ft is found from Eq. (22-18).

$$h_x = \frac{0.332k}{x}(\mathbf{Re}_x)^{1/2}\mathbf{Pr}^{1/3}$$

$$= \frac{(0.332)(0.0164)}{(2.1)}(5 \times 10^5)^{1/2}(0.72)^{1/3}$$

$$= 1.65 \text{ Btu/(h)(ft}^2\text{)(°F)}$$

The mean effective heat-transfer coefficient for the section of the plate between the leading edge and the point where the laminar boundary layer breaks down

($x = 2.1$ ft) is twice the local coefficient at $x = 2.1$; thus $h_m = 2h_x = 3.3$ Btu/(h)(ft^2)(°F).

Heat transfer to a fluid entering a pipe A fluid usually enters a pipe with uniform velocity and temperature profiles. If the pipe wall is heated or cooled over its entire length, then both thermal and hydrodynamic velocity profiles begin developing at the pipe inlet. As in the case of flow parallel to a flat plate, the local coefficient of heat transfer has a value of infinity at the inlet and diminishes toward an asymptotic value far down the pipe.

The problem of the developing hydrodynamic boundary layer has been solved for isothermal flow and has already been discussed in Chap. 11. The problem in which the hydrodynamic and thermal boundary layers develop simultaneously has been studied by Kays.[1] His approach was to combine Langhaar's results for the developing velocity profile with a numerical solution of the differential energy balance. He obtained solutions, limited to fluids with **Pr** = 0.7, for boundary conditions of constant wall temperature, uniform heat flux from the wall, and constant temperature difference between the wall and the fluid.

Natural convection from a vertical plate An important system to the engineer is that in which heat is transferred from a vertical plate to a fluid moving parallel to it by natural convection. In any system in which the density of the fluid varies with position, natural convection is present. However, when the fluid is also moving as a result of forced convection, the natural-convection effects are not often important. If the flow due to forced convection is reduced, a region is entered in which both mechanisms are important. The example we shall consider in this section is the limiting case in which flow results entirely from heating.

This problem can be solved by using the energy and continuity equations given in the foregoing analysis for forced laminar flow past a heated plate. However, the differential momentum balance must include a body-force term due to the effect of gravity on the heated fluid. The resulting momentum balance is written as

$$\rho\left(u_x \frac{\partial u_x}{\partial x} + u_y \frac{\partial u_x}{\partial y}\right) = g\rho\beta(t - t_\infty) + \mu \frac{\partial^2 u_x}{\partial y^2} \qquad (22\text{-}24)$$

Although the body force caused by the variation of density is important in this equation, the effects of compressibility on the other terms in the differential balances are not important, so that they may be used with the assumption of constant physical properties. This set of equations has been solved[2] for the case of natural-convection

[1] W. M. Kays, *Trans. Am. Soc. Mech. Eng.*, 77:1265 (1955).

[2] Pohlhausen's solution is outlined in standard references such as M. Jakob, "Heat Transfer," vol. 1, p. 444, John Wiley & Sons, Inc., New York, 1949.

heating of ideal diatomic gases adjacent to a vertical plate at constant temperature. The solution expressed in dimensionless groups is

$$\mathbf{Nu}_m = 0.478(\mathbf{Gr})^{1/4} \tag{22-25}$$

where the mean Nusselt number $\mathbf{Nu}_m = h_m L/k$ and the Grashof number $\mathbf{Gr} = [gL^3(t_s - t_\infty)]/v^2 T$.

The heat-transfer coefficient h_m is the mean coefficient applicable between the lower edge of the plate and any point at a distance L above the lower edge. It is used with a temperature difference which is the wall temperature minus the fluid temperature outside the thermal boundary layer, t_∞ or t_o. The temperature in the denominator of the Grashof number must be expressed in absolute units because the analytical solution contains the assumption that the coefficient of volumetric expansion β is equal to the reciprocal of the absolute temperature T. This is true only for ideal gases. The Grashof number is found in most correlations of natural-convection heat transfer. It serves in much the same role as the Reynolds number in forced convection heat transfer. The relationship between the two groups is made more evident if the Grashof number is broken down as follows:

$$\mathbf{Gr} = \frac{gL^3\beta\,\Delta t}{v^2}$$

$$= \left(\frac{g\beta\,\Delta t}{uv/L^2}\right)\left(\frac{Lu^2}{uv}\right)$$

$$\sim \left(\frac{\text{buoyant force}}{\text{viscous drag}}\right)\left(\frac{\text{inertial force}}{\text{viscous drag}}\right)$$

$$\sim \left(\frac{\text{buoyant force}}{\text{viscous drag}}\right)(\text{Reynolds number})$$

The velocity u introduced above may be regarded as any characteristic velocity of the system.

Equation (22-25) is valid only for systems in which the flow is completely laminar. If the vertical plate for which it applies is of sufficient length, the boundary layer will become turbulent. The transition from a natural-convection laminar boundary layer to a natural-convection turbulent boundary layer has been observed experimentally to depend on the value of the product **Pr Gr**. This combination of dimensionless groups constitutes a new dimensionless group known as the *Rayleigh number* **Ra**. When $\mathbf{Ra} < 10^9$, the boundary layer is entirely laminar and Eq. (22-25) applies. When $\mathbf{Ra} > 10^9$, the upper portion of the boundary layer becomes turbulent, and equations to be presented in subsequent chapters are required if the entire system is to be dealt with.

In the start-up of a process, the portions of the system heating or cooling by natural convection will experience variations of temperature with time. A case of particular significance is the nuclear reactor in which the removal of control rods in the absence of forced circulation could give rise to excessively high temperatures during the interval in which natural circulation is becoming established. Electrical components constitute another example of a system which, in most circumstances, is expected to cool by natural convection and for which transient behavior may be significant. Most studies of transient natural convection involve numerical solutions of the basic energy equation applied to simple systems. An extensive summary of work in this area is provided by Gebhart.[1]

Other systems Engineers are interested in laminar flow in other heat-transfer systems with developing velocity profiles. One system is the flat plate in which heating is begun at some distance from the leading edge. The hydrodynamic boundary layer in this case is already partially developed when the thermal boundary layer starts to develop.

The problem of flow normal to a heated cylinder or sphere is of considerable importance. At low Reynolds numbers, for which there is laminar flow with no separation, it is possible to obtain numerical solutions of the differential balances. However, for most practical cases the fluid separates from the sides of the cylinder, and empirical methods must be used. This problem will be considered more extensively in Chap. 23, "Heat Transfer with Turbulent Flow."

Heat Transfer with a Developed Velocity Distribution in a Pipe

The heat-transfer problems most frequently encountered have to do with the heating and cooling of fluids in pipes. Although the entrance effects which we have just discussed affect overall performance significantly in short pieces of equipment, we are often concerned with the performance of equipment in which the entrance effect is negligible.

The velocity profile in developed, isothermal, laminar flow has the shape of a parabola. The equation for this profile was derived first in this book in Chap. 2 from a force balance, and subsequently in Chap. 10 from the differential mass and momentum balances. It is

$$u_x = 2u_b\left[1 - \left(\frac{r}{r_i}\right)^2\right] \tag{22-26}$$

[1] B. Gebhart, "Heat Transfer," 2d ed., p. 356, McGraw-Hill Book Company, 1971.

If the fluid is heated or cooled, the velocity profile can be greatly altered because of the effect of temperature on viscosity. The complications resulting in the heat-transfer problems are so great that only approximate solutions have been obtained. Graetz[1] has provided solutions for two cases. In one the distortion of the velocity profile is assumed to be negligible and the parabolic profile is maintained. In the other solution, the distortion is assumed to be so great that the velocity profile is flat over the pipe cross section. This type of flow is called *rodlike*, or *plug*, flow. Such a condition might be approximately achieved if a liquid in laminar flow were heated from the pipe wall. The two Graetz solutions are discussed below.

Parabolic velocity profile This solution is obtained by making the assumption that Eq. (22-26) represents the axial fluid velocity at all points in the system. As boundary conditions, the wall temperature of the tube is assumed to be uniform and the inlet fluid temperature is assumed to be constant over the pipe cross section.

The differential energy balance, Eq. (8-11), can be expressed in cylindrical coordinates and, if applied to a steady-state system with axial symmetry, reduces to

$$u_x \frac{\partial t}{\partial x} = \alpha \left(\frac{\partial^2 t}{\partial r^2} + \frac{1}{r} \frac{\partial t}{\partial r} + \frac{\partial^2 t}{\partial x^2} \right) \tag{22-27}$$

If the further assumption is made that heat conduction in the direction of flow is negligible in comparison with the other transport terms, then $\partial^2 t/\partial x^2$ can be dropped from Eq. (22-27). If Eq. (22-26) is combined with this simplified energy equation, we obtain

$$\frac{\partial^2 t}{\partial r^2} + \frac{1}{r} \frac{\partial t}{\partial r} = \frac{2u_b}{\alpha} \left[1 - \left(\frac{r}{r_i} \right)^2 \right] \frac{\partial t}{\partial x} \tag{22-28}$$

This partial differential equation can be solved by the technique of separation of variables, illustrated in Chap. 19. The temperature variable is assumed to be the product of two quantities, one of which is a function of radial distance r and the other a function of axial distance x. Two ordinary differential equations are obtained. The solution of the equation relating t and x is a simple exponential relation. The solution of the equation relating t and r is a series. The combined solutions give the equation

$$\frac{t_s - t}{t_s - t_o} = \sum_{n=0}^{n=\infty} B_n \phi_n e^{\beta_n^2 (x/r_i)/\mathbf{Re}\,\mathbf{Pr}} \tag{22-29}$$

The values of the constants B_n and β_n have been provided for the first 10 terms in the series by Sellars, Tribus, and Klein.[2] The same reference contains equations for

[1] L. Graetz, *Ann. Phys.*, N.F., **18**:79 (1883) and **25**:737 (1885).

[2] J. R. Sellars, M. Tribus, and J. S. Klein, *Trans. Am. Soc. Mech. Eng.*, **78**:441 (1956).

determining values of ϕ_n, which is a function of r/r_i. Details of the Graetz solution are given by Drew[1] and by Jakob.[2] The bulk temperature of the fluid at any axial position can be obtained by combining Eq. (22-29), which defines temperature, and Eq. (22-26), for velocity, in the following expression:

$$t_b = \frac{1}{\pi r_i^2 u_b C_p \rho} \int_{r=0}^{r=r_i} (\rho C_p t u_x)(2\pi r)\, dr$$

which simplifies to

$$t_b = \frac{2}{r_i^2 u_b} \int_{r=0}^{r=r_i} t u_x r\, dr \tag{22-30}$$

The local coefficient of heat transfer h_x can be found by writing a heat balance on a differential length of pipe dx.

$$h_x(\pi D\, dx)(t_s - t_b) = wC_p\, dt_b$$

This equation reduces to

$$h_x = \frac{wC_p}{\pi D} \frac{dt_b}{dx} \frac{1}{t_s - t_b} \tag{22-31}$$

This expression for the local coefficient of heat transfer h_x is converted into an equation for the local Nusselt number $h_x D/k$ by multiplying both sides by D/k.

Thus the manipulation of Eqs. (22-29) to (22-31) will lead finally to a complicated equation expressing the group $\mathbf{Nu}_x$ as a function of wC_p/kx, which is known as the *Graetz number*. This can be expanded and written as $\mathbf{Pe}(\pi/4)(D/x)$, where the quantity **Pe**, called the *Peclet number*, is the product **Re Pr**. The results of this analysis are shown in Fig. 22-3.

Flat velocity profile If the flow is assumed to be rodlike, $u_x = u_b$ at all points and the combined mass, energy, and momentum equations simplify to

$$\frac{\partial^2 t}{\partial r^2} + \frac{1}{r}\frac{\partial t}{\partial r} = \frac{u_b}{\alpha}\frac{\partial t}{\partial x} \tag{22-32}$$

The solution of this equation is somewhat simpler than the solution of the equation involving the parabolic profile. This occurs because the partial differential equation, when the variables are separated, gives two ordinary differential equations which can be solved by standard methods. One solution is a simple exponential equation, and the other involves a Bessel function. The final solution, when

[1] T. B. Drew, *Trans. Am. Inst. Chem. Eng.*, **26**:26 (1931).

[2] Op. cit., p. 451.

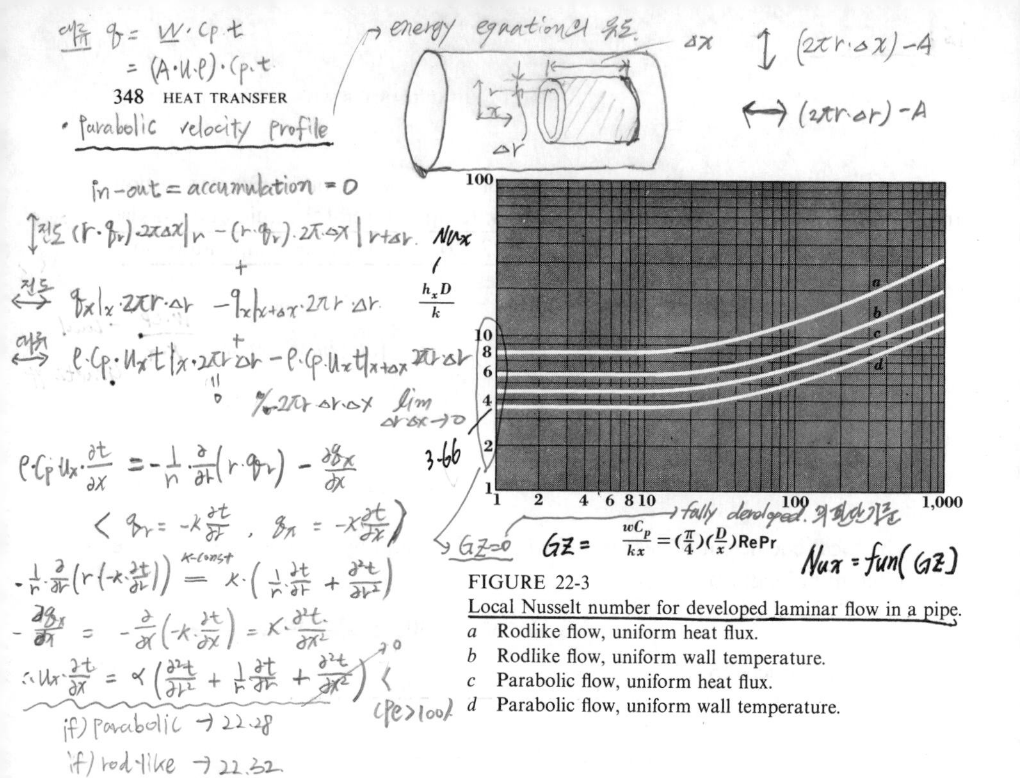

FIGURE 22-3
Local Nusselt number for developed laminar flow in a pipe.
a Rodlike flow, uniform heat flux.
b Rodlike flow, uniform wall temperature.
c Parabolic flow, uniform heat flux.
d Parabolic flow, uniform wall temperature.

combined with the boundary conditions of uniform wall temperature and uniform inlet temperature, is as follows.

$$\frac{t_s - t}{t_s - t_o} = \sum_{n=1}^{n=\infty} \frac{2}{a_n J_1(a_n)} J_0\left(\frac{a_n r}{r_i}\right) e^{-2a_n^2(x/r_i)/\text{Re Pr}} \tag{22-33}$$

The quantity a_n is the nth root of the expression $J_0(a_n) = 0$. J_0 and J_1 are Bessel functions of zero and first order, which can be found in standard tabulations. Details of this solution are given in the references previously cited for the problem of flow with a parabolic profile.

The equation for the bulk temperature, Eq. (22-30), is simplified when applied to rodlike flow because $u_x = u_b$. This gives

$$t_b = \frac{2}{r_i^2} \int_{r=0}^{r=r_i} tr\, dr \tag{22-34}$$

The local coefficient of heat transfer h_x can be found from t_b by employing Eq. (22-31). This coefficient is then used to calculate the local Nusselt number $h_x D/k$. The equation for this useful group is

$$\frac{h_x D}{k} = \frac{\sum_{n=1}^{n=\infty} e^{-2a_n^2(x/r_i)/\text{Re Pr}}}{\sum_{n=1}^{n=\infty} a_n^{-2} e^{-2a_n^2(x/r_i)/\text{Re Pr}}} \tag{22-35}$$

This relation is shown graphically in Fig. 22-3.

Uniform heat flux Both flow systems assumed in the preceding pages give different solutions if the boundary condition of constant wall temperature is replaced with the condition of uniform heat flux from the wall. The behavior of the local Nusselt number with this boundary condition is shown in Fig. 22-3.

The local Nusselt number can be seen to approach asymptotic values for long tubes for all cases shown in Fig. 22-3. The asymptotic values, in the case of uniform heat flux, can be predicted without the complete solution of the partial differential equations, as illustrated in Example 22-2.

Example 22-2 Show that the local Nusselt number for uniform heat flux to a fluid in rodlike flow approaches a value of 8 for long pipes.

The combined mass, energy, and momentum equations for this system reduce to

$$\frac{\partial^2 t}{\partial r^2} + \frac{1}{r}\frac{\partial t}{\partial r} = \frac{u_b}{\alpha}\frac{\partial t}{\partial x} \tag{22-32}$$

If heat is added at a constant rate per unit length of pipe, the temperature profile will approach some constant shape at great distances from the pipe inlet. In this region $\partial t/\partial x$ is constant at all radial and axial points, so that Eq. (22-32) can be written as

$$\frac{\partial^2 t}{\partial r^2} + \frac{1}{r}\frac{\partial t}{\partial r} = \frac{u_b}{\alpha}\frac{\partial t}{\partial x} = a \tag{1}$$

in which a is a constant.

Because the flow is rodlike,

$$\frac{\partial t}{\partial x} = \frac{dt_b}{dx} = \frac{a\alpha}{u_b} \tag{2}$$

From Eq. (22-31),

$$h_x = \frac{wC_p}{\pi D}\frac{a\alpha}{u_b}\frac{1}{t_s - t_b} \tag{3}$$

The flow rate w is

$$w = \rho u_b \frac{\pi D^2}{4}$$

We substitute this in Eq. (3) to get

$$h_x = \left(\rho u_b \frac{\pi D^2}{4}\right)\frac{C_p}{\pi D}\frac{ak}{\rho C_p u_b}\frac{1}{t_s - t_b}$$

$$= \frac{akD}{4(t_s - t_b)}$$

This is rearranged to give the local Nusselt number:

$$\frac{h_x D}{k} = \frac{aD^2}{4(t_s - t_b)} \tag{4}$$

The temperature at any radial point can be found in terms of the constant a by integrating Eq. (1). This is done by the standard method of substituting a new variable for $\partial t/\partial r$ and reducing Eq. (1) to a first-order equation, which can then be integrated directly. The integrated equation for temperature is

$$t = \frac{ar^2}{4} + C_1 \ln r + C_2 \tag{5}$$

The boundary conditions $\partial t/\partial r = 0$ at $r = 0$, and $t = t_s$ at $r = D/2$, are substituted to obtain values for the arbitrary constants C_1 and C_2.

The solution is

$$t = a\left(\frac{r^2}{4} - \frac{D^2}{16}\right) + t_s \tag{6}$$

In Eq. (6) t_s is a function of x, so that $t = f(r,x)$.

When this equation is integrated according to Eq. (22-34), the bulk temperature is found to be

$$t_b = -\frac{aD^2}{32} + t_s \tag{7}$$

This is rearranged to give the temperature difference

$$t_s - t_b = \frac{aD^2}{32}$$

which is substituted in Eq. (4) to give

$$\frac{h_x D}{k} = \frac{aD^2}{4}\frac{32}{aD^2} = 8 \tag{8}$$

Average coefficients It is convenient for design purposes to know the average heat-transfer coefficient for an entire length of pipe. If the surface temperature t_s is constant and t_{b2} represents the bulk temperature at the pipe discharge, an arithmetic average coefficient h_a over the area A can be defined by the following equation:

$$q = h_a A\frac{(t_s - t_o) + (t_s - t_{b2})}{2} = h_a A(t_s - t_b)_a \tag{22-36}$$

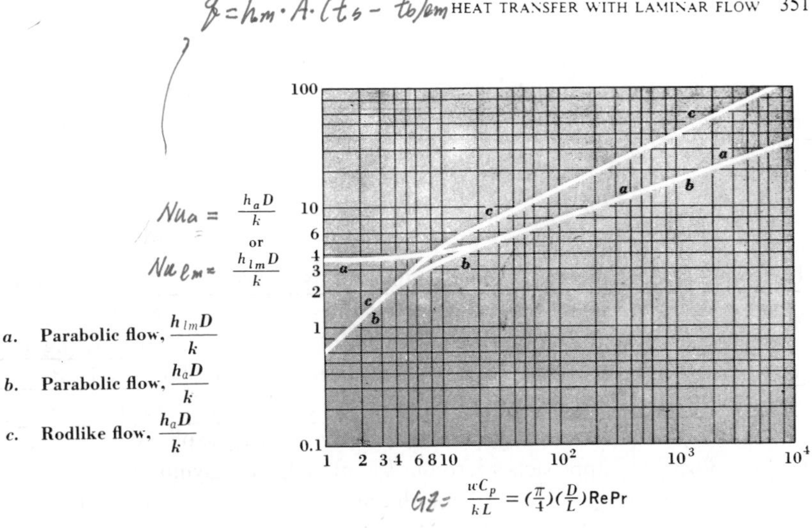

FIGURE 22-4
Average Nusselt number for developed laminar flow in a pipe with uniform wall temperature.

Two of the quantities in this equation are

$$q = wC_p(t_{b2} - t_o)$$

and

$$A = \pi DL$$

We substitute these in Eq. (22-36) and rearrange to get the Nusselt number based on the arithmetic average temperature difference.

$$\frac{h_a D}{k} = \frac{2wC_p}{\pi k L} \frac{t_{b2} - t_o}{(t_s - t_o) + (t_s - t_{b2})} \tag{22-37}$$

A value of t_{b2} can be obtained by substituting in Eq. (22-30), the results of one of the solutions for t as a function of x and r [Eq. (22-29) or (22-33)]. This value of t_{b2} can then be put into Eq. (22-37) in order to calculate $h_a D/k$. Some results of this procedure are shown in Fig. 22-4. An equivalent way of defining h_a is to write an equation similar to (22-20):

$$\frac{q}{b} = h_a L(t_s - t_b)_a = \int_0^L h_x(t_s - t_b)\, dx \tag{22-38}$$

This would give the same results as Eq. (22-37).

For long pipes and low flow rates the discharge bulk temperature t_{b2} is approximately equal to the surface temperature t_s, so that Eq. (22-37) simplifies to

$$\frac{h_a D}{k} = \frac{2wC_p}{\pi k L} \tag{22-39}$$

This simplification applies to both rodlike and parabolic flow. The result is that curves b and c in Fig. 22-4 merge for low values of wC_p/kL and have a slope of 1.

Frequently, an equation similar to (22-36) is written with the driving force as the logarithmic mean of $t_s - t_o$ and $t_s - t_{b2}$. The behavior of the two kinds of mean Nusselt number $h_a D/k$ and $h_{lm} D/k$ as functions of wC_p/kL is shown in Fig. 22-4.

The logarithmic-mean temperature difference gives the physically correct result for long pipes; Δt_{lm} approaches zero but h approaches an asymptotic value, given for example by curve d of Fig. 22-3 for the limit of large x. For short pipes with parabolic flow, h_a and h_{lm} are the same. Therefore, curve a of Fig. (22-4) is of the most practical interest, and the following equations are given in terms of $h_{lm} D/k$, written as $h_m D/k$ or $\mathbf{Nu}_m$.

At values of $wC_p/kL > 20$, the parabolic-flow equation for the average Nusselt number $h_m D/k$ can be represented adequately by the empirical expression

$$\frac{h_m D}{k} = 1.62\left(\frac{4wC_p}{\pi k L}\right)^{1/3} \tag{22-40}$$

The quantity $4wC_p/\pi kL$ is equal to $(Du_b \rho/\mu)(C_p \mu/k)(D/L)$, so that Eq. (22-40) can be written as

$$\mathbf{Nu}_m = 1.62\,\mathbf{Re}^{1/3}\,\mathbf{Pr}^{1/3}\left(\frac{D}{L}\right)^{1/3} \tag{22-41}$$

This equation is the basis of the well-known heat-transfer correlation of Sieder and Tate,[1] who proposed the following equation:

$$\mathbf{Nu}_m = 1.86\,\mathbf{Re}^{1/3}\,\mathbf{Pr}^{1/3}\left(\frac{D}{L}\right)^{1/3}\left(\frac{\mu}{\mu_s}\right)^{0.14} \tag{22-42}$$

where μ = fluid viscosity at average bulk temperature
μ_s = viscosity at wall temperature

The term μ/μ_s is an empirical correction for the distortion of the velocity profile which results from the effect of temperature on viscosity.

[1] E. N. Sieder and G. E. Tate, *Ind. Eng. Chem.*, **28**:1929 (1936).

Another representation found quite commonly is obtained by dividing both sides of Eq. (22-42) by the product **Re Pr**. This gives

$$\frac{h_m}{C_p G}\left(\frac{C_p \mu}{k}\right)^{2/3}\left(\frac{\mu_s}{\mu}\right)^{0.14} = 1.86\left(\frac{D u_b \rho}{\mu}\right)^{-2/3}\left(\frac{D}{L}\right)^{1/3} \tag{22-43}$$

The reliability of this equation is discussed by McAdams,[1] who points out that some experimental data deviate as much as 250 percent above the values predicted by Eq. (22-43). One probable reason for this is that the equation is derived from an approximation which assumes developed flow in short tubes, and in short tubes the entrance effects which we discussed earlier may be quite significant. The assumption of developed flow gives lower heat-transfer coefficients than those actually obtained with entrance effects, so that designs based on developed flow are likely to have higher heat-transfer rates than predicted.

PROBLEMS

22-1 Lubricating oil is being heated in a double-pipe heat exchanger. The oil enters a 1-in, schedule-40 inner steel pipe with a velocity of 0.3 m/s at a temperature of 40°C. It leaves at 70°C. The annular space between the pipes contains steam condensing at 105°C. The heat-transfer coefficient for the condensing steam is 11,500 $W/m^2 \cdot K$. Find the length of pipe required for the exchanger. The oil has the following average physical properties: density, 900 kg/m^3; specific heat, 2000 $J/kg \cdot K$; thermal conductivity, 0.12 $W/m \cdot K$.

22-2 A solution to the problem of heat transfer to a fluid in laminar flow in a pipe with constant wall temperature is given by Lévêque.[2] An equation derived for heat transfer from a flat plate is adapted by using for the velocity gradient at the wall a value obtained from the equation describing the parabolic velocity profile in a circular tube. The temperature difference driving force used is $t_s - t_o$ where t_o is both the fluid temperature at the edge of the thermal boundary layer for flow over a flat plate and the inlet fluid temperature for flow in a tube. The resulting equation for the local Nusselt number is:

$$\mathbf{Nu}_x = 1.077\left(\frac{D}{x}\right)^{1/3} \mathbf{Re}^{1/3}\,\mathbf{Pr}^{1/3}$$

This equation gives the same value for the Nusselt number in the region **Re Pr** $(D/x) > 100$ as is obtained from the more complicated Graetz solution for developed parabolic flow shown in Fig. 22-3. Since the Lévêque equation is not valid beyond the length where the thermal boundary layer reaches the center of the pipe, the region in which it coincides with the solution by Graetz is a rough measure of the length of the thermal boundary layer or thermal entrance effect. Calculate this length in

[1] W. H. McAdams, "Heat Transmission," 3d ed., p. 237, McGraw-Hill Book Company, New York, 1954.

[2] J. Lévêque, *Ann. Mines*, **13**(12):201, 305, 381 (1928). See also Knudsen and Katz, op. cit., p. 363.

terms of pipe diameters for Reynolds numbers of 100, 1000, and 2000 for air, water, light oil, and mercury at 200°F.

Fluid	Pr
Air	0.72
Water	1.88
Light oil	62.00
Mercury	0.016

22-3 Equation (22-19) does not apply for fluids with a Prandtl number less than 0.6. Find the local Nusselt number at the critical Reynolds number $\mathbf{Re}_x = 500{,}000$ for mercury flowing past a flat horizontal plate at 200°F. At this temperature the Prandtl number for mercury is 0.016.

22-4 The partial differential equation describing the temperature distribution in a fluid being heated while flowing in a parabolic laminar flow is given earlier in the chapter as Eq. (22-28). Show by a procedure similar to that used in Example 22-2 that for a very long pipe with uniform heat flux the local Nusselt number has an asymptotic value of $\frac{48}{11}$.

22-5 A fluid with uniform inlet temperature t_o is flowing at a steady mass velocity G lb/(h)(ft^2) in laminar motion through a closed duct of rectangular cross section, the surfaces of which are kept at a uniform temperature t_s at all points. Because of the effect of temperature on the viscosity of the fluid, the velocity profile may be assumed to be flat as in plug-type flow. The width of the duct is much greater than the depth, so that heat transfer at the vertical sides may be neglected. Longitudinal conduction is also negligible, and the thermal properties of the fluid can be assumed constant.

Derive an expression for the mixing-cup, or bulk, temperature of the liquid in terms of the dimensions of the system, the physical properties of the liquid, and the distance x from the inlet. Similarities between this problem and Example 19-2, which illustrated unsteady-state heat conduction, should be noted.

22-6 A certain fluid flowing in a circular pipe has a temperature profile described by the equation

$$\frac{t - t_s}{t_m - t_s} = 1 - \frac{r^2}{r_i^2}$$

in which t = fluid temperature at radius r, t_s = wall temperature, t_m = mixing-cup temperature, r_i = inside radius of pipe. Prove that for this system the Nusselt number has a value of 4.

22-7 A viscous oil is flowing at $\mathbf{Re} = 100$ through a copper tube, 1-in ID and 20 ft long, arranged as a double-pipe heat exchanger. Water flows in the annular space, countercurrent to the oil, and we assume that all the resistance to heat transfer is on the oil side. If the water enters at 80°F and leaves at 120°F, and the oil enters at 150°F, what is the approximate heat-transfer rate in the exchanger, Btu/h? For the oil, use the following constant properties: bulk viscosity, 50 cP; density, 50 lb/ft^3; thermal conductivity, 0.070 Btu/(h)(ft)(°F); viscosity at interface, 30 cP; specific heat, 0.50 Btu/(lb)(°F).

23

HEAT TRANSFER WITH TURBULENT FLOW

Forced-convection heat transfer to a fluid flowing in turbulent motion in a pipe may be the commonest heat-transfer system in industry. Although forced convection may be associated with laminar flow and natural convection with turbulent flow, these are cases of secondary importance. Heat-transfer coefficients are higher with turbulent flow than with laminar flow, and heat-transfer equipment is usually designed to take advantage of this fact.

The state of knowledge concerning heat transfer with turbulent flow is necessarily limited by our knowledge of isothermal turbulent flow. We saw in Chap. 12 that the use of the Navier-Stokes equations in the analysis of isothermal turbulent flow is complicated because of the fluctuations in the velocity components. The use of the differential energy balance in the analysis of nonisothermal turbulent flow is difficult for the same reason. Heat is transferred in most turbulent streams primarily by the movement of numerous macroscopic elements of fluid (eddies) between regions at different temperatures. We cannot predict the behavior of these eddies with time, but even if we could, the expressions describing this behavior would probably be so complicated that the combined solution of the equations of energy and momentum would not be possible. Nevertheless, answers must be found to these problems. In this chapter, we consider some theoretical developments in turbulent heat transfer which are of use to the engineer, and in the next chapter we shall examine some design equations. The theory tells us why the design equations work and what some of their limitations are.

Entrance Effects

In the preceding chapter on heat transfer with laminar flow, the first part of the discussion was concerned with heat transfer in developing flow and the latter part with heat transfer in developed flow. Entrance effects which are associated with developing flow exist also in turbulent systems and may have a significant effect on overall performance in flow through short pipes ($L/D < 60$). Once this entrance region has been passed, the heat-transfer coefficients in developed turbulent flow remain essentially constant.

Entrance effects in a pipe Numerous possible combinations of thermal and hydrodynamic entrance conditions exist. In the following examples we assume that the fluid enters with a uniform temperature and that the pipe wall is at some uniform temperature higher than that of the entering fluid. Certain flow conditions at the entrance will be considered, and their effects on the local heat-transfer coefficients deduced qualitatively from the knowledge we have gained thus far of hydrodynamic and heat-transfer theory.

1 The fluid enters the pipe with a uniform velocity profile at a rate such that **Re** < 2100. Under these conditions a laminar boundary layer will build up, starting from the leading edge, until it fills the pipe at some distance downstream. The heat-transfer coefficient will be infinite at the inlet but will continue to diminish even after the point of developed flow has been reached.

2 The fluid enters the pipe in laminar flow with a uniform velocity profile at a rate such that **Re** > 2100. This condition can be achieved with a rounded entrance. At the leading edge a laminar boundary layer develops which, at a critical distance, breaks down into a turbulent boundary layer in a manner already described in Chap. 11 with reference to flow past a flat plate. This turbulent boundary layer increases in thickness with increasing downstream distance until it fills the pipe with a turbulent core and a laminar sublayer at the wall. Downstream from this point the system is identical in all respects with the system that would develop if the flow had been turbulent from the entrance. This flow behavior is reflected in the values of the local heat-transfer coefficient, which decrease from infinity at the inlet to some minimum value at the critical point where the laminar boundary layer changes into a turbulent boundary layer. Near this point the heat-transfer coefficient increases in

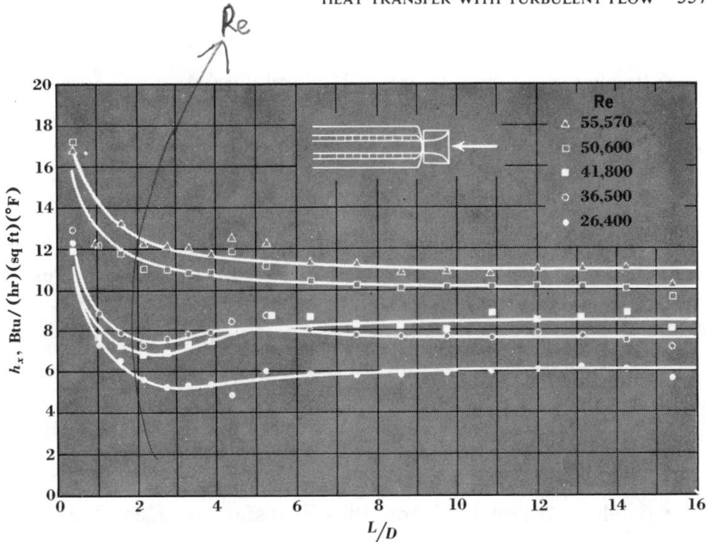

FIGURE 23-1
Local coefficients of heat transfer near the inlet of a tube with a bellmouth entrance. *(From L. M. K. Boelter, G. Young, and H. W. Iverson, Natl. Adv. Comm. Aeronaut. Tech. Mem. 1451, 1948.)*

magnitude for a short distance, but then continues to decrease downstream until the turbulent boundary layers meet at the center of the pipe. Experimental values of h_x are shown in Fig. 23-1 for air flowing into a tube with a bellmouth entrance. The point of minimum h_x moves toward the entrance with increasing values of **Re**.

3 The fluid enters the pipe in a state of turbulence with a nonuniform velocity profile at a rate such that **Re** > 2100. The velocity profile at the entrance may be caused by the presence of a sudden contraction or a pipe bend immediately upstream. A thermal boundary layer will build up, starting from the beginning of the heated length, and will fill the pipe at some downstream point. However, this entrance length for nonuniform turbulent flow usually does not have any significant effect on the local heat-transfer coefficients beyond 10 pipe diameters, whereas entrance effects in laminar flow often persist for 50 or more pipe diameters. The local heat-transfer coefficient at the beginning of the heated length is infinite in turbulent flow because of the temperature discontinuity, just as it is for laminar flow. However, it quickly falls to some constant value, as described above.

Boelter, Young, and Iverson[1] measured heat-transfer coefficients for air

[1] L. M. K. Boelter, G. Young, and H. W. Iverson, *Natl. Adv. Comm. Aeronaut. Tech. Mem.* 1451, 1948.

flowing into a pipe (1.785-in ID) and correlated the average coefficient h_m in terms of the coefficient far downstream h_∞ by the equation

$$h_m = h_\infty\left(1 + \frac{K}{L/D}\right) \qquad (23\text{-}1)$$

Values of K depend on the entrance condition as shown in Table 23-1. At distances less than 5 pipe diameters from the inlet, the results could not be correlated.

The experiments with entrances equipped with orifices (and for a simple sharp-edged entrance as well) showed that peaks existed in plots h_x versus L/D at values of L/D between 1 and 3. These are caused by the turbulence surrounding the vena contracta.

The Analogy between Momentum Transfer and Heat Transfer

The similarity of heat and momentum transport was noted by Osborne Reynolds in 1874; his work has led to useful, simple equations relating the friction factor, the heat-transfer coefficient, and the mass-transfer coefficient. Improvements were made in these equations by Prandtl (1910) and Taylor (1916); we shall study a derivation which leads to the Prandtl-Taylor equation. Murphree (1932) and von Kármán (1939) extended this work still further; an equation bears von Kármán's name. Subsequently, further modifications have been made, including work by Reichardt (1940), Boelter, Martinelli, and Jonassen (1941), Martinelli (1947), Lyon (1951), and Deissler (1954).[1]

Table 23-1 VALUES OF K FOR Eq. (23-1)

Type of entrance	K
Bellmouth	0.7
Bellmouth with one screen	1.2
Short calming section ($L/D = 2.8$) with sharp-edged entrance	~ 3
Long calming section ($L/D = 11.2$) with sharp-edged entrance	1.4
45° angle-bend entrance	~ 5
90° angle-bend entrance	~ 7
1-in square-edged orifice, located 1 in upstream from entrance	~ 16
1.4-in square-edged orifice, located 1 in upstream from entrance	~ 7

[1] These equations are discussed by M. Jakob, "Heat Transfer," vol. 1, chap. 24, John Wiley & Sons, Inc., New York, 1949; and by J. G. Knudsen and D. L. Katz, "Fluid Dynamics and Heat Transfer," chap. 15, McGraw-Hill Book Company, New York, 1958.

In this book we shall not consider in detail derivations of equations more complex than the Prandtl-Taylor equation. The equations for heat transfer can be extended easily to cover mass transfer; this is done in Chap. 33, which is devoted to mass transfer in turbulent systems.

Reynolds analogy Reynolds stated that in the transport of heat or momentum between a fluid and a solid surface, two mechanisms contributed to the transport process:

"*1* The natural internal diffusion of the fluid when at rest.

2 The eddies caused by visible motion which mixes the fluid up and continually brings fresh particles into contact with the surface."[1]

The first cause he recognized to be dependent on the nature of the fluid, and the second cause to be a function of the velocity of the fluid past the surface. The combined effects of these causes led to a heat-transfer equation, which he wrote as

$$H = At + B\rho vt$$

where t = difference in temperature between surface and fluid
ρ = density
v = bulk velocity
A and B = constants
H = heat transmitted per unit of area of surface per unit of time

The resistance R to motion offered by friction in the fluid he wrote as

$$R = A'v + B'\rho v^2$$

in which A' and B' are also constants. Various considerations, which he did not name specifically, led to the supposition that A and B were proportional to A' and B'. In our present nomenclature we recognize that the quantity Reynolds wrote as $B\rho v$ is equivalent for highly turbulent systems to the coefficient of heat transfer h, and the quantity B' is proportional to the friction factor f. Thus Reynolds said, in effect, that h was proportional to f.

Reynolds' ideas have been extended by others and placed in mathematical terms to give an equation relating h and f. We shall consider two principal methods of doing this, one based on a proportionality and the other on equations of flux of heat and momentum.

The proportionality of heat and momentum transfer can be stated in terms of four quantities, which we shall define with reference to a fluid at a bulk temperature

[1] O. Reynolds, *Proc. Manchester Lit. Phil. Soc.*, **14**:7 (1874). See also O. Reynolds, "Papers on the Mechanical and Physical Subjects," vol. 1, p. 81, Cambridge University Press, London, 1900.

t_b flowing through a pipe and losing heat to the wall of the pipe, which is at temperature t_s. The four quantities are:

(a) The heat flux from the fluid to the pipe wall, $h(t_b - t_s)$ Btu/(h)(ft^2)
(b) The momentum flux at the wall, $\tau_s g_c$ (lb)(ft/h)/(h)(ft^2)
(c) The rate at which energy available for transfer as heat is transported parallel to the pipe wall, $wC_p(t_b - t_s)$ Btu/h
(d) The rate at which momentum is transported parallel to the pipe wall, wu_b (lb)(ft/h)/h

It is then postulated that these four quantities can be described by the following proportionality:

$$\frac{(a)}{(c)} = \frac{(b)}{(d)}$$

From this the heat-transfer coefficient can be found as

$$h = \frac{\tau_s g_c C_p}{u_b} \tag{23-2}$$

It has been shown in Eq. (12-67) that the shear stress at the wall τ_s is related to the friction factor by the equation

$$\tau_s g_c = \frac{f u_b^2 \rho}{2} \tag{23-3}$$

Therefore we combine Eqs. (23-2) and (23-3) to obtain an equation which, though not actually derived by Reynolds, is known as the *Reynolds analogy*:

$$h = \frac{f u_b \rho C_p}{2} \tag{23-4}$$

The second method of deriving the equation has a semitheoretical basis and will be examined after we have discussed the application of mixing-length theory in the next section.

Eddy thermal diffusivity and mixing length In Chap. 12 on turbulent flow, we wrote Eq. (12-37), relating total shear stress $\bar{\tau}_{yx}{}'$ to velocity gradient $d\bar{u}_x/dy$:

$$\bar{\tau}_{yx}{}' = \frac{\rho}{g_c}(\nu + \nu_e)\frac{d\bar{u}_x}{dy} \tag{12-37}$$

In this equation ν, the kinematic viscosity, depends only on the molecular properties of the fluid, whereas ν_e, the eddy kinematic viscosity, depends on the motion of the fluid. The quantity ν_e was shown to be related to the Prandtl mixing length ℓ and

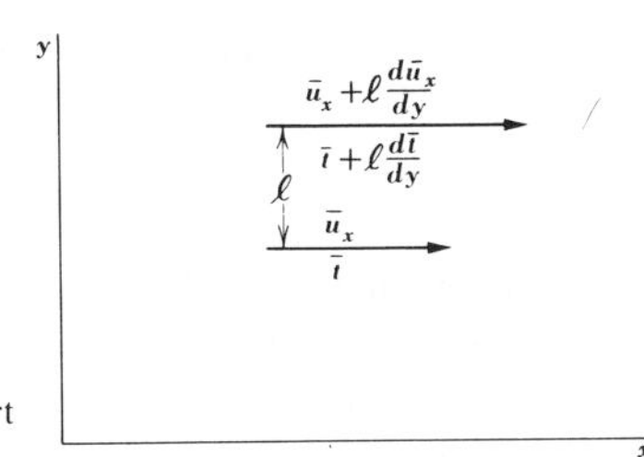

FIGURE 23-2
Effect of Prandtl mixing length in the transport of heat in a turbulent stream.

the velocity gradient by the equation

$$\nu_e = \ell^2 \left| \frac{d\bar{u}_x}{dy} \right| \tag{12-39}$$

A similar quantity known as the *eddy thermal diffusivity* can be derived. Consider a fluid flowing under the conditions represented in Fig. 23-2. Velocities and temperatures are shown at two planes separated by a distance equal to the Prandtl mixing length ℓ. Fluid is assumed to be transported between the planes with a velocity equal to the time-average magnitude of the fluctuating velocity component $\overline{|u'_y|}$. Energy is transported with the packets of fluid at a rate per unit area which equals the mass flux times the product of specific heat and temperature difference, $\overline{|u'_y|}\rho C_p[\ell(d\bar{t}/dy)]$. In Chap. 12 it was indicated that the fluctuating velocity components at a point u'_y and u'_x had the same time-average magnitude, and a relation was given for these quantities in terms of the Prandtl mixing length:

$$\overline{|u'_y|} = \overline{|u'_x|} = \ell \left| \frac{d\bar{u}_x}{dy} \right| \tag{12-40}$$

Therefore the heat flux due to the turbulent motion of the fluid becomes $\rho C_p \ell^2 |d\bar{u}_x/dy| \, d\bar{t}/dy$. This expression contains the quantity $\ell^2 |d\bar{u}_x/dy|$, which was defined as the eddy kinematic viscosity ν_e in Eq. (12-39). When the same group appears in an equation for turbulent heat transfer it is referred to as the eddy thermal diffusivity and will be written as α_e.

The turbulent heat flux can therefore be written $\rho C_p \alpha_e(d\bar{t}/dy)$. The Fourier equation for heat conduction, Eq. (17-1), could have been written in terms of molecular thermal diffusivity:

$$\frac{q}{A} = -\rho C_p \alpha \frac{d\bar{t}}{dy} \tag{23-5}$$

If an equation is written to represent heat transferred in a fluid by the combined mechanisms of conduction and turbulent diffusion, we obtain

$$\frac{q}{A} = -\rho C_p(\alpha + \alpha_e)\frac{d\bar{t}}{dy} \tag{23-6}$$

which is similar to Eq. (12-37) given on p. 352.

On the basis of our derivations, the eddy kinematic viscosity ν_e and the eddy thermal diffusivity α_e are equal. Although experimental evidence indicates that this is only approximately true, the assumption of equality is often made in theoretical work. Derivations for both terms are based on the assumption that packets of fluid travel a distance equal to the Prandtl mixing length and then give up their excess heat and axially directed momentum. This is most unlikely to occur in practice, but it is not unreasonable to suppose that the actual transport of heat and momentum differs from the idealized representation by about the same amount.

Although the subject of mass transfer will be considered later, it is appropriate at this point to write an equation for mass transfer of a component A in a binary-flow system:

$$J_A = -(D_{AB} + D_{ABe})\frac{d\bar{\rho}_A}{dy} \tag{23-7}$$

This equation contains the molecular diffusivity D_{AB} and the eddy diffusivity D_{ABe}. A derivation similar to that outlined above for α_e will be given later to show that $D_{ABe} = \ell^2 \left| d\bar{u}_x/dy \right|$. Thus D_{ABe}, α_e, and ν_e are equal, according to the mixing-length theory.

Derivation of Reynolds analogy from eddy transport theory Equation (23-4) relating h and f was derived from a proportionality based on intuition. It can also be derived from the equations of turbulent flux of heat and momentum, and this derivation is, perhaps, more revealing because the assumptions that will have to be made reveal the shortcomings of the final equation.

Equation (23-6) is written for a fluid being cooled in a pipe with heat being transferred radially. The distance y is measured from the wall, so Eq. (23-6) becomes

$$\frac{q}{A} = \rho C_p(\alpha + \alpha_e)\frac{dt}{dy} \tag{23-8}$$

All the quantities in this equation are on a time mean basis, and a similar equation is written for momentum flux, in which only time mean quantities are referred to. Total shear on the face of a cylindrical element in the x direction is

$$\tau g_c = \rho(\nu + \nu_e)\frac{du}{dy} \tag{23-9}$$

If the molecular transport coefficients α and ν are assumed to be negligible when compared with the turbulent transport coefficients, Eqs. (23-8) and (23-9) can be written in terms of dy.

$$dy = \frac{\rho C_p \alpha_e \, dt}{q/A} = \frac{\rho \nu_e \, du}{\tau g_c}$$

If α_e and ν_e are assumed to be identical, this expression may be written

$$du = \frac{C_p \tau g_c}{q/A} \, dt \qquad (23\text{-}10)$$

The variation of τ with radial position has already been shown for both laminar and turbulent flow to follow the relation

$$\frac{\tau}{\tau_s} = 1 - \frac{y}{r_i}$$

If heat flux in a turbulent system is analogous to momentum flux, then it is assumed to follow a similar equation.

$$\frac{q/A}{(q/A)_s} = 1 - \frac{y}{r_i} \qquad (23\text{-}11)$$

From these two equations we see that $\tau g_c/(q/A)$ is constant for all radial positions. It is most convenient to write it at the wall, so we shall designate it $\tau_s g_c/[(q/A)_s]$, although in most convection equations the quantity q/A is implicitly understood as referring to flux at the wall. Because the ratio is constant, Eq. (23-10) can be integrated with $C_p \tau_s g_c/[(q/A)_s]$ outside the integral sign. We integrate between conditions at the wall and at some radial position in the fluid at which the local time mean velocity is the same as the bulk velocity and assume that the fluid temperature at this point is the same as the bulk, or mixing-cup, temperature.

$$u_b - u_s = \frac{C_p \tau_s g_c}{(q/A)_s}(t_b - t_s)$$

We know that the velocity at the wall u_s is zero. Furthermore, we have

$$\left(\frac{q}{A}\right)_s = h(t_b - t_s)$$

and

$$\tau_s g_c = \frac{f u_b^2 \rho}{2}$$

Therefore we get

$$u_b = C_p \frac{f u_b^2 \rho}{2} \frac{t_b - t_s}{h(t_b - t_s)} \qquad (23\text{-}12)$$

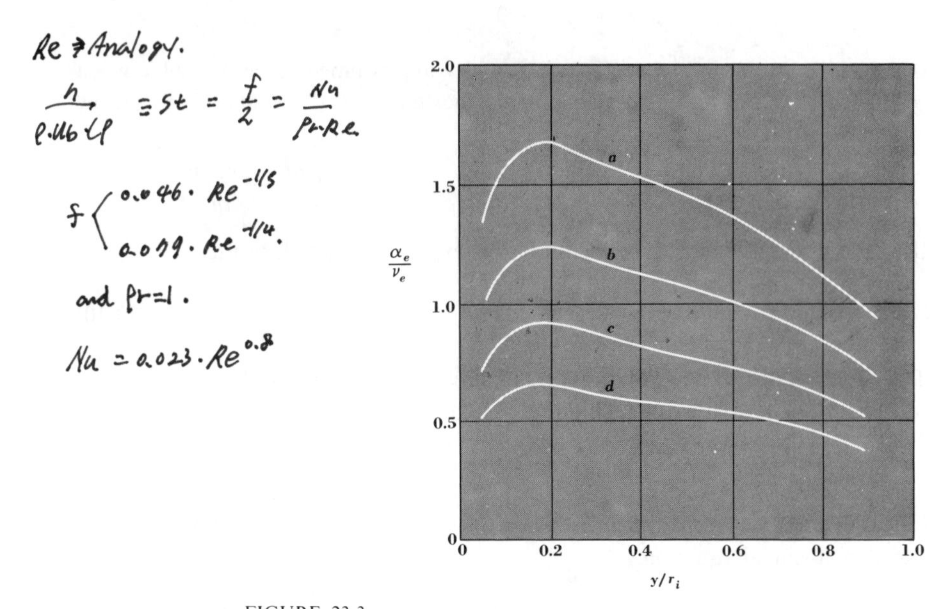

FIGURE 23-3
Ratio of turbulent transport coefficients for mercury in a vertical tube. (*From S. E. Isakoff and T. B. Drew, " Proceedings of the General Discussion on Heat Transfer," p. 479, Inst. Mech. Eng. and Am. Soc. Mech. Eng., New York, 1951.*)
a **Re** = 400,000.
b **Re** = 200,000.
c **Re** = 100,000.
d **Re** = 50,000.

which reduces to

$$h = \frac{f u_b \rho C_p}{2} \qquad (23\text{-}4)$$

One of the major assumptions in this derivation was that α and ν were negligible compared with α_e and ν_e. The ratio ν/α is equal to the Prandtl number, so for fluids with **Pr** = 1, $\nu = \alpha$. Therefore, in this special case, $\nu + \nu_e = \alpha + \alpha_e$, and Eq. (23-4) can be obtained without disregarding the molecular transport terms. It is significant that Eq. (23-4) gives best results when applied to gases, for which, as we know, **Pr** ~ 1.

The equality of α_e and ν_e has been challenged by many investigators. These quantities can be calculated from measurements of temperature and velocity profiles. Most results indicate that $0.5 < \alpha_e/\nu_e < 2$. The variation of this ratio with distance from the wall and with **Re** is shown in Fig. 23-3.

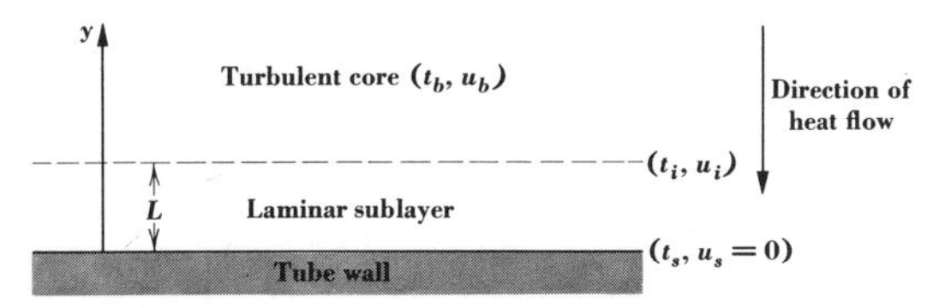

FIGURE 23-4
Flow model for the derivation of the Prandtl-Taylor equation.

The Prandtl-Taylor analysis One of the deficiencies of the Reynolds analogy was that, for fluids with Prandtl numbers other than unity, the molecular coefficients of thermal diffusivity and kinematic viscosity were neglected. This is often a valid assumption in the turbulent core of fluid in a pipe. However, the laminar sublayer has an important effect on heat transfer, and in this region, where the eddy coefficients diminish to zero, the molecular coefficients α and ν cannot be discarded.

The Prandtl-Taylor analysis is an attempt to overcome this shortcoming. Basically, the method employed is to write a conduction equation for the laminar sublayer and a Reynoids-analogy equation for the turbulent core. These equations are then combined in the customary method of combining resistances in series to produce a single equation for the overall resistance.

A flow system in a circular pipe is represented in Fig. 23-4. The fluid is considered to be in only two states of motion, as indicated.

In the laminar sublayer the velocity is assumed to be so small that all the heat which enters from the turbulent zone is conducted to the wall. Thus the heat flux in this layer is

$$\frac{q}{A} = \frac{k(t_i - t_s)}{L} \tag{23-13}$$

The momentum flux in the layer is

$$\tau g_c = \mu \frac{du}{dy} \tag{23-14}$$

The shear-stress distribution in a circular pipe is

$$\tau = \tau_s\left(1 - \frac{y}{r_i}\right)$$

In the laminar sublayer, which is extremely thin, $y/r_i \ll 1$ and $\tau \sim \tau_s$. Therefore we rewrite Eq. (23-14) as

$$du = \frac{\tau_s g_c}{\mu} dy$$

which can be integrated over the thickness L of the sublayer to give

$$u_i = \frac{\tau_s g_c}{\mu} L$$

which determines the thickness of the layer as

$$L = \frac{\mu u_i}{\tau_s g_c} \tag{23-15}$$

This is combined with Eq. (23-3) to eliminate τ_s in favor of f.

$$L = \frac{2\mu u_i}{f u_b{}^2 \rho}$$

This expression for L is substituted in Eq. (23-13) to give an equation for the heat flux in the sublayer.

$$\frac{q}{A} = \frac{k f u_b{}^2 \rho}{2\mu u_i}(t_i - t_s) \tag{23-16}$$

Heat and momentum transfer in the core are assumed to follow the Reynolds analogy. The rate at which momentum is transported parallel to the pipe wall is $w u_b$ when the turbulent zone extends to the wall, as in the derivation which led to Eq. (23-4). However, if a laminar sublayer surrounds the core, the rate at which momentum is transported parallel to the edge of the turbulent core is $w(u_b - u_i)$. Very little fluid is flowing in the laminar sublayer, so that the bulk velocity u_b and mass-flow rate w in the core are very nearly the same as the total bulk velocity and mass-flow rate. The heat-transfer coefficient h in Eq. (23-4) becomes the heat-transfer coefficient for only the turbulent core and will be designated as h'. Thus the proportionality which leads to a relation for h' becomes

$$\frac{h'(t_b - t_i)}{w C_p (t_b - t_i)} = \frac{\tau_s g_c}{w(u_b - u_i)} \tag{23-17}$$

from which we get

$$h' = \frac{\tau_s g_c C_p}{u_b - u_i}$$

and, substituting for $\tau_s g_c$ with Eq. (25-3), we obtain

$$h' = \frac{f u_b{}^2 \rho C_p}{2(u_b - u_i)} \tag{23-18}$$

Therefore the heat flux

$$\frac{q}{A} = h'(t_b - t_i)$$

becomes

$$\frac{q}{A} = \frac{fu_b{}^2\rho C_p}{2(u_b - u_i)}(t_b - t_i) \qquad (23\text{-}19)$$

Equations (23-16) and (23-19) are next arranged in tems of driving force.

$$t_i - t_s = \frac{q}{A}\frac{2\mu u_i}{kfu_b{}^2\rho}$$

$$t_b - t_i = \frac{q}{A}\frac{2(u_b - u_i)}{fu_b{}^2\rho C_p}$$

These are added to give

$$t_b - t_s = \frac{q}{A}\left[\frac{2\mu u_i}{kfu_b{}^2\rho} + \frac{2(u_b - u_i)}{fu_b{}^2\rho C_p}\right] \qquad (23\text{-}20)$$

Since the conventional equation for convection is

$$\frac{q}{A} = h(t_b - t_s)$$

we see that an equation for h can be written as

$$h = \frac{1}{2\mu u_i/kfu_b{}^2\rho + [2(u_b - u_i)]/fu_b{}^2\rho C_p} \qquad (23\text{-}21)$$

If the right-hand side of this equation is multiplied by $fu_b C_p \rho/2$ in both the numerator and denominator, we obtain

$$h = \frac{fu_b\rho C_p/2}{\dfrac{u_i}{u_b}\dfrac{C_p\mu}{k} + 1 - \dfrac{u_i}{u_b}} \qquad (23\text{-}22)$$

For fluids with **Pr** = 1, this equation reduces to Eq. (23-4).

In our discussions of the universal velocity profile in Chap. 12 we learned that for the laminar sublayer the dimensionless velocity and distance variables u^+ and y^+ are equal. This equality can be seen in Fig. 12-6. It persists up to $u^+ = y^+ = 5$ and provides a basis for determining the velocity u_i in Eq. (23-22). The relation between u and u^+ was shown to be

$$u^+ = \frac{u}{u^*} \qquad (12\text{-}46)$$

The term u^* was shown to be related to the friction factor as follows:

$$u^* = u_b\sqrt{\frac{f}{2}} \qquad (12\text{-}68)$$

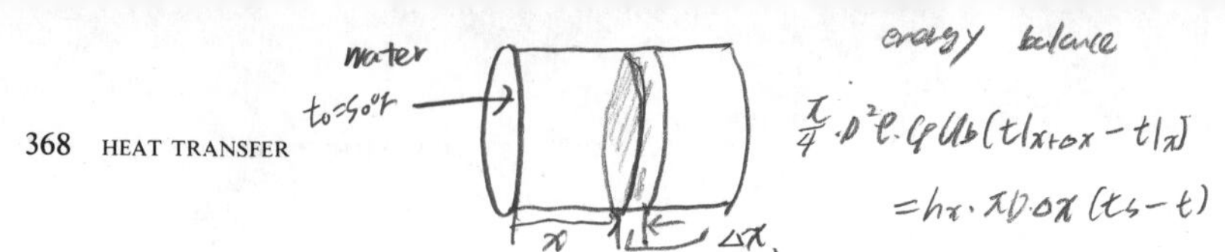

Therefore we can write

$$u = u^+ u_b \sqrt{\frac{f}{2}}$$

and at the edge of the laminar sublayer, where $u^+ = 5$, we have

$$u_i = 5u_b \sqrt{\frac{f}{2}} \tag{23-23}$$

which can be substituted in Eq. (23-22) to give

$$h = \frac{f u_b \rho C_p / 2}{1 + 5\sqrt{f/2}(\mathbf{Pr} - 1)} \tag{23-24}$$

This may also be written in terms of the Nusselt number if we multiply both sides by D/k to give

$$\mathbf{Nu} = \frac{\mathbf{Re}\,\mathbf{Pr}\,(f/2)}{1 + 5\sqrt{f/2}(\mathbf{Pr} - 1)} \tag{23-25}$$

Several other expressions have been offered for u_i/u_b in place of the quantity $5\sqrt{f/2}$ shown above. The difference is due to the presence of a buffer region between the laminar sublayer and the turbulent core, as shown in Fig. 12-6. If the equations representing the laminar sublayer and the turbulent core are extrapolated, they meet in the buffer region at $u^+ = y^+ = 11.6$. This quantity is therefore suggested by some authors in place of the quantity 5 in the denominator of Eq. (23-25).

Other analogies A major improvement in the Prandtl-Taylor equation was made by von Kármán, who considered the resistance to heat transfer to be composed of three parts. These correspond to the laminar sublayer, the buffer zone, and the turbulent core and are shown in the plot of the universal velocity profile, Fig. 12-6.

The derivation of von Kármán's equation is similar to the Prandtl-Taylor analysis. Equations are written for the laminar sublayer ($y^+ \leq 5$), in which the molecular thermal diffusivity and kinematic viscosity are considered but the eddy terms are neglected. This leads to an expression for the temperature drop over this layer. In the analysis of the buffer layer ($5 \leq y^+ \leq 30$), both the molecular and eddy transport terms are considered. An empirical equation [$u^+ = 5 + 5 \ln (y^+/5)$], representing the velocity in the buffer layer, is used to obtain an equation for the eddy kinematic viscosity, which is then substituted for the eddy thermal diffusivity in the heat-flux equation. Integration of this equation with both eddy and molecular thermal diffusivities gives an equation for the temperature drop over the buffer layer. Finally, the Reynolds analogy is applied to the turbulent core ($y^+ \geq 30$) to supply an equation for the temperature drop over this resistance. The three temperature drops are then added, as in the Prandtl-Taylor analysis, to give an equation for the overall temperature drop. The heat flux q/A divided by the overall temperature drop gives the heat-transfer coefficient shown in the following equation:

$$h = \frac{f u_b \rho C_p/2}{1 + 5\sqrt{f/2}\{\mathbf{Pr} - 1 + \ln[(1 + 5\,\mathbf{Pr})/6]\}} \tag{23-26}$$

Like the Prandtl-Taylor equation, this also reduces to Reynolds analogy, Eq. (23-4), for fluids with $\mathbf{Pr} = 1$.

The results of most analogies can be put in general form, as illustrated in Eqs. (23-24) and (23-26), in which the numerator of the right-hand side is the simple Reynolds analogy, Eq. (23-4), and the denominator is a complex group of terms which serves as a correction. Some modifications include terms which account for the difference between bulk velocity and centerline velocity; others correct for the variation of $\tau g_c/(q/A)$ with radial position. In the work of Deissler, the variation of viscosity with temperature is taken into account.

No further discussion of the theoretical analogies will be given. The references mentioned earlier in this chapter contain summaries of most of the important equations and references to others. One final analogy worth discussing, however, is the empirical j-factor relation proposed by Colburn.[1] This analogy may be obtained by replacing the entire denominator in either Eq. (23-24) or (23-26) by the expression $\mathbf{Pr}^{2/3}$. The equation is then rearranged to present it in its customary form.

$$\frac{h}{u_b \rho C_p}\mathbf{Pr}^{2/3} = \frac{f}{2} = j_H \tag{23-27}$$

The term j_H is called the *j factor for heat transfer*. As with the other analogies, the j-factor equation reduces to Reynolds analogy for fluids with $\mathbf{Pr} = 1$.

We shall see that a similar expression exists for mass-transfer coefficients, in which the Reynolds analogy for mass transfer is altered in the same way as for heat transfer; however, the Schmidt number replaces the Prandtl number. The j factor obtained for mass transfer is written j_M.

The j-factor relation has been shown experimentally to have considerable merit in correlating heat- and mass-transfer data. It has been used for many systems other than pipes and flat plates. In general, it has been found that $j_H = j_M$. However, in systems in which the fluid flows around spheres and other bluff objects, the frictional loss is often due more to form drag than to frictional drag. In these cases j_H still equals j_M but neither is equal to $f/2$, which is usually determined from overall pressure-loss measurements. In systems in which form drag is absent, e.g., pipes and flat plates, the relation $j_H = j_M = f/2$ holds.

[1] A. P. Colburn, *Trans. Am. Inst. Chem. Eng.*, **29**:174 (1933).

Equations for Heat-transfer Coefficients Based on the Analogies

Equations for the prediction of friction factors can be combined with the analogy equations to provide a direct method of predicting heat-transfer coefficients. For example, Reynolds analogy gives the equation

$$h = \frac{f u_b \rho C_p}{2} \qquad (23\text{-}4)$$

which can be expressed as

$$\mathbf{Nu} = \frac{f}{2}\,\mathbf{Re}\,\mathbf{Pr} \qquad (23\text{-}28)$$

In Chap. 12 we saw that a convenient empirical equation for the friction factor for flow in a pipe was

$$f = 0.046\,\mathbf{Re}^{-0.2} \qquad (12\text{-}72)$$

Equations (23-28) and (12-72) can be combined to give

$$\mathbf{Nu} = 0.023\,\mathbf{Re}^{0.8}\,\mathbf{Pr} \qquad (23\text{-}29)$$

Because this equation is based on (23-4), which applies only when $\mathbf{Pr} = 1$, we write Eq. (23-29) as

$$\mathbf{Nu} = 0.023\,\mathbf{Re}^{0.8} \qquad (23\text{-}30)$$

This, as we shall see, is identical with the well-known Dittus-Boelter equation when that equation is used for fluids with $\mathbf{Pr} = 1$. The Prandtl-Taylor analogy and the von Kármán analogy can be combined with equations for the friction factor to produce equations for the Nusselt number as a function of **Re** and **Pr**. The empirical j-factor analogy of Colburn can also be used. If the form of the Colburn analogy used in Eq. (23-27) is rearranged in terms of dimensionless groups, we obtain

$$j_H = \frac{f}{2} = \frac{\mathbf{Nu}}{\mathbf{Re}\,\mathbf{Pr}}\,\mathbf{Pr}^{0.67} \qquad (23\text{-}31)$$

Equation (12-72) is substituted for f, and Eq. (23-31) reduces to

$$0.023\,\mathbf{Re}^{-0.2} = \frac{\mathbf{Nu}}{\mathbf{Re}\,\mathbf{Pr}^{0.33}}$$

or

$$\mathbf{Nu} = 0.023\,\mathbf{Re}^{0.8}\,\mathbf{Pr}^{0.33} \qquad (23\text{-}32)$$

which is a common form.

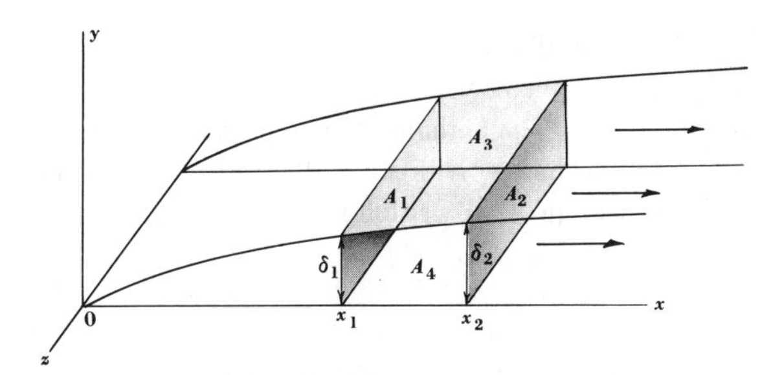

FIGURE 23-5
Control volume for analysis of heat transfer in a turbulent boundary layer.

Turbulent Flow Parallel to a Flat Plate

Heat-transfer coefficients in the region of the laminar boundary layer on a flat plate were analyzed in the previous chapter, starting with the basic differential balances. The problem could also have been solved in a semiempirical fashion by what is known as the *von Kármán integral method.* In this chapter we are dealing with turbulent flow and are unable to use the Navier-Stokes equations, so we shall use the integral method to obtain an approximate solution to the problem.

The use of the integral method for isothermal flow has already been illustrated in Chap. 12. An empirical equation was written for the velocity distribution in the boundary layer, and a force-momentum balance made on a segment of the layer. This subsequently yielded an equation for the drag coefficient at the surface of the plate. The equation for the velocity distribution was

$$\frac{u_x}{u_o} = \left(\frac{y}{\delta}\right)^{1/7} \qquad (12\text{-}96)$$

In this problem we shall assume that a similar equation describes the temperature distribution.

$$\frac{t_s - t}{t_s - t_o} = \left(\frac{y}{\delta_{th}}\right)^{1/7} \qquad (23\text{-}33)$$

The Prandtl number is assumed to be 1, so that $\delta = \delta_{th}$; the boundary of the control volume shown in Fig. 23-5 is the boundary of both the hydrodynamic and thermal boundary layers. Viscous dissipation and conduction in the x and z directions are neglected.

Heat is carried into this system by fluid transport through faces A_1 and A_3 and out through face A_2. It is also conducted into the system from the section of the plate of area A_4.

The overall energy balance, Eq. (4-12), is written for the control volume shown in Fig. 23-5. We neglect the variation of the kinetic energy $u^2/2g_c$ and the potential energy gz/g_c. The enthalpy H is represented by $C_p t$; the flow is steady, so $\partial \breve{E}/\partial \theta$ is zero; and no shaft work is done, $\dot{W}_s = 0$. The heat-conduction rate into the control volume through A_4 is written as

$$q = \iint_{A_4} h_x(t_s - t_o)\, dA$$

Thus the overall energy balance becomes

$$\iint_{A_1} \rho C_p tu \cos\alpha\, dA + \iint_{A_2} \rho C_p tu \cos\alpha\, dA + \iint_{A_3} \rho C_p tu \cos\alpha\, dA$$
$$= \iint_{A_4} h_x(t_s - t_o)\, dA \qquad (23\text{-}34)$$

At face A_1, $u \cos\alpha = -u_x$, and at face A_2, $u \cos\alpha = u_x$, so we can write

$$\iint_{A_2} \rho C_p tu_x\, dA - \iint_{A_1} \rho C_p tu_x\, dA + \iint_{A_3} \rho C_p tu \cos\alpha\, dA$$
$$= \iint_{A_4} h_x(t_s - t_o)\, dA \qquad (23\text{-}35)$$

As a further simplification, the mass flow through face A_3 is related to the mass flow through faces A_1 and A_2 by the following equation, derived in Chap. 12 for an incompressible fluid:

$$\iint_{A_3} \rho u \cos\alpha\, dA = -\iint_{A_2} \rho u_x\, dA + \iint_{A_1} \rho u_x\, dA \qquad (12\text{-}81)$$

Since the temperature of the fluid at the upper edge of the boundary layer (face A_3) is constant at the free-stream temperature t_o, we multiply Eq. (12-81) by t_o and substitute in Eq. (23-35) for the integral over A_3 to get

$$\iint_{A_2} \rho C_p(t - t_o)u_x\, dA - \iint_{A_1} \rho C_p(t - t_o)u_x\, dA = \iint_{A_4} h_x(t_s - t_o)\, dA \qquad (23\text{-}36)$$

Because all the areas have the same width and there is no variation of any quantity in the z direction, we can write Eq. (23-36) with single integrals and insert suitable limits of integration.

$$\int_0^{\delta_2} \rho C_p(t - t_o)u_x\, dy - \int_0^{\delta_1} \rho C_p(t - t_o)u_x\, dy = \int_{x_1}^{x_2} h_x(t_s - t_o)\, dx \qquad (23\text{-}37)$$

If we now consider the length $x_2 - x_1$ to approach zero, the left-hand side of Eq. (23-37) becomes a differential quantity and the right-hand side becomes the heat-transfer rate from a segment of finite width but differential length. Therefore we have

$$d\left[\int_0^\delta \rho C_p(t - t_o)u_x \, dy\right] = h_x(t_s - t_o)\, dx \qquad (23\text{-}38)$$

from which we obtain

$$h_x = \rho C_p \frac{d}{dx}\left(\int_0^\delta \frac{t - t_o}{t_s - t_o} u_x \, dy\right) \qquad (23\text{-}39)$$

The dimensionless temperature function under the integral sign can be rearranged to give

$$\frac{t - t_o}{t_s - t_o} = 1 - \frac{t_s - t}{t_s - t_o}$$

$$= 1 - \left(\frac{y}{\delta}\right)^{1/7} \qquad (23\text{-}40)$$

Furthermore, we have

$$u_x = u_o\left(\frac{y}{\delta}\right)^{1/7} \qquad (23\text{-}41)$$

We substitute Eqs. (23-40) and (23-41) in (23-39), change the variable y in the differential to y/δ, and rewrite the upper limit of integration to get

$$h_x = \rho C_p u_o \frac{d}{dx}\left\{\delta \int_0^1 \left[1 - \left(\frac{y}{\delta}\right)^{1/7}\right]\left(\frac{y}{\delta}\right)^{1/7} d\frac{y}{\delta}\right\} \qquad (23\text{-}42)$$

The integral can be obtained directly; it has a numerical value of $\frac{7}{72}$. Therefore the local coefficient of heat transfer is

$$h_x = \rho C_p u_o \frac{7}{72}\frac{d\delta}{dx} \qquad (23\text{-}43)$$

The thickness of the boundary layer δ was found in Chap. 12 to be represented by the equation

$$\frac{\delta}{x} = 0.376(\mathbf{Re}_x)^{-1/5} \qquad (12\text{-}102)$$

Differentiating this equation, we get

$$\frac{d\delta}{dx} = 0.376\left(\frac{\mu}{u_o \rho}\right)^{1/5}\left(\frac{4}{5}\right)x^{-1/5}$$

$$= 0.301(\mathbf{Re}_x)^{-1/5} \qquad (23\text{-}44)$$

The derivative is substituted in Eq. (23-43) to give

$$\frac{h_x}{\rho C_p u_o} = 0.0292(\mathbf{Re}_x)^{-1/5} \qquad (23\text{-}45)$$

If both sides of this equation are multiplied by $xu_o \rho/\mu$ and $C_p \mu/k$, we get the equation in terms of the local Nusselt number $h_x x/k$,

$$\mathbf{Nu}_x = 0.0292(\mathbf{Re}_x)^{4/5}\,\mathbf{Pr}$$

Since $\mathbf{Pr} = 1$, the final form of the equation is

$$\mathbf{Nu}_x = 0.0292(\mathbf{Re}_x)^{4/5} \qquad (23\text{-}46)$$

If the expression for h_x is integrated from the leading edge of the plate to a distance $x = L$, as was illustrated in Chap. 22, we obtain the mean Nusselt number,

$$\mathbf{Nu}_m = 0.0365(\mathbf{Re}_L)^{4/5} \qquad (23\text{-}47)$$

This derivation for $\mathbf{Nu}_m$ has been based on the assumption that the turbulent boundary layer starts at the leading edge. However, a laminar boundary layer precedes the turbulent boundary layer, which commences at a critical value of approximately $\mathbf{Re}_x = 500{,}000$. The local coefficients can be found by applying Eq. (22-19) from the leading edge to the critical distance and Eq. (23-46) from the critical point downstream. In this method of using this equation, x must always be the distance from the leading edge of the plate, not the distance from the point where turbulence starts. The direct use of Eq. (23-47) to obtain the mean coefficient when a significant portion of the plate is covered by a laminar boundary layer is wrong. Instead, the mean coefficient is found by integrating, using Eq. (22-19) from $x = 0$ to the end of the laminar zone, and further integration from this point to the end of the plate, using Eq. (23-46).

The results derived above are, as mentioned, restricted to use for fluids which have a Prandtl number at (or close to) unity and for which we can thus assume that the momentum and thermal boundary layers are of equal thickness. For fluids outside this category a number of approaches have been suggested. One procedure is based on the following simple relationship between turbulent boundary-layer thicknesses;

$$\delta/\delta_{th} = \mathbf{Pr}^n \qquad (23\text{-}48)$$

It will be recalled that a similar relation (with $n = 1/3$) was used for the analysis of heat transfer in laminar boundary layers. If the upper limits of integration in Eqs. (23-37) to (23-39) are taken as the edge of the thermal boundary layer, and the substitution suggested in Eq. (23-48) is used to convert values of the momentum boundary-layer thickness to thermal boundary-layer thickness, an expression similar to Eq. (23-45), which, in addition, contains the Prandtl number, is obtained.

$$\frac{h_x}{\rho C_p u_0} = 0.0292(\mathbf{Re}_x)^{-1/5}\,\mathbf{Pr}^{-8n/7} \qquad (23\text{-}49)$$

If a value of $n = \frac{1}{3}$ is taken, the exponent on the Prandtl number is -0.38 and Eqs. (23-46) and (23-47) become

$$\mathbf{Nu}_x = 0.0292(\mathbf{Re}_x)^{4/5}\,\mathbf{Pr}^{0.62} \qquad (23\text{-}50)$$

and

$$\mathbf{Nu}_m = 0.0365(\mathbf{Re}_L)^{4/5}\,\mathbf{Pr}^{0.62} \qquad (23\text{-}51)$$

This problem has been the subject of both theoretical and experimental study. Hanna and Myers,[1] utilizing a considerably more detailed integral analysis than that shown above, arrived at a value of 0.50 for the Prandtl number exponent in Eqs. (23-50) and (23-51). Extension of the Colburn *j*-factor analogy from pipes to plates indicates that the Prandtl number exponent should be 0.33. Experimental data reported by Zhukauskas and Ambrazyavichyus[2] for turbulent heat transfer from a flat plate for the Prandtl number range $0.7 < \mathbf{Pr} < 380$ indicate the value for the Prandtl number exponent to be 0.43.

Cellular Convection

When a thin, horizontal layer of fluid is uniformly heated from below, the heated fluid rises because of decreased density and is replaced by cooler fluid from the upper surface. If the thickness of the fluid is small in comparison to the expanse of its surface, the fluid will tend to circulate in a series of cells known as *Bénard cells*. Bénard[3] observed that heating of a thin layer of oil caused the circulation to take place in a lattice of polygons which eventually formed a honeycomb pattern of regular hexagons. It has since been shown that an important criterion for the development of cellular convection is the value of the Rayleigh number ($\mathbf{Ra} = \mathbf{Gr}\,\mathbf{Pr}$), a dimensionless group already seen to be of importance in natural convection flows. At values of $\mathbf{Ra} > 1700$, circulation patterns exist, either of the hexagonal kind observed by Bénard or as parallel rows in a kind of two-dimensional roll structure. Recent work indicates that substantial differences in surface tension, either as a result of temperature or concentration differences, are necessary for development of the hexagonal pattern. In the absence of a surface-tension driving

[1] O. T. Hanna and J. E. Myers, *Engineering Experiment Station Bulletin No. 148*, Purdue University, 1962.

[2] A. A. Zhukauskas and A. B. Ambrazyavichyus, *Int. J. Heat and Mass Transfer*, **3**:305 (1961).

[3] H. Bénard, *Ann. Chim. et Phys.*, **23**:62 (1901).

force the two-dimensional roll structures result. A review of the subject by Whitehead[1] indicates that in addition to the small-scale manifestations of cellular convection seen in the laboratory and the chemical plant, the phenomenon is also evidently occurring (1) inside the earth where cells rising from the molten interior might be generating the continental drift motions on the crust; (2) in the upper layers of the ocean; (3) in the atmosphere, as indicated frequently in well-ordered cumulus cloud patterns; (4) on the surface of the sun, described by astronomers as having "granulation" patterns.

Cellular convection can also occur in polymer coatings causing undesirable effects, such as pigment flocculation and segregation. Reviews of this problem with suggestions for practical remedies to suppress the Bénard cells have been provided by Hansen and Pierce.[2]

PROBLEMS

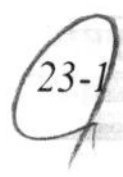

23-1 Air is heated by passing it through a series of eight sections of straight steel pipe which extend through a rectangular tank containing hot water. The pipe is 1-in, schedule-40, and the immersed lengths are 2 ft long. The return bends are outside the water tank. The bulk air velocity is 50 ft/s, and the air pressure can be assumed to be constant at 1 atm. If the air enters at 60°F, find its exit temperature. The inside-pipe-wall temperature can be assumed to be constant at the temperature of the water, which is 210°F.

23-2 Derive Eq. (23-47) for the average Nusselt number for a plate of length L from Eq. (23-46) for the local Nusselt number for turbulent flow.

23-3 Air at 60°F is moving with a free-stream velocity of 100 ft/s across a flat horizontal plate with a uniform surface temperature of 200°F. The plate is 2 ft wide and 4 ft long. Find the total heat-transfer rate in both the portion of the plate covered by the laminar boundary layer and the portion covered by the turbulent boundary layer.

23-4 Derive an equation for flow in a pipe relating the heat-transfer coefficient h and the friction factor f for the case where $\alpha_e \ll \alpha$ and $\nu_e \ll \nu$.

23-5 Derive the Prandtl-Taylor equation by integrating the flux equations (23-8) and (23-9).

23-6 Using the Reynolds analogy, calculate the heat-transfer coefficient in W/m$^2 \cdot$K for a fluid flowing in a pipe with a pressure drop due to friction of 23 kPa per 100 m of pipe. The bulk velocity is 1.5 m/s and the pipe has an ID of 0.05 m. Specific heat of the fluid is 2100 J/kg · K.

23-7 Water flows in turbulent motion through a straight section of 1-in, 16-gauge stainless steel tubing at a rate such that the pressure drop over a given length is 100 kPa. Steam is condensing on the outside of the tube at a constant average temperature of 105°C while inside the tube the water is flowing at a constant average temperature of 20°C. If

[1] J. A. Whitehead, Jr., *Am. Sci.*, **59**:444 (1971).

[2] C. M. Hansen and P. E. Pierce, *Ind. Eng. Chem. Prod. Res. Develop.*, **12**:67 (1973); **13**:218 (1974).

the heat-transfer rate under these conditions is 6600 W/m of tubing, find the heat-transfer rate if the flow is increased to a value that gives a pressure drop of 200 kPa. Assume that the friction factor does not change and that the heat transfer coefficient for the steam side is constant at $h = 5700\ \mathrm{W/m^2 \cdot K}$. The temperatures may be taken as constant at the values given above.

23-1 Air $t_0 = 60$, $u_b = 50$ ft/s; OD, ID, Δx, t_2, $t_s = 210°F$

OD: 1.315 in, ID: 1.049 in

$L: 2\,ft \times 8 = 16\,ft$

$\mu = 0.018\,cp \times \frac{kg\,m^{-1}s^{-1}/10^3}{cp} \times \frac{1\,lb}{0.453\,kg} \times \frac{0.3048\,m}{1\,ft} = 12.11 \times 10^{-6}\ lb\cdot ft^{-1}s^{-1}$

$C_p = 0.25\ Btu \cdot lb^{-1} °F^{-1}$

$Re = \frac{\rho u_b D}{\mu} = \frac{0.0709 \cdot 50 \cdot (1.409/12)}{(12.11 \times 10^{-6})} = 25898$ (Turbulent)

$\rho = \frac{PM}{RT} = \frac{1\,atm \times 29\,lb/lbmol}{0.73\,ft^3 atm/lbmol\,°R \times 560°R} = 0.0709\ lb \cdot ft^{-3}$

energy balance, $\frac{\pi}{4}D^2 \rho C_p u_b (t|_{x+\Delta x} - t|_x) = h_x \pi D \Delta x (t_s - t)$ $/\Delta x$, $\lim \Delta x \to 0$.

G.E: $\frac{dt}{dx} + \frac{1}{\rho C_p u_b} \cdot \frac{4}{D} \cdot h_x (t - t_s) = 0 \Rightarrow \ln\left(\frac{t_2 - t_s}{t_0 - t_s}\right) = -\frac{4 h_a L}{\rho C_p u_b D}$..

$\to t_2 = t_s + (t_0 - t_s) \cdot \exp\left(-\frac{4 h_a L}{\rho C_p u_b D}\right)$

Air $u_b = 100$ ft/s, $t_0 = 60°F$; 2 ft; 4 ft

- properties of air, arithmetic mean of bulk temp.

$\frac{200 + 60}{2} = 130°F$

$\mu = 12.78 \times 10^{-6}\ lb/ft\cdot s$

$C_p = 0.285\ Btu/lb \cdot °F$

$\rho = 0.0673\ lb/ft^3$

$k = 0.0160\ Btu/h\,ft \cdot °F$

$Pr = 0.732$.

- laminar leading edge

$Re_L = 5 \times 10^5 = \frac{\rho u_b L}{\mu} \Rightarrow L = 0.95\,ft$.

1) at Laminar Region. ($x < 0.95\,ft$)

$h_m = 0.664 \times \frac{k}{L} \times (5 \times 10^5)^{1/2} Pr^{1/3} = 7.13\ Btu/h\,ft^2\,°F$.

$q = h_m \cdot A \cdot \Delta t = (7.13)(0.95 \times 2) \times (200 - 60) = 1896.56\ Btu/h$.

2) at turbulent Region ($x > 0.95\,ft$)

$Re_L = \frac{\rho u_b L}{\mu} = \frac{0.0673 \times 100 \times 4}{12.78 \times 10^{-6}} = 2.106 \times 10^6$.

$h_m = \frac{1}{L}\int h_x\,dx = \frac{1}{(4 - 0.95)} \int_{0.95}^{4} \frac{k}{x}\, 0.0292 \cdot \left(\frac{\rho u_b x}{\mu}\right)^{4/5} Pr^{0.62}$. $\to \int x^{-1/5} = \frac{4}{5} x^{4/5}$

$= 12.36\ Btu/h\,ft^2\,°F$.

$q = h_m \cdot A \cdot \Delta t = (12.36) \times (3.05 \times 2) \times (200 - 60)$

$= 10554.47\ Btu/h$.

$q_{tot} = 12451.03\ Btu/h$

24

SOME DESIGN EQUATIONS FOR CONVECTIVE HEAT TRANSFER

The Application of Dimensional Analysis to Heat Transfer

In the preceding chapters we examined three methods of obtaining equations for predicting heat-transfer coefficients in laminar- and turbulent-flow systems. The use of the combined momentum, energy, and continuity equations gave solutions to heat-transfer problems in laminar flow but not in turbulent flow. A second method, the von Kármán integral method, was used to obtain heat-transfer coefficients for turbulent flow over a flat plate. The third method, based on the analogy between heat and momentum transfer, gave solutions to heat-transfer problems with turbulent flow in a pipe.

A fourth method of obtaining heat-transfer equations, for either laminar or turbulent flow, is the use of dimensional analysis. As an illustration, we shall apply this procedure, outlined in Chap. 13, to a problem of natural-convection heat transfer. Natural-convection heat transfer from a plate to a fluid has already been discussed briefly in the chapter on laminar flow. However, laminar flow persists only to a certain point above the leading edge and then breaks down into turbulence. The answer obtained by dimensional analysis is not limited to one type of flow.

Example 24-1 It will be assumed that the heat-transfer coefficient in natural convection in a heated pipe can be related in a dimensionless equation to the physical variables which are listed below. Find the dimensionless groups which form the basis of such an equation.

Six fundamental dimensions are chosen: length $[L]$, mass $[M]$, time $[\theta]$, temperature $[T]$, heat $[H]$, and force $[F]$. Because heat and force are capable of being expressed in terms of the other four dimensions, we include in our list of variables two dimensional constants, the proportionality factor g_c from Newton's second law, and the mechanical equivalent of heat J.

Variable	Symbol	Dimension
Length of heated section	L	$[L]$
Fluid density	ρ	$[M/L^3]$
Fluid viscosity	μ	$[M/L\theta]$
Fluid thermal conductivity	k	$[H/L\theta T]$
Dimensional constant	g_c	$[ML/F\theta^2]$
	J	$[FL/H]$
Mean heat-transfer coefficient	h_m	$[H/\theta TL^2]$
Temperature difference, $t_s - t_o$	Δt	$[T]$
Coefficient of thermal expansion	β	$[T^{-1}]$
Specific heat of fluid	C_p	$[H/MT]$
Gravitational acceleration	g	$[L/\theta^2]$
Bulk velocity of fluid	u_b	$[L/\theta]$
Diameter of pipe	D	$[L]$

The total number of variables is 13, and we have chosen to express these in terms of six dimensions. As is often the case, the maximum number of variables which will not form a dimensionless group is equal to the number of fundamental dimensions. Thus, according to Buckingham's theorem, we should obtain seven dimensionless groups. The variables which we choose to be common to all groups are the first six in the table above. Each of the remaining seven variables will, in turn, be added to the first six, to give the seven groups. The first group is

$$\pi_1 = L^a \rho^b \mu^c k^d g_c{}^e J^f h_m{}^g$$

We substitute the dimensions and obtain

$$[1] = [L]^a \left[\frac{M}{L^3}\right]^b \left[\frac{M}{L\theta}\right]^c \left[\frac{H}{L\theta T}\right]^d \left[\frac{ML}{F\theta^2}\right]^e \left[\frac{FL}{H}\right]^f \left[\frac{H}{\theta TL^2}\right]^g$$

which yields the following simultaneous equations:

$$
\begin{array}{ll}
[L] & 0 = a - 3b - c - d + e + f - 2g \\
[M] & 0 = b + c + e \\
[\theta] & 0 = -c - d - 2e - g \\
[T] & 0 = -d - g \\
[H] & 0 = d - f + g \\
[F] & 0 = -e + f
\end{array}
$$

These can be solved to give $b = c = e = f = 0$, $a = g$, and $d = -g$. Therefore π_1 reduces to $(h_m L/k)^g$. The exponent g is of no significance and could have been assumed to be unity throughout the analysis.

The second group, π_2, is found in a similar fashion. The first six variables are combined with the variable Δt, which represents the temperature-difference driving force.

$$\pi_2 = L^a \rho^b \mu^c k^d g_c{}^e J^f (\Delta t)^g$$

which gives

$$1 = [L]^a \left[\frac{M}{L^3}\right]^b \left[\frac{M}{L\theta}\right]^c \left[\frac{H}{L\theta T}\right]^d \left[\frac{ML}{F\theta^2}\right]^e \left[\frac{FL}{H}\right]^f [T]^g$$

from which we obtain the equations

$$
\begin{array}{ll}
[L] & 0 = a - 3b - c - d + e + f \\
[M] & 0 = b + c + e \\
[\theta] & 0 = -c - d - 2e \\
[T] & 0 = -d + g \\
[H] & 0 = d - f \\
[F] & 0 = -e + f
\end{array}
$$

These equations can be solved to give $a = b = 2g$, $c = -3g$, and $d = e = f = g$. Taking the exponent g as unity, the group becomes

$$\pi_2 = \frac{L^2 \rho^2 k g_c J \, \Delta t}{\mu^3}$$

In the analysis for π_3, the additional variable is β, replacing Δt. Because β has the reciprocal of the dimension possessed by Δt, the analysis for π_3 will yield

$$\pi_3 = \frac{\mu^3 \beta}{L^2 \rho^2 k g_c J}$$

The remaining groups, obtained by the procedures already illustrated, are

$$\pi_4 = \frac{C_p \mu}{k}$$

$$\pi_5 = \frac{L^3 \rho^2 g}{\mu^2}$$

$$\pi_6 = \frac{L \rho u_b}{\mu}$$

$$\pi_7 = \frac{D}{L}$$ ////

The seven groups obtained in Example 24-1 can be used to correlate the results of experiments in natural convection. One common way of doing this is to write an equation of the form

$$\pi_1 = \alpha \pi_2{}^{\beta} \pi_3{}^{\gamma} \pi_4{}^{\delta} \pi_5{}^{\varepsilon} \pi_6{}^{\zeta} \pi_7{}^{\eta} \tag{24-1}$$

The constants represented by the Greek letters can be determined by a statistical analysis of the experimental results, often the least-squares method. Groups which have no effect on h_m can easily be located because the exponents on these groups will be very small. Two groups may have the same exponent, indicating that the variables they contain may be combined in a single group. As an example, if the analysis of Example 24-1 is used as a basis for correlating natural-convection heat transfer from a flat, vertical plate (i.e., a tube of infinite diameter), it is found that the combined groups, $\pi_2 \pi_3 \pi_5$, give a single dimensionless group, $L^3 \rho^2 g \beta \, \Delta t / \mu^2$, known as the *Grashof number*, which together with π_1, the Nusselt number, and π_4, the Prandtl number, are adequate to correlate the results. The groups π_6 and π_7 are meaningless in this analysis because $u_b = 0$ and $D = \infty$. The correlating equation is

$$\pi_1 = \alpha(\pi_2 \pi_3 \pi_5)^{\beta} (\pi_4)^{\delta} \tag{24-2}$$

The problem of natural-convection heat transfer with laminar flow of gases has been seen in a previous chapter to be represented by the expression

$$\frac{h_m L}{k} = 0.478 \left(\frac{L^3 \rho^2 g \, \Delta t}{\mu^2 T} \right)^{0.25} \tag{22-25}$$

The Prandtl number is approximately constant at unity for most gases, so the absence of that group would be expected. The Grashof group in Eq. (22-25) differs from the product $\pi_2 \pi_3 \pi_5$ only in that for an ideal gas, β equals $1/T$, where T is the absolute temperature. Equation (24-2) is the basis of a number of correlations for natural-convection heat transfer from a flat plate for gases and liquids in both laminar and turbulent systems.

Equation (24-1) can also serve as the basis for the analysis of forced-convection heat transfer. If the flow is turbulent, the length of the pipe has little effect on the heat-transfer coefficient ($h_m \approx h$), so groups π_1 and π_7 are combined to give the Nusselt number hD/k. Groups π_2, π_3, and π_5, which have to do with gravitational forces in a variable-density field, have negligible effects; π_4, the Prandtl number, remains, and π_6 and π_7 combine to give the Reynolds number $Du_b\rho/\mu$. Thus an equation is obtained of the form

$$\frac{hD}{k} = \alpha\left(\frac{Du_b\rho}{\mu}\right)^{\beta}\left(\frac{C_p\mu}{k}\right)^{\gamma} \tag{24-3}$$

Although most equations which are obtained by the procedures of dimensional analysis are of the form $\pi_1 = \alpha\pi_2{}^{\beta}\pi_3{}^{\gamma}\cdots$, there is no necessity that this be so. In fact we shall see shortly that heat-transfer coefficients from spheres may be correlated by an equation of the general form $\pi_1 = \alpha + \beta\pi_2{}^{\gamma}\pi_3{}^{\delta}$.

Summary of Equations Used for Predicting Convective Heat-Transfer Coefficients

The equations used by engineers for predicting heat-transfer coefficients have various origins, ranging from fundamental derivations to empirical correlations of dimensionless groups. The fact that most equations are expressed in terms of dimensionless groups should not be taken to mean that they all have their origin in dimensional analysis. We have seen that such common groups as the Nusselt, Reynolds, Prandtl, and Grashof numbers also occur in equations obtained analytically.

The equations presented below have been obtained from a variety of sources. Since most of them have been reprinted in numerous textbooks, we shall not attempt to give the original references.[1]

Convection in Circular Pipes

Laminar flow The theoretical equations proposed in the chapter on heat transfer with laminar flow have been used with success where the temperature gradient has not been so large as to affect fluid properties significantly. The results

[1] An excellent source of heat-transfer equations and data is the textbook by W. H. McAdams, "Heat Transmission," 3d ed., McGraw-Hill Book Company, New York, 1954. See also the review article by S. Whitaker, Forced Convection Heat Transfer Correlations for Flow in Pipes, Past Flat Plates, Single Cylinders, Single Spheres, and for Flow in Packed Beds and Tube Bundles, *AIChE J.*, **18**:361 (1972).

of these derivations have been plotted (Fig. 22-3) in terms of the local Nusselt number $\mathbf{Nu}_x$, which contains the local heat-transfer coefficient h_x. Mean heat-transfer coefficients based on arithmetic and logarithmic mean temperature-difference driving force have been calculated from the local coefficients and are plotted in Fig. 22-4. For the case of constant wall temperature, if the Graetz number wC_p/kL is less than 5 (perhaps as a result of a low flow rate or a long pipe), the Nusselt number $h_m D/k$ is approximately equal to 3.6. If wC_p/kL is greater than 10, the Nusselt number can be approximated by the equation

$$\frac{h_m D}{k} = 1.62\left(\frac{4wC_p}{\pi kL}\right)^{1/3} \tag{22-40}$$

Sieder and Tate proposed a modification of Eq. (22-40) which accounts for the effect of temperature on viscosity by the use of a factor $(\mu/\mu_s)^{0.14}$. The quantity μ_s is the viscosity at the wall temperature. The other properties, including μ, are evaluated at the arithmetic average bulk temperature. The Sieder and Tate equation expressed in dimensionless groups is

$$\mathbf{Nu}_m = 1.86\ \mathbf{Re}^{1/3}\mathbf{Pr}^{1/3}\left(\frac{D}{L}\right)^{1/3}\left(\frac{\mu}{\mu_s}\right)^{0.14} \tag{22-42}$$

which is limited by the same restrictions as Eq. (22-40).

The equations given above apply when the pipe-wall temperature is uniform. This condition is often approximately achieved in condensers. In Chap. 22 we saw that laminar-flow systems with uniform heat flux at the wall have also been analyzed. This condition may be achieved in heat exchangers when the overall temperature difference is constant along the exchanger. The data in Fig. 22-3 can be used for this case.

In all forced-convection systems undergoing heat transfer, natural-convection flow effects contribute to the development of the flow pattern. However, these effects are usually significant only when the forced-convection flow is laminar and the temperature-difference driving force is large. In vertical tubes the natural-convection effects can either augment or oppose the forced-convection flow. When the two effects are additive, the consequence is to cause higher heat-transfer coefficients than would be predicted by design equations such as Eq. (22-41). When the flows are opposing, the heat-transfer coefficients are lower than those predicted by Eq. (22-41). Opposing fluid motions in vertical tubes may cause a complete breakdown of the forced-convection laminar profile.

Natural-convection effects contribute to the flow pattern and thus to the heat-transfer coefficients in horizontal tubes as well as in vertical tubes. The consequence of natural convection in horizontal tubes is an increase of the heat-transfer coefficient, but the increase is likely to be smaller than for vertical tubes. Equations and graphs for estimating the effects of natural convection on laminar flow in both vertical and horizontal tubes are available.[1]

Turbulent flow Three well-known equations for correlating heat-transfer data are presented. All apply to the fully developed turbulent flow of fluids for which **Pr** > 0.7. In pipes for which $L/D > 60$, the entrance effects are negligible. The equations can, however, be used for shorter pipes if the entrance region is treated separately by the use of Eq. (23-1) and Table 23-1.

The first equation considered is that of Dittus and Boelter, for which all fluid properties are evaluated at the arithmetic mean bulk temperature.

$$\frac{hD}{k} = 0.023\left(\frac{Du_b\rho}{\mu}\right)^{0.8}\left(\frac{C_p\mu}{k}\right)^{0.3 \text{ or } 0.4} \tag{24-4}$$

This equation correlates data for heating of the fluid when the exponent on the Prandtl number is 0.4 and for cooling of the fluid when the exponent is 0.3. The difference appears to be the result of temperature effects on viscosity in the laminar sublayer. Equation (23-15) shows that the thickness of this layer is proportional to fluid viscosity. If two turbulent systems of the same fluid at the same bulk temperature are compared in heating and cooling experiments, the system being heated will have a higher temperature in the laminar sublayer than the system being cooled. Consequently, if the fluid is a liquid, the sublayer of the heated system will be thinner, and the heat-transfer coefficient will therefore be higher than for the system being cooled. Since the Prandtl number is greater than 1 for most liquids, raising the exponent in Eq. (24-4) has the effect of raising h. For most gases the Prandtl number is near unity and approximately independent of temperature, so that the value of the exponent on **Pr** is of minor consequence. An understanding of this lack of effect for gases may be gained from the fact that viscosity and thermal conductivity increase at almost the same rate with temperature, so that the thermal resistance of the sublayer is approximately constant.

Another well-known equation is that of Colburn. If the exponent on the Prandtl number in Eq. (24-4) is taken as 0.33, the equation can be rearranged by dividing both sides by **Pr Re** to read

$$\frac{h}{u_b\rho C_p}\left(\frac{C_p\mu}{k}\right)^{2/3} = 0.023\left(\frac{Du_b\rho}{\mu}\right)^{-0.2} \tag{24-5}$$

[1] Engineering Sciences Data (Chemical Engineering Series), Item no. 68006, Institution of Chemical Engineers, London, 1968.

Each side of this equation is equal to the quantity previously defined as j_H. The fluid properties are evaluated at the arithmetic mean of the wall and bulk fluid temperatures.

Sieder and Tate have presented an equation for turbulent-flow heat-transfer coefficients which has a viscosity correction similar to the correction they proposed for laminar-flow coefficients. Because it takes account of the effect of temperature on viscosity more directly than either the Dittus-Boelter or Colburn equations, it is valid for fluids whose Prandtl numbers are as high as 10^4. It is written

$$\frac{h}{u_b \rho C_p}\left(\frac{C_p \mu}{k}\right)^{2/3}\left(\frac{\mu_s}{\mu}\right)^{0.14} = 0.023\left(\frac{D u_b \rho}{\mu}\right)^{-0.2} \tag{24-6}$$

All fluid properties are evaluated at the arithmetic average bulk temperature except the quantity μ_s, which is evaluated at the average wall temperature.

An analysis[1] of Eq. (24-4) for accuracy in predicting experimental results shows an overall root-mean-square deviation of 13 percent for 651 data points reported in a dozen studies. All points represent conditions of developed flow and minimal radial variation of fluid properties. The distribution of errors shown in Table 24-1 shows

Table 24-1 **DISTRIBUTION OF ERRORS FOR Eq. (24-4)**

Pr	Re				
	$4 \times 10^3 - 10^4$	$10^4 - 3.16 \times 10^4$	$3.16 \times 10^4 - 10^5$	$10^5 - 3.16 \times 10^5$	$3.16 \times 10^5 - 10^6$
	71	61	44	29	22
0.32–1	−11.06	−13.54	−11.09	−10.15	−7.43
	15.35	17.27	13.24	12.76	9.61
		31	43	82	16
1–3.16		4.12	−1.48	−5.48	−6.18
		6.43	6.96	10.29	11.24
	15	45	58	5	
3.16–10	6.35	10.97	−1.94	5.55	
	12.13	16.66	9.46	13.82	
	18	46	8		
10–31.6	4.21	3.24	−4.31		
	12.14	15.18	5.19		
	6	27	6		
31.6–100	−6.65	1.10	−2.28		
	10.24	6.99	7.58		
	5	13			
100–316	−28.39	−21.15			
	28.55	21.72			

SOURCE: Engineering Sciences Data (Chemical Engineering Series), Item no. 67016, Institution of Chemical Engineers, London, 1967.

[1] Engineering Sciences Data (Chemical Engineering Series), Item no. 67016, Institution of Chemical Engineers, London, 1967.

no obvious trends except for the rather large deviations at **Pr** > 100. The same report contains the following least-squares correlation of the 651 data points:

$$\frac{h}{u_b \rho C_p} = \exp\left[-3.796 - 0.205 \ln \mathbf{Re} - 0.505 \ln \mathbf{Pr} - 0.0225(\ln \mathbf{Pr})^2\right] \qquad (24\text{-}7)$$

This equation represents the 651 data points, with an overall root-mean-square error of 10.2 percent. The distribution of errors using Eq. (24-7) shown in Table 24-2 indicates significant improvement over Eq. (24-4), particularly in the Prandtl number ranges of 0.32–1 and 100–316.

Tables 24-1 and 24-2 show in descending order the number of data points, the average error, and the root-mean-square error for each range of Reynolds number and Prandtl number.

Example 24-2 In a heat exchanger water flows through a long 1-in, 16-gauge copper tube at a bulk velocity of 7 ft/s and is heated by steam condensing at 300°F on the outside of the tube. The water enters at 60°F and leaves at 140°F. Find the heat-transfer coefficient for the water.

Table 24-2 DISTRIBUTION OF ERRORS FOR Eq. (24-7)

Pr	Re				
	$4 \times 10^3 - 10^4$	$10^4 - 3.16 \times 10^4$	$3.16 \times 10^4 - 10^5$	$10^5 - 3.16 \times 10^5$	$3.16 \times 10^5 - 10^6$
0.32–1	71 −0.25 9.62	61 −1.75 9.78	44 0.97 6.52	29 2.14 7.02	22 5.17 7.45
1–3.16		31 4.54 6.62	43 −0.58 6.49	82 −0.65 10.73	16 1.08 9.30
3.16–10	15 4.11 10.94	45 9.48 15.76	58 −3.19 9.84	5 4.86 13.02	
10–31.6	18 3.30 11.95	46 2.25 15.32	8 −5.51 6.38		
31.6–100	6 −2.98 8.50	27 5.29 8.61	6 2.81 7.17		
100–316	5 −5.31 5.86	13 −4.58 7.23			

SOURCE: Engineering Sciences Data (Chemical Engineering Series), Item no. 67016, Institution of Chemical Engineers, London, 1967.

The physical properties of water at 100°F are

$$\rho = 62.0 \text{ lb/ft}^3$$
$$C_p = 0.998 \text{ Btu/(lb)(°F)}$$
$$\mu = 0.000458 \text{ lb/(ft)(s)}$$
$$k = 0.364 \text{ Btu/(h)(ft)(°F)}$$

The ID of the tubing is 0.870 in.

$$\mathbf{Re} = \frac{(0.870/12)(7)(62.0)}{0.000458}$$
$$= 68{,}600$$
$$\mathbf{Pr} = \frac{(0.998)(0.000458)(3600)}{0.364}$$
$$= 4.53$$

Because the flow is turbulent and the tube is quite long, we have a choice of calculating the heat-transfer coefficient from any one of the four standard equations (24-4) to (24-7).

(*a*) Using the Dittus-Boelter equation (24-4), we get

$$h = \frac{(0.023)(0.364)}{(0.870/12)}(68{,}600)^{0.8}(4.53)^{0.4}$$
$$= 1560 \text{ Btu/(h)(ft}^2\text{)(°F)}$$

The exponent on **Pr** is 0.4 because the fluid is being heated.

(*b*) In the Colburn equation (24-5), the physical properties must be evaluated at the arithmetic mean of the wall and bulk-fluid temperatures. The arithmetic mean bulk temperature is 100°F. The temperature of the condensing steam is 300°F. To calculate the wall temperature we need to know the coefficient of heat transfer for the condensing steam. The calculation of coefficients for condensation will be explained in a later chapter. For the present problem we assume that the average coefficient for condensation is of the same magnitude as the heat-transfer coefficient for water. If we neglect the temperature drop over the copper wall, the average wall temperature is

200°F. Therefore, to use the Colburn equation, we must know the physical properties of water at 150°F. These are

$$\rho = 61.2 \text{ lb/ft}^3$$
$$C_p = 1.00 \text{ Btu/(lb)(°F)}$$
$$\mu = 0.000292 \text{ lb/(ft)(s)}$$
$$k = 0.384 \text{ Btu/(h)(ft)(°F)}$$

$$\mathbf{Re} = \frac{(0.870/12)(7)(61.2)}{0.000292}$$

$$= 107{,}000$$

$$\mathbf{Pr} = \frac{(1.00)(0.000292)(3600)}{0.384}$$

$$= 2.73$$

Substituting these values in Eq. (24-5) gives

$$h = (0.023)(7)(3600)(61.2)(1.0)(107{,}000)^{-0.2}(2.73)^{-2/3}$$

$$= \frac{35{,}600}{(10.1)(1.95)}$$

$$= 1800 \text{ Btu/(h)(ft}^2\text{)(°F)}$$

(*c*) The Sieder-Tate equation (24-6) contains the physical properties evaluated at the bulk temperature. It can be rearranged to read

$$\frac{hD}{k} = 0.023\ \mathbf{Re}^{0.8}\ \mathbf{Pr}^{0.33}\left(\frac{\mu}{\mu_s}\right)^{0.14}$$

$\mathbf{Re} = 68{,}000$, and $\mathbf{Pr} = 4.53$ as in part (*a*); $\mu = 0.000458$ lb/(ft)(s); μ_s is the viscosity at the average wall temperature, which we have already estimated to be 200°F, so that $\mu_s = 0.000205$ lb/(ft)(s). Therefore we get

$$h = \frac{(0.023)(0.364)}{(0.870/12)}(68{,}600)^{0.80}(4.53)^{0.33}\left(\frac{0.000458}{0.000205}\right)^{0.14}$$

$$= (0.115)(7400)(1.65)(1.12)$$
$$= 1580 \text{ Btu/(h)(ft}^2\text{)(°F)}$$

(*d*) Equation (24-7) provides still another means of determining h. We use the same values of the physical properties employed in solving the Dittus-Boelter equation in part (*a*).

$$\frac{h}{u_b \rho C_p} = \exp\left[-3.796 - 0.205 \ln 68{,}600 - 0.505 \ln 4.53 - 0.0225(\ln 4.53)^2\right]$$

$$= e^{-6.89}$$

$$= 0.00105$$

$$h = (0.00105)(62.0)(7)(3600)(0.998)$$

$$= 1640 \text{ Btu/(h)(ft}^2\text{)(°F)}$$

Tables 24-1 and 24-2 show that, for the system specified in this example, Eqs. (24-4) and (24-7) might be expected to predict values of the heat-transfer coefficient within 10 percent of experimental values. ////

Convection in Noncircular Conduits

A heat-transfer system which is commonly encountered is that in which fluids at different temperatures flow in concentric pipes. The heat-transfer coefficient of the fluid in the annular space can be predicted from equations similar to those which apply for circular pipes; however, the equivalent diameter, as defined in Chap. 14, must be used. For the annular space this equivalent diameter D_{eq} is the ID of the outer pipe D_2 minus the OD of the inner pipe D_1. Wiegand has proposed an equation for the heat-transfer coefficient on the outer wall of the inner pipe, which is

$$\frac{hD_{eq}}{k} = 0.023\left(\frac{D_{eq} u_b \rho}{\mu}\right)^{0.8}\left(\frac{C_p \mu}{k}\right)^{0.4}\left(\frac{D_2}{D_1}\right)^{0.45} \tag{24-8}$$

For the inner wall of the outer pipe, Eq. (24-6) is recommended, with the substitution of D_{eq} for D. Equations (24-6) and (24-8) apply only to turbulent flow. The use of the equivalent diameter in conventional heat-transfer equations has been found to be valid for other configurations, such as triangular and rectangular ducts. Equations (24-4) to (24-7), when used with D_{eq}, give reasonably reliable results.

Convection Normal to a Cylinder

Numerous investigations have been made of heat transfer between a cylinder and a fluid flowing normal to the axis of the cylinder. For Reynolds numbers less than 1.0, the equations of continuity, energy, and momentum have been solved numerically to yield local heat-transfer coefficients. At higher flow rates experimental investigations have provided information on local heat-transfer coefficients. Figure 24-1 shows measured values of the local Nusselt number plotted as a function of angular position at various values of **Re**. High heat-transfer coefficients occur at

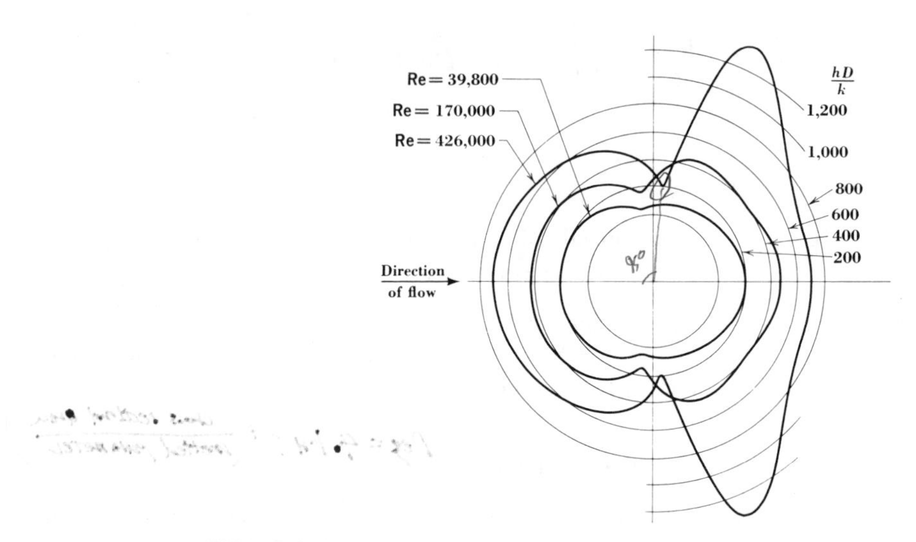

FIGURE 24-1
Local values of Nusselt number for air flow normal to a cylinder. [*From E. Schmidt and K. Wenner, Forsch. Gebiete Ingenieurw.*, **12**:*65 (1941); Trans. Natl. Adv. Comm. Aeronaut. Tech. Memor. 1050 (1943)*].

the forward stagnation point and decrease as the laminar boundary layer thickens. At values of **Re** from 10^3 to 10^5 separation of the laminar boundary layer occurs at slightly more than 80° from the stagnation point. The velocity gradient at the cylinder surface at this point is nearly zero, and the heat-transfer coefficient has a minimum value, as shown on the curve for **Re** = 39,800 in Fig. 24-1. Beyond this separation point the surface is exposed to a turbulent wake, which causes the heat-transfer coefficient to increase.

At higher values of **Re** the boundary layer changes from laminar to turbulent at approximately 95°. When this happens in flow over a flat plate or inside a tube near the entrance, a minimum occurs, followed by a sudden increase of the heat-transfer coefficient, as we have already seen. Thus a minimum heat-transfer coefficient is observed at the transition point of 95° for a cylinder, followed by a sharp increase in the heat-transfer coefficient for that portion of the surface covered by the turbulent boundary layer. The turbulent boundary layer then separates from the cylinder at a position between 130 and 150° from the stagnation point, causing a second minimum to occur in the curve for the heat-transfer coefficient. This behavior is seen in the curve for **Re** = 426,000 in Fig. 24-1. The values of the local heat-transfer coefficients and the location of the minima are strongly dependent on the intensity of turbulence in the free stream.

Natural convection.

Average coefficients for the entire surface of the cylinder can be determined from plots of local coefficients, but are more often found from correlations of average coefficients. Most writers recommend an equation of the type

$$\frac{hD}{k} = b\left(\frac{Du_o\rho}{\mu}\right)^n \tag{24-9}$$

in which the constants b and n depend on **Re**, as shown in Table 24-3. The values in this table are for the calculation of the heat-transfer coefficients of gases. For liquids the same constants are used, but in addition the right-hand side of Eq. (24-9) must be multiplied by 1.1 $\mathbf{Pr}^{1/3}$. All fluid properties are evaluated at the arithmetic mean of the surface and free-stream temperatures.

Data on natural convection from horizontal cylinders are of value in estimating heat losses from pipes. A suitable correlation for the average Nusselt number for a single horizontal cylinder when the product **Pr Gr** ranges from 10^3 to 10^9 is

$$\frac{hD}{k} = 0.53\left(\frac{D^3\rho^2 g\beta\,\Delta t}{\mu^2}\right)^{1/4}\left(\frac{C_p\mu}{k}\right)^{1/4} \tag{24-10}$$

The presence of the Grashof number $D^3\rho^2 g\beta\,\Delta t/\mu^2$ would be expected in any correlation of natural-convection heat transfer, as was pointed out earlier in this chapter. The physical properties are evaluated at the arithmetic average of the surface temperature and the temperature of the fluid outside the thermal boundary layer.

Example 24-3 A pipe contains steam at a temperature such that the temperature of the outer surface of the insulation around the pipe is 56°C. The OD of the insulation is 0.1 m, and the line is in a room in which the air is at a temperature of 20°C. Find the convective heat-transfer coefficient between the surface of the insulation and the air.

Table 24-3 CONSTANTS FOR USE IN Eq. (24-9)

$Du_o\rho/\mu$	n	b
1–4	0.330	0.891
4–40	0.385	0.821
40–4000	0.466	0.615
4000–40,000	0.618	0.174
40,000–250,000	0.805	0.0239

SOURCE: Taken from W. H. McAdams, "Heat Transmission," 3d ed., p. 260, McGraw-Hill Book Company, New York, 1954.

Equation (24-10) is used, so the physical properties must be evaluated at 38°C. They are

$$k = 0.0266 \text{ W/m} \cdot \text{K}$$
$$\rho = 1.14 \text{ kg/m}^3$$
$$\beta = 0.00322 \text{ K}^{-1}$$
$$\mu = 1.92 \times 10^{-5} \text{ Pa} \cdot \text{s}$$
$$C_p = 1000 \text{ J/kg} \cdot \text{K}$$
$$\mathbf{Gr} = \frac{(0.1)^3(1.14)^2(9.8)(0.00322)(36)}{(1.92 \times 10^{-5})^2}$$
$$= 4.00 \times 10^6$$
$$\mathbf{Pr} = \frac{(1000)(1.92 \times 10^{-5})}{0.0266}$$
$$= 0.722$$

Therefore, using Eq. (24-10), we get

$$h = \frac{(0.53)(0.0266)}{0.1}(4 \times 10^6)^{1/4}(0.722)^{1/4}$$
$$= 5.81 \text{ W/m}^2 \cdot \text{K}$$

The product **Pr Gr** is obviously within 10^3 to 10^9, the range of applicability of the equation. ////

Convection Normal to a Bank of Circular Tubes

The effect of one tube on another is to increase the heat-transfer coefficient of the downstream tube. Numerous correlations have been proposed for heat-transfer coefficients of a fluid flowing across a bundle of tubes. One of the best-known correlations is that of Grimison, who wrote a conventional equation for gases,

$$\frac{hD}{k} = b\left(\frac{DG_{\max}}{\mu}\right)^n \qquad (24\text{-}11)$$

and supplied values of the constants b and n, from which mean coefficients can be obtained for banks of tubes 10 or more rows deep. $G_{\max}$ is the product of density times the velocity at the minimum cross section. Fluid properties are evaluated at the arithmetic mean of the air and tube-wall temperatures, as recommended for Eq.

(24-9). A table of the constants b and n is given by McAdams.[1] As an example, tubes spaced in a pattern of equilateral triangles in which the center-to-center distance between tubes is twice the tube diameter are represented by the values $b = 0.482$ and $n = 0.556$. When the tubes are located on a square grid with the same center-to-center spacing, $b = 0.229$, $n = 0.632$. When used for fluids other than gases, the right-hand side of Eq. (24-11) should be multiplied by 1.1 $\mathbf{Pr}^{1/3}$.

For tube bundles less than 10 rows deep, the average heat-transfer coefficient for the bundle can be calculated from Eq. (24-11) and the ratios given in Table 24-4.

Example 24-4 Air is being heated by blowing it over a bank of five rows of 1-in, 14-gauge copper tubes 4 ft long. The tubes are spaced with their centers at the corners of equilateral triangles; the center-to-center distance is 2 tube diameters. There are five tubes per row, and there is a space of $\frac{1}{2}$ in between the outside tubes and the duct. Steam is condensing at 227°F inside the tubes, and the resistance to heat transfer of the condensing steam and the tube walls can be neglected. If 500 ft^3/min of air enters the heater at 1 atm pressure and 60°F, find the mean heat-transfer coefficient for the air. Assume that the exit air temperature is 100°F.

The mean heat-transfer coefficient for the tube bank will be found using Grimison's equation (24-11).

The exit temperature of the air is 100°F, so the mean air temperature is 80°F. The physical properties of air will be taken at $(80 + 227)/2 = 153$°F. They are $k = 0.0164$ Btu/(h)(ft)(°F), and $\mu = 1.36 \times 10^{-5}$ lb/(ft)(s). The mass velocity is based on air at 60°F. Thus we use $\rho_{60°F} = 0.076$ lb/ft^3.

The minimum free-cross-sectional area for flow in the heater is (4)(4) ×

Table 24-4 RATIO OF MEAN HEAT-TRANSFER COEFFICIENT FOR BANK OF TUBES N ROWS DEEP TO THE COEFFICIENT FOR TUBES IN A SINGLE ROW

	N									
	1	2	3	4	5	6	7	8	9	10
Triangular grid	1	1.10	1.22	1.31	1.35	1.40	1.42	1.44	1.46	1.47
Square grid	1	1.25	1.36	1.41	1.44	1.47	1.50	1.53	1.55	1.56

[1] *Ibid.*

$(\frac{1}{12}) + (2)(4)(0.5/12) = 1.667\ \text{ft}^2$. Therefore the maximum mass velocity is

$$G_{max} = \frac{(500)(0.076)}{(1.667)(60)}$$

$$= 0.380\ \text{lb/(s)(ft}^2)$$

and

$$\frac{DG_{max}}{\mu} = \frac{(\frac{1}{12})(0.380)}{1.36 \times 10^{-5}}$$

$$= 2330$$

From the discussion following Eq. (24-11), we obtain the constants for this system, which are $b = 0.482$, $n = 0.556$. Thus

$$h = \frac{(0.482)(0.0164)}{\frac{1}{12}}(2330)^{0.556}$$

$$= 7.1\ \text{Btu/(h)(ft}^2)(^\circ\text{F})$$

Grimison's equation applies to banks of 10 rows of tubes or more. In this problem the bank contains only 5 rows. From Table 24-4 we see that the ratio of the mean coefficient for a bank of 5 rows to the mean coefficient for a bank of 10 rows is 1.35/1.47. Therefore, for the bank of 5 rows, the mean heat-transfer coefficient is

$$h = (7.1)(1.35/1.47)$$

$$= 6.5\ \text{Btu/(h)(ft}^2)(^\circ\text{F}) \qquad ////$$

Convection from Spheres

Both local and average heat-transfer coefficients have been measured for single spheres. In a qualitative way the local coefficients behave in the same manner for spheres as for cylinders. Heat-transfer coefficients diminish from the forward stagnation point to the point of separation of the laminar boundary layer and then increase. When the transition to a turbulent boundary layer is followed by separation, the two points are indicated by sudden trends of increasing h_x in the region 90 to 120° from stagnation.

Many of the data used for predicting average heat-transfer coefficients for spheres have been obtained by measurements of mass-transfer rates. A generally accepted correlation is that of Froessling, which is written

$$\frac{hD}{k} = 2.0 + 0.6\left(\frac{C_p\mu}{k}\right)^{1/3}\left(\frac{Du_o\rho}{\mu}\right)^{1/2} \qquad (24\text{-}12)$$

The quantity 2.0 is the value obtained for the Nusselt number for heat transfer by conduction from a sphere of diameter D in an infinite stagnant medium. This constant can be obtained by writing an equation for steady-state, spherical conduction as follows:

$$q = -(k)(4\pi r^2)\frac{dt}{dr} = \text{const} \qquad (24\text{-}13)$$

We integrate this equation from the surface of the sphere ($r = D/2$, $t = t_s$) to some point a great distance away ($r = \infty$, $t = t_o$). The heat-transfer rate q is constant not only with time, but also with radial position. Equation (24-13) is therefore written as

$$-\frac{q}{4\pi k}\int_{D/2}^{\infty}\frac{dr}{r^2} = \int_{t_s}^{t_o} dt \qquad (24\text{-}14)$$

which gives

$$-\frac{q}{4\pi k}\left(-\frac{1}{\infty} + \frac{1}{D/2}\right) = t_o - t_s \qquad (24\text{-}15)$$

This equation is rearranged to give

$$q = 2\pi Dk(t_s - t_o) \qquad (24\text{-}16)$$

However, we write by convention

$$q = (h)(\pi D^2)(t_s - t_o) \qquad (24\text{-}17)$$

Thus, combining Eqs. (24-16) and (24-17), we obtain

$$\frac{hD}{k} = 2 \qquad (24\text{-}18)$$

Convective Heat Transfer between a Fluid and a Packed Bed

One method of solution of this problem has been suggested by Ranz.[1] Although the particles of a bed may not be spherical, an analysis based on flow in a regularly packed bed of spheres gives a good approximation for the heat-transfer coefficient.

The spheres are assumed to be packed in the closest possible arrangement, called *rhombohedral packing*. The bed consists of parallel layers of spheres in planes normal to the direction of flow. The spheres in successive layers are directly centered over the orifices formed by the openings between the spheres in the previous layer. The spheres in any layer are centered directly above those in the third layer

[1] W. E. Ranz, *Chem. Eng. Progr.*, **48**:247 (1952).

FIGURE 24-2
Section through a packed bed of spheres.

below. In this arrangement, the centers of any four adjacent spheres are at the corners of a tetrahedron.

The method is based on the assumption that the heat-transfer coefficient for any sphere in the bed is the same as it would be for flow past a single sphere at the velocity through one of the orifices between the spheres. The jet from any orifice impinges directly on the sphere in the next layer.

It remains to calculate the relation of the velocity through the space shown in Fig. 24-2 to the superficial velocity. The ratio of the total bed cross-sectional area to the area of the openings must be found. If the diameter of the spheres is small compared with the overall width or diameter of the bed, the ratio desired is equal to the ratio of area of the triangle shown in Fig. 24-2 to the area of the opening between the arcs of the circles in one triangle. The side of the equilateral triangle is D, so the area is $\sqrt{3}(D^2/4)$. The area of the sum of the three sectors is $\pi D^2/8$, so the ratio of the area of the triangle to the area of the orifice is $2\sqrt{3}/(2\sqrt{3} - \pi)$, or 10.73.

The method is reduced to the simple procedure of multiplying the superficial velocity by 10.73 and substituting the result for u_o in Eq. (24-12). The coefficients predicted by this method are about 10 percent higher than those measured for a bed of randomly packed, nearly spherical particles. The empirical correlation of such data usually consists of an equation relating the j factor to the Reynolds number for a certain range of **Re**. A typical equation, representing the results of Bradshaw,[1] is

$$j_H = \frac{h}{u_{bs}\rho C_p}\mathbf{Pr}^{2/3} = 2.50\left[\frac{Du_{bs}\rho}{\mu(1-\varepsilon)}\right]^{-1/2} \tag{24-19}$$

The form of the Reynolds number used in this equation is discussed in Chap. 14; it is designated $\mathbf{Re}_p$.

[1] R. D. Bradshaw, *AIChE J.*, **9**:590 (1963). Eq. (24-19) is recommended for values of $\mathbf{Re}_p$ ranging from 400 to 10,000.

Convection from a Plane Surface

Heat transfer between plane surfaces and a fluid in forced convection has been discussed in the two preceding chapters on laminar and turbulent flow. A laminar boundary layer builds up, starting from the leading edge of the plate, and extends a distance limited by the critical Reynolds number, which is approximately $\mathbf{Re}_x = 500{,}000$. Beyond this point the boundary layer is turbulent. Equations (22-19) and (23-46) can be used to predict local heat-transfer coefficients in the two zones. An overall average coefficient can be obtained by integrating $h_x\,dx$ over the entire length of the plate.

If the fluid is moving past the plate as the result of natural convection, the heat-transfer coefficients are still dependent on whether the boundary layer is laminar or turbulent. It was pointed out in the section on dimensional analysis earlier in this chapter that correlations of natural-convection heat-transfer coefficients usually involve the Grashof number, $(L^3\rho^2 g\beta\,\Delta t)/\mu^2$. Equation (22-25) for ideal gases heated by laminar natural convection from a vertical plate contains the Grashof number:

$$\mathbf{Nu}_m = 0.478\ \mathbf{Gr}^{1/4} \qquad (22\text{-}25)$$

McAdams gives a more general equation for laminar natural convection,

$$\mathbf{Nu}_m = 0.59\ \mathbf{Gr}^{1/4}\ \mathbf{Pr}^{1/4} \qquad (24\text{-}20)$$

and for combined laminar and turbulent natural convection,

$$\mathbf{Nu}_m = 0.13\ \mathbf{Gr}^{1/3}\ \mathbf{Pr}^{1/3} \qquad (24\text{-}21)$$

Equation (24-20) applies when the product $\mathbf{Gr\ Pr}$ lies between 10^4 and 10^9, and Eq. (24-21) applies for $\mathbf{Gr\ Pr} > 10^9$. The fluid properties are evaluated at the arithmetic average temperature $(t_s + t_o)/2$.

Natural-convection equations are most frequently used to determine heat loss to the atmosphere. For this reason it is convenient to have simplified forms of Eqs. (24-20) and (24-21) in which values of the physical properties of air at typical ambient conditions have already been substituted. Perry[1] offers the following useful dimensional formulas:

(1) For all surfaces, $\mathbf{Gr\ Pr} > 10^9$

$$h = 0.18(\Delta t)^{1/3} \qquad (24\text{-}22)$$

(2) For a horizontal cylinder, $10^3 < \mathbf{Gr\ Pr} < 10^9$

$$h = 0.50(\Delta t/D'_0)^{1/4} \qquad (24\text{-}23)$$

[1] J. H. Perry, "Chemical Engineers' Handbook," 4th ed., p. 10-11, McGraw-Hill Book Company, New York, 1963. See also 5th ed; p. 10-11.

(3) For vertical plates, $10^3 < \mathbf{Gr\,Pr} < 10^9$

$$h = 0.28(\Delta t/L)^{1/4} \qquad (24\text{-}24)$$

(4) For heated horizontal plates facing upward, $10^5 < \mathbf{Gr\,Pr} < 10^7$

$$h = 0.38(\Delta t)^{1/4} \qquad (24\text{-}25)$$

(5) For heated horizontal plates facing downward, $10^5 < \mathbf{Gr\,Pr} < 10^7$

$$h = 0.20(\Delta t)^{1/4} \qquad (24\text{-}26)$$

The variables in the above equations are expressed in the customary units except for D'_0, the diameter of a horizontal cylinder, which is in inches; L, the height of a vertical plate, is expressed in feet. Perry[1] also offers reduced dimensional equations for natural convection in water and organic liquids.

Heat Transfer to Liquid Metals

The use of liquid metals as heat-transfer fluids is best known in the field of nuclear engineering. Liquid metals can be heated over a wide range of temperature without the necessity of keeping the system under great pressure. However, a disadvantage is that they are often somewhat difficult to handle.

Another advantage of the use of liquid metals is that they have high heat-transfer coefficients compared with many liquids and, in addition, have high heat capacities per unit volume. These properties make them useful for removing heat from nuclear reactors, where large amounts of heat must be removed quickly and economically.

Although liquid metals have about the same viscosity as many liquids, they have thermal conductivities about 100 times greater, and hence have Prandtl numbers much smaller than 1. Because of the high thermal conductivity, the ratio of the heat-transfer rate by the molecular mechanism to the rate by the eddy mechanism is much larger than for most other fluids. As a result, most standard correlations of heat-transfer coefficients cannot be applied to liquid metals, and separate correlations have been developed.

The equations derived by Lyon and by Martinelli using the analogy between heat and momentum transfer have been mentioned in the previous chapter. Their results indicate that molecular conduction of heat is significant in the core of a turbulent stream as well as in the laminar sublayer. Lyon[2] employed velocity-distribution data in his equation to obtain values for the Nusselt number for liquid metals. His results can be represented by the following approximate equation,

[1] Perry, p. 10-11.
[2] R. N. Lyon, *Chem. Eng. Progr.*, **47**:75 (1951).

which gives the average heat-transfer coefficient for circular pipes for which $L/D > 60$.

$$\frac{hD}{k} = 7 + 0.025\left(\frac{Du_b\rho}{\mu}\right)^{0.8}\left(\frac{C_p\mu}{k}\right)^{0.8} \tag{24-27}$$

In this equation, which applies where there is uniform heat flux from the wall, the physical properties are evaluated at the bulk temperature of the fluid.

Seban and Shimazaki[1] have proposed a similar equation for liquid metals in circular tubes with a constant wall temperature. It differs from Eq. (24-27) only in that the constant 7 is replaced by 5. Both equations apply when **Re Pr** is greater than 100 and, obviously, become identical at high values of **Re Pr**. Dwyer[2] has noted that this form of the equation represents approximately 90 percent of the available experimental data within ± 20 percent. Accurate experimental measurements of heat transfer in liquid metals are difficult to make because the heat-transfer rates are often substantially higher than for ordinary fluids. Such effects as nonwetting and depositing of scale-forming impurities at the heat-transfer surface are believed to be responsible for substantial scatter in many experimental results. Axial-conduction and natural-convection effects are frequently overlooked, though they also may be the cause of some of the scatter.

Boundary-layer flow of liquid metals on flat plates has been considered by Grosh and Cess.[3] They suggested that, because liquid metals have a much thicker thermal boundary layer than momentum boundary layer, the free-stream velocity can be assumed to represent the actual velocity at all values of y in the temperature field. This simplifies the differential energy balance to

$$u_0 \frac{\partial t}{\partial x} = \alpha \frac{\partial^2 t}{\partial y^2} \tag{24-28}$$

If x/u_0 in this equation is equivalent to the time variable θ, it is seen that the equation becomes

$$\frac{\partial t}{\partial \theta} = \alpha \frac{\partial^2 t}{\partial y^2} \tag{19-1}$$

which describes one-dimensional, unsteady-state conduction into a plate of infinite thickness. Thus Eq. (19-13), the solution to the unsteady-state conduction problem, is also a solution to Eq. (24-28). The abbreviated form of this solution can be written as

$$\frac{t_s - t}{t_s - t_0} = \operatorname{erf} \frac{y}{\sqrt{4\alpha x/u_0}} \tag{24-29}$$

[1] R. A. Seban and T. T. Shimazaki, *Trans. Am. Soc. Mech. Eng.*, **73**:803 (1951).
[2] O. E. Dwyer, *Atom. Energy Rev.*, **4**:3 (1966).
[3] R. J. Grosh and R. D. Cess, *Trans. Am. Soc. Mech. Eng.*, **80**:667 (1958).

The error function can be evaluated either from tabulated results or from the convergent series

$$\operatorname{erf} n = \frac{2}{\sqrt{\pi}}\left(n - \frac{n^3}{3 \cdot 1!} + \frac{n^5}{5 \cdot 2!} - \frac{n^7}{7 \cdot 3!} + \cdots\right) \tag{24-30}$$

For small values of n (i.e., as y approaches zero) the first term of this series can be seen to provide an adequate solution to the heat transfer problem:

$$\frac{t_s - t}{t_s - t_0} = \frac{2}{\sqrt{\pi}} \frac{y}{\sqrt{4\alpha x/u_0}} = y\sqrt{\frac{u_0}{\pi \alpha x}} \tag{24-31}$$

Therefore, the local heat-transfer coefficient can be written for this system as

$$h_x = k\left\{\frac{d(t_s - t)/(t_s - t_0)}{dy}\right\}_{y=0} \tag{22-11}$$

$$= k\sqrt{\frac{u_0}{\pi \alpha x}} \tag{24-32}$$

which can be rearranged to read

$$\frac{h_x x}{k} = 0.564\left(\frac{x u_0 \rho}{\mu}\right)^{1/2}\left(\frac{C_p \mu}{k}\right)^{1/2} \tag{24-33}$$

The results expressed in dimensionless numbers are as follows:

$$\mathbf{Nu}_x = \sqrt{\frac{\mathbf{Re}_x\ \mathbf{Pr}}{\pi}} \tag{24-34}$$

This equation was checked by Grosh and Cess with numerical solutions of the differential energy balance and found to give values approximately 10 percent too low in the range $0.005 < \mathbf{Pr} < 0.025$.

Heat Transfer to Nonnewtonian Fluids

The properties of nonnewtonian fluids and, in particular, their flow characteristics were discussed in Chap. 2. Because of the high apparent viscosities encountered in these systems, turbulent flow is encountered less often than in systems containing newtonian fluids. Thus, most heat-transfer studies involving nonnewtonian fluids are limited to the laminar flow regime. The basic differential energy-balance equations derived earlier remain applicable. The principal innovations result from the velocity profile equations substituted in the energy-balance equations before integration. A summary of both analytical and empirical solutions for the various classes of nonnewtonian fluids is provided by Skelland.[1]

Effect of Surface Roughness on Heat-transfer Coefficients

Surface roughness has an effect on the friction factor in turbulent flow, as has been seen in Chaps. 12 and 14. However, in laminar flow the effect is negligible. A similar situation exists in heat transfer. Except for a slight increase in the heat-transfer area, the roughness found on commercial heat-transfer surfaces has little effect on heat transfer with laminar flow. When the flow is turbulent, the roughness may or may not have a significant effect. If the roughness elements do not protrude from the laminar sublayer, then the heat-transfer coefficient will not be affected. On the other hand, protuberances that extend into the buffer zone or the turbulent core will cause an increase both in the loss of energy due to friction and in the heat-transfer coefficient.

It has been known for some time that if surfaces with transverse fins are used, the increases in the heat-transfer area and in the heat-transfer coefficient are usually offset by an increase in the power required to pump the fluid past the heat-transfer surface.

In a study of the effect of roughness on both the friction factor and the heat-transfer coefficient, Brouillette[2] cut V-shaped grooves in thick-walled copper tubing and measured both heat-transfer and pressure-drop data for water. He found that a roughness ratio of 0.05 (protuberance height/tubing diameter) caused the heat-transfer coefficient to double; however, this effect was accompanied by a fourfold increase in the friction factor. The inside heat-transfer area was increased by the presence of the grooves, but even the combined effects of increased h and A were not generally sufficient to give as high a ratio of heat flux to pumping power as was obtained for a smooth tube.

The disproportionate increase in friction factor is probably caused by a substantial contribution of form drag to the total frictional loss. With flow past plates and through smooth pipes, in which form drag is absent, the analogy between heat and momentum transfer is quite good. However, with flow across cylinders, form drag constitutes a major portion of the total resistance to flow, and the friction factor (as calculated from total pressure drop) is no longer related to the heat-transfer coefficient by a direct proportionality such as the j-factor relationship. The

[1] A. H. P. Skelland, "Non-Newtonian Flow and Heat Transfer," John Wiley & Sons, Inc., New York, 1967.

[2] E. C. Brouillette, M.S. thesis in chemical engineering, Purdue University, Lafayette, Ind., 1955.

same situation seems to exist in flow systems in which the roughness elements on a solid surface extend beyond the laminar sublayer into the turbulent core.

Savage[1], in an extensive study of heat transfer and pressure drop in tubes with rectangular, internal fins normal to the direction of flow, found wide variations in the relative contributions of form drag and skin drag to the total pressure loss. The proportional variation of the mean heat-transfer coefficient was much less. For nearly all fin heights and spacings, the additional turbulence caused by the fins increased the mean heat-transfer coefficient by 75 to 100 percent. Thus, for close spacings, the increase in heat-transfer rate was due as much to an increase in the area of the heat-transfer surface as to the increase in the heat-transfer coefficient.

PROBLEMS

24-1 A horizontal 2-in, schedule-40 steel pipe is carrying water at a bulk velocity of 5 ft/s at an average temperature of 200°F. The pipe is insulated with a layer of 85 percent magnesia 1 in thick. The air surrounding the insulation is at 90°F. Find the temperature at the outer surfaces of the insulation and the pipe and the heat loss by convection in Btu/(h)(lin ft). Neglect radiation loss.

24-2 Air at 36°C flows at a velocity of 5.5 m/s normal to a $1\frac{1}{2}$-in, schedule-40 steel pipe. The pipe contains saturated steam at 135°C and is 3 m long. Assuming that the resistance to heat transfer in the condensing steam and in the steel pipe is negligible compared with the resistance of the air, find the heat loss from the pipe.

24-3 It is necessary to heat 14,300 lb/h of oil from 95 to 190°F in a 1-in, schedule-40 steel pipe jacketed by a second pipe in which steam is condensing at 10 psig. The oil has an average specific heat of 0.46 Btu/(lb)(°F), a specific gravity of 0.82, and a thermal conductivity of 0.073 Btu/(h)(ft)(°F). The following table shows the viscosity of the oil as a function of temperature:

Temperature, °F	95	110	130	160	190
Viscosity, cP	13.5	10.8	8.1	5.5	4.8

The heat-transfer coefficient for the condensing steam can be taken as 1500 Btu/(h)(ft^2)(°F). Find the length of pipe required and discuss the assumptions made in your solution.

24-4 The inside vertical wall of a furnace 2.45 m high and 2.45 m wide has a surface temperature of 1093°C. The wall is composed of 0.1 m of firebrick, 0.1 m of Silocel brick, and 0.2 m of building brick. The atmospheric temperature outside the furnace is 21°C. Find the heat loss from this surface in watts by convection.

24-5 A steel ball $\frac{1}{2}$ in in diameter is quenched from 800°F by dropping it in a stream of oil at 60°F moving past the ball with a velocity of 5 ft/s. Because the resistance to conduction within the ball is much less than the resistance to heat transfer in the oil, the

[1] D. W. Savage, *AIChE J.*, 9:694 (1963).

ball can be assumed to be at a uniform temperature at any instant. Find the time for the ball to cool to 200°F. The average physical properties of the oil are $k = 0.073$ Btu/(h)(ft)(°F); $\rho = 52$ lb/ft^3; $C_p = 0.54$ Btu/(lb)(°F); $\mu = 0.00083$ lb/(ft)(s).

24-6 A spherical drop of water is moving through air with a constant velocity of 80 ft/s. The air temperature is 80°F, and the drop temperature is constant at the wet-bulb temperature of 63°F. If the drop has an initial diameter of $\frac{1}{4}$ in, find the time for it to evaporate completely.

24-7 An electric conductor, which is a solid cylindrical copper wire, is covered by a layer of insulation. The copper wire has a diameter of 0.01 ft. The insulation, which has maximum permissible temperature of 200°F, has a thermal conductivity $k = 0.04$ Btu/(h)(ft)(°F). The air around the wire has a bulk temperature of 100°F.

The addition of insulation to a cylinder usually decreases the heat loss. However, for very small cylinders, the addition of insulation may increase the heat loss, and this is of considerable importance in the design of electrical systems where heat dissipation is desired.

For the system described above, find the thickness of insulation which would assure a maximum heat loss.

24-8 A sphere of solid, radioactive material with a fixed radius r_i generates heat at the fixed rate of q W. It has been proposed that by covering the sphere with some suitable material, the temperature t_i (at the interface between the radioactive material and the covering material) might be decreased to a minimum. Assuming fixed values for q, r_i, outside convective coefficient h, and outside bulk temperature of the surroundings t_o, find the equation for the thickness of the coating which gives the minimum value for the interfacial temperature t_i.

24-9 Orange juice is being concentrated by evaporation of moisture from a film of juice flowing down the inside wall of a large glass tube. The tube is heated on the outside by steam at 105°C. The orange juice flavor is easily affected by overheating; therefore, no portion of the juice can have a temperature exceeding 70°C for even a short period of time. Find the limiting bulk velocity for the case where the bulk or mixing-cup temperature of the orange juice is 60°C.

Data

1 Convective coefficient for steam, $h = 6000$ W/m$^2 \cdot$ K.

2 Convective coefficient for orange juice, $h = 1500\ u_b^{1/2}$ where u_b is bulk velocity of film in m/s and h is in W/m$^2 \cdot$ K.

3 Thermal conductivity of glass = 0.9 W/m $\cdot$ K.

4 Thickness of glass wall = 0.006 m.

24-10 A window pane, consisting of a single layer of glass, separates air outdoors at 10°F from air indoors at 70°F. Neglecting the end effects produced by the window frame, sketch the shapes of the boundary layer that will be produced by natural convection on each side of the window pane. Also, sketch qualitatively the curves of the local heat-transfer coefficients for each side of the pane and indicate how the variations of the relative heat-transfer resistances cause certain parts of the window to be covered either with ice or condensate, or to remain dry.

24-11 Gasoline flowing inside the tubes of a heat exchanger is being cooled from 95 to 45°C by pentane boiling outside the tubes at 35°C. Assume all the resistance to heat transfer is on the gasoline side; the flow is turbulent. If the flow rate of gasoline is doubled, what is its new outlet temperature?

24-12 Ammonia is evaporated continuously at −15°C in a refrigeration plant by cooling 0.12 kg/s of brine from 10 to −6°C. The overall heat-transfer coefficient based on outside area is 300 $W/m^2 \cdot K$. If the specific heat of the brine is 2800 J/kg·K, calculate the outside heat-transfer area required.

24-13 Alcohol entering at 115°F flows in the inner pipe of a double-pipe heat exchanger and is cooled with water flowing in the jacket. The water enters countercurrent to the alcohol at 50°F and leaves at 100°F. The inner pipe is 1-in, schedule-40 steel pipe; k for steel is 26 Btu/(h)(ft)(°F). The individual coefficients are $h_i = 180$; $h_o = 300$ Btu/(h)(ft^2)(°F). In addition it was found that the heat exchanger had developed scale deposits after a few weeks of operation; the dirt-deposit coefficients are $h_{di} = 1000$; $h_{do} = 500$.

The flow rate of water is 1000 lb/h. The flow rate of alcohol is 2200 lb/h and its specific heat is 0.50 Btu/(lb)(°F). What length of pipe is needed for the exchanger?

24-14 Benzene is cooled from 130 to 70°F in the inner pipe of a double-pipe heat exchanger. Water flows countercurrently to the benzene in the jacket entering at 50°F and leaving at 85°F. The exchanger consists of $\frac{7}{8}$-in, 16-BWG copper tubing jacketed by a $1\frac{1}{2}$-in, schedule-40 steel pipe. The velocity of the water is 4 ft/s. Neglect wall resistance and any scale deposits and assume the pipes are sufficiently long to permit the neglect of end effects. Compute the heat-transfer coefficients of the benzene and water and the overall coefficient based on the outside area of the inner pipe. Assume that viscosity is constant at the bulk stream temperature.

24-15 A 60 percent sucrose solution is heated from 10 to 15°C in a double-pipe heat exchanger by water flowing countercurrently and cooling from 49 to 27°C. The solution flows inside a 1-in, schedule-40 steel pipe at 4 kg/s. How long must the pipe be? Neglect the pipe wall and water side resistances.

Data:
$C_p = 3980$ J/kg·K
$k = 0.26$ W/m·K
$\mu = 0.094$ Pa·s

24-16 Air at 70°F is to be heated by passing it through a 2-ft ID vertical pipe filled to a depth of 12 ft with $\frac{1}{4}$-in-long by $\frac{1}{4}$-in-diameter cylindrical particles. The particles are radioactive and they all generate heat at a rate of 50 Btu/(h)(ft^3). The porosity of the bed of particles is 0.41, and the air is blown through the bed at 1000 lb/h.

(*a*) What is the outlet temperature of the air?

(*b*) Find the temperature profile of the pellets in the bed.

The heat-transfer coefficient between the pellets and the air, which we assume is everywhere constant, is given by

$$j_H = 0.91\ \mathbf{Re}^{-0.51}\psi \qquad (\mathbf{Re} < 50) \tag{1}$$

$$j_H = 0.61\ \mathbf{Re}^{-0.41}\psi \qquad (\mathbf{Re} > 50) \tag{2}$$

where

$$j_H = \frac{h}{C_p G_o}\left(\frac{C_p \mu}{k}\right)^{2/3} \tag{3}$$

and

$$\mathbf{Re} = \frac{G_o}{a\mu\psi} \tag{4}$$

In these formulas, a is the surface area of cylinders per unit volume of bed, G_o is the superficial mass velocity, and ψ is an empirical shape factor, equal to 0.91 for cylinders. The properties of the fluid (air) are evaluated at a mean temperature defined as for Eq. (24-5).

24-17 Oil is heated from 38 to 93°C by flowing inside the 1-in, 16-BWG tubes of a heat exchanger. Steam condenses outside the tubes at 100°C. The flow rate of oil is 1.26 kg/s and the inside area of the heat exchanger tubes is 18.60 m^2; the specific heat of the oil is 3140 J/kg · K.

(*a*) What is the overall heat-transfer coefficient based on the inside area?

(*b*) If the steam-side coefficient is 2990 W/m^2 · K, what is the oil-side coefficient? Neglect the resistance of the tube wall.

(*c*) In order to increase the heating capacity, the steam pressure is increased so that it condenses at 149°C. What flow rate of oil is now obtained if the oil is still heated from 38 to 93°C?

Assume that the steam-side coefficient remains 2990 W/m^2 · K and that the properties of the oil are not changed by the new steam-side temperature.

24-18 (*a*) Water flows through a 16-gauge, $\frac{1}{2}$-in copper tube at the rate of 0.233 lb/s. The tube has no insulation and extends through the air space under a house. The air is at 60°F and the natural convection coefficient for the outside of the tube $h_o = 2.5$ Btu/(h)(ft^2)(°F). The convection coefficient for the inner surface of the tube $h_i = 500$ Btu/(h)(ft^2)(°F). If the tubing runs from a hot water tank in which the water is held at a constant temperature of 140°F, find the temperature of the water at a point 60 ft downstream where it enters a faucet.

(*b*) It is suggested that it might be economical to insulate the hot water line. If insulation $\frac{1}{2}$ in thick is wrapped around the pipe at a cost of 50c/lin ft, how many hours of continuous operation are required to pay for the insulation? Assume that the thermal conductivity of the insulation is zero and the cost of heat is $2.50 per million Btu.

24-19 Water is flowing through a $\frac{1}{2}$-in, 16-gauge copper tube at a bulk velocity of 5 ft/s. The water enters the tube at a temperature of 40°F. The tube is horizontal, 1000 ft long, and suspended in a room in which the air temperature is 80°F. What is the exit temperature of the water? Assume radiant heat-transfer is zero and that the heat-transfer resistance of the tube wall and the water inside the tube are negligible.

25

BOILING AND CONDENSATION

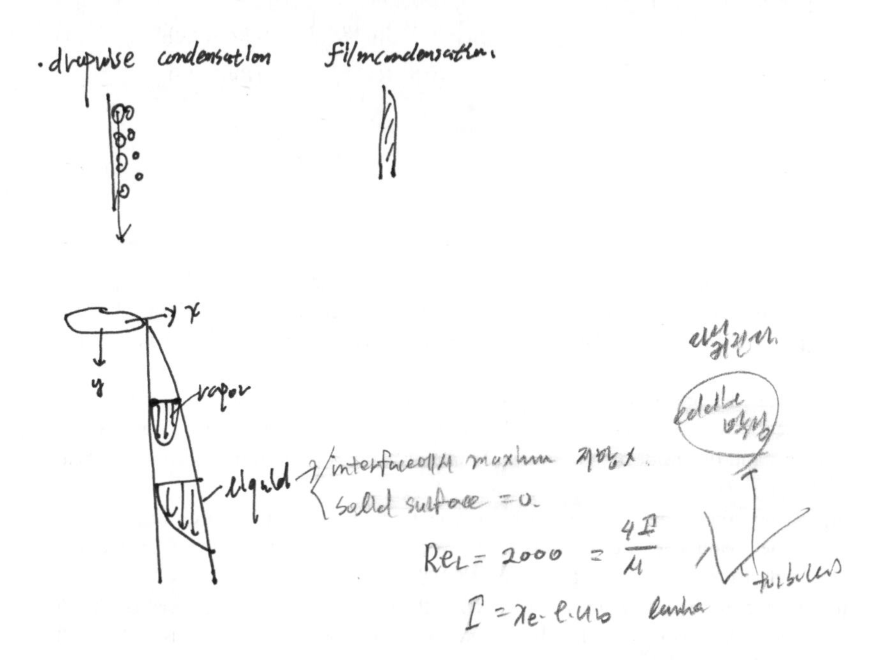

The transfer of heat which accompanies a change of phase is often distinguished by high rates. Heat flux as high as 50 million Btu/(h)(ft^2) has been obtained in boiling systems. This mechanism of transferring heat has become of importance in rocket technology and nuclear reactor design, where large quantities of heat are produced in small spaces. Although condensation rates have not reached a similar magnitude, coefficients of heat transfer for condensation as high as 20,000 Btu/(h)(ft^2)(°F) have been reported. Condensing coefficients in ordinary commercial equipment are generally higher than in systems where a change of phase does not occur.

BOILING

Mechanisms

Among the various heat-transfer mechanisms, boiling is perhaps the most complex. For example, in the quenching of very hot metals in liquids, the heat-transfer rate often increases as the temperature difference between the metal and liquid decreases. A similar occurrence, known as the *Leidenfrost* phenomenon, is observed when

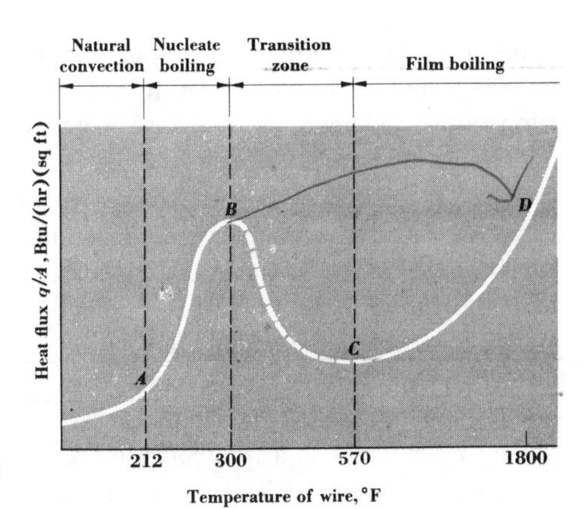

FIGURE 25-1
Boiling regimes for water at atmospheric pressure.

drops of liquid fall on a very hot surface. Although the drops bounce vigorously on the surface, they may take several seconds to evaporate. However, when drops fall on a much cooler surface of the same material, the bouncing does not occur. Instead, the drops wet the surface, spread out, and evaporate in a second or less. The explanation for both the paradoxical quenching behavior and the Leidenfrost phenomenon lies in the fact that boiling heat transfer occurs by several different mechanisms and that the mechanism is often more important in determining heat flux than the temperature-difference driving force.

An early investigation of these mechanisms was made by Nukiyama[1] in 1934. He immersed a platinum wire in water at 212°F and heated the wire electrically. The heat flux from the wire was calculated from the power input, and the temperature of the wire was determined from its electric resistance. His results and those of subsequent investigators were of the form shown in Fig. 25-1.

Nukiyama found that as the temperature of the wire increased above 212°F, the rate of heat transfer increased smoothly, until at 300°F the temperature of the wire suddenly jumped to about 1800°F. However, further increases in the power input caused only regular increases in the wire temperature. A decrease in the power input from this point caused the wire temperature to decrease, as shown by the smooth curve below 1800 down to 570°F, whereupon the temperature dropped suddenly to below 300°F. Below this point, it followed the curve already established for increasing heat flux.

Nukiyama concluded that at least three types of boiling existed, and subsequent work has proved him correct. The first mechanism encountered is now

[1] Shiro Nukiyama, Maximum and Minimum Values of Heat Transmitted from Metal to Boiling Water under Atmospheric Pressure, *J. Soc. Mech. Eng., Japan*, **37**(206):367–374 (1934).

known as *nucleate boiling*. At surface temperatures only a few degrees above 212°F, bubbles form and rise in columns from discrete places on the heater surface. As the temperature difference is increased, more sites on the surface begin producing bubbles, so that more columns of bubbles are available for removing heat from the surface. If the temperature difference becomes sufficiently high, the concentration of bubble columns reaches some maximum, indicated as point B on Fig. 25-1, at which liquid can no longer reach the heater surface at a sufficient rate to form the required amount of vapor. At this point the mechanism changes to what is known as *film boiling*, and the temperature of the wire rises at once (even though the heat flux remains constant) to the value corresponding to point D. The name film boiling applies because the heater surface is now in contact with only the vapor phase, which exists as a film between the heater and the liquid. Vaporization of liquid takes place at the liquid-vapor interface, causing the film thickness to increase until the vapor breaks off from the film in irregularly shaped bubbles at random locations. The wire melts if the point D occurs at such a high temperature that the metal is above its melting point. Nukiyama found this burnout to occur with copper wires, but not with platinum wires, which have a much higher melting point.

Film boiling may be stable at points both above and below D. However, if the heat flux is decreased starting at D, the film breaks down in the vicinity of point C. The mechanism of nucleate boiling is then reestablished at the same heat flux but on the segment of the curve between A and B.

The transition zone from B to C has received only minor attention. Because of the difficulty of controlling the temperature of the wire, electric heaters are of little use for experimental work in this region. With all heaters the surface is partially blanketed with vapor in the transition zone, but the thickness of the vapor varies greatly. Because of the instability of operation in this region and because increases in the driving force cause a decrease in heat flux, this mechanism is of little interest at present in industrial operations.

The portion of the curve to the left of point A on Fig. 25-1 does not represent boiling, but rather heating by natural convection. The fluid may be at its boiling point, but if the heater surface is no more than a few degrees hotter, no bubbles appear on the surface. The portion of the curve near point A is a region in which the first few bubble columns have formed; the heat load is shared between the mechanisms of nucleate boiling and natural convection.

Nucleate Boiling

No very satisfactory mechanism has yet been proposed for nucleate boiling. Early experimental work was largely devoted to measuring boiling performance; the results were correlated using empirical equations composed of dimensionless groups.

However, contradictions in the experimental results of the different investigators have brought most of their equations into disrepute. Recently, equations based partly on mechanistic concepts have been proposed and will be discussed later in this section. Many aspects of boiling are still unexplained, and predictions of boiling coefficients without reference to experimental data may easily be in error by 100 percent.

Nucleation It is observed in chemistry laboratories that liquids in glass containers often do not begin to boil at the saturation temperature corresponding to atmospheric pressure. In fact, superheating in excess of 100°F can be obtained if care is taken to use pure liquids and clean glassware. However, this state of supersaturation can be quickly ended, and violent ebullition of vapor occurs if impurities are added to the system.

Although the phenomenon of superheating is most spectacularly demonstrated in the nonboiling, unstable system described above, it also occurs in all boiling systems. The temperature of a boiling liquid, measured some distance from the heated surface, is often as much as a degree higher than the temperature of the vapor above the liquid; the vapor is at the saturation temperature. Adjacent to the heated surface the liquid superheat may be as high as 30 to 50°F. This zone is extremely thin; most of the temperature drop usually occurs less than 1 mm from the heater.

The explanation for the presence of superheat in boiling systems is that the saturation temperatures and pressures ordinarily used by engineers (e.g., as found in steam tables) apply to equilibrium at a flat interface between vapor and liquid. However, if a force balance is written at the equator of a spherical vapor bubble of radius r, the internal pressure p_g exceeds the surrounding liquid pressure p_ℓ by an amount equal to the effect of the surface force, which tends to contract the bubble. Thus we write

$$p_g(\pi r^2) = p_\ell(\pi r^2) + 2\pi r\sigma \qquad (25\text{-}1)$$

from which we get

$$p_g - p_\ell = \frac{2\sigma}{r} \qquad (25\text{-}2)$$

To achieve the excess pressure $p_g - p_\ell$, which may be very high in small bubbles, the liquid which forms the vapor nucleus must be at a temperature much higher than the flat-surface saturation temperature. If nuclei with large radii of curvature are available, the required excess pressure will be smaller and boiling will commence at a lower superheat. A perfectly pure liquid in contact with a flat heated surface would, in theory, require an infinite amount of superheat. A liquid is not, however, a continuum, and clusters of molecules could serve as nuclei. Furthermore, the

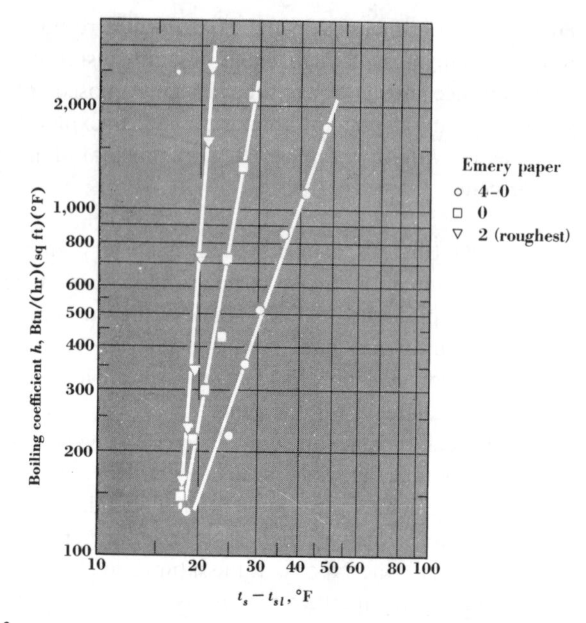

FIGURE 25-2
Coefficients of heat transfer for n-hexane boiling on a flat plate polished with three grades of emerypaper. [*From H. Kurihara, AIChE J.*, **6**(1):83 (1960).]

conditions of perfect purity and flatness are unobtainable, so that the highest superheats recorded are of the order of 100 to 200 F. The effect of solid impurities in preventing bumping can be explained on the basis of the probable large radius of curvature of the nuclei they provide. However, it is likely that boiling does not require the presence of foreign matter, but can be initiated on the heating surface. Numerous experiments have shown that metal surfaces polished with emery paper have higher coefficients of heat transfer when rough paper is used to polish the surface than when fine paper is used. Sample results are shown in Fig. 25-2.

One way in which surface conditions are believed to affect boiling is by the trapping of vapor in surface cavities before boiling begins. The pits on the heater surface may contain air when boiling begins, but after a short time all the air is dissipated and only molecules of the substance being boiled remain in the vapor space. If the pits are round-bottomed, the liquid will fill them completely when boiling stops, and they will not serve as sources of bubbles when boiling is resumed. However, if the pits have angular bottoms and the liquid-vapor system has a sufficiently large contact angle, the liquid will not completely fill the pits when boiling stops. The liquid-vapor interface will bulge downward into the pit, but in this case the surface forces act with the pressure in the vapor space to balance the liquid pressure. The liquid advances only until the radius of curvature of the interface, which is now negative, is small enough so that $p_g - 2\sigma/r$ is equal to p_ℓ and a stable gas pocket remains. Upon further heating, bubbles are once again emitted from this site.

Experiments have shown that the heat-transfer coefficient is proportional to some power of the concentration of bubble sites ranging from 0.25 to 0.46, although no theoretical reasons for this relation are known. The number of sites on a heater surface increases with an increase of superheat, but little is known about this relationship other than that it probably depends on the roughness of the surface and the physical properties of the boiling fluid. The accurate prediction of boiling coefficients may be possible when these two mechanisms are understood.

Bubble behavior The formation, growth, and release of bubbles is an extremely rapid sequence of events. Photographic studies have shown that a typical sequence[1] at a nucleation site involves a growth time for the bubble of perhaps 0.01 s; following departure of the bubble from the site, there is a delay time of 0.01 to 0.06 s before a new bubble appears and repeats the growth cycle. A further complication is that the active production of bubbles from a site may either continue for a long period of time or may be terminated after a short burst of bubbles has been produced. This erratic behavior could be due to temperature fluctuations of the surface on which boiling takes place. As indicated earlier, this surface must first be superheated with respect to the saturation temperature of the liquid for any boiling to take place. Recent studies[2] have shown that wide fluctuations of temperature are constantly occurring at all points on the boiling surface. Thus, when research workers describing their experiments specify a "surface temperature," it is some kind of average temperature, meaningful only in terms of the way in which it has been measured.

The concept of fluctuating surface temperature is consistent with the mechanism of boiling proposed by Moore and Mesler,[3] who found that the surface temperature beneath a bubble site was lower than the surrounding surface temperatures. They suggested that vaporization of a thin layer of liquid (a "microlayer") between the growing bubble and the heating surface caused removal of heat from the surface, thus lowering the surface temperature at that location. This mechanism is presumably supplemented by vaporization of liquid into the bubble from other parts of the vapor-liquid interface.

[1] J. B. Roll, *AIChE J.*, **10**:530 (1964).
[2] T. Raad, *AIChE J.*, **17**:1260 (1971).
[3] F. D. Moore and R. B. Mesler, *AIChE J.*, **7**:620 (1961).

Correlations of nucleate-boiling heat-transfer coefficients Correlations by Rohsenow[1] and by Forster and Zuber[2] have met with some success in the prediction of boiling coefficients. Both correlations have some basis in theory, but in the end resort to a combination of dimensionless groups for which no mechanistic explanation is offered. Both equations are of the form

$$\mathbf{Nu} = a\,\mathbf{Re}^b\,\mathbf{Pr}^c \qquad (25\text{-}3)$$

The characteristic velocity used by Rohsenow in **Re** is fnV_b, where f is the frequency of bubble formation at a site (bubbles per second), n is the number of active sites per unit area, and V_b is the volume of a bubble. The quantity fnV_b can be seen to have the units of velocity. The Forster and Zuber correlation, on the other hand, employs a velocity which is the radial velocity of the interface of the growing bubble. They found that the product of radial velocity and bubble radius is independent of time; thus the Reynolds number, which contains this product, is also independent of time and can be calculated from the properties of the system.

Although the two equations discussed above seem to have firmer bases than the many completely empirical correlations which preceded them, they do not offer entirely reliable methods of predicting boiling coefficients. One difficulty is that the effects of none of the individual variables are known. In boiling from electrically heated wires, the wire diameter seems to be a variable. In boiling from horizontal tubes, the average boiling coefficient varies with tube size while the local boiling coefficient may vary as much as 100 percent from the bottom to the top of the tube, as shown in Fig. 25-3.

The effect of surface tension on boiling coefficients is unknown. Westwater,[3] in a thorough review of the state of knowledge of boiling, shows that if the effect of surface tension on the boiling coefficient were represented by the equation

$$h = \text{const}\ \sigma^n \qquad (25\text{-}4)$$

the values predicted by various experimenters for the exponent n would range from -2.5 to 1.275.

The effect of degree of superheat on the boiling coefficient is variable. Many commercial surfaces give data which can be represented approximately by the equation[4]

$$h = \text{const}\ (\Delta t)^{2.5} \qquad (25\text{-}5)$$

[1] W. M. Rohsenow, *Trans. Am. Soc. Mech. Eng.*, **74**:969 (1952).

[2] H. K. Forster and N. Zuber, *AIChE J.*, **1**:531 (1955).

[3] J. W. Westwater, "Advances in Chemical Engineering," vol. 1, chap. 1, Academic Press, Inc., New York, 1956.

[4] The temperature-difference driving force is defined by some writers as the temperature of the heater surface minus the bulk temperature of the liquid. Others employ the temperature of the heater surface minus the saturation temperature of the liquid. The latter definition gives slightly greater values of Δt than the former and appears to be in more general use.

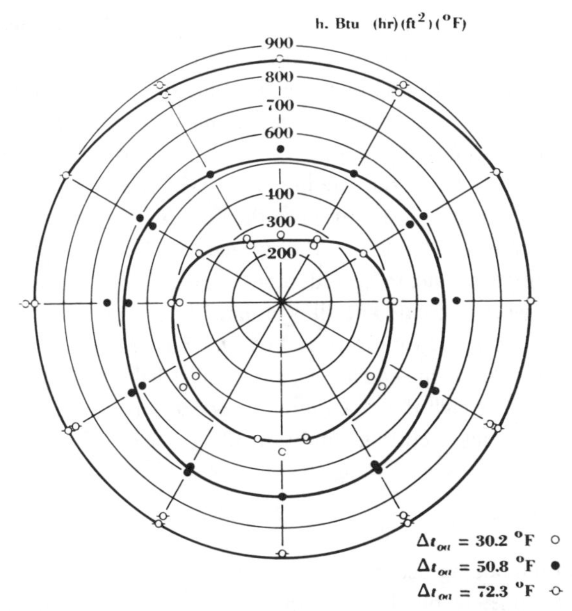

FIGURE 25-3
Boiling coefficient for methanol around $1\frac{1}{4}$ in horizontal copper tube: Δt_{oa} = average value of overall temperature-difference driving force. [*From R. Lance, AIChE J.*, **4**:75 *(1958)*.]

However, values of the exponent as low as 1.4 have been reported on copper tubes, whereas flat surfaces polished with coarse emery paper may give values as high as 25.

The conclusion that one must draw from the present state of knowledge of nucleate boiling is that although we have some understanding of boiling performance, we are unable to predict reliable values for the boiling coefficients of heat transfer at the present time. In these circumstances the engineer has no alternative but to seek experimental data.

Maximum Heat Flux

As the superheat of a boiling liquid is increased, the concentration of active centers on the heating surface increases and the heat flux also increases. The mass rate of the vapor rising from the surface must at steady state be equal to the mass rate of liquid proceeding toward the surface. As the boiling rate increases, the rate of liquid influx must increase; since the area available for liquid flow decreases with increasing number of bubble columns, the liquid velocity must increase very sharply. A limiting condition is reached because the drag exerted by each phase on the other prevents indefinite increases in velocity. At this limiting condition, the liquid flow toward the heated surface cannot increase and the surface becomes largely blanketed with vapor. If the heat input to the surface is held constant (as with electrical heating), the surface temperature will rise quickly to some much higher value, at which heat is transmitted to the fluid by the mechanism of film boiling. If the heat input is a variable and the surface temperature is slowly increased (as in the case of boiling on steam-heated tubes), the boiling mechanism enters the transition zone between nucleate and film boiling.

The maximum heat flux at which nucleate boiling occurs is of considerable importance in industrial operations. Film boiling has few advantages; the superheat at which the maximum flux occurs with nucleate boiling is generally regarded as a limiting operating condition.

Film Boiling

Film boiling occurs when the superheat is sufficiently high to keep the heater surface completely blanketed with vapor. Heat is transmitted through the gas film by conduction, convection, and radiation. High heat fluxes are often obtained with film boiling at high surface temperatures because of the contribution of radiation. In addition, high heat fluxes are obtained if the vapor in the film is turbulent, as, for example, in film boiling from a long vertical surface.

Film boiling is important in rocket technology when liquefied gases such as hydrogen and oxygen are used to cool the rocket engine. Extremely high superheats occur in these systems, causing heat to be transferred by film boiling. A correlation for predicting film-boiling heat-transfer coefficients has been provided by Bromley.[1]

CONDENSATION

Mechanisms

Heat transfer by the condensation of vapors on a cold surface is a common industrial process. Condensation occurs usually on the outer surfaces of tubes in a shell-and-tube heat exchanger, and in most condensers the liquid inside the tubes which receives the heat is water.

[1] L. A. Bromley, *Chem. Eng. Progr.*, **46**:221 (1950).

The vapor being condensed may consist of only one substance, as in the case of steam being condensed in a power plant, or it may be a mixture of substances. The mixture may consist of some substances which are condensable and some which are not, as in the condensation of moisture from air in a dehumidifier. Finally, the mixture may consist of substances which are all condensable at the temperature of the cold surface, as in the case of vapor from a distillation column. In this chapter we shall consider primarily the problem of the condensation of a pure substance.

Vapor-liquid equilibrium is influenced by pressure as well as by temperature, so the pressure in a condensing system is an important variable. Because the cooling fluid in large condensers is usually water at near-atmospheric temperature, the approximate surface temperature of the condenser is usually fixed within a fairly narrow range. Condensation of the vapors is then made to occur by the choice of a suitable operating pressure. It is in this manner that the operating pressure is determined for many industrial separations.

Condensation on solid surfaces occurs by two methods. One is known as *film* condensation. If the condensate wets the surface readily, a liquid film forms on which further condensation occurs. The latent heat of condensation is transferred through the liquid film to the surface. When a vapor consisting of a single component condenses in this fashion, nearly the entire resistance to heat transfer from the vapor to the surface is found in the liquid film. The second kind of condensation, known as *dropwise* condensation, occurs when the liquid wets the surface sparingly so that the condensate coalesces into discrete drops attached to the surface. The drops grow by further condensation on their surfaces and by coalescing with adjacent drops. If the solid surface is not horizontal, the drops grow until they reach a sufficient size to cause them to run down the surface to its lowest point and fall off. In flowing downward, the drops sweep clear a path, so that a substantial portion of the surface is free from condensate at all times. This unwetted surface offers no resistance to heat transfer, and for this reason coefficients for dropwise condensation reach extremely high values. McAdams[1] reports values of h_m measured by various workers which range from 4000 to 75,000 Btu/(h)(ft^2)(°F). Unfortunately, dropwise condensation appears to be possible only when the surface is coated with a "promoter"—for example, benzyl mercaptan, which prevents wetting of the surface.

A comparison of experimentally determined performance curves for dropwise and film condensation is shown in Fig. 25-4. The curve for film condensation was obtained with steam condensing on a polished $\frac{3}{4}$-in copper surface. Results for dropwise condensation were obtained using the same surface coated with montan wax, a mixture of long-chain fatty acids and paraffin hydrocarbons.

[1] W. H. McAdams, "Heat Transmission," 3d ed., chap. 13, McGraw-Hill Book Company, New York, 1954.

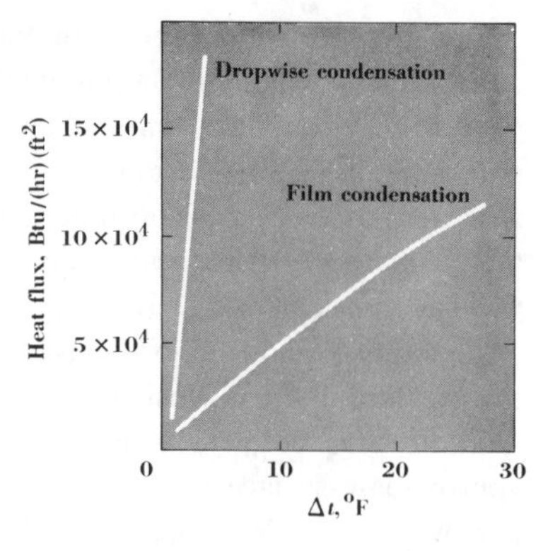

FIGURE 25-4
Comparison of dropwise and film condensation on a vertical surface. [*From D. W. Tanner, C. J. Potter, D. Pope and D. West, Int. J. Heat and Mass Transfer*, **8**:*419* (*1965*).]

Condensers treated with the various promoters usually give dropwise condensation for only a short time because the promoter is dissipated. For this reason industrial condensers are designed on the assumption that only film condensation occurs, and we shall consider only this mechanism in the remainder of this chapter.

The local heat-transfer coefficient for condensation usually varies with position; it varies along the length of a vertical tube and around the perimeter of a horizontal tube. The derivation we shall examine in the next section is for a mean condensing coefficient with a constant temperature difference. This temperature difference, as was mentioned earlier in Chap. 21, serves in part to define the heat-transfer coefficient. In condensing it is taken as the temperature at the vapor-liquid interface minus the temperature at the liquid-solid interface. If the vapor is a pure substance, its saturation temperature will be the temperature at the vapor-liquid interface. If the vapor is superheated or is a mixture, its temperature will not be the same as the vapor-liquid interface temperature. Nevertheless, this interface temperature minus the liquid-solid interface temperature is the driving force used in defining condensing coefficients.

Condensation on Vertical Tubes

Nusselt[1] in 1916 presented theoretical relations for the mean coefficient of heat transfer for a pure saturated vapor condensing on vertical and horizontal tubes. It was assumed in both cases that the flow of liquid was entirely viscous, that there was negligible drag at the vapor-liquid interface, and that the contribution of inertia forces to the downward flow of liquid was negligible. It has been shown since that the last two assumptions are usually valid but that in many situations the liquid flow is turbulent. This usually occurs with condensation on a long vertical tube. Near

[1] W. Nusselt, *Z. dtsch. Ing.*, **60**:541, 569 (1916).

the top of the tube the flow is laminar, but when the condensate film attains a certain thickness, the flow becomes turbulent and Nusselt's equation no longer applies. The criterion for transition between the two types of flow is the value of Reynolds number $\mathbf{Re}_L$. For a film of thickness x_e flowing down a vertical tube of diameter D with a bulk velocity u_b, the hydraulic radius is found by its customary definition to be simply x_e. Thus, at any distance L from the top of the tube where the film thickness is x_e, we write

$$\mathbf{Re}_L = \frac{4x_e u_b \rho}{\mu} \tag{25-6}$$

The mass-flow rate is $x_e u_b \rho \pi D$, and the mass flow per unit of tube perimeter is $x_e u_b \rho$. This quantity, designated as Γ, increases from a value of zero at the top of the tube to a maximum at the bottom. Using it in the Reynolds number, we obtain

$$\mathbf{Re}_L = \frac{4\Gamma}{\mu} \tag{25-7}$$

Laminar flow exists when the Reynolds number is less than 2000, and turbulent flow when it is greater. This is therefore a limitation on the validity of the Nusselt equation.

The tube diameter is usually much greater than the thickness of the film, so the cylindrical system may be represented approximately by a rectangular coordinate system. The origin is at the top of the tube, with the x axis directed outward normal to the tube surface and the y axis directed downward.

The following Navier-Stokes equation is the basis of the hydrodynamic analysis, which, as usual, precedes the heat-transfer analysis.

$$u_x \frac{\partial u_y}{\partial x} + u_y \frac{\partial u_y}{\partial y} + u_z \frac{\partial u_y}{\partial z} + \frac{\partial u_y}{\partial \theta} = g_c Y - \frac{g_c}{\rho}\frac{\partial p}{\partial y} + \nu\left(\frac{\partial^2 u_y}{\partial x^2} + \frac{\partial^2 u_y}{\partial y^2} + \frac{\partial^2 u_y}{\partial z^2}\right) \tag{9-53}$$

The left side of this equation is zero because the flow is steady, and u_x, u_z, and $\partial u_y/\partial y$ are assumed to be zero. On the right side, $\partial p/\partial y$ is zero, and the variation of u_y in both y and z directions is neglected, so that Eq. (9-53) becomes

$$\frac{d^2 u_y}{dx^2} = \frac{-g_c Y}{\nu} \tag{25-8}$$

The body force Y reduces to g/g_c, because the y axis is vertical, so that Eq. (25-8) becomes

$$\frac{d^2 u_y}{dx^2} = \frac{-g}{\nu} \tag{25-9}$$

Equation (25-9) is integrated twice to give

$$u_y = \frac{-gx^2}{2\nu} + C_1 x + C_2 \tag{25-10}$$

The two arbitrary constants are found by the application of two boundary conditions: $u_y = 0$ at $x = 0$, and $du_y/dx = 0$ at the vapor-liquid interface, where $x = x_e$. This latter condition is true if drag at the vapor-liquid interface is negligible. The use of these values gives

$$u_y = \frac{g}{\nu}\left(x_e x - \frac{x^2}{2}\right) \tag{25-11}$$

The mass rate can be found by integrating over the film thickness. Neglecting the change in perimeter with increase in x, we obtain

$$w = \int_0^{x_e} \frac{g}{\nu}\left(x_e x - \frac{x^2}{2}\right)\rho\pi D\,dx = \frac{\rho\pi D g}{\nu}\frac{x_e^3}{3} \tag{25-12}$$

From this we see that the mass rate per unit of tube perimeter, which we defined earlier as Γ, is

$$\Gamma = \frac{\rho g x_e^3}{3\nu} \tag{25-13}$$

To find the mean heat-transfer coefficient, a heat balance is written for a differential length of tube dy located at a distance y from the top of the tube. The heat-flow rate is

$$dq = h\pi D\,dy(t_{sv} - t_s) \tag{25-14}$$

The temperature gradient in the y direction is assumed to be negligible, so all the heat of condensation released at the vapor-liquid interface is conducted horizontally to the tube surface. If the film has a thickness x_e, we can write

$$dq = k\pi D\,dy\frac{t_{sv} - t_s}{x_e} \tag{25-15}$$

Combining Eqs. (25-14) and (25-15), we obtain

$$h = \frac{k}{x_e} \tag{25-16}$$

By definition, the local heat-transfer coefficient is

$$h = \frac{dq}{dA(t_{sv} - t_s)}$$

$$= \frac{\lambda\,dw}{dA(t_{sv} - t_s)} \tag{25-17}$$

where dw = condensation rate, lb/h, at the differential section and λ = latent heat of vaporization. However, the local condensation rate can also be written as

$$\frac{dw}{dA} = \frac{dw}{\pi D\, dy} = \frac{d\Gamma}{dy} \tag{25-18}$$

Combining Eqs. (25-16) to (25-18) gives

$$h = \frac{k}{x_e} = \frac{\lambda}{t_{sv} - t_s}\frac{d\Gamma}{dy} \tag{25-19}$$

or

$$t_{sv} - t_s = \frac{x_e \lambda}{k}\frac{d\Gamma}{dy} \tag{25-20}$$

For the entire length L of the tube on which laminar flow exists, we write

$$q = h_m A(t_{sv} - t_s) = \lambda w_L \tag{25-21}$$

This equation is solved for the temperature difference

$$t_{sv} - t_s = \frac{\lambda w_L}{h_m A} = \frac{\lambda \Gamma_L}{h_m L} \tag{25-22}$$

which is substituted in Eq. (25-20) and rearranged to give

$$dy = \frac{x_e h_m L}{k\Gamma_L}\, d\Gamma \tag{25-23}$$

From Eq. (25-13) the film thickness is

$$x_e = \left(\frac{3\nu\Gamma}{\rho g}\right)^{1/3} \tag{25-24}$$

This result is substituted in Eq. (25-23):

$$dy = \frac{h_m L}{k\Gamma_L}\left(\frac{3\nu}{\rho g}\right)^{1/3} \Gamma^{1/3}\, d\Gamma \tag{25-25}$$

This equation is integrated from $y = 0$ to $y = L$ and from $\Gamma = 0$ to $\Gamma = \Gamma_L$ and rearranged to give the mean heat-transfer coefficient for the entire laminar region

$$h_m = \tfrac{4}{3}k\left(\frac{\rho g}{3\nu\Gamma_L}\right)^{1/3}$$

$$= 0.925\left(\frac{k^3 \rho g}{\nu \Gamma_L}\right)^{1/3} \tag{25-26}$$

If Eq. (25-22) is rearranged, the quantity Γ_L can be expressed as

$$\Gamma_L = \frac{h_m L(t_{sv} - t_s)}{\lambda} \tag{25-27}$$

Substituting Eq. (25-27) in Eq. (25-26), we obtain

$$h_m = 0.925\left[\frac{k^3 \rho g \lambda}{\nu h_m L(t_{sv} - t_s)}\right]^{1/3} \tag{25-28}$$

The mean heat-transfer coefficient can be readily extracted by cubing both sides of Eq. (25-28), solving for h_m, and taking the fourth root. This procedure gives

$$h_m = 0.942\left[\frac{k^3 \rho g \lambda}{\nu L(t_{sv} - t_s)}\right]^{1/4} \tag{25-29}$$

The mean coefficient can also be expressed in terms of $\mathbf{Re}_L$, as defined in Eq. (25-7). To achieve this result we rearrange Eq. (25-26) to give

$$h_m\left(\frac{\nu^2}{k^3 g}\right)^{1/3} = 1.47\left(\frac{4\Gamma_L}{\mu}\right)^{-1/3}$$

$$= 1.47(\mathbf{Re}_L)^{-1/3} \tag{25-30}$$

in which $\mathbf{Re}_L$ is the Reynolds number at the bottom of the laminar-flow region on the tube.

Equation (25-29) shows a dependence of h_m on the temperature drop over the condensate film. In the operation of a condenser the cold fluid flowing through the inside of the tube increases in temperature, so that the assumption that $t_{sv} - t_s$ is constant is not valid. Nevertheless, the relation between temperature drop and h_m is only by the power of $\frac{1}{4}$, so the variation of h_m because of this is usually small.

The physical properties of the condensate vary with temperature, but the only property affected significantly is the viscosity. It has been shown that the viscosity should be evaluated at a temperature equal to the temperature of the saturated vapor minus $\frac{3}{4}(t_{sv} - t_s)$. The other properties can also be used at this temperature. A minor correction may also be made to the latent heat λ, because the total heat flux is actually the sum of the latent heat plus the sensible heat given up by subcooling the liquid in the condensate film. An accepted procedure is to multiply λ by the factor $[1 + (0.4C_p(t_{sv} - t_s)/\lambda)]^2$ before using it in Eq. (25-29).

Equation (25-29) has been compared with experimental data and found to be generally too conservative. It has been suggested that this is because ripples on the surface of the condensate film give a certain amount of mixing. Experimental values of h_m range as high as 50 percent greater than the values predicted by Eq. (25-29). Consequently, the recommended procedure is to multiply the theoretical values of h_m obtained from Eq. (25-29) by a factor of 1.2. Both the theoretical and recommended relations are shown in Fig. 25-5.

If the tube is sufficiently long, the film thickness increases and the flow becomes turbulent. A theoretical solution is not available for this situation. However, the

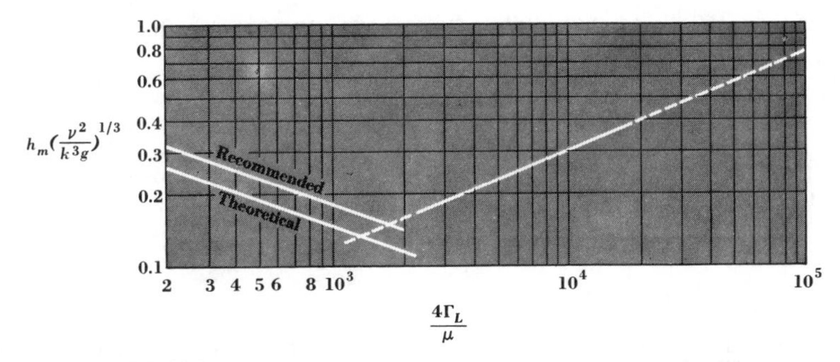

FIGURE 25-5
Condensing coefficients on a vertical tube. (*From W. H. McAdams, "Heat Transmission," 3d ed., p. 335, McGraw-Hill Book Company, New York, 1954.*)

mean coefficient has been measured experimentally in numerous investigations. McAdams[1] recommends the following empirical correlation:

$$h_m\left(\frac{\nu^2}{k^3 g}\right)^{1/3} = 0.0077\left(\frac{4\Gamma_L}{\mu}\right)^{0.4} \tag{25-31}$$

This relation, which is recommended for values of the Reynolds number $4\Gamma_L/\mu$ greater than 2000, is shown in Fig. 25-5, in addition to the curves for laminar flow. The mean coefficient obtained from Eq. (25-31) applies to the entire tube, i.e., both the laminar and turbulent portions.

Condensation on Horizontal Tubes

The local coefficient of heat transfer for a vapor condensing on a horizontal tube is a function of position, just as it is for a vertical tube. However, for a horizontal tube the variation occurs around the periphery of the tube rather than along the axis. The highest coefficient is on the top of the tube, where the condensate film is thinnest, and the lowest is on the underside.

Nusselt's derivation for condensation on a horizontal tube is similar to that for a vertical tube and will not be given here. It leads to the following equation for the mean heat-transfer coefficient:

$$h_m = 0.725\left|\frac{k^3 \rho g \lambda}{\nu D(t_{sv} - t_s)}\right|^{1/4} \tag{25-32}$$

[1] Loc. cit.

This expression, which is similar to Eq. (25-29) for the vertical tube, can be rearranged, using the substitutions shown below, to give h_m as a function of $\mathbf{Re}_L$.

$$h_m\left(\frac{\nu^2}{k^3 g}\right)^{1/3} = 1.20(\mathbf{Re}_L)^{-1/3} \qquad (25\text{-}33)$$

As before, the Reynolds number is expressed as

$$\mathbf{Re} = \frac{(4)(\text{hydraulic radius})(u_b \rho)}{\mu} \qquad (25\text{-}34)$$

The hydraulic radius is found by the customary procedure and is equal to the thickness of the film x_e at any point. The quantity $x_e u_b \rho$ is the mass rate of condensate per lineal foot of tubing at a certain peripheral position on each side of the tube. Equation (25-34) can be written as follows:

$$\mathbf{Re} = \frac{4x_e u_b \rho}{\mu} \qquad (25\text{-}35)$$

The Reynolds number at the base of the tube, which we shall write as $\mathbf{Re}_L$, is the one used for defining h_m in Eq. (25-33). It can be written in terms of the total mass rate per foot of tubing as

$$\mathbf{Re}_L = \frac{4}{\mu}\frac{\Gamma_L}{2} = \frac{2\Gamma_L}{\mu} \qquad (25\text{-}36)$$

where Γ_L is total rate of condensate formation on the tube per foot of length, and $\Gamma_L/2$, the rate for each side, is the mass rate just before the streams of condensate join at the bottom of the tube. Equation (25-36) therefore defines $\mathbf{Re}_L$ as used in going from Eq. (25-32) to Eq. (25-33).

If a number of horizontal tubes are arranged in a vertical row so that all the condensate from each tube falls on the tube beneath, the mean coefficients on successive tubes diminish. The derivation which leads to Eq. (25-32) can be extended theoretically[1] to include the effect of N tubes in a vertical row to give a mean coefficient for the entire bank of tubes as

$$(h_m)_N = 0.725\left[\frac{k^3 \rho g \lambda}{\nu N D(t_{sv} - t_s)}\right]^{1/4} \qquad (25\text{-}37)$$

This equation has been found to give satisfactory predictions of mean coefficients for banks of tubes, provided the mass-flow rate of condensate on the lowest tube is such that $\mathbf{Re}_L < 2000$. There will, of course, always be a certain amount of turbulence on any tube below the top, because of the condensate dripping from the tube above. This has the effect of increasing $(h_m)_N$. On a single horizontal tube it is

[1] M. Jakob, "Heat Transfer," vol. 1, chap. 30, John Wiley & Sons, Inc., New York, 1949.

unlikely that condensation will occur at a sufficient rate to cause turbulent flow. However, the flow of vapor over the condensing film may cause ripples to occur, with the result that the coefficients will be higher than those predicted from Eq. (25-32).

Superheated Vapors

If the vapor being condensed enters the condenser in a superheated condition, this superheat must be removed along with the latent heat by the cooler fluid inside the tubes. The mechanism of condensation is coupled with desuperheating so that an analysis of the process might be expected to require the consideration of both steps. It has been found experimentally, however, that the resistance offered by desuperheating is negligible. Even with superheat the temperature of the liquid at the vapor-liquid interface is the saturation temperature, so that the usual condensing coefficient is employed with the driving force $t_{sv} - t_s$, as in the case of a saturated vapor. The extremely small resistance for desuperheating will be more readily understood when we have considered simultaneous heat and mass transfer in Chap. 36.

Effect of Noncondensable Gases

This phenomenon is familiar in the form of condensation of atmospheric moisture to give dew. It also occurs in industrial equipment in which liquids containing dissolved air are boiled and later condensed.

Condensation of a constituent will occur only if the partial pressure of that constituent in the gas exceeds the vapor pressure at the temperature of the cold surface. If this partial-pressure difference exists, the condensable constituents will diffuse through the gas phase to the cold surface and condense there. In addition to the latent heat given up by condensation at the surface, sensible heat will be transferred by conduction or convection to the interface from the bulk of the gas phase.

This situation, like the condensation of superheated vapor, is a coupled process. However, in this case the diffusion step is not usually negligible; in fact, the rate of diffusion may be so low as to become the controlling step. The analysis of this process requires the simultaneous application of equations for heat and mass transfer and will not be discussed at this point.

In some processes the gas contains only a small portion of condensable material and flows through the condenser continuously. In others, however, the gas entering the condenser is nearly all condensable. These total condensers must contain provision for continuous or periodic venting; otherwise the noncondensable gases will accumulate and the capacity of the condenser will diminish to zero.

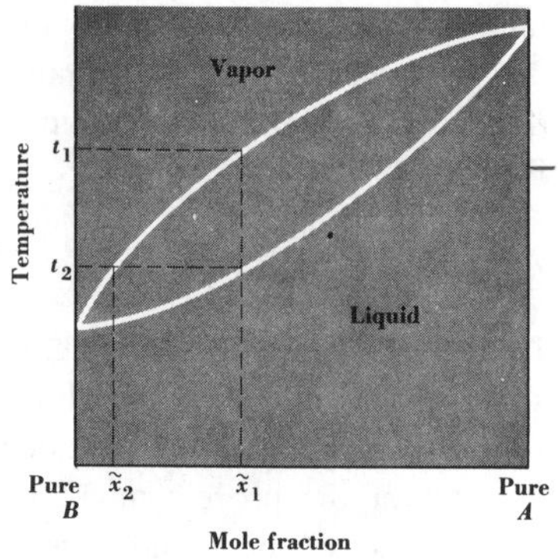

FIGURE 25-6
Vapor-liquid equilibrium diagram illustrating equilibrium in condensation of mixtures.

Condensation of Mixtures

Condensers for the condensation of mixtures may operate in a variety of ways. If all the vapor entering is to be condensed and opportunity is provided for the establishment of equilibrium conditions, the bulk of the vapor phase in the condenser at any instant will acquire an equilibrium composition different from that of the liquid. The liquid which leaves must necessarily have the same composition as the vapor entering, but the bulk vapor phase will be richer than either of these in the most volatile components. This is illustrated in Fig. 25-6, which shows the vapor-liquid equilibrium for a binary system of components A and B at constant pressure. A saturated vapor of composition x_1 enters the condenser continuously at temperature t_1. The liquid leaving is of the same composition but is assumed to be at the liquid-saturation temperature t_2. The bulk of the vapor in the condenser is at the same temperature as the liquid t_2 but is of a different composition x_2.

The determination of the heat load in a condenser is rather simple under these conditions. The vapor-liquid interface temperature is merely the saturation temperature t_2 of the liquid at the condenser pressure. The equations derived for predicting condensing coefficients for pure substances are used with the physical properties of the liquid condensate. The temperature difference is, as before, the vapor-liquid interface temperature minus the solid-liquid interface temperature.

The assumptions made above regarding equilibrium between the bulk gas phase and the liquid are not always valid. Another way of analyzing the problem would be to assume that the bulk gas phase in the condenser has the same composition as the entering gas. Under these conditions, the vapor-liquid interfacial temperature and the liquid interfacial composition are determined by trial-and-error procedures and are functions of the solid-liquid interfacial temperature.

PROBLEMS

25-1 A shell-and-tube condenser contains four rows with four copper tubes per row, arranged in square pitch. The tubes are $\frac{3}{4}$-in 16 BWG and are 3 ft long. Water is flowing through the tubes at a rate such that the convection heat-transfer coefficient for the water is $h = 1000$ Btu/(h)(ft^2)(°F). The water rises only slightly in temperature and may be assumed to be at a constant temperature of 60°F. Pure, saturated steam at 5 psig is condensing on the shell side. Determine the capacity of the condenser in Btu/h under these conditions when it is (*a*) in a vertical position and (*b*) in a horizontal position.

25-2 A condenser contains horizontal, finned tubes which are 0.013-m-OD copper tubes to which a continuous spiral fin is attached. The fin has an overall diameter of 0.038 m, a thickness of 0.0005 m and a pitch such that it spirals around the tube 630 times per meter of tube length. Estimate the condensing coefficient for the top tube in the condenser in W/K · m^2 of total condensing area. Assume that the metal surface is at 60°C at all points and that the condensing vapor is saturated steam at 99°C. The fins may be considered to have an effective height equal to the area of a fin face divided by the outside fin diameter.

25-3 The condensation of a superheated vapor is a coupled process of desuperheating and condensation. If no condensation occurs, the desuperheating process has a resistance to heat transfer which may be 100 times greater than the resistance offered when condensation occurs. Nevertheless, when desuperheating and condensation are combined, the resistance of the desuperheating step is negligible compared with the condensing resistance. Explain this paradox.

25-4 If the bubbles formed in water at a particular surface have an average diameter of 2μ, what would be the approximate temperature of the boiling liquid at atmospheric pressure? Take the surface tension as 58 dyn/cm.

25-5 If the value of h_m found in Prob. 25-1*a* is 900 Btu/(h)(ft^2)(°F), what would the value be if the steam pressure were changed to 20 psig?

T°C	μ, cP
30	0.801
40	0.656
50	0.549
60	0.467
70	0.406
80	0.356

Assume k and ρ are not changed by temperature.

25-6 It has been found that a relation of the type

$$\mathbf{Nu} = f(\mathbf{Gr}, \mathbf{Pr})$$

fails for correlating boiling heat-transfer coefficients. Explain why this may be so.

26

RADIANT HEAT TRANSFER

The Nature of Thermal Radiation

Radiant energy of the form called thermal radiation is emitted by every body having a temperature greater than absolute zero. This does not mean, however, that the amount of thermal radiation emitted is always significant. Its importance in a heat-transfer process depends on the amount of heat being transferred simultaneously by other mechanisms. In systems at or below room temperature thermal radiation is likely to be negligible. However, at temperatures of "red heat" ($\sim 1000°F$) and higher, radiant transmission is often the principal mechanism of heat transfer.

Bodies may emit radiant energy of other forms in addition to thermal radiation. Bombardment of a substance by electrons produces radiation which we call x rays. Exposure to one form of radiation often causes a body to emit another, or secondary, radiation; e.g., certain minerals fluoresce under ultraviolet light. In fact, there is a continuous spectrum of electromagnetic radiation in which the various arbitrary divisions are referred to by names reflecting the method of origin or some characteristic quality. All forms have the same velocity of propagation but differ in wavelength and origin. All forms produce heat when absorbed. Nevertheless, it is only the electromagnetic radiation produced by virtue of the temperature of the emitter that we call thermal radiation. Some of this thermal radiation we also call visible light. Most of it, however, falls outside the spectrum of visible light and is ordinarily included in what we call infrared radiation. Table 26-1 gives the approximate ranges of wavelength of some forms of radiation.

The amount of thermal radiation emitted by a body depends on its temperature and surface condition. At any temperature the radiation emitted extends over a spectrum, as shown in Fig. 26-1. As the temperature changes, both the total emissive power and the distribution of energy change. It can be seen that the wavelength at which the maximum emissive power occurs decreases as the temperature of the emitter increases. The spectrum of visible light occurs on the low-wavelength side of the thermal-radiation spectrum, so increasing the temperature causes increasing amounts of the thermal energy to appear as light. In addition, the distribution of the energy in the visible range alters. At low temperatures, where visible light is just emitted (e.g., 1000°F), the light is red because of the encroachment of the thermal spectrum onto the visible spectrum from the red side. As the temperature increases, more energy is emitted at the lower wavelengths and the light becomes more nearly white.

The sun emits energy at 10,000°F, and so its spectral distribution curve is shifted substantially to the left of curve *a* in Fig. 26-1. In fact, the solar spectrum straddles the so-called visible range of 3500 to 7800 Å, which is more than a coincidence.

Typically, the solar energy reaching the earth's surface contains about 5 percent of its energy in the ultraviolet range, 40 percent in the visible range, and 55 percent in the infrared range. The light we receive has passed through the earth's atmosphere which absorbs energy in amounts that differ with wavelength of the radiation, length of the path, and composition of the atmosphere. The effect of these factors on the

Table 26-1 CHARACTERISTIC WAVELENGTHS OF RADIATION

	Wavelength range, cm
Cosmic rays	$<0.001 \times 10^{-8}$
Gamma rays	$0.01\text{–}0.15(\times 10^{-8})$
X rays	$0.06\text{–}1000(\times 10^{-8})$
Ultraviolet	$100\text{–}3500(\times 10^{-8})$
Electric lamps	$2000\text{–}50{,}000(\times 10^{-8})$
Visible	$3500\text{–}7800(\times 10^{-8})$
Infrared	$7800 \times 10^{-8}\text{–}0.04$
Radio	$0.04\text{–}10^{7}$

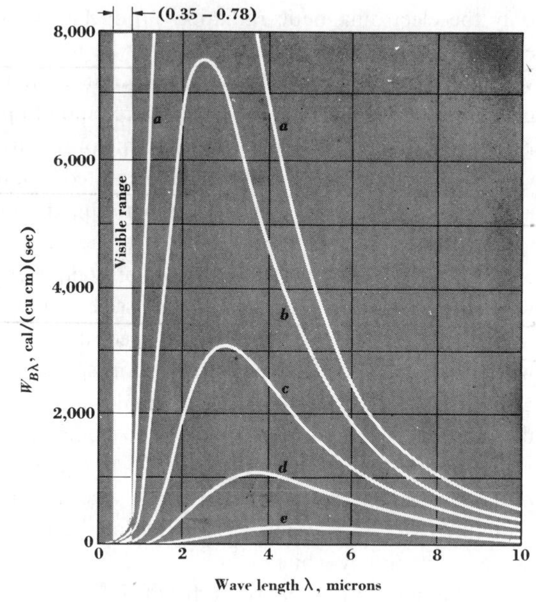

FIGURE 26-1
Spectral distribution of total energy emitted by a black body. (*From M. Jakob, "Heat Transfer," vol. I, John Wiley & Sons, Inc., New York, 1950.*)
a $T = 1400°K$.
b 1200°K.
c 1000°K.
d 800°K.
e 600°K.

quality of sunlight as we receive it is evident in the changing character of the light as the length of its path changes with time of day and amount of atmospheric pollution.

We might expect another object at the same temperature as the sun to emit light of the same quality as sunlight. However, surface conditions also affect the emission at various wavelengths and so both surface conditions and temperature would have to be duplicated. The practical difficulty of operating a filament lamp at solar temperature to produce solarlike radiation is evident. Most filament lamps operate with filament temperatures below 5000°F; hence, the light they emit is deficient in ultraviolet and oversupplied with infrared radiation, as compared with solar radiation. To more nearly match solar radiation it is necessary to change from filament lamps, which produce light by thermal radiation, to fluorescent lamps, in which an electric discharge produces radiation which is then transformed by an appropriate blend of fluorescent phosphors into a spectrum approximating the solar spectrum.

An aspect of solar radiation which has acquired engineering importance is the effect of radiation pressure on the paths of vehicles in space flight. Maxwell in 1865 proposed that radiation impinging on a surface exerted a force on that surface. Einstein's equation relating mass, energy, and the velocity of light can be used to express the radiation as an equivalent mass flux. Equations presented in Chap. 5 can be used to calculate the equivalent force or radiation pressure as demonstrated in Example 26-1.

Example 26-1 Find the radiation pressure on the earth produced by solar radiation assuming the incident radiation rate to be 442 Btu/(h)(ft^2).

In the Einstein equation $E = mc^2/g_c$

where

$$E = 442\ \text{Btu/(h)(ft}^2\text{)}$$

$$c = 9.84 \times 10^8\ \text{ft/s}$$

$$g_c = 32.2\ \text{(lb)(ft)/(lb}_f\text{)(s}^2\text{)}$$

Thus,

$$m = \frac{(32.2)(442)(778)}{(9.84 \times 10^8)^2(3600)}$$

$$= 3.17 \times 10^{-15}\ \text{lb/(s)(ft}^2\text{)}$$

We write a modified form of Eq. (5-7) to obtain the following expression:

$$\frac{\text{Radiation force}}{\text{area}} = \frac{1}{g_c}\,\Delta(wu_x)$$

$$= \frac{mc}{g_c}$$

$$= \frac{(3.17 \times 10^{-15})(9.84 \times 10^8)}{32.2}$$

$$= 9.7 \times 10^{-8}\ \text{lb}_f/\text{ft}^2$$

To express the answer in metric units we multiply by the conversion factor 4.79×10^6 and obtain a value for the radiation pressure of 0.46 dyn/m^2.

The result obtained above assumes no reflection of solar radiation; complete reflection would, of course, double the answer. All solar radiation not reflected is absorbed and ultimately re-emitted. However, rotation of the earth causes it to be re-emitted approximately uniformly in all directions, thus producing no resulting force effect. ////

Absorption, Reflection, and Transmission[1]

Radiant energy striking matter may be absorbed, transmitted, or reflected. The fraction of the total energy which is absorbed is called the *absorptivity* α; the fraction transmitted is the *transmissivity* τ; and the fraction reflected is the *reflectivity* ρ. These definitions lead to the equation

$$\alpha + \tau + \rho = 1 \qquad (26\text{-}1)$$

A body having an absorptivity of unity is called a *black body*. If absorptivity plus reflectivity equals 1 for a substance, it is said to be opaque. Certain solids are only partially opaque in very thin sections, but the absorption of energy in most opaque solids is accomplished essentially at the surface. Some solids, of course, are nearly transparent, as are most liquids and gases.

Absorption does not occur equally for radiation of all wavelengths; since the spectral distribution of energy is a function of the temperature of the emitter, the absorptivity of a receiver depends on the temperature of the emitter. The variation of absorptivity of a solid with wavelength is shown in Fig. 26-2. Selective absorption is especially significant in absorption by gases, which are transparent to radiation of some wavelengths but highly absorbent for others. The absorptivity of a material also depends on its own temperature, although any variation because of this effect is usually less significant than the variation resulting from temperature changes of the emitter. Thus the specification of α for a particular surface requires the designation of two temperatures. In solving many problems, however, it is necessary to assume that the absorptivity of a surface is not a function of the spectral distribution; the surface in these circumstances is spoken of as a "gray" surface.

Eye sensitivity to light is obviously a function of wavelength. Not so obvious is the fact that even in the visible range the eye sensitivity varies. Relative response in an average person follows an error-function-type curve which has a maximum midway in the visible range and approaches zero asymptotically at the boundaries of the visible range. The response curve differs somewhat for different people; pronounced aberrations in the curve are a measure of color blindness.

Reflectivity and transmissivity are characteristics which we experience in the everyday world. The criteria for predicting these quantities for visible light also apply to thermal radiation. Polished metallic surfaces have high reflectivities, and granular surfaces have low reflectivities. Gases and liquids usually transmit most of the radiation incident upon them, although liquids are capable of reflecting substantial amounts of energy. Because of the relation between α, τ, and ρ in Eq. (26-1), the remarks made above concerning variation of absorptivity with temperature obviously apply also to the sum of transmissivity and reflectivity.

[1] A useful reference is R. Siegel and J. R. Howell, "Thermal Radiation Heat Transfer," McGraw-Hill Book Company, New York, 1972.

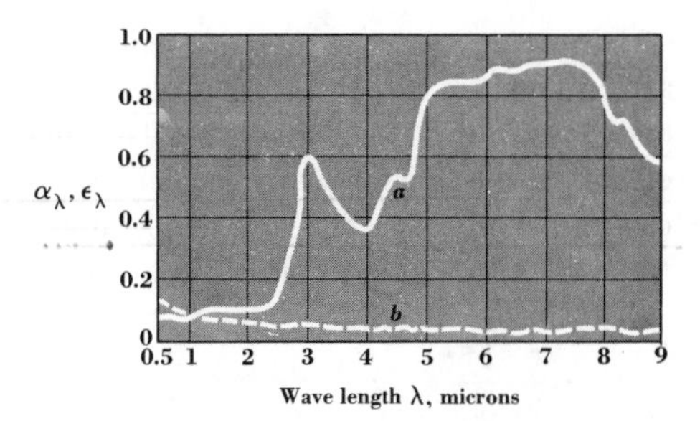

FIGURE 26-2
Variation of absorptivity and emissivity with wavelength. [*From W. Sieber, Z. Tech. Phys.*, **22**:*130* (*1941*).]
a White fireclay.
b Polished copper.

Reflection from a macroscopic section of surface depends greatly on the character of the surface. If the surface is very smooth, the angles of incidence and reflection are the same. However, most surfaces encountered in engineering are sufficiently rough so that some reflection occurs in all directions. If the total reflection is independent of the angle of incidence, it is said to be "diffuse." This assumption is made in solving most engineering problems.

Emissivity

If one or more small non-heat-generating bodies are contained within a large evacuated enclosure, they will eventually reach a state of thermal equilibrium at which each receives radiant heat at the same rate at which it loses it. If W_1 is the total rate of emission of energy per unit area for a body of area A_1, W_o is the flux of radiant energy from the enclosure which strikes area A_1, and α_1 is the absorptivity of surface A_1, then we can write

$$W_1 A_1 = W_o \alpha_1 A_1 \qquad (26\text{-}2)$$

from which

$$\frac{W_1}{\alpha_1} = W_o \qquad (26\text{-}3)$$

For additional small bodies W_o will remain constant, so we can write

$$\frac{W_1}{\alpha_1} = \frac{W_2}{\alpha_2} = \cdots = \frac{W_n}{\alpha_n} \qquad (26\text{-}4)$$

Thus in a system at thermal equilibrium, all bodies will have the same temperature and all bodies will have the same ratio of total emissive power to absorptivity. This principle is called *Kirchhoff's law*.

If one of the surfaces in the enclosure absorbs all the radiation incident upon it, it has an absorptivity of 1 and, as mentioned earlier, is referred to as a black body. Because W/α is fixed, this places an upper limit on the emissive power of that surface, or any other at the same temperature, so that a black body is referred to as an ideal emitter as well as a perfect absorber. Thus the emissive power of a black body [in Btu/(h)(ft^2 of surface area)] is a function only of its own temperature. By definition, the emissivity of a real surface is the ratio of its emissive power W_1 to that of a black body W_2 at the same temperature. This emissivity is designated as ε_1. If we rewrite Eq. (26-4) to include this definition, we have

$$\frac{W_1}{W_2} = \frac{\alpha_1}{\alpha_2} = \varepsilon_1$$

in which object 2 is chosen to be a black body. Because α_2 is unity, we obtain for the nonblack body the relation

$$\alpha_1 = \varepsilon_1 \qquad (26\text{-}5)$$

To generalize, we may state that in a system at thermal equilibrium the emissivity of any object in the system is the same as its absorptivity. In a system where various portions are at different temperatures, this is not true, but is often assumed to be true so that a solution can be obtained to a problem. As mentioned earlier, the absorptivity of a surface actually varies with the wavelength of the incident radiation; sometimes, however, the surface is assumed to be "gray," and α is assumed to be constant. In these circumstances, α for a surface is evaluated by determining the emissivity, not at the actual surface temperature, but rather at the temperature of the source of the radiation, because this is the temperature the absorbing surface would have if it were at thermal equilibrium with the emitter. It is a fact that the temperature of the absorber has some effect on absorptivity as well, but the effect of the temperature of the emitter is usually more important.

Emissivity, like absorptivity, is low for polished metal surfaces and is moderately high for oxidized metal surfaces. It is also high for most nonmetallic substances. Table 26-2 contains some typical values of emissivity. Emissivity, like absorptivity, varies with wavelength and also with the angle between the emitted beam and the emitting surface. The variation with wavelength is shown in Fig. 26-2. The values in Table 26-2 are for the total emissivity, which includes radiation of all wavelengths normal to the emitting surface. The variation of emissivity with angle often is not large; in such circumstances the variation is neglected and the values of Table 26-2 are taken as being total hemispherical emissivities. Some of the values in this table are so close to unity that the surfaces may be considered to be black bodies.

One way of achieving nearly perfect black-body conditions is by absorbing or emitting radiation through a small hole in a hollow object. This is illustrated in Fig.

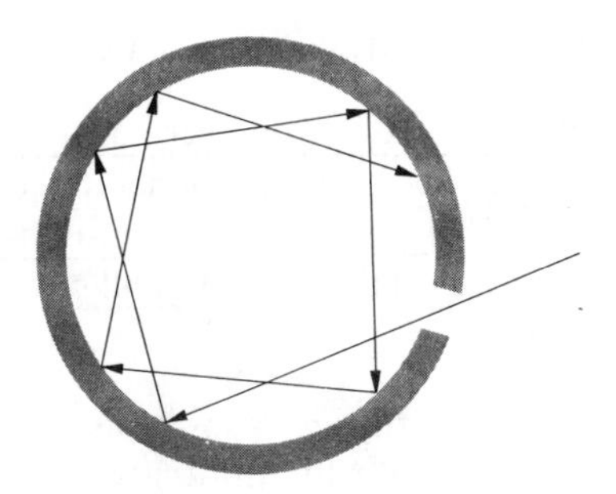

FIGURE 26-3
Absorption of radiation in a hollow container.

26-3. Radiation entering the hole is likely to be reflected numerous times before finally being reflected out of the hole by which it entered. Each time it impinges on the surface, a fraction α is absorbed, so that after a few reflections, nearly all the radiation which entered is absorbed. In effect, the hole in the container is equivalent to a surface of the same area with an absorptivity of 1. Similarly, in its emissive characteristics, the effect of the hole is that of a black body of the same area at a temperature equal to the temperature of the enclosure. This principle is employed when making temperature measurements by a method which requires that the emitting surface have an emissivity of 1. The instrument, a pyrometer, is sighted on a cavity or wedge-shaped hole in the walls of the system.

Table 26-2 TOTAL EMISSIVITIES OF SOME SURFACES

Surface	Temperature, °F	Emissivity
Polished aluminum	73	0.040
Polished copper	242	0.023
Polished iron	800–1800	0.144–0.377
Cast iron, newly turned	72	0.435
Oxidized iron	212	0.736
Asbestos board	74	0.96
Red brick	70	0.93
Sixteen different oil paints, all colors	212	0.92–0.96
Water	32–212	0.95–0.963

SOURCE: J. H. Perry, "Chemical Engineers' Handbook," 5th ed., p. 10-46, McGraw-Hill Book Company, New York, 1973. See also "Thermophysical Properties of Matter," The Thermophysical Properties Research Center Data Series, vols. 7–9, Plenum Press, New York, 1972; and G. G. Gubareff, J. E. Jansen, and R. H. Torborg, "Thermal Radiation Properties Survey," 2d ed., Honeywell Research Center, Minneapolis, 1960.

The Stefan-Boltzmann Law

It can be shown from thermodynamic considerations that the total emissive power of a black body is proportional to the fourth power of the absolute temperature. This relation, known as the Stefan-Boltzmann law, applies to the total radiant energy of all wavelengths emitted in all directions and is written[1]

$$W_B = \sigma T^4 \qquad (26\text{-}6)$$

The quantity σ, known as the *Stefan-Boltzmann constant*, has a value of 0.1714×10^{-8} Btu/(h)(ft^2)(°R)^4, which is equivalent to 5.676×10^{-8} W/m$^2 \cdot$K^4. If the emissivity of a real surface is expressed as a total emissivity ε, the total emissive power of the real body in Btu/(h)(ft^2) is

$$W = \varepsilon \sigma T^4 \qquad (26\text{-}7)$$

An expression for the monochromatic emissive power of a black body $W_{B\lambda}$ as a function of the wavelength was derived from quantum theory by Planck in 1900. The equation,[2] known as *Planck's law*, is

$$W_{B\lambda} = \frac{C_1 \lambda^{-5}}{e^{C_2/\lambda T} - 1} \qquad (26\text{-}8)$$

The term $W_{B\lambda}$ has the dimensions of energy/(area)(time)(unit increment of wavelength). It is expressed in units of ergs/(s)(cm^3) when the constants C_1 and C_2 are 3.74×10^{-5} (erg)(cm^2)/s and 2.59 (cm)(°R), respectively. This function is plotted in Fig. 26-1. It can be differentiated to obtain $dW_{B\lambda}/d\lambda$, and if this derivative is equated to zero, it yields the following expression for the wavelength at which the maximum emission of energy occurs:

$$5\lambda T(1 - e^{-C_2/\lambda T}) = C_2 \qquad (26\text{-}9)$$

This equation can be solved by trial and error for λT to give

$$\lambda_{\max} = \frac{0.5216}{T} \qquad (26\text{-}10)$$

in which the constant 0.5216 has units of (cm)(°R). Equation (26-10) is known as *Wien's displacement law*; it predicts a decreasing wavelength at peak emissive power for increasing temperature. It is of interest that the radiation from the sun has a maximum intensity at a wavelength of approximately 5×10^{-5} cm. According to Eq. (26-10), the surface temperature of the sun is 10,400°R.

[1] We use T to indicate absolute temperature (Kelvin or degrees Rankine) in this book and t for degrees Fahrenheit or degrees Celsius.

[2] Another common form of Planck's law gives the intensity of the monochromatic radiation *normal* to the radiating surface. Equation (26-8), however, gives the monochromatic emissive power, which represents all the energy of wavelength λ emitted through a *hemispherical* solid angle. This is obtained by multiplying the normal intensity by a factor π.

Equation (26-8) for the monochromatic emissive power can be integrated over the entire spectrum to give the total emissive power. Thus the Stefan-Boltzmann constant from Eq. (26-6) becomes

$$\sigma = \frac{1}{T^4}\int_0^\infty W_{B\lambda}\, d\lambda \qquad (26\text{-}11)$$

Substituting Planck's law, Eq. (26-8), in Eq. (26-11) yields

$$\sigma = \frac{C_1}{T^4}\int_0^\infty \frac{\lambda^{-5}}{e^{C_2/\lambda T} - 1}\, d\lambda$$

$$\sigma = \frac{C_1}{C_2{}^4}\int_0^\infty \frac{(C_2/\lambda T)^5}{e^{C_2/\lambda T} - 1}\, d\frac{\lambda T}{C_2}$$

which is rearranged to give

$$\sigma = \frac{C_1}{C_2{}^4}\int_0^\infty \frac{(C_2/\lambda T)^3}{e^{C_2/\lambda T} - 1}\, d\frac{C_2}{\lambda T} \qquad (26\text{-}12)$$

The value of σ obtained by integrating Eq. (26-12) is $\dfrac{C_1}{15}\left(\dfrac{\pi}{C_2}\right)^4$. If the values of C_1 and C_2 given above are substituted in the expression and the answer converted to engineering units, we obtain a numerical value for σ of 0.1714×10^{-8} Btu/(h)(ft^2)(°R)^4. According to Jakob,[1] this value is about $1\frac{1}{2}$ percent smaller than the best experimental measurements.

Radiant Heat Exchange between Black Surfaces

If two surfaces at different temperatures are arranged so that radiant energy can be exchanged between them, a net flow of heat will occur from the hotter to the colder surface. Perhaps the simplest system to consider is one in which the surfaces are flat, parallel planes of infinite length and breadth. Most systems, however, do not have simple geometry, and the net heat-flow rate depends on the spatial arrangement of the two (or more) surfaces. To simplify the discussion, we shall assume in this section that we are dealing only with black surfaces, so that all the energy incident upon any surface is absorbed. We consider a system in which radiant energy is exchanged between two elements, 1 and 2, with areas dA_1 and dA_2 shown in Fig. 26-4. The line joining these elements has a length r, and the angles between this line and normals to the two surfaces are β_1 and β_2. The rate of transfer of radiant heat from 1 to 2 is proportional to the product of the apparent areas of the elements and

[1] M. Jakob, "Heat Transfer," vol. 1, p. 39, John Wiley & Sons, Inc., New York, 1949.

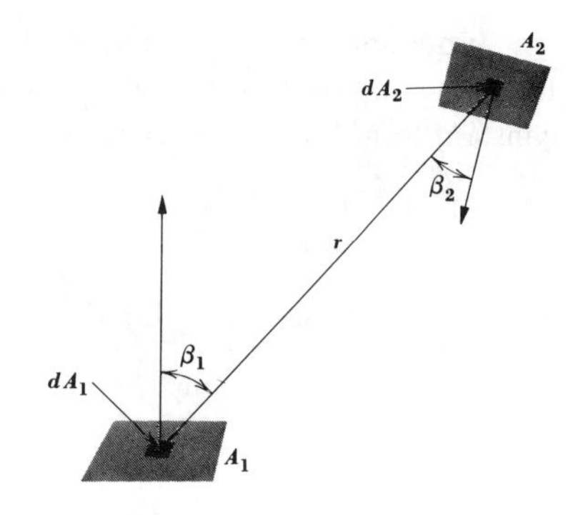

FIGURE 26-4
Radiant heat exchange between two area elements.

inversely proportional to the square of the distance between them. The apparent areas, i.e., the areas normal to a beam of radiation, are $dA_1 \cos \beta_1$ and $dA_2 \cos \beta_2$, so that the rate of radiation from 1 to 2 is

$$dq_{1\to2} = \frac{I_1\, dA_1 \cos \beta_1\, dA_2 \cos \beta_2}{r^2} \tag{26-13}$$

The quantity I_1 is a proportionality constant called the *intensity of radiation* from surface 1. We use the following procedure to relate it to the total emissive power of surface 1, which we have seen to be $\sigma T_1^4\, dA_1$. The quantity $dA_2 \cos \beta_2/r^2$ is the differential solid angle subtended from the center of 1 to the boundaries of 2 and is written $d\omega$. We integrate[1] Eq. (26-13) over the entire hemispherical angle 2π.

$$dq_{1\to2} = I_1\, dA_1 \int_0^{2\pi} \cos \beta_1\, d\omega$$

$$= \pi I_1\, dA_1$$

[1] The integration may be illustrated with reference to the accompanying diagram, Fig. 26-5. The apparent area of dA_2, which is the differential area normal to the beam, is $dA_2 \cos \beta_2$. It is shown here as a portion of the surface of a hemisphere which has dA_1 at the center. The circumference of the segment is $2\pi r \sin \beta_1$, and the width is $r\, d\beta_1$. The normal area of the circumferential segment, therefore, is $2\pi r^2 \sin \beta_1\, d\beta_1$. We integrate, starting with the general form, Eq. (26-13).

$$dq_{1\to2} = I_1\, dA_1 \int_0^{\pi/2} \frac{\cos \beta_1 \cos \beta_2\, dA_2}{r^2}$$

We substitute in this for $\cos \beta_2\, dA_2$ and get

$$dq_{1\to2} = I_1\, dA_1 \int_0^{\pi/2} \frac{(\cos \beta_1)(2\pi r^2 \sin \beta_1\, d\beta_1)}{r^2}$$

$$= 2\pi I_1\, dA_1 \int_0^{\pi/2} \cos \beta_1 \sin \beta_1\, d\beta_1$$

$$= \pi I_1\, dA_1$$

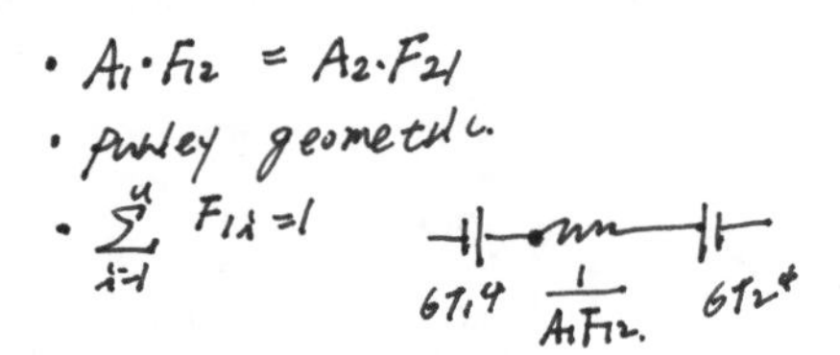

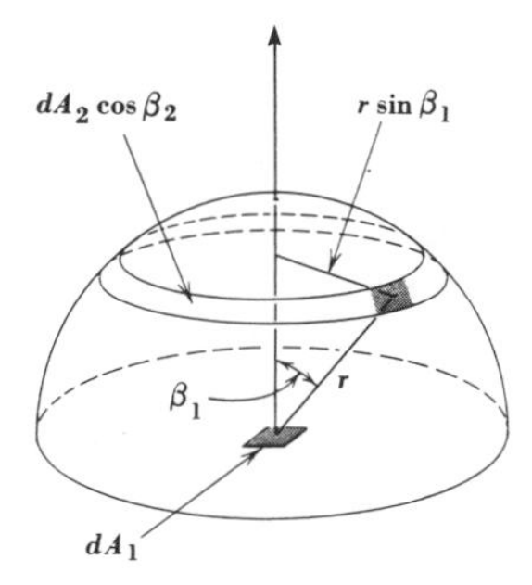

FIGURE 26-5
Hemispherical model for determining intensity of radiation.

By equating this quantity with the result from the Stefan-Boltzmann equation we see that the intensity I_1 of radiation from surface 1 is $\sigma T_1{}^4/\pi$.

Thus we rewrite Eq. (26-13) as

$$dq_{1\to 2} = \frac{\sigma T_1{}^4 \, dA_1 \cos \beta_1 \, dA_2 \cos \beta_2}{\pi r^2} \qquad (26\text{-}14)$$

A consideration of radiant-heat flow from surface 2 to surface 1 leads to a similar expression.

$$dq_{2\to 1} = \frac{\sigma T_2{}^4 \, dA_1 \cos \beta_1 \, dA_2 \cos \beta_2}{\pi r^2} \qquad (26\text{-}15)$$

The net flow of heat is found by taking the difference

$$dq_{12} = \frac{\sigma \, dA_1 \cos \beta_1 \, dA_2 \cos \beta_2}{\pi r^2}(T_1{}^4 - T_2{}^4) \qquad (26\text{-}16)$$

If the two surfaces have finite areas, A_1 and A_2, Eqs. (26-14) and (26-15) are evaluated by performing the indicated double integrations in each case.

In engineering it is convenient to use simpler formulas to represent heat flow, so the integrated form of Eq. (26-14) is represented as

$$q_{1\to 2} = \sigma A_1 F_{12} T_1{}^4 \qquad (26\text{-}17)$$

and Eq. (26-15) as

$$q_{2\to 1} = \sigma A_2 F_{21} T_2{}^4 \qquad (26\text{-}18)$$

The quantity F_{12} designates the fraction of the total radiation leaving surface 1 which strikes surface 2, and the quantity F_{21} represents the fractional impingement of radiation from 2 on 1. F_{12} and F_{21} are known as *view factors*. If the two surfaces are at the same temperature, $q_{1\to 2}$ and $q_{2\to 1}$ must be equal, from which it follows that $A_1 F_{12} = A_2 F_{21}$. Therefore we can write either of the following

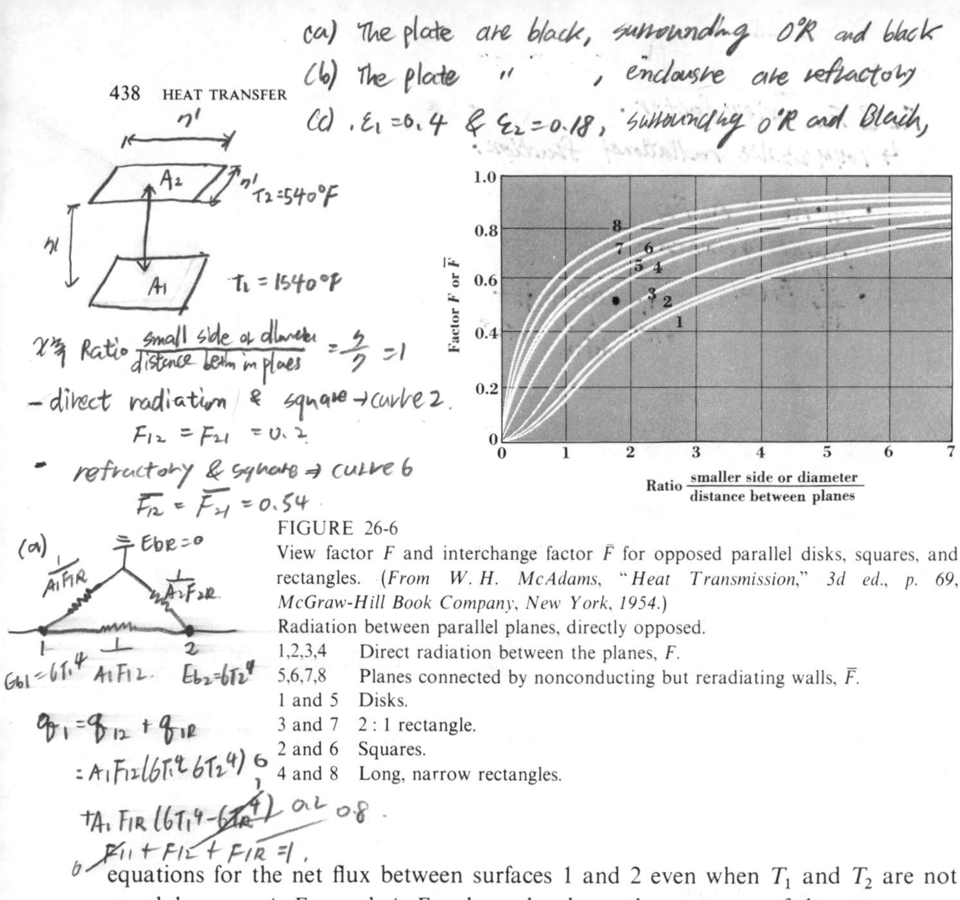

FIGURE 26-6
View factor F and interchange factor $\bar{F}$ for opposed parallel disks, squares, and rectangles. (*From W. H. McAdams, "Heat Transmission," 3d ed., p. 69, McGraw-Hill Book Company, New York, 1954.*)
Radiation between parallel planes, directly opposed.

1,2,3,4	Direct radiation between the planes, F.
5,6,7,8	Planes connected by nonconducting but reradiating walls, $\bar{F}$.
1 and 5	Disks.
3 and 7	2 : 1 rectangle.
2 and 6	Squares.
4 and 8	Long, narrow rectangles.

equations for the net flux between surfaces 1 and 2 even when T_1 and T_2 are not equal, because $A_1 F_{12}$ and $A_2 F_{21}$ depend only on the geometry of the system.

$$q_{12} = \sigma A_1 F_{12}(T_1{}^4 - T_2{}^4) \tag{26-19}$$

$$q_{12} = \sigma A_2 F_{21}(T_1{}^4 - T_2{}^4) \tag{26-20}$$

Although the equality of $A_1 F_{12}$ and $A_2 F_{21}$ was established considering black surfaces, it is equally valid for nonblack surfaces.

The view factor F_{12} can be evaluated by combining Eq. (26-17) with the integrated form of Eq. (26-14). By this procedure it is found that

$$F_{12} = \frac{1}{A_1} \int_{A_2} \int_{A_1} \frac{dA_1 \cos \beta_1 \, dA_2 \cos \beta_2}{\pi r^2} \tag{26-21}$$

Values of the view factor have been calculated for a number of spatial arrangements shown in Fig. 26-6. Other values are given in Perry, sec. 10. The method of determining these factors is illustrated in Example 26-2.

Example 26-2 Determine the view factor for radiant-heat transfer between a small disk of area A_1 and a parallel large disk of area A_2. The disks are assumed to be directly opposed; i.e., a line joining their centers is normal to

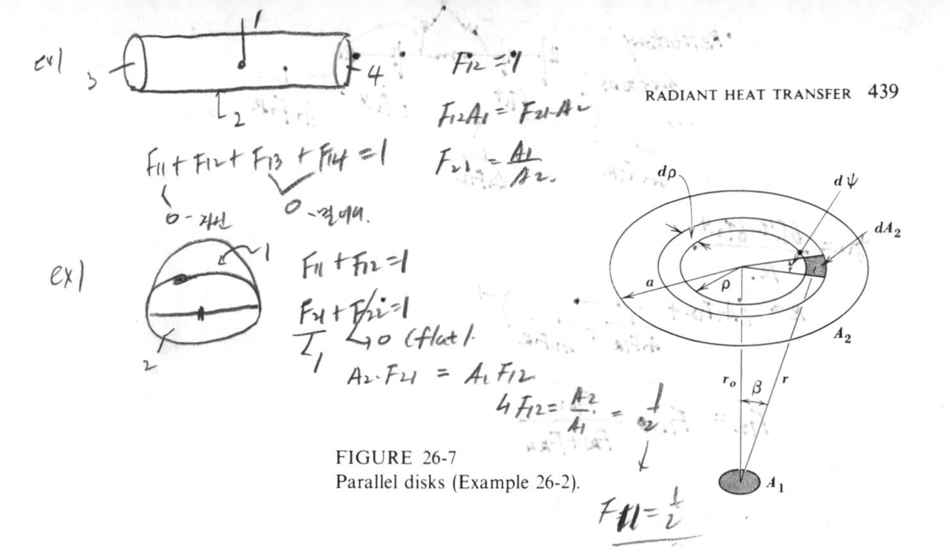

FIGURE 26-7
Parallel disks (Example 26-2).

both disks. The large disk has a radius a, and the distance between centers is r_o. This system is shown in Fig. 26-7. Equation (26-21) is

$$F_{12} = \frac{1}{A_1} \int_{A_2} \int_{A_1} \frac{dA_1 \cos \beta_1 \, dA_2 \cos \beta_2}{\pi r^2}$$

In this problem A_1 is small compared with A_2, so for the integration over A_1, the other quantities under the integral sign can be considered constant. In addition, the angles β_1 and β_2 are equal; they will be designated simply as β. Therefore the equation for the view factor reduces to

$$F_{12} = \int_{A_2} \frac{\cos^2 \beta \, dA_2}{\pi r^2}$$

The differential element dA_2 is shown in the figure. The area of the element is

$$dA_2 = \rho \, d\psi \, d\rho$$

It can also be seen from the construction that $\cos \beta = r_o/r$. Thus the view factor may be written as

$$F_{12} = \int_0^a \int_0^{2\pi} \frac{r_o^2}{\pi r^4} \rho \, d\psi \, d\rho$$

The system is symmetrical, so the first integration produces the result

$$F_{12} = \int_0^a \frac{2r_o^2}{r^4} \rho \, d\rho$$

The quantity r is a function of ρ as follows:

$$r^2 = \rho^2 + r_o^2$$

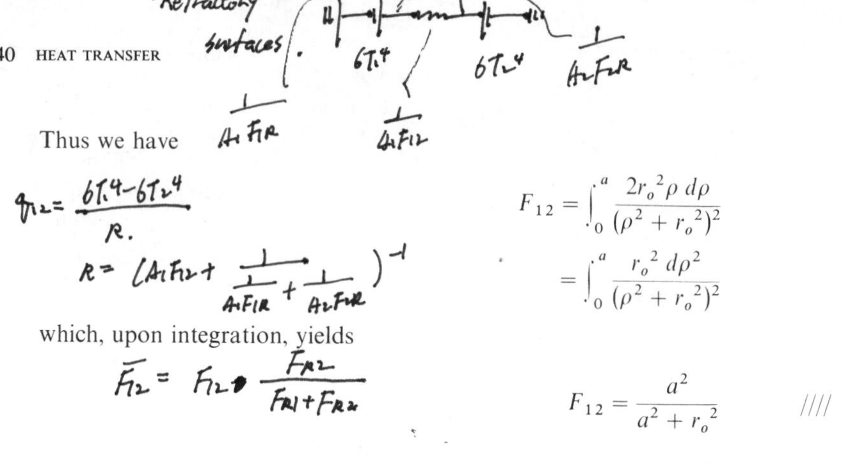

Thus we have

$$F_{12} = \int_0^a \frac{2r_o^2 \rho \, d\rho}{(\rho^2 + r_o^2)^2}$$

$$= \int_0^a \frac{r_o^2 \, d\rho^2}{(\rho^2 + r_o^2)^2}$$

which, upon integration, yields

$$F_{12} = \frac{a^2}{a^2 + r_o^2} \qquad ////$$

Multiple surfaces In a system enclosed by i surfaces, it follows that

$$F_{11} + F_{12} + F_{13} + \cdots + F_{1i} = 1 \qquad (26\text{-}22)$$

If surface 1 cannot "see" itself, $F_{11} = 0$. If the surfaces are all black, the net heat flow between surface 1 and all other surfaces is

$$q_{1,\,\text{net}} = \sigma A_1 F_{12} T_1^4 - \sigma A_2 F_{21} T_2^4 + \sigma A_1 F_{13} T_1^4 - \sigma A_3 F_{31} T_3^4 + \cdots \qquad (26\text{-}23)$$

The terms containing $\sigma A_1 T_1^4$ can be collected and expressed as a single term using the summation expressed in Eq. (26-22). Equation (26-23) can then be written as

$$q_{1,\,\text{net}} = \sigma A_1 T_1^4 - \sigma(A_2 F_{21} T_2^4 + A_3 F_{31} T_3^4 + \cdots + A_i F_{i1} T_i^4) \qquad (26\text{-}24)$$

Refractory surfaces In many heat-transfer systems the high- and low-temperature surfaces which serve as "source" and "sink" for radiant heat are connected by a wall of some refractory material through which only a negligible amount of heat flows. Radiant energy is absorbed by these refractory surfaces, but at steady operation it is emitted at the same rate if there is no heat loss through the refractory. Hence there is no net heat flux associated with the radiant interchange at the refractory surfaces.

If we consider two black surfaces 1 and 2 connected by a refractory surface R, the heat transfer from surface 1 to surface 2 must include not only the direct transfer represented by the view factor F_{12}, but also the energy from surface 1 which is absorbed by the refractory surface and reradiated to surface 2. A factor $\bar{F}_{12}$, called the *interchange* factor, is defined to represent the sum of these two effects. The refractory surface radiates energy[1] to both surfaces 1 and 2 (F_{R1} and F_{R2}) and also to itself, if it can see itself (F_{RR}). However, all the energy radiated to itself is ultimately reradiated to surface 1 or surface 2, and the fraction reradiated to surface 2 is

[1] It is assumed that this reradiation (plus any reflection) is equal in all directions, or "diffuse," and is therefore independent of the direction from which it arrived.

$F_{R2}/(F_{R1} + F_{R2})$. The interchange factor, therefore, is the sum of the fraction radiated directly plus the fraction reradiated.

$$\bar{F}_{12} = F_{12} + F_{1R}\frac{F_{R2}}{F_{R1} + F_{R2}} \tag{26-25}$$

If both sides of Eq. (26-25) are multiplied by A_1 and the right side rearranged, we obtain

$$A_1\bar{F}_{12} = A_1F_{12} + \frac{1}{F_{R1}/A_1F_{1R}F_{R2} + 1/A_1F_{1R}}$$

For any two surfaces, even if one is a refractory, the relations $A_1F_{1R} = A_RF_{R1}$ and $A_RF_{R2} = A_2F_{2R}$ are valid, since they depend only on spatial considerations. We then use these two relations to obtain

$$A_1\bar{F}_{12} = A_1F_{12} + \frac{1}{1/A_2F_{2R} + 1/A_1F_{1R}} \tag{26-26}$$

Therefore, from a knowledge of A_1, A_2, and the view factors F_{12}, F_{1R}, and F_{2R}, the interchange factor $\bar{F}_{12}$ can be calculated. The interchange factors for a number of geometrical arrangements are shown in Fig. 26-6.

The net radiant exchange between two surfaces which are connected by refractory reradiating surfaces is given by the equation

$$q_{12} = \sigma A_1\bar{F}_{12}T_1^4 - \sigma A_2\bar{F}_{21}T_2^4 \tag{26-27}$$

If this equation is applied to two surfaces at the same temperature, it follows that the quantities $A_1\bar{F}_{12}$ and $A_2\bar{F}_{21}$ must be equal. Since they depend entirely on geometric considerations, it follows that they are equal when not at the same temperature. Thus we rewrite Eq. (26-27) as

$$q_{12} = \sigma A_1\bar{F}_{12}(T_1^4 - T_2^4) \tag{26-28}$$

Gray Surfaces

If the surfaces from which there is a net heat flux are not black bodies, the emissive power of each of the surfaces is determined by using the Stefan-Boltzmann equation multiplied by the emissivity. Furthermore, the absorption of energy on each surface is not complete, but equals the amount of energy incident on the surface multiplied by the absorptivity. These two factors make an accurate analysis of radiant transfer between real surfaces much more complicated than between black bodies. To simplify the problem, it is often assumed that the surfaces are gray. As mentioned earlier, this assumption is equivalent to assuming that absorptivity is independent of

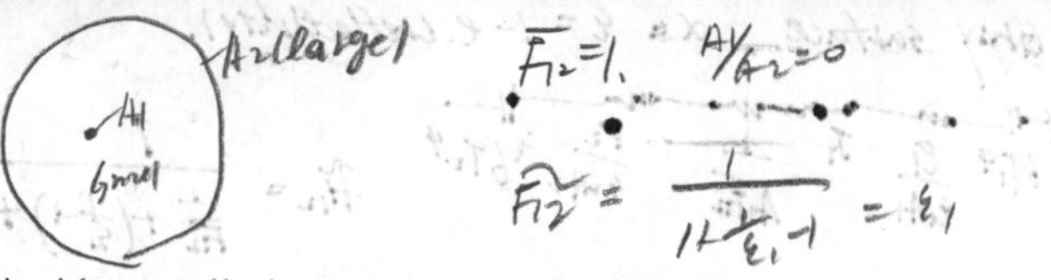

the wavelength of the incident radiation, and thus of the temperature and other characteristics of the emitter. In effect, the emissivity and absorptivity of a surface are assumed to be the same. Hottel[1] has shown that for the case of two surfaces 1 and 2, between which there is net heat flux, connected by any number of refractory reradiating zones, the net flux can be represented by the equation

$$q_{12} = \sigma A_1 \mathscr{F}_{12}(T_1{}^4 - T_2{}^4) \qquad (26\text{-}29)$$

in which the factor $\mathscr{F}_{12}$ is

$$\mathscr{F}_{12} = \frac{1}{(1/\bar{F}_{12}) + [(1/\varepsilon_1) - 1] + (A_1/A_2)[(1/\varepsilon_2) - 1]} \qquad (26\text{-}30)$$

The derivation of this equation is rather complex; for details, the reader is referred to the work of Hottel.[1,2] Equation (26-30) is readily applied to a number of simple but useful systems. For the case of a small body 1 in a large enclosure 2, e.g., a pipe in a room, $\bar{F}_{12} = 1$ and $A_1/A_2 = 0$. Therefore $\mathscr{F}_{12} = \varepsilon_1$. When radiant energy is exchanged between two large parallel planes, the interchange factor $\bar{F}_{12}$ is unity and the areas are equal, so we get

$$\mathscr{F}_{12} = \frac{1}{1/\varepsilon_1 + 1/\varepsilon_2 - 1} \qquad (26\text{-}31)$$

These two systems can be analyzed directly without the aid of Eq. (26-30); the same results as shown above are obtained.

The Radiation Heat-Transfer Coefficient

Unless radiation from a surface occurs in a vacuum, it is usually accompanied by convective heat transfer. If the radiating surfaces are at uniform temperatures, we calculate the heat transfer by convection by the usual methods. The heat transfer by radiation is computed from the Stefan-Boltzmann equation with the appropriate coefficients. The total rate of heat transfer is the sum of the contributions of the two mechanisms. However, in dealing with both mechanisms simultaneously, it is often convenient to define a heat-transfer coefficient for radiation in the manner of a convective coefficient. This matter has already been discussed in Chap. 21, and a coefficient defined by the following equation:

$$q = h_r A(t_{s1} - t_{s2}) \qquad (21\text{-}11)$$

[1] H. C. Hottel, chap. 4 in W. H. McAdams, "Heat Transmission," 3d ed., McGraw-Hill Book Company, New York, 1954.

[2] H. C. Hottel and A. F. Sarofim, "Radiative Transfer," McGraw-Hill Book Company, New York, 1967.

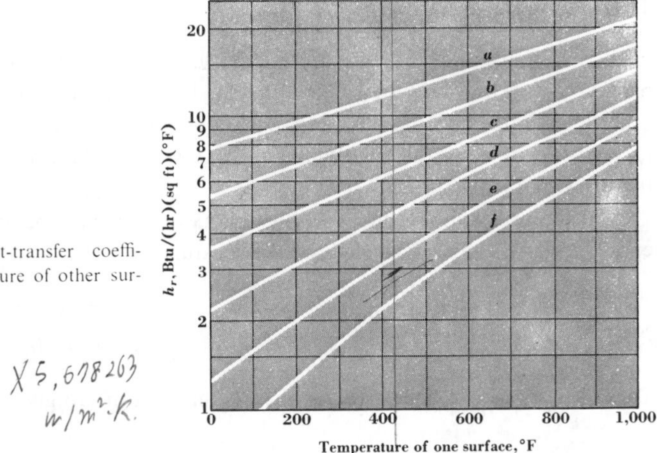

FIGURE 26-8
The radiant-heat-transfer coefficient. Temperature of other surface:
a 1000°F.
b 800°F.
c 600°F.
d 400°F.
e 200°F.
f 0°F.

The temperatures t_{s1} and t_{s2} are the surface temperatures of the source and sink, respectively. To obtain an expression for h_r, we write Eq. (21-11) in terms of the absolute temperatures of the surfaces:

$$q = h_r A(T_1 - T_2) \qquad (26\text{-}32)$$

The equation for radiant-heat flux between black surfaces with a view factor of 1 is

$$q = \sigma A(T_1{}^4 - T_2{}^4) \qquad (26\text{-}33)$$

By combining Eqs. (26-32) and (26-33), we obtain the definition of h_r:

$$h_r = \sigma \frac{T_1{}^4 - T_2{}^4}{T_1 - T_2} \qquad (26\text{-}34)$$

The strong dependence of h_r on temperature is seen in this equation. Nevertheless, the use of h_r is as rigorous as the use of the Stefan-Boltzmann equation, provided this variation in h_r is taken into account. Values of h_r calculated from Eq. (26-34) are shown in Fig. 26-8. The ratio in Eq. (26-34), if divided out, gives the definition of h_r as

$$h_r = \sigma(T_1{}^3 + T_1{}^2 T_2 + T_1 T_2{}^2 + T_2{}^3) \qquad (26\text{-}35)$$

It is evidently of no importance in the determination of h_r which subscript, 1 or 2, represents the source and the sink.

In radiation between gray surfaces or between surfaces with view factors other than 1, it is necessary to modify Eq. (26-32) by multiplying the right-hand side by $\mathscr{F}_{12}$, just as would be done with Eq. (26-33).

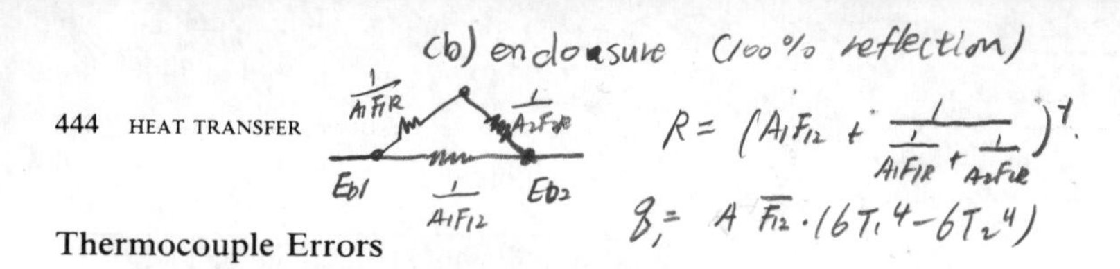

Thermocouple Errors

A thermocouple used to measure a fluid temperature in a container may give a reading significantly different from the fluid temperature if the walls are at some other temperature. This difference occurs because the thermocouple tip maintains its temperature constant by exchanging heat with the fluid by convection and with the walls by radiation.

If the wall temperature t_s is higher than the gas temperature t_g, the thermocouple indicates some intermediate temperature t_c. The heat-flow equations are

$$\begin{aligned} q &= h_r A_c \mathscr{F}_{cs}(t_s - t_c) \\ &= h A_c (t_c - t_g) \end{aligned}$$

These equations can be solved for the gas temperature to give

$$t_g = t_c - \frac{h_r \mathscr{F}_{cs}}{h}(t_s - t_c) \tag{26-36}$$

A calculation showing the magnitude of this effect is given in Example 26-3.

Example 26-3 A glass thermometer ($\varepsilon = 0.96$) is inserted in a large circular duct at right angles to the wall. The thermometer bulb, 0.65 cm in diameter, is at the center of the duct. Air flows through the duct at a rate such that the convective heat-transfer coefficient between the thermometer and the air is 100 $W/m^2 \cdot K$.

(*a*) If the walls of the duct are at 425°C and the thermometer reads 150°C, what is the temperature of the air in the duct?

(*b*) If silver foil ($\varepsilon = 0.03$) is wrapped around the thermometer, what will be the new thermometer reading?

(*a*) The thermometer can see only the duct, so its view factor is unity. The area of the thermometer is so much less than the area of the duct that $(A_c/A_s)(1/\varepsilon_s - 1)$ is negligible compared with the other terms in the denominator of Eq. (26-30). Hence, for this problem, Eq. (26-30) becomes

$$\mathscr{F}_{cs} = \frac{1}{\frac{1}{1} + (1/\varepsilon_c - 1) + 0} = \varepsilon_c = 0.96$$

The radiation heat-transfer coefficient is found using Fig. 26-8 to be 43 $W/m^2 \cdot K$. The gas temperature is found using Eq. (26-36).

$$t_g = 150 - \frac{(43)(0.96)(425 - 150)}{100} = 36°\text{C}$$

(*b*) The thermometer reading is now found by trial and error from Eq. (26-36).

$$36 = t_c - \frac{(h_r)(0.03)}{100}(425 - t_c)$$

Assume $t_c = 40°C$; using Fig. 26-8, $h_r = 34$ W m$^2 \cdot$ K.

$$36 = 40 - \frac{(34)(0.03)(425 - 40)}{100}$$

$$= 40 - 3.9$$

$$= 36$$

It is evident that the thermometer would read approximately 40°C.

Example 26-3 shows that the error in a temperature reading caused by radiation can be reduced by covering the surface of the temperature-sensing device with some material which has a low emissivity and absorptivity. The error can also be reduced by enclosing the device in a shield such as a hollow tube. The gas is free to pass between the thermocouple and the tube, but the thermocouple can see little more than the shield. In these circumstances the shield exchanges heat primarily with the gas and the duct and is at some intermediate temperature. The thermocouple, however, exchanges heat with the shield and the gas and is at some temperature closer to the gas temperature than it would have been if the shield were not used.

Two other factors may affect the reading of a thermocouple in low-velocity gas systems. If the gas is air, it neither emits nor absorbs important amounts of radiation itself, but has merely the convective effect already discussed. However, if the gas contains appreciable quantities of triatomic or higher polyatomic molecules, these constituents absorb and emit significant amounts of radiation. This radiant transfer with the gas affects the thermocouple reading but causes the thermocouple to give a more nearly correct reading. A fourth mechanism results from the fact that the thermocouple is probably at the wall temperature at the point where it emerges from the duct. The resulting gradient between the temperature at the wall and that at the tip affects the heat balance on the tip and may give rise to significant errors if the thermocouple is short.

Radiation from Gases

Although solids and liquids emit radiation over a continuous spectrum, gases do not. In fact, most monatomic and diatomic gases, like those contained in air, emit scarcely any radiation. Other gases such as water vapor and carbon dioxide emit radiation only within certain spectral bands. Only gases with a dipole moment triatomic and higher polyatomic (most gases) emit radiation in significant amounts. The gases which radiate energy also absorb it, but only within the same bands in which it is emitted. Gases which do not emit radiation do not absorb it.

The width of the emission (or absorption) bands for a particular gas depends on both temperature and pressure. Furthermore, the intensity of emission at any one set of conditions varies with wavelength over the bandwidth. An approximate representation for water vapor gives bands from 2.2 to 3.3, 4.8 to 8.5, and 12 to 25 μm. For carbon dioxide the bands are from 2.4 to 3.0, 4.0 to 4.8, and 12.5 to 16.5 μm. The higher-wavelength bands are usually not of great significance in absorption because only a small fraction of the total thermal radiation from solids falls in these ranges. Visible light (0.35 to 0.78 μm) is not absorbed by either water vapor or carbon dioxide.

The absorption of radiant energy by a particular gas depends on the number of molecules in the path of the beam. The extent of absorption in a gaseous medium can be increased by increasing the length of the path or by increasing the partial pressure of the absorbing constituent. Let us call $I_{\lambda L}$ the intensity of radiation at a particular wavelength after having passed through a thickness L of gas, and $I_{\lambda o}$ the intensity of the radiation at the same wavelength before it enters the gas. These intensities are related by the exponential equation

$$I_{\lambda L} = I_{\lambda o} e^{-m_\lambda L} \qquad (26\text{-}37)$$

The quantity m_λ is a characteristic of the particular gas, its partial pressure, and the wavelength of the radiation considered. It can be seen from Eq. (26-37) that if a gas layer is infinitely thick, the absorption will be complete at wavelengths for which m_λ is greater than zero, i.e., within the bands characteristic of that gas. Since m_λ is a function of wavelength, the absorptivity of a finite thickness of gas will vary with λ.

The rigorous treatment of problems of gas radiation is complex even when the geometrical shapes are simple. Consequently, the total emission or absorption of radiation associated with a gaseous mass of finite dimensions is usually analyzed by the approximate methods of Hottel,[1] which yield results of engineering value. Hottel has determined for a number of gases what is known as the emissivity as a function of temperature and pressure by considering a hemispherical mass of gas of radius L radiating to the midpoint of the base of the hemisphere. From the midpoint the beam length is the same in all directions. The gas emissivity ε_g is defined as the ratio of the energy radiated from the hemispherical body of gas to an element of surface at the midpoint divided by the energy radiated to the same element by a black hemispherical surface at the gas temperature. The rate of radiation from the gas is therefore $\sigma \varepsilon_g T_g^4$ per unit area of receiving surface. If the

[1] Ibid.

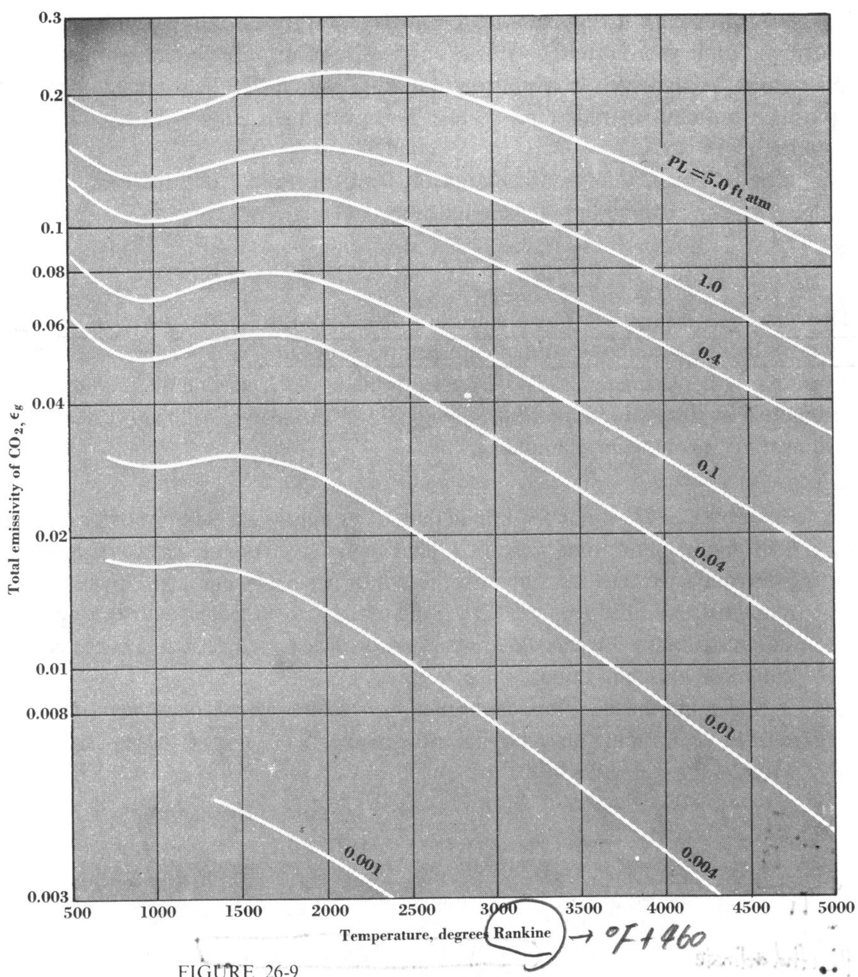

FIGURE 26-9
Total emissivity of carbon dioxide. (*From W. H. McAdams, "Heat Transmission," 3d ed., p. 83, McGraw-Hill Book Company, New York, 1954.*)

surface element at the midpoint is radiating heat back to the gas, the absorption rate by the gas will be $\sigma\alpha_g T_1{}^4$, in which α_g is the absorptivity of the gas for black-body radiation from a surface at T_1. The net rate of radiant heat transfer is the difference, $\sigma(\varepsilon_g T_g{}^4 - \alpha_g T_1{}^4)$. Values of the emissivity are plotted in a series of graphs, such as Fig. 26-9, for carbon dioxide in a system at a total pressure of 1 atm. Absorptivities are obtained using the same charts.

The parameter in Fig. 26-9 is PL, in which P is the partial pressure of carbon dioxide in atmospheres and L is the beam length in feet. Shapes other than hemispheres can be analyzed by employing the beam length of a hemisphere giving an equivalent amount of radiation. Table 26-3 gives these lengths for a number of systems.

The radiant transfer between a gas at temperature T_g and a black surface of finite area A at temperature T_1 is found from the equation

$$q = \sigma A(\varepsilon_g T_g{}^4 - \alpha_g T_1{}^4) \qquad (26\text{-}38)$$

The emissivity of the gas ε_g is evaluated at the temperature of the gas. The absorptivity of the gas is evaluated at the temperature of the surface. As empirical corrections to α_g, the parameter PLT_1/T_g is used in place of PL in finding a value on the ordinate axis; this value is multiplied by $(T_g/T_1)^{0.65}$ to give α_g. The procedures are outlined in the following example.

Example 26-4 A stack 3 ft in diameter contains a gas with 5 percent CO_2 at 2000°F and 1 atm total pressure. The flow of gas past the surface of the stack is at rate such that the convective heat-transfer coefficient from the gas to the refractory is 1.5 Btu/(h)(ft^2)(°F). If the refractory surface is at 1900°F and has an emissivity of 1.0, calculate the amounts of heat transferred from the gas by radiation and convection.

To find the heat flux due to radiation we first find the parameter PL to be used in determining absorptivity and emissivity. The mean beam length L is found using Table 26-3.

$$PL = (0.05)(0.90)(3)$$
$$= 0.135$$

Table 26-3 EQUIVALENT BEAM LENGTHS FOR GAS RADIATION†

Shape	Equivalent beam length L
Sphere	0.60 × diameter
Cylinder of infinite length	0.90 × diameter
Space between infinite parallel planes	2.0 × distance between planes
Space outside an infinite bank of tubes with centers on equilateral triangles; tube diameter equals clearance	2.8 × clearance

† From H. C. Hottel, chap. 4 in W. H. McAdams, "Heat Transmission," 3d ed., McGraw-Hill Book Company, New York, 1954. See also H. C. Hottel and A. F. Sarofim, "Radiative Transfer," chap. 7, McGraw-Hill Book Company, New York, 1967.

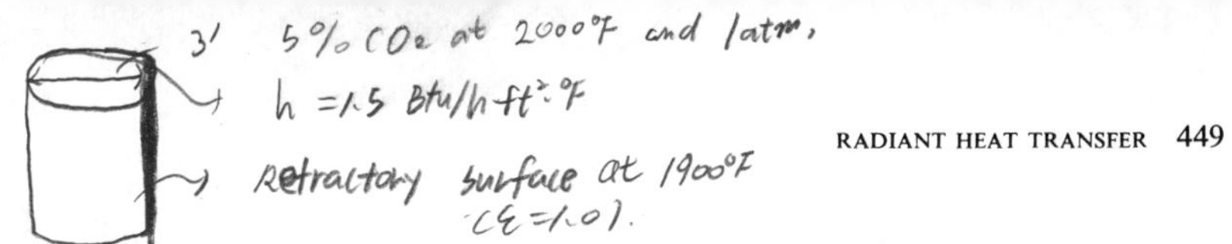

The gas temperature T_g is 2460°R. Therefore we find, from Fig. 26-9, that the gas emissivity ε_g is 0.07.

The parameter for finding gas absorptivity is

$$\frac{PLT_1}{T_g} = 0.135\left(\tfrac{2360}{2460}\right)$$

$$= 0.130$$

The ordinate value on Fig. 26-9 at this value of the parameter and at $T = 2360$ is 0.072, which we use in calculating the value of α_g.

$$\alpha_g = 0.072\left(\tfrac{2460}{2360}\right)^{0.65}$$

$$= 0.074$$

The net radiant transfer of energy is

$$\frac{q}{A} = 0.171\left[0.070\left(\tfrac{2460}{100}\right)^4 - 0.074\left(\tfrac{2360}{100}\right)^4\right]$$

$$= 0.171(25{,}700 - 22{,}900)$$

$$= 480 \text{ Btu/(h)(ft}^2\text{)}$$

The heat flux by convection is

$$\frac{q}{A} = (1.5)(100)$$

$$= 150 \text{ Btu/(h)(ft}^2\text{)} \qquad ////$$

If the enclosure is not black, some of the radiation striking it is reflected back into the gas and to other walls. It has been found that if the emissivity of the enclosure is greater than 0.7, it is usually sufficiently accurate to use an effective emissivity ε', which is equal to $(\varepsilon + 1)/2$, in which ε is the actual emissivity of the enclosure. Equation (26-38) is therefore modified empirically to give

$$q = \sigma A \varepsilon'(\varepsilon_g T_g{}^4 - \alpha_g T_1{}^4) \qquad (26\text{-}39)$$

For systems containing water vapor, the methods are similar to those outlined above for carbon dioxide. For mixtures of water vapor and carbon dioxide, there is a slight overlap of the emission bands, so the emissivities are not additive. The reader is referred to the works of Hottel mentioned above for methods of solving problems involving water vapor, mixtures of emitting gases, and gases containing suspended luminous particles.

Solar Radiation

Solar radiation is responsible for the maintenance of the surface temperature on the earth. The solar constant of 442 Btu/(h)(ft^2) represents the rate at which solar energy reaches the earth's atmosphere per square foot of area normal to the beam. If this rate is multiplied by the normal receiving area of the earth, $\pi D^2/4$, in which D is the earth's diameter, we obtain the total rate of receipt of solar energy. This rate may be equated to $\sigma \pi D^2 T_m{}^4$, which represents the total emission rate from the earth at some mean surface temperature T_m, assuming an emissivity of 1.0. This calculation gives an approximate mean surface temperature of 504°R, or 44°F. The rate of conduction from the earth's core as mentioned in Chap. 16 is approximately 0.01 Btu/(h)(ft^2) and is obviously negligible compared with the radiant flux.

We are made aware of the effects of radiation in the formation of frost. When the sun is down, some thermal radiation is still received on the earth's surface from space, but the amount is small. Consequently, there is a net loss of radiant energy from the ground at night. This loss may be offset by convection from the earth's atmosphere, but if the air is very still, the surface temperature may drop below 32°F. Thus frost will form, even though the air temperature is above freezing. It is well known that frost is avoided in such circumstances either by a slight breeze or by cloud cover. Loss by radiation is sometimes reduced artificially by the creation of smoke clouds; the prevention of freezing is effected more by the reduction of radiant-energy loss and by convective mixing than by the heat of combustion released in forming the smoke. The success of motor-driven propellers in preventing frost in citrus groves indicates that the effects of mixing may be the more significant of these two mechanisms.

It is evident that the energy reaching the surface of the earth is somewhat depleted, from having passed through the atmosphere—a thickness of roughly 90 mi of air, dust, water vapor, and carbon dioxide. The dust, water vapor, and carbon dioxide are all capable of absorbing radiant energy, and some of this energy is reradiated directly back into space instead of to the earth. However, the absorptive effect of the gases is diminished by the fact that most of the solar radiation, emitted at 10,000°F, is at wavelengths below the absorptive bands of carbon dioxide and water, so that the gases of the atmosphere are nearly transparent to solar radiation. Radiation emitted from the earth is affected, however. The emission temperatures for this terrestrial radiation are much lower than the sun's temperature, so terrestrial radiation is emitted at higher wavelengths. These overlap the absorption bands of carbon dioxide and water vapor, so that much of the terrestrial radiation is absorbed by the atmosphere and reradiated to the earth. This phenomenon is referred to as a *greenhouse* effect, because the transmission of solar energy through glass and subsequent trapping by the glass of reradiation from within the greenhouse assist in keeping a greenhouse at the proper temperature in cold weather.

The atmospheric effects depend on the concentration of the absorbing gases in the atmosphere. It has been suggested that the increasing concentration of carbon dioxide in the atmosphere in the last two centuries as the result of combustion may upset the balance between rates at which radiation is received and emitted by the earth. Any such change would supposedly be a decrease in the rate at which energy is lost from the earth, with the result that the mean temperature of the earth's surface would rise.

It has been proposed that long-term cyclical variations in the earth's climate have been the result of fluctuations of carbon dioxide concentration in the atmosphere in past geologic eras. Periods of high atmospheric carbon dioxide concentration supposedly caused increased temperatures and melting of polar ice sheets which thus increased the volume of the oceans. This decreased the concentration of carbon dioxide in the oceans and resulted in increased absorption of carbon dioxide from the atmosphere. The subsequent decrease of carbon dioxide in the atmosphere then allowed greater radiation loss from the earth, subsequent cooling, reestablishment of the polar icecaps, and net loss of carbon dioxide from the ocean to the atmosphere as a result of the shrinking volume of sea water.

This cycle, which is periodically brought to public attention, is often oversimplified. Some of the complexity of the problem is indicated by the quantities shown in Table 26-4, most of which need to be included in any meaningful

Table 26-4 ESTIMATED CARBON DIOXIDE BALANCE ON EARTH'S ATMOSPHERE

Losses from atmosphere	
Absorption by oceans	100×10^9 tons/yr
To photosynthesis	60×10^9 tons/yr
Weathering of igneous rocks to carbonates	0.1×10^9 tons/yr
Gains by atmosphere	
Diffusion from oceans	100×10^9 tons/yr
Respiration and decay	60×10^9 tons/yr
Combustion in homes and factories	6×10^9 tons/yr
Release from soil of recently cleared farmland	2×10^9 tons/yr
Release by hot springs and volcanoes	0.1×10^9 tons/yr
Amounts presently held	
In oceans	$130{,}000 \times 10^9$ tons
In fossil fuels	$40{,}000 \times 10^9$ tons
In atmosphere	$2{,}300 \times 10^9$ tons

SOURCE: G. N. Plass, *Sci. Am.*, **201**:41 (1959)

analysis. The most important effect usually overlooked is that of gaseous transport to and from the oceans, which serves as a substantial damper on fluctuations of atmospheric carbon dioxide content.[1]

A phenomenon similar to the greenhouse effect is that of the presumably increasing amounts of particulate material suspended in the atmosphere. These materials, in contrast to carbon dioxide, absorb both incoming solar radiation and radiation from the earth and re-emit radiation in all directions. The net effect would thus be to prevent much solar radiation from ever reaching the earth. This presumably would have the result of lowering the mean temperature of the earth, in contrast to the alleged effect of the carbon dioxide.

Solar Heating

Although the direct use of solar radiation as a source of thermal energy is of considerable current interest, evidence of attempts to utilize solar energy in heating homes extends back at least 2500 years.[2] Studies of construction in ancient Greece show the use of techniques to maximize solar heating such as building the northern part of a house two stories and the southern part one story. Protruding eaves were used to block the sun from south-facing rooms in summer while letting winter sunlight, arriving from a lower angle in the sky, flow into the rooms. The Romans also left evidence of having taken solar energy into account in the design of buildings. More recently, a French scientist of the eighteenth century, Buffon, experimented with multiple-mirror solar furnaces, while others of that century, including Priestley and Lavoisier, used lenses to concentrate sunlight for the high temperatures needed in their experiments. Nineteenth century developments brought the first solar-powered steam engines and the first large-scale solar units for distilling fresh water from salt water. Extensive work has been done in solar experimentation and construction throughout the twentieth century, including the development and marketing of solar water heaters in Florida in the 1920s.

One way to utilize the sun's energy is to make direct use of the high-energy photons as in photosynthesis in plant leaves or by generation of electricity from photovoltaic cells. A simpler way is merely to absorb the radiation at a surface and use the resulting thermal energy for heat or power production. Considering only this second way, the great variety of collection devices and techniques can be divided into two categories:

1 Passive systems. These rely on transfer of heat by conduction, radiation, and natural convection.

[1] For scientific studies of various aspects of this problem, see papers by Eriksson, Möller, Lieth, Bolen and Keeling, and Kanwisher in *J. Geophys. Res.*, **68**:13 (1963).

[2] For an account of historical developments in the utilization of solar energy see P. M. Cheremisinoff and T. C. Regino, "Principles and Applications of Solar Energy," Ann Arbor Science Publishers, Ann Arbor, 1978.

2 *Active systems.* These utilize all the heat-transfer mechanisms of passive systems plus forced convection.

Passive systems are usually an integral part of a building and in effect represent the application of principles of heat transfer in the design of the building to minimize the cost of heating and cooling. Because these systems operate without pumping costs and generally have low maintenance costs, they are attractive economically if capital costs are not too high. However, because natural convection heat-transfer coefficients are usually of low magnitude, the heat-transfer rates associated with passive solar-heating systems are also likely to be low. Active systems have the advantage that the velocity of the heat-transfer fluid can be controlled independently of the heat flux so that much higher heat-transfer coefficients can be obtained and more efficient utilization of the entrapped solar energy can result.

Flat-plate collectors The commonest active device for receiving solar energy is the flat-plate collector. This is simply a flat surface (commonly 4 ft wide and 8 ft long in current commercial modules) which is heated by the sun on its upper surface and insulated on its lower surface. The heat absorbed by the plate is carried away by a fluid circulating through tubes which are either attached to the surface or are formed in the interior of the plate during its manufacture. Collectors made of plastic usually have tubes which are an integral part of the sheet while metallic collectors are generally made of aluminum sheet with copper tubes brazed, soldered, or mechanically fastened to the aluminum surface. The choice of metals is obviously based on their high thermal conductivities. Plastic collectors may cost only one-third as much as the metallic collectors but have the disadvantages of shorter life and lower rates of heat conduction.

Flat-plate collector performance is described by the equation

$$q = \alpha S - U(t_s - t_a) \qquad (26\text{-}39)$$

in which q is the net absorbed heat flux in Btu/(h)(ft^2), α is the absorptivity of the collector surface, S is the solar intensity or insolation in Btu/h per square foot of surface of the collector, U is the overall heat-transfer coefficient describing loss of heat from the surface to the surroundings, t_s is the temperature of the surface, and t_a is the ambient temperature. If Eq. (26-39) is divided by S, the quantity q/S defines the collection efficiency.

A detailed examination of Eq. (26-39) reveals some of the problems associated with solar collection. Although the solar constant is 442 Btu/(h)(ft^2), this represents the solar energy flux before entering the earth's atmosphere. The energy arriving on a square foot of the earth's surface, even in summer and at midday, may be only half the amount striking the earth's atmosphere. Many factors affect this quantity, including absorption by the atmosphere, additional absorption by atmospheric pollutants and cloud cover, the latitude of the collector location, day of the year, and time of day. The collector is generally not placed horizontally but preferably is mounted facing south at an angle from horizontal equal to 15° plus the latitude of the location. In this position the surface is generally considered to be in the optimum, time-average position for receiving solar radiation. Kreider and Kreith[1] give extensive tables showing values of the insolation on surfaces as a function of the variables described above. For example, on June 21 at 40° north latitude the solar insolation on a south-facing surface mounted at 50° to horizontal is quoted as 2 Btu/(h)(ft^2) at 5 A.M. and 7 P.M., 59 Btu/(h)(ft^2) at 7 A.M. and 5 P.M., 179 Btu/(h)(ft^2) at 9 A.M. and 3 P.M., and 263 Btu/(h)(ft^2) at noon. The total for the entire day is 1974 Btu per square foot of collector surface.

The absorptivity of the surface is an important quantity and it is desirable to have α as high as possible. Unfortunately, heat losses from the collector include losses by re-radiation and to minimize these it is desirable to keep the emissivity ε of the surface as low as possible. Although α and ε are the same in a system at thermal equilibrium, they are not the same for the surface of a solar collector. The incoming radiation, having been emitted from a much hotter source, will be at much lower wavelengths than the energy re-radiated from the collector surface. Much research has been done to find surface coatings with high absorptivities for radiation at wavelengths below 2 μm while also having low emissivities for radiation emitted at temperatures below 200°F. This is known as selectivity. For example, copper oxide coated on aluminum has been found to have values of $\alpha = 0.93$ and $\varepsilon = 0.11$ for these conditions. Unfortunately, coatings with good selectivity are still quite costly. Durability of the coating is also important in the sense that the selectivity must be retained for long periods of time.

The term $U(t_s - t_a)$ in Eq. (26-39) represents the sum of all upward heat losses from the collector. Losses from the back and the edges are usually negligible if the unit is properly insulated. Thus, the overall coefficient U is a composite term covering losses from the surface due both to re-radiation and convection. In an effort to minimize convective losses, some collectors are provided with permanent transparent coverings of glass or plastic using either a single sheet or parallel sheets. Usually the air gap between these sheets or between the sheet and the collector surface is about $\frac{1}{2}$ in to minimize natural convection. The collector covers can be highly effective in reducing losses but of course also add to the cost of the apparatus. The

[1] J. F. Kreider and F. Kreith, "Solar Heating and Cooling," McGraw-Hill Book Co., New York, 1975. See also F. Kreith and J. F. Kreider, "Principles of Solar Engineering," McGraw-Hill Book Co., New York, 1978.

covers also decrease the amount of solar radiation striking the actual surface of the collector. They help, on the other hand, by absorbing some of the energy radiated from the collector surface and re-radiating it back to the collector.

In actual operation, the surface temperature t_s of the collector varies with position. If water is flowing through parallel, closely spaced tubes from the bottom to the top of the collector, t_s will increase along the collector as the water temperature increases. At the water inlet end, the surface temperature can be found by equating the net heat absorbed at the surface of a differential segment of collector with the gain in heat content of the water flowing through that element. This involves a trial-and-error solution for t_s and also determination of the temperature of the water leaving the differential element. A series of iterations starting from the inlet and proceeding to the outlet will produce the entire temperature profile of the surface (which is of no great value) and also the exit temperature of the water. From this the total heat gain of the collector is readily determined. A simpler, more empirical procedure is to replace t_s in Eq. (26-39) with the inlet water temperature and multiply the right-hand side of the equation by an empirical coefficient known as the *heat-removal efficiency*. This coefficient is specific for each collector design and water flow rate.

Flat-plate solar collectors are now being produced commercially in a variety of designs. Unfortunately, there is more involved in the successful application of these collectors than the design factors discussed above. Solar radiation is not an intensive source of heat in comparison with more familiar heat sources and is highly variable with time of day, season, and latitude. The obvious advantage is that the source of the energy is free. However, if a large capital investment is required to collect this energy, the true cost of solar heating can be high. It is thought that if costs of conventional fuels continue to rise, a point will be reached at which solar collectors will become economical. However, this overlooks the fact that the manufacture of the collector modules requires substantial amounts of electricity (in the case of aluminum) or hydrocarbons (in the case of plastic materials) for their fabrication. Thus, as energy costs rise, materials costs and the costs of the collectors also rise. As a rule of thumb, the cost of a solar installation is considered to be due in almost equal parts to three factors: cost of the collector modules, cost of the piping and controls, and labor charges. Obviously, all three kinds of charges are tied to inflation and rise as energy costs rise. Any meaningful economic analysis of a proposed solar-collector system must consider the interest charges on the money to pay for the system, the cost of pumping the collector fluid, maintenance costs of the system, the lifetime that can be reasonably anticipated for the apparatus, and any tax incentives that may be available. Only by doing a quantitative evaluation of all these factors along with a heat-transfer analysis to determine fuel saving, can the economic feasibility of a solar-collector system be determined.

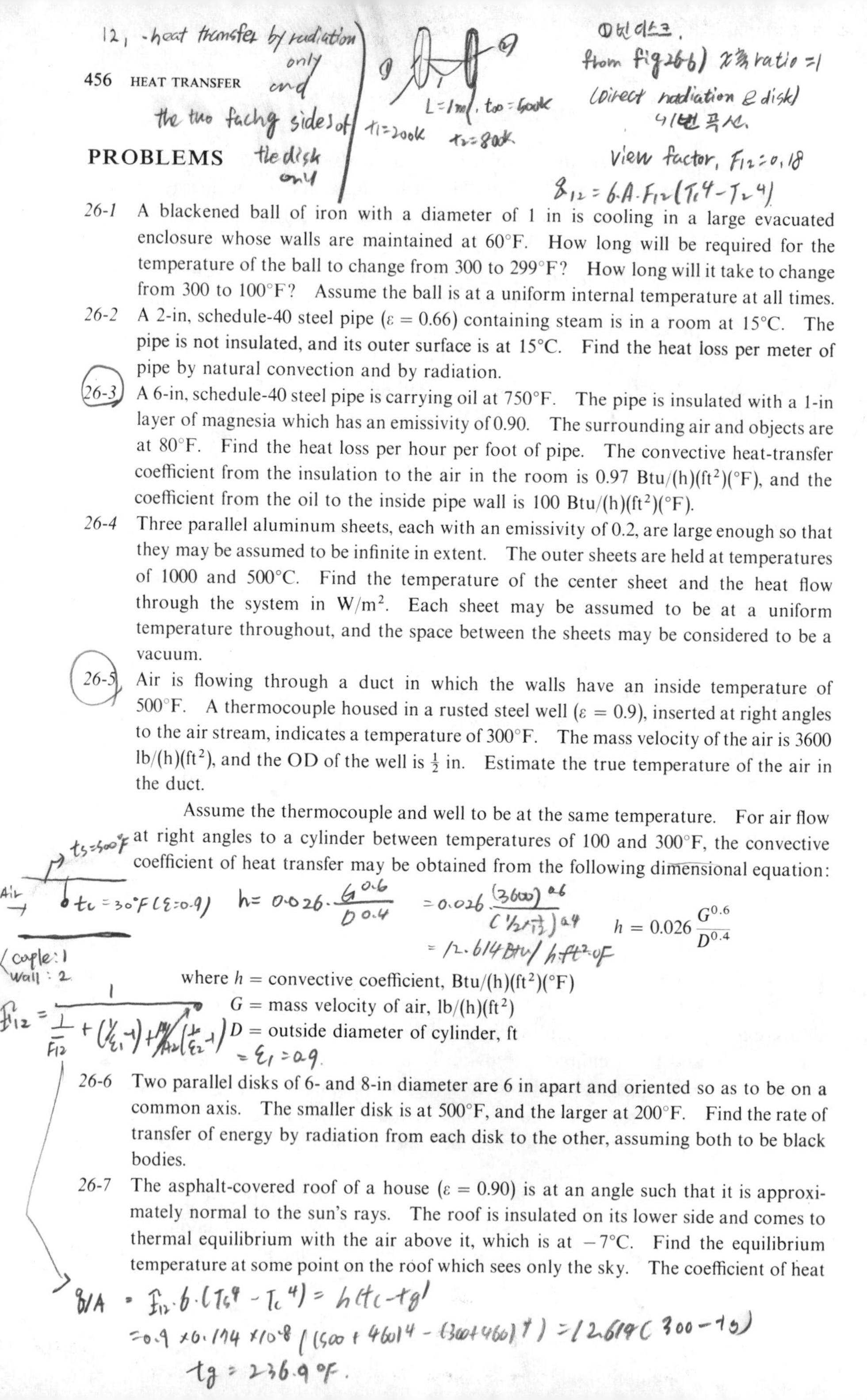

PROBLEMS

26-1 A blackened ball of iron with a diameter of 1 in is cooling in a large evacuated enclosure whose walls are maintained at 60°F. How long will be required for the temperature of the ball to change from 300 to 299°F? How long will it take to change from 300 to 100°F? Assume the ball is at a uniform internal temperature at all times.

26-2 A 2-in, schedule-40 steel pipe ($\varepsilon = 0.66$) containing steam is in a room at 15°C. The pipe is not insulated, and its outer surface is at 15°C. Find the heat loss per meter of pipe by natural convection and by radiation.

26-3 A 6-in, schedule-40 steel pipe is carrying oil at 750°F. The pipe is insulated with a 1-in layer of magnesia which has an emissivity of 0.90. The surrounding air and objects are at 80°F. Find the heat loss per hour per foot of pipe. The convective heat-transfer coefficient from the insulation to the air in the room is 0.97 Btu/(h)(ft^2)(°F), and the coefficient from the oil to the inside pipe wall is 100 Btu/(h)(ft^2)(°F).

26-4 Three parallel aluminum sheets, each with an emissivity of 0.2, are large enough so that they may be assumed to be infinite in extent. The outer sheets are held at temperatures of 1000 and 500°C. Find the temperature of the center sheet and the heat flow through the system in W/m^2. Each sheet may be assumed to be at a uniform temperature throughout, and the space between the sheets may be considered to be a vacuum.

26-5 Air is flowing through a duct in which the walls have an inside temperature of 500°F. A thermocouple housed in a rusted steel well ($\varepsilon = 0.9$), inserted at right angles to the air stream, indicates a temperature of 300°F. The mass velocity of the air is 3600 lb/(h)(ft^2), and the OD of the well is $\frac{1}{2}$ in. Estimate the true temperature of the air in the duct.

Assume the thermocouple and well to be at the same temperature. For air flow at right angles to a cylinder between temperatures of 100 and 300°F, the convective coefficient of heat transfer may be obtained from the following dimensional equation:

$$h = 0.026 \frac{G^{0.6}}{D^{0.4}}$$

where h = convective coefficient, Btu/(h)(ft^2)(°F)
G = mass velocity of air, lb/(h)(ft^2)
D = outside diameter of cylinder, ft

26-6 Two parallel disks of 6- and 8-in diameter are 6 in apart and oriented so as to be on a common axis. The smaller disk is at 500°F, and the larger at 200°F. Find the rate of transfer of energy by radiation from each disk to the other, assuming both to be black bodies.

26-7 The asphalt-covered roof of a house ($\varepsilon = 0.90$) is at an angle such that it is approximately normal to the sun's rays. The roof is insulated on its lower side and comes to thermal equilibrium with the air above it, which is at -7°C. Find the equilibrium temperature at some point on the roof which sees only the sky. The coefficient of heat transfer for natural convection between the roof and the air is 8.5 $W/m^2 \cdot K$. It can be assumed that the transmissivity of solar energy by the earth's atmosphere is 0.70 on this particular day.

26-8 A furnace stack 6 ft in diameter contains hot gas at 2000°F. The stack has brick walls 6 in thick. The gas in the stack is at 1 atm pressure and contains 10 percent by volume of carbon dioxide; the remainder is oxygen and nitrogen. The bulk velocity of the gases is such that the convective coefficient of heat transfer inside the stack is 5 Btu/(h)(ft²)(°F). Find the heat flux through the walls of the stack in Btu/(h)(ft²). The thermal conductivity of the brick is 0.7 Btu/(h)(ft)(°F), and its emissivity is 0.90. The atmospheric temperature is 70°F, and the mean combined convective and radiative coefficient of heat transfer outside the stack is 6 Btu/(h)(ft²)(°F).

26-9 (*a*) A 60-W light bulb has a cylindrical tungsten filament 8×10^{-4} m in diameter and 0.04 m in length. Assuming that conduction and convection losses and reradiation from the surroundings are all negligible, find the filament operating temperature. Assume this temperature to be uniform at all points in the filament at any given time.

(*b*) If the lower limit of visible radiation from the filament occurs at a temperature of 1000°C, how long does it take for the light to "go out"?

Data for the filament:

1 Emissivity = 0.39.
2 Density = 19,000 kg/m^3.
3 Specific heat = 155 $J/kg \cdot K$.

26-10 Sunlight shines through a vertical glass window for which the transmissivity is 0.8, the absorptivity 0.18, and the reflectivity 0.02. The solar constant is 442 Btu/(h)(ft² normal to radiation) and of this energy 50 percent may be assumed to reach the earth's surface and strike the window. On one side of the window the bulk air temperature is 80°F and on the other side the bulk air temperature is 30°F. Neglecting all thermal effects due to reradiation, determine the steady-state temperature of the window when the sunlight strikes the window at an angle of 45°. The convection coefficients for the window may each be determined from the simplified equation $h = 0.5\,(\Delta t)^{0.5}$ where Δt is the temperature difference (°F) between the window and the surrounding bulk air and h is the convection coefficient in Btu/(h)(ft²)(°F). Assume the window glass has the same temperature at all points.

26-11 A small black area of 0.01 ft² is radiating toward a plane mirror, which reflects all the radiation away except that which falls on a vertical black strip 10 ft high and 0.1 ft wide opposite the source. The strip is 10 ft away from the source, placed vertically with the bottom of the strip at the same level as the source. Find F_{12}.

26-12 Two vertical black disks, each 1 m in diameter, are placed parallel to each other on their center line at a separation of 1 m. The left-hand disk is held at 2000 K by electric heating, and the right-hand disk is held at 800 K by cooling water circulated behind it. Consider heat transfer by radiation only and from the two facing sides of the disks only. The disks radiate to each other, and to the surroundings at 500 K. At what rate is heat removed by the cooling water from the right-hand disk?

26-13 Two vertical black disks, each 3 ft in diameter, are placed parallel to each other on their center line at a separation of 1 ft. The left-hand disk is heated by electrical resistance to 1040°F, and each disk is perfectly insulated on all surfaces except the two faces which see each other. The disks are in space, and the surroundings can be assumed to emit negligible radiation.

(*a*) Calculate the steady-state temperature of the right-hand disk.

(*b*) What is the rate of electrical heat input to the left-hand disk?

26-14 A metal tube of 0.025-m OD, 3 m long, contains steam condensing at 95°C. The rate of heat loss from the pipe to the surroundings is 200 W. If the air and walls of the room are at 40°C, calculate the emissivity of the metal pipe. The resistance to heat transfer of the condensing film and tube wall may be assumed negligible.

26-15 Hot oil at 500°F flows through a 1-in, 16-gauge, horizontal ceramic tube. The heat-transfer coefficient of the oil is $h = 100$ Btu/(h)(ft^2)(°F), the pipe is in a room in which the air and walls of the room are at 80°F, and the emissivity of the pipe surface is 0.90. What is the temperature of the outer surface of the pipe? The thermal conductivity of the ceramic material in the tube is $k = 0.63$ Btu/(h)(ft)(°F).

26-16 Air at a bulk temperature of 200°F is passing through a large duct whose inside wall is at 600°F and has an emissivity of 0.9. Located at the centerline of the duct, running parallel to the duct axis, is a 1-in, 16-gauge aluminum tube ($\varepsilon = 0.1$) which is open at both ends so that air flows both outside and inside the tube. Assuming that the metal in the aluminum tube is everywhere at the same temperature, what is that temperature? The heat-transfer coefficient between tube surface and air is $h = 0.80$ Btu/(h)(ft^2)(°F).

26-17 A heat pipe constructed of thin-walled, polished, copper tubing ($\frac{1}{4}$-in OD) is to be designed to remove heat at the rate of 5 Btu/h from an enclosure containing an electric circuit. The heat pipe contains a pure organic fluid which, at the pressure inside the pipe, boils and condenses at 150°F. One end of the pipe is to be inside the enclosure housing the circuit, and the other end will protrude horizontally into the room which surrounds the enclosure. The room is at 70°F. How long must the section of the heat pipe be which extends into the room? List and justify all assumptions.

26-18 The water in a flat-plate solar collector is at an average temperature of 70°F and the air outside is at 60°F. The convective heat-transfer coefficient between the surface of the collector and the water is $h_w = 35$ Btu/(h)(ft^2)(°F), and the convective heat-transfer coefficient between the surface of the collector and the air is $h_a = 0.44$ Btu (h)(ft^2)(°F). Assume the collector surface has negligible resistance to heat conduction and that the emissivity and absorptivity of the surface are both unity. Find the heat flow rate to the water, the convective loss to the air, and the radiation loss to the surroundings, all in Btu/(h)(ft^2), when the insolation rate is 200 Btu/(h)(ft^2). Also determine the plate temperature. Heat losses from the back and sides are negligible.

26-19 A solar collector 4 ft wide and 8 ft long has a cross-sectional area for flow of 0.361 ft^2, which is the sum of the areas of the 208 rectangular flow channels (0.125 in × 0.200 in). Water entering the base of this collector is at 70°F and flows at the rate of 4 gal/min. The collector surface has an emissivity and an absorptivity both equal to 0.92. The convective heat-transfer coefficient between the surface of the collector and the water is $h_w = 118$ Btu/(h)(ft^2)(°F). The convective heat-transfer coefficient between the surface and the air is $h_a = 0.44$ Btu/(h)(ft^2)(°F) but because this is so low the convective heat loss to the air may be neglected. The conductivity of the plate material is large and the thickness is small, so the temperature drop over this thickness may also be neglected. Calculate the exit temperature of the water when the insolation rate is 200 Btu/(h)(ft^2) at the surface of the collector. Heat losses from the back and the sides are negligible.

3.
6.065''
6.625''
8.625''

- K(steel) = 26 Btu/h·ft·°F
- K(magnesia) = 0.034 Btu/h·ft·°F
- $\varepsilon_{mag} = 0.9$

the oil) - 750°F, h_i = 100 Btu/hft²·°F

Air) - 80°F, h = 0.97 Btu/hft²·°F

Pipe → Air) - heat tr = convection + radiation.
- Pipe length is infinite.
- st. st.

◦ 파이프 내부 저항

$\frac{1}{h_i A_1}$ (R_1), $\frac{\Delta r}{K A_{lm1}}$ (R_2), $\frac{\Delta r}{K A_{lm2}}$ (R_3)

$$R_1 = \frac{1}{h_i A_1} = \frac{1}{100\cdot\pi(6.065/12)} \approx 6.30\times10^{-3}$$

$$R_2 = \frac{\Delta r}{K A_{lm1}} = \frac{(6.625-6.065)/2/12}{26\times\frac{\pi(6.625-6.065)/12}{\ln(6.625/6.065)}} = 5.41\times10^{-4}$$

$$R_3 = \frac{\Delta r}{K\cdot A_{lm2}} = \frac{(8.625-6.625)/24}{0.034\times\frac{\pi(8.625-6.625)/12}{\ln(8.625/6.625)}} = 1.23$$

파이프 → 공기
q = 대류 + 복사.

대류 q $= h_o\cdot A_o(t_s - 80°F) = (0.97)(\pi\times 8.625/12)(t_s - 80°F) = 2.19(t_s - 80°F)$

복사 q $= A_o\varepsilon\sigma\{(t_s+460)^4 - (80+460)^4\} = 0.9\times 0.1714\times10^{-8}\{(t_s+460)^4 - 540^4\}\times\pi\times(8.625/12)$

$= 1.54\times10^{-9}\{(t_s+460)^4 - 540^4\}\times 2.26$

$q_{tot} = 2.19(t_s-80) + 1.54\times10^{-9}\{(t_s+460)^4 - 540^4\}$

$$q_1 = q_2 \rightarrow \frac{750 - t_s}{6.30\times10^{-3} + 5.41\times10^{-4} + 1.23} = 2.19(t_s-80) + 1.54\times10^{-9}\{(t_s+460)^4 - 540^4\}\times 2.26$$

$t_s = 173.0°F$

$$q/L = \frac{750 - 173}{6.30\times10^{-3} + 5.41\times10^{-4} + 1.23}$$

= 466.5 Btu/h·ft.

27

HEAT-EXCHANGE EQUIPMENT

In this chapter we shall consider the application of heat-transfer theory to the design and operation of certain types of heat-exchange equipment. We have covered a number of such applications in previous chapters, but the following topics are best discussed after all the principal mechanisms of heat transfer have been studied.

HEAT EXCHANGERS

Shell-and-Tube Heat Exchangers

The heat exchangers found most commonly in industry contain a number of parallel tubes enclosed in a single shell and for that reason are called shell-and-tube exchangers. They are used when large quantities of fluid must be heated or cooled and double-pipe exchangers of sufficient capacity would be too bulky and expensive. As in the case of double-pipe exchangers, the fluid streams may be flowing in the same or opposite directions. In Fig. 27-1, one fluid enters the tubes from a header and, leaving the tubes, flows into another header from which it leaves the unit. The other fluid flows countercurrently through the shell side.

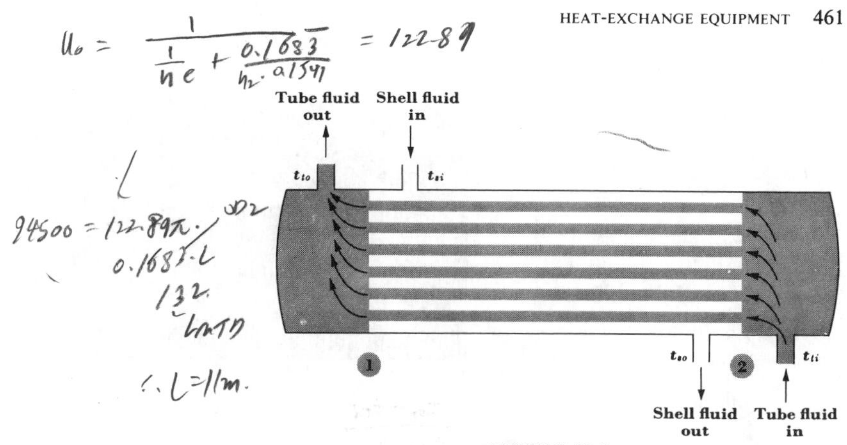

FIGURE 27-1
Single-pass, shell-and-tube heat exchanger.

The temperature behavior of the fluids in the exchanger shown in Fig. 27-1 is analogous to that in the double-pipe exchanger shown in Fig. 21-7. Furthermore, the equations for determining the heat-transfer area in the two cases are identical. If U_o can be assumed constant, we use the relation

$$q = U_o A_o \Delta t_{lm} \tag{21-29}$$

If U_o is a linear function of Δt, we use Eq. (21-34); if the variation is not linear, then graphical or numerical integration of Eq. (21-35) is carried out. In all these instances A_o is the total outside area of all the tubes in the shell-and-tube exchanger.

Examination of Fig. 27-1 shows that not all of the flow on the shell side is parallel to the tube axes; the method by which the fluid flows into and out of the shell side obviously causes a certain amount of cross flow in addition to the flow parallel to the tubes. In practical circumstances heat exchangers generally have segmental or cross-flow baffles (not shown in Fig. 27-1) installed in the shell which cause the shell-side fluid to make a number of cross-flow passes over the tubes while moving from the inlet to the outlet of the shell. The main purpose of the baffles is to minimize channeling by which some of the fluid flows preferentially in certain paths that have little contact with the heat-transfer surface. Each baffle may extend to half of the shell cross section; they are spaced at distances as close as one-fifth of the shell diameter. In addition to providing more uniform flow and heat-transfer characteristics, the baffles also help to support the tubes. In heat exchangers in which condensation or boiling is taking place on the shell side, baffles are not necessary; hence, tube supports are installed. More detailed information on heat exchanger design is provided by Perry,[1] Kern,[2] and Taborek.[3]

[1] Perry, sec. 11.
[2] D. Q. Kern, "Process Heat Transfer," McGraw-Hill Book Company, New York, 1950.
[3] J. Taborek, *Heat Transfer Eng.*, **1**:15 (1979).

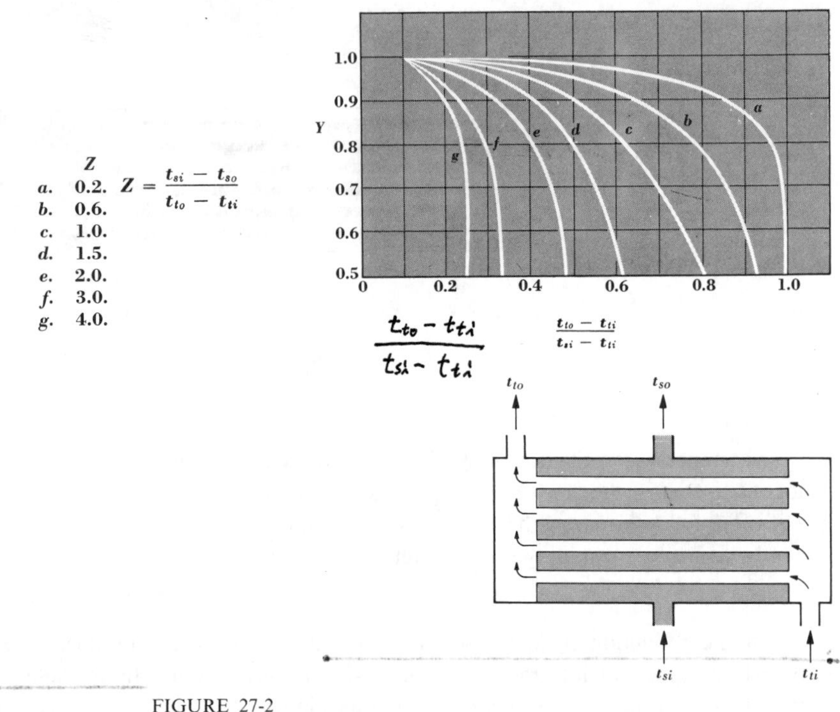

FIGURE 27-2
Correction factor for a cross-flow heat exchanger. [*From R. W. Bowman, A. C. Mueller, and W. M. Nagle, Trans. Am. Soc. Mech. Eng.*, **62**:*283* (*1940*).]

Although the heat exchanger in Fig. 27-1 would, in practice, have baffles causing a number of cross-flow passes of the shell-side fluid, it would nevertheless be designed on the assumption that only countercurrent flow existed. However, if the exchanger were quite short, it is evident that the assumption of fluids flowing in parallel concurrent or countercurrent flow would not be realized. Instead, the principal path of the fluid in the shell would be normal to the tube bundle. In these circumstances the analogy with the double-pipe exchanger breaks down. An analysis assuming complete cross flow has been carried out by Bowman, Mueller, and Nagle, who integrated the theoretical equations for this system. They compared their equation for q with a modified form of Eq. (21-29)

$$q = U_o A_o Y \frac{\Delta t_2 - \Delta t_1}{\ln (\Delta t_2/\Delta t_1)} \tag{27-1}$$

in which Δt_1 and Δt_2 are the terminal temperature differences taken as if the fluids in

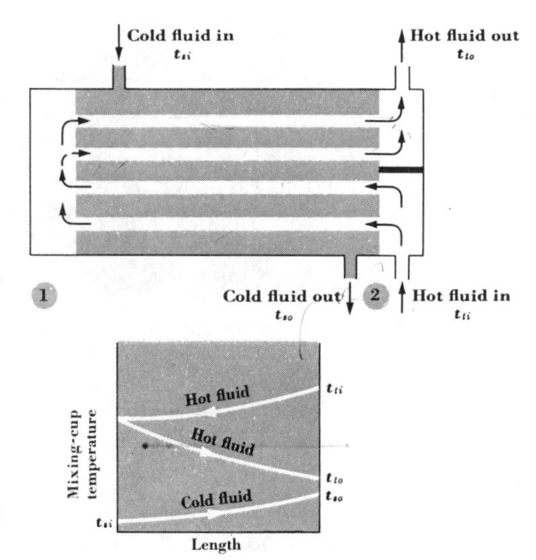

FIGURE 27-3
Heat exchanger with one shell pass and two tube passes.

the cross-flow exchanger were in countercurrent flow. The factor Y was then obtained by equating the two expressions. In designing cross-flow exchangers, values of the factor Y may be read from Fig. 27-2 and used with the terminal temperature differences for counterflow to obtain A_o.

Multipass Heat Exchangers

Shell-and-tube heat exchangers are sometimes modified by putting a partition in one of the headers so that the tube-side fluid makes two passes through the shell. This arrangement is shown schematically in Fig. 27-3.

In part of the exchanger the fluids in both the tube and shell sides are in concurrent flow, but in the remainder of the exchanger counterflow exists. This case has also been analyzed by Bowman, Mueller, and Nagle, and the necessity of using a cumbersome equation has been eliminated by providing a chart of values for Y to be used in conjunction with Eq. (27-1). As before, the values of Δt in Eq. (27-1) are taken as though counterflow existed in the entire exchanger.

Many other combinations have been analyzed, including systems with more than one shell pass. Multiple passes on the shell side are achieved by placing longitudinal partitions in the shell. Heat-transfer areas are calculated for these systems by the use of a correction factor as applied to Eq. (27-1) in the same manner as those already discussed. A number of charts for determining these factors are

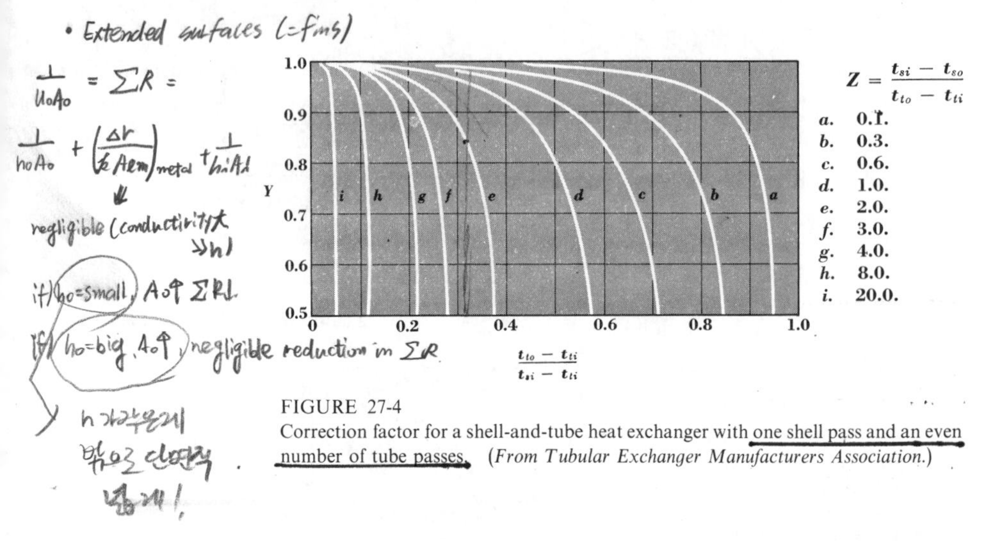

FIGURE 27-4
Correction factor for a shell-and-tube heat exchanger with one shell pass and an even number of tube passes. (*From Tubular Exchanger Manufacturers Association.*)

given in Perry's Handbook, sec. 10. Figure 27-4 gives correction factors for a shell-and-tube exchanger with one shell pass and two, four, six, eight, etc., tube passes.

In all systems it is assumed in the analysis that the fluid in any one pass of the shell of the exchanger is completely mixed at any cross section normal to flow. This is not actually achieved in the shell, but is often approached because the baffles in the shell cause mixing at various positions along the axis of the exchanger.

In some condensers and reboilers the temperature of the shell-side fluid is constant throughout the exchanger. For these systems, the correction factor Y has a value of unity and the logarithmic mean temperature difference is used, as in Eq. (21-29).

Example 27-1 In a tubular heater with eight tube passes and a single shell pass, hot gas flows through the shell side and a liquid flows through the tubes. The gas enters at 160°F and leaves at 102°F; the liquid enters at 52°F and leaves at 87°F. If the overall coefficient of heat transfer U_o is 5.5 Btu/(h)(ft^2)(°F) and the total heat-transfer rate is 100,000 Btu/h, find the total heat-transfer area.

The parameters for determining the factor Y are

$$\frac{t_{to} - t_{ti}}{t_{si} - t_{ti}} = \frac{87 - 52}{160 - 52} = 0.32$$

$$Z = \frac{t_{si} - t_{so}}{t_{to} - t_{ti}} = \frac{160 - 102}{87 - 52} = 1.66$$

From Fig. 27-4, $Y = 0.89$. The terminal temperature differences for counterflow are $\Delta t_1 = 50°$ and $\Delta t_2 = 73°$. The total heat-transfer area is

$$A_o = \frac{100{,}000}{(5.5)(0.89)[(73 - 50)/\ln(\frac{73}{50})]}$$

$$= 336 \text{ ft}^2 \qquad ////$$

Extended Surfaces[1]

The use of extended surfaces, often called *fins*, to give high heat-transfer rates in heat-exchange devices has been common for many years. An automobile radiator consists of a bank of tubes through which hot water from the engine circulates and loses heat to the surrounding air. The tubes are inserted through holes in a series of parallel plates which receive heat from the tube walls and transmit it to the air by the mechanism of forced convection. Some heat flows directly through the tube walls to the air, but most of it is dissipated through the plates. The amount dissipated by radiation is a small fraction of the total heat loss, so the name "radiator" is not descriptive of the device. On a much larger scale, the use of air coolers to remove heat in power plants and oil refineries has become important in regions where ample cooling water is not available.

Other forms of extended surfaces attached to tubes are circular or spiral ribbon-type fins; longitudinal fins, which are plates attached to the outer surface running lengthwise along the tube; and pin fins, which are small pins or strips attached by one end to the outer surface of the tube in the manner of bristles. Fins may be welded or brazed to the tube and may be of the same or a different kind of metal. Some fins are formed directly from the metal of a thick-walled tube. Fins on tubes are usually found only on the outer surface.

Extended surfaces do not increase the heat-transfer rate significantly in all circumstances. This can be shown by the following considerations. The sum of the resistances to heat transfer is equal to the reciprocal of the product of area and the overall coefficient of heat transfer defined in Chap. 21. In a system with two fluid resistances and a metal-wall resistance, the relation is

$$\frac{1}{U_o A_o} = \sum R = \frac{1}{h_o A_o} + \left(\frac{\Delta r}{k A_{lm}}\right)_{\text{metal}} + \frac{1}{h_i A_i} \qquad (27\text{-}2)$$

The quantity A_{lm} in the metal-resistance term is valid only for concentric, cylindrical surfaces; however, the metal resistance is usually not great and for the moment will

[1] Good general references are D. Q. Kern and A. D. Kraus, "Extended Surface Heat Transfer," McGraw-Hill Book Company, 1972; and W. M. Kays and A. L. London, "Compact Heat Exchangers," 2d ed., McGraw-Hill Book Company, New York, 1964.

be ignored. It can be seen that the effect of the presence of the fins on the outer surface will be to increase A_o and consequently to reduce the heat-transfer resistance $1/h_o A_o$ of the fluid outside the tube. If the outside coefficient of heat transfer h_o is so small that the outside resistance is much greater than either the metal resistance or the inside resistance, the overall resistance will be reduced almost in proportion to the amount by which A_o is increased. Thus it is possible to double the outside heat-transfer area by the use of fins and reduce the total resistance by almost one-half. An example of an application of this principle is the condensation of steam inside a copper tube as air is heated in passing over the finned outer surface. Conversely, if the coefficient of heat transfer on the outer surface h_o is high and the coefficient on the inner surface h_i is low, the outer resistance will constitute a small part of the total resistance. The effect of doubling the outside area is merely to cut in half an already negligible resistance so that the overall resistance is reduced only slightly. This would be the situation if the foregoing system were reversed so that steam condensed on the outer, finned surfaces of tubes while air or some other gas passed through the tube. An analysis of these two situations is shown in Example 27-2.

If fouling of a tube surface occurs, the use of fins would not appear to be beneficial because of the possibility of trapping foreign particles between the fins. However, flow between fins is often at a fairly high velocity, and if the deposit is not likely to plug the space between the fins entirely, their presence may still be beneficial. The fouling resistance is $1/h_d A_o$, and if h_d is small compared with the other coefficients in the system, any increase in the surface area A_o on which fouling occurs cuts down the total resistance by a significant amount.[1]

Example 27-2 Air is to be heated by the condensation of steam in a heat exchanger where the heat-transfer coefficient of the air is 60 $W/m^2 \cdot K$ and the coefficient for the condensing steam is 6000 $W/m^2 \cdot K$. Find the heat-transfer rate in W/lin m of tubing for the following systems:

(*a*) Steam is condensing inside $\frac{3}{4}$-in, 18-gauge copper tubing, and air is flowing over the outside. The tubes are not finned.

(*b*) Same as (*a*), except that the outer surfaces of the tubes are finned.

(*c*) Steam is condensing on the outside of the finned tubes, and air is passing through the inside.

We shall assume that the overall temperature difference is fixed at 55°C and that the resistance of the metal wall to heat transfer is negligible for all three systems. The outside area of the smooth tube is 0.060 m^2/lin m, and the inside area is 0.052 m^2/lin m. The finned tube has 600 circular fins/m of

[1] For design data on heat transfer to finned tubes see the series of articles by D. L. Katz and associates in *Pet. Refiner*, vols. 33, 34(1954-55). See also R. L. Webb, *Heat Transfer Eng.*, **1**:33 (1980).

0.00132 m height. The total outside area of tubing and fins is 0.161 m^2/m. The heat-transfer coefficients given above will be assumed to be constant for each fluid in all systems.

(*a*) Smooth tubing; steam inside, air outside:

$$\text{Steam resistance} = \frac{1}{(6000)(0.052)} = 0.00321 \text{ K/W}$$

$$\text{Air resistance} = \frac{1}{(60)(0.060)} = 0.278 \text{ K/W}$$

$$\sum R = 0.281 \text{ K/W}$$

$$q = \frac{\Delta t}{\sum R} = \frac{55}{0.281} = 196 \text{ W}$$

(*b*) Finned tubing; steam inside, air outside:

$$\text{Steam resistance} = 0.00321 \text{ K/W}$$

$$\text{Air resistance} = \frac{1}{(60)(0.161)} = 0.104 \text{ K/W}$$

$$\sum R = 0.107 \text{ K/W}$$

$$q = \frac{55}{0.107} = 514 \text{ W}$$

(*c*) Finned tubing; air inside, steam outside:

$$\text{Steam resistance} = \frac{1}{(6000)(0.161)} = 0.00104 \text{ K/W}$$

$$\text{Air resistance} = \frac{1}{(60)(0.052)} = 0.321 \text{ K/W}$$

$$\sum R = 0.322 \text{ K/W}$$

$$q = \frac{55}{0.322} = 171 \text{ W}$$

In system (*b*) the effect of the fins in reducing the major resistance is to cause a significant increase in the heat-transfer rate over the rate in system (*a*). In system (*c*), however, the effect of the fins is merely to reduce further the resistance of the steam, which was negligible even on an unfinned surface. The major resistance, which is that of the air stream, is at the inside surface of the tube, which is smaller in area than the outside of the smooth tube. Thus the total resistance in system (*c*) with the finned tubes is greater than in system (*a*), when smooth tubes were used. ////

Fin efficiencies Although the resistance to heat transfer of the metal wall was neglected in Example 27-2, this procedure is not always justifiable. Fins as high as 1 in and only a few hundredths of an inch in thickness are not uncommon, and even when made of copper, such fins may offer a significant resistance to heat transfer. This resistance is, of course, more likely to be significant when the fluid resistances are small.

A quantitative method of describing the effect of the fin resistance is by the use of fin efficiency. This quantity is defined as the actual heat-transfer rate from a fin divided by the rate that would be obtained if the entire fin were at the temperature of the base of the fin, i.e., the outer cylindrical surface of the tube. The fin efficiency can be predicted mathematically, as shown in Example 27-3.

The heat flow from a tube surface to a fluid can be written as

$$q = hA_f\eta_f \,\Delta t + hA_t \,\Delta t \qquad (27\text{-}3)$$

where A_f = fin area

A_t = area of tube surface between fins

η_f = fin efficiency

h = heat-transfer coefficient, assumed constant at all points on fin and tube surface

Δt = temperature difference between base of fin and bulk-fluid phase

When written in Ohm's law form, Eq. (27-3) becomes

$$q = \frac{\Delta t}{1/[h(A_f\eta_f + A_t)]} \qquad (27\text{-}4)$$

in which the denominator has the dimensions of resistance. When an overall coefficient of heat transfer U_o based on the outside area is written, the term $1/h_o$, which appears in Eq. (21-15), is replaced by $A_o/h_o(A_f\eta_f + A_t)$, where A_o is the sum of A_f and A_t.

Fin efficiences have been calculated for a number of configurations. A simple system is examined in Example 27-3, and a chart for determining the efficiency of a circular fin is given in Fig. 27-5.

Example 27-3 A longitudinal steel fin 1 in high and $\frac{1}{8}$ in thick is exposed to an air stream at 70°F. The air moving past the fin has a uniform convection coefficient $h = 15$ Btu/(h)(ft^2)(°F). The fin has a thermal conductivity of 25 Btu/(h)(ft)(°F) and a base temperature of 250°F. Calculate the fin efficiency and the heat flow per lineal foot assuming that all temperature gradients in planes parallel to the base of the fin are negligible. A section of the fin is shown in Fig. 27-6.

First a heat balance will be written for the differential element of width w ($\frac{1}{96}$ ft), height dx, and length 1 ft.

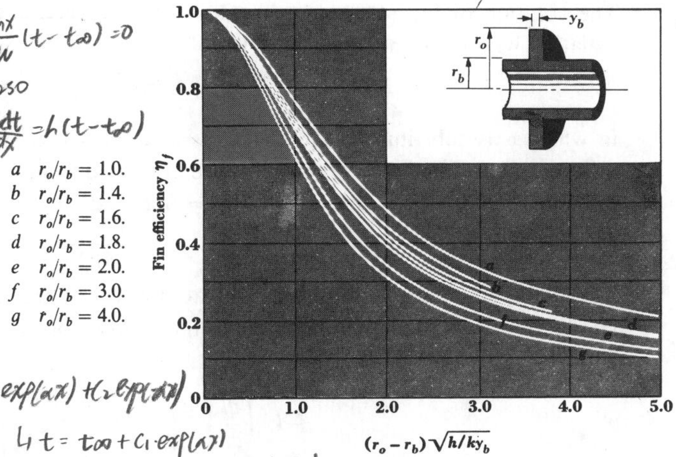

FIGURE 27-5
Efficiency of circular fins of constant thickness. [*From K. A. Gardner, Trans. Am. Soc. Mech. Eng.*, **67**:*621* (*1945*).]

$$\text{Rate of heat conduction in} = -kw\frac{dt}{dx} \tag{1}$$

$$\text{Rate of heat conduction out} = -kw\left(\frac{dt}{dx} + \frac{d^2t}{dx^2}\,dx\right) \tag{2}$$

$$\text{Rate of heat convection out} = h(2\,dx)(t - 70) \tag{3}$$

At steady state, (1) = (2) + (3), which reduces to the following differential equation:

$$\frac{d^2t}{dx^2} - \frac{2h}{kw}(t - 70) = 0 \tag{4}$$

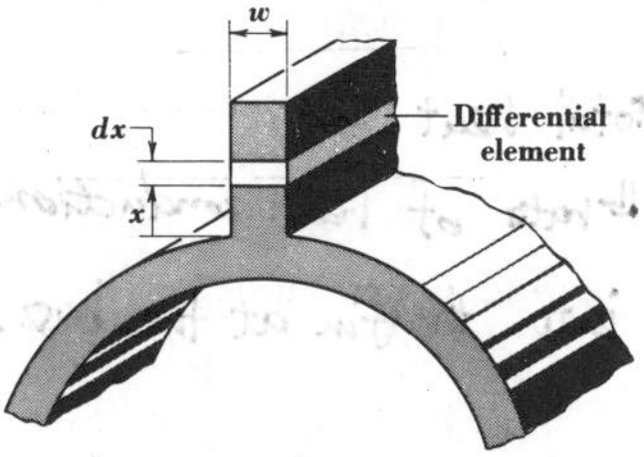

FIGURE 27-6
Longitudinal fin on a section of a circular tube (Example 27-3).

The solution of Eq. (4), which can be written down directly for this linear equation with constant coefficients, is

$$t - 70 = C_1 e^{\alpha x} + C_2 e^{-\alpha x} \tag{5}$$

in which α is substituted for $\sqrt{2h/kw}$ as a convenience, and C_1 and C_2 are arbitrary constants. These constants are evaluated from a knowledge of two boundary conditions. One is that at $x = 0$, $t = 250°F$. Therefore

$$180 = C_1 + C_2 \tag{6}$$

The other boundary condition is that at the top of the fin ($x = \frac{1}{12}$ ft) the conduction rate to the surface equals the convection rate away from the surface. This equality is written as

$$-kw\left(\frac{dt}{dx}\right)_{x=1/12} = hw(t_{x=1/12} - 70) \tag{7}$$

Substituting the proper values from Eq. (5) into Eq. (7) gives

$$-k(C_1 \alpha e^{\alpha x} - C_2 \alpha e^{-\alpha x}) = h(C_1 e^{\alpha x} + C_2 e^{-\alpha x})$$

The constant α is

$$\alpha = \left[\frac{(2)(15)}{(25)(\frac{1}{96})}\right]^{1/2} = 10.7 \text{ ft}^{-1}$$

At $x = \frac{1}{12}$, $\alpha x = 0.895$.

The second equation for C_1 and C_2 is thus

$$-(25)(10.7)(C_1 e^{0.895} - C_2 e^{-0.895}) = 15(C_1 e^{0.895} + C_2 e^{-0.895}) \tag{8}$$

Equations (6) and (8) are solved to give

$$C_1 = 23.4$$
$$C_2 = 156.6$$

The heat flow from the fin may be found by combining Eqs. (3) and (5) to obtain an equation for the heat loss from a differential area and integrating over the entire fin area. It is easier, however, to find the rate of heat conduction into the fin at the base, which is also equal to the total heat loss. Thus the total heat loss is

$$q = -kA\left(\frac{dt}{dx}\right)_{x=0}$$
$$= -(25)(\tfrac{1}{96})(10.7)(23.4 - 156.6)$$
$$= 370 \text{ Btu/(h)(lin ft)}$$

The total heat loss if the fin were at 250°F throughout would be

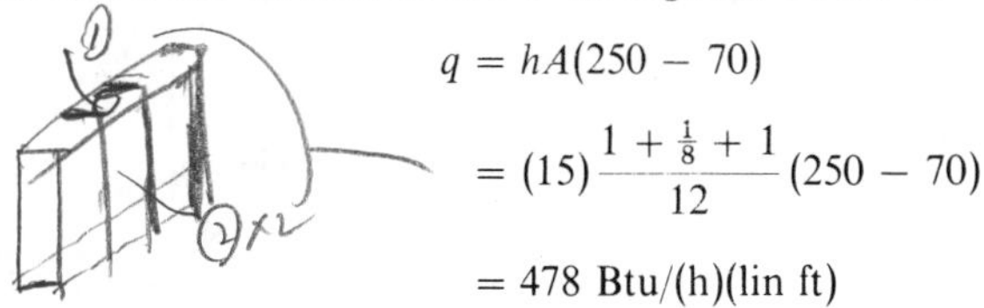

$$q = hA(250 - 70)$$

$$= (15)\frac{1 + \frac{1}{8} + 1}{12}(250 - 70)$$

$$= 478 \text{ Btu/(h)(lin ft)}$$

The fin efficiency η_f is thus $\frac{370}{478}$, or 0.78. The temperature at the top of the fin may be found by substituting $x = \frac{1}{12}$ in Eq. (5); it is 191°F. ////

One of the assumptions made in Example 27-3 was that the heat-transfer coefficient between the fin and the air was the same at all points on the fin. This is probably justifiable for developed flow parallel to a longitudinal fin. It is less likely to be valid for the more typical case of flow normal to circular fins of the type shown in Fig. 27-5. Stynes[1] has shown experimentally that the local transfer coefficient on a fin may be six times larger near the tip than near the base. Furthermore, he has found that the mean transfer coefficient is usually somewhat higher on the upstream face of the fin than on the downstream face.

Wilson's Method of Analysis

A useful method of determining the individual convective heat-transfer coefficients in a heat exchanger is that of Wilson.[2] If a single-phase fluid is flowing in developed turbulent flow inside the tubes of an exchanger, the convective coefficient h_i can be estimated by the Dittus-Boelter equation

$$\frac{h_i D}{k} = 0.023\left(\frac{Du_b\rho}{\mu}\right)^{0.8}\left(\frac{C_p\mu}{k}\right)^{0.3} \qquad (24\text{-}4)$$

This equation shows that h_i is proportional to $u_b^{0.8}$ if everything else is held constant. However, we shall find it desirable to change the average temperature of the fluid as u_b changes; the effect of the temperature if the fluid is water is represented by the term $1 + 0.011t$ in the equation

$$h_i = au_b^{0.8}(1 + 0.011t) \qquad (27\text{-}5)$$

The constant a could be calculated by equating (24-4) and (27-5), but it is usually determined from the results of the Wilson-line experiment described below. The temperature t is the average water temperature in degrees Fahrenheit.

[1] S. K. Stynes, *AIChE J.*, **10**:437 (1964).
[2] E. E. Wilson, *Trans. Am. Soc. Mech. Eng.*, **37**:47 (1915).

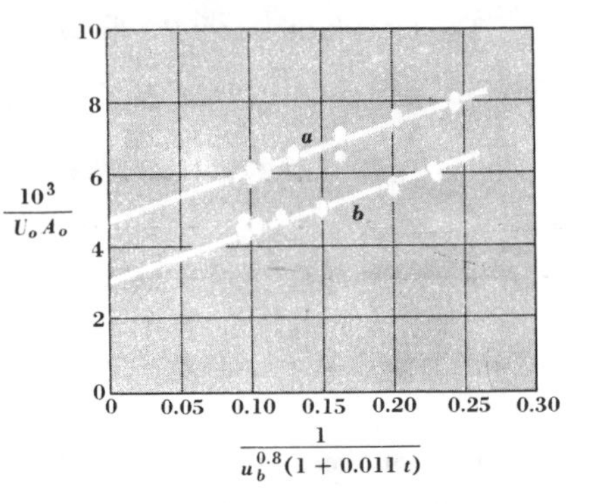

FIGURE 27-7
Wilson plot for boiling methanol on a $1\frac{1}{4}$-in tube. [*From R. P. Lance, AIChE J.*, **4**:*75 (1958)*.]
a $\Delta t_o = 50°F$.
b $\Delta t_o = 70°F$.

If a series of runs is conducted in which the velocity of the water in the tubes is varied, the overall coefficient can be represented by the equation

$$\frac{1}{U_o A_o} = \frac{1}{h_o A_o} + \frac{\Delta r}{k A_{lm}} + \frac{1}{a u_b^{0.8}(1 + 0.011t)A_i} \tag{27-6}$$

The first two terms on the right-hand side are the resistances of the outside fluid and the tube wall, respectively. If they are held constant for all runs, a linear relation exists between $1/U_o A_o$ and $1/u_b^{0.8}(1 + 0.011t)$. Such a plot is shown in Fig. 27-7, in which the overall coefficients are for the boiling of methanol outside a $1\frac{1}{4}$-in copper tube through which hot water is flowing. It can be seen in Fig. 27-7 that the predicted linear form is realized.

If the straight lines through the points are extrapolated to the ordinate at which $1/u_b^{0.8}(1 + 0.011t)$ equals zero, the velocity u_b at this point is infinite, and the resistance of the fluid inside the tube to heat transfer is zero. Therefore the intercepts of the straight lines on the ordinate axis of Fig. 27-7 give values of $1/U_o A_o$ equal to the sum of the tube wall and outside fluid resistances.

$$\left(\frac{1}{U_o A_o}\right)_{u_b = \infty} = \frac{1}{h_o A_o} + \frac{\Delta r}{k A_{lm}} \tag{27-7}$$

The quantity $\Delta r / k A_{lm}$ can be readily calculated, and the value of the convective coefficient for the outside fluid h_o obtained.

It is essential that all resistances other than the inside resistance be held as nearly constant as possible during runs at different velocities. This means that the outside coefficient h_o must be held constant. Because h_o is a function of temperature drop for such systems as boiling, condensation, and natural convection, care must be taken that the temperature drop over the fluid outside the tube is held constant for successive runs. This is achieved by altering the temperature level of the fluid inside the tube each time its velocity is changed, so that the inclusion of the factor $(1 + 0.011t)$ is essential for accurate results. Conditions are usually chosen so that the temperature change of the water is small from inlet to outlet.

Fouling coefficients can be estimated by the Wilson method if the outside fluid coefficient h_o can be predicted or is negligible. The fouling resistance $1/h_{do}A_o$ is a part of the intercept value.

In some pieces of equipment the value of the exponent on u_b in Eq. (27-5) may not be known. Upstream disturbances may cause it to differ from the value of 0.8 used for developed turbulent flow in circular pipes. If an incorrect exponent is used in the plot (for example, 0.8 when it should be 0.7), the experimental data should still extrapolate to the proper value of the ordinate representing all the other resistances. However, the line will no longer be quite straight, and extrapolation is more difficult. Nevertheless, the curvature is often not great enough to cause serious errors, and the method remains, in spite of this deficiency, a useful technique.

EVAPORATION

Equipment Types

The vaporization of a liquid for the purpose of concentrating a solution is a common step in chemical processing and is done in many ways. The simplest device is an open pan or kettle which receives heat from a coil or jacket or by direct firing underneath the pan. A somewhat more complicated apparatus is the horizontal-tube evaporator in which a liquid in the shell side of a closed, vertical, cylindrical vessel is evaporated by passing steam or other hot gas through a bundle of horizontal tubes contained in the lower part of the vessel. In both devices nucleate boiling occurs in the solution at the heated surface, and any circulation of the liquid that occurs is a result of the heating. The liquid level in the horizontal-tube evaporator is usually at less than half the height of the vessel; the empty space permits disengagement of entrained liquid from the vapor passing overhead.

A more efficient device is the vertical-tube evaporator. This consists of a vertical, cylindrical vessel with a bundle of vertical, steam-heated tubes at the base of the vessel. However, the bundle of tubes operates with steam around the tubes and the liquid to be evaporated within the tubes. Essentially, the bundle acts as a vertical shell-and-tube heat exchanger in which steam is admitted to the shell side and the evaporating liquid is on the tube side. Boiling within the tubes causes upward flow of the liquid-vapor mixture. After disengagement, the vapor is removed from the top of the vessel and the liquid remaining flows down in an annular space between the tube bundle and the wall of the vessel to the space beneath

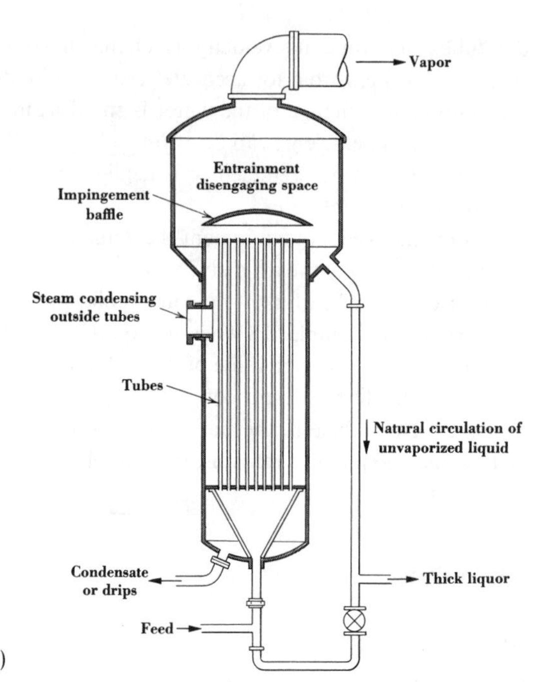

FIGURE 27-8
Long-tube, natural circulation evaporator. (*Swenson Evaporator Co.*)

the tube bundle, where it begins the cycle once again. These evaporators are fed continuously, and the concentrated liquor is removed continuously from the bottom of the vessel. The apparatus described above is often referred to as a short-vertical-tube, basket-type evaporator and is one of the most common types. The flow is caused entirely by natural convection.

An evaporator which works in similar fashion is the long-tube, natural-circulation evaporator. Operating on somewhat the same principle as a smoke-stack, the long-tube evaporator gives higher fluid velocities than the short-tube evaporator. This high velocity is desirable because it promotes higher heat-transfer coefficients between the evaporating fluid and the tube walls. The method of operation is shown schematically in Fig. 27-8. A further improvement is sometimes made by pumping the circulating liquid around to give even higher velocities and higher heat-transfer coefficients than can be obtained by natural convection. These forced-circulation evaporators are especially useful for evaporating viscous liquids for which natural-circulation rates would be low.

Diagrams of various types of evaporator are given in Perry, sec. 11. Among the most important of recent designs is the falling-film evaporator, in which a film of solution flows down the surface of a tube and receives heat through the tube wall. The short exposure time in the falling-film evaporator makes it useful for concentrating solutions of heat-sensitive materials. Long-tube evaporators, in which the tubes are filled with liquid, may also be used for heat-sensitive solutions by not permitting recirculation. This type is sometimes referred to as a *Kestner evaporator*.

Evaporator operation The evaporator is merely a special kind of heat exchanger, and the heat-transfer rate is predicted from the equation

$$q = UA\,\Delta t \qquad (27\text{-}8)$$

The area in Eq. (27-8) may be either the area on which steam is condensing or the area of the surface adjacent to the liquid solution. The reciprocal of the product of the overall coefficient and the area, $1/UA$, is equal to the sum of the individual resistances of the condensing steam, the tube wall, the boiling fluid, and any fouling present. The resistances of both the condensing steam and the boiling fluid vary with position, so that $1/UA$ also varies.

In the vertical-tube evaporator the condensing coefficient is infinite at the top of the tube and decreases as the condensate film thickness increases down the tube. The heat-transfer coefficient for the evaporating solution also changes with position along the tube. At the bottom of the tube the fluid is a single phase, and if the fluid is not at the saturation temperature, the heat-transfer coefficient is predictable from the usual equations describing heat transfer in circular tubes. When boiling begins, however, the heat-transfer coefficient increases because of the contribution of the growing bubbles to turbulence in the laminar sublayer. In addition, the presence of the vapor bubbles decreases the bulk density of the fluid in the tube and causes an increase in velocity in accordance with the continuity equation. This effect further increases the heat-transfer coefficient and causes more rapid vapor formation. The effect is autocatalytic and gives a marked increase in the heat-transfer coefficient at successively higher positions. If sufficient vapor is formed, however, the heat-transfer coefficient is adversely affected because parts of the walls are in contact with only vapor.

As a result of the variation of both the condensing and boiling heat-transfer coefficients, it can be seen that the determination of the overall coefficient U from the individual resistances is likely to be extremely difficult, if not impossible. In fact, overall coefficients for evaporators are usually predicted from the performance of other evaporators operating under conditions as nearly similar as possible.

A further difficulty encountered in the prediction of evaporator performance is the choice of a significant overall temperature difference. The temperature of the liquid entering an evaporator tube may be below the saturation temperature for the pressure at that point. As the fluid moves up the tube, its temperature increases because of heating and its pressure decreases as in any fluid flowing in a tube of constant cross section. Thus the approach to saturation temperature is rapid. When boiling begins, the decreasing hydrostatic head and the increase in velocity as a result of vapor formation are additional factors causing a rapid decrease of pressure, so that the saturation temperature decreases markedly along the tube. Consequently, the local value of the overall temperature difference between the condensing steam and the boiling fluid is most difficult to predict. Although recent work has been done to provide a more rigorous approach to the problem, most evaporator design is done with an overall temperature difference which is the temperature of the saturated steam minus the saturation temperature of the strong solution at the pressure existing in the vapor space of the evaporator. Overall coefficients are computed from evaporator performance using this Δt, and the same method of defining the temperature difference is employed for the design of other systems.

The boiling point of a solution at a certain pressure is a function of the concentration of the solution. The calculation of the boiling-point rise due to dissolved solids is discussed in most textbooks of physical chemistry. In the design of evaporators it is common practice to use a representation known as a *Dühring plot* for calculating the boiling point of solutions. Such a plot is shown in Fig. 27-9. The boiling point of pure water at the pressure in the vapor space of the evaporator is determined using steam tables. The boiling point of the solution is then found on the Dühring plot using the appropriate value of the concentration parameter.

Multiple-Effect Operation

One of the principal costs of evaporation is the cost of steam for heating. A considerable reduction in operating costs is achieved by operating a battery of evaporators in which the overhead vapor from one evaporator (or "effect") is condensed in the steam chest of the next evaporator, thus saving both the cost of condensing the vapor from the first unit and supplying heat for the second. Several evaporators may operate in a battery in this fashion. If the vapor from one unit is to heat an adjacent unit, it must condense at a higher temperature than the boiling liquid in that effect. This temperature difference is achieved by operating succeeding units at lower pressures. The vapor from the last unit is usually sent to a condenser operating under a vacuum.

The method of operation shown in Fig. 27-10 is called *forward feeding*. By this method the dilute feed solution is supplied to the same effect that receives the high-pressure steam. Subsequent effects are at successively lower pressures. Another method of operation is known as *backward feeding*. With this sytem the high-pressure steam is supplied to the effect receiving the most concentrated liquid,

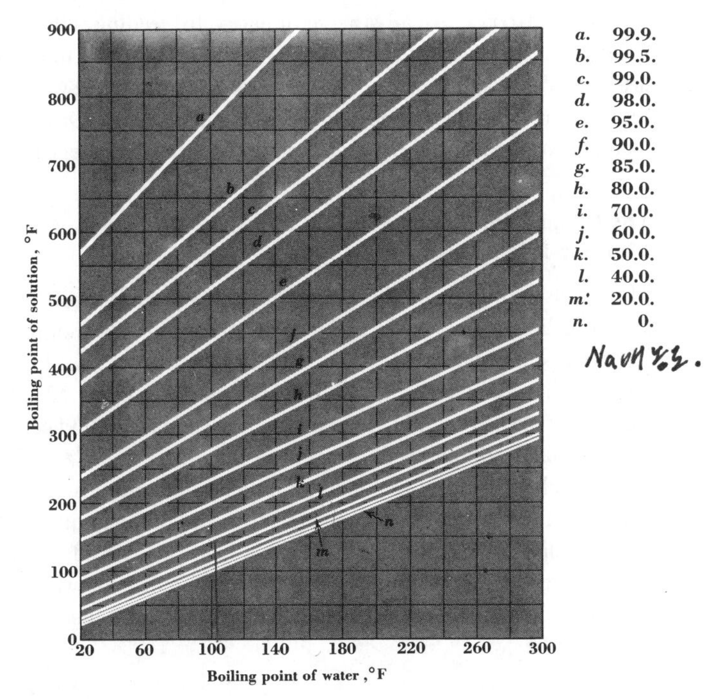

FIGURE 27-9
Boiling points of solutions of sodium hydroxide (percent by weight of NaOH). (*From G. G. Brown et al., "Unit Operations," John Wiley & Sons, Inc., New York, 1950.*)

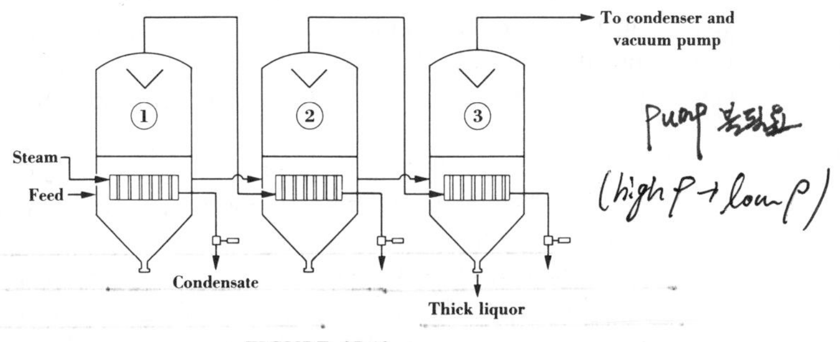

FIGURE 27-10
Forward feeding of multiple-effect evaporator battery.

whereas the dilute feed enters the effect supplied with steam at the lowest pressure in the entire battery. An advantage of backward feeding is that the most concentrated liquor is at the highest temperature, tending to give a more equal distribution of heat-transfer coefficients in the various effects. However, interstage pumps are required in this system and add to equipment costs. Other methods of operation include *parallel feeding*, in which fresh feed enters each effect of the system.

The Calculation of the Performance of Multiple-Effect Evaporator Systems

The precise solution of the theoretical equations for the design of an evaporator system is limited by the accuracy of the predicted values of the overall heat-transfer coefficients. However, it is necessary to estimate the heat-transfer area, the operating conditions of each unit, and the steam consumption. One basic consideration is that the areas are usually required to be identical in all effects of a battery, so as to simplify operation and reduce capital costs.

The largest items by far in an enthalpy balance around a single effect are the latent heat terms for the condensing steam and boiling liquid. The contributions to the enthalpy balance from the sensible heat of the feed, the thick liquor, and the condensate are much smaller, and if the differences between effects resulting from these liquid streams are neglected, the heat load on all effects is the same. Therefore we can write

$$U_1 \, \Delta t_1 = U_2 \, \Delta t_2 = U_3 \, \Delta t_3 \qquad (27\text{-}9)$$

This equation serves as a basis for estimating the temperature drop over each effect.

If the overall coefficients for the various effects are identical, we find

$$q = UA(\Delta t_1 + \Delta t_2 + \Delta t_3) \qquad (27\text{-}10)$$

which shows that the total rate of evaporation from the multiple-effect system is no greater than the rate that would be obtained from a single effect operating with a temperature difference equal to $\Delta t_1 + \Delta t_2 + \Delta t_3$. Thus it is evident that the advantage of multiple-effect operation is not in increasing capacity, but rather in steam and cooling-water economy.

Example 27-4 A triple-effect evaporator operating with backward feed is to be used to concentrate 45 tons/h of a solution of 10 mass percent NaOH in water to 50 mass percent. Saturated steam at 140 psia is supplied to the steam chest of the first effect, and a pressure of 1.5 psia is to be maintained in the vapor space of the last effect. The feed is supplied at 120°F. The overall heat-transfer coefficient for each effect is estimated to be 500 Btu/(h)(ft^2)(°F). Find the heat-transfer area of each effect.

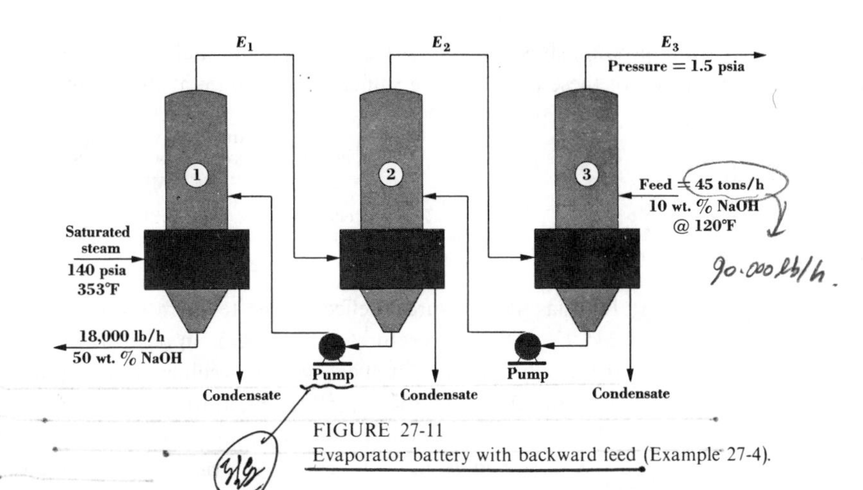

FIGURE 27-11
Evaporator battery with backward feed (Example 27-4).

The system is shown schematically in Fig. 27-11. First, some elements of the material balance are considered:

$$\text{Amount of concentrated liquor from effect 1} = \frac{(45)(2000)(0.10)}{0.50}$$

$$= 18{,}000 \text{ lb/h}$$

$$\text{Total evaporation rate} = 90{,}000 - 18{,}000$$

$$= 72{,}000 \text{ lb/h}$$

If equal evaporation rates are assumed for each effect, we have

$$E_1 = E_2 = E_3 = 24{,}000 \text{ lb/h}$$

The amount and concentration of the thick-liquor stream from each effect are found by a material balance. For example, considering effect 3:

$$\text{Amount of thick liquor} = 90{,}000 - 24{,}000$$

$$= 66{,}000$$

$$\text{Concentration of thick liquor} = \frac{(90{,}000)(0.10)}{66{,}000}$$

$$= 0.136$$

From Fig. 27-9 the boiling-point rise due to the dissolved sodium hydroxide in an aqueous solution of 13.6 mass percent concentration is 7°F at a pressure of 1.5 psia (boiling point of pure water equal to 115°F). The material balances for each of the three effects yield the information in the following table. The boiling-point rises for effects 1 and 2 depend slightly on pressure. Rough guesses of these pressures are sufficiently accurate at this point.

Thick liquor	Amount, lb/h	Concentration, mass % NaOH	BP rise, °F
From effect 1	18,000	50	76
From effect 2	42,000	21.4	14
From effect 3	66,000	13.6	7

At 1.5 psia, the pressure in effect 3, the temperature of saturated water vapor is 115°F. The temperature of the steam supplied to the first effect is 353°F. If the solution in each effect were not subject to a boiling-point rise due to the dissolved NaOH, the overall temperature-difference driving force to be shared between the three effects would be 353 − 115 = 238°F. However, effect of the dissolved solid is to cause a series of boiling-point rises of 76 + 14 + 7 = 97°F. As we have seen in Chap. 25, the superheat in a vapor stream does not contribute to the overall temperature-difference driving force. Thus 97° must be subtracted from the total of 238° to obtain the sum of effective Δt's in the three evaporators. This gives

$$\sum \Delta t = 238 - 97$$
$$= 141$$

Assuming $q_1 = q_2 = q_3$, $U_1 = U_2 = U_3$, and $A_1 = A_2 = A_3$, we obtain $\Delta t_1 = \Delta t_2 = \Delta t_3$; hence $\Delta t = \frac{141}{3} = 47°F$ for each effect. We can now specify the temperatures of all the streams, as shown in Fig. 27-12.

Initially, we assumed that $E_1 = E_2 = E_3 = 24{,}000$ lb/h, and this has permitted us to estimate roughly the temperatures and compositions of all streams. Using this information, shown in Fig. 27-12, we write enthalpy balances around each effect, using enthalpy information from the accompanying enthalpy chart, Fig. 27-13, for NaOH solutions, and from the steam tables.

Enthalpy balances:

For effect 1
$$(1193.0 - 324.8)S + 130(90{,}000 - E_2 - E_3) = 1194E_1 + 305(18{,}000)$$

For effect 2
$$(1194 - 199)E_1 + 78(90{,}000 - E_3) = 1140E_2 + 130(90{,}000 - E_2 - E_3)$$

For effect 3
$$(1140 - 137)E_2 + 78(90{,}000) = 1115E_3 + 78(90{,}000 - E_3)$$

In addition,

$$E_1 + E_2 + E_3 = 72{,}000$$

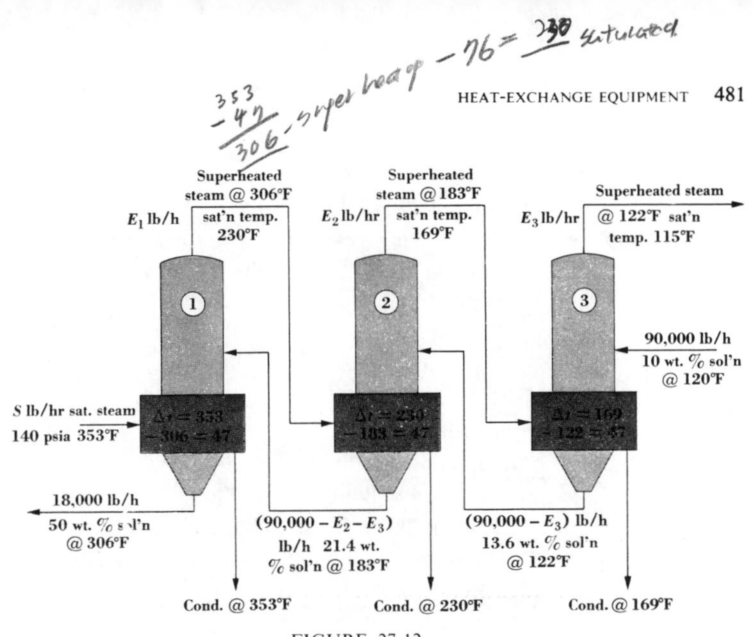

FIGURE 27-12
Temperatures and flow rates, first trial (Example 27-4).

These four equations contain four unknown quantities E_1, E_2, E_3, and S and may be solved simultaneously to give

$$E_1 = 26{,}200 \text{ lb/h}$$
$$E_2 = 23{,}100 \text{ lb/h}$$
$$E_3 = 22{,}700 \text{ lb/h}$$
$$S = 35{,}400 \text{ lb/h}$$

The areas of the effects can now be calculated, taking q as the total enthalpy change of the condensing vapor in each effect.

$$A_1 = \frac{(35{,}400)(1194.1 - 330.5)}{(500)(47)} = 1300 \text{ ft}^2$$

$$A_2 = \frac{(26{,}200)(1194 - 199)}{(500)(47)} = 1110 \text{ ft}^2$$

$$A_3 = \frac{(23{,}100)(1140 - 137)}{(500)(47)} = 987 \text{ ft}^2$$

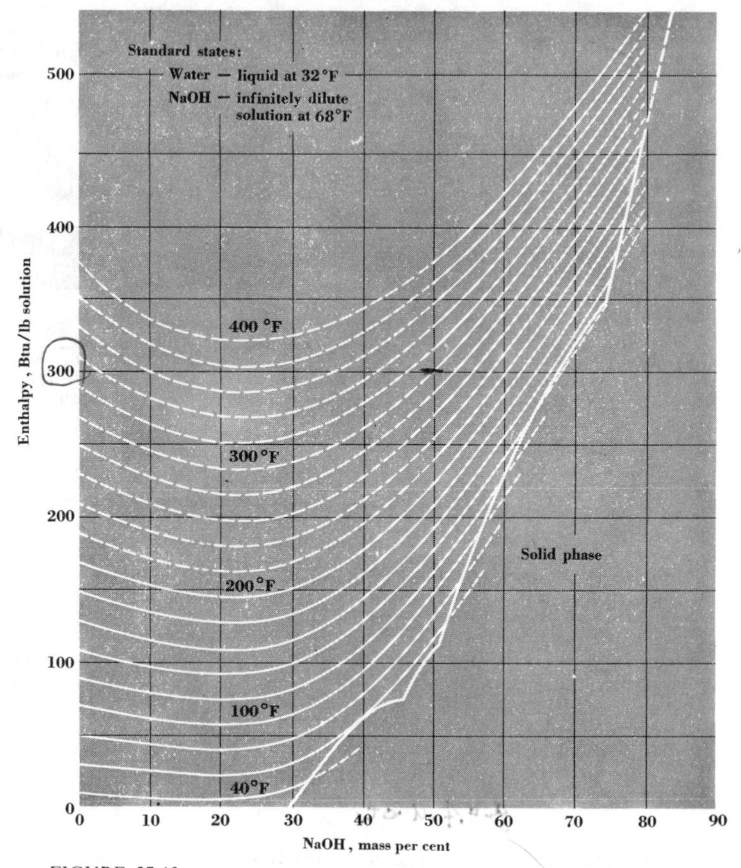

FIGURE 27-13
Enthalpy-concentration diagram for aqueous solutions of NaOH. (*From G. G. Brown et al., "Unit Operations," John Wiley & Sons, Inc., New York, 1950.*)

The arithmetic average area is 1130 ft^2. In view of the accuracy of U, this is probably an adequate answer. However, the calculations can be refined by modifying each value of Δt by the ratio of the calculated area to the average area as follows:

$$\Delta t_1 = \frac{(47)(1300)}{1130} = 54$$

$$\Delta t_2 = \frac{(47)(1110)}{1130} = 46$$

$$\Delta t_3 = \frac{(47)(987)}{1130} = 41$$

With these figures the temperature distribution previously estimated can be revised and the four enthalpy balances solved once more for E_1, E_2, E_3, and S. The results are

$$E_1 = 26{,}100 \text{ lb/h}$$
$$E_2 = 23{,}200 \text{ lb/h}$$
$$E_3 = 22{,}700 \text{ lb/h}$$
$$S = 33{,}000 \text{ lb/h}$$

from which we recalculate the areas and obtain

$$A_1 = 1080 \text{ ft}^2$$
$$A_2 = 1140 \text{ ft}^2$$
$$A_3 = 1120 \text{ ft}^2$$

The arithmetic average area is 1110 ft^2. The total rate of evaporation expressed as pounds of vapor removed from the feed solution per pound of steam supplied to effect 1 is 72,000/33,000, or 2.2. This ratio is known as the *steam economy*. For a triple-effect evaporator battery it might be expected to be roughly $\frac{3}{1}$. However, a large amount of heat is lost in the concentrated solution from the first effect, and only 26,100 lb/h of vapor is produced from 33,000 lb/h of steam. In subsequent effects the quantities are more nearly equal.

PROBLEMS

27-1 A double-pipe heat exchanger is constructed of 200 ft of 2-in, schedule-40 steel pipe jacketed with 3-in, schedule-40 steel pipe. Hot water at the rate of 30,000 lb/h under pressure enters the inside pipe at a temperature of 300°F, and 22,300 lb/h of cold water enters the jacket at 60°F. Calculate the temperatures of the two streams leaving the exchanger and the total heat-transfer rates when the streams are countercurrent and when they are concurrent. The convective coefficients of heat transfer inside and outside the clean tube may be taken as 1200 and 1175, respectively.

27-2 A multipass heat exchanger has two passes on the shell side and four passes on the tube side. It is cooling a petroleum oil inside the tubes from 135 to 52°C. Water is the coolant in the shell, with inlet and outlet temperatures of 13 and 31°C. The individual heat-transfer coefficients are h (oil) = 270, h (water) = 965, h (scale on water side) = 2840 W/m$^2\cdot$K. Neglect the metal-wall resistance. The exchanger has 120 1-in-OD, 16 BWG tubes, each 2 m long. Calculate the heat-transfer rate in W.

27-3 A lubricating oil is being heated in a concentric-pipe heat exchanger. The oil enters the inner pipe with a velocity of 3.5 ft/s at a temperature of 100°F. The inner pipe is a 1-in, schedule-40 steel pipe. The oil leaves at a temperature of 160°F. The annular space between the inner and outer pipes contains steam condensing at a uniform temperature of 220°F. The steam-side coefficient is 2000 Btu/(h)(ft^2)(°F). Calculate the length of the heat exchanger, assuming that the heat-transfer coefficient for the oil is constant at the arithmetic average temperature. The oil has the following properties at 130°F: viscosity, 17.0 cP; specific heat, 0.47 Btu/(lb)(°F); density, 55.9 lb/ft^3; thermal conductivity, 0.070 Btu/(h)(ft)(°F). The viscosity of the oil at 220°F is 4.2 cP.

27-4 One kilogram per second of an organic liquid is heated from 15 to 60°C in a concentric-pipe heat exchanger. Saturated steam at a uniform temperature of 171°C is condensing in the outer pipe, which is 8-in, schedule-40 steel pipe. The exchanger is 12 m long, and the inner section is 4-in, schedule-40 steel pipe.

If the inside pipe is replaced by a 6-in, schedule-40 pipe, calculate the length of exchanger required to give the same exit temperature of 60°C.

Data:

1 Specific heat of organic liquid = 2100 J/kg · K.

2 Condensing steam coefficient = 8500 W/m^2 · K—assume constant.

3 Developed turbulent flow may be assumed to exist in the inside pipe under both sets of conditions.

4 The steam condensate leaves the exchanger as a saturated liquid.

5 The pipe-wall resistance to heat transfer may be assumed to be negligible.

27-5 Air at 60°F is being heated by blowing it over a bank of five rows of 1-in, 14-gauge vertical copper tubes 4 ft high. The tubes are staggered and spaced on equilateral triangles 2 tube diameters on a side; there is $\frac{1}{2}$ in between the side tubes and the duct. There are five tubes per row. Steam at 5 psig is condensing inside the tubes; the condensate is not subcooled. The steam-side heat-transfer coefficient may be taken as 2000 Btu/(h)(ft^2)(°F), and the resistance of the tube wall may be neglected.

It is desired to heat 500 ft^3/min of air measured at 1 atm and 60°F. What will be the air temperature at the discharge end of the exchanger?

27-6 One kilogram per second of an organic liquid at 15°C is being heated on the tube side of a single-pass shell-and-tube heat exchanger by saturated steam condensing at 110°C on the shell side. The condensate is not subcooled. The exchanger consists of twenty $\frac{3}{8}$-in, 20-gauge copper tubes, each 6 m long. When the exchanger is first placed in operation, it is found that the organic liquid is heated to 90°C as it leaves the exchanger. After a certain period of operation, it is found that the liquid exit temperature is 80°C.

(*a*) What is the fouling coefficient under these conditions?

(*b*) At what temperature must saturated steam now be used to heat 1 kg of liquid per second to 90°C?

Assume that the effect of temperature on the inside heat-transfer coefficient can be neglected and that the resistances of the tube wall and the condensing steam are small compared with the convective resistance of the organic liquid. The specific heat of the organic liquid is 2300 J/kg · K.

27-7 A heat exchanger for chilling water is to be designed using $\frac{3}{4}$-in, 18-gauge copper tubes. Water at the rate of 40 gal/min is to enter the tube side of the single-pass shell-and-tube exchanger at 70°F and is to be cooled to 50°F. Heat is removed on the shell side by Freon-12 boiling at 47°F. A bulk water velocity of 5 ft/s in the tubes will be adequate, and under these circumstances the average inside heat-transfer coefficient h is 980 Btu/(h)(ft^2)(°F). The boiling coefficient is not constant, but is a function of the temperature difference over the boiling film. Boiling coefficients may be obtained by plotting the data given below on a log-log plot of h versus Δt. Determine the total length of tubing, the number of tubes required, and the length per tube, by the following methods:

(*a*) Assume U is constant at the value obtained for that point in the exchanger where the water is at 60°F.

(*b*) Assume U is a linear function of Δt.

(*c*) Use the function U actually obtained from the data.

Data:

Boiling coefficient, h Btu/(h)(ft^2)(°F)	400	600	800	1000	1400
Temperature drop over boiling film, Δt°F	6.8	8.6	10.5	12.2	14.9

27-8 Find the fin efficiency of a pin fin attached by one end to a tube surface which is at 200°F. Air flowing past the fin is at 60°F and has a convective coefficient $h = 10$ Btu/(h)(ft^2)(°F), which may be assumed to be constant at all points on the fin. The fin is a solid steel cylinder $\frac{1}{8}$ in in diameter and 1 in high. Temperature gradients parallel to the base of the fin may be neglected.

27-9 An evaporator is supplied with an aqueous feed solution containing 5 mass percent solids at a rate of 5 kg/s and a temperature of 93°C. The thick liquor leaves the evaporator at a temperature of 66°C, containing 45 mass percent solids. The saturation temperature of the vapor from the evaporator is 52°C. Saturated steam at 121°C is condensed in the steam chest and leaves as condensate at the same temperature. Assuming the specific heats of all solutions are equal to 4184 J/kg·K and the overall coefficient of heat transfer is 3600 W/m^2·K, find the heat-transfer area and the steam economy.

27-10 An aqueous glue solution is to be concentrated from 4 to 50 wt percent solids in a double-effect evaporator. Twenty tons of feed enters the system per hour, preheated to the boiling point of the effect into which it is introduced. Steam is available at 10 psig, and condenser facilities are capable of producing an absolute pressure of 4 inHg in the last effect. Assume that there is no rise in boiling point due to dissolved solids and that the specific heats of all solutions are constant at 1.2 Btu/(lb)(°F). All condensate streams leave at the condensing temperatures.

(*a*) Using forward feed, calculate the area of the heating surface required, the steam consumption, and the steam economy for each effect. The areas in each effect are to be equal. The overall heat-transfer coefficients are $U_1 = 400$ and $U_2 = 350$ Btu/(h)(ft^2)(°F).

(*b*) Repeat part (*a*), using backward feed and the coefficients given.

(*c*) Repeat part (*a*), using parallel feed and the coefficients given.

27-11 The tubes on which the data of Fig. 27-7 were taken had an OD of 1.25 in, a wall thickness of 0.065 in, and a length of 25 in. Find the boiling coefficient of heat transfer for each set of runs, assuming that no fouling was present.

27-12 A shell-and-tube heat exchanger being used to heat 7.6 kg/s of oil from 27 to 150°C is made of 300 1-in, 16 BWG copper tubes which are 4 m long. There are two tube passes and one shell pass. The oil is heated by steam condensing on the shell side at 177°C and 930.8 kPa abs. Calculate the oil-side heat-transfer coefficient. Neglect the resistance of the tube wall. The specific heat of the oil is 3350 J/kg·K and the steam-side coefficient is 4540 W/m²·K.

27-13 Oil A is heated from 70 to 136°F on the tube side of a shell-and-tube heat exchanger. The flow rate of the oil is 300,000 lb/h. The fluid B on the shell side is cooled from 200°F to some undetermined temperature. The heat exchanger available has one shell pass and four tube passes. The shell ID is 39 in and contains 1024 $\frac{3}{4}$-in, 14 BWG, copper tubes, each 12 ft long on 1-in square pitch. The h of the shell side can be assumed to be 160 Btu/(h)(ft²)(°F). Calculate T_2 of the shell-side fluid and its flow rate in lb/h.

Data (oil A properties constant from 130°F to 220°F):

$Cp = 0.475$ Btu/(lb)(°F)
$\mu = 2.9$ cP
$k = 0.0789$ Btu/(ft²)(h)(°F)/ft
$\rho = 51.5$ lb/ft³
Cp of shell-side fluid = 0.475 Btu/(lb)(°F).

27-14 Light oil is heated from 70 to 200°F as it flows through the inner pipe of a double-pipe heat exchanger. Saturated steam condenses in the annular space at a uniform temperature of 250°F. The total exchanger length is 100 ft, the inner section consists of 1-in, schedule-40 steel pipe, and the oil flow rate is 26.9 gal/min. If the inner pipe is replaced with an equal length of $1\frac{1}{2}$-in, schedule-40 steel pipe, and the flow rate of the oil is still 26.9 gal/min, what is the new exit temperature of the oil? Assume turbulent flow at all times and neglect the heat-transfer resistances of condensing steam, pipe wall, and fouling. The specific heat of the oil is 0.50 Btu/(lb)(°F) and the density is 40 lb/(ft³).

27-15 A countercurrent, double-pipe heat exchanger with a tube surface area A_o of 10 ft² cools 1000 lb/h of hot oil [specific heat = 0.40 Btu/(lb)(°F)] from 150 to 100°F using cooling water which increases in temperature from 60 to 90°F through the exchanger. If an emergency should occur which would cause the inlet temperature of the available cooling water to be 70°F instead of 60°F, the water flow rate would have to be increased to keep the exit oil temperature at 100°F. Find the new exit temperature of the water and the new water flow rate. Assume the overall coefficient of heat transfer is not significantly changed by increasing the water flow rate.

27-16 Air is being heated by blowing it over a bank of seven rows of 1-in, 14-gauge copper tubes 2 ft long. The tubes are placed on a square grid with center-to-center distance of 2 in. There are seven tubes per row, and there is a space of 1 in between the outside tubes and the duct. The air ($p = 14.7$ psia) is to be heated from 60 to 100°F by hot water (under pressure) which flows in the tubes, entering at 210°F and leaving at 170°F. Assume that all the heat-transfer resistance is on the air side. Find the flow rate (lb/h) of air which is being heated. Properties of the air are: thermal conductivity = 0.0156 Btu/(h)(ft)(°F); viscosity = 0.0183 cP; specific heat = 0.24 Btu/(lb)(°F). This kind of exchanger is called a *crossflow exchanger*, and it has just one shell pass, with seven parallel rows of tubes.

27.1.

cold water 60°F 22300 lb/h
hot water 300°F, 30000 lb/h $h_i = 1200$
$h_o = 1175$
2.067 / 2.375 in.
200 ft

steel pipe ⇒ $k = 26$ Btu/h ft °F
※ assumption
→ steady-state.

$\frac{1}{h_oA_o} = \frac{1}{1175 \cdot \pi \cdot D_o L} = \frac{1}{1175 \cdot \pi \cdot \frac{2.375}{12} \cdot 1}$

$\frac{\Delta r}{KA_{em}} = \frac{1/24(2.375-2.067)}{26 \cdot \frac{\pi 1/12(2.375-2.067)}{\ln(2.375/2.067)}} = 0.000850249$

$\frac{1}{h_iA_i} = \frac{1}{1200 \cdot \pi \cdot D_i L} = \frac{1}{1200 \cdot \pi \cdot \frac{2.067}{12} \cdot 1}$

i) counter current

300°F, t_1, t_2, 60°F, q

$\frac{1}{h_oA_o} \quad \frac{\Delta r}{KA_{em}} \quad \frac{1}{h_iA_i} = \frac{\Sigma R}{L}$

• $W_h \cdot C_{p,h} \cdot \Delta t_h = W_c \cdot C_{p,c} \cdot \Delta t_c$ 같은 water. $C_{ph} = C_{pc} = 1$ Btu/lb°F

⇒ $30000(300-t_1) = 22300(t_2-60)$

→ $t_2 = 1/223(103380 - 300t_1)$, $W_h C_{ph} \Delta t_h = q = \frac{\Delta t_{lm}}{\Sigma R/L}$

$= \dfrac{\dfrac{(300-t_2)-(t_1-60)}{\ln[(300-t_2)/(t_1-60)]} \cdot 200}{\dfrac{1}{1175 \cdot \pi \cdot \frac{2.375}{12}} + 0.000850249 + \dfrac{1}{1200 \cdot \pi \cdot \frac{2.067}{12}}}$ ⇒ $30000 \cdot 1 \cdot (300-t_1)$

$= \dfrac{300-t_1-t_2}{\ln\left(\frac{300-t_2}{t_1-60}\right)} \cdot 226.03 \cdot 200$

t_2에 ① 대입하면 $t_1 = 163.17$°F

$t_2 = \frac{1}{223}(103380 - 300t_1) = 244.08$°F

$q = 30000(300 - 163.17) = 4104900$ Btu/h

ii) ~~counter current~~ concurrent
— counter current 와 마찬가지로.

300, 60, t_1, t_2, q

$t_2 = \frac{1}{223}(103380 - 300t_1)$

$q = 30000(300 - t_1)$

$= \dfrac{(300-60)-(t_1-t_2)}{\ln\left(\frac{300-60}{t_1-t_2}\right)} \cdot 266.03 \cdot 200$

$t_1 = 199.27$°F $\quad \mathcal{E} = \dfrac{240 - t_1 + \frac{1}{223}(103380 - 300t_1)}{\ln\left(240 / t_1 - \frac{1}{223}(103380 - 300t_1)\right)} \cdot 266.03 \cdot 200$

$t_2 = \frac{1}{223}(103380 - 300t_1) = 195.51$°F

∴ $q = 30000(300 - 199.27) = 3021900$ Btu/h

· ch 29). 3) $D_{AB} = 7.4 \times 10^{-8} \frac{(x \cdot M_B)^{1/2}}{V_A^{0.6}} \cdot \frac{T}{\mu_B}$

-batch polymerization, water temp 일정. x, M_B, T, μ_B - const.

$$D_{AB} \propto \frac{1}{V_A^{0.6}}$$

$$\Rightarrow \frac{D_{AB \cdot 1000Å}}{D_{AB \cdot 10Å}} = \frac{(V_{A \cdot 10Å})^{0.6}}{(V_{A \cdot 1000Å})^{0.6}} \Rightarrow D_{AB \cdot 100Å} = \left(\frac{\frac{\pi}{6} \times (10Å)^3}{\frac{\pi}{6}(1000Å)^3}\right)^{0.6} \cdot D_{AB \cdot 10Å}$$

$$\Rightarrow D_{AB \cdot 100Å} = \left(\frac{10Å}{1000Å}\right)^{3 \times 0.6} \cdot 10^{-6} = 2.512 \times 10^{-10} cm^2/s.$$

4. $D_{AB} = 7.4 \times 10^{-8} \cdot \frac{(x M_B)^{1/2}}{V_A^{0.6}} \cdot \frac{T}{\mu_B}$, B = water, $x = 2.6$, $M_B = 18$

$\mu_B = $ 1.1404 cp (at 15°C), 1.0050 cp (at 20°C)

i) methanol (density : 17.912 kmol/m³,

$= 55.828\ cm^3/mol$ ↑↓ V_A

5. i) 1atm, 293K, ⟨Air-Benzene system⟩

$$D_{AB} = \frac{1 \times 10^{-3} T^{1.75} [(1/M_A) + (1/M_B)]^{1/2}}{P[(\Sigma_A V_i)^{1/3} + (\Sigma_B V_i)^{1/3}]^2}$$

$$= \frac{1 \times 10^{-3} \cdot (293)^{1.75} \cdot [(1/29) + (1/78)]^{1/2}}{1 \cdot [(20.1)^{1/3} + (6(16.5 + 1.98) - 20.2)^{1/3}]^2} = 0.744\ cm^2/s$$

ii) from table 29-3,

$$D_{AB} \propto T^{1.75}.$$

$$D_{AB}(293K) = \left(\frac{293k}{298k}\right)^{1.75} \cdot 0.0962.$$

$$= 0.0934\ cm^2/s$$

page8.) Notation을 봐라.

1. Diffusivity.

Solid: $D_{AB} = D_0 \cdot \exp(-E/RT)$, D_{AB} [=] cm^2/s. $10^{-8} \sim 10^{-10}$

liquid: $D_{AB} \propto \frac{T}{\mu_{AB}}$ $\sim 10^{-5} \sim 10^{-6}$

(Stokes-Einstein)

gases: $D_{AB} \sim \frac{T^{1.5\sim1.8}}{P}$ $\sim 0.1 \sim 1$

• Fourier Number: $\frac{D_{AB}\cdot\theta}{L^2} \sim 1$: 정강강에. $\begin{cases} D_{AB}\cdot\theta = L^2 \\ \sqrt{D_{AB}\cdot\theta} = L \end{cases}$

$L \propto \sqrt{\nu\theta}$ (유체) → kinematic viscosit

$L \propto \sqrt{\alpha\theta}$ (열) → thermal diffusivity

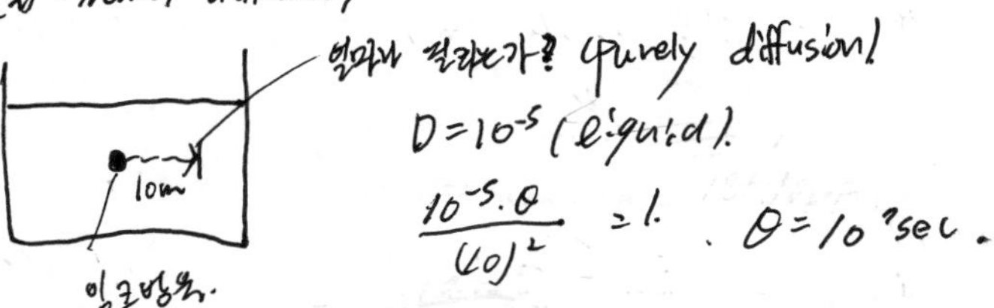

• Knudsen Number; $K_n = \frac{\lambda}{L} = 1$

$L = \lambda$ 일때 이동.

활성탄. 촉매, L = pore size. λ = mean free path.

보통 diffusion과 다른 knudsen diffusion.

- 벽의 영향을 받는데.

$K_n = \frac{\lambda}{L}$:

$D_{KA} = \frac{\bar{V}_A \cdot L}{3}$

$\bar{V}_A = \left(\frac{8RT}{\pi M_A}\right)^{1/2}$

PART THREE

Mass Transfer

ex 30.1) Air

Iodine 1g at 85°C

Air - Stagnat, $\rightarrow 0$

0.08m

1.5×10^{-3}m

$$\tilde{N}_A = -\frac{1}{A} \cdot \frac{1}{M} \cdot \frac{dm}{d\theta} \quad ①$$

$A = \frac{\pi}{4}D^2 = \frac{\pi}{4}(1.5\times10^{-3}m)^2 = 1.767\times10^{-6}m^2$

$M = 126.9$ g/mol.

at stedy state diffusion)

$$\tilde{N}_A = \frac{D_{AB}\cdot \tilde{c}(\tilde{x}_{A0} - \tilde{x}_{AL})}{L\cdot \tilde{x}_{B\cdot lm}} \quad ②$$

$$D_{AB} = 0.07\,cm^2/s \cdot \left(\frac{358.15}{273.15}\right)^{1.5} = 0.105\,cm^2/s \times 10^{-4}\,m^2/s$$

L = 0.08 m.

$$\tilde{c} = \frac{P}{RT} = \frac{1atm}{(0.082)(273+85)} = 0.03406\,\frac{mol}{L} \times \frac{10^3}{m^3} = \frac{kmol}{m^3}$$

$$\tilde{x}_{A\cdot 0} = \frac{26.78\,mmHg}{760\,mmHg} = 0.0352. \quad \tilde{x}_{B0} = (1-0.0352)$$

(증기압)

$\tilde{x}_{A\cdot L} = 0,$ $\quad \tilde{x}_{B\cdot L} = 1,$

$$\tilde{x}_{B\cdot lm} = \frac{\tilde{x}_{BL} - \tilde{x}_{B\cdot 0}}{\ln(\tilde{x}_{BL}/\tilde{x}_{B\cdot 0})} = \frac{1-(1-0.0352)}{\ln\left(1/(1-0.0352)\right)} = 0.9823$$

① = ②

$$-\frac{1}{A\cdot M}\frac{dm}{d\theta} = \frac{D_{AB}\cdot \tilde{c}}{L\cdot \tilde{x}_{B\cdot lm}} \cdot (\tilde{x}_{A0} - \tilde{x}_{AL})$$

$\theta = 0 \quad m = m$

$\theta = \theta \quad m = 0$

$$\theta = \frac{m}{A\cdot M} \cdot \frac{L\cdot \tilde{x}_{B\cdot lm}}{D_{AB}\cdot \tilde{c}} \cdot \frac{1}{(\tilde{x}_{A0} - \tilde{x}_{AL})}$$

$$= \frac{0.001\,kg}{(1.767\times10^{-6}m^2)(126.9\,kg/kmol)} \cdot \frac{0.08 \cdot 0.9823}{(0.105\times10^{-4}\,m^2/s)\times 0.03406 \cdot 0.352}$$

$= 2.78\times10^{7}$s $= 322.2$ days.

28

INTRODUCTION TO MASS TRANSFER

General Remarks

Some of the most typical chemical engineering problems lie in the field of mass transfer. As mentioned at the beginning of this book, a distinguishing mark of the chemical engineer is his ability to design and operate equipment in which reactants are prepared, chemical reactions take place, and separations of the resulting products are made. This ability rests largely on a proficiency in the science of mass transfer. Applications of the principles of momentum and heat transfer are common in many branches of engineering, but the application of mass transfer has traditionally been largely limited to chemical engineering. Other important applications occur in metallurgical processes, in problems of high-speed flight, and in waste treatment and pollution-control processes.

By mass transfer is meant the tendency of a component in a mixture to travel from a region of high concentration to one of low concentration. For example, if an open test tube with some water in the bottom is placed in a room in which the air is relatively dry, water vapor will diffuse out through the column of air in the test tube. There is a mass transfer of water from a place where its concentration is high (just above the liquid surface) to a place where its concentration is low (at the outlet of the tube). If the gas mixture in the tube is stagnant, the transfer occurs by molecular diffusion. If there is a bulk mixing of the layers of gas in the tube by mechanical stirring or because of a density gradient, mass transfer occurs primarily by the mechanism of forced or natural convection. These mechanisms are analogous to the transfer of heat by conduction and by convection; there is, however, no counterpart in mass transfer for thermal radiation.

The analogy between momentum and energy transfer has already been studied in some detail, and it is now possible to extend the analogy to include mass transfer. This matter has been mentioned in earlier chapters and will be studied in detail in the chapters to follow.

In discussing the fundamentals of mass transfer we shall consider mainly binary mixtures, although multicomponent mixtures are important in industrial applications. Some of these more complicated situations will be discussed after the basic principles have been illustrated in terms of binary mixtures.

Molecular Diffusion

Molecular diffusion occurs in a gas as a result of the random motion of the molecules. This motion is sometimes referred to as a *random walk*. Across a plane normal to the direction of the concentration gradient (or any other plane), there are fluxes of molecules in both directions. The direction of movement for any one molecule is independent of the concentration in dilute solutions. Consequently, in a system in which there is a concentration gradient, the fraction of molecules of a particular species (referred to as species A) which will move across a plane normal to the gradient is the same for both the high- and low-concentration sides of the plane. Because the total number of molecules of A on the high-concentration side is greater than on the low-concentration side, there is therefore a net movement of A in the direction in which the concentration of A is lower. If there are no counteracting effects, the concentrations throughout the mixture tend to become the same. In the analogous transfer of heat in a gas by conduction, the distribution of hotter molecules (those which have a higher degree of random molecular motion) tends to be evened out by random mixing on a molecular scale. Similarly, if there is a gradient of directed velocity (as distinguished from random velocity) across the plane, the velocity distribution tends toward uniformity as a result of the random molecular mixing. There is a transfer of momentum, which is proportional to the viscosity of the gas.

The above remarks apply only in an approximate and qualitative way. The quantitative prediction of the diffusivity, thermal conductivity, and viscosity of a gas from a knowledge of molecular properties can be quite complicated. The consideration of such relations forms an important part of the subject of statistical mechanics.

Molecular diffusion also occurs in liquids and solids. Crystals in an unsaturated solution dissolve, with subsequent diffusion away from the solid-liquid interface. Diffusion in solids is of importance in metallurgical operations. When iron which is unsaturated with respect to carbon is heated in a bed of coke, the concentration of the carbon near the surface is increased by inward diffusion of carbon atoms.

Eddy Diffusion

Just as momentum and energy can be transferred by the motion of finite parcels of fluid, so mass can be transferred. We have seen that the rate of these transfer operations, caused by bulk mixing in a fluid, can be expressed in terms of the eddy kinematic viscosity, the eddy thermal diffusivity, and the eddy diffusivity. This latter quantity can be related to a mixing length which is the same as that defined in connection with momentum and energy transfer. In fact, the analogy between heat and mass transfer is so straightforward that equations developed for the former are often found to apply to the latter by a mere change in the meaning of the symbols. The reader is referred to the discussions in Chaps. 2 and 23.

Eddy diffusion is apparent in the dissipation of smoke from a smokestack. Turbulence causes mixing and transfer of the smoke to the surrounding atmosphere. In certain locations where atmospheric turbulence is lacking, smoke originating at the surface of the earth is dissipated largely by molecular diffusion. This causes serious pollution problems because mass is transferred less rapidly by molecular diffusion than by eddy diffusion.

Convective Mass-Transfer Coefficients

In the study of heat transfer we found that the solution of the differential energy balance was sometimes cumbersome or impossible, and it was convenient to express the rate of heat flow in terms of a convective heat-transfer coefficient by an equation like

$$\frac{q}{A} = h(t_s - t_m) \qquad (21\text{-}1)$$

The analogous situation in mass transfer is handled by an equation of the form

$$N_A = k_\rho(\rho_{As} - \rho_{Am}) \qquad (28\text{-}1)$$

The mass flux N_A is measured relative to a set of axes fixed in space. The driving force is the difference between the concentration at the phase boundary (a solid surface or a fluid interface) and the concentration at some arbitrarily defined point in the fluid medium. The convective coefficient k_ρ may apply to forced or natural convection; there are no mass-transfer counterparts for boiling, condensation, or radiation heat-transfer coefficients. The value of k_ρ is a function of the geometry of the system and the velocity and properties of the fluid, just as was the coefficient h.

Separation Operations

The size of separation equipment depends on the rate at which a component is transferred from one phase to another, and this rate is in turn dependent on how fast the component is transferred to the interphase boundary. As the two individual heat-transfer coefficients influence the size of a heat exchanger, so the gas- and liquid-phase mass-transfer coefficients affect the size of a gas-absorption column.

For instance, ammonia can be separated from a mixture with air by passing the gas up through an absorption tower. Water is fed to the top of the tower and flows downward countercurrent to the rising gas in such a way that there is a large area of interphase contact. At any point in the tower the concentration of ammonia in the gas phase exceeds the concentration which would be in equilibrium with the water phase. As a result, the ammonia travels through the gas toward the surface of the water, is absorbed, and travels away from the interface into the interior of the water phase. The height of the tower is largely determined by the rate of mass transfer of ammonia from one phase to the other, and this rate will be expressed in terms of concentration differences and mass-transfer coefficients.

If the tower is sufficiently high, almost all the ammonia is removed from the air before it leaves at the top. To separate the ammonia from the water, the solution leaving the bottom of the tower is subjected to fractional distillation.

Although the rate of mass transfer is the dominant factor in the design of a gas-absorption tower or other separation device, the overall mass and energy balances must also be satisfied. The capacity of a tower is determined by its diameter, and this dimension is calculated by the use of the principles of fluid dynamics to compute the permissible velocities of the fluid phases in the tower. The principles of thermodynamics assist in the prediction of the phase-equilibrium relations which govern the maximum driving forces available.

In the rest of this book we shall study, first, the principles of mass transfer through a single phase by the several possible mechanisms, and then we shall apply the results to interphase transfer as encountered in some of the common separation operations.

29

MOLECULAR DIFFUSION AND DIFFUSIVITY

Fick's Law

The rate of mass transfer is given by Fick's (first) law as[1]

$$\tilde{I}_A = -D_{AB}\tilde{\rho}\frac{d\tilde{x}_A}{dy} \qquad (29\text{-}1)$$

for ordinary molecular diffusion in the y direction. The quantity $\tilde{I}_A$ is the flux of constituent A, measured in lb mol/(h)(ft^2) with respect to axes moving with the molar average velocity $\tilde{u}$ of the fluid mixture. The driving force for diffusion is the gradient $d\tilde{x}_A/dy$, measured in ft^{-1}, and the proportionality factor D_{AB} represents the diffusivity, or coefficient of diffusion (ft^2/h) for a binary mixture of A and B. This equation is the form which naturally arises from the kinetic theory of (ideal) gases; the molecular flux $\tilde{I}_A$ is proportional to the gradient in terms of molecules per (cm^3)(cm).

The use of the molar flux $\tilde{I}_A$ has come about because of its convenient relation to common laboratory experiments for the measurement of diffusivities. By suitable methods, two different gas compositions are established in two well-defined,

[1] A similar equation was first given by A. Fick, *Ann. Phys.* (Leipzig), **170**:59 (1855).

isolated regions in a constant-volume apparatus. The two regions, both at identical pressure and temperature, are then put into communication in such a way that $\tilde{I}_A$ and $d\tilde{x}_A/dy$ can be measured to give D_{AB} by Eq. (29-1). For ideal gases, the molar concentration $\tilde{\rho}$ is everywhere the same, so that the molar average velocity $\tilde{u}$ is zero. The observed flux $\tilde{N}_A$, measured with respect to the apparatus fixed in space, is identical to $\tilde{I}_A$. This is a consequence of equimolar counterdiffusion. For nonideal systems the numerical value of D_{AB} depends on the choice of reference frame; in this book, for gaseous systems, we limit our discussion to ideal gases, for which $\tilde{\rho}$ is constant. For diffusion in open systems there may be a flux because of bulk flow with respect to a plane fixed in space ($\tilde{u} \neq 0$); then $\tilde{I}_A$ and $\tilde{N}_A$ would no longer be equal. This matter is discussed further in Chap. 30.

Since $\tilde{I}_A$ and $\tilde{I}_B$ represent molar fluxes with respect to the center of moles,[1] it is always necessary that their sum be zero:

$$\tilde{I}_A + \tilde{I}_B = 0 \qquad (29\text{-}2)$$

For liquids we base our definition of D_{AB} on the flux J_A in lb/(h)(ft^2) with respect to the center of mass. The appropriate form of Fick's law is then

$$J_A = -D_{AB}\rho \frac{dx_A}{dy} \qquad (29\text{-}3)$$

For an interdiffusing system of constant density the flux J_A is identical to N_A, the flux with respect to a stationary closed apparatus. In such a system the center of mass is stationary, so the mass average velocity u is zero. It is customary to use Eq. (29-3) for liquids, because the molar density may vary sharply with composition for a liquid; the density is often nearly constant.[2] By definition, we have

$$J_A + J_B = 0 \qquad (29\text{-}4)$$

which is analogous to Eq. (29-2). In Chap. 30 we shall discuss in more detail the relations among the fluxes $\tilde{I}_A$, J_A, N_A, and $\tilde{N}_A$.

Equations (29-1) and (29-3) are but two of many possible forms of Fick's law. For instance, the principles of irreversible thermodynamics[3] indicate that the proper driving force for nonideal systems is the gradient of the chemical potential $d\mu_A/dy$. Without entering into the theoretical basis of this idea, we can understand its plausibility by analogy to momentum and energy transfer. In momentum transfer the shear stress approaches zero as the velocity gradient decreases, and two

[1] The center of moles moves with the molar average velocity $\tilde{u}$.

[2] As an example, the densities of aqueous ethyl alcohol solutions range from 62.4 to 49.9 lb/ft^3, whereas the molar densities range from 3.0 to 1.0 lb mol/ft^3.

[3] A basic reference is S. R. de Groot and P. Mazur, "Nonequilibrium Thermodynamics," North-Holland, Amsterdam, 1962. A simple treatment is given by D. D. Fitts, "Nonequilibrium Thermodynamics," McGraw-Hill Book Company, New York, 1962.

parts of a system come to thermal equilibrium ($q/A = 0$) when the temperatures in the two parts become equal. Thus it is logical to assume that the chemical potential is the driving force in mass transfer, since a system comes to chemical or composition equilibrium as the chemical potential of each component becomes equal throughout the system. We therefore write

$$\tilde{I}_A = -L_{AB}\frac{d\mu_A}{dy} \qquad (29\text{-}5)$$

where L_{AB} is a proportionality factor (phenomenological coefficient) related to the diffusivity. The chemical potential μ_A (Btu/lb mol) can be found in terms of concentrations.

From thermodynamics, the chemical potential is related to the activity by

$$d\mu_A = RT\, d \ln a_A \qquad (29\text{-}6)$$

The activity is in turn related to the mole fraction $\tilde{x}_A$ by

$$a_A = \frac{\gamma_A \tilde{\rho}_A}{\tilde{\rho}} = \gamma_A \tilde{x}_A \qquad (29\text{-}7)$$

in which γ_A is called the *activity coefficient*. The standard state, for which the activity is unity, is here chosen as pure A ($\tilde{x}_A = 1.0$). The activity coefficient with this standard state always approaches unity as the mole fraction approaches unity, and for ideal solutions, γ_A is unity for all values of $\tilde{x}_A$.

Equations (29-6) and (29-7) may be used to express the gradient of chemical potential as

$$\frac{d\mu_A}{dy} = RT\left(\frac{d \ln \gamma_A}{dy} + \frac{d \ln \tilde{x}_A}{dy}\right) \qquad (29\text{-}8)$$

or

$$\frac{d\mu_A}{dy} = \frac{RT}{\tilde{x}_A}\left(\tilde{x}_A \frac{d \ln \gamma_A}{d\tilde{x}_A} + 1\right)\frac{d\tilde{x}_A}{dy} \qquad (29\text{-}9)$$

When Eq. (29-9) is substituted in Eq. (29-5), we get

$$-\tilde{I}_A = \frac{L_{AB}RT}{\tilde{x}_A}\left(\tilde{x}_A \frac{d \ln \gamma_A}{d\tilde{x}_A} + 1\right)\frac{d\tilde{x}_A}{dy} \qquad (29\text{-}10)$$

Since the relation between γ_A and $\tilde{x}_A$ can be found for a nonideal system by the usual methods of classical thermodynamics, the quantity in parentheses in Eq. (29-10) can be evaluated as a function of composition at a given pressure and temperature.

In ideal systems $d \ln \gamma_A / d\tilde{x}_A$ is always zero; thus Eq. (29-10) becomes

$$\tilde{I}_A = -\frac{L_{AB}RT}{\tilde{x}_A}\frac{d\tilde{x}_A}{dy} \qquad (29\text{-}11)$$

By comparing this equation with our first form of Fick's law, Eq. (29-1), we obtain

$$L_{AB} = \frac{D_{AB}\tilde{\rho}_A}{RT} \qquad (29\text{-}12)$$

where we have used the relation

$$\tilde{\rho}_A = \tilde{\rho}\tilde{x}_A$$

For a system of constant $\tilde{\rho}$, Eq. (29-1) can be written as

$$\tilde{I}_A = -D_{AB}\frac{d\tilde{\rho}_A}{dy} \qquad (29\text{-}13)$$

and for a system of constant ρ, Eq. (29-3) also can be written as

$$J_A = -D_{AB}\frac{d\rho_A}{dy} \qquad (29\text{-}14)$$

which is the same as Eq. (2-19). These forms of Fick's laws are those frequently found in classical texts.

The preceding discussion has been presented here because the phenomenological coefficient L_{AB} often appears in the literature, but it is evident that it is merely an alternative to the use of D_{AB}. For example, for ideal gases it is known from theory and experiment that D_{AB} varies only slightly with composition. However, from Eq. (29-12), we see that L_{AB} is proportional to the mole fraction for an ideal gas, so that L_{AB} is not useful for this situation. For nonideal gases, liquids, and solids, no form of Fick's law has been found that assures a reasonably constant diffusion coefficient.[1]

Since all such equations are empirical in basis, we choose the simple forms, Eqs. (29-1) and (29-3), as the most practical, accepting the fact that D_{AB} often varies widely with composition. Of course, D_{AB} is also a function of pressure and temperature, as are the analogous coefficients, the viscosity and the thermal conductivity.

The expressions for the fluxes $\tilde{I}_A$ or J_A must be combined with the mass balances of Chaps. 3 or 7 in the solution of mass-transfer problems. Since these balances were made in terms of mass and not moles, J_A has appeared therein, e.g., Eq. (7-18). We shall apply these equations only to situations where it is reasonable to assume the diffusivity and the density are constant,[2] leading to Eq. (2-17). As already mentioned, the flux $\tilde{I}_A$ is particularly useful for gases, and its use will be developed in the following chapters.

[1] The variation of D_{AB} with composition for binary liquid systems has been discussed by J. Leffler and H. T. Cullinan, Jr., *Ind. Eng. Chem., Fundam.*, **9**:84 (1970), and previously by A. Vignes, *Ind. Eng. Chem., Fundam.*, **5**:189 (1966).

[2] A useful paper which discusses the forms of Fick's law which may be convenient for various situations is H. T. Cullinan, Jr., *Ind. Eng. Chem. Fundam.*, **4**:133 (1965).

Diffusivities

Solids One component in a solid will diffuse through the other at a measurable rate if there is a suitable concentration gradient and the temperature is high enough. The effects of diffusion in solids are very important in metallurgy. As mentioned earlier, the depth to which carbon will penetrate in a given time from the surface of steel being carburized (case-hardened) is governed by the laws of diffusion. The rate of reaction in some chemical processes is determined by solid diffusion, but the number of applications of importance to the chemical engineer is much less than the number concerned with diffusion in liquids and gases.

The atoms in a crystalline solid are localized at certain positions defined by the appropriate lattice, such as face-centered cubic. An atom vibrates about its equilibrium position, and occasionally a vibration becomes strong enough to cause the atom to escape its original position and jump to a new place. Its energy is low at the lattice position but higher (unfavorable) in between; the atom needs a certain activation energy E to jump over this energy barrier.

Let us now consider the movement of an impurity atom which is similar to the solvent atoms so that it fits into the host lattice positions. In order for the atom to be able to jump to an adjacent site, the latter must be vacant. Such vacancies do exist, and they also wander about in the lattice owing to the jumping of the solvent atoms. Any atom may jump in any possible lattice direction; diffusion occurs simply because there are more atoms jumping where their concentration is higher. The distance an atom diffuses in the direction of the concentration gradient is of course only a tiny fraction of the total distance it has travelled during its jumps. The effects of the activation energy, the vacancy concentration, and the statistical nature of the process can be developed into a quantitative theory.[1] The results can also be represented by the empirical equation

$$D_{AB} = D_o e^{-E/RT} \tag{29-15}$$

which is in the typical form which describes an activated process.

Of other possible ways in which a solid may diffuse we mention also the interstitial mechanism. If the impurity atom is small enough to fit into spaces between the lattice sites, it can jump to places that are determined mainly by the concentration of the solute atoms. The process is inherently faster than that by the vacancy mechanism. The spaces between interstitial sites may not be big enough for the solute atoms to pass except at times when holes open up because of the vibration of the lattice atoms. Interstitial diffusion is also an activated process and follows Eq. (29-15).

[1] A lucid treatment is given by P. G. Shewmon, "Diffusion in Solids," McGraw-Hill Book Company, New York, 1963.

Diffusivities for a few dilute solid systems are given in Table 29-1. In general, D_{AB} is of the order of 10^{-8} to 10^{-10} cm^2/s. For concentrated systems the diffusivity is a strong function of composition, and the theory is more involved. Shewmon[1] discusses this subject also.

Liquids Diffusion in liquids is important in many separation operations, notably liquid-liquid extraction, gas absorption, and distillation. An inspection of Table 29-2 shows that for all the systems listed, the values of D_{AB} are in the general range 10^{-5} to 10^{-6} cm^2/s. The diffusivity is a function of composition.

There are two prominent theories for the prediction of approximate diffusivities in liquids. In the Eyring theory the molecules of a liquid are pictured as forming a quasi-crystalline lattice and the analysis is performed more or less as it was for diffusion in a solid. In the hydrodynamical theory the diffusivity is first related to the force which acts on a sphere moving in a continuum. This force can be evaluated in terms of Stokes' law, and the resulting expression is called the *Stokes-Einstein equation.* The results of both theories can be put into the form

$$\frac{D_{AB}\mu_B}{T} = F(V) \qquad (29\text{-}16)$$

in which $F(V)$ represents a function which depends on the molecular volume V of the mixture. For the Stokes-Einstein equation the right-hand side of Eq. (29-16) is simply inversely proportional to the radius of the diffusing molecule A.

Equation (29-16) has been used by Wilke[2] and later by Wilke and Chang[3] as a

Table 29-1 DIFFUSIVITIES IN SOLIDS

System	Parameters in Eq. (29-15) D_0, cm^2/s	E, cal/g atom
Bi in Pb	7.7×10^{-3}	18,600
Hg in Pb	3.6×10^{-1}	19,000
Cu in Au	5.8×10^{-4}	27,400
Sb in Au	5.3×10^{-5}	21,700
Al in Cu	1.2×10^{-2}	37,500
Zn in Cu	8×10^{-1}	38,000
Cd in Cu	3.5×10^{-9}	8,200

SOURCE: R. M. Barrer, "Diffusion in and through Solids," pp. 141, 275, The Macmillan Company, New York, 1941.

[1] Ibid.
[2] C. R. Wilke, *Chem. Eng. Progr.*, **45**:218 (1949).
[3] C. R. Wilke and P. Chang, *AIChE J.*, **1**:270 (1955).

basis of a method for the correlation of liquid diffusivities. Their relation is the following dimensional equation:

$$\frac{D_{AB}\mu_B}{T} = 7.4 \times 10^{-8} \frac{(\chi M_B)^{1/2}}{V_A^{0.6}} \tag{29-17}$$

where D_{AB} = diffusion coefficient, cm²/s, for a dilute solution
μ_B = viscosity of (dilute) solution, cP
T = temperature, K
M_B = molecular weight of solvent
V_A = molar volume of solute, cm³/g mole, at normal boiling point
χ = association parameter of the solvent

Values of χ are: water, 2.6; methanol, 1.9; ethanol, 1.5; benzene, 1.0; ethane, 1.0; heptane, 1.0; other nonassociated solvents, 1.0. Values of V_A, if not available experimentally, can be estimated from the Le Bas volumes.[1]

Table 29-2 DIFFUSIVITIES IN LIQUIDS
Substance B (solvent) = water

Solute	Temperature, °C	Concentration	$D_{AB} \times 10^5$, cm²/s
Hydrogen	25	$\tilde{x}_A = 0$ (infinite dilution)	4.80
Oxygen	25		2.41
Carbon dioxide	25		2.00
Ammonia	12		1.64
Chlorine	25		1.25
Methanol	15		1.26
Acetic acid	20		1.19
Benzoic acid	25		1.21
Acetone	15		1.22
Ethanol	25	$\tilde{x}_A = 0.05$	1.13
		0.10	0.90
		0.275	0.41
		0.50	0.90
		0.70	1.40
		0.95	2.20
n-Butanol	30	$\tilde{x}_A = 0.446$	0.267
		0.546	0.437
		0.642	0.560
		0.778	0.920
		0.869	1.24

SOURCE: R. C. Reid, J. M. Prausnitz and T. K. Sherwood, "The Properties of Gases and Liquids," 3d ed., McGraw-Hill Book Company, New York, 1977; and P. A. Johnson and A. L. Babb, *Chem. Rev.*, **56**:387–453 (1956).

[1] Perry, p. 3-229.

Gases Experimental values of the diffusivity of a number of binary gaseous systems are given in Table 29-3. The values fall approximately in the range 0.1 to 1.0 cm^2/s. The diffusivities can be considered independent of concentration for gases at pressures near 1 atm.

Fuller et al.[1] give a large collection of data for many gas pairs and also compare various methods of prediction. The following equation is recommended:

$$D_{AB} = \frac{1 \times 10^{-3} T^{1.75}[(1/M_A) + (1/M_B)]^{1/2}}{p[(\sum_A v_i)^{1/3} + (\sum_B v_i)^{1/3}]^2} \tag{29-18}$$

in which M_A and M_B are the molecular weights, T is in °K, p is in atm, and D_{AB} is in cm^2/s. The volume of each molecule is made up for common organic vapors from the chemical formula from the values given in the upper part of Table 29-4. For the common gases the volumes in the lower part of Table 29-4 should be used. These volumes are 10 to 15 percent lower than the Le Bas volumes given in Perry.[2]

The form of Eq. (29-18) is based on the kinetic theory of gases, but it has been altered somewhat to obtain a good statistical fit with the existing experimental data. An equation which is more directly related to theory is

$$D_{AB} = \frac{BT^{3/2}[(1/M_A) + (1/M_B)]^{1/2}}{p\, r_{AB}^2\, I_D} \tag{29-19}$$

In this equation,[3] r_{AB} is the collision diameter (Å) which is calculated from values of

Table 29-3 DIFFUSIVITIES IN GASES
(Pressure about 1 atm)

System	Temperature, K	D_{AB}, cm^2/s
Air–chlorine	293	0.124
Air–carbon dioxide	276	0.142
Air–sulfur dioxide	293	0.122
Air–water	298	0.260
Air–ammonia	295	0.247
Air–benzene	298	0.0962
Air–ethanol	298	0.135
Hydrogen–nitrogen	294	0.763
Hydrogen–helium	298	1.320

SOURCE: E. N. Fuller, P. D. Schettler, and J. C. Giddings, *Ind. Eng. Chem.*, **58**(5):19 (1966).

[1] E. N. Fuller, P. D. Schettler, and J. C. Giddings, *Ind. Eng. Chem.*, **58**(5):19 (1966).
[2] Perry, p. 3-229.
[3] C. R. Wilke and C. Y. Lee, *Ind. Eng. Chem.* **47**:1253 (1955); Perry, p. 3–234.

r_A and r_B for the pure components by the relation

$$r_{AB} = \frac{r_A + r_B}{2} \tag{29-20}$$

The collision integral I_D is a function of r_{AB} and also $(\varepsilon/kT)_{AB}$, where

$$\left(\frac{\varepsilon}{k}\right)_{AB} = \sqrt{\left(\frac{\varepsilon}{k}\right)_A \left(\frac{\varepsilon}{k}\right)_B} \tag{29-21}$$

The parameters r_i and $(\varepsilon/k)_i$ characterize the intermolecular potential energy and are important even for gases which follow the ideal-gas law, i.e., most gases at ambient conditions. The necessary values of r_i, $(\varepsilon/k)_i$, I_D, and B are given elsewhere.[1] The values are usually obtained experimentally from measurements of the viscosity.

We have given here only a qualitative description of the theories for the diffusivities of gases and liquids. A good introduction to the theoretical procedures to be used in estimating the diffusivities in gases and liquids is given by Bird, Stewart,

Table 29-4 SPECIAL ATOMIC DIFFUSION VOLUMES

Atomic and Structural Diffusion Volume Increments

C	16.5	(Cl)	19.5
H	1.98	(S)	17.0
O	5.48	Aromatic or Heterocyclic rings	−20.2
N	5.69		

Diffusion Volumes of Simple Molecules

H_2	7.07	CO_2	26.9
D_2	6.70	N_2O	35.9
He	2.88	NH_3	14.9
N_2	17.9	H_2O	12.7
O_2	16.6	(CCl_2F_2)	114.8
Air	20.1	(SF_6)	69.7
Ne	5.59		
Ar	16.1	(Cl_2)	37.7
Kr	22.8	(Br_2)	67.2
(Xe)	37.9	(SO_2)	41.1
CO	18.9		

( ) indicates that listed value is based on only a few data points.

[1] Wilke and Lee, ibid.; and Perry, pp. 3-230 to 3-235.

and Lightfoot;[1] more empirically oriented discussions are given by Treybal.[2] Perry's Handbook contains experimental data; Reid and Sherwood[3] contains experimental data and a review of the methods for predicting the diffusivity.

Diffusion in Porous Solids[4]

In some chemical operations, such as heterogeneous catalysis, an important factor affecting the rate of reaction is the diffusion of a gaseous component through a porous solid. The effective diffusivity in the solid is reduced below what it would be in a free fluid, for two reasons. First, the tortuous nature of the path increases the distance which a molecule must travel to advance a given distance in the solid. Second, the free cross-sectional area is restricted. For many catalyst pellets, the effective diffusivity of a gaseous component is of the order of a tenth of its value in a free gas.

If the pressure is low enough and the pores are small enough, the mechanism of diffusion is basically changed; there exists what is called *Knudsen diffusion*. This type of diffusion occurs when the size of the pores approaches the mean free path of the gaseous molecules. A typical molecule now collides predominantly with the walls of the pores rather than with other molecules.

From the kinetic theory of gases it can be shown that for Knudsen diffusion the coefficient is given by

$$D_{KA} = \frac{\bar{v}_A \bar{d}}{3} \qquad (29\text{-}22)$$

when $\bar{d}$ is the average pore diameter and $\bar{v}$ is the mean molecular speed given by

$$\bar{v}_A = \left(\frac{8RT}{\pi M_A}\right)^{1/2} \qquad (29\text{-}23)$$

Since the flux given by

$$\tilde{N}_A = -\tilde{\rho} D_{KA} \frac{d\tilde{x}_A}{dy} \qquad (29\text{-}24)$$

is determined only by collisions with the wall, the flux of A is independent of that for B. Note also that D_{KA} is independent of pressure.

[1] R. B. Bird, W. E. Stewart, and E. N. Lightfoot, "Transport Phenomena," sec. 16.4, John Wiley & Sons, Inc., New York, 1961.

[2] R. E. Treybal, "Mass Transfer Operations," 3d ed., chap. 2, McGraw-Hill Book Company, New York, 1980.

[3] R. C. Reid, J. M. Prausnitz and T. K. Sherwood, "The Properties of Gases and Liquids," 3d ed., chap. 11, McGraw-Hill Book Company, New York, 1977.

[4] C. N. Satterfield, "Mass Transfer in Heterogeneous Catalysis," The M.I.T. Press, Cambridge, Mass., 1970.

For diffusion under conditions where both ordinary and Knudsen diffusion are important, the Bosanquet relation

$$\frac{1}{D_A} = \frac{1}{D_{AB}} + \frac{1}{D_{KA}} \qquad (29\text{-}25)$$

applies for equimolar countercurrent diffusion. The coefficient D_A is defined by

$$\tilde{N}_A = -\tilde{\rho} D_A \frac{d\tilde{x}_A}{dy} \qquad (29\text{-}26)$$

PROBLEMS

29-1 The activity coefficient of a binary solution of A in B can sometimes be correlated by the simple equation

$$\ln \gamma_A = A(1 - \tilde{x}_A)^2$$

Make a plot of L_{AB}/D_{AB} versus $\tilde{x}_A$ at 20°C for $\tilde{\rho} = 50$ kg mol/m^3 and $A = 0.406$.

29-2 What is the ratio of the diffusivity of cadmium in copper to that of aluminum in copper at 20°C? At about what temperature would the diffusivities be equal? The solutions are dilute.

29-3 Consider a batch polymerization taking place in a dilute benzene solution at 40°C. Early in the reaction, when the average diameter of the polymer molecule is 10 Å, the diffusivity of the polymer in benzene is 10^{-6} cm^2/s. Later on, when the polymer has grown to an average size of 1000 Å (assume it is spherical), what will be its diffusivity?

29-4 Check the method of Wilke and Chang by calculating the diffusivities of methanol, acetic acid, and acetone in water and comparing with the values given in Table 29-2.

29-5 Estimate the diffusivity in the air-benzene system at 1 atm and 293 K and at 1 atm and 1000 K. Calculate the latter value completely from Eq. (29-18). Calculate it also from the value in Table 29-3, using Eq. (29-18) only to correct this value for the effect of temperature.

29-6 Use the equation of Fuller et al. to calculate D_{AB} for chlorine in air at 20°C and 1 atm, and at 500°C and 1 atm.

29-7 Repeat Prob. 29-6 for ethanol in air at 21°C and 100°C.

29-8 Use Eq. (29-19) to solve Probs. 29-5 and 29-6, referring to Perry, pp. 3-230 to 3-235 for details.

29-9 For equimolar countercurrent diffusion of $H_2(A)$ and $N_2(B)$ at 25°C, make a plot of $1/D_A$ as a function of pressure for a porous solid of an average pore size of 10,000 Å. Let the pressure vary from 10^{-6} torr to 10^3 torr.

29-10 For equimolar countercurrent diffusion of $H_2(A)$ and $N_2(B)$ at 25°C and 1 atm abs, make a plot of $1/D_A$ as a function of pore diameter in the range 1 to 1000 nm.

29-11 A copper object is plated with gold to a thickness of 10 μm. If it is desired to keep the penetration depth of the copper into the gold below 1 μm during 10 yr of operation at constant temperature, what is the maximum such temperature?

The unsteady-state diffusion equation can be solved for the penetration depth δ of A into a thick layer of B to give $\delta = 4\sqrt{D_{AB}\theta}$.

29-12 Nitrogen is flowing through a tube at 110 kPa and 300 K; the atmosphere around the tube is air at 300 K and atmospheric pressure. The mean free paths of N_2 and O_2 can be estimated as 10^{-7} m.

If there is a tiny orifice in the tube wall, $\bar{d} = 10^{-8}$ m, estimate the flux of oxygen into the pure nitrogen in the tube.

If the hole in the tube is larger ($\bar{d} = 10^{-5}$), what would be the flux of O_2 into the N_2?

Estimate also the fluxes of nitrogen out of the two leaks of different size.

30.3. 0.003m, → A porous sphere -- saturated with methanol.
↳ drop - large tank of pure water at 25°C
↳ center $\tilde{x}_A = 0.01$.

$$Y = \frac{\tilde{x}_{AS} - \tilde{x}_A}{\tilde{x}_{AS} - \tilde{x}_{A0}} = \frac{0 - 0.01}{0 - 1}$$

$\tilde{x}_{AS} = 0$ (pure water ...), $\tilde{x}_{A0} = 1$ (초기-saturated), $\tilde{x}_A = 0.01$.

$$n = \frac{r}{r_m} = 0. \quad m = \frac{k}{h \cdot r_m} \rightarrow \frac{D_{AB}}{k_{\tilde{x}} \cdot r_m} = 0$$

(- interdiffusion process -- -- $k_{\tilde{x}} \gg D$).

X 값 → from fig 19-5, = 0.55,

$$X = Fo \Rightarrow \frac{\alpha \cdot \theta}{r^2} \rightarrow \frac{D_{AB} \cdot \theta}{r_m^2} = 0.55.$$

D_{AB} (methanol) = 1.26×10^{-5} cm²/s.

$$D_{eff} = \frac{1}{16} \times D_{AB} = 1.26 \times 10^{-10} \text{ m}^2/\text{s}.$$

$$\theta = \frac{r^2 \cdot (0.55)}{D_{eff}} = \frac{0.55 \times (0.0015 \text{ m})^2}{1.26 \times 10^{-10} \text{ m}^2/\text{s}}$$

$= 9.82 \times 10^3$ h = 2.728 hr

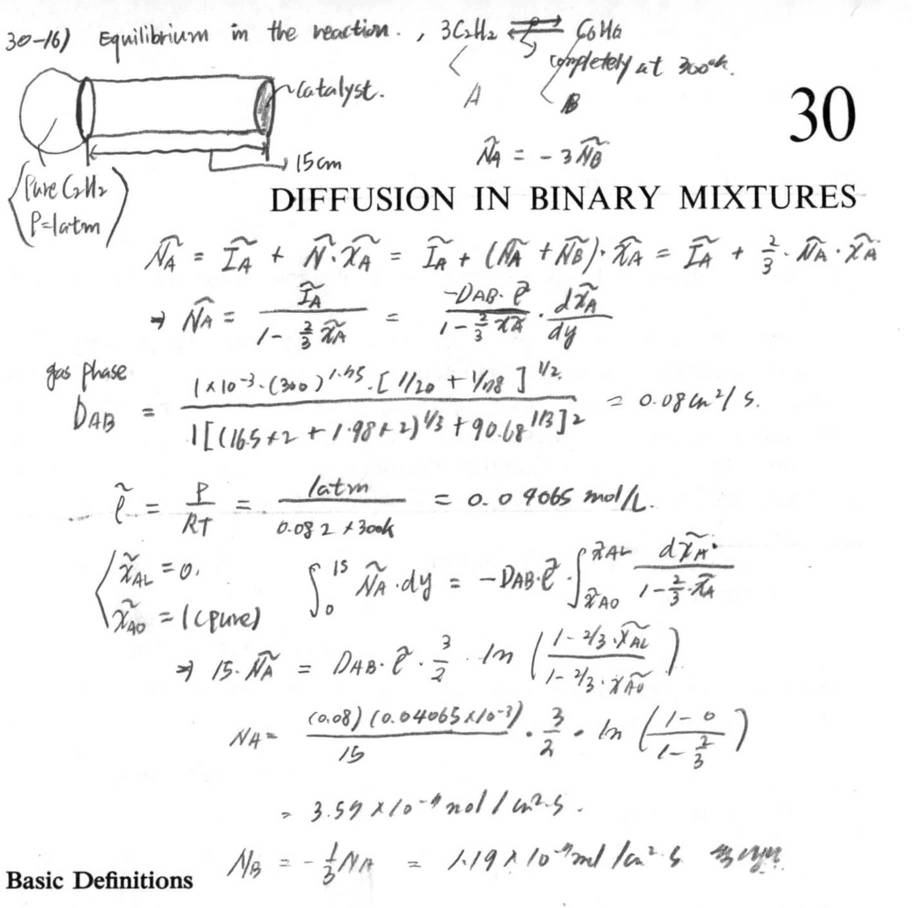

30

DIFFUSION IN BINARY MIXTURES

Basic Definitions

In Chap. 29 we mentioned two forms of Fick's law, one for a molar flux with respect to the molar average velocity, Eq. (29-1), and one for a mass flux with respect to the mass average velocity, Eq. (29-3). Furthermore, in many problems we wish to know the flux with respect to axes fixed in space. In what follows we develop the relations among the various fluxes. Let us recall, first, the relations between mole fraction and mass fraction:

$$x_A = \frac{\rho_A}{\rho} = \frac{\tilde{x}_A M_A}{\tilde{x}_A M_A + \tilde{x}_B M_B} \tag{30-1}$$

$$\tilde{x}_A = \frac{\tilde{\rho}_A}{\tilde{\rho}} = \frac{x_A/M_A}{x_A/M_A + x_B/M_B} \tag{30-2}$$

By definition, we also have

$$\rho_A + \rho_B = \rho \tag{30-3}$$

$$\tilde{\rho}_A + \tilde{\rho}_B = \tilde{\rho} \tag{30-4}$$

and

$$\rho_A = \tilde{\rho}_A M_A \tag{30-5}$$

The mass flux N of a mixture is now simply defined as

$$N = N_A + N_B \qquad (30\text{-}6)$$

The flux when designated as N is the value of the flux relative to a set of axes fixed in space.

The velocity of a single object is intuitively apparent. The velocity of an aggregate of particles which move but retain the same positions relative to each other is obviously the same as the velocity of any individual particle. The velocity of an aggregate is not apparent, however, when the individual particles move with different velocities. In this case we define the velocity of the aggregate, say of particles of molecular species A, as the mass flux of A divided by the concentration of A ($u_A = N_A/\rho_A$).

For a mixture of A and B, the velocity of the mixture is defined by substituting for the fluxes in Eq. (30-6).

$$u\rho = u_A \rho_A + u_B \rho_B \qquad (30\text{-}7)$$

The quantity u is called the *mass average velocity* of the mixture. If we divide Eq. (30-7) by the density of the mixture, we obtain

$$u = x_A u_A + x_B u_B \qquad (30\text{-}8)$$

The molar flux is

$$\tilde{N} = \tilde{N}_A + \tilde{N}_B \qquad (30\text{-}9)$$

Because $u_A = N_A/\rho_A = \tilde{N}_A/\tilde{\rho}_A$ and $u_B = N_B/\rho_B = \tilde{N}_B/\tilde{\rho}_B$, we can substitute for $\tilde{N}_A$ and $\tilde{N}_B$ in Eq. (30-9) to obtain

$$\tilde{N} = u_A \tilde{\rho}_A + u_B \tilde{\rho}_B \qquad (30\text{-}10)$$

The flux $\tilde{N}$ can be expressed in terms of a molar average velocity $\tilde{u}$, which is defined as $\tilde{N}/\tilde{\rho}$, by writing

$$\tilde{u}\tilde{\rho} = u_A \tilde{\rho}_A + u_B \tilde{\rho}_B$$

This can also be written as

$$\tilde{u} = \tilde{x}_A u_A + \tilde{x}_B u_B \qquad (30\text{-}11)$$

In any of the equations derived in previous chapters for the integral or differential balances which apply to mixtures, the velocity used has been u, the mass average velocity.

An inspection of Eqs. (30-6) to (30-11) shows that the mass flux N is *not* equal to the molar flux $\tilde{N}$ times the average molecular weight, because u and $\tilde{u}$ are different. In the flow of a uniform mixture through a pipe, there is no interdiffusion effect, u_A and u_B are identical, and the mass flux *is* then equal to the molar flux multiplied by the average molecular weight. However, in cases for which there is a concentration gradient in the direction of flow, u_A and u_B differ, with the effect noted above. In many mass-transfer situations, the total flux is of the same order of magnitude as the fluxes caused by diffusion, so the above distinction is not merely academic.

We now proceed to develop the relations among the various diffusional fluxes. The symbols J_A and $\tilde{J}_A$ represent, respectively, the mass flux and molar flux [lb or lb mol/(h)(ft^2)] of component A with respect to a set of axes moving at the mass average velocity u of the fluid. The symbols I_A and $\tilde{I}_A$ represent the fluxes with respect to a set of axes which moves at the molar average velocity $\tilde{u}$ of the fluid. The following equations hold:

$$J_A = \tilde{J}_A M_A \qquad (30\text{-}12)$$

$$I_A = \tilde{I}_A M_A \qquad (30\text{-}13)$$

The relations between J_A and $\tilde{I}_A$ and between $\tilde{J}_A$ and I_A are not so obvious and will be derived presently.

In chemical engineering it is usually desired to relate the flux to a fixed surface such as a fluid interface rather than to an average velocity. Thus we need to use the mass and the molar flux referred to a set of axes fixed in space. These fluxes have been defined in Eqs. (30-6) to (30-11) as N_A and $\tilde{N}_A$, respectively.

In most work in ordinary diffusion in gases, it has become customary to use the flux $\tilde{I}_A$ defined by Eq. (29-1). However, the relations for thermal diffusion have been expressed in terms of a flux such as J_A. These two fluxes must sometimes be added algebraically, so we need to be able to convert the flux $\tilde{I}_A$ into J_A. We shall presently develop the relation between $\tilde{I}_A$ and J_A, based on the definitions already given. We begin the study of the relations among the various fluxes by considering those defined on the mass basis.

The flux N_A is the sum of the diffusive flux J_A with respect to the center of mass plus the contribution to the flux of A caused by the bulk mass movement N:

$$N_A = J_A + N x_A \qquad (30\text{-}14)$$

which is expressed in terms of velocity as

$$u_A \rho_A = J_A + u\rho \frac{\rho_A}{\rho}$$

This is rearranged to give

$$J_A = \rho_A(u_A - u) \qquad (30\text{-}15)$$

which indicates the basis of the statement that J_A is the flux of A relative to axes moving at the mass average velocity.

Equation (30-14) can also be written for the second component B as

$$N_B = J_B + Nx_B \tag{30-16}$$

The sum of Eqs. (30-14) and (30-16) implies that $J_A = -J_B$, as already given by Eq. (29-4). For component B we can write

$$J_B = -D_{BA}\rho \frac{dx_B}{dy}$$

$$= D_{BA}\rho \frac{dx_A}{dy} \tag{30-17}$$

since the sum of the mass fractions must be unity. From the basic relations

$$J_A = -D_{AB}\rho \frac{dx_A}{dy} \tag{29-3}$$

and

$$J_A = -J_B \tag{29-4}$$

we find that

$$D_{AB} = D_{BA} \tag{30-18}$$

for a binary system.

The equation connecting the fluxes $\tilde{N}_A$ and $\tilde{I}_A$ is obtained next. The molar flux of A with respect to fixed axes is the sum of $\tilde{I}_A$, the flux with respect to the molar average velocity, and the flux caused by the bulk flow related to $\tilde{u}$, that is, the flux $\tilde{N}$. The equation expressing these relations is

$$\tilde{N}_A = \tilde{I}_A + \tilde{N}\tilde{x}_A \tag{30-19}$$

This may be written in terms of velocity as

$$u_A\tilde{\rho}_A = \tilde{I}_A + \tilde{u}\tilde{\rho}\frac{\tilde{\rho}_A}{\tilde{\rho}}$$

from which we obtain

$$\tilde{I}_A = \tilde{\rho}_A(u_A - \tilde{u}) \tag{30-20}$$

This equation reveals the basis of the statement that $\tilde{I}_A$ is the molar flux with respect to axes moving at the molar average velocity. Equation (30-19) can be written for B and added to the equation for A to show that $\tilde{I}_A = -\tilde{I}_B$, as asserted in Eq. (29-2). We can also show that $D_{AB} = D_{BA}$ for this case.

Next, $\tilde{J}_A$ and $\tilde{I}_A$ are related by the following reasoning. The molar flux $\tilde{J}_A$ with respect to the mass average velocity is equal to the molar flux with respect to the molar average velocity plus the flux of A caused by the bulk flux of moles with respect to the mass average velocity. This analysis is parallel to that leading to Eq. (30-19) and gives

$$\tilde{J}_A = \tilde{I}_A + (\tilde{J}_A + \tilde{J}_B)\tilde{x}_A \tag{30-21}$$

Equation (30-21) can be further simplified, for from Eqs. (30-12) and (29-4), we obtain

$$M_A \tilde{J}_A + M_B \tilde{J}_B = 0 \tag{30-22}$$

$$\tilde{J}_B = -\frac{M_A}{M_B}\tilde{J}_A \tag{30-23}$$

Equation (30-21) then becomes, with some rearrangement,

$$\tilde{I}_A = \tilde{J}_A\left(1 - \tilde{x}_A + \tilde{x}_A \frac{M_A}{M_B}\right) \tag{30-24}$$

or

$$\tilde{I}_A = \frac{\tilde{J}_A}{M_B}(M_B \tilde{x}_B + M_A \tilde{x}_A) \tag{30-25}$$

Since the quantity in parentheses is the average molecular weight, or $\rho/\tilde{\rho}$, the final general relation between $\tilde{I}_A$ and $\tilde{J}_A$ is

$$\tilde{I}_A = \frac{\tilde{J}_A \rho}{M_B \tilde{\rho}} \tag{30-26}$$

The derivations of the other relations among the fluxes follow those already given, and their development is left as an exercise for the reader.

The Differential Mass Balance

Mass fluxes The equations derived so far in this chapter apply only to steady one-dimensional flow with no generation. For more complicated situations the differential mass balances derived in Chap. 7 must be used. Let us now review these equations and relate them to the various kinds of fluxes we have defined.

In obtaining Eqs. (7-17) and (7-18), a diffusional flux J_A was considered superimposed upon the flux caused by the bulk flow, which we now know is represented by the mass average velocity. In Eq. (7-18),

$$\rho_A\left(\frac{\partial u_x}{\partial x} + \frac{\partial u_y}{\partial y} + \frac{\partial u_z}{\partial z}\right) + \frac{D\rho_A}{D\theta} + \frac{\partial J_{Ax}}{\partial x} + \frac{\partial J_{Ay}}{\partial y} + \frac{\partial J_{Az}}{\partial z} - r_A = 0$$

the quantities u_x, u_y, and u_z are the components of the mass average velocity. Equation (7-18) is convenient for use in systems of constant density, for which the term in parentheses is zero and for which we have

$$J_{Ax} = -D_{AB}\frac{\partial \rho_A}{\partial x} \qquad (7\text{-}19)$$

Equations similar to Eq. (7-19) hold for the other two coordinate directions. For a fluid of constant density, Eq. (7-18) becomes

$$\frac{D\rho_A}{D\theta} = D_{AB}\left(\frac{\partial^2 \rho_A}{\partial x^2} + \frac{\partial^2 \rho_A}{\partial y^2} + \frac{\partial^2 \rho_A}{\partial z^2}\right) + r_A \qquad (7\text{-}22)$$

if D_{AB} is constant.

For steady one-dimensional flow in the y direction, with no generation term, Eq. (7-22) becomes

$$u\frac{d\rho_A}{dy} = D_{AB}\frac{d}{dy}\frac{d\rho_A}{dy} \qquad (30\text{-}27)$$

This equation can be integrated immediately to yield

$$u\rho_A = D_{AB}\frac{d\rho_A}{dy} + \text{const} \qquad (30\text{-}28)$$

The quantity $u\rho_A$ is the flux of A caused by the bulk flow u, and it equals Nx_A, or $(N_A + N_B)x_A$. The diffusion term is $-J_A$, so we can write Eq. (30-28) as

$$\text{const} = J_A + (N_A + N_B)x_A \qquad (30\text{-}29)$$

The right-hand side of Eq. (30-29) represents the sum of the flux caused by diffusion and that caused by bulk flow, so it is thus equal to the total flux of A with respect to fixed axes, that is, N_A. The constant of integration is therefore N_A, and Eq. (30-29) is identical with Eq. (30-14).

Molar fluxes For isothermal gaseous systems near atmospheric pressure the molar density of the mixture is constant. The derivation of the equation of continuity at the beginning of Chap. 7 can be repeated in terms of moles, and the result, similar to Eq. (7-12), is, in the absence of chemical reaction,

$$\frac{\partial \tilde{u}_x}{\partial x} + \frac{\partial \tilde{u}_y}{\partial y} + \frac{\partial \tilde{u}_z}{\partial z} + \frac{1}{\tilde{\rho}}\frac{D\tilde{\rho}}{D\theta} = 0 \qquad (30\text{-}30)$$

The quantities $\tilde{u}_x$, $\tilde{u}_y$, $\tilde{u}_z$ are the components of the molar average velocity $\tilde{u}$. For a system of constant molar density, $D\tilde{\rho}/D\theta$ is zero, so we have

$$\frac{\partial \tilde{u}_x}{\partial x} + \frac{\partial \tilde{u}_y}{\partial y} + \frac{\partial \tilde{u}_z}{\partial z} = 0 \qquad (30\text{-}31)$$

For a molar balance on component A, Eq. (7-16) becomes

$$d(\tilde{u}_x \tilde{\rho}_A + \tilde{I}_{Ax})\, dy\, dz = \left[\frac{\partial(\tilde{u}_x \tilde{\rho}_A)}{\partial x} + \frac{\partial \tilde{I}_{Ax}}{\partial x}\right] dx\, dy\, dz \qquad (30\text{-}32)$$

The analysis leads to

$$\tilde{\rho}_A\left(\frac{\partial \tilde{u}_x}{\partial x} + \frac{\partial \tilde{u}_y}{\partial y} + \frac{\partial \tilde{u}_z}{\partial z}\right) + \frac{D\tilde{\rho}_A}{D\theta} + \frac{\partial \tilde{I}_{Ax}}{\partial x} + \frac{\partial \tilde{I}_{Ay}}{\partial y} + \frac{\partial \tilde{I}_{Az}}{\partial z} = 0 \qquad (30\text{-}33)$$

This equation is analogous to Eq. (7-18), but the generation term $\tilde{r}_A$ has been omitted. Equation (30-30) applies only to nonreacting systems, for $\sum \tilde{r}_i$ is not in general zero; $\sum r_i$ is always zero. If $\tilde{\rho}$ is constant, Eq. (30-31) holds, and we also have

$$\tilde{I}_{Ax} = -D_{AB}\frac{\partial \tilde{\rho}_A}{\partial x} \qquad (30\text{-}34)$$

This equation is the equivalent of Eq. (29-1); similar relations hold for the other two coordinate directions. For a system of constant molar density, with no reaction occurring, Eq. (30-33) can be written as

$$\frac{D\tilde{\rho}_A}{D\theta} = D_{AB}\left(\frac{\partial^2 \tilde{\rho}_A}{\partial x^2} + \frac{\partial^2 \tilde{\rho}_A}{\partial y^2} + \frac{\partial^2 \tilde{\rho}_A}{\partial z^2}\right) \qquad (30\text{-}35)$$

which is analogous to Eq. (7-22).

For one-dimensional steady flow in the y direction, Eq. (30-35) becomes

$$\tilde{u}\frac{d\tilde{\rho}_A}{dy} = D_{AB}\frac{d}{dy}\frac{d\tilde{\rho}_A}{dy} \qquad (30\text{-}36)$$

Integration yields

$$\tilde{u}\tilde{\rho}_A = D_{AB}\frac{d\tilde{\rho}_A}{dy} + \text{const} \qquad (30\text{-}37)$$

The constant is equal to the molar flux $\tilde{N}_A$, so that Eq. (30-37) is equivalent to Eq. (30-19).

Application of the differential balances The simultaneous solution of the differential mass, energy, and momentum equations for a multicomponent system can be exceedingly complicated. For instance, in a gaseous system Eq. (30-35) might be useful, but the Navier-Stokes equations are expressed in terms of mass units, not molar units. It would be necessary to use Eq. (7-18), for variable ρ, with equations similar to Eq. (9-50), rather than the Navier-Stokes equations for constant ρ, Eqs. (9-52) to (9-54). Fortunately, in most practical situations the solutions of the Navier-Stokes equations which hold when no mass transfer is taking place are not

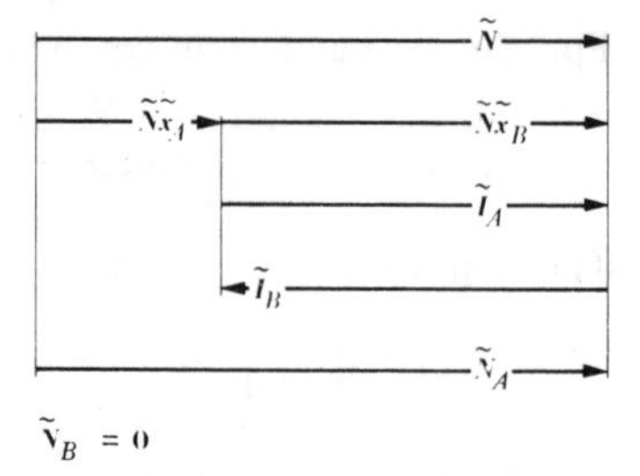

FIGURE 30-1
Fluxes for diffusion of A through stagnant B at a point where $\tilde{x}_A = 0.33$.

appreciably affected by the occurrence of mass transfer. For instance, the parabolic velocity profile associated with laminar flow in a tube is little changed if the walls of the tube are made of some soluble material which diffuses toward the centerline. For mass transfer, Eq. (7-22) can be used with solutions of gases for which the variation of concentrations never becomes large enough to have an appreciable effect on the density. However, in an analysis of a flowing gaseous reaction system involving large changes in concentrations, appreciable errors may be made if the secondary effects omitted in simpler situations are not considered.

Similar remarks can be made about the use of Eqs. (8-11) and (8-14) for systems in which mass transfer is taking place. Fluxes of energy caused by diffusion have been omitted from these equations, but these effects can be ignored in many common situations. Such effects will, however, be studied in Chap. 36. For a binary system, the velocities appearing refer to the mass average velocity.

Diffusion of *A* through Stagnant *B*

In the situation illustrated in Fig. 30-2, water vapor is diffusing at a constant rate from the liquid surface up through the layer of stagnant air in the tube. The air is not moving with respect to fixed axes, so $N_B = \tilde{N}_B = 0$. Since we are dealing with gases for which $\tilde{\rho}$ is constant, we use Eq. (30-19) and write it with $\tilde{N}_B = 0$:

$$\tilde{N}_A = \tilde{I}_A + \tilde{N}_A \tilde{x}_A \qquad (30\text{-}38)$$

$$\tilde{N}_A = \frac{\tilde{I}_A}{1 - \tilde{x}_A} \qquad (30\text{-}39)$$

The relations among the fluxes can be understood more clearly by reference to Fig. 30-1, in which the length of the arrows (vectors) is proportional to the fluxes represented.

Equation (29-1) is next substituted into Eq. (30-39) to give

$$\tilde{N}_A = -\frac{D_{AB}\tilde{\rho}}{1 - \tilde{x}_A}\frac{d\tilde{x}_A}{dy} \qquad (30\text{-}40)$$

This equation can be integrated between 0 and L and $\tilde{x}_{Ao}$ and $\tilde{x}_{AL}$ to give

$$\tilde{N}_A \int_0^L dy = -D_{AB}\tilde{\rho} \int_{\tilde{x}_{Ao}}^{\tilde{x}_{AL}} \frac{d\tilde{x}_A}{1 - \tilde{x}_A} \tag{30-41}$$

and

$$\tilde{N}_A = \frac{D_{AB}\tilde{\rho}}{L} \ln \frac{1 - \tilde{x}_{AL}}{1 - \tilde{x}_{Ao}} \tag{30-42}$$

or

$$\tilde{N}_A = \frac{D_{AB}\tilde{\rho}}{L} \ln \frac{\tilde{x}_{BL}}{\tilde{x}_{Bo}} \tag{30-43}$$

An equivalent form of Eq. (30-43) is

$$\tilde{N}_A = \frac{D_{AB}\tilde{\rho} \ln (\tilde{x}_{BL}/\tilde{x}_{Bo})}{L(\tilde{x}_{BL} - \tilde{x}_{Bo})} (\tilde{x}_{Ao} - \tilde{x}_{AL}) \tag{30-44}$$

A logarithmic mean mole fraction is defined by

$$\tilde{x}_{B,\,lm} = \frac{\tilde{x}_{BL} - \tilde{x}_{Bo}}{\ln (\tilde{x}_{BL}/\tilde{x}_{Bo})} \tag{30-45}$$

so that Eq. (30-44) can be written as

$$\tilde{N}_A = \frac{D_{AB}\tilde{\rho}}{L\tilde{x}_{B,\,lm}} (\tilde{x}_{Ao} - \tilde{x}_{AL}) \tag{30-46}$$

This equation has the familiar form in which a flux equals a coefficient multiplied by a concentration difference. A mass-transfer coefficient $k_{\tilde{x}}$ can be defined by

$$\tilde{N}_A = k_{\tilde{x}}(\tilde{x}_{Ao} - \tilde{x}_{AL}) \tag{30-47}$$

and

$$k_{\tilde{x}} = \frac{D_{AB}\tilde{\rho}}{L\tilde{x}_{B,\,lm}} \tag{30-48}$$

For dilute solutions of A, $\tilde{x}_{B,\,lm}$ is approximately unity. Since $D_{AB}\tilde{\rho}$ does not vary with pressure for ideal gases, $k_{\tilde{x}}$ is independent of pressure. The preceding derivation can also be done in terms of partial pressures, beginning with Eq. (29-1), in the form

$$\tilde{I}_A = -\frac{D_{AB}\tilde{\rho}}{p}\frac{d\bar{p}_A}{dy} = \frac{-D_{AB}}{RT}\frac{d\bar{p}_A}{dy} \tag{30-49}$$

It is left to the reader to obtain equations analogous to Eqs. (30-40) to (30-48) in terms of partial pressures.

Gas-phase diffusivities have been measured experimentally using an arrangement similar to the one shown in Fig. 30-2, called a *Stefan experiment*. We shall refer again to Eq. (30-47) in Chap. 31 in connection with the definition of mass-transfer coefficients.

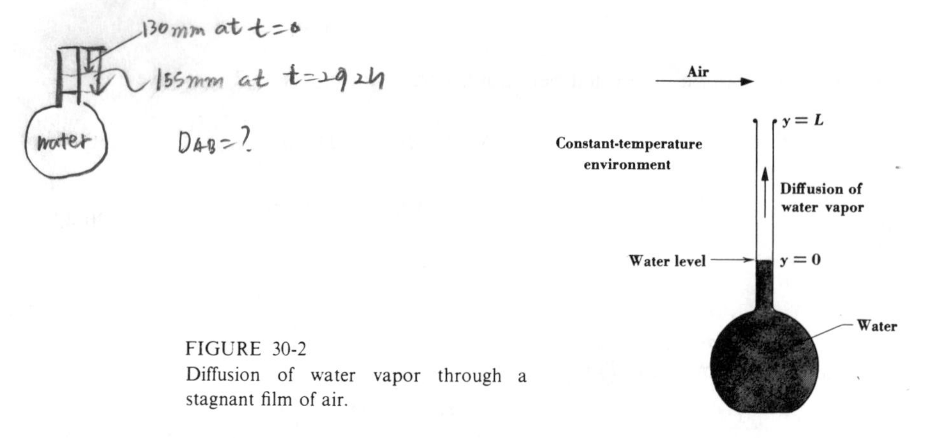

FIGURE 30-2
Diffusion of water vapor through a stagnant film of air.

Example 30-1 An apparatus of the type shown in Fig. 30-2 is used for the experimental determination of the diffusivity of a vapor through gas. For the air-water system, calculate the value of D_{AB} from the following information.

The entire apparatus is enclosed in a constant-temperature environment at 54°C and 1 atm, and the air is circulated over a desiccant so that the concentration of water vapor in the air is zero. It is assumed that there is no convective mixing in the tube above the level of the water. If 292 h is required for the level to fall from 130 to 155 mm below the top of the tube, what is the diffusivity in the water-air system at 54°C?

The molar flux $\tilde{N}_A$ is related to the rate at which the water level falls by

$$\tilde{N}_A = \frac{\tilde{\rho}_{AL}\, dL}{d\theta} \tag{1}$$

where $\tilde{\rho}_{AL}$ = molar density of liquid water

L = length of diffusion path

The flux is also given by Eq. (30-46), which was derived for steady-state diffusion. The path for diffusion increases in length by only 20 percent in 292 h, so steady-state diffusion can be assumed. Thus we write

$$\tilde{N}_A = \tilde{\rho}_{AL} \frac{dL}{d\theta} = \frac{D_{AB}\tilde{\rho}(\tilde{x}_{Ao} - \tilde{x}_{AL})}{L\tilde{x}_{B,\,lm}} \tag{2}$$

In this equation $\tilde{x}_{Ao}$ is the mole fraction water vapor in the gas just above the water level, and $\tilde{x}_{AL}$ is measured at the outlet of the tube. Equation (2) is now integrated from $\theta = 0$ to $\theta = \theta$ and from $L = L_o$ to $L = L$:

$$\int_0^\theta d\theta = \frac{\tilde{\rho}_{AL}\tilde{x}_{B,\,lm}}{\tilde{\rho} D_{AB}(\tilde{x}_{Ao} - \tilde{x}_{AL})} \int_{L_o}^{L} L\, dL \tag{3}$$

This equation is solved for D_{AB} to give

$$D_{AB} = \frac{\tilde{\rho}_{AL}\tilde{x}_{B,\,lm}}{\theta\tilde{\rho}(\tilde{x}_{Ao} - \tilde{x}_{AL})} \frac{L^2 - L_o^{\,2}}{2} \tag{4}$$

In the present problem we have

$$\tilde{\rho}_{AL} = 969/18 = 54.8 \text{ kg mol/m}^3$$

$$\tilde{\rho} = \frac{p}{RT} = \frac{1}{(0.082)(327)} = 0.0373 \text{ kg mol/m}^3$$

$$\tilde{x}_{AL} = 0$$

The vapor pressure of water at 54°C is 15.4 kPa. Hence

$$\tilde{x}_{Ao} = 15.4/101.3 = 0.152$$

$$\tilde{x}_{B,\,lm} = \frac{1 - 0.848}{\ln(1/0.848)} = 0.922$$

$$L = 0.155 \text{ m}$$

$$L_o = 0.13 \text{ m}$$

$$\theta = 1.05 \times 10^6 \text{ s}$$

When these quantities are substituted into Eq. (4), it is found that D_{AB} equals 3.03×10^{-5} m²/s. ////

Example 30-2 We consider the diffusion of salt (NaCl) through water in an apparatus similar to the one shown in Fig. 30-2. The equipment is maintained at 68°F, and the bulb contains a quantity of salt crystals. It is assumed that the liquid in the bulb is well mixed and that there is no eddy mixing in the diffusion tube. Thus the concentration at the lower end of the tube is constant; it is a saturated solution of salt in water at 68°F. The water surrounding the tube contains a negligible amount of salt.

The mechanism of the diffusion of an electrolyte in water is complicated and has been extensively studied. Although the different ions may tend to travel at different rates, the requirement of electrical neutrality makes it possible to regard the diffusion of a single salt as a diffusion of salt molecules. The diffusivity is a function of concentration; Reid and Sherwood[1] give a curve of the diffusivity of salt in water versus concentration. The variation is rather small; we choose a value of D_{AB} at 68°F as 1.35×10^{-5} cm²/s, or 5.22×10^{-5} ft²/h.

[1] R. C. Reid, J. M. Prausnitz, and T. K. Sherwood, "The Properties of Gases and Liquids," 3d ed., p. 593, McGraw-Hill Book Company, New York, 1977.

As the solid salt goes into solution to replace the salt which has diffused out of the tube, the volume of the solid phase decreases. There must be an inward flow of solution to fill this volume. Thus at the tube outlet there is a net outward flow of dissolved salt and a net inward flow of water. The solution of the problem is started by a consideration of the various mass balances, which will furnish the necessary relation between N_A (salt) and N_B (water).

We first make an overall mass balance on the bulb, which contains solid salt (M_S) and saturated solution (M_L; $x_{Ao} = 0.265$ at 68°F):

$$M_S + M_L = M \tag{1}$$

These masses all vary with time; however, the volume of the bulb V is constant:

$$\frac{M_S}{\rho_S} + \frac{M_L}{\rho_L} = V \tag{2}$$

and

$$-\frac{1}{\rho_S}\frac{dM_S}{d\theta} = \frac{1}{\rho_L}\frac{dM_L}{d\theta} \tag{3}$$

A balance on the salt yields

$$M_S + M_L x_{Ao} = M_A \tag{4}$$

Equations (1) and (4) are differentiated to give the rate of change of the masses of the various components in the bulb:

$$\frac{dM_S}{d\theta} + \frac{dM_L}{d\theta} = \frac{dM}{d\theta} \tag{5}$$

$$\frac{dM_S}{d\theta} + x_{Ao}\frac{dM_L}{d\theta} = \frac{dM_A}{d\theta} \tag{6}$$

Equation (3) is then used to eliminate $dM_S/d\theta$ in these equations to give

$$\left(1 - \frac{\rho_S}{\rho_L}\right)\frac{dM_L}{d\theta} = \frac{dM}{d\theta} \tag{7}$$

$$\left(x_{Ao} - \frac{\rho_S}{\rho_L}\right)\frac{dM_L}{d\theta} = \frac{dM_A}{d\theta} \tag{8}$$

The ratio of $dM_A/d\theta$ to $dM/d\theta$ must be the same as N_A/N, the flux of salt out over the flux of salt plus water out. Dividing Eq. (8) by (7) gives

$$\frac{N_A}{N} = \frac{x_{Ao} - (\rho_S/\rho_L)}{1 - (\rho_S/\rho_L)} \tag{9}$$

We call this constant ratio $1/K$, so that Eq. (30-16) becomes, for this problem,

$$N_A = J_A + KN_A x_A \tag{10}$$

Fick's law in the form of Eq. (29-3) is substituted to give

$$N_A = -D_{AB} \frac{\rho\, dx_A}{dy} + KN_A x_A \tag{11}$$

This equation is rearranged and integrated along the tube:

$$N_A \int_o^L dy = -D_{AB} \int_{x_{Ao}}^{x_{AL}} \frac{\rho\, dx_A}{1 - Kx_A} \tag{12}$$

and

$$N_A = -\frac{D_{AB}}{L} \int_{x_{Ao}}^{x_{AL}} \frac{\rho\, dx_A}{1 - Kx_A} \tag{13}$$

The known experimental relation between the solution density ρ and the salt mass fraction x_A permits the evaluation of the integral by graphical or numerical methods. However, the specific gravity of NaCl solutions (Perry, p. 3-78) varies from 1.0 only to about 1.20 for a saturated solution, so that we can approximate the integral by using a constant value, say $\rho_{av} = 1.10$ g/cm^3. The integral can then be evaluated to yield

$$N_A = \frac{D_{AB}\rho_{av}}{KL} \ln \frac{1 - Kx_{AL}}{1 - Kx_{Ao}} \tag{14}$$

The flux of salt out of the bulb can now be calculated for the following values of the parameters:

$$\rho_L = 1.20 \text{ g/cm}^3 \text{ (Perry, p. 3-78)}$$
$$\rho_S = 2.16 \text{ g/cm}^3 \text{ (Perry, p. 3-21)}$$
$$x_{Ao} = 0.265 \text{ (Perry, p. 3-94)}$$
$$K = 0.521$$
$$\rho_{av} = 1.10(62.4) = 68.6 \text{ lb/ft}^3$$
$$D_{AB} = 5.22 \times 10^{-5} \text{ ft}^2\text{/h}$$
$$L = 0.5 \text{ ft}$$
$$x_{AL} = 0$$

The result is

$$N_A = 2.03 \times 10^{-3} \text{ lb/(h)(ft}^2)$$

and also

$$N = 1.06 \times 10^{-3} \text{ lb/(h)(ft}^2)$$
$$N_B = -0.97 \times 10^{-3} \text{ lb/(h)(ft}^2)$$

The fluxes can be illustrated by vectors as

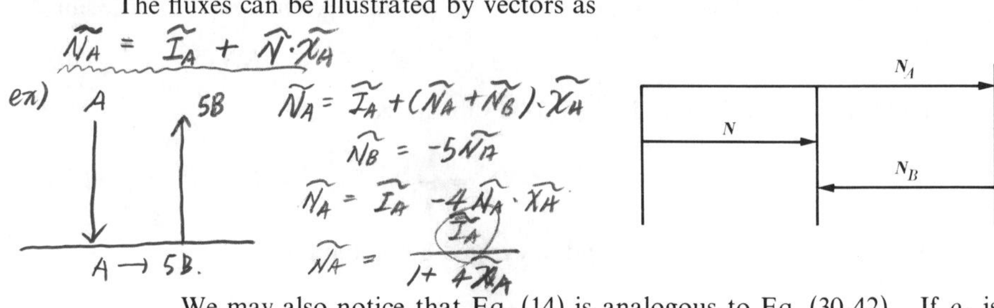

We may also notice that Eq. (14) is analogous to Eq. (30-42). If ρ_S is much greater than ρ_L (e.g., sublimation of a solid), then $K = 1$ and $N = N_A$, $N_B = 0$; the shrinkage of the solid causes a negligible flux of B. ////

Equimolar Counterdiffusion

If a mole of B diffuses to the left for every mole of A that diffuses to the right, $\tilde{N}_A = -\tilde{N}_B$ and Eq. (30-19) becomes simply

$$\tilde{N}_A = \tilde{I}_A = -D_{AB}\tilde{\rho}\frac{d\tilde{x}_A}{dy} \tag{30-50}$$

For steady state this equation can be integrated to yield

$$\tilde{N}_A = \frac{D_{AB}\tilde{\rho}}{L}(\tilde{x}_{Ao} - \tilde{x}_{AL}) \tag{30-51}$$

We shall see in a later chapter that this equation is useful in describing the rate of mass transfer in distillation problems. An energy balance often requires that 1 mol of A diffuse to a liquid surface, condense, and liberate just enough heat to evaporate 1 mol of B, which diffuses away from the liquid surface.

Another case in which equimolar counterdiffusion occurs is in the interdiffusion of two gases. Consider a cylinder with a transverse partition at the center dividing the vessel into two parts. The space on either side of the partition is occupied by a different gas at the same pressure and temperature. If the partition is now removed, the two gases will interdiffuse at equal molar rates, for the molar density $\tilde{\rho}$ remains constant. Equation (30-50) applies at any instant.

If the two interdiffusing components were liquids of essentially equal densities, we should have $N_A = -N_B$ and also

$$N_A = J_A = -D_{AB}\rho\frac{dx_A}{dy} \tag{30-52}$$

Equation (30-50) or (30-52) cannot be integrated directly for this situation, since N_A and dx_A/dy are functions of both position and time. Rather, the differential balance must be used for this unsteady-state problem. Equation (30-35) reduces to

$$\frac{\partial \tilde{\rho}_A}{\partial \theta} = D_{AB} \frac{\partial^2 \tilde{\rho}_A}{\partial y^2} \tag{30-53}$$

for the gaseous system, and Eq. (7-22) reduces to

$$\frac{\partial \rho_A}{\partial \theta} = D_{AB} \frac{\partial^2 \rho_A}{\partial y^2} \tag{30-54}$$

for the liquid system. The boundary values pertinent to this problem can be specified, and the solution obtained by the method of the separation of the variables illustrated in Example 19-2. The interdiffusion of two fluids described above is analogous to the heat-transfer problem of calculating the temperature distribution as a function of position and time in two conducting plates, initially at different uniform temperatures and suddenly brought into contact.

In order to study a problem in unsteady-state diffusion without introducing any new mathematics, let us consider the following example, which is the analog of Example 19-2.

Example 30-3 A slab of porous solid 1 cm thick is soaked in pure ethanol. The void space in the solid occupies 50 percent of its volume. The pores are fine, so that molecular diffusion can take place through the liquid in the passages; there is no convective mixing. The effective diffusivity of the system ethanol-water in the pores is one-tenth that in the free liquid.

If the slab is placed in a large well-agitated reservoir of pure water at 25°C how long will it take for the mass fraction of ethanol at the center of the slab to fall to 0.009? Assume that there is no resistance to mass transfer in the water phase and that the concentration of ethanol in the water, and thus at the surface of the slab, is constant at zero.

Since the densities of alcohol and water differ by only 20 percent, we shall assume that ρ is constant; it is sufficiently accurate to use Eq. (30-54):

$$\frac{\partial \rho_A}{\partial \theta} = D_{AB} \frac{\partial^2 \rho_A}{\partial y^2} \tag{1}$$

or

$$\frac{\partial x_A}{\partial \theta} = D_{AB} \frac{\partial^2 x_A}{\partial y^2} \tag{2}$$

These equations imply that D_{AB} is constant, whereas Table 29-2 shows that it varies greatly with concentration. In order to avoid mathematical complication we shall assume D_{AB} is constant at average value of 1.0×10^{-9} m^2/s. During most of the time spent in achieving a low ethanol concentration at the center of the slab, the ethanol concentration in the slab is quite low. Therefore an average value of D_{AB} corresponding to a small x_A has been chosen.

We now make the following substitutions:

$$\tau = \frac{D_{AB}\theta}{y_o{}^2}$$

$$Y = x_A$$

$$n = \frac{y}{y_o}$$

The distance y is measured normal to the center of the slab; y_o is half the thickness of the slab.

By substitution, Eq. (2) becomes

$$\frac{\partial Y}{\partial \tau} = \frac{\partial^2 Y}{\partial n^2} \tag{3}$$

This is the same as Eq. (2) of Example 19-2. The boundary conditions on $Y(n,\tau)$ are as follows:

$$\frac{\partial Y}{\partial n}(0,\tau) = 0$$

$$Y(1,\tau) = 0$$

$$Y(n,0) = 1.0$$

These are identical with the boundary conditions of Example 19-2, so we can write the solution immediately from the results of that example, as follows:

$$Y = \sum_{i=1}^{i=\infty} \frac{-2(-1)^i}{\frac{2i-1}{2}\pi} \exp\left[\frac{-(2i-1)^2\pi^2\tau}{4}\right] \cos\left(\frac{2i-1}{2}\right)\pi n \tag{4}$$

For $n = 0$ and $Y = 0.009$, it is found that $\tau = 2.01$. Thus we have

$$\theta = \frac{\tau y_o{}^2}{D_{AB}}$$

$$= \frac{(2.01)(0.005)^2}{(1/10)(10^{-9})}$$

$$= 502{,}500 \text{ s } (139.58 \text{ h})$$

The concentration profiles at various times can be calculated from Eq. (4). Their appearance is the same as the curves shown in Fig. 19-2, if the ordinate is interpreted as the mass fraction water.

A great variety of problems in heat conduction have been solved in the literature.[1] The analogy to molecular diffusion is such that the solutions of most of these problems can be used to find immediately the solution of the corresponding diffusion problems. Solutions to many problems in molecular diffusion are given by Crank.[2] ////

Thermal Diffusion

It has been mentioned that in addition to ordinary molecular diffusion there exist thermal diffusion, pressure diffusion, and forced diffusion. If two regions in a mixture are maintained at different temperatures so that there is a flux of heat, it has been found that a concentration gradient is set up. In a binary mixture, one kind of molecule tends to travel toward the hot region and the other kind toward the cold region. This is called the *Soret effect*. Usually this effect has a negligible influence on mass transfer, but it is useful in the separation of certain mixtures. An analogous phenomenon, called the *Dufour effect*, is the tendency toward the generation of a temperature gradient in conjunction with mass transfer arising from a concentration gradient.

Pressure diffusion occurs when there is a gradient of pressure in a fluid mixture, as in a deep closed well or in a closed tube which is rotated around an axis perpendicular to the axis of the tube. The lighter component tends to travel toward the region of lower pressure, i.e., the top of the well or the end of the tube near the axis of rotation. The ultracentrifuge operates according to the principles of pressure diffusion.

Forced diffusion is caused by the action of an external force other than gravity on the molecules. This external force must be such that it acts to a different degree on different species. An example is the diffusion of the ions in an electrolyte in an electric field.

As a result of the concentration gradient generated by one of the above phenomena, ordinary diffusion occurs in a direction opposite that of the other kind of diffusion. In many situations a steady state is reached at which the flux by ordinary diffusion just balances that by one of the types described above, so that the properties of the system at a given point remain constant with time. As an example

[1] See, for example, H. S. Carslaw and J. C. Jaeger, "Conduction of Heat in Solids," 2d ed., Clarendon Press, Oxford, 1959.

[2] J. Crank, "Mathematics of Diffusion," Clarendon Press, Oxford, 1956.

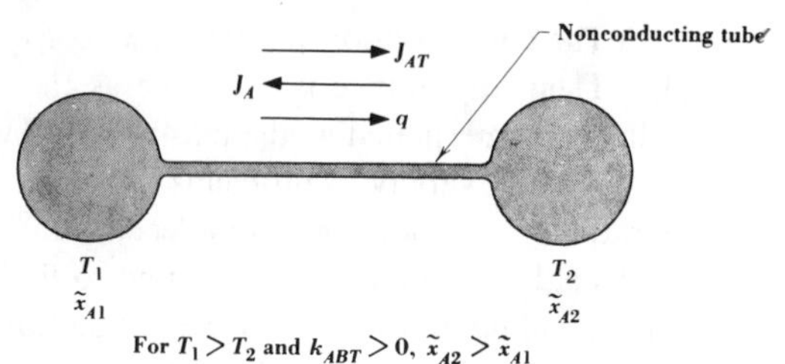

FIGURE 30-3
Thermal diffusion in a binary mixture at steady state.

of such a case, we consider the following example of thermal diffusion.

The two bulbs shown in Fig. 30-3 are connected by a capillary tube made of nonconducting material. The gas in the tube is stagnant, i.e., there is no convective mixing, whereas the gas in the bulbs is well mixed by convection. The flux caused by thermal diffusion in a binary mixture is obtained from irreversible thermodynamics[1] as

$$J_{AT} = -D_{ABT} \frac{1}{T} \frac{dT}{dy} \tag{30-55}$$

Although there is a temperature gradient from one bulb to another, at steady state the temperature at any point is constant. There is also no net flow of A or B from one bulb to the other, so

$$N_A = 0 \tag{30-56}$$

or

$$J_{AT} + J_A = 0 \tag{30-57}$$

Thus we have

$$D_{ABT} \frac{1}{T} \frac{dT}{dy} = -D_{AB} \rho \frac{dx_A}{dy} \tag{30-58}$$

However, it is customary to express the composition in terms of mole fractions rather than mass fractions, so we use Eqs. (30-26), (30-12), and (29-1) to obtain

$$D_{ABT} \frac{1}{T} \frac{dT}{dy} = \frac{-D_{AB} \tilde{\rho}^2 M_A M_B}{\rho} \frac{d\tilde{x}_A}{dy} \tag{30-59}$$

A thermal-diffusion ratio K_{ABT} is defined by

[1] A concise, readable introduction to this subject is the book by K. A. Denbigh, "Thermodynamics of the Steady State," Methuen & Co., Ltd., London, 1951. See also the references on p. 482.

$$K_{ABT} = \frac{D_{ABT}\rho}{D_{AB}\tilde{\rho}^2 M_A M_B} \qquad (30\text{-}60)$$

so that Eq. (30-59) can be written as

$$\frac{d\tilde{x}_A}{dy} = -K_{ABT}\frac{1}{T}\frac{dT}{dy} \qquad (30\text{-}61)$$

Although K_{ABT} is actually a complicated function of temperature, an estimate of the relation between the temperature and concentration profiles can be obtained by the integration of Eq. (30-61) with K_{ABT} constant. The result is

$$\tilde{x}_{A2} - \tilde{x}_{A1} = K_{ABT} \ln \frac{T_1}{T_2} \qquad (30\text{-}62)$$

The thermal-diffusion ratio may be evaluated at a mean temperature in this approximate equation. The quantity K_{ABT} can be either positive or negative, depending on the particular system under consideration. A positive value of K_{ABT} means that component A tends to accumulate in the colder part of the system. The thermal-diffusion ratio is also a function of composition, but $\tilde{x}_{A2} - \tilde{x}_{A1}$ is usually so small that the variation can be ignored in the integration of Eq. (30-61). A few values of K_{ABT} are listed in Table 30-1.

Table 30–1 SOME VALUES OF THE THERMAL-DIFFUSION RATIO K_{ABT}

System A-B	Temperature °C	Composition $\tilde{x}_A$	$100K_{ABT}$
He-Ne	−68	0.538	7.65
	57	0.538	7.82
	92	0.538	7.83
He-Ar	57	0.10	2.50
	57	0.30	6.60
	57	0.50	9.31
H_2-He	57	0.50	−4.81
He-N_2	−13	0.345	8.31
H_2-Ar	−15	0.47	−6.35
H_2-N_2	− 9	0.42	7.49
H_2-CO	−27	0.53	7.38
H_2-CO_2	27	0.53	6.89

SOURCE: J. O. Hirschfelder, C. F. Curtiss, and R. B. Bird, "Molecular Theory of Gases and Liquids," pp. 584–585, John Wiley & Sons, Inc., New York, 1954.

PROBLEMS

30-1 One gram of iodine is placed in a bulb like that illustrated in Fig. 30-2. If the tube through which diffusion takes place is 0.08 m long and 1.5×10^{-3} m in diameter, how long will it take the solid to disappear at 85°C? The air in the bulb is assumed to be saturated with iodine, and the iodine concentration in the surrounding atmosphere is zero.

30-2 If the bulb of Prob. 30-1 is initially filled with methanol to the outlet of the tube, how long will it take for the methanol level to fall to the bottom of the tube, i.e., 0.08 m? The surrounding fluid is air at 25°C.

30-3 A porous sphere of 0.003-m diameter is saturated with methanol. It is dropped into a large tank of pure water at 25°C. If it is assumed that the interdiffusion process is controlled by the rate of diffusion within the sphere, calculate the time required for the mole fraction at the center of the sphere to fall to $\tilde{x}_A = 0.01$. The effective diffusivity through the porous sphere is one-tenth of the ordinary molecular diffusivity. The solution may be obtained by the use of the graphs in Chap. 19.

30-4 The vapor pressure of naphthalene at 15°C is 4.84 Pa. Find the rate of evaporation of naphthalene from a sphere 0.01 m in diameter at 15°C if the rate is so slow that the sphere diameter can be assumed to be constant. The sphere will be considered to be surrounded by an infinite stagnant mass of air at the same temperature. Express your answer in mass percent evaporated per day with reference to the mass of the sphere when its diameter is 0.01 m. The density of naphthalene is 1152 kg/m^3.

30-5 Consider the sphere of Prob. 30-3 at the same initial conditions. Calculate the time required to remove all but 1 percent of the methanol initially within the sphere. Do you expect this time to be longer or shorter than that found in Prob. 30-3? Figure 14-1 of Perry is useful.

30-6 From relations between J_A and $\tilde{I}_A$ and between x_A and $\tilde{x}_A$, show that the coefficients D_{AB} which appear in Eqs. (29-1) and (29-3) are identical.

30-7 A Ney-Armistead diffusion cell consists of two bulbs of equal volume connected by a capillary as shown in Fig. 30-3. For the measurement of the coefficient of ordinary diffusion, one cell is initially filled with pure helium and the other with pure argon, all at the same pressure and temperature. Show how to calculate D_{AB} from data on the mole fraction of helium in one bulb as a function of time, and the dimensions of the apparatus. The volume of the capillary is small compared to that of the bulb.

30-8 When $x_{He} = 0.7$ in the left-hand bulb (originally pure helium) of the apparatus described in the previous problem, calculate the fluxes $\tilde{N}_A$, $\tilde{N}_B$, $\tilde{N}$, $\tilde{I}_A$, J_A, J_B, N_A, N_B, and N at the exit of the bulb. The temperature is 30°C and the pressure is 101.325 kPa. Each bulb has a volume of 2.75×10^{-4} m^3 and the capillary is 0.1 m long and has a 0.005-m ID.

30-9 A gas A is passed over a catalyst surface on which the dimerization reaction $2A \rightarrow B$ occurs. There is no reaction in the fluid phase. The product B which is produced diffuses away from the surface into the fluid stream. If the mass transfer of A to the surface is represented by diffusion through a stationary film, find the relation analogous to Eq. (30-42) which applies to this problem.

30-10 A thin layer of palladium metal is highly permeable to hydrogen but not to other gases such as nitrogen. Thus hydrogen can be separated from a mixture of 50 mol percent hydrogen in nitrogen by passing the mixture over a palladium surface. On the other side of the palladium the hydrogen pressure is adjusted so that at the surface of the palladium next to the mixture $\tilde{y}_{H_2} = 0.20$. If resistance to mass transfer is represented by a stagnant film 0.001 m thick, calculate the flux of hydrogen through the palladium in $kmol/m^2 \cdot s$. The process takes place at 20°C and 101.325 kPa.

30-11 Pure CO_2 is contained in a glass bulb from which a capillary leads to the atmosphere. How long will it take for the mole fraction of CO_2 in the bulb to fall to 0.5 at 20°C? The bulb contains 1.0×10^{-4} m^3 and the capillary has a 0.001-m ID and a length of 0.1 m.

30-12 In order to test an automotive-exhaust-emission-control device, we want to make a gaseous mixture of 5.0 ppm (vol) (i.e., parts per million on a volumetric or molar basis) of a hydrocarbon in air. The mixture is made by allowing the hydrocarbon to diffuse out of a small tube into a stream of air. The hydrocarbon liquid level is held constant in the tube by a suitable reservoir. For an air flow rate of 1.67×10^{-5} m^3/s, what diameter of tube is needed if the liquid level is 0.1 m below the end of the tube exposed to the flowing air?

Data:

1 diffusion coefficient for hydrocarbon = 1.0×10^{-5} m^2/s
2 vapor pressure = 10132.5 Pa
3 total pressure = 101325 Pa
4 temperature = 25°C

30-13 A modification of the Stefan experiment is to be used to measure the diffusivity of A in B. The diffusion takes place in a tube. The inside of the closed end of the tube is coated with a catalyst which is so active that equilibrium in the reaction ($A \rightleftarrows 2B$) is maintained at the closed end of the diffusion tube; the equilibrium constant K is known. The other end of the tube is open to a large reservoir of pure A. The whole apparatus is kept at constant temperature. Find the equation to calculate the diffusivity from the measured flux of A into the tube and the other experimental parameters. Steady state may be assumed.

30-14 If the thermal diffusion cell of Fig. 30-3 is initially filled at constant temperature and 101.325 kPa with an equimolar mixture of hydrogen and helium, what will be the steady-state compositions in the bulbs when one is heated to 100°C and the other cooled to 0°C?

30-15 A Wicke-Kallenbach apparatus is sometimes used to measure diffusion coefficients. It is convenient to analyze because it operates at steady state. Consider a capillary 10 cm long and 100 μm in diameter which is connected to the sides of larger tubes at each of its ends. The configuration forms a letter H; pure helium flows through the left vertical, and pure argon through the right vertical. The flows are adjusted so that there is no pressure difference from one end of the capillary to the other. The flows of the two pure gases are large compared to the diffusion fluxes through the capillary, the horizontal bar in the H. The pressure is 10 Pa and the temperature 298 K. Calculate $\tilde{N}$ and N through the capillary, and specify the direction of flow.

30-16 Equilibrium in the reaction

$$3C_2H_2 \rightleftarrows C_6H_6$$

is completely toward benzene at 300 K. A disk of catalyst is placed at the end of a tube 15 cm long; the opposite end of the tube is exposed to pure acetylene at 1 atm, 300 K. The catalyst is so active that the concentration of acetylene is zero adjacent to the disk. Find the production rate of benzene at steady state.

30-17 Two large bulbs are connected by a capillary tube of 2-mm ID which is 30 cm long. One bulb is filled with pure hydrogen and the other with pure nitrogen. Find the magnitude and direction of the bulk flux of mass shortly after the bulbs are connected together by the capillary.

30-18 A bulb like that of Fig. 30-2 contains diethyl ether at 17.9°C, at which temperature its vapor pressure is 400 torr. For a stagnant vapor column 5 cm high, what is the initial flux (g mol/cm$^2 \cdot$s) of ether out of the tube?

30-19 Pure chlorine gas is contained in a 100-mL bulb from which a 2-mm ID capillary 10 cm long leads to the atmosphere. How long will it take for the mole fraction of Cl_2 in the bulb to fall to 0.05 at 20°C?

18) 5cm

diethyl ether at 17.9°C, - vapor pressure = 400 torr.
- stagnant vapor column, 5cm high,
what is the initial flux (g mol/cm²·s) of ether out of the tube.

$$\tilde{N}_A = \frac{D_{AB} \cdot \tilde{c}}{L \cdot \tilde{x}_{B.lm}} (\tilde{x}_{A.0} - \tilde{x}_{A.L}), \quad \tilde{c} = \frac{P}{RT} = \frac{1 atm}{--} = 4.2 \times 10^{-5} mol/cm^3$$

$$x_{A0} = \frac{400 mmHg}{760 mmHg} = 0.5263. \quad x_{B.0} = 0.4737$$

$$x_{AL} = 0, \quad x_{B.L} = 1.$$

$$\tilde{N}_A = 5.58 \times 10^{-7} mol/cm^2 \cdot s.$$

table A-18, 298k, $D_{AB} = 0.093$.

31

CONVECTIVE MASS-TRANSFER COEFFICIENTS

Mass Transfer between Phases

Mass transfer may occur across a single phase, as, for example, in a gaseous system in which a substance evaporates at one interface and condenses at another interface. Mass transfer may also occur by a coupled process of transfer of a substance from the interior of one phase to the phase boundary and then into the interior of a second phase. If interfacial resistances are neglected, the total resistance to mass transfer in the first example is the resistance of the single gaseous phase. In the second example the total resistance is the sum of the resistances of the two phases which are in series. These phases are usually either two immiscible liquids or a gas and a liquid.

To solve mass-transfer problems the boundary conditions must be known. The boundaries of the mass-transfer systems are either the phase interfaces or arbitrarily chosen average conditions within the phases. Regardless of how the mathematical boundaries for solving the problem are chosen, phase equilibrium exists at the interface.[1] Although a concentration difference of the diffusing component exists across every interface, the difference is not necessarily a driving force for mass transfer. The compositions at an interface where equilibrium exists are not independent variables, but are fixed in accordance with the principles of physical equilibrium. The analogous situation in heat transfer is much easier to visualize. Thermal equilibrium between adjoining phases is implicitly assumed when we say that adjacent phases have the same temperature at the interface. In momentum transfer we usually assume that two phases have the same velocity at the interface.

The rate equations are written with composition difference as the driving force, and no difficulties are encountered if the rate equation is written for a single phase. However, if an overall rate equation is desired, the difference in the actual compositions of two phases will not be an adequate expression for the driving force. Such a difference usually occurs even when the phases are in equilibrium. To avoid this difficulty, the composition of one of the phases is expressed in terms of an equilibrium composition, i.e., the composition the other phase would have if it were in equilibrium with the first. Thus the overall driving force is expressed in units of composition based on one phase only.

Phase Equilibrium

Equilibrium in a multiphase system is subject to the restrictions of the phase rule. These restrictions are expressed in the equation

$$P + V = C + 2 \tag{31-1}$$

where P = number of phases at equilibrium
V = number of variants or degrees of freedom of system

If no chemical reactions are occurring in the system, the quantity C is simply the number of chemical compounds or free elements present. If chemical reactions do occur, then C is the number of chemical compounds and free elements minus the number of chemical equilibria or other restrictions placed on the chemical behavior of the system.

The phase rule can be regarded as analogous to an algebraic system of variables and independent equations. The number of variables minus the number of equations determines how many variables may be assigned arbitrary values. This is comparable with the number of variants in a physical system. In an algebraic system when the number of variables and the number of equations are equal, no

[1] At unusually high mass fluxes there is evidence that the adjacent phases are not in equilibrium. This effect is equivalent to an interfacial resistance to mass transfer. An interfacial resistance may also be caused by the presence of impurities which may be segregated at the interface.

variable can be arbitrarily specified. An example of this in phase equilibrium is the occurrence of a "triple point," i.e., a situation in which a single pure compound exists as three phases in physical equilibrium. According to the phase rule, V has a value of zero in this instance, which means that there can be only one set of conditions at which these three phases coexist. No variable (such as pressure or temperature) may be altered without causing the disappearance of one of the phases.

The phase rule does not tell us the numerical values of any of the variables, nor does it tell us the forms of any of the relations which exist among variables. Nevertheless, it is of great value to the engineer and scientist in planning experiments in which numerical values are to be measured and analytical relations deduced. It is of equal value in chemical-plant design in that it indicates the number of process variables that can be controlled independently in any system at equilibrium. For example, in gas-absorption operations a constituent is usually absorbed from a gas stream by a liquid. Such a system, if it consists of three chemical compounds and exists as two phases, may be seen from Eq. (31-1) to have three variables at equilibrium. In process terms this means that pressure, temperature, and the concentration in one phase of the diffusing component may be arbitrarily specified.

Distillation operations may involve only binary mixtures, or there may be many components. In a binary, two-phase system only two variables exist. This means that if the pressure and temperature are fixed, the compositions of the equilibrium phases may not be varied. It is common in binary-distillation calculations to specify pressure and one composition variable. In such circumstances, the temperature is fixed and may be found by performing appropriate calculations.

In humidification operations air may be considered as a single chemical constituent since it enters no chemical reactions and always retains the same proportions of its various constituents in such operations. Therefore, in humidification, there are, in effect, two chemical constituents to be considered—air and water. Because there are also two phases to be considered, the number of variables is two. Most such systems operate at atmospheric pressure so that only a single variable remains to be controlled. Thus a choice of temperature automatically fixes the concentration of water in the gas phase (the humidity) when equilibrium exists.

Equilibrium Relations

It is desirable, at this point, to review some of the quantitative equations for calculating equilibrium compositions. These and many other useful equilibrium equations can be found in books on physical chemistry. It is our intention in this chapter to discuss only certain simple equations relating to vapor-liquid equilibrium and gas absorption. Other relations will be reviewed in later chapters as needed.

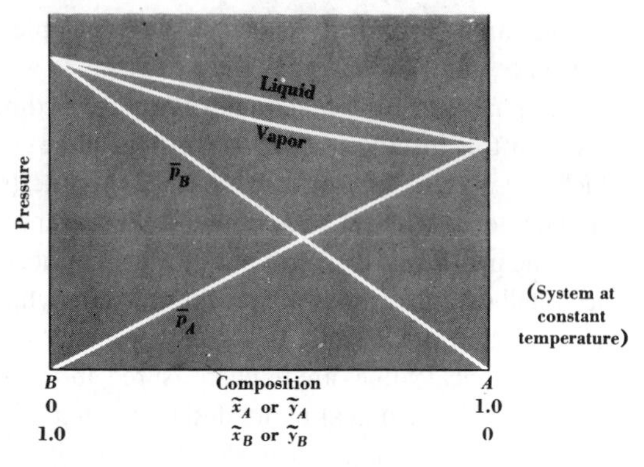

FIGURE 31-1
Partial pressures and equilibrium compositions for a system containing an ideal gas and an ideal liquid solution.

The partial pressure of a component in a gas phase is given by definition as

$$\bar{p}_A = \tilde{y}_A p \qquad (31\text{-}2)$$

where $\bar{p}_A$ = partial pressure of component A
$\tilde{y}_A$ = mole fraction of component A in gas phase[1]
p = total pressure on system

This definition applies whether the gas is ideal or nonideal. Because the sum of the mole fractions must be unity, it follows that the sum of the partial pressures always equals the total pressure.

Raoult's law is a relation often used for determining the partial pressure of a constituent in a liquid solution. It is expressed as

$$\bar{p}_A = \tilde{x}_A p_A \qquad (31\text{-}3)$$

where $\tilde{x}_A$ = mole fraction of component A in liquid
p_A = vapor pressure of substance A at temperature of system

This equation is valid when the gas and liquid phases behave as an ideal gas and an ideal solution, respectively. The concept of the ideal gas is familiar to most students, but the ideal solution is less so. It will suffice here to say that mixtures of nonpolar molecules of the same chemical type and approximately the same size approach ideal-solution behavior; an example is the adjacent members of a homologous series. The heat effect and volume change of mixing are zero for an ideal

[1] The symbols x_A and $\tilde{x}_A$ have until now been used to express composition in any phase. Henceforth, they will refer specifically to the liquid phase, and y_A and $\tilde{y}_A$ will refer to the vapor phase.

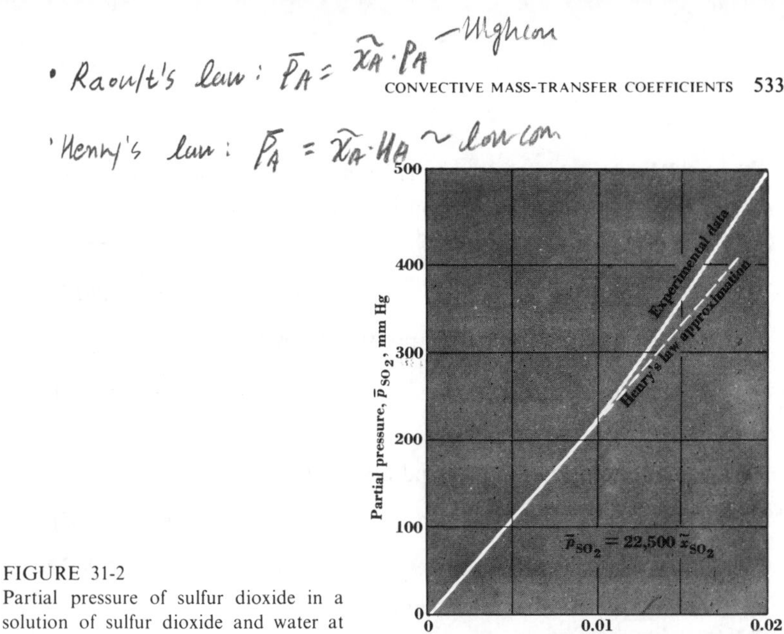

FIGURE 31-2
Partial pressure of sulfur dioxide in a solution of sulfur dioxide and water at 20°C.

solution. If the partial pressures of both components are plotted as a function of mole fraction at constant temperature, as shown in Fig. 31-1, they are represented by straight lines with a slope equal to the vapor pressure. Raoult's law usually matches experimental data best for components present in high concentrations.

Henry's law is a general form of which Raoult's law is a special case. It is expressed as

$$\bar{p}_A = \tilde{x}_A H_A \tag{31-4}$$

which simply expresses a linear relation between partial pressure and composition. The quantity H_A is a constant which is usually found experimentally. In the special case in which Raoult's law applies, the Henry's law constant H_A becomes the vapor pressure of component A. Henry's law usually applies to the low-concentration range, as is illustrated by the experimental data in Fig. 31-2. This law is of value in the design of gas-absorption systems because many of them operate with dilute solutions. Liquid-liquid phase equilibrium is discussed in Chap. 38.

Example 31-1 A mixture of benzene and toluene exists as a gas-liquid system in equilibrium at 80°C. The gas phase contains 65 mole percent benzene and 35 mol percent toluene. Find the total pressure on the system and the composition of the liquid phase, assuming the liquid to be an ideal solution and the gas to be ideal. The vapor pressures of benzene and toluene at 80°C are 756 and 287 mmHg, respectively.

We designate benzene as component A and toluene as component B. The partial pressures, from Raoult's law, are

$$\bar{p}_A = 756\tilde{x}_A \tag{1}$$

$$\bar{p}_B = 287\tilde{x}_B \tag{2}$$

From the definition of partial pressure we write

$$\bar{p}_A = 0.65p \tag{3}$$

$$\bar{p}_B = 0.35p \tag{4}$$

Combining Eqs. (1) and (3) and (2) and (4), we obtain

$$756\tilde{x}_A = 0.65p \tag{5}$$

$$287\tilde{x}_B = 0.35p \tag{6}$$

In addition, we know the equation

$$\tilde{x}_A + \tilde{x}_B = 1 \tag{7}$$

Equations (5) to (7) are solved for the three variables:

$$\tilde{x}_A = 0.415$$

$$\tilde{x}_B = 0.585$$

$$p = 483 \text{ mmHg} \qquad ////$$

Example 31-2 Derive equations for the equilibrium compositions of the gas and liquid in a system at constant temperature in which the two phases behave as an ideal gas and an ideal solution, respectively.

The two components are designated A and B. We write the partial pressures using Raoult's law.

$$\bar{p}_A = \tilde{x}_A p_A$$

$$\bar{p}_B = \tilde{x}_B p_B$$

The total pressure is the sum of the partial pressures.

$$p = \bar{p}_A + \bar{p}_B = \tilde{x}_A p_A + \tilde{x}_B p_B$$
$$= \tilde{x}_A p_A + (1 - \tilde{x}_A)p_B$$

Solving for $\tilde{x}_A$, we obtain

$$\tilde{x}_A = \frac{p - p_B}{p_A - p_B}$$

At a constant temperature the vapor pressures p_A and p_B are fixed. The liquid composition $\tilde{x}_A$ is therefore a linear function of the total pressure, as shown in

Fig. 31-1; the total pressure ranges from $p = p_B$ at $\tilde{x}_A = 0$ to $p = p_A$ at $\tilde{x}_A = 1$. The vapor composition is found from the definition of partial pressure.

$$\bar{p}_A = \tilde{y}_A p = \tilde{x}_A p_A$$

from which we obtain

$$\tilde{y}_A = \frac{\tilde{x}_A p_A}{p}$$

$$= \frac{p_A}{p} \frac{p - p_B}{p_A - p_B} \qquad ////$$

Individual Mass-Transfer Coefficients

The concept of the mass-transfer coefficient has already been discussed in Chap. 28 and coefficients defined in Eqs. (28-1) and (30-47). In this section we shall consider the mass-transfer coefficient in more detail.

Although the analogy between heat transfer and mass transfer has been useful for the solution of many mass-transfer problems, some basic differences often prevent us from making use of our knowledge of heat-transfer coefficients for predicting mass-transfer coefficients. One of these has to do with boundary conditions. In most heat-transfer systems where convective effects exist, heat is being transferred between a single-phase fluid and a solid. In most convective mass-transfer systems, however, two fluid phases are in contact and mass transfer occurs between the fluid phases. In the heat-transfer system the fluid has zero velocity at the interface between the fluid and the solid and an adjacent viscous sublayer exists; in the mass-transfer system the velocity at the interface between the fluids is probably not zero and the existence of a viscous sublayer is often unlikely. The difference in flow behavior has the practical result that much of the experimental information available on heat transfer is of little value in mass transfer, and vice versa. An additional complication arises in the analysis of transfer rates between two fluids in direct contact, for the average interfacial area is a function of flow conditions and is difficult or impossible to measure. There are some applications, however, in which heat and mass transfer occur in similar systems, and some in which they occur simultaneously. An example of the latter is spray drying, an operation in which a liquid spray is injected into a hot-air stream; heat and mass are transferred between adjacent fluid phases. The operation of a catalytic reactor represents a system in which heat and mass are transferred simultaneously between a fluid phase and a solid surface. When possible, in succeeding discussions on mass transfer, the reader will be referred to a previous discussion of heat-transfer coefficients. Frequently, however, mass-transfer systems will be discussed which have no counterpart of interest in heat transfer.

If a mass-transfer system consists of a solute dissolving at a steady rate from a solid surface and diffusing into a liquid stream, we define the mass-transfer coefficient $k_{\tilde{x}}$ by the following equation, which is similar to Eq. (30-47):

$$\tilde{N}_A = k_{\tilde{x}}(\tilde{x}_{As} - \tilde{x}_A) \qquad (31\text{-}5)$$

In this equation $\tilde{N}_A$ is the moles of solute A leaving the interface per unit time per unit area of interface. The mole fraction of solute in the liquid at the interface $\tilde{x}_{As}$ represents the composition the solution would have if it were in equilibrium with the solid solute at the temperature of the system. The quantity $\tilde{x}_A$ represents the composition expressed in terms of the mole fraction of solute at some point within the fluid phase. If the flow is of the type known as boundary-layer flow, $\tilde{x}_A$ would be chosen as the mole fraction of solute at the edge of the concentration boundary layer and would be written $\tilde{x}_{Ao}$. If, however, the flow were in a closed conduit, the value $\tilde{x}_A$ would be chosen as the mole fraction of solute at the mixing-cup concentration. (The mixing-cup concentration is the concentration if the stream were uniform. It is defined in a manner similar to the mixing-cup temperature.) The mass-transfer coefficient $k_{\tilde{x}}$ is usually expressed in the units lb mol/(h)(ft^2) or kmol/s$\cdot$m^2. The subscript $\tilde{x}$ signifies that it applies to a liquid phase and must be used in conjunction with a driving force expressed in terms of mole fraction in the liquid phase.

Mass transfer at a steady rate from a solid interface to a gas stream is described by the equation

$$\tilde{N}_A = k_{\tilde{y}}(\tilde{y}_{As} - \tilde{y}_A) \qquad (31\text{-}6)$$

The flux $\tilde{N}_A$ has the same units as in Eq. (31-5). The mole fraction of solute in the gas phase adjacent to the interface is $\tilde{y}_{As}$. This is assumed to be the composition that would occur at equilibrium at the same temperature and pressure. The quantity $\tilde{y}_A$ is usually either the mole fraction of solute at the edge of the concentration boundary layer or the mole fraction at the mixing-cup concentration. The mass-transfer coefficient $k_{\tilde{y}}$ has the same units as $k_{\tilde{x}}$, but as the subscript indicates, it must be used with a driving force expressed as mole fraction in the gas phase.

A system in which a solute is being transferred at a steady rate from a gas phase to a liquid phase can be described by two equations:

$$\tilde{N}_A = k_{\tilde{y}}(\tilde{y}_A - \tilde{y}_{As}) \qquad (31\text{-}7)$$

$$\tilde{N}_A = k_{\tilde{x}}(\tilde{x}_{As} - \tilde{x}_A) \qquad (31\text{-}8)$$

In a system of adjacent gas and liquid phases the values of $\tilde{y}_A$ and $\tilde{x}_A$ are most likely to be mole fractions at the mixing-cup concentrations of the respective phases.

Overall Mass-Transfer Coefficients

For interphase transfer it is often convenient to use an overall mass-transfer coefficient, just as an overall heat-transfer coefficient was employed in heat transfer. The overall mass-transfer coefficient may be defined by either of the following equations:

$$\tilde{N}_A = K_{\tilde{y}}(\tilde{y}_A - \tilde{y}_A^*) \qquad (31\text{-}9)$$

$$\tilde{N}_A = K_{\tilde{x}}(\tilde{x}_A^* - \tilde{x}_A) \qquad (31\text{-}10)$$

The quantities $\tilde{y}_A^*$ and $\tilde{x}_A^*$ represent equilibrium concentrations. The quantity $\tilde{y}_A^*$ is the mole fraction of solute in a gas at equilibrium with a liquid of composition $\tilde{x}_A$. Likewise, $\tilde{x}_A^*$ is the mole fraction of solute in the liquid at equilibrium with a gas of composition $\tilde{y}_A$. Therefore $\tilde{y}_A - \tilde{y}_A^*$ and $\tilde{x}_A^* - \tilde{x}_A$ are expressions representing the overall driving force for mass transfer between two fluid phases.

An equation relating the overall mass-transfer coefficients to the coefficients for the individual phases can be found by rewriting Eqs. (31-7) and (31-8) as follows:

$$\tilde{y}_A - \tilde{y}_{As} = \frac{\tilde{N}_A}{k_{\tilde{y}}} \qquad (31\text{-}11)$$

$$\tilde{x}_{As} - \tilde{x}_A = \frac{\tilde{N}_A}{k_{\tilde{x}}} \qquad (31\text{-}12)$$

Equilibrium between the gas and liquid phases is assumed to follow Henry's law. If the partial pressure found from Eq. (31-4) is divided by the total pressure, we obtain

$$\tilde{y}_A = \frac{\tilde{x}_A H_A}{p} \qquad (31\text{-}13)$$

The quantity H_A/p is designated as m, and Eq. (31-13) is written as

$$\tilde{y}_A = m\tilde{x}_A \qquad (31\text{-}14)$$

At an interface $\tilde{y}_A$ and $\tilde{x}_A$ in Eq. (31-14) are written $\tilde{y}_{As}$ and $\tilde{x}_{As}$, and we assume that equilibrium exists at the interface. Therefore we write Eq. (31-12) as

$$\frac{\tilde{y}_{As}}{m} - \tilde{x}_A = \frac{\tilde{N}_A}{k_{\tilde{x}}} \qquad (31\text{-}15)$$

and multiply by m to obtain

$$\tilde{y}_{As} - m\tilde{x}_A = \frac{\tilde{N}_A m}{k_{\tilde{x}}} \qquad (31\text{-}16)$$

This expression is added to Eq. (31-11) to give

$$\tilde{y}_A - m\tilde{x}_A = \tilde{N}_A\left(\frac{1}{k_{\tilde{y}}} + \frac{m}{k_{\tilde{x}}}\right) \qquad (31\text{-}17)$$

The term $m\tilde{x}_A$ represents the mole fraction of solute in a gas at equilibrium with a liquid of mole fraction $\tilde{x}_A$. Thus we recognize that $m\tilde{x}_A$ is identical with $\tilde{y}_A^*$. Making this substitution, we write Eq. (31-17) as

$$\tilde{y}_A - \tilde{y}_A^* = \tilde{N}_A\left(\frac{1}{k_{\tilde{y}}} + \frac{m}{k_{\tilde{x}}}\right) \qquad (31\text{-}18)$$

The defining expression for $K_{\tilde{y}}$, Eq. (31-9), is written in similar fashion.

$$\tilde{y}_A - \tilde{y}_A^* = \frac{\tilde{N}_A}{K_{\tilde{y}}} \qquad (31\text{-}19)$$

By comparing Eqs. (31-18) and (31-19) we see that $K_{\tilde{y}}$ can be defined in terms of the coefficients for the individual phases as

$$K_{\tilde{y}} = \frac{1}{1/k_{\tilde{y}} + m/k_{\tilde{x}}} \qquad (31\text{-}20)$$

By similar procedures we find an expression for $K_{\tilde{x}}$.

$$K_{\tilde{x}} = \frac{1}{1/k_{\tilde{x}} + 1/mk_{\tilde{y}}} \qquad (31\text{-}21)$$

Even though the individual coefficients $k_{\tilde{y}}$ and $k_{\tilde{x}}$ may be constant throughout a system, the overall coefficient may vary as a result of variation of m. The quantity m, as we have already seen, is a function of concentration, temperature, and total pressure; it is usually constant only for dilute solutions at a constant temperature in a column in which the fractional change in pressure is negligible. As we shall show below, Eqs. (31-20) and (31-21) are not correct if the equilibrium line is curved, i.e., if m is not constant.

The discussion above is illustrated by Fig. 31-3. From an overall mass balance on a piece of contacting equipment, such as a gas-absorption tower, we shall find that there is an algebraic relation between the bulk-gas composition and the bulk-liquid composition at any horizontal section in a tower, i.e., at any value of $\tilde{x}_A$. This relation is represented in Fig. 31-3 by the upper curve, called an *operating line*. The position of this line depends on the compositions at the ends of the tower and on the flow rates of the two phases.

At the point P in Fig. 31-3, the gas and liquid compositions are $\tilde{y}_A$ and $\tilde{x}_A$,

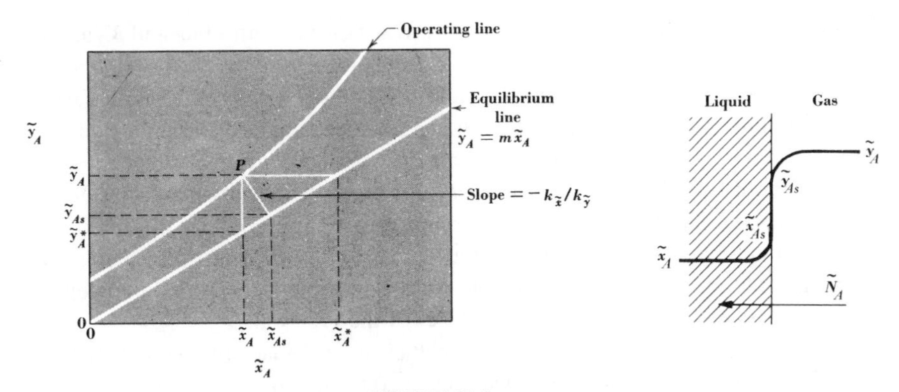

FIGURE 31-3
Graphs showing compositions at an interface between phases.

respectively. The compositions at the interface, $\tilde{y}_{As}$ and $\tilde{x}_{As}$, are found by combining Eqs. (31-7) and (31-8) to give

$$-\frac{k_{\tilde{x}}}{k_{\tilde{y}}} = \frac{\tilde{y}_A - \tilde{y}_{As}}{\tilde{x}_A - \tilde{x}_{As}} \tag{31-22}$$

This equation is represented by a straight line of slope $-k_{\tilde{x}}/k_{\tilde{y}}$ passing through $\tilde{x}_A$, $\tilde{y}_A$ and $\tilde{x}_{As}$, $\tilde{y}_{As}$. Since the latter point is on the equilibrium line, the values of $\tilde{x}_{As}$ and $\tilde{y}_{As}$ are found by the construction shown in the figure. Also given in Fig. 31-3 are the composition $\tilde{y}_A^*$, which is in equilibrium with $\tilde{x}_A$, and the composition $\tilde{x}_A^*$, which is in equilibrium with $\tilde{y}_A$.

In the derivation of Eq. (31-20), $\tilde{x}_{As}$ was replaced by $\tilde{y}_{As}/m$ in Eq. (31-15). Thus the m appearing in Eqs. (31-16) *et seq.* is the value of m corresponding to $\tilde{x}_{As}$. In going from Eq. (31-17) to (31-18), $m\tilde{x}_A$ was replaced by $\tilde{y}_A^*$; this substitution is valid only if the m at $\tilde{x}_{As}$ is the same as the m at $\tilde{x}_A$. For a curved equilibrium line, these m's are not identical, and so Eqs. (31-20) and (31-21) are not correct.[1]

The above restrictions on the use of the overall coefficient become irrelevant if it is approximately equal to one of the individual coefficients. For example, if $m/k_{\tilde{x}}$ is negligible in comparison with $1/k_{\tilde{y}}$, it is said that the gas-phase resistance is controlling and $K_{\tilde{y}}$ is the same as $k_{\tilde{y}}$. Note that it is not sufficient for m to be small or $k_{\tilde{x}}$ to be large; their *ratio* must be small compared with $1/k_{\tilde{y}}$.† Similar remarks

[1] This matter is discussed in some detail in T. K. Sherwood, R. L. Pigford, and C. R. Wilke, "Mass Transfer," p. 459, McGraw-Hill Book Company, New York, 1975.

† In many systems a chemical reaction occurs in the liquid phase. This does not, however, make m zero even when the equilibrium partial pressure of the solute is zero. The reason for this is that the solute may have to diffuse a significant distance into the liquid before reacting.

apply to a situation for which the liquid-phase resistance is controlling and $K_{\tilde{x}}$ and $k_{\tilde{x}}$ are the same.

Although an overall coefficient should not be used for systems with a curved equilibrium line, reference to the literature shows that such a coefficient is often the only one reported. It is almost impossible in two-fluid contactors to measure $\tilde{x}_{As}$ and $\tilde{y}_{As}$ experimentally and to calculate $k_{\tilde{y}}$, for example, directly from data on the rate of mass transfer between phases. Usually only $\tilde{x}_A$ and $\tilde{y}_A$ can be measured, and the individual resistances must be estimated indirectly. In heat transfer in a heat exchanger, the situation is more tractable, for it is possible to measure the analogous temperatures, which are those at the surfaces of the tube walls.

A justification for the use of overall coefficients is the lack of precision of most data on mass transfer. Standard deviations of 20 to 30 percent are not uncommon, so that minor errors introduced by the use of an overall coefficient may not be significant. This lack of precision is not caused by carelessness, but by the difficulty of obtaining reproducible data on compositions and velocities in a device such as a packed absorption tower.

Mass-Transfer Coefficients and Bulk Flow

The characteristics of the mass-transfer coefficients which we have defined depend to some extent on whether they apply to a case for which $\tilde{N}_A = -\tilde{N}_B$ or to a case for which $\tilde{N}_B = 0$. The former case is encountered in adiabatic binary distillation, where an energy balance often results in the requirement that a mole of A be transferred in one direction for every mole of B transferred in the opposite direction. The latter case is encountered in gas absorption and liquid-liquid extraction, where the solute is transferred toward or away from the interface through the second component of the phase, which does not dissolve.

In order to discuss this situation clearly, we must define another mass-transfer coefficient $k_{\tilde{x}}^o$ (or $k_{\tilde{y}}^o$, k_ρ^o, etc.). This is the value which $k_{\tilde{x}}$ would have for the given fluids and the given hydrodynamic conditions if the net rate of mass transfer (the bulk flow, $\tilde{N}_A + \tilde{N}_B$) were zero. This coefficient is different from $k_{\tilde{x}}$ for appreciable rates of mass transfer when $\tilde{N}_B$ is zero; for small transfer rates or for $\tilde{N}_B = -\tilde{N}_A$, $k_{\tilde{x}}^o$ and $k_{\tilde{x}}$ are indistinguishable.

The relation between $k_{\tilde{x}}^o$ and $k_{\tilde{x}}$ can be obtained rigorously only for certain cases of laminar flow. For most systems, including turbulent flow, such a relation is not known; the following analysis is at best a good approximation. The flux for $\tilde{N}_A = -\tilde{N}_B$ across a layer of fluid may be expressed by the equation

$$\tilde{N}_A = \frac{\tilde{\rho} D_{AB}}{\delta_m} (\tilde{x}_{As} - \tilde{x}_A) \qquad (31\text{-}23)$$

which has been obtained from Eq. (30-51). The path length for diffusion L has been replaced by a fictitious "film" thickness δ_m. This thickness is chosen so as to give the transfer rate for the actual system in terms of the resistance of a stagnant layer of thickness δ_m which would offer the same resistance. We compare Eqs. (31-8) and (31-23) and find that for equimolar counterdiffusion the coefficient $k_{\tilde{x}}$ can be expressed as

$$k_{\tilde{x}} = \frac{\tilde{\rho} D_{AB}}{\delta_m} \tag{31-24}$$

Because $k_{\tilde{x}}$ and $k_{\tilde{x}}{}^o$ are equal in this special case, Eq. (31-24) also serves as the definition of $k_{\tilde{x}}{}^o$.

If we have $\tilde{N}_B = 0$, the analysis is based on Eqs. (30-46) to (30-48), with the result

$$k_{\tilde{x}} = \frac{\tilde{\rho} D_{AB}}{\delta_m (1 - \tilde{x}_A)_{lm}} \tag{31-25}$$

which is found by combining Eqs. (31-8) and (30-46). We have already defined $k_{\tilde{x}}{}^o$ as $\tilde{\rho} D_{AB}/\delta_m$; hence, for the special case of $\tilde{N}_B = 0$, we obtain

$$k_{\tilde{x}}{}^o = k_{\tilde{x}} (1 - \tilde{x}_A)_{lm}. \tag{31-26}$$

We shall use Eq. (31-26) to estimate the effect of composition on $k_{\tilde{x}}$. It is assumed that the coefficient $k_{\tilde{x}}{}^o$ is not affected by composition, but is fixed by the fluid properties and velocities and the dimensions of the system. For example, correlations will be obtained involving a Nusselt-like group for mass transfer, which is called the *Sherwood number*. This group is defined in terms of $k_{\tilde{x}}{}^o$ or $k_\rho{}^o$ as

$$\mathbf{Sh} = \frac{k_{\tilde{x}}{}^o D}{\tilde{\rho} D_{AB}} \quad \text{or} \quad \frac{k_\rho{}^o D}{D_{AB}} \tag{31-27}$$

If it is desired to use the value of $k_{\tilde{x}}{}^o$ obtained from such a correlation for $\tilde{N}_B = 0$ and concentrated solutions, $k_{\tilde{x}}{}^o$ must be replaced by $k_{\tilde{x}}(1 - \tilde{x}_A)_{lm}$.

Application to a Packed Tower—Dilute Solutions

The convective mass-transfer coefficients have been arbitrarily defined just as heat-transfer coefficients were defined. To obtain the transfer rate, the flux must be multiplied by the area across which the transfer is occurring. This computation is possible in a heat exchanger or in mass transfer between a solid surface and a fluid. However, the interfacial area between a liquid flowing down over the packing in an absorption tower and the gas rising in the tower is difficult to measure or compute and so is often unknown. This is also true of spray towers and liquid-liquid extraction devices.

To assist in analyzing situations in which the magnitude of the interfacial area is not available, we define the quantity a, the interfacial area per unit volume of contacting device. The rate of mass transfer per unit volume of tower is then given as

$$\frac{\text{Rate of mass transfer}}{\text{Volume}} = k_{\tilde{x}}a(\tilde{x}_{As} - \tilde{x}_A) \qquad (31\text{-}28)$$

Since the rate of transfer per unit volume and the driving force are readily measurable, coefficients obtained from experiments on packed towers and similar devices are usually given on a volumetric basis; one finds data for $k_{\tilde{y}}a$, $k_{\tilde{x}}a$, $K_{\tilde{y}}a$, and $K_{\tilde{x}}a$, in units such as lb mol/(h)(ft^3).

In a packed gas-absorption tower the quantity a is not the same as the surface area of the solid packing. The interfacial contact area between gas and liquid per unit volume of tower is a function of the gas and liquid flow rates, and it does not necessarily vary in the same way as the convective coefficient.[1]

In most applications of the mass-transfer coefficients, Eq. (31-28) applies at any point along the phase interface and must be integrated over the area contained in the whole equipment in order to obtain a useful result. We illustrate this procedure by a consideration of the absorption from a gas of a soluble gas A by water in a countercurrent contacting device, the packed tower of Fig. 31-4. A mixture of air and a solute A, which is soluble in water, is blown into the bottom of the tower and rises at a rate $\tilde{G}$, measured in lb mol/(h)(ft^2) of cross section of empty tower. The tower is filled with a packing made of ceramic rings. Water, which absorbs component A, flows down over the packing at $\tilde{L}$ lb mol/(h)(ft^2) of cross section of empty tower.

Although the concentration in each phase diminishes with distance z up the tower, mass transfer in the z direction is usually assumed to be negligible. The solute is transferred in the direction of decreasing concentration from the bulk of the gas phase to the liquid interface and then to the bulk of the liquid phase. Transfer occurs by both eddy and molecular diffusion and is described by the various convective mass-transfer coefficients. If the tower is very high, almost all the solute is removed from the gas leaving at the top.

Any one of the rate equations (31-7) to (31-10) applies at a point in the tower. The total amount of A transferred per unit time in the tower can be expressed by an equation like

$$(\tilde{N}_A)_{\text{av}} A_i = K_{\tilde{y},\,\text{av}}(\tilde{y}_A - \tilde{y}_A^*)_{\text{av}} A_i \qquad (31\text{-}29)$$

[1] H. L. Shulman, C. F. Aldrich, A. Z. Proulx, and J. O. Zimmerman, *AIChE J.*, **1**:253 (1955); typical values of a are of the order of 10 ft^{-1}. See also P. V. Danckwerts, "Gas-Liquid Reactions," chap. 9, McGraw-Hill Book Company, New York, 1970.

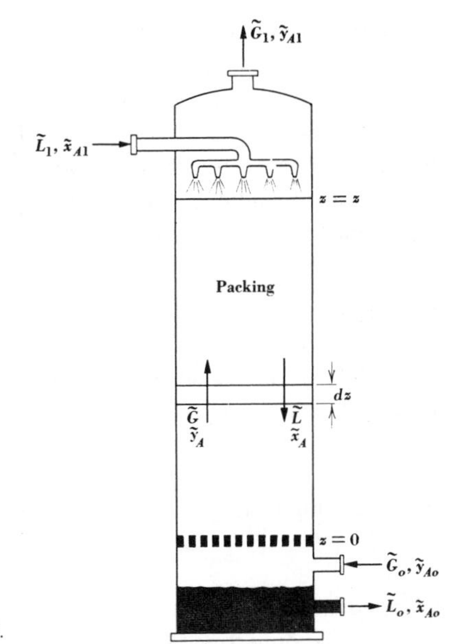

FIGURE 31-4
Packed gas-absorption tower.

in which A_i is the total interfacial area. The right-hand side of Eq. (31-29) contains the average concentration difference. In general, this quantity must be found by integration, but in some simplified cases we shall see that the logarithmic mean of the terminal concentration differences applies, just as the logarithmic mean temperature difference applied to certain heat exchangers. The left-hand side of Eq. (31-29) is equal to the change from the top to the bottom of the tower of the product of the flow rate of one of the phases and its concentration; this quantity is analogous to the total heat-transfer rate in an exchanger.

We now proceed to the derivation of an equation giving the tower height in terms of the flow rates, coefficients, and terminal concentrations. At any height z above the bottom of the tower, we assume that the concentrations are uniform over the cross section of the tower. Since conditions change with z, Eq. (31-7), for example, applies only for a differential element of interfacial area dA_i of height dz. It is customary to express A_i in terms of the quantity a (the interfacial area per unit volume of tower), the height z, and the cross-sectional area of the empty tower A:

$$A_i = aAz \qquad (31\text{-}30)$$

or if a is assumed constant,

$$dA_i = aA\,dz \qquad (31\text{-}31)$$

We then write Eq. (31-9) as

$$\tilde{N}_A \, dA_i = K_{\tilde{y}}{}^o a(\tilde{y}_A - \tilde{y}_A^*)A \, dz \qquad (31\text{-}32)$$

The left-hand side of Eq. (31-32) is the mass transferred in the height dz; for dilute solutions we assume that $\tilde{L}$ and $\tilde{G}$ are not appreciably changed by the addition or removal of solute A. We write $K_y{}^o$, where the superscript reminds us that the coefficient applies to a dilute solution. A mass balance then gives

$$\begin{aligned} -\tilde{N}_A \, dA_i &= A\tilde{G} \, d\tilde{y}_A \\ &= A\tilde{L} d\tilde{x}_A \end{aligned} \qquad (31\text{-}33)$$

Integration of this equation from the dilute (1) end of the tower to any point within the tower yields

$$\tilde{G}(\tilde{y}_A - \tilde{y}_{A1}) = \tilde{L}(\tilde{x}_A - \tilde{x}_{A1}) \qquad (31\text{-}34)$$

or, integrating to the rich (o) end,

$$\tilde{G}(\tilde{y}_{Ao} - \tilde{y}_{A1}) = \tilde{L}(\tilde{x}_{Ao} - \tilde{x}_{A1}) \qquad (31\text{-}35)$$

These equations indicate that $\tilde{y}_A$ and $\tilde{x}_A$ at any point in the tower must fall on a straight line connecting $(\tilde{x}_{A1}, \tilde{y}_{A1})$ and $(\tilde{x}_{Ao}, \tilde{y}_{Ao})$. This line is the operating line and is shown in Fig. 31-3. The lower part of this line, which corresponds to dilute solutions, is straight. For such solutions, Henry's law is a good approximation, and the equilibrium line is also straight, as shown in Fig. 31-3. The location of the various concentrations on a $\tilde{y} - \tilde{x}$ graph is further illustrated in Fig. 31-5.

As already discussed, the use of the overall coefficient and Eq. (31-32) are appropriate for these dilute solutions, for which the operating and equilibrium lines are straight. Equations (31-32) and (31-33) are now combined and written as

$$\int_o^z dz = z = \frac{\tilde{G}}{K_{\tilde{y}}{}^o a} \int_{\tilde{y}_{A1}}^{\tilde{y}_{Ao}} \frac{d\tilde{y}_A}{\tilde{y}_A - \tilde{y}_A^*} \qquad (31\text{-}36)$$

The right-hand side of this equation is evaluated by reasoning which is analogous to that leading to the logarithmic mean-temperature-difference driving force. Since $\tilde{y}_A$ and $\tilde{y}_A^*$ are both linear functions of $\tilde{x}_A$ or $\tilde{y}_A$, their difference $\tilde{y}_A - \tilde{y}_A^*$, or Δ, must also be a linear function of $\tilde{y}_A$. Thus we obtain

$$\frac{d\Delta}{d\tilde{y}_A} = \text{const} = \frac{\Delta_o - \Delta_1}{\tilde{y}_{Ao} - \tilde{y}_{A1}} \qquad (31\text{-}37)$$

where Δ is $\tilde{y}_A - \tilde{y}_A^*$. The differential $d\tilde{y}_A$ in Eq. (31-36) is replaced using Eq. (31-37), and we get

$$z = \frac{\tilde{G}}{K_{\tilde{y}}{}^o a} \frac{\tilde{y}_{Ao} - \tilde{y}_{A1}}{\Delta_o - \Delta_1} \int_{\Delta_1}^{\Delta_o} \frac{d\Delta}{\Delta} \qquad (31\text{-}38)$$

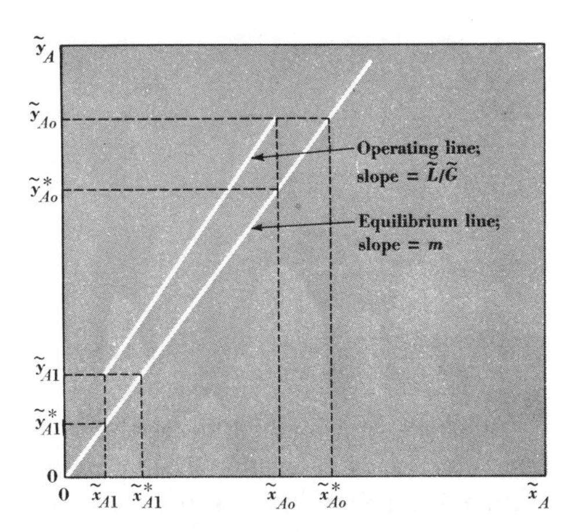

FIGURE 31-5
Graph of contacting dilute solutions.

Integration yields

$$z = \frac{\tilde{G}}{K_{\tilde{y}}{}^o a} \frac{\tilde{y}_{Ao} - \tilde{y}_{A1}}{(\tilde{y}_A - \tilde{y}^*_A)_{lm}} \tag{31-39}$$

where

$$(\tilde{y}_A - \tilde{y}^*_A)_{lm} = \frac{(\tilde{y}_A - \tilde{y}^*_A)_o - (\tilde{y}_A - \tilde{y}^*_A)_1}{\ln[(\tilde{y}_A - \tilde{y}^*_A)_o/(\tilde{y}_A - \tilde{y}^*_A)_1]} \tag{31-40}$$

A similar development can be done in terms of $k_{\tilde{y}}{}^o a$, $K_{\tilde{x}}{}^o a$, or $k_{\tilde{x}}{}^o a$ to obtain

$$z = \frac{\tilde{G}}{k_{\tilde{y}}{}^o a} \frac{\tilde{y}_{Ao} - \tilde{y}_{A1}}{(\tilde{y}_A - \tilde{y}_{As})_{lm}} \tag{31-41}$$

$$z = \frac{\tilde{L}}{K_{\tilde{x}}{}^o a} \frac{\tilde{x}_{Ao} - \tilde{x}_{A1}}{(\tilde{x}^*_A - \tilde{x}_A)_{lm}} \tag{31-42}$$

$$z = \frac{\tilde{L}}{k_{\tilde{x}}{}^o a} \frac{\tilde{x}_{Ao} - \tilde{x}_{A1}}{(\tilde{x}_{As} - \tilde{x}_A)_{lm}} \tag{31-43}$$

Each of these four equations for z can be written as the product of a height, called the *height of a transfer unit* (HTU) and a pure number, called the *numbers of transfer units* (NTU). Thus Eqs. (31-39) and (31-41) to (31-43) become, respectively,

$$z = H_{OG} n_{OG} \tag{31-44}$$

$$z = H_G n_G \tag{31-45}$$

$$z = H_{OL} n_{OL} \tag{31-46}$$

$$z = H_L n_L \tag{31-47}$$

in which
$$H_{OG} = \frac{\tilde{G}}{K_{\tilde{y}}{}^{o}a} \tag{31-48}$$

$$H_G = \frac{\tilde{G}}{k_{\tilde{y}}{}^{o}a} \tag{31-49}$$

$$H_{OL} = \frac{\tilde{L}}{K_{\tilde{x}}{}^{o}a} \tag{31-50}$$

$$H_L = \frac{\tilde{L}}{k_{\tilde{x}}{}^{o}a} \tag{31-51}$$

One reason why the transfer-unit system is useful is that HTU always has the dimension simply of length. The forms given are used for calculations when mole fractions are used for the driving force. However, there is a permanence of form, so that if concentrations are used for the driving force, we have

$$H_G = \frac{u_G}{k_\rho{}^{o}a} = \frac{u_G}{k_{\tilde{\rho}}{}^{o}a} \tag{31-52}$$

in which u_G is the superficial velocity of the gas phase, equal to G/ρ or $\tilde{G}/\tilde{\rho}$; $k_\rho{}^o$ is identical to $k_{\tilde{\rho}}{}^o$. Equations similar to (31-52) can be written for H_L, H_{OG}, and H_{OL}.

In much of the literature, the gas-phase mass-transfer coefficient is expressed in terms of a driving force of partial pressure.

$$\tilde{N}_A = k_G(\bar{p}_A - \bar{p}_{As}) \tag{31-53}$$

This leads to the definition of H_G as $\tilde{G}/k_G{}^o ap$ because $k_G{}^o p$ is identical to $k_{\tilde{y}}{}^o$.

The equations given in this chapter can be applied to gas absorption, distillation, liquid-liquid extraction, and other separation operations. The symbols y and G have referred to the gas phase, and x and L to the liquid phase; if both phases are liquids, however, the equations apply equally well. We merely designate one of the phases as the y phase and the other as the x phase. For example, in liquid-liquid extraction we usually refer to the extract phase as the y phase and refer to the raffinate phase as the x phase.

In liquid-liquid extraction and in leaching, it is common to express compositions in mass fractions rather than mole fractions. All our previous equations are easily adapted to this basis. For example, the y-phase HTU is

$$H_G = \frac{G}{k_y{}^{o}a} \tag{31-54}$$

where
$$N_A = k_y(y_{As} - y_A) \tag{31-55}$$

and
$$k_y{}^o = k_y(1 - y_A)_{lm} \tag{31-56}$$

For situations in which mass-transfer resistances are given in terms of HTUs, it is convenient to use these quantities directly to find $\tilde{x}_{As}$ and $\tilde{y}_{As}$. Equation (31-22) becomes

$$-\frac{H_G \tilde{L}(1-\tilde{y}_A)_{lm}}{H_L \tilde{G}(1-\tilde{x}_A)_{lm}} = \frac{\tilde{y}_A - \tilde{y}_{As}}{\tilde{x}_A - \tilde{x}_{As}} \qquad \tilde{N}_B = 0 \tag{31-57}$$

Equations (31-20) and (31-21) can be expressed in terms of HTUs as

$$H_{OG} = H_G + \frac{m\tilde{G}}{\tilde{L}} H_L = \frac{m\tilde{G}}{\tilde{L}} H_{OL} \tag{31-58}$$

All the equations developed in this section apply to dilute solutions, for which both operating and equilibrium lines are straight and for which one uses overall coefficients. In general, solutions of mole fraction A less than 0.05 can be considered dilute. Overall coefficients are easier to use than the individual coefficients, for the interfacial compositions need not be computed. For the overall gas-phase resistance, the driving force $\tilde{y}_A - \tilde{y}_A^*$ is simply $\tilde{y}_A - m\tilde{x}_A$, the vertical distance between the operating and equilibrium lines. Similarly, for the liquid side, the driving force is $\tilde{x}_A^* - \tilde{x}_A$ or $(\tilde{y}_A/m) - \tilde{x}_A$, the horizontal distance between the lines. If the overall coefficients are not directly available, they are easily calculated from the relations which have been given above.

In Chaps. 32 and 33 we shall consider applications of the differential and overall balances to the laminar and then to the turbulent boundary layer. The flow interactions will be similar to those studied in the analogous chapters on momentum transfer and on energy transfer, and it will usually be convenient to express the results in terms of convective mass-transfer coefficients. However, in Chap. 34, where some coefficients useful in the design of packed towers and other devices will be given, it will prove more convenient to use HTUs. It will be found that HTUs for packed gas-absorption towers are of the order of 0.3 to 3 ft; for liquid-liquid extraction, as high as 60 to 100 ft; and for gas-solid adsorption, as low as 0.01 ft. In Chap. 35 we shall study the calculation of NTUs for concentrated solutions, where the operating and equilibrium lines may be curved, and in subsequent chapters a number of separation operations will be discussed.

PROBLEMS

31-1 A gas contains 70 mol percent benzene (A) and 30 mol percent toluene (B) at 80°C. The pressure on the system is increased isothermally until condensation begins. Find this pressure and the composition of the first drop of liquid condensate.

If the pressure is further increased until all the vapor is condensed, find the composition of the last bubble of vapor to condense. Sketch a graph of $\tilde{x}_A$ or $\tilde{y}_A$ versus $\tilde{L}/\tilde{V}$, the ratio of liquid to vapor.

31-2 Find the value of m (for $\tilde{x}_A \to 0$) for the following gases dissolved in water. Use the data of Appendix Tables A-20, A-21, and A-22.
(*a*) Chlorine at 25°C and 1 atm
(*b*) Ammonia at 30°C and 1 atm
(*c*) Carbon dioxide at 40°C and 100 atm

31-3 Make a $\tilde{y}_A$ versus $\tilde{x}_A$ graph for CO_2 in air at 1 atm dissolved in 15.3 weight percent monoethanolamine at 40°C. Use data from table 14-37A of Perry (4th ed.).

31-4 A gaseous mixture at 1 atm pressure contains 10 mol percent *n*-heptane, 70 mol percent *n*-hexane, and 20 mol percent water vapor. It is cooled at constant pressure until it is completely condensed.
(*a*) Find the temperature at which condensation begins and the composition of the first drop of condensate.
(*b*) Find the temperature at which condensation is completed and the composition of the last bubble of vapor before it condenses.

Assume that *n*-heptane and *n*-hexane form an ideal-liquid solution which is immiscible with water. The following table contains vapor-pressure data for the pure compounds in millimeters of mercury.

	Temperature, °C				
Compound	50	60	70	80	90
Water	90	150	230	355	525
n-Heptane	150	220	310	435	590
n-Hexane	410	584	815	1115	1420

31-5 A gas-absorption column operates at a temperature of 20°C and an average pressure of 1 atm. Water flows down through the column and absorbs sulfur dioxide from a gas stream flowing up. At the top of the column the water entering is essentially free from sulfur dioxide and the gas leaving contains 10 mol percent sulfur dioxide in air. At the bottom of the column the water contains 0.7 mol percent sulfur dioxide and the gas contains 30 mol percent sulfur dioxide. The individual mass-transfer coefficients $k_{\tilde{x}}$ and $k_{\tilde{y}}$ are constant throughout the column at 4.0 and 0.3 lb mol/(h)(ft^2), respectively. Find the overall mass-transfer coefficients $K_{\tilde{x}}$ and $K_{\tilde{y}}$ and the interfacial compositions at both the top and bottom of the column.

31-6 An absorption tower has been built to remove 80 percent of the soluble gas A fed into the bottom of the tower at a concentration of $\tilde{y}_{A_0} = 0.01$. The absorbing liquid is pure water and the liquid flow rate used is 1.5 times the minimum possible.
(*a*) Make a sketch of the operating and equilibrium lines ($m = 1$).
(*b*) A new law is passed so that 95 percent of the A must be removed; the feed gas composition and flow rate remain the same. We want to use the same tower. How can it be done? Calculate the new operating conditions and show the new

operating line. Assume that H_{OG} does not change and that there is no problem with flooding.

31-7 Sulfur dioxide is to be removed from a stream of air in which it is present at a mole fraction of 0.0296. A packed absorption tower will be used of such a diameter that the gas flux is 9.22×10^{-3} kmol/s · m². Pure water will be fed to the top of the tower and it will descend at 0.437 kmol/s · m² countercurrent to the gas which enters the bottom of the tower, as in Fig. 31-4. The mole fraction of SO_2 in the effluent gas is to be reduced to 0.0030. For such a tower, operating at 20°C, we have the following data on the coefficients and the equilibrium:

$$k_{\tilde{x}}a = 0.943 \text{ kmol/s} \cdot \text{m}^3$$

$$k_{\tilde{y}}a = 0.0496 \text{ kmol/s} \cdot \text{m}^3$$

$$m = 29.6 \quad \text{(from Fig. 31-2)}$$

Calculate the height of tower needed.

31-8 A gas containing 0.03 mol fraction acetone (A) in air is fed to the bottom of a packed absorption tower, and the gas from the top of the tower is to contain only 0.002 mol fraction A. The solvent, pure water, is fed to the top of the tower at such a rate that the water leaving the bottom of the tower contains 0.01 mol fraction A. If the overall gas-phase HTU is 2 ft, what tower height is needed? The distribution coefficient m is 2.50.

31-9 Solute A is present at 0.02 mol fraction in the gas fed to the bottom of a 6-m-high packed tower, and the gas leaving the top of the tower contains 0.001 mol fraction A. The solvent is pure water, and it leaves the bottom of the tower with 0.02 mol fraction A. The slope of the straight equilibrium curve is 0.80.

Calculate H_{OG}. What is H_G, if we know from an independent correlation that $H_L = 0.35$ m?

31-10 Air containing 0.03 mol fraction A is to be scrubbed in a countercurrent packed absorption tower by pure water. The water leaving the bottom of the tower contains 0.010 mol fraction A. The equilibrium line slope is 20, the tower height is 6 m, and H_{OG} is 1 m. What percentage of the entering A is removed in the tower?

$\tilde{N}_A = k_{\tilde{x}}(\tilde{x}_A^* - \tilde{x}_A)$. at equil $m\tilde{x}_A = \tilde{y}_A^*$

$$K_{\tilde{y}} = \frac{1}{1/k_{\tilde{y}} + m/k_{\tilde{x}}}$$

$$K_{\tilde{x}} = \frac{1}{1/k_{\tilde{x}} + \frac{1}{m \cdot k_{\tilde{y}}}}$$

32

MASS TRANSFER WITH LAMINAR FLOW

·Laminar flow parallel to a flat plate.

heat

$$\frac{d^2\left(\frac{t_s-t}{t_s-t_0}\right)}{d\eta^2} + \frac{Pr[f(\eta)]}{2}\cdot\frac{d\left(\frac{t_s-t}{t_s-t_0}\right)}{d\eta} = 0$$

$y=0,\ \frac{t_s-t}{t_s-t_0}=0.\ \frac{u_x}{u_0}=0$

$y\to\infty\ \frac{t_s-t}{t_s-t_0}=1\ \ \frac{u_x}{u_0}=1$

$x=0\ \frac{t_s-t}{t_s-t_0}=1\ \ " =1$

mass

$$\frac{d^2\left(\frac{\rho_{AS}-\rho_A}{\rho_{AS}-\rho_{A0}}\right)}{d\eta^2} + \frac{Sc\, f(\eta)}{2}\cdot\frac{d\left(\frac{\rho_{AS}-\rho_A}{\rho_{AS}-\rho_{A0}}\right)}{d\eta} = 0.$$

$y=0,\ \frac{\rho_{AS}-\rho_A}{\rho_{AS}-\rho_{A0}}=0.\ \frac{df(\eta)}{d\eta}=0$

$y\to\infty\ \ " =1\ \ " =1$

$u_{ys}\neq 0.$

$$N_f = k_\rho\cdot(\rho_{AS}-\rho_{A0}) = -D_{AB}\cdot\left(\frac{\partial\rho_A}{\partial y}\right)\Big|_{y=0} + \rho_{AS}\cdot u_{y\cdot s}$$

$$= k'_\rho\cdot(\rho_{AS}-\rho_{A0}) + \rho_{AS}\cdot u_{y\cdot s}$$

where $k_\rho' = \frac{-D_{AB}}{\rho_{AS}-\rho_{A0}}\left(\frac{\partial\rho_A}{\partial y}\right)_{y=0}$

Mass transfer occurs in only a few systems in which the fluid is in laminar flow. The principal reason for this is that most mass-transfer systems contain more than one fluid phase, so that stable laminar boundary layers cannot be built up in that part of the system in which mass transfer is important. There are a few systems, however, in which mass transfer occurs between a solid and a fluid, and in such systems we have not only a laminar boundary layer adjacent to the solid surface, but also the possibility of developed laminar flow throughout the fluid phase.

The problem of mass transfer between a fluid and a solid is similar to the problem of heat transfer between a fluid and a solid. Although the systems are usually not completely analogous, simplifications can often be made so that the heat-transfer analysis is applicable to the mass-transfer systems. The present chapter covers only mass transfer in a fluid in laminar flow and corresponds to Chap. 22, "Heat Transfer with Laminar Flow."

Laminar Flow Parallel to a Flat Plate

A fluid flowing over a flat plate contains a momentum boundary layer in which the velocity of the fluid changes from zero at the plate to the free-stream velocity at the edge of the boundary layer. If mass is being transferred between the fluid and the plate, a concentration boundary layer also exists in which the concentration of the solute changes from the equilibrium value at the solid-fluid interface to the free-stream concentration in the fluid. Just as thermal and momentum boundary layers are often of different thickness in one system, so the concentration and momentum boundary layers may be of different thickness. In fact, it is not uncommon to have heat and mass transfer occurring in the same flow system, so that the three kinds of boundary layer occur simultaneously, and all may be of different thickness.

If the flow is laminar, mass transfer occurs normal to the bulk motion of the fluid only by molecular diffusion and the associated convective flow discussed in Chap. 30. The differential mass balance for a multicomponent system in which molecular diffusion occurs was derived in Chap. 7. If Eq. (7-18) is written for steady, two-dimensional flow of a binary mixture of constant density, we obtain

$$u_x \frac{\partial \rho_A}{\partial x} + u_y \frac{\partial \rho_A}{\partial y} = D_{AB}\left(\frac{\partial^2 \rho_A}{\partial x^2} + \frac{\partial^2 \rho_A}{\partial y^2}\right) + r_A \qquad (32\text{-}1)$$

If no generation of component A occurs and diffusion in the x direction is neglected, Eq. (32-1) becomes

$$u_x \frac{\partial \rho_A}{\partial x} + u_y \frac{\partial \rho_A}{\partial y} = D_{AB} \frac{\partial^2 \rho_A}{\partial y^2} \qquad (32\text{-}2)$$

This equation is similar to the simplified form of the differential momentum balance for two-dimensional flow of a fluid of constant density and viscosity:

$$u_x \frac{\partial u_x}{\partial x} + u_y \frac{\partial u_x}{\partial y} = \nu \frac{\partial^2 u_x}{\partial y^2} \qquad (32\text{-}3)$$

A solution for this latter equation is presented graphically in Fig. 11-8, in which the dimensionless velocity u_x/u_o is plotted against the function $y(u_o/\nu x)^{1/2}$ (usually designated as η).

If the diffusivity D_{AB} and the kinematic viscosity ν are equal, Eqs. (32-2) and (32-3) become identical, except for the dependent variables ρ_A and u_x. These may be replaced in the two differential equations by dimensionless quantities, so that the boundary values are similar, the new variables being u_x/u_o and $(\rho_{As} - \rho_A)/(\rho_{As} - \rho_{Ao})$. The concentrations ρ_{As} and ρ_{Ao} at the plate and in the free stream, respectively, are assumed constant. The boundary conditions for boundary-layer flow over the flat plate with diffusion from the plate into the stream are

$$\text{At } y = 0 \qquad \frac{\rho_{As} - \rho_A}{\rho_{As} - \rho_{Ao}} = 0, \quad \frac{u_x}{u_o} = 0$$

$$\text{At } y = \infty \qquad \frac{\rho_{As} - \rho_A}{\rho_{As} - \rho_{Ao}} = 1, \quad \frac{u_x}{u_o} = 1$$

$$\text{At } x = 0 \qquad \frac{\rho_{As} - \rho_A}{\rho_{As} - \rho_{Ao}} = 1, \quad \frac{u_x}{u_o} = 1$$

Consequently, in this specific case the solutions of the diffusion and momentum equations become identical. From this we conclude that the dimensionless concentration and velocity profiles in the boundary layer are the same; therefore the concentration and momentum boundary layers have an equal thickness.

It is significant that one requirement which leads to the identical solutions described above is that D_{AB} and ν be identical. The quotient ν/D_{AB} is a dimensionless group known as the *Schmidt number* and must be equal to unity in this special case. The Schmidt number indicates the same relation between mass and momentum transfer that the Prandtl number does between heat and momentum transfer. It will be recalled that in Chap. 22 the energy equation was compared with the momentum equation and the two found to be identical when the Prandtl number was unity. Equations (32-2) and (32-3) are analogous to Eq. (22-1); the boundary conditions above may be compared with those given on page 328.

As mentioned earlier, a solution for Eq. (32-3) has been presented graphically in Fig. 11-8. This graphical solution is valid for Eq. (32-2) for a Schmidt number of unity, but only in the limiting case in which the diffusion rate approaches zero. This is because the solution presented in Fig. 11-8 was derived on the assumption that the normal velocity at the plate u_{ys} was zero. If diffusion is occurring at a steady rate from the plate into the stream, the velocity component u_y cannot be zero at the plate, and only in the limiting case when u_{ys} approaches zero is the solution presented in Fig. 11-8 valid for mass transfer.

The solution to Eqs. (32-2) and (32-3) when u_y at the plate is not zero is found by transforming the partial differential equations into ordinary differential equations by means of the similarity transformations

$$\eta = y\sqrt{\frac{u_o}{\nu x}} \qquad (11\text{-}15)$$

$$f(\eta) = \frac{\psi}{\sqrt{x\nu u_o}} \qquad (11\text{-}16)$$

The transformation converts Eq. (32-2) into the form

$$\frac{d^2\left(\dfrac{\rho_{As}-\rho_A}{\rho_{As}-\rho_{Ao}}\right)}{d\eta^2}+\frac{\mathbf{Sc}\, f(\eta)}{2}\,\frac{d\left(\dfrac{\rho_{As}-\rho_A}{\rho_{As}-\rho_{Ao}}\right)}{d\eta}=0 \tag{32-4}$$

which is analogous to Eq. (22-3).

For the boundary condition for which $y = 0$ ($\eta = 0$), the requirement that $(\rho_{As} - \rho_A)/(\rho_{As} - \rho_{Ao})$ be zero has already been given. The requirement that u_x/u_o be zero leads to $df(\eta)/d\eta = 0$, as may be seen from Eq. (11-19). In Chaps. 11 and 22, u_{ys} has been zero, so Eq. (11-20) has led to a requirement that $f(\eta)$ also be zero. In the present case, Eq. (11-20) leads to $f(\eta) = -2u_{ys}x/\sqrt{x\nu u_o}$. Since $f'(\eta)$ is zero, $f(\eta)$ must be constant, so u_{ys} must vary as $1/\sqrt{x}$ in order for the transformation being used to be valid.

At $y = \infty$, we have seen that $(\rho_{As} - \rho_A)/(\rho_{As} - \rho_{Ao}) = 1.0$. Since $u_x/u_o = 1.0$, Eq. (11-19) shows that $df(\eta)/d\eta = 1.0$. The boundary conditions discussed above are summarized by

$$\text{At } y=0 \qquad \frac{\rho_{As}-\rho_A}{\rho_{As}-\rho_{Ao}}=0,\quad \frac{df(\eta)}{d\eta}=0,\quad \frac{f(\eta)}{-2}=\frac{u_{ys}x}{\sqrt{x\nu u_o}}$$

$$=\frac{u_{ys}}{u_o}(\mathbf{Re}_x)^{1/2}=\text{const}$$

$$\text{At } y=\infty \qquad \frac{\rho_{As}-\rho_A}{\rho_{As}-\rho_{Ao}}=1,\quad \frac{df(\eta)}{d\eta}=1$$

In the first boundary condition, the constant contains the factor -2. The velocity at the surface must vary according to

$$\frac{u_{ys}}{u_o}=\frac{\text{const}}{(\mathbf{Re}_x)^{1/2}} \tag{32-5}$$

It has been shown by Eckert and Drake[1] that this condition is achieved with the condition of constant temperature and concentration at the wall.

A graph of the solution of Eq. (32-4) is shown in Fig. 32-1 for a binary fluid with a Schmidt number of unity. Both positive and negative values of the parameter $(u_{ys}/u_o)(\mathbf{Re}_x)^{1/2}$ are represented; the positive values apply when mass transfer occurs from the plate to the fluid, and the negative values when mass transfer is from the fluid to the plate. The line for which the parameter is zero represents a system in which the mass-transfer rate is negligible compared with the velocity of the free

[1] E. R. G. Eckert and R. M. Drake, Jr., "Heat and Mass Transfer," 2d ed., chap. 16, McGraw-Hill Book Company, New York, 1959.

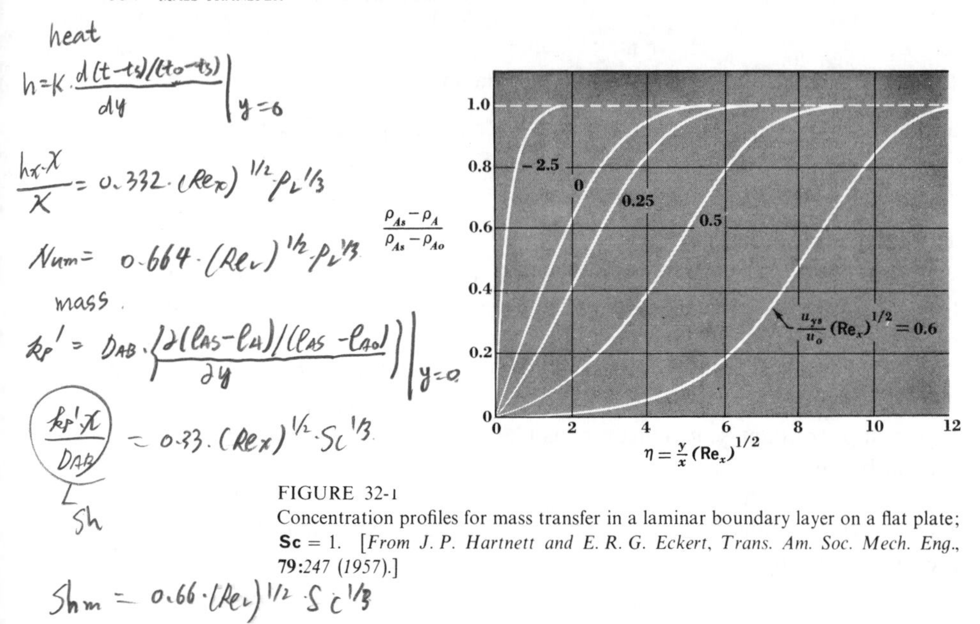

FIGURE 32-1
Concentration profiles for mass transfer in a laminar boundary layer on a flat plate; **Sc** = 1. [*From J. P. Hartnett and E. R. G. Eckert, Trans. Am. Soc. Mech. Eng.*, **79**:*247* (*1957*).]

stream. This line is identical with the dimensionless velocity profile in Fig. 11-8. It is also identical with the line for **Pr** = 1 in Fig. 22-1. If heat and mass transfer are occurring simultaneously in boundary-layer flow, the identity of the differential energy and diffusion equations and their boundary conditions leads to identical temperature and composition profiles when **Sc**/**Pr** is unity. This ratio of dimensionless groups readily reduces to the ratio of thermal to molecular diffusivity, which is known as the *Lewis number*.

The results shown in Fig. 32-1 can be presented in terms of a mass-transfer coefficient k'_ρ. The transfer rate between the fluid and plate may be written by equating the values of N_A found by substituting the appropriate quantities in Eqs. (28-1) and (30-14).

$$N_A = k_\rho(\rho_{As} - \rho_{Ao}) = -D_{AB}\left(\frac{\partial \rho_A}{\partial y}\right)_{y=0} + \rho_{As}u_{ys}$$

$$= k'_\rho(\rho_{As} - \rho_{Ao}) + \rho_{As}u_{ys} \qquad (32\text{-}6)$$

The coefficient k'_ρ thus defined can be written as

$$k'_\rho = \frac{-D_{AB}}{\rho_{As} - \rho_{Ao}}\left(\frac{\partial \rho_A}{\partial y}\right)_{y=0}$$

or as

$$k'_\rho = D_{AB}\left\{\frac{\partial[(\rho_{As} - \rho_A)/(\rho_{As} - \rho_{Ao})]}{\partial y}\right\}_{y=0} \tag{32-7}$$

The relation of k'_ρ to k_ρ in some special cases is discussed in Chap. 36.

The behavior of the mass-transfer coefficient can be inferred qualitatively by a comparison of Eq. (32-7) with the curves of Fig. 32-1. Systems for which the parameter is negative (indicating mass transfer from the fluid to the plate) have large slopes at the origin ($y = 0$, $\eta = 0$), indicating high mass-transfer coefficients. Systems in which the parameter is large and positive have low slopes at the origin, indicating much lower mass-transfer coefficients. The systems represented by a positive parameter are those in which mass is transferred away from the plate.

A numerical value of the local mass-transfer coefficient can be computed for a fluid with **Sc** = 1 by measuring the slope at the origin of the appropriate curve in Fig. 32-1 and using Eq. (32-7) to compute k'_ρ. For the limiting case of $u_{ys} = 0$, $k_\rho = k'_\rho = k_\rho{}^o$ and the slope at the origin is 0.33. Consequently, we can write

$$\left\{\frac{\partial[(\rho_{As} - \rho_A)/(\rho_{As} - \rho_{Ao})]}{\partial y}\right\}_{y=0} = \frac{0.33}{x}(\mathbf{Re}_x)^{1/2}$$

Inserting this in Eq. (32-7), we obtain

$$\frac{k_\rho{}^o x}{D_{AB}} = 0.33\,(\mathbf{Re}_x)^{1/2} \tag{32-8}$$

For a fluid with **Sc** other than unity, curves similar to those in Fig. 32-1 could be obtained from a solution of Eq. (32-4). For the case of $u_{ys} = 0$, the curves given in Fig. 22-1 apply to mass transfer if the ordinate is interpreted as $(\rho_{As} - \rho_A)/(\rho_{As} - \rho_{Ao})$ and the parameter **Pr** is interpreted as **Sc**. An approximate expression for the mass-transfer coefficient corresponding to the concentration profiles of Fig. 22-1 is

$$\frac{k_\rho{}^o x}{D_{AB}} = 0.33\,(\mathbf{Re}_x)^{1/2}\,\mathbf{Sc}^{1/3} \tag{32-9}$$

The dimensionless group $k_\rho{}^o x/D_{AB}$ is known as the *Sherwood number* and corresponds to the Nusselt number in heat transfer. The procedures followed and assumptions used in obtaining Eq. (32-9) are similar to those employed in the analysis of heat transfer in Chap. 22, which resulted in Eq. (22-8). As for Eq. (22-8), Eq. (32-9) is valid for $\mathbf{Sc} > 0.6$; in addition, both equations apply only if u_{ys} is negligible.

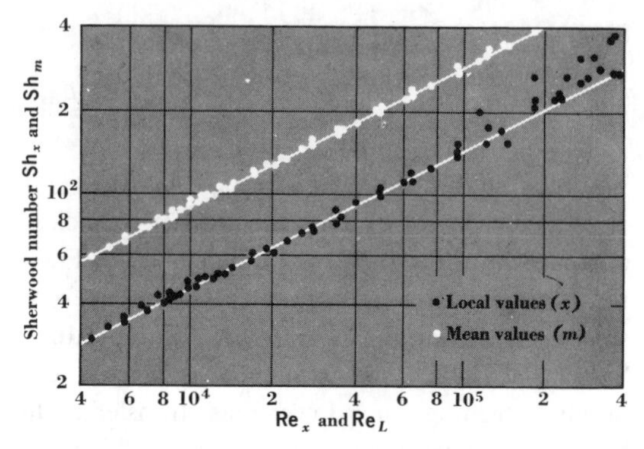

FIGURE 32-2
Local and mean Sherwood numbers for sublimation of naphthalene into air in a laminar boundary layer. [*From W. J. Christian and S. P. Kezios, AIChE J.*, **5**:*61 (1959).*]

Mean mass-transfer coefficients for a plate of finite length can be found by writing Eq. (32-9) to obtain $k_\rho{}^o$ for a differential element of surface and integrating over the length of the plate. This procedure was illustrated in Chap. 22 in the determination of mean heat-transfer coefficients for a flat plate from local coefficients. As in the heat-transfer analysis, the equation giving the mean coefficient differs from the equation for the local coefficient by a factor of 2. Thus the equation for the mean coefficient is

$$\mathbf{Sh}_m = 0.66\,(\mathbf{Re}_L)^{1/2}\,\mathbf{Sc}^{1/3} \qquad (32\text{-}10)$$

in which the mean Sherwood number contains the mean mass-transfer coefficient for a plate of length L.

Very few experimental investigations of mass transfer from a flat plate have been reported; this is in contrast to the many investigations of heat transfer. Christian and Kezios[1] describe measurements of local and mean mass-transfer coefficients for naphthalene subliming outward into an air stream flowing parallel to the axis of a hollow cylinder. Their results are shown in Fig. 32-2. The straight lines drawn through the data are represented by equations differing from Eqs. (32-9) and (32-10) only in the coefficients, which are 0.339 and 0.678, respectively. These results might be expected to differ somewhat from the flat-plate equations because of the curvature of the surface; however, for the cylinders chosen (0.75- and 1-in diameter), the radius of curvature was so large compared with the boundary-layer thickness that flat-plate

[1] W. J. Christian and S. P. Kezios, *AIChE J.*, **5**(1):61 (1959).

behavior was obtained. The Schmidt number for the air-naphthalene system was 2.40.

In boundary-layer systems in which combined heat and mass transfer occur and **Pr** = **Sc** = 1.0, the dimensionless temperature profiles are also represented by the curves of Fig. 32-1. The reasoning concerning mass-transfer coefficients can be extended to heat-transfer coefficients. The heat-transfer rate from a hot gas stream flowing over a flat plate is reduced by the transfer of mass into the boundary layer from the surface of the plate. One method of achieving this effect is to force a second gas through the porous flat plate into the boundary layer. In high-speed flow, sublimation of the plate itself may occur, with the effect not only of absorbing the latent heat of sublimation, but also of reducing the heat-transfer coefficient as a result of the mass-transfer effect just discussed. The effect is achieved more simply by evaporation of a liquid from the surface of the plate. The opposite effect of a high heat-transfer coefficient is illustrated by the high heat-transfer coefficients encountered in desuperheating of steam near a surface on which condensation occurs.

Mass Transfer with Developed Velocity Distribution in a Pipe

The development of a concentration boundary layer on a flat plate is similar to the development of a concentration profile near the entrance of a pipe. When the boundary layer fills the pipe, the flow is said to be developed. If the partial pressure of the diffusing component is constant at the wall, the concentration profile in the pipe will eventually become flat. This condition might be achieved in a pipe whose inner surface is coated with some material soluble in the fluid stream. It might also be achieved in a very long wetted-wall column operated in such a way that the fluid flowing down the column (or pipe) is at a constant temperature throughout and has, therefore, a constant vapor pressure. A different boundary condition is achieved if the walls of the pipe are porous and the solute is forced through the walls at a constant rate per unit area at all points along the pipe.

The two methods of operation described above are analogous to the boundary conditions of constant wall temperature and constant heat flux in heat-transfer systems. We have seen earlier in this chapter that the differential equations for molecular diffusion of mass and for conduction of heat are similar. We have also seen that in mass-transfer systems in which the normal velocity at the wall is small compared with the free-stream velocity, the mass-transfer results are analogous to heat-transfer results in which no mass transfer is taking place. The fortunate consequence of this is that the results for heat transfer in a pipe presented in Chap. 22 can be used to predict mass-transfer coefficients by merely substituting the Sherwood number for the Nusselt number and the Schmidt number for the Prandtl number in the heat-transfer solutions. Solutions for the local Sherwood number can be obtained from Fig. 22-3 for either uniform flux or uniform concentration at the wall for fluid streams with flat and parabolic profiles. Solutions for arithmetic average and log-mean driving force can be obtained from Fig. 22-4.

Other Systems

Mass transfer in a fluid may cause density differences which will result in natural convective flow similar to natural convection by heat transfer. The convective mass-transfer coefficients can be predicted from the equations used in heat transfer by substituting the Sherwood and Schmidt numbers for the Nusselt and Prandtl numbers, respectively. The Grashof number for mass transfer is $gL^3(\rho_o - \rho_s)/v^2\rho_s$.

Mass transfer also occurs by molecular diffusion inside drops. Various conditions may prevail within the drops, ranging from complete stagnation to complete mixing. An intermediate condition is one in which viscous circulation occurs within the drops. An analysis of this special case of mass transfer with laminar flow will be deferred until Chap. 34, when all the fluid-flow patterns and their effects on mass transfer inside drops will be considered simultaneously.

PROBLEMS

32-1 A thin sheet of naphthalene 0.1 in thick and 4 in square is placed in an air stream parallel to the direction of flow. The air is at 32°F and 1 atm and is moving in laminar flow with a uniform velocity of 50 ft/s. How long must the sheet be exposed to the air before one-quarter of its mass will have sublimed? Assume that the upper and lower surfaces remain flat at all times. The molecular diffusivity for the air-naphthalene system is 0.199 ft^2/h, and the Schmidt number is 2.57 under the conditions given in the problem. The vapor pressure of naphthalene at 32°F is 0.0059 mmHg. The temperature drop between the air and the naphthalene as a result of the vaporization of naphthalene can be assumed to be negligible. Neglect evaporation from the edges.

32-2 Water is flowing at 14°C through a smooth metal tube of 0.02-m ID in fully developed laminar flow at **Re** = 1000. From $x = 0$ the metal is replaced by a tube cast of benzoic acid, for which **Sc** = 1850 for solutions in water at 14°C. The solubility of the solid is 2.39 kg/m^3. Use Fig. 22-4 to calculate the length of soluble wall required to achieve an average concentration of acid in water of 1.19 kg/m^3. Can Eq. (22-41) be used to obtain this result?

32-3 A vertical tube of 0.03-m ID is arranged as a wetted-wall tower. A mixture of 1 percent (volume) toluene in air enters the bottom of the tube in fully developed laminar flow at 0.25 m/s and 25°C. Toluene is so strongly absorbed by the descending liquid that the toluene concentration next to the surface is zero. How tall must the tower be to remove 70 percent of the entering toluene? Assume the surface of the liquid has no ripples and moves slowly so that the gas behaves as if it were flowing in a tube with solid walls.

32-2)

ρ_{A0} ρ_{Ab}

$$Gz = \frac{\pi}{4} \cdot \frac{D}{L} \cdot Re \cdot Sc = \frac{291 \times 10^4}{L}$$

Assume $Gz > 0$

$$Sh_m = 1.62 \cdot Re^{1/3} \cdot Sc^{1/3} \cdot \left(\frac{D}{L}\right)^{\frac{1}{3}}$$

$\frac{k_\rho \cdot D}{D_{AB}}$

$$k_\rho = 1.38 \times 10^{-6} / L^{\frac{1}{3}}$$

$Sc = \frac{\nu}{D_{AB}}$

$$N_{A,total} = \frac{\pi D^2}{4} \cdot u_b \cdot (\rho_{Ab} - \rho_{A0})$$

(2.39 − 0) (2.39 − 1.191)

$$= k_\rho \cdot \pi \cdot DL \cdot \frac{(\rho_s - \rho_{A0}) + (\rho_{As} - \rho_{Ab})}{2}$$

or LMCD

$$L^{\frac{2}{3}} = 1857.$$

$\frac{D \cdot \rho \cdot u_b}{\mu} = 1000$, $u_B = 0.0585$ m/s

$L = 1132.3$

$D_{AB} = 1.21 \times 10^{-9}$ (table 29-1)

$\sim \times \frac{287}{298}$ 절대온도의 비율 공식 보정

33

MASS TRANSFER WITH TURBULENT FLOW

Integral method.

$$\left(\frac{\rho_{AS}-\rho_A}{\rho_{AS}-\rho_{A0}}\right) = \left(\frac{y}{\delta}\right)^{1/7}$$

$$\frac{k_\rho^\circ x}{D_{AB}} = 0.0292 (Re_x)^{4/5} Sc$$

// (eddy property theory, Prandtl mixing length)

Analogy

Reynolds ⇒ $k_x^\circ M_m = \frac{f u_b \rho}{2}$

↳ mean molecular weight

Prandtl $Sh = \frac{Re \cdot Sc (f/2)}{1 + 5\sqrt{f/2}(Sc-1)}$

Colburn $j_m = \frac{f}{2} = \frac{k_x^\circ M_m}{u_b \rho} Sc^{2/3}$

Much of our understanding of the mechanism of mass transfer with turbulent flow has been gained from analogous studies of heat transfer with turbulent flow. The analogy between heat and mass transfer is very useful; however, its limitations must frequently be considered, and this is particularly true in applications involving turbulent flow.

In this chapter we shall consider mass transfer with turbulent flow in somewhat the same order as we considered heat transfer with turbulent flow in Chap. 23. It is desirable to consider mass transfer in a turbulent boundary layer on a flat plate because of the small but increasing number of direct applications of this theory; a more important reason is that this study leads to an understanding of mass transfer with flow over surfaces of more complex geometry. We shall also consider the classical analogies among heat, mass, and momentum transfer between a fluid and the inner wall of a pipe. Finally, we shall examine the penetration theory, which is of importance as a mechanistic picture. One unique feature of the penetration theory is that it describes mass transfer (and heat transfer) between two fluid phases. This is in contrast to most transport theories, which are more or less restricted in applicability to transport between a fluid and a solid phase.

Turbulent Flow Parallel to a Flat Plate

The equations for mass transfer in a turbulent boundary layer cannot be solved analytically. However, like the problems of heat and momentum transfer in a turbulent boundary layer, the mass-transfer problem can be attacked using the von Kármán integral method. The treatment is similar to that in Chap. 23, but will be given in some detail so that the assumptions will be evident.

The velocity in the turbulent boundary layer will be represented by the well-known empirical equation

$$\frac{u_x}{u_o} = \left(\frac{y}{\delta}\right)^{1/7} \qquad (12\text{-}95)$$

This expression has been established for turbulent-boundary-layer flow without mass transfer, so that its use in this derivation is probably limited to systems with low concentrations of the diffusing component.

The composition profile of the boundary layer is assumed to follow an analogous equation.

$$\frac{\rho_{As} - \rho_A}{\rho_{As} - \rho_{Ao}} = \left(\frac{y}{\delta_c}\right)^{1/7} \qquad (33\text{-}1)$$

We assume that the Schmidt number is unity, so that the momentum-boundary-layer thickness δ and the concentration-boundary-layer thickness δ_c are the same at any distance x from the leading edge of the plate. The concentration ρ_{As} of the diffusing component at the surface of the plate will be taken as constant at all points on the plate, and it will be assumed that there is no flow of component B from the plate into the boundary layer.

A material balance for component A is written for the control volume shown in Fig. 33-1. Diffusion in the x direction is assumed to be negligible compared with the transport due to fluid motion in this direction. Velocity and concentration gradients do not exist in the z direction.

The net inflow of solute A through the lower surface A_4 equals the net outflow through the upper surface A_3 and through surfaces A_1 and A_2 at steady state. The flows are written using Eq. (3-10).

$$\iint_{A_1} \rho_A u \cos\alpha \, dA + \iint_{A_2} \rho_A u \cos\alpha \, dA + \iint_{A_3} \rho_{Ao} u \cos\alpha \, dA$$

$$= \iint_{A_4} k_\rho(\rho_{As} - \rho_{Ao}) \, dA \qquad (33\text{-}2)$$

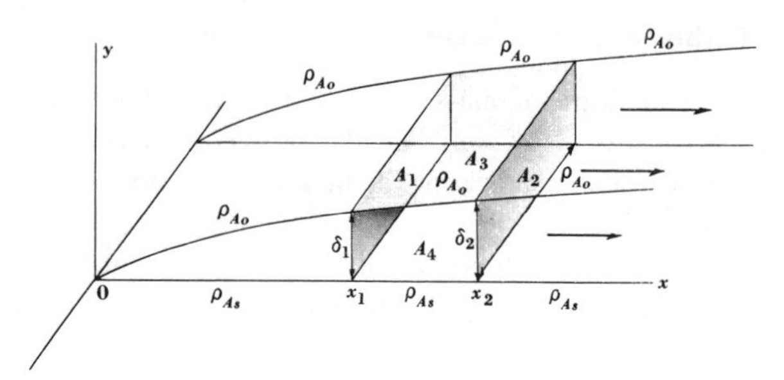

FIGURE 33-1
Mass transfer from a flat plate to a turbulent boundary layer.

At surface A_1, $u \cos \alpha = -u_x$, and at A_2, $u \cos \alpha = u_x$, so that we write

$$\iint_{A_2} \rho_A u_x \, dA - \iint_{A_1} \rho_A u_x \, dA + \iint_{A_3} \rho_{Ao} u \cos \alpha \, dA$$

$$= \iint_{A_4} k_\rho(\rho_{As} - \rho_{Ao}) \, dA \qquad (33\text{-}3)$$

To eliminate the term for flow of solute through surface A_3, we write a total mass balance on the control volume.

$$\iint_{A_2} \rho u_x \, dA - \iint_{A_1} \rho u_x \, dA + \iint_{A_3} \rho u \cos \alpha \, dA = \iint_{A_4} k_\rho(\rho_{As} - \rho_{Ao}) \, dA \qquad (33\text{-}4)$$

We restrict our analysis to dilute solutions for which the concentration of the diffusing component is low so that we can assume that the total density of the fluid is uniform throughout the boundary layer. Therefore we can divide each term in Eq. (33-4) by the total density ρ and multiply by ρ_{Ao}, which is also constant. The result of making this change is shown in Eq. (33-5), which has been solved for the flow through A_3.

$$\iint_{A_3} \rho_{Ao} u \cos \alpha \, dA = \iint_{A_1} \rho_{Ao} u_x \, dA - \iint_{A_2} \rho_{Ao} u_x \, dA$$

$$+ \iint_{A_4} k_\rho(\rho_{As} - \rho_{Ao}) \frac{\rho_{Ao}}{\rho} \, dA \qquad (33\text{-}5)$$

Equation (33-5) is then substituted in Eq. (33-3) to give

$$\iint_{A_2} (\rho_A - \rho_{Ao})u_x \, dA - \iint_{A_1} (\rho_A - \rho_{Ao})u_x \, dA$$

$$= \iint_{A_4} [k_\rho(\rho_{As} - \rho_{Ao})]\left(1 - \frac{\rho_{Ao}}{\rho}\right) dA \qquad (33\text{-}6)$$

All surfaces have the same width in the z direction; none of the variables is a function of z, so that Eq. (33-6) can be written as a sum of single integrals.

$$\int_0^{\delta_2} (\rho_A - \rho_{Ao})u_x \, dy - \int_0^{\delta_1} (\rho_A - \rho_{Ao})u_x \, dy$$

$$= \int_{x_1}^{x_2} [k_\rho(\rho_{As} - \rho_{Ao})]\left(1 - \frac{\rho_{Ao}}{\rho}\right) dx \qquad (33\text{-}7)$$

As the length $x_2 - x_1$ approaches zero, the left-hand side of Eq. (33-7) approaches differential magnitude. The right-hand side applies to a segment of finite width in the z direction, but differential width dx in the x direction. Therefore we write

$$d\left[\int_0^{\delta} (\rho_A - \rho_{Ao})u_x \, dy\right] = [k_\rho(\rho_{As} - \rho_{Ao})]\left(1 - \frac{\rho_{Ao}}{\rho}\right) dx \qquad (33\text{-}8)$$

from which we obtain

$$k_\rho = \frac{d}{dx}\left(\int_0^{\delta} \frac{\rho_A - \rho_{Ao}}{\rho_{As} - \rho_{Ao}} u_x \, dy\right) \frac{\rho}{\rho - \rho_{Ao}} \qquad (33\text{-}9)$$

The concentration term under the integral sign can be rewritten as

$$\frac{\rho_A - \rho_{Ao}}{\rho_{As} - \rho_{Ao}} = 1 - \frac{\rho_{As} - \rho_A}{\rho_{As} - \rho_{Ao}}$$

$$= 1 - \left(\frac{y}{\delta}\right)^{1/7} \qquad (33\text{-}10)$$

The local velocity is

$$u_x = u_o\left(\frac{y}{\delta}\right)^{1/7} \qquad (33\text{-}11)$$

Equations (33-10) and (33-11) are substituted in Eq. (33-9) and integrated as in Chap. 23. The result is

$$k_\rho = \frac{7u_o}{72} \frac{d\delta}{dx} \frac{\rho}{\rho - \rho_{Ao}} \qquad (33\text{-}12)$$

The thickness of the boundary layer was found in Chap. 12 to be

$$\delta = 0.376x(\mathbf{Re}_x)^{-1/5} \qquad (12\text{-}102)$$

This equation is differentiated to give

$$\frac{d\delta}{dx} = 0.301(\mathbf{Re}_x)^{-1/5} \qquad (33\text{-}13)$$

which is substituted in Eq. (33-12) to give

$$k_\rho = 0.0292u_o(\mathbf{Re}_x)^{-1/5}\frac{\rho}{\rho - \rho_{Ao}} \qquad (33\text{-}14)$$

The quantity $\rho/(\rho - \rho_{Ao})$ in Eq. (33-14) will make only a slight contribution to k_ρ in dilute systems. If the diffusing component is not present in the free stream, the quantity has a value of unity.

If bulk transport from the plate is negligible, $\rho/(\rho - \rho_{Ao})$ does not appear in Eq. (33-14). The value obtained for k_ρ is designated $k_\rho{}^o$ for this special case, as mentioned in Chap. 31. We obtain a solution similar to the heat-transfer solution of Chap. 23. Both sides of the equation are multiplied by $xu_o\rho/\mu$ and $\mu/\rho D_{AB}$ to give

$$\frac{k_\rho{}^o x}{D_{AB}} = 0.0292(\mathbf{Re}_x)^{4/5}\,\mathbf{Sc} \qquad (33\text{-}15)$$

It was assumed initially that the Schmidt number was unity, so that the concentration and momentum boundary layer could be assumed to be of equal thickness. Thus Eq. (33-15) is written as

$$\frac{k_\rho{}^o x}{D_{AB}} = 0.0292(\mathbf{Re}_x)^{4/5} \qquad (33\text{-}16)$$

which is analogous to Eq. (23-46) for the local coefficient of heat transfer in turbulent flow over a flat plate.

Eddy Mass Diffusivity and Mixing Length

The eddy kinematic viscosity ν_e and the eddy thermal diffusivity α_e have been defined in Eqs. (12-37) and (23-6). In the discussions leading to these definitions, it was indicated that a similar quantity known as the eddy mass diffusivity D_{ABe} could be defined in equations for mass transfer with turbulent flow.[1] The derivation for D_{ABe} is based on a flow system as represented in Fig. 33-2. Velocities and concentrations

[1] The eddy mass diffusivity is often referred to simply as the eddy diffusivity.

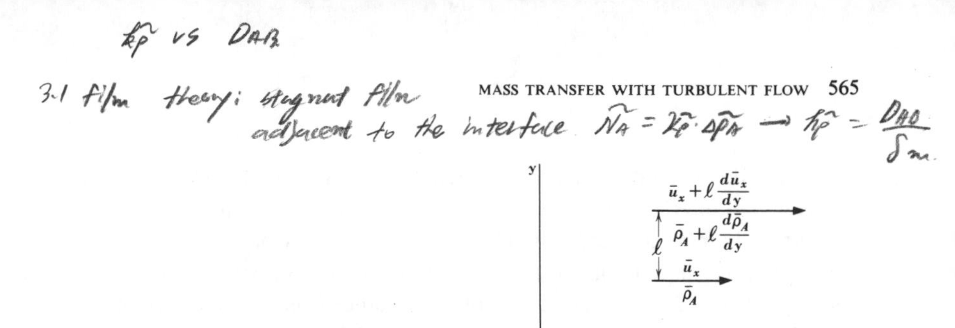

FIGURE 33-2
Eddy transport of mass.

are shown at two points separated by a distance equal to the Prandtl mixing length ℓ. Fluid is transported between the two points with a velocity equal to $\overline{|u_y'|}$, whereas component A is transported with a mass rate per unit area equal to the product of velocity and concentration difference, $-\overline{|u_y'|}\,\ell(d\bar{\rho}_A/dy)$. This concept has already been employed in Chaps. 12 and 23. In Chap. 12 it was shown that the fluctuating velocity component $\overline{|u_y'|}$ could be expressed in terms of the Prandtl mixing length and the velocity gradient as

$$\overline{|u_y'|} = \ell\left|\frac{d\bar{u}_x}{dy}\right| \tag{12-40}$$

Therefore the mass-transfer rate due to turbulent transport is

$$N_A = -\ell^2\left|\frac{d\bar{u}_x}{dy}\right|\frac{d\bar{\rho}_A}{dy} \tag{33-17}$$

The quantity $\ell^2\,|d\bar{u}_x/dy|$ has previously been shown to be equal to the eddy kinematic viscosity and also to the eddy thermal diffusivity. The derivation given above shows that it is also a proportionality constant in the equation for turbulent mass transfer. Consequently, it will be referred to as the eddy mass diffusivity, and by analogy to the molecular mass diffusivity D_{AB}, it will be designated D_{ABe}. A phenomenological equation for the mass flux due to the combined mechanisms of molecular and eddy diffusion is

$$N_A = -(D_{AB} + D_{ABe})\frac{d\bar{\rho}_A}{dy} \tag{33-18}$$

The eddy mass diffusivity in a system bounded by a solid wall is a function of the distance from the wall because the wall affects the character of turbulence in the adjacent fluid. In the laminar sublayer the Prandtl mixing length and D_{ABe} are ordinarily taken as zero. Mass transfer in this region is a function of the molecular diffusivity D_{AB}. Far from the wall, however, D_{ABe} is usually much greater than D_{AB}, so that in this region the molecular diffusivity can often be neglected.

The eddy mass diffusivity in a system with a transport across a free surface, e.g., a gas-liquid interface, is also believed to increase with the distance from the interface. Lamourelle and Sandall[1] report the results of mass-transfer measurements in falling liquid films in accord with the notions of previous workers that the eddy diffusivity varies as the square of the distance from the free surface.

The assumption that D_{ABe}, α_e, and ν_e are identical is often made in theoretical derivations, though the model which is the basis of these definitions is vastly oversimplified. Experimental work indicates that the terms are of the same magnitude but are not equal.

Sherwood and Woertz[2] measured diffusion of water vapor from one plane wall to another across streams of air, helium, and carbon dioxide. Values of D_{ABe} and ν_e were determined from measurements of the concentration and velocity profiles at the center of the rectangular duct. As mentioned in earlier chapters, comparisons of heat and momentum transfer indicate a range of values for α_e/ν_e between 0.5 and 2.0; the value of this quotient is shown in Fig. 23-3 to vary with distance from the wall and also to be a function of **Re**. Thus it would seem likely that D_{ABe}/ν_e would behave in similar fashion. In the work of Sherwood and Woertz the ratio D_{ABe}/ν_e was approximately constant at 1.6 for $2000 < \mathbf{Re} < 120{,}000$. No values were reported for locations other than at the centerline, but it is likely that some variation with position would occur.

An interesting application of turbulent-flow theory is found in the effect of wire grids on eddy diffusivity. As Hinze[3] points out, the presence of a grid decreases the scale of turbulence in the fluid downstream from the grid but causes an increase in the intensity of the turbulence. If the grid openings are sufficiently small, the net effect is to decrease the eddy diffusivity. The decrease in the scale of turbulence apparently has a greater effect in this case in diminishing D_{ABe} than the increase in intensity has in raising it. Thus a grid inserted in a fluid stream for the purpose of increasing the rate of mixing may, in fact, cause a decrease in the rate of mixing.

The Analogy between Momentum Transfer and Mass Transfer

Although Reynolds was concerned only with the analogy between heat and momentum transfer, the equations which are referred to as the Reynolds analogy can be readily extended to cover mass transfer. This is also true for the equations of Prandtl and Taylor, von Kármán, etc., which were derived or referred to in Chap. 23. In this section we shall consider primarily relations between mass and momentum transfer. The relations between heat and mass transfer, if desired, can be

[1] A. P. Lamourelle and O. C. Sandall, *Chem. Eng. Sci.*, **27**:1035 (1972).
[2] T. K. Sherwood and B. B. Woertz, *Ind. Eng. Chem.*, **31**:1034 (1939).
[3] J. O. Hinze, "Turbulence: An Introduction to Its Mechanism and Theory," p. 341, McGraw-Hill Book Company, New York, 1959.

obtained by combining the mass-transfer equations from the present chapter with the heat-transfer equations from Chap. 23.

The proportionality of mass and momentum transfer which leads to the Reynolds analogy can be arrived at by considering a fluid flowing in a pipe. The fluid is transferring both momentum and mass from the stream to the wall of the pipe. The following flux and rate expressions are considered:

(a) The mass flux from the fluid to the pipe wall for negligible bulk radial flow, $k_{\tilde{x}}^{o}(\tilde{x}_A - \tilde{x}_{As})$, lb mol/(h)(ft²)
(b) The momentum flux at the wall, $\tau_s g_c$, (lb)(ft/h)/(h)(ft²)
(c) The rate at which mass available for radial transfer is transported parallel to the pipe wall, $(w/M_m)(\tilde{x}_A - \tilde{x}_{As})$, lb mol/h
(d) The rate at which momentum is transported parallel to the pipe wall, wu_b, (lb)(ft/h)/h

It is postulated that these quantities are related by the proportionality

$$\frac{a}{c} = \frac{b}{d}$$

from which it follows that

$$k_{\tilde{x}}^{o} M_m = \frac{\tau_s g_c}{u_b} \tag{33-19}$$

It was shown in Chap. 12 that the shear stress at the wall of a pipe is related to the friction factor by the equation

$$\tau_s g_c = \frac{f u_b^2 \rho}{2} \tag{12-67}$$

If Eq. (12-67) is substituted in Eq. (33-19), we obtain

$$k_{\tilde{x}}^{o} M_m = \frac{f u_b \rho}{2} \tag{33-20}$$

in which M_m is the mean (or mixing-cup) molecular weight of the fluid in the pipe. By combining Eqs. (23-4) and (33-20), we relate the heat- and mass-transfer coefficients.

$$\frac{h}{C_p k_{\tilde{x}}^{o} M_m} = \frac{h}{\tilde{C}_p k_{\tilde{x}}^{o}} = \frac{h}{C_p k_x^{o}} = 1 \tag{33-21}$$

This expression is sometimes called the *Lewis equation.*

Equations (33-20) and (33-21) can also be derived analytically. The analytical procedure was presented in Chap. 23 as an a ternative to the intuitive statement of proportionality relating heat and momentum transfer. A similar treatment with similar assumptions will yield the equations given above, relating the friction factor and the heat- and mass-transfer coefficients.

The value of the analytical approach is that it bares the assumptions which must be true for the answer to have validity. One of these assumptions is that the mass flux is described by the same radial function as shear stress:

$$\frac{\tau}{\tau_s} = \frac{\tilde{N}_A}{\tilde{N}_{As}} = 1 - \frac{y}{r_i} \tag{33-22}$$

It can be seen that if D_{ABe}/ν_e is a function of radial position, the first two parts of Eq. (33-22) are not likely to be equal.

Another assumption in the analytical procedure is that the molecular diffusivities D_{AB} and ν are negligible compared with the eddy diffusivities. This limits the use of the Reynolds analogy to the turbulent core except for the case where the Schmidt number of the fluid is unity. The Schmidt number is by definition ν/D_{AB}, and at a value of $\mathbf{Sc} = 1$, ν equals D_{AB}. If in addition ν_e is equal to D_{ABe}, Eq. (33-20) can be obtained without neglecting the molecular-transport coefficients. For a fluid with $\mathbf{Sc} = 1$, therefore, the Reynolds analogy applies not only in the turbulent core, but also in the laminar sublayer. Similarly, if $\alpha = D_{AB}$, Eq. (33-21) can be derived for a system containing both laminar- and turbulent-flow regions.

The more complex forms of the analogy between mass and momentum transfer also parallel the forms presented in Chap. 23. The derivations are similar, so few details will be given here. In the Prandtl-Taylor analysis molecular transport of mass and momentum are assumed to be the only mechanisms of importance in the laminar sublayer. In the turbulent core, Reynolds' analogy is applied. The two sets of equations are then combined to give an equation for the overall mass flux, from which the following equation is obtained:

$$k_{\tilde{x}}{}^{o}M_m = \frac{fu_b\rho/2}{1 + 5\sqrt{f/2}(\mathbf{Sc} - 1)} \tag{33-23}$$

If both sides of this equation are multiplied by $D/\rho D_{AB}$, the equation may be rearranged to give

$$\frac{k_{\tilde{x}}{}^{o}M_m D}{\rho D_{AB}} = \frac{(f/2)(Du_b\rho/\mu)(\mu/\rho D_{AB})}{1 + 5\sqrt{f/2}(\mathbf{Sc} - 1)} \tag{33-24}$$

which is expressed in terms of dimensionless groups as

$$\mathbf{Sh} = \frac{(f/2)\mathbf{Re}\ \mathbf{Sc}}{1 + 5\sqrt{f/2}(\mathbf{Sc} - 1)} \tag{33-25}$$

The quantity D is the pipe diameter, so **Re** is the Reynolds number for the pipe; the density ρ is the mixing-cup density of the fluid in pounds per cubic foot. The quantity $k_{\tilde{x}}{}^{o}M_m/\rho$ can be expressed as $k_{\tilde{x}}{}^{o}RT/p$, if the fluid is an ideal gas; it has the dimensions of length/time, as does $k_\rho{}^{o}$.

The von Kármán analogy can also be expressed in terms of mass and momentum transfer. The equation comparable with Eq. (23-26) is

$$\mathbf{Sh} = \frac{(f/2)\mathbf{Re}\,\mathbf{Sc}}{1 + 5\sqrt{f/2}\{\mathbf{Sc} - 1 + \ln[(1 + 5\mathbf{Sc})/6]\}} \tag{33-26}$$

The dimensionless groups in Eq. (33-26) contain exactly the same variables as are found in Eq. (33-25).

One final form of the analogy is the empirical *j*-factor relation proposed by Colburn, which was also discussed in Chap. 23. This analogy may be written for mass transfer as

$$\frac{k_{\tilde{x}}{}^{o}M_m}{u_b\rho}\mathbf{Sc}^{2/3} = j_M = \frac{f}{2} \tag{33-27}$$

or, in terms of dimensionless groups, as

$$\frac{\mathbf{Sh}}{\mathbf{Re}\,\mathbf{Sc}^{1/3}} = j_M = \frac{f}{2} \tag{33-28}$$

The term j_M is known as the *j factor for mass transfer*. As pointed out in Chap. 23, for systems in which form drag is absent, it is generally found that $j_M = j_H = f/2$; if form drag is present, the equality is between only j_M and j_H and is only approximate. The essence of the *j*-factor relation is that the term $\mathbf{Sc}^{2/3}$ is suggested as a simple empirical replacement for the complex denominators on the right-hand side of the Prandtl-Taylor equation (33-25) and the von Kármán equation (33-26). As in the case of the heat-transfer equations, all forms of the analogy shown here reduce to the simple Reynolds analogy for $\mathbf{Sc} = 1$.

The Penetration Theory

The penetration theory of mass transfer was first suggested by Higbie[1] as a mechanistic picture of mass transfer from a gas-liquid interface into a liquid. The methods of analysis which we have presented to this point are adequate for describing transfer across a stationary interface. The penetration theory applies, in addition, to transfer across a fluid-fluid interface which may not be stationary.

Previous notions of the mechanism of interphase mass transfer were derived from the concept of a stagnant film in each fluid adjacent to the interface. Although it was known that a stable film was not actually present in most mass-transfer systems, the concept of a somewhat flexible film of indefinite thickness comparable

[1] R. Higbie, *Trans. Am. Inst. Chem. Eng.*, **31**:365 (1935).

with the viscous sublayer in boundary-layer flow was the basis of most mass-transfer models. Mass was visualized as being transferred through this film by molecular diffusion in accordance with the equations for steady-state mass transfer. This theory led to the definition of transfer coefficients in terms of the diffusivity and the thickness of the film. In this book we have almost always presented the individual phase-transfer coefficients for turbulent flow as empirical quantities without reference to the film theory. In most systems, such as flow over a flat plate, we have seen that there is no stagnant film; the quantity transferred from the plate into the boundary layer flows normal to the plate by diffusion and parallel to the plate by the motion of the fluid. The film theory has been used briefly, however, in Chap. 31 to obtain a relation between $k_{\tilde{x}}$ and $k_{\tilde{x}}^o$ for turbulent flow [refer to Eqs. (31-23) to (31-26)].

Higbie proposed that the principal mechanism of mass transfer involved the motion of turbulent eddies from the core of the fluid to the interface, followed by a short interval of unsteady-state molecular diffusion from the interface into the fluid before it was displaced from the surface by subsequent eddies. The average rate of mass transfer with this model depends on the time of exposure of an eddy at the interface and the total amount of the diffusing component that is transferred from the interface into the eddy during this time.

Higbie assumed that all eddies which reach the interface have the same exposure time. During this time, diffusion into the eddy is expressed by an equation which we recognize as a simplified form of Eq. (7-19).

$$\frac{\partial \rho_A}{\partial \theta} = D_{AB} \frac{\partial^2 \rho_A}{\partial y^2} \tag{33-29}$$

The boundary conditions are

$$
\begin{array}{lll}
1 & \rho_A = \rho_{Ao} & \theta = 0 \\
2 & \rho_A = \rho_{As} & \theta > 0, y = 0 \\
3 & \rho_A = \rho_{Ao} & y = \infty
\end{array}
$$

The surface concentration ρ_{As} and the concentration at a great distance from the surface ρ_{Ao} are constant. Equation (33-29) is a familiar form which has already been solved in Chap. 19 for the case of one-dimensional unsteady-state heat conduction into a medium of infinite thickness. The boundary conditions for the two problems are the same so the solution to the heat-transfer problem, Eq. (19-13), can be written in terms of mass-transfer variables as

$$\frac{\rho_{As} - \rho_A}{\rho_{As} - \rho_{Ao}} = \operatorname{erf} \frac{y}{2\sqrt{\theta D_{AB}}} \tag{33-30}$$

where

$$\operatorname{erf} \frac{y}{2\sqrt{\theta D_{AB}}} = \frac{2}{\sqrt{\pi}} \int_0^{y/2\sqrt{\theta D_{AB}}} e^{-n^2}\, dn$$

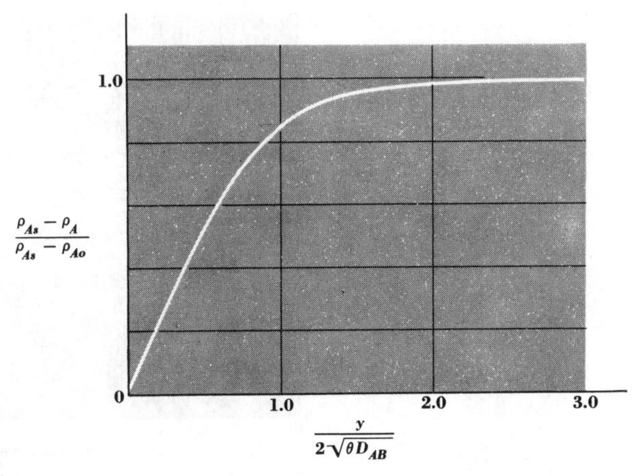

FIGURE 33-3
The Gauss error-integral function representing unsteady-state diffusion, Eq. (33-30).

The right-hand side is a tabulated function known as the *Gauss error integral*, for which values may be found in most sets of mathematical tables. A graphical representation of the function is shown in Fig. 33-3. A schematic representation of the concentration gradients obtained by this method is shown in Fig. 33-4 for a single eddy.

The rate of diffusion from the interface is

$$N_A = -D_{AB}\left(\frac{\partial \rho_A}{\partial y}\right)_{y=0} \tag{33-31}$$

if we neglect the convective-transport term. The function in Eq. (33-30), when differentiated and inserted in Eq. (33-31), gives

$$N_A = (\rho_{As} - \rho_{Ao})\sqrt{\frac{D_{AB}}{\pi\theta}} \tag{33-32}$$

The total amount of solute transferred per unit of interfacial area in a contact time θ_c is found by integrating, as in the expression

$$\int_0^{\theta_c} N_A \, d\theta = (\rho_{As} - \rho_{Ao})\left(\frac{D_{AB}}{\pi}\right)^{1/2} \int_0^{\theta_c} \frac{d\theta}{\theta^{1/2}}$$

$$= 2(\rho_{As} - \rho_{Ao})\left(\frac{\theta_c D_{AB}}{\pi}\right)^{1/2} \tag{33-33}$$

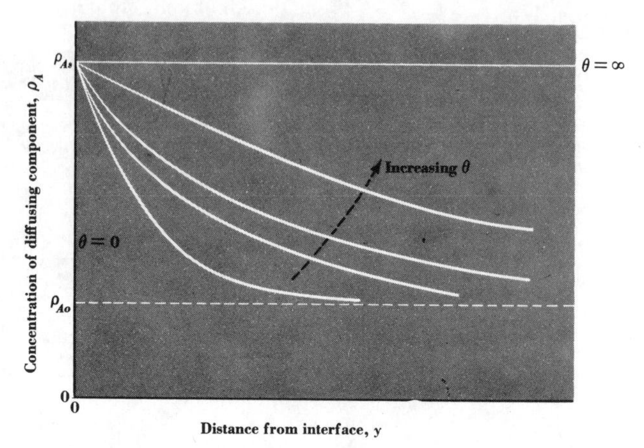

FIGURE 33-4
Schematic representation of concentration gradients for a single eddy.

The average rate over the time θ_c is found by dividing the total amount of solute transferred by θ_c to give

$$N_{A,\,\mathrm{av}} = 2(\rho_{As} - \rho_{Ao})\left(\frac{D_{AB}}{\pi\theta_c}\right)^{1/2} \tag{33-34}$$

If Eq. (33-34) is compared with Eq. (32-6), defining the mass-transfer coefficient in a steady-state system, we obtain

$$k_\rho(\rho_{As} - \rho_{Ao}) = 2(\rho_{As} - \rho_{Ao})\left(\frac{D_{AB}}{\pi\theta_c}\right)^{1/2}$$

from which

$$k_\rho = 2\left(\frac{D_{AB}}{\pi\theta_c}\right)^{1/2} \tag{33-35}$$

This equation is of little value as a method of calculating k_ρ because the average contact time of an eddy at the interface θ_c is not usually known. The most valuable conclusion that can be drawn from Eq. (33-35) is that if the penetration theory is correct, the mass-transfer coefficient must be proportional to the square root of the molecular diffusivity. Most mass-transfer correlations are of the type

$$\mathbf{Sh} = a\mathbf{Re}^b\,\mathbf{Sc}^c \tag{33-36}$$

The molecular diffusivity D_{AB} appears to the first power in the denominator of the Sherwood number. Because D_{AB} also appears in the denominator of the Schmidt number to the first power, the exponent c must have a value of 0.5 if k_ρ is to be proportional to $D_{AB}^{1/2}$.

If the concept of the stagnant film as the controlling resistance to mass transfer were realized, the mass-transfer coefficient would be related to D_{AB} by the equation[1]

$$k_\rho = \frac{D_{AB}}{\delta_m} \tag{33-37}$$

in which δ_m is the equivalent thickness of the film. This equation is equivalent to Eq. (31-24). The first power dependence of k_ρ on D_{AB} gives a value of zero for the exponent c in Eq. (33-36). Most experimental work indicates a value of c closer to 0.5 than to zero, thus supporting the belief that the penetration theory is a more accurate mechanism for representing interphase mass transfer than the stagnant-film theory.

The penetration theory has been modified by Danckwerts,[2] who proposed that the times of exposure of the surface elements are not the same and that the average rate of mass transfer depends on the distribution of the elements of surface among the various "age groups." The average absorption rate is found by multiplying the fraction of surface which has age θ by the instantaneous absorption rate for a surface of that age and then summing this expression for all elements of surface. Danckwerts proposed a distribution function $\phi(\theta)$ such that the fraction of surface having terminal ages between θ and $\theta + d\theta$ is $\phi(\theta)\, d\theta$. The instantaneous rate of absorption per element is given in Eq. (33-32), so that the instantaneous rate for all elements of age between θ and $\theta + d\theta$ is

$$\phi(\theta)(\rho_{As} - \rho_{Ao})\left(\frac{D_{AB}}{\pi\theta}\right)^{1/2} d\theta$$

Summing over all ages between $\theta = 0$ and $\theta = \infty$, we obtain, for the instantaneous rate per unit of interfacial area,

$$N_A = \int_0^\infty \phi(\theta)(\rho_{As} - \rho_{Ao})\left(\frac{D_{AB}}{\pi\theta}\right)^{1/2} d\theta \tag{33-38}$$

Danckwerts considered it likely that there is no relation between the age of an element and the probability that the element may be replaced. This distribution of ages can be represented by a function $\phi(\theta)$ of the form

$$\phi(\theta) = se^{-s\theta} \tag{33-39}$$

[1] In this discussion of the penetration theory the mass transferred by bulk flow is not included, so that k_ρ and $k_\rho{}^o$ are identical.

[2] P. V. Danckwerts, *Ind. Eng. Chem.*, **43**:1460 (1951).

in which s is the fractional rate of renewal of surface (the dimension is reciprocal time). This expression is inserted in Eq. (33-38) to give

$$N_A = (\rho_{As} - \rho_{Ao})\left(\frac{D_{AB}}{\pi}\right)^{1/2} \int_0^\infty \frac{se^{-s\theta}}{\theta^{1/2}}\,d\theta = (\rho_{As} - \rho_{Ao})(sD_{AB})^{1/2} \qquad (33\text{-}40)$$

This equation may be combined with the definition of the mass-transfer coefficient k_ρ to give

$$k_\rho(\rho_{As} - \rho_{Ao}) = (\rho_{As} - \rho_{Ao})(sD_{AB})^{1/2}$$

from which we get

$$k_\rho = (sD_{AB})^{1/2} \qquad (33\text{-}41)$$

The constant s must be found experimentally, so that Eq. (33-41) does not provide a direct means of predicting k_ρ. However, it does confirm the square-root relationship between k_ρ and D_{AB} obtained by Higbie.

Although the initial applications suggested for the penetration theory were limited to mass transfer at a gas-liquid interface, it is an equally valid picture as a mechanism for transporting heat.[1] It has also been suggested that the mechanism represents transfer in a fluid adjacent to a solid-liquid interface. Johnson and Huang[2] measured rates of solution of organic solids in an agitated vessel and found that the mass-transfer coefficient was related to the diffusivity by an equation similar to Eq. (33-36), in which values of the exponent c ranged from 0.422 to 0.526. Hanratty[3] has also shown that the penetration model represents mass transfer in the vicinity of a solid surface. Figure 33-5 shows a comparison of the dimensionless concentration profiles plotted by Hanratty for mass transfer from a fluid to a solid. The dashed line represents the equation of Higbie, which is based on equal ages for all elements. Danckwerts' results, assuming an age distribution, are shown by the solid line. Figure 33-5 indicates that the concentration profile is not particularly sensitive to the choice of distribution function used to represent surface age. A number of other authors have also proposed distribution functions containing a variable representing surface lifetime of a fluid element. Harriott,[4] however, has proposed a two-variable distribution function in which the variables are the age of the element and the thickness of the zone to be penetrated.[5] A summary of the models which have been suggested is presented by Scriven, along

[1] H. L. Toor and J. M. Marchello, *AIChE J.*, **4**:97 (1958).
[2] A. I. Johnson and Chen Huang, *AIChE J.*, **2**:412 (1956).
[3] T. J. Hanratty, *AIChE J.*, **2**:359 (1956).
[4] P. Harriott, *Chem. Eng. Sci.*, **17**:149 (1962).
[5] L. E. Scriven, *Chem. Eng. Educ.*, **2**:150 (1968); **3**:26 (1969); **3**:94 (1969).

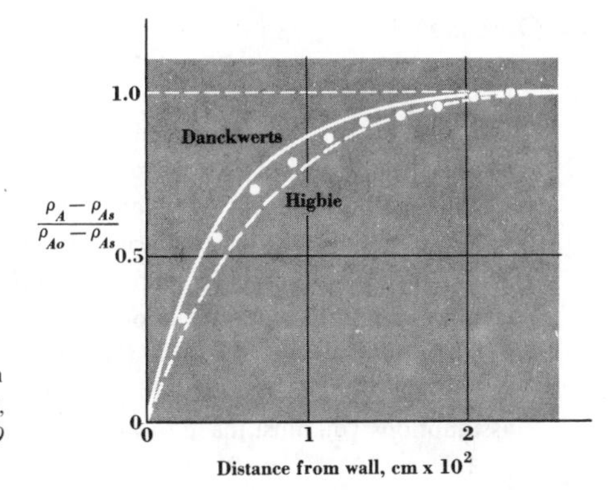

FIGURE 33-5
Concentration profile compared with predictions from penetration theory, [*From T. J. Hanratty, AIChE J.*, **2**:*359* (*1956*).]

with an analysis that proposes that the correct mechanistic view of mass transfer is one in which a variety of small-scale, nearly laminar flows are considered to carry the solute to or from the fiuid interface. This concept, designated by Scriven as *the stagnation-flow model*, leads to an equation for instantaneous mass flux consisting of the right side of Eq. (33-32) multiplied by a coefficient dependent on the direction and strength of the convective flow normal to the surface, i.e., the stagnation flow.

Interfacial Turbulence

Although mass transfer of a solute between stagnant, immiscible liquids might be expected to proceed entirely by molecular diffusion, spontaneous interfacial turbulence has been found to occur in a number of systems. The effect of this turbulence is to give much higher mass-transfer rates than would be obtained solely by molecular diffusion. A paper by Sternling and Scriven[1] summarizes a large number of reported instances of this effect. For example, a solution of 10 percent methanol in toluene placed quietly in water causes a turbid emulsion of water droplets to appear in the organic phase. The water phase, however, remains clear. Numerous other examples are cited, and the various forms of observed interfacial turbulence are described. It is noted that no effect is observed at a liquid-liquid interface when no solute is present. Sternling and Scriven attribute the turbulence to hydrodynamic instability caused by fluctuations in interfacial tension associated with mass transfer across the interface.

[1] C. V. Sternling and L. E. Scriven, *AIChE J.*, **5**:514 (1959).

PROBLEMS

33-1 Derive an equation for the mean Sherwood number for a flat plate on which the turbulent boundary layer extends from the leading edge.

33-2 Repeat Prob. 32-1 for a plate 0.1 m wide and 0.6 m long. Assume that the transition from a laminar to a turbulent boundary layer occurs at $\mathbf{Re}_x = 3 \times 10^5$.

33-3 A pan containing water is placed in a position where it is exposed to a 10 mi/h wind. The pan contains a uniform depth of $\frac{1}{2}$ in of water, is quite wide, and has a length of 8 ft in the wind direction. The water in the pan is at a constant temperature of 60°F; the moisture content of the air far from the pan is 0.005 lb water/lb air. Find the length of time required for all the water in the pan to evaporate. List any assumptions you must make to work this problem. The kinematic viscosity of the air is 1.70×10^{-4} ft²/s; $D_{AB} = 2.79 \times 10^{-4}$ ft²/s. The transition to turbulent flow occurs at $\mathbf{Re}_x = 3 \times 10^5$.

33-4 Develop an equation for the overall coefficient $K_{\tilde{x}}$ based on a constant coefficient $k_{\tilde{y}}$ on the gas side and a coefficient $k_{\tilde{x}}$ on the liquid side determined by the penetration theory with the distribution function of Higbie. Note that while an element of liquid resides at the surface, $\tilde{y}_{As}$ or $m\tilde{x}_{As}$ varies, although $\tilde{y}_A$ in the bulk of the gas is considered constant.

33-5 Water is flowing in a smooth tube of 0.02-m ID at a Reynolds number of 10,000. Compare the values of the Sherwood number for the dissolution of cinnamic acid at 6.6°C ($\mathbf{Sc} = 2920$) predicted by the analogies of Reynolds, von Kármán, and Colburn.

"dilute"

$$N_A = -\frac{1}{A}\cdot\frac{dm}{d\theta} \quad \left(\begin{array}{l} m: \text{질량 mass} \\ A: \text{면적} \end{array}\right)$$

$$= -\left(\frac{1}{\pi D^2}\right)\frac{d\left(\frac{\rho\pi D^3}{6}\right)}{d\theta}$$

$$= -\frac{\rho}{2}\cdot\frac{dD}{d\theta}, \qquad N_A = k_\rho(\rho_{AS} - \rho_{A\infty})$$

$\rho_{A\infty} \to 0$. Sh = 2 이므로 $k_\rho = \frac{2D_{AB}}{D}$

"PSSA" $$-\frac{\rho}{2}\cdot\frac{dD}{d\theta} = \frac{2D_{AB}}{D}\cdot\rho_{AS}$$

$$\Rightarrow -\int_{D=1/24}^{D=0} D\,dD = \frac{4D_{AB}\rho_{AS}\cdot\theta}{\rho}$$

$$\left[\frac{D^2}{2}\right]_0^{1/24} = \frac{4D_{AB}\cdot\rho_{AS}\cdot\theta}{\rho}$$

$$\theta = \left(\frac{1}{24}\right)^2\left(\frac{1.145\times 62.4}{(2\times 4)\times 0.2\times(9.5\times 10^{-6})}\right)$$

$$= 8160\text{ h} = 340\text{ days}$$

34

SOME DESIGN EQUATIONS FOR CONVECTIVE MASS TRANSFER

The Application of Dimensional Analysis to Mass Transfer

In the two preceding chapters we have derived a number of equations for predicting convective-mass-transfer coefficients in laminar- and turbulent-flow systems. As in the study of heat transfer, we have seen that these equations could be obtained by three methods: (1) the analytical solutions of the basic differential equations, (2) the von Kármán integral method, and (3) the classical Reynolds analogy and subsequent improvements based on a knowledge of the flow regimes adjacent to a solid surface. In addition, we examined the penetration theory, which is based on the principle of surface renewal and seems especially appropriate for describing mass transfer between two fluid phases.

One more method of obtaining equations for predicting mass-transfer coefficients is the use of dimensional analysis. Because this method can be used for both laminar and turbulent systems, it is considered in the present chapter rather than in either of the previous two chapters.

The principles and methods of dimensional analysis have been discussed extensively in Chap. 13 and illustrated by examples in Chaps. 13 and 24. The following example is presented as an illustration of the methods of dimensional analysis applied to mass transfer.

Example 34-1 The principles of dimensional analysis are to be used to find the dimensionless groups which might appear in an equation for predicting the mass-transfer coefficient k_ρ for a fluid in a vertical pipe. Mass will be assumed to be transferred between the fluid and the pipe wall as the result of a concentration difference $\Delta\rho_A$. The fluid in the pipe may be moving as the result of natural convection associated with density gradients, or as the result of forced convection.

The variables chosen are listed below. They contain only three fundamental dimensions: length $[L]$, mass $[M]$, and time $[\theta]$. Because the variables do not contain any dimensions which can be derived from these three, e.g., force or energy, it is not necessary to include in the list of variables any dimensional constants such as the mechanical equivalent of heat J or the proportionality constant g_c from Newton's second law.

Variable	Symbol	Dimension
Diameter of pipe	D	$[L]$
Bulk-fluid density	ρ	$[M/L^3]$
Bulk-fluid viscosity	μ	$[M/L\theta]$
Mass-transfer coefficient	$k_\rho{}^o$	$[L/\theta]$
Concentration difference	$\Delta\rho_A$	$[M/L^3]$
Gravitational acceleration	g	$[L/\theta^2]$
Bulk velocity	u_b	$[L/\theta]$
Molecular diffusivity	D_{AB}	$[L^2/\theta]$
Length of pipe	L	$[L]$

The list contains nine variables expressed in terms of three dimensions. It can be determined by manipulating the variables that the maximum number of variables which will not form a dimensional group in this situation equals the number of fundamental dimensions, which is three. If an additional variable is added however, it becomes possible to form dimensionless groups so that these various groups will be constructed from three common variables combined successively with each of the remaining six variables. The result will be six dimensionless groups.

The first three variables in the list are chosen as common to all groups. The first group is

$$\pi_1 = D^a \rho^b \mu^c (k_\rho{}^o)^d$$

The dimensions are substituted to give

$$1 = [L]^a \left[\frac{M}{L^3}\right]^b \left[\frac{M}{L\theta}\right]^c \left[\frac{L}{\theta}\right]^d$$

The exponents on the dimensions are summed to give the following equations:

$$[L] \qquad 0 = a - 3b - c + d$$
$$[M] \qquad 0 = b + c$$
$$[\theta] \qquad 0 = -c - d$$

These equations can be solved to give $c = -a = -b = -d$, so that the group π_1 is found to be $(\mu/D\rho k_\rho{}^o)^c$.

The second group π_2 is formed by writing

$$\pi_2 = D^a \rho^b \mu^c (\Delta\rho_A)^d$$

which gives the following relation among dimensions:

$$1 = [L]^a \left[\frac{M}{L^3}\right]^b \left[\frac{M}{L\theta}\right]^c \left[\frac{M}{L^3}\right]^d$$

The sums of the exponents give the equations

$$[L] \qquad 0 = a - 3b - c - 3d$$
$$[M] \qquad 0 = b + c + d$$
$$[\theta] \qquad 0 = c$$

From the solution of these equations we find $a = c = 0$ and $b = -d$. Thus $\pi_2 = (\Delta\rho_A/\rho)^d$.

Four additional groups are formed by combining D, ρ, and μ with each of the four remaining variables. These groups are found to be

$$\pi_3 = \frac{D^3 \rho^2 g}{\mu^2}$$

$$\pi_4 = \frac{D u_b \rho}{\mu}$$

$$\pi_5 = \frac{\mu}{\rho D_{AB}}$$

$$\pi_6 = \frac{L}{D} \qquad ////$$

The six groups found in the preceding example contain at least two groups which have already been found by analytical procedures to affect mass transfer. These are π_4, the Reynolds number, and π_5, the Schmidt number. Additional groups can be recognized by forming suitable combinations. For example, π_5/π_1 reduces to $k_\rho{}^o D/D_{AB}$, which we call the Sherwood number. The product $\pi_2 \pi_3 \pi_6{}^3$ is the Grashof number for mass transfer, $gL^3\rho\ \Delta\rho/\mu^2$, which was mentioned in Chap. 32.

The combinations of dimensionless groups which describe a series of experimental results (and can be used for predicting future performance) are found by writing equations of the form

$$\pi_1 = \alpha\pi_2^{\beta}\pi_3^{\gamma}\pi_4^{\delta}\pi_5^{\varepsilon}\pi_6^{\zeta}\cdots \qquad (34\text{-}1)$$

and determining the constants by statistical analysis. If the system of Example 34-1 were limited to natural convection, it is likely that the exponent on Reynolds number in Eq. (34-1) would be very close to zero. The other groups would probably have exponents which could be combined to give an equation of the form

$$\frac{\pi_5}{\pi_1} = \alpha(\pi_2\pi_3\pi_6^{\ 3})^{\beta}\pi_5^{\ \gamma} \qquad (34\text{-}2)$$

which we recognize in more familiar notation as

$$\mathbf{Sh} = \alpha\ \mathbf{Gr}^{\beta}\ \mathbf{Sc}^{\gamma} \qquad (34\text{-}3)$$

If the fluid in the pipe were moving as the result of forced convection, the groups which might influence natural convection (π_2 and π_3) would be found to have small exponents. Hence they would not appear in the final correlation, which would be of the form

$$\frac{\pi_5}{\pi_1} = \alpha\pi_4^{\ \beta}\pi_5^{\ \gamma}\pi_6^{\ \delta} \qquad (34\text{-}4)$$

The group π_6, L/D, is a measure of the importance of the entrance region for turbulent systems and would likely disappear from most correlations for long pipes, so that the final correlation would be of the form

$$\frac{\pi_5}{\pi_1} = \alpha\pi_4^{\ \beta}\pi_5^{\ \gamma} \qquad (34\text{-}5)$$

which is more familiar when written as

$$\mathbf{Sh} = \alpha\ \mathbf{Re}^{\beta}\ \mathbf{Sc}^{\gamma} \qquad (34\text{-}6)$$

The group obtained by taking the product $\pi_4\pi_5$, i.e., **Re Sc**, is known as the *Peclet number for mass transfer* (**Pe** $= Du_b/D_{AB}$). An analogous form (**Pe** $= Du_b/\alpha$), also known as the Peclet number, was used in the analysis of heat transfer to fluids in laminar flow in Chap. 22.

Mass-Transfer Coefficients for Flow in Pipes

Most data on mass transfer between a fluid and a pipe wall have been obtained using wetted-wall columns. This device consists basically of a vertical section of circular pipe in which a gas is usually flowing upward. A volatile liquid flows down the inside surface of the pipe and evaporates into the gas stream. The principal reason for the use of wetted-wall columns for studying mass transfer is that the interfacial area is known. This makes it possible to determine the actual mass-transfer coefficient, for example, k_ρ, rather than the product, $k_\rho a$. The gas is usually only slightly soluble in the liquid, so that the liquid at the interface is almost pure. The liquid may be supplied at the adiabatic-saturation temperature.[1] If the column operates adiabatically, the liquid remains at this temperature as it falls through the section, so that the concentration ρ_{As} of the diffusing component in the gas phase at the interface is constant. The wetted-wall column can be used with the gas phase in either laminar or turbulent flow, but care must be taken to avoid the formation of ripples at the vapor-liquid interface since these make it difficult to determine the interfacial area.

It would seem that the correct velocity to be used in the Reynolds number describing the gas flow would be the velocity of the gas relative to that of the liquid, i.e., the sum of the bulk gas velocity and the gas-liquid interfacial velocity (when flow is countercurrent). However, since the interfacial velocity is usually not known, it is customary to use simply the bulk velocity of the gas relative to the wall of column.

Another method of obtaining data on mass-transfer coefficients in pipes is to use pipes cast from some soluble material and to pass a suitable solvent through the pipe. Alternatively, one might cast the pipe from a volatile substance and pass a gas through the pipe. In both cases solid materials which dissipate very slowly must be used to prevent significant increases in the pipe diameter during an experimental run. Liquid-phase systems usually yield results at higher Schmidt numbers (**Sc** $\sim$ 2000) than gases (**Sc** $\sim$ 1) because the molecular diffusivity of liquids is of the order of 10^{-5} times the molecular diffusivity of gases.

Laminar flow In Chap. 32 it was indicated that predictions of mass transfer for laminar flow in a pipe could be made using the results presented in Chap. 22 on heat transfer with laminar flow in a pipe. Heat-transfer solutions were presented for systems with flat and parabolic velocity profiles for both constant wall temperature and constant heat flux. The solutions for constant wall concentration, which are analogous to the solutions for constant wall temperature, will be considered here; the solutions for a constant mass flux from the wall will not be considered.

The integrated Graetz solutions for the mixing-cup concentration ρ_{Ab} with constant wall concentration and parabolic and flat velocity profiles are shown in Fig. 34-1. The solutions are obtained as follows. The line representing the parabolic profile is found by inserting the mass-transfer solution, which is equivalent to the Graetz solution for heat transfer, Eq. (22-29), into an integral expression for mixing-cup

[1] This term is defined in Chap. 36.

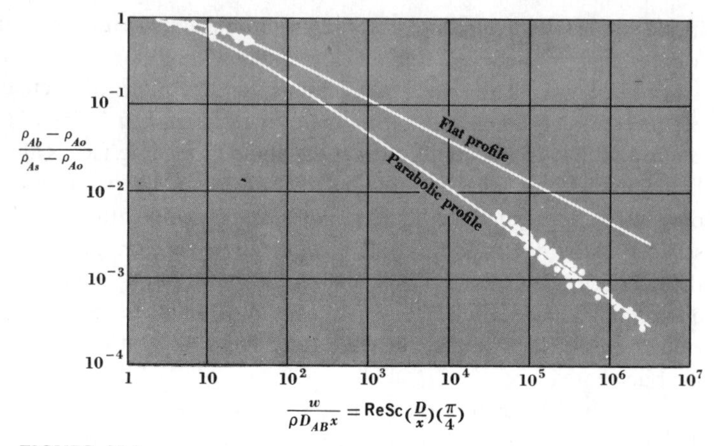

FIGURE 34-1
Exit concentration of solute for laminar flow in a pipe. [*From W. H. Linton, Jr., and T. K. Sherwood, Chem. Eng. Progr.*, **46**:*258 (1950)*.]

concentration, similar to Eq. (22-30), and integrating over the pipe cross section. The equivalent mass-transfer solution used is simply Eq. (22-29), with the dependent variable taken as $(\rho_{As} - \rho_A)/(\rho_{As} - \rho_{Ao})$ instead of $(t_s - t)/(t_s - t_o)$. The independent variable in Eq. (22-29) is **Re Pr** (r_i/x). The quantity represented on the abscissa in Fig. 34-1 differs from this by the factor $\pi/2$; it is written $w/\rho D_{AB}x$, which we recognize as **Re Sc** $(D/x)(\pi/4)$.

The line for the parabolic profile at values of $w/\rho D_{AB}x$ above 40 can be seen to be linear and is described by the equation

$$\frac{\rho_{Ab} - \rho_{Ao}}{\rho_{As} - \rho_{Ao}} = 5.5\left(\frac{w}{\rho D_{AB}x}\right)^{-2/3} \tag{34-7}$$

The solution as in the case of heat transfer applies only for pipe lengths sufficiently short so that the concentration boundary layers do not meet at the center of the pipe.

The solution for a flat velocity profile is also shown in Fig. 34-1. This curve can be calculated from Eq. (22-23) by making the same changes of variables as indicated for the parabolic system. For short tubes (high values of $w/\rho D_{AB}x$) this function can be represented approximately by the equation

$$\frac{\rho_{Ab} - \rho_{Ao}}{\rho_{As} - \rho_{Ao}} = 4.0\left(\frac{w}{\rho D_{AB}x}\right)^{-1/2} \tag{34-8}$$

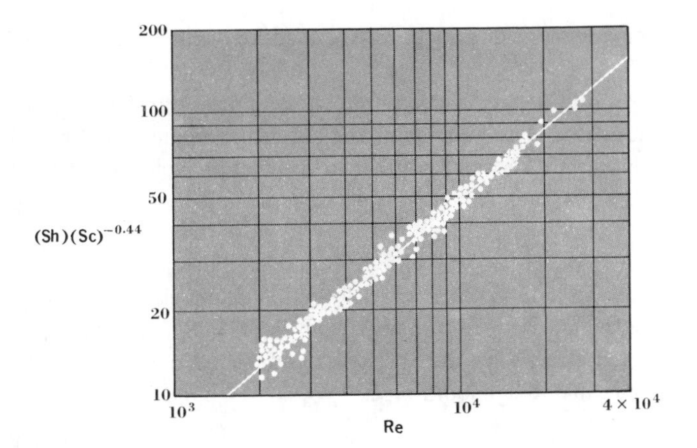

FIGURE 34-2
Mass-transfer coefficients for vaporization of liquids with turbulent gas flow in a wetted-wall column. [*From E. R. Gilliland and T. K. Sherwood, Ind. Eng. Chem.*, **26**:*516* (*1934*).]

The data in Fig. 34-1 at high values of $w/\rho D_{AB}x$ were obtained by the use of soluble pipe walls, as described earlier. This technique was used with Schmidt numbers as high as 3000. It can be seen that the data fit the theoretical line for the parabolic profile quite well. At low values of $w/\rho D_{AB}x$ the results appear to be closer to the line based on the assumption of a flat velocity profile. These data were obtained by the operation of a wetted-wall column in which liquids were vaporized into a gas stream. For these systems the Schmidt number was approximately unity. The fact that the values lie above the line representing the parabolic profile has been attributed to natural-convection effects.

Turbulent flow Extensive data on mass transfer with turbulent flow in wetted-wall columns have been obtained by Gilliland and Sherwood.[1] Results for nine liquids evaporating into air are shown in Fig. 34-2. The best-fitting straight line is represented by the equation

$$\frac{k_\rho^o D}{D_{AB}} = 0.023\left(\frac{Du_b\rho}{\mu}\right)^{0.83}\left(\frac{\mu}{\rho D_{AB}}\right)^{0.44} \tag{34-9}$$

The Sherwood number in this equation is often written as $(k_\rho D/D_{AB})(p_{B,lm}/p)$ when applied to gases. In these groups $k_\rho p_{B,lm}/p$ is identical with $k_\rho(1 - \tilde{x}_A)_{lm}$ if the gas phase is ideal. The group $k_\rho(1 - \tilde{x}_A)_{lm}$ represents a form of the mass-transfer

[1] E. R. Gilliland and T. K. Sherwood, *Ind. Eng. Chem.*, **26**:516 (1934).

coefficient which is independent of concentration effects and is defined in Chap. 31 as $k_\rho{}^o$

Additional equations for turbulent mass transfer in a pipe can be obtained from the analogies between mass and momentum transfer in Chap. 33. Equation (33-28) relates the Sherwood number to the friction factor as follows:

$$\frac{\mathbf{Sh}}{\mathbf{Re}\ \mathbf{Sc}^{1/3}} = \frac{f}{2} \tag{33-28}$$

This equation is based on Colburn's j-factor relation. If an empirical equation for the friction factor is combined with Eq. (33-28), we obtain an equation which can be solved directly for the Sherwood number. Equation (12-72) is a convenient form, applicable to a smooth pipe.

$$f = 0.046\ \mathbf{Re}^{-0.20} \tag{12-72}$$

If this expression is combined with Eq. (33-28), we obtain

$$\mathbf{Sh} = 0.023\ \mathbf{Re}^{0.80}\ \mathbf{Sc}^{0.33} \tag{34-10}$$

This expression gives results quite close to those predicted by Eq. (34-9), differing chiefly in the exponent on the Schmidt number. The experimental results which led to Eq. (34-9) were obtained for systems in which the Schmidt number varied only from 0.6 to 2.5, so the exponent 0.44 is questionable. In the work of Linton and Sherwood, referred to in the section on laminar flow, some runs were made in the turbulent-flow regime with Schmidt numbers between 1000 and 2200. These results, when combined with the wetted-wall data of Gilliland, indicate that the correct exponent on the Schmidt number is 0.33. Consequently, Eq. (34-10) is recommended as being applicable over a wider range than Eq. (34-9).

The Prandtl-Taylor and von Kármán analogies can also be combined with Eq. (12-72) to give equations for predicting mass-transfer coefficients. However, the Colburn analogy has been found to give generally satisfactory results and has the advantage of simplicity.

The mass-transfer coefficients for turbulent flow in Eqs. (34-9) and (34-10) can be applied either to point conditions in the region of developed flow or to the entire pipe surface if the pipe length is greater than 50 diameters. The mass-transfer coefficient at the pipe entrance is infinite, just as is the heat-transfer coefficient. However, with turbulent flow the region of developed flow is usually attained in 10 pipe diameters or less, after which the coefficient is substantially constant at the value predicted from the correlation. When the pipe length exceeds 50 diameters, the high values of the coefficient near the entrance contribute so little to the total mass transfer that the mean coefficient for the entire pipe is approximately the same as the local coefficient in the developed flow region.

Example 34-2 Air passes through a naphthalene tube which has an ID of 1 in and a length of 6 ft at bulk velocities of (*a*) 2 ft/s and (*b*) 50 ft/s. The air is at 50°F and has an average pressure of 1 atm. Assuming that the change in pressure along the tube is negligible and that the naphthalene surface is at 50°F, find the percent saturation of the discharge air and the sublimation rate of naphthalene in the tube in pounds per hour.

Properties of air at 50°F, 1 atm	$\rho = 0.078$ lb/ft^3 $\mu = 1.20 \times 10^{-5}$ lb/(ft)(s)
Properties of naphthalene at 50°F	Vapor pressure = 0.0209 mmHg Molecular diffusivity in air = 0.200 ft^2/h Molecular weight = 128.2

(*a*)

$$u_b = 2 \text{ ft/s}$$

$$\mathbf{Re} = \frac{(\frac{1}{12})(2)(0.078)}{1.20 \times 10^{-5}}$$

$$= 1080 \qquad \text{(hence laminar flow)}$$

$$\mathbf{Sc} = \frac{(1.20 \times 10^{-5})(3600)}{(0.078)(0.200)}$$

$$= 2.77$$

$$\mathbf{Re\ Sc}\frac{D}{x}\frac{\pi}{4} = (1080)(2.77)(\tfrac{1}{72})\left(\frac{\pi}{4}\right)$$

$$= 32.6$$

From Fig. 34-1, using the curve for the parabolic profile, we find

$$\frac{\rho_{Ab} - \rho_{Ao}}{\rho_{As} - \rho_{Ao}} = 0.42$$

The air entering the tube is assumed to be free from naphthalene, so that $\rho_{Ao} = 0$. The concentration ρ_{As} of naphthalene at the interface is found from the vapor pressure by the following approximation (valid for dilute mixtures):

$$\rho_{As} = \frac{(0.0209)(128)(0.078)}{(760)(29)}$$

$$= 9.5 \times 10^{-6} \text{ lb/ft}^3$$

Therefore the exit concentration of naphthalene is

$$\rho_{Ab} = (0.42)(9.5 \times 10^{-6})$$
$$= 4.0 \times 10^{-6}\ \text{lb/ft}^3$$

$$\text{Total evaporation rate} = \left(\frac{\pi}{4}\right)\left(\frac{1}{12}\right)^2 (2)(4.0 \times 10^{-6})(3600)$$
$$= 1.6 \times 10^{-4}\ \text{lb/h}$$

The exit air is 42 percent saturated, as was shown by the reading from Fig. 34-1.

(*b*)

$$u_b = 50\ \text{ft/s}$$

$$\mathbf{Re} = \frac{(\frac{1}{12})(50)(0.078)}{1.20 \times 10^{-5}}$$
$$= 27{,}000$$

For turbulent flow we use Eq. (34-10) to find the mass-transfer coefficient.

$$\mathbf{Sh} = 0.023(27.000)^{0.8}(2.77)^{0.33}$$
$$= (0.023)(3500)(1.40)$$
$$= 113$$

The mole fraction of naphthalene is so low that $(1 - \tilde{x}_A)_{lm} \sim 1$. Hence $k_\rho{}^o = k_\rho$.

$$k_\rho = \frac{(113)(0.200)}{(\frac{1}{12})}$$
$$= 271\ \text{ft/h}$$

A naphthalene balance is written for a differential length of tube. Because of the low vaporization rate, the total volumetric rate of flow can be assumed to be independent of the distance from the tube inlet.

$$\frac{\pi}{4}\left(\frac{1}{12}\right)^2 (50)\, d\rho_{Ab} = k_\rho \frac{\pi}{12} dx(\rho_{As} - \rho_{Ab})$$

This is rearranged to give

$$\int_0^{\rho_{Ab}} \frac{d\rho_{Ab}}{\rho_{As} - \rho_{Ab}} = \frac{48}{50} k_\rho \int_0^6 dx$$

This equation is integrated and gives the expression

$$-\ln \frac{\rho_{As} - \rho_{Ab}}{\rho_{As}} = \frac{(48)(271)(6)}{(50)(3600)}$$
$$= 0.43$$

from which we obtain

$$\frac{\rho_{As}}{\rho_{As} - \rho_{Ab}} = 1.54$$

This is solved to give the exit mixing-cup concentration.

$$\begin{aligned}\rho_{Ab} &= 0.35\rho_{As} \\ &= (0.35)(9.5 \times 10^{-6}) \\ &= 3.33 \times 10^{-6} \text{ lb/ft}^3\end{aligned}$$

$$\text{Total evaporation rate} = \frac{\pi}{4}\left(\frac{1}{12}\right)^2 (50)(3.33 \times 10^{-6})(3600)$$

$$= 3.3 \times 10^{-3} \text{ lb/h}$$

The exit air is 35 percent saturated. ////

Mass-Transfer Coefficients for Two-Phase Flow in Packed Beds

The measurement of mass-transfer coefficients for packed beds has been the subject of much research for the past thirty years. In spite of this effort, the complexity of the problem is so great that the reliability of predicted mass-transfer rates is still far from satisfactory. In the correlations for packed beds to be presented in this chapter, experimental results frequently deviate by as much as 25 percent from the recommended correlations, and in a number of systems the deviation is even greater.

Some of the reasons for these deviations have been indicated in preceding chapters. The principal difficulty seems to be in the determination of the interfacial area. Another difficulty arises from the fact that most measurements are for overall coefficients representing a pair of resistances in series. To determine single-phase coefficients, the experimental system must be arranged so that one resistance is either negligible or can be calculated. A third difficulty is that very little is known regarding entrance effects in packed beds. Most experimental systems involve much shorter beds than are found in industry. As a consequence, end effects may exert a greater influence in the experimental systems than in the industrial systems.

As a result of the inadequacy of theoretical methods for predicting mass-transfer coefficients, the designers of mass-transfer equipment lean heavily on performance data whenever possible. Perry's Handbook gives several pages (pp. 18-41 to 18-48) of data on HTUs (these were defined in Chap. 31) for both single- and two-phase systems for a variety of fluid systems and packings. A summary of articles containing data on HTUs is given by Cornell, Knapp, and Fair.[1] In addition, correlations to be given below can be used for the calculation of individual coefficients or HTUs.

[1] D. Cornell, W. G. Knapp, and J. R. Fair, *Chem. Eng. Progr.*, **56**:68 (1960).

Gases The experimental results for gas-phase resistance have been measured in a variety of ways. One way has been to absorb a gas in a liquid in which it has a very low vapor pressure or in which a rapid reaction occurs between the absorbed solute and some nonvolatile component in the liquid. The resistance to mass transfer in the liquid phase is assumed to be negligible, so that the overall coefficient is identical with the gas-phase coefficient. This technique must be used with discretion; in the past it has sometimes been used in systems in which the resistance in the liquid phase has not been negligible. A second method has been to vaporize a pure liquid flowing down the tower into an insoluble gas passing the liquid. As a third method, the resistance of the liquid phase may be predicted from appropriate correlations and subtracted from the total resistance to give the gas-phase resistance. This latter method has been used by Fellinger for the determination of gas-phase coefficients in systems in which ammonia is absorbed from air into water flowing over various kinds of packing. A substantial amount of Fellinger's data is reproduced in the section of Perry referred to above. Curves for both H_G and H_{OG} are shown to depend on both gas and liquid rates. The data for H_G, expressed in feet, have been correlated by an empirical equation of the form

$$H_G = \alpha G^\beta L^\gamma \, \mathbf{Sc}^{0.5} \qquad (34\text{-}11)$$

in which gas and liquid mass velocities are expressed in lb/(h)(ft^2). The use of the Schmidt number is to permit generalization of the results to systems other than air and ammonia. The exponent on the Schmidt number of 0.5 is justifiable on the basis of the penetration theory, in which the mass-transfer coefficient k_ρ was shown to be a function of $D_{AB}^{0.5}$. The exponents in Eq. (34-11) are not really constants, but

Table 34-1 CONSTANTS FOR DETERMINING H_G IN Eq. (34-11)

Type of packing	α	β	γ	Range of values, lb/(h)(ft²) G	L
Raschig rings:					
$\frac{3}{8}$ in	2.32	0.45	−0.47	200–500	500–1500
1 in	7.00	0.39	−0.58	200–800	400–500
	6.41	0.32	−0.51	200–600	500–4500
2 in	3.82	0.41	−0.45	200–800	500–4500
Berl saddles:					
$\frac{1}{2}$ in	32.4	0.30	−0.74	200–700	500–1500
	0.811	0.30	−0.24	200–700	1500–4500
1 in	1.97	0.36	−0.40	200–800	400–4500
$1\frac{1}{2}$ in	5.05	0.32	−0.45	200–1000	400–4500

SOURCE: R. E. Treybal, "Mass-transfer Operations," p. 239, McGraw-Hill Book Company, New York, 1955.

depend on both the packing and the gas and liquid rates. A table of these values is given by Treybal; sample results are shown in Table 34-1.

A similar correlation for gas-phase coefficients has been suggested by Sherwood[1] as representative of the data of a number of investigators. This correlation, when expressed in the form of Eq. (34-11), proposes values of 1.01, 0.31, and -0.33 for the constants α, β, and γ, respectively. Within the limits of accuracy of the data, there is no distinction associated with packing size.

Liquids Most experimental results for liquid-phase mass-transfer coefficients have been obtained by the absorption or desorption in water of gases of slight solubility. Typical gases used have been oxygen, hydrogen, and carbon dioxide. For these systems the slope of the line representing Henry's law is quite large, so that the average overall mass-transfer coefficient (which is measured from terminal conditions) and the liquid-phase mass-transfer coefficient are approximately equal, as can be deduced from Eq. (31-21).

An extensive investigation of liquid-phase mass transfer has been described by Sherwood and Holloway. Their results are correlated by the following empirical equation:

$$H_L = \phi\left(\frac{L}{\mu}\right)^{\eta} \mathbf{Sc}^{0.5} \qquad (34\text{-}12)$$

The height of a transfer unit H_L is expressed in feet, the liquid rate L in lb/(h)(ft^2), and the viscosity μ of the liquid phase in lb/(ft)(h). The Schmidt number applies to the liquid phase. The constants ϕ and η depend on the type of packing. Typical values recommended by Sherwood and Holloway are given in Table 34-2.

Table 34-2 CONSTANTS FOR DETERMINING H_L IN Eq. (34-12)

Type of packing	ϕ	η	Range of L
Raschig rings:			
$\frac{3}{8}$ in	0.0018	0.46	400–15,000
1 in	0.010	0.22	400–15,000
2 in	0.012	0.22	400–15,000
Berl saddles:			
$\frac{1}{2}$ in	0.0067	0.28	400–15,000
1 in	0.0059	0.28	400–15,000
$1\frac{1}{2}$ in	0.0062	0.28	400–15,000

SOURCE: T. K. Sherwood and F. A. L. Holloway, *Trans. Am. Inst. Chem. Eng.*, **36**:39 (1940).

[1] T. K. Sherwood and R. L. Pigford, "Absorption and Extraction," 2d ed., p. 285, McGraw-Hill Book Company, New York, 1952.

Most experimental results show that H_L is independent of the gas rate except at very high gas velocities, so that Eq. (34-12) does not contain the term G. The definition of H_L is given as

$$H_L = \frac{\tilde{L}}{k_{\tilde{x}}{}^o a} \qquad (31\text{-}51)$$

Thus it can be seen that H_L varies directly with the liquid rate and also as a result of whatever effect the liquid rate has on $k_{\tilde{x}}{}^o$ and a. The net effect of these three factors is expressed in the empirical constant η in Eq. (34-12). As for the dependence of H_L on the gas rate, it can be seen from Eq. (31-51) that if the gas rate affects either $k_{\tilde{x}}{}^o$ or a, then H_L should be dependent on G. Investigations by Shulman et al.[1] indicate that $k_{\tilde{x}}{}^o$ is independent of the gas rate and that the interfacial area a is almost independent of gas velocity for Raschig rings and Berl saddles with a size of 1 in. and greater. For smaller sizes, however, a significant increase occurs and should be reflected in any correlation such as Eq. (34-12).

The same investigation by Shulman and coauthors indicates the reason for the strong dependence of H_G on both gas and liquid rates. H_G is defined as

$$H_G = \frac{\tilde{G}}{k_{\tilde{y}}{}^o a} \qquad (31\text{-}49)$$

In this equation the mass-transfer coefficient $k_{\tilde{y}}{}^o$ and the gas rate $\tilde{G}$ establish a dependence of H_G on the gas velocity while the strong effect of liquid rate on the interfacial area a makes H_G also dependent on L. Hence both G and L enter Eq. (34-11) for calculating H_G. Figure 34-3 shows typical data for the interfacial area as a function of G and L to illustrate the basis for these assertions.

Mass Transfer from Spheres and Cylinders

Extensive data have been taken on both heat and mass transfer from single spheres. Much of the mass-transfer work has been done using liquid spheres evaporating into a passing air stream. Other investigations have employed spheres of solid material which have either sublimed into passing air streams or dissolved into a passing stream of liquid. As might be expected, the results of both heat- and mass-transfer research are represented by similar equations. Equation (24-12) for heat transfer from single spheres describes mass transfer if the Nusselt and Prandtl numbers are replaced by the Sherwood and Schmidt numbers. For mass transfer it is written as

$$\frac{k_\rho{}^o D}{D_{AB}} = 2 + 0.6\left(\frac{\mu}{\rho D_{AB}}\right)^{1/3}\left(\frac{D u_o \rho}{\mu}\right)^{1/2} \qquad (34\text{-}13)$$

[1] H. L. Shulman, C. F. Ulrich, A. Z. Proulx, and J. O. Zimmerman, *AIChE J.*, **1**:253 (1955).

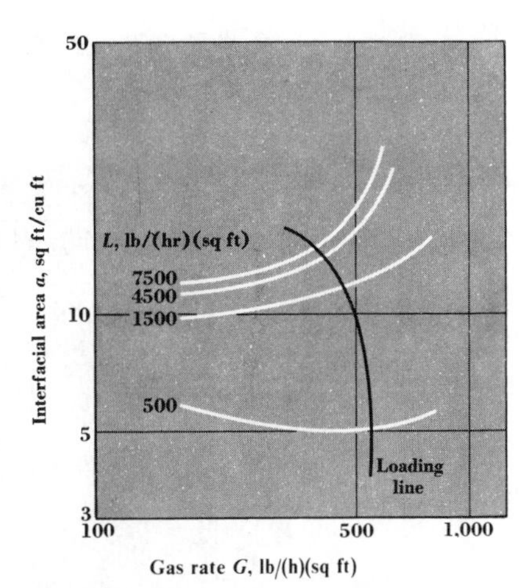

FIGURE 34-3
Effective interfacial area for 0.5-in Raschig rings. [*From H. L. Shulman, C. F. Ulrich, A. Z. Proulx, and J. O. Zimmerman, AIChE J.*, **1**:*253 (1955).*]

The constant 2 in this equation can be derived in the manner suggested for the analogous situation in heat transfer in Chap. 24. The analysis is based on a system in which molecular diffusion of a single component occurs at steady state outward from a spherical surface into an infinite, stagnant medium. The second term on the right-hand side of Eq. (34-13) represents the contribution of the fluid motion to mass transfer from the sphere.

Mass-transfer measurements on a cylinder of naphthalene in a wind tunnel are reported by Sandall and Mellichamp. Local values of the Sherwood number, plotted as a function of peripheral position, are shown in Fig. 34-4. The decrease

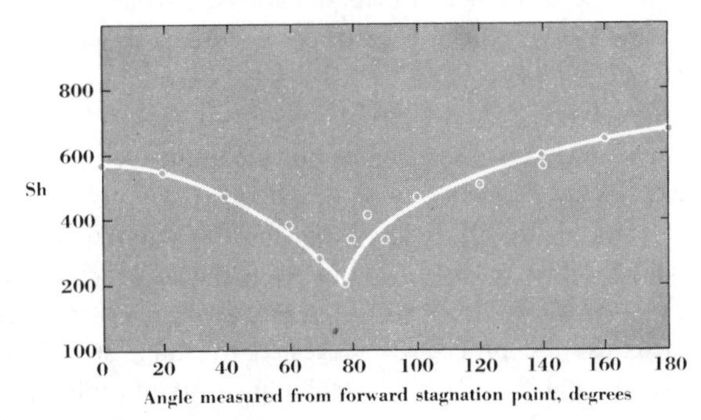

FIGURE 34-4
Local values of the Sherwood number for air flow around a naphthalene cylinder at **Re** = 110,000. [*From O. C. Sandall and D. A. Mellichamp, Chem. Eng. Educ.*, **5**:*134 (1971).*]

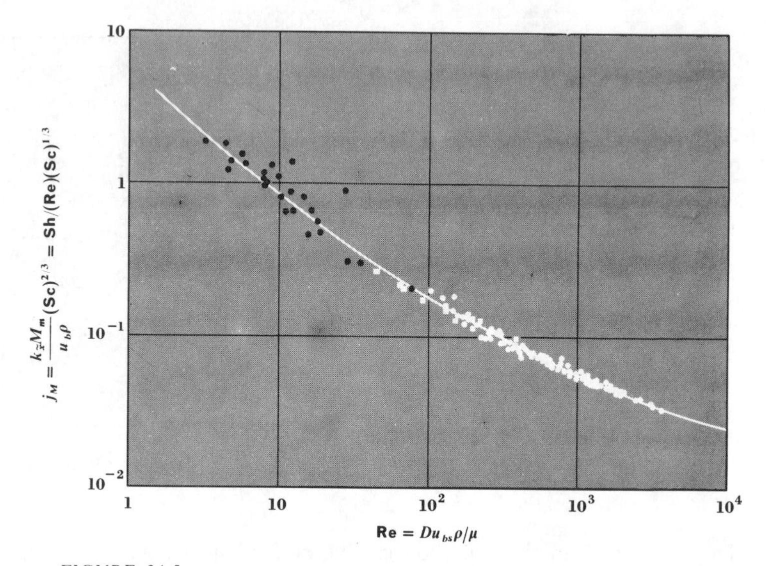

FIGURE 34-5
Average mass-transfer coefficients for single-phase flow in packed beds. [*From M. Hobson and G. Thodos, Chem. Eng. Progr.*, **45**:517 (1949).]

from a high mass-transfer rate at the forward stagnation point to a minimum value at 80° from stagnation where the laminar boundary layer separates confirms similar studies in heat transfer reported in Chap. 24. Mean transfer coefficients were found to be well-correlated by the mass-transfer analog of Eq. (24-9).

Mass transfer from beds of spheres and other solids is important in the analysis of the performance of beds of catalyst particles. Most of the investigations in this field have been with beds of porous solids in which water is evaporated into an air stream. Hobson and Thodos[1] have extended the correlations of previous workers by measuring mass-transfer rates from packed beds to liquid streams. Figure 34-5 shows their results, with the line recommended as fitting the data. The correlation applies to cylinders as well as spheres; the equivalent diameter of a cylinder is taken as $(D_c x_e + D_c{}^2/2)^{1/2}$, in which D_c is the actual diameter of the cylinder and x_e is the length. The velocity term in **Re** is the superficial velocity.

An alternative method of estimating mass-transfer coefficients for a packed bed is to use the procedure suggested by Ranz, which is described in detail for heat transfer in Chap. 24. The technique is to multiply the superficial fluid velocity in the bed by 10.73 and to use this figure for the velocity in **Re** in Eq. (34-13). The value obtained for the mass-transfer coefficient as a result of this procedure is about 10

[1] M.. Hobson and G. Thodos, *Chem. Eng. Progr.*, **45**:517 (1949).

percent higher than the average mass-transfer coefficient for a bed of randomly packed, nearly spherical particles. The data used by Ranz to test this hypothesis are those of Gamson, Thodos, and Hougen, which are shown in Fig. 34-5. Since these data are represented by the correlations of both Ranz and Hobson and Thodos, it is obvious that the two methods are equivalent over the range tested. Additional experimental results of Hobson and Thodos in Fig. 34-5, however, cover a much wider range of **Re** than has been tested by Ranz.

Mass transfer in both fixed and fluidized beds has been studied by Bradshaw.[1] Experimental measurements were made of the rates of evaporation of moisture from beds of porous spheres and pellets. The results for fixed beds check closely those of Hobson and Thodos and other investigators. The results for fixed beds extend as low as $\mathbf{Re}_p = 400$ and for fluidized beds as high as $\mathbf{Re}_p = 12{,}000$. The mass-transfer coefficients for both fixed and fluidized beds were correlated for the entire range of Reynolds number by the equation

$$j_M = 2.25(\mathbf{Re}_p)^{-0.50} \qquad (34\text{-}14)$$

The standard deviation of this correlation was 23 percent for the five different kinds of particles used. A similar equation was found to correlate heat-transfer coefficients in fixed and fluidized beds. The Reynolds number in Eq. (34-14) is $Du_{bs}\rho/\mu(1-\epsilon)$.

It is likely that some of the scatter in mass-transfer correlations is the result of attempts to make them too nearly all-inclusive. Karabelas et al.[2] tabulate the correlations presented in 24 papers and claim that greater success in correlation would have been obtained if separate correlations had been proposed for the various flow regimes defined by extreme values of such dimensionless groups as **Gr**, **Re**, **Sc**, etc.

Mass Transfer in Spheres, Drops, and Bubbles

A number of operations involve mass transfer between a continuous fluid and a discontinuous phase. The discontinuous phase may be solid spheres or other particles which are being dried; it may consist of liquid drops from which a solute is being extracted, or it may consist of bubbles in which transfer of a less volatile substance is occurring to the vapor-liquid interface, with simultaneous transfer of a more volatile component back into the gaseous phase. In the preceding section of this chapter we have considered mass transfer in a continuous phase. We shall now examine mass transfer in the discontinuous phase.

If the discontinuous phase consists of solid material or of stagnant gas or liquid, the only mechanism of transport within this phase is molecular diffusion. The

[1] Robert D. Bradshaw, *AIChE J.*, **9**:590 (1963).

[2] A. J. Karabelas, T. W. Wegner, and T. J. Hanratty, *Chem. Eng. Sci.*, **26**:1581 (1971). See also G. A. Hughmark, *Ind. Eng. Chem. Fundam.*, **19**:198 (1980).

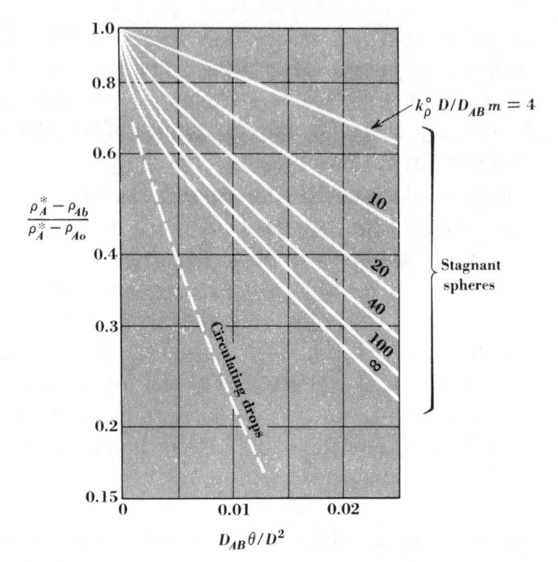

FIGURE 34-6
Bulk concentration change in stagnant and circulating drops.

partial differential equations for unsteady-state diffusion may be written in spherical coordinates and solved if the discontinuous phase has spherical symmetry. The solution for the concentration of diffusing component at any radial position as a function of time may then be integrated from the center to the surface to obtain the bulk, or mixing-cup, concentration as a function of time. A solution to this problem is shown in Fig. 34-6 for a sphere of uniform initial concentration ρ_{Ao} immersed in a continuous phase of equivalent bulk concentration ρ_A^*, where $\rho_A^* = m\rho_A{}^C$, m is the distribution coefficient, and $\rho_A{}^C$ is the concentration in the continuous phase. The bulk concentration ρ_{Ab} in the sphere is shown as a function of molecular diffusivity inside the sphere D_{AB}, sphere diameter D, and time of contact θ.

The solution for the rigid sphere referred to above is found in a number of advanced textbooks on heat transfer and diffusion.[1] The family of curves in Fig. 34-6 is applicable both to systems in which the interfacial concentration is constant and to systems in which the interfacial concentration varies because of a resistance to mass transfer in the continuous phase. The resistance to mass transfer in the continuous phase determines the value of the parameter $k_\rho{}^o D/D_{AB}\,m$, a Biot number. The mass-transfer coefficient $k_\rho{}^o$ for the continuous phase can be estimated from Eq. (34-13). In this equation one uses the diffusivity D_{AC}, the molecular diffusivity of

[1] J. Crank, "The Mathematics of Diffusion," p. 91, Clarendon Press, Oxford, 1956.

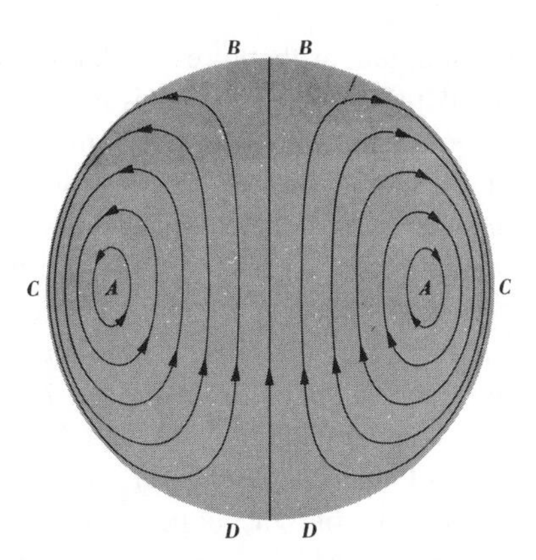

FIGURE 34-7
Circulating pattern for a rising drop or bubble.

the solute with respect to the second constituent in the continuous phase, where D is the sphere diameter; $k_\rho{}^o D/D_{AC}$ is a Sherwood group.

If the discontinuous phase is a fairly nonviscous liquid and if the sphere is a liquid drop greater than 1 or 2 mm in diameter, it is likely that the drop will not be stagnant. It has been shown by a number of investigators that viscous circulation often occurs within spherical drops in the manner shown in Fig. 34-7. The liquid is stagnant only along a circular ring which cuts the horizontal plane of the drop shown here at points A. On the outer surface of the drop there is circulation from B to C to D.

An analysis of mass transfer in circulating drops has been made by Kronig and Brink,[1] who assumed that the streamlines shown in Fig. 34-7 are lines of constant concentration and that unsteady-state molecular diffusion occurs normal to the streamlines. Their solution is shown graphically as the dashed line in Fig. 34-6. Unlike the solution for the stagnant drops shown on the same figure, the solution for the circulating drops applies only for negligible resistance to mass transfer in the continuous phase. In this way it corresponds to the bottom line of the family of curves for stagnant drops.

Other possible mechanisms of mass transfer within drops include natural convection due to concentration gradients. This mechanism might be superimposed on the circulation mechanism described above. Circulation is also affected if dirt or other foreign material collects at the interface between the drop and the bulk phase. A final possibility is that a state of turbulent mixing may exist within the drop so that its concentration may be nearly uniform at any instant. This would be equivalent to assuming negligible resistance to mass transfer within the drop. Such

[1] R. Kronig and J. C. Brink, *Appl. Sci. Res.*, **A2**:142 (1950). See also R. Streicher and K. Schügerl, *Chem. Eng. Sci.*, **32**:23 (1977).

a condition probably exists for most large gas bubbles. In these circumstances nearly the entire resistance is in the continuous phase.

A condition of complete mixing may be approached for liquid drops which rise only a short distance. During formation and immediately afterward, the drops may undergo considerable oscillation, which causes mixing. For a short time of rise, this oscillation may persist for the entire period of immersion, with the result that the dispersed phase will offer a negligible resistance to mass transfer. Under these circumstances the outside resistance will also be decreased from the value predicted for a perfect sphere from Eq. (34-13).

The contacting of two fluid phases in the manner described above is done in vessels known as *spray towers*, which are little more than empty shells. An extensive discussion of the operation of this type of equipment is given by Treybal.[1] Performance data are also presented by Treybal for predicting flow capacities and extraction rates. The extraction-rate data are of limited value because of uncertainty regarding the size of drops produced. The effect of drop size on mass-transfer rate is very great, as can be seen in Fig. 34-6.

PROBLEMS

34-1 A cylindrical tube with an ID of 1 in is cast from benzoic acid. Water at 57°F flows through the tube at a bulk velocity of 0.1 ft/s. How long must the tube be if the solution leaving the tube is to be 1 percent saturated with benzoic acid? What is the mean mass-transfer coefficient for this system based on an arithmetic average driving force? The saturation concentration of benzoic acid in water at 57°F is 2.36 g/L, and the Schmidt number is 1870.

34-2 Repeat Prob. 34-1 for a bulk velocity of 3 m/s using SI units.

34-3 A stream of air at 25°C containing 5 mol percent ethanol is passed upward at the rate of 400 lb/(h)(ft^2) through a tower packed with 1-in Raschig rings. The ethanol is to be absorbed by a countercurrent stream of water at 25°C flowing at a rate of 1000 lb/(h)(ft^2). Find H_L, H_G, H_{OL}, and H_{OG}. How would you account for the relative magnitude of the gas- and the liquid-phase resistances?

34-4 A naphthalene sphere with a diameter of $\frac{1}{2}$ in is placed in a closed room at 50°F. Find the length of time required for the sphere to disappear completely, assuming the air in the room to be an infinite, stagnant medium. The unsteady-state process of evaporation is so slow that sublimation from the surface of the sphere at any instant can be treated as though it were occurring from a sphere of constant diameter. The physical properties of air and naphthalene can be found in Example 34-2.

34-5 The naphthalene sphere of Prob. 34-4 is suspended in an air stream at 10°C moving with a velocity of 9 m/s. Find the length of time required for the sphere to sublime

[1] R. E. Treybal, "Liquid Extraction," 2d ed., chap. 11, McGraw-Hill Book Company, New York, 1963.

completely, assuming it to remain spherical. Comment on the shape the particle will actually acquire.

34-6 Acetic acid is to be removed from an aqueous solution by allowing droplets of acetic acid–water solution to fall through a column of benzene. The initial concentration of acetic acid is 0.01 weight percent, and this is to be reduced to 0.005 weight percent. Find the distance the drops must fall if the drops have a diameter of 5 mm and are assumed to be (*a*) stagnant, (*b*) circulating, (*c*) completely mixed. Both fluid phases are at 25°C. At this temperature the diffusivity of acetic acid in water is 0.88×10^{-5} cm^2/s, and the diffusivity of acetic acid in benzene is 1.92×10^{-5} cm^2/s. The drops may be assumed to achieve their terminal velocity very quickly. This velocity can be estimated using a drag coefficient taken from Fig. 14-6. For part (*b*) it is necessary to neglect the resistance to mass transfer in the continuous phase. The distribution coefficient m is 43.5, concentration of acetic acid in water divided by that in benzene.

34-7 A solute A in a dilute mixture with air is to be absorbed in water in a tower packed with 1-in Raschig rings. The liquid flow will be such that L is greater than 500 lb/(h)(ft²). Flow rates will be chosen so that $\tilde{L}/\tilde{G}$ is 3.0 in all cases.

At 50°C Eqs. (34-11) and (34-12) give $H_G = 0.35$ m and $H_L = 0.35$ m; m is 2.0. At 30°C, m is 1.4. Estimate H_{OG} at 50 and 30°C.

34-8 Substance A is converted to P (isomerization) at 100°C only in the presence of a catalyst. The catalyst is in the form of 3-mm diameter nonporous spheres, and it is contained in a 2-cm ID metal tube. The reaction occurs on the outside surface of the spheres so fast that the surface concentration of A is zero. Thus the apparent rate of reaction is completely controlled by the rate of mass transfer of A to the surface of the catalyst.

Pure A (molecular weight = 50) is forced at 1 atm and 100°C through the packed bed of catalyst spheres at a velocity which gives a Reynolds number $\mathbf{Re}_p$ of 100. What length of catalyst bed is required to obtain a 99 percent conversion of A to P? For the gas (both A and P) the Schmidt number is 5 and the viscosity is 0.02 cP. The spheres form a bed which has a void fraction of 0.4.

34-9 Pure air at 25°C and 1 atm is fed through a tube made of naphthalene of 1-cm ID and 1.2 m long. The flow rate is 1.4 m/s. Find j_M, **Sh**, and the percent saturation with naphthalene of the air leaving the tube.

35

CONTINUOUS CONTACTING OF IMMISCIBLE PHASES

Mass-transfer coefficients and transfer units have been defined and applied to a number of flow situations in Chaps. 31 to 34. The material in these chapters is analogous to that in Chaps. 21 to 24 on convective heat-transfer coefficients. Chapters 25 and 26 have no counterpart in mass transfer, but Chap. 27, which discusses the integration of the heat-transfer-rate equation over the area of a heat exchanger, has many points of similarity to this chapter. We shall discuss here the integration of the mass-transfer-rate equation over the area of interphase contact in equipment such as a gas-absorption tower or a liquid-liquid-extraction tower.

In the study of convective heat transfer we considered the flow of heat from one fluid to another through a conducting wall. The wall prevented direct contact between the fluids. In mass transfer there is no wall and the fluids are in direct contact. However, if the fluids are immiscible, the mass-transfer area is found by methods similar to the methods used for finding the heat-transfer area as shown in Chap. 27. In the present chapter we study mass transfer between immiscible fluids for simple cases for which heat-transfer effects are not important. In Chap. 36 we shall study the simultaneous transfer of heat and mass between immiscible phases.

The study of the application of the principles of mass transfer to the separation operations is complicated in comparison with heat transfer by two additional

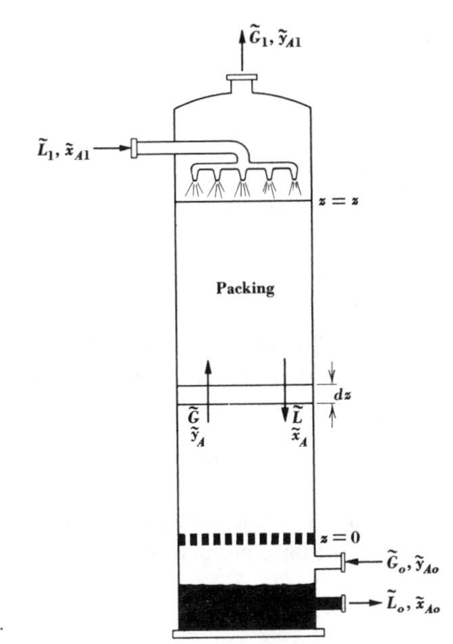

FIGURE 35-1
Packed gas-absorption tower.

factors. First, the interphase contacting is often performed by stages, rather than continuously, as in heat exchangers. Second, the phases in direct contact may be partially miscible, and this situation causes complications in the solutions of the mass balances for the ternary or multicomponent systems which must be considered. Operations involving these additional factors are studied in Chaps. 37 *et seq.*

Integrated Equation for a Tower—Concentrated Solutions

In Chap. 31 we found the equations which represent the mass balance and the rate of transfer between phases for the differential element shown in Fig. 35-1. These equations were then integrated with the assumption of dilute solutions, and we found that the average driving force over a tower is the logarithmic mean, as it is for the analogous heat-transfer problem. We now consider the more general case of concentrated solutions. Here the use of overall coefficients is not appropriate, so we use the gas-phase coefficient and replace the rate expression Eq. (31-32) by

$$\tilde{N}_A \, dA_i = k_{\tilde{y}}a(\tilde{y}_A - \tilde{y}_{As})A \, dz \qquad (35\text{-}1)$$

The coefficient $k_{\tilde{y}}a$ occurs in this relation; it is a function of composition according to

$$k_{\tilde{y}}a = \frac{k_{\tilde{y}}{}^{o}a}{(1 - \tilde{y}_A)_{lm}} \tag{35-2}$$

which follows from Eq. (31-26). The quantity $k_{\tilde{y}}{}^{o}a$ is a function of L, G, **Sc**, etc., as discussed in the previous chapter. The mass balance on the differential element becomes

$$-\tilde{N}_A\, dA_i = Ad(\tilde{G}\tilde{y}_A) = Ad(\tilde{L}\tilde{x}_A) \tag{35-3}$$

where $\tilde{G}$ and $\tilde{L}$ are now variables because of the appreciable amounts of A which may be transferred. This equation can be integrated over the tower between the concentration limits shown in Fig. 35-1 to yield

$$(\tilde{N}_A)_{\text{av}}\, A_i = A(\tilde{G}_o\, \tilde{y}_{Ao} - \tilde{G}_1 \tilde{y}_{A1}) = A(\tilde{L}_o\, \tilde{x}_{Ao} - \tilde{L}_1 \tilde{x}_{A1}) \tag{35-4}$$

Alternatively, the mass balances can be integrated from the bottom of the tower to any intermediate point where the concentrations are $\tilde{x}_A$ and $\tilde{y}_A$.

$$\tilde{G}_o\, \tilde{y}_{Ao} - \tilde{G}\tilde{y}_A = \tilde{L}_o\, \tilde{x}_{Ao} - \tilde{L}\tilde{x}_A \tag{35-5}$$

or

$$\tilde{y}_A = \frac{\tilde{L}}{\tilde{G}}\, \tilde{x}_A + \frac{1}{\tilde{G}}\, (\tilde{G}_o \tilde{y}_{Ao} - \tilde{L}_o\, \tilde{x}_{Ao}) \tag{35-6}$$

The concentrations of contiguous phases in a tower are given by Eqs. (35-5) and (35-6). The line representing Eq. (35-6) on a graph of $\tilde{y}_A$ versus $\tilde{x}_A$ is an operating line. With the units used here, the operating line is in general curved; $\tilde{G}$ and $\tilde{L}$ are functions of $\tilde{x}_A$. However, for dilute solutions, the operating line becomes straight, as already discussed.

Equations (35-1) and (35-3) are now combined to give

$$-dz = \frac{d(\tilde{G}\tilde{y}_A)}{k_{\tilde{y}}a(\tilde{y}_A - \tilde{y}_{As})} \tag{35-7}$$

Since the phases are immiscible, the molar mass velocity of inert fluid in the z direction, $\tilde{G}_B$, is a constant. Thus an equation for $\tilde{G}$ is

$$\tilde{G} = \frac{\tilde{G}_B}{1 - \tilde{y}_A} \tag{35-8}$$

This expression for $\tilde{G}$ is substituted in Eq. (35-7), and the indicated differential is taken. We obtain, after some algebra,

$$-dz = \frac{\tilde{G}\, d\tilde{y}_A}{k_{\tilde{y}}a(1 - \tilde{y}_A)(\tilde{y}_A - \tilde{y}_{As})} \tag{35-9}$$

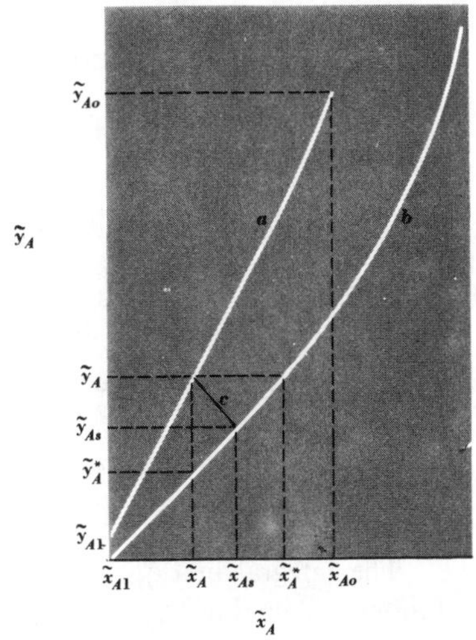

FIGURE 35-2
Graph of compositions in tower of Fig. 35-1.
a Operating line.
b Equilibrium line.
c Slope $= -\dfrac{H_G \tilde{L}(1 - \tilde{y}_A)_{lm}}{H_L \tilde{G}(1 - \tilde{x}_A)_{lm}}$.

The right-hand side of Eq. (35-9) appears difficult to integrate, for $\tilde{G}$ varies, and $k_{\tilde{y}}a$ is a function of $\tilde{G}$ and $(1 - \tilde{y}_A)_{lm}$. However, the coefficient $k_{\tilde{y}}{}^o a$ is by definition independent of $(1 - \tilde{y}_A)_{lm}$ [Eq. (35-2)]. We therefore write Eq. (35-9) as

$$-dz = \frac{\tilde{G}}{k_{\tilde{y}}{}^o a} \frac{(1 - \tilde{y}_A)_{lm}\, d\tilde{y}_A}{(1 - \tilde{y}_A)(\tilde{y}_A - \tilde{y}_{As})} \qquad (35\text{-}10)$$

We recognize the group $\tilde{G}/k_{\tilde{y}}{}^o a$ as the height of the transfer unit H_G [Eq. (31-49)]. Since H_G is roughly proportional to $\tilde{G}^{0.3}$ (Table 34-1), a 20 percent change in $\tilde{G}$ produces a change of only about 6 percent in H_G. Since $\tilde{G}$ rarely changes by more than 20 percent, it is permissible to neglect the variation of H_G over a tower. The limits of integration can be deduced from Figs. 35-1 and 35-2, so that Eq. (35-10) becomes

$$z = H_G \int_{\tilde{y}_{A1}}^{\tilde{y}_{Ao}} \frac{(1 - \tilde{y}_A)_{lm}\, d\tilde{y}_A}{(1 - \tilde{y}_A)(\tilde{y}_A - \tilde{y}_{As})} \qquad (35\text{-}11)$$

The number of y-phase transfer units is then

$$n_G = \int_{\tilde{y}_{A1}}^{\tilde{y}_{Ao}} \frac{(1 - \tilde{y}_A)_{lm}\, d\tilde{y}_A}{(1 - \tilde{y}_A)(\tilde{y}_A - \tilde{y}_{As})} \qquad (35\text{-}12)$$

which is found from Eq. (35-11) and the definition of n_G:

$$z = H_G n_G \qquad (31\text{-}45)$$

Given the precision of most mass-transfer data, it is permissible to replace the logarithmic mean mole fraction of inert fluid by the arithmetic mean:

$$(1 - \tilde{y}_A)_{lm} = \frac{(1 - \tilde{y}_A) + (1 - \tilde{y}_{As})}{2} \tag{35-13}$$

When this equation is substituted in Eq. (35-12), there results

$$n_G = \frac{1}{2}\int_{\tilde{y}_{A1}}^{\tilde{y}_{Ao}} \frac{d\tilde{y}_A}{\tilde{y}_A - \tilde{y}_{As}} + \frac{1}{2}\int_{\tilde{y}_{A1}}^{\tilde{y}_{Ao}} \frac{(1 - \tilde{y}_{As})\, d\tilde{y}_A}{(1 - \tilde{y}_A)(\tilde{y}_A - \tilde{y}_{As})} \tag{35-14}$$

The second integral can be evaluated using the method of partial fractions, and Eq. (35-14) reduces to

$$n_G = \int_{\tilde{y}_{A1}}^{\tilde{y}_{Ao}} \frac{d\tilde{y}_A}{\tilde{y}_A - \tilde{y}_{As}} + \frac{1}{2}\ln\frac{1 - \tilde{y}_{A1}}{1 - \tilde{y}_{Ao}} \tag{35-15}$$

The integral in Eq. (35-15) must be evaluated graphically for concentrated solutions. Values of $1/(\tilde{y}_A - \tilde{y}_{As})$ are obtained as a function of $\tilde{y}_A$ from a graph such as Fig. 35-2. At a number of values of $\tilde{y}_A$ between $\tilde{y}_{Ao}$ and $\tilde{y}_{A1}$, $\tilde{y}_{As}$ is determined from Eq. (31-55) by drawing a line through $\tilde{x}_A$, $\tilde{y}_A$ (i.e., a point on the operating line), with a slope $-H_G \tilde{L}(1 - \tilde{y}_A)_{lm}/H_L \tilde{G}(1 - \tilde{x}_A)_{lm}$. This slope may vary considerably through the tower. It is thus evident that n_G can be determined only if *both* H_G and H_L are known.

The above analysis can also be done on the basis of the liquid, or x-phase, resistance. The resulting equations are

$$z = H_L n_L \tag{35-16}$$

and

$$n_L = \int_{\tilde{x}_{A1}}^{\tilde{x}_{Ao}} \frac{d\tilde{x}_A}{\tilde{x}_{As} - \tilde{x}_A} - \frac{1}{2}\ln\frac{1 - \tilde{x}_{A1}}{1 - \tilde{x}_{Ao}} \tag{35-17}$$

Overall coefficients or transfer units should not be used for concentrated solutions unless no data on the individual resistances are available.

The equations given above apply to all transfer operations for which $\tilde{N}_B$ is zero. The proper forms to use with compositions expressed in mass fractions, concentrations, or partial pressures can be derived by the reader.

If we have $\tilde{N}_A = -\tilde{N}_B$, as is usually assumed in binary distillation, the second terms in Eqs. (35-15) and (35-17) are omitted. In addition, Eq. (31-55) does not hold, but rather $k_{\tilde{x}} = k_{\tilde{x}}^o$.

Simplification for Dilute Solutions

General remarks We have seen that the only rigorously correct way to treat the countercurrent contacting of concentrated solutions is by the use of the individual coefficients or HTUs. The interfacial compositions must be computed and the

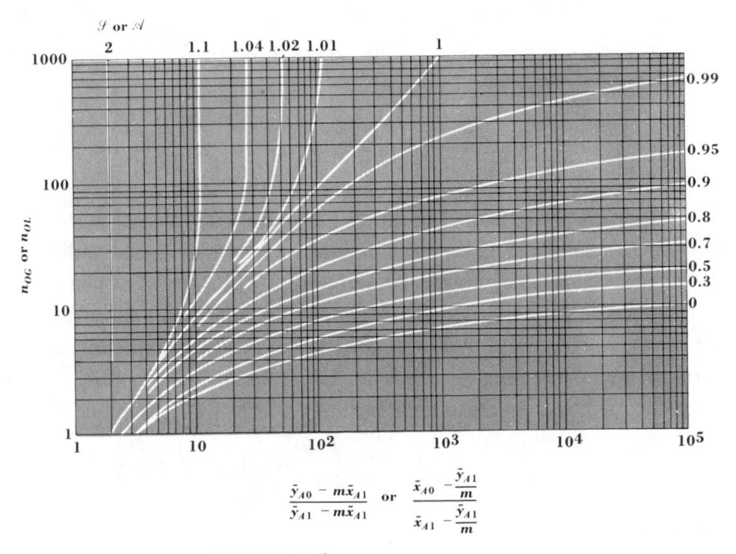

FIGURE 35-3
Graph for NTUs with straight operating and equilibrium lines.

NTUs found from a graphical integration. The operating and equilibrium lines are generally curved, but if the solutions are dilute, both lines are approximately straight. The distinction between dilute and concentrated solutions is arbitrary, for the curvature of the equilibrium line depends on the particular system involved. In any event, the curvature of the operating line becomes appreciable as mole fractions rise above 0.05. Thus, by dilute solutions, we mean situations for which $\tilde{L}$, $\tilde{G}$, and m can be considered constant.

Since the quantities $1 - \tilde{y}_{A1}$ and $1 - \tilde{y}_{Ao}$ are both approximately unity, Eq. (35-15) can be adapted to give the number of overall gas-transfer units as

$$n_{OG} = \int_{\tilde{y}_{A1}}^{\tilde{y}_{Ao}} \frac{d\tilde{y}_A}{\tilde{y}_A - \tilde{y}_A^*} \tag{35-18}$$

In a similar way, we obtain from Eq. (35-17)

$$n_{OL} = \int_{\tilde{x}_{A1}}^{\tilde{x}_{Ao}} \frac{d\tilde{x}_A}{\tilde{x}_A^* - \tilde{x}_A} \tag{35-19}$$

Logarithmic mean driving force In Chap. 31 we found that the logarithmic mean driving force is appropriate for dilute solutions. From the definitions of H_{OG} and n_{OG} it is clear that Eq. (31-36) is equivalent to Eq. (35-18) above. We find that, since $\tilde{y}_A - \tilde{y}_A^*$ is a linear function of $\tilde{y}_A$, the NTUs can be expressed as

$$n_{OG} = \frac{\tilde{y}_{Ao} - \tilde{y}_{A1}}{(\tilde{y}_A - \tilde{y}_A^*)_{lm}} \tag{35-20}$$

A similar analysis gives

$$n_{OL} = \frac{\tilde{x}_{Ao} - \tilde{x}_{A1}}{(\tilde{x}_A^* - \tilde{x}_A)_{lm}} \tag{35-21}$$

Colburn's equation for the NTUs An alternative method of integrating Eq. (35-18) has been given by Colburn.[1] For a straight equilibrium line which passes through the origin, $\tilde{y}_A^*$ is equal to $m\tilde{x}_A$. In turn, $\tilde{x}_A$ can be expressed in terms of $\tilde{y}_A$ by Eq. (35-6), which is linear for a straight operating line; that is, $\tilde{L}/\tilde{G} = \text{const.}$ By these two substitutions Eq. (35-18) is rearranged so that the only variable under the integral sign is $\tilde{y}_A$. The details of the integration are not given here; the resulting equation for the number of overall gas-transfer units is

$$n_{OG} = \frac{\ln \{(1 - \mathcal{S})[(\tilde{y}_{Ao} - m\tilde{x}_{A1})/(\tilde{y}_{A1} - m\tilde{x}_{A1})] + \mathcal{S}\}}{1 - \mathcal{S}} \tag{35-22}$$

in which

$$\mathcal{S} = \frac{m\tilde{G}}{\tilde{L}} = \frac{1}{\mathcal{A}} \tag{35-23}$$

The quantity $\mathcal{S}$ is called the *stripping factor*, and $\mathcal{A}$ is the *absorption factor*. For most separations $\mathcal{A}$ has a value near unity; for simple gas absorption an economic optimal value is about 1.4.

A similar equation can be derived for n_{OL}. The result is

$$n_{OL} = \frac{\ln \left\{(1 - \mathcal{A})\left[\left(\tilde{x}_{Ao} - \frac{\tilde{y}_{A1}}{m}\right)\Big/\left(\tilde{x}_{A1} - \frac{\tilde{y}_{A1}}{m}\right)\right] + \mathcal{A}\right\}}{1 - \mathcal{A}} \tag{35-24}$$

A graph for doing calculations by Eqs. (35-22) and (35-24) is given in Fig. 35-3. The equation for n_{OG} is to be used when the operating line lies above the equilibrium line, as in gas absorption. Here $\mathcal{S}$ is usually less than one; if it is greater than one, the possible separation that can be obtained is limited, no matter how many transfer units there are, as shown by the vertical asymptotes in Fig. 35-3. The equation for n_{OL} is to be used when the operating line is below the equilibrium line, as in gas desorption (stripping) or liquid extraction. Here $\mathcal{A}$ is usually less than 1 for an economic process.

There is no basic reason not to use n_{OL} for absorption or n_{OG} for stripping, and indeed this can be done if the log mean formulas are used (dilute solution) or a graphical solution is done (general case). However, the particular form of Eqs.

[1] A. P. Colburn, *Trans. Am. Inst. Chem. Eng.*, **35**:211 (1939).

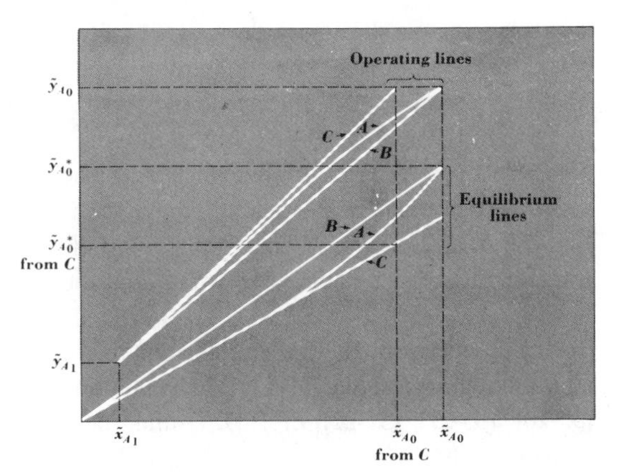

FIGURE 35-4
Errors arising from the assumption of dilute solutions.

(35-22) and (35-24) limit their use as described above. Many other forms are possible, some of which do permit the calculation of n_{OL} for gas absorption, for example.[1, 2]

Applicability of the simplified equations It may sometimes be expedient to use Eq. (35-20) or (35-22) when the operating and equilibrium lines are actually slightly curved, and it is useful to have an idea of the effect of the simplifying assumptions on the accuracy of the result. Figure 35-4 shows the operating and equilibrium lines which might apply to a calculation of n_{OG} from given values of $\tilde{y}_{A1}$, $\tilde{y}_{Ao}$, $\tilde{x}_{A1}$, and $m\tilde{G}/\tilde{L}$. The ratio of the slopes of the equilibrium and operating lines is taken as the value at the dilute (1) end of the tower. These values of $\tilde{L}$ and $\tilde{G}$ are, to a good approximation, the mass velocities of the solute-free streams, and m is based on the Henry's law constant at extreme dilution.

The data given can be used to calculate $\tilde{x}_{Ao}$ by Eq. (35-4), so that the ends of the operating line are fixed. Since the slope of the line is known at its dilute end, the actual operating line A can be sketched with little error with no further computation. If actual solubility data are known, the equilibrium line A in Fig. 35-4 can also be drawn as shown. The number of overall gas-transfer units, for example, could then be obtained with little error from these lines and an integration of Eq. (35-18).

The use of Eq. (35-20), which contains the log mean driving force, corresponds to the assumption that the ends of the actual curved operating and equilibrium lines A are connected by the straight lines B. For the lines shown in Fig. 35-4, Eq. (35-20) would give slightly more than the correct number of overall y-phase transfer units,

[1] Stanley Hartland, "Counter-Current Extraction: An Introduction to the Design and Operation of Counter-Current Extractors," Pergamon Press, New York, 1970.

[2] All previous equations for NTUs, Eqs. (35-18) to (35-21), apply as written for the operating line *above* the equilibrium line.

since the average driving force $\tilde{y}_A - \tilde{y}_A^*$ obtained from the lines B is less than the correct value obtained from the lines A.

The use of Eq. (35-22) corresponds to the assumption that a straight operating line passes through $\tilde{x}_{A1}$, $\tilde{y}_{A1}$ with a slope $\tilde{L}_1/\tilde{G}_1$. The straight equilibrium line passes through the origin with a slope m. These lines are labeled C in Fig. 35-4, and it can be seen that the average driving force $\tilde{y}_A - \tilde{y}_A^*$ obtained from them is greater than the actual average driving force. Thus the value of n_{OG} calculated in this way is less than the correct value. Note also that the value of $\tilde{x}_{Ao}$ found from the lines C is in error.

The curvature of the equilibrium line can have the opposite sign from the curvature shown in Fig. 35-4. The evaluation of the error made by the use of Eqs. (35-20), (35-21), (35-22), or (35-24) must be based on the behavior of the specific system involved.

Modes of Operation

Generality of equations It has been mentioned several times that the equations of this chapter and of Chap. 31 can be applied to a number of separation operations. The specific example of gas absorption has been used, so that our explanation would not be too abstract. However, the symbols G and L and y_A and x_A can apply to two liquid phases or to a fluid and a solid phase as well as to the gas and liquid phases in gas absorption or distillation. The subscript o always refers to the rich end of the countercurrent contacting device, and the subscript 1 to the dilute end. This convention means that $z = 0$ is sometimes at the top of the actual tower. Either mass or molar units can be used.

Gas absorption For gas absorption the operating line on a $\tilde{y}_A$ versus $\tilde{x}_A$ diagram is placed above the equilibrium line; molar units are generally used. The solute A is transferred from a mixture with a gas B into a liquid C; B and C are immiscible. The appropriate driving force is $\tilde{y}_A - \tilde{y}_{As}$, $\tilde{y}_A - \tilde{y}_A^*$, $\tilde{x}_{As} - \tilde{x}_A$, or $\tilde{x}_A^* - \tilde{x}_A$. Figure 35-5 shows a typical $\tilde{y}_A$ versus $\tilde{x}_A$ diagram for dilute solutions. Equation (35-20) shows that the number of transfer units increases as the average distance between the operating and equilibrium lines is decreased. In the most common situation, we desire to reduce the concentration of A in a known quantity of gas to a certain value, using liquid of a known composition. Thus $\tilde{y}_{A1}$, $\tilde{y}_{Ao}$, $\tilde{x}_{A1}$, and $\tilde{G}$ are fixed. The lower end of the operating line is fixed, so that as the amount of absorbing liquid $\tilde{L}$ is varied, the slope of the line is changed, and its upper end moves along the horizontal line $\tilde{y}_A = \tilde{y}_{Ao}$. The minimum value of $\tilde{L}$ is one which causes the driving force to be zero at any point in the tower. For the straight lines shown in Fig. 35-5, the point of touching, or pinch point, occurs at the rich end where

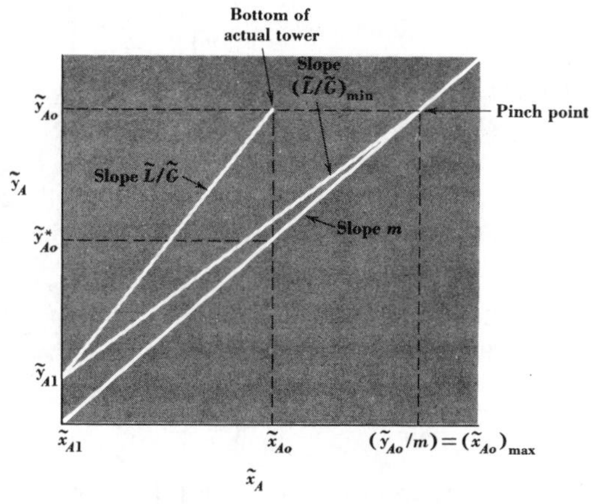

FIGURE 35-5
Graph for gas absorption showing $(\tilde{L}/\tilde{G})_{min}$.

$\tilde{x}_{Ao}$ equals $\tilde{y}_{Ao}/m$. At a pinch point the driving force is zero, and from Eq. (35-18) it is evident that the NTU is infinite.

The value of the slope $\tilde{L}/\tilde{G}$ which causes a pinch point is the minimum possible, and the actual operating conditions are often defined with reference to $(\tilde{L}/\tilde{G})_{min}$. For example, it is common to operate gas-absorption equipment at $\tilde{L}/\tilde{G}$ equal to about $1.5(\tilde{L}/\tilde{G})_{min}$. There is an economic optimum value of $\tilde{L}/\tilde{G}$. Too low a value results in a very high tower, and too large a value requires a large amount of absorbing liquid; this increases the cost of recovering the solute from this liquid, and the diameter of the absorption tower must be increased to accommodate the large liquid flow.

Desorption, or stripping The operation of transferring a volatile solute A from a liquid C to a gas B is called *desorption*, or *stripping*. The gas phase can be regarded as a solvent which removes the solute from the liquid. The rich liquid feed is introduced to the top of a contacting tower, and a gas is blown into the bottom. The rich gas leaves the top of the tower, and the stripped liquid leaves the bottom. Mass is transferred in the opposite direction from that for gas absorption, and the appropriate driving force is $\tilde{y}_{As} - \tilde{y}_A$, $\tilde{y}_A^* - \tilde{y}_A$, $\tilde{x}_A - \tilde{x}_{As}$, or $\tilde{x}_A - \tilde{x}_A^*$.

The operating line is below the equilibrium line, as shown in Fig. 35-6. In the usual case the lower end of the operating line is fixed, and various amounts of gas may be used to treat the liquid feed of concentration $\tilde{x}_{Ao}$. The minimum gas rate corresponds to a maximum slope $\tilde{L}/\tilde{G}$.

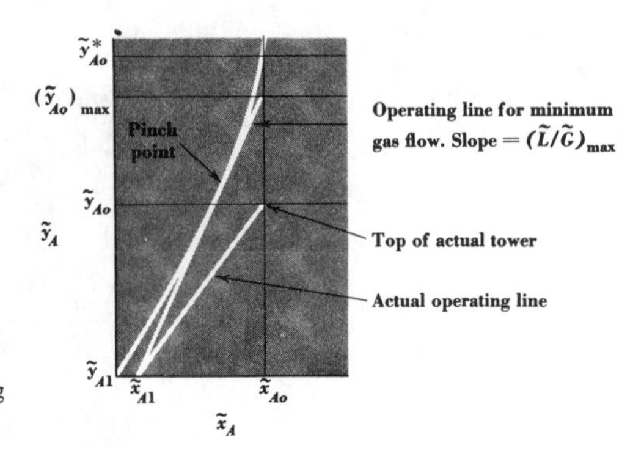

FIGURE 35-6
Graph for gas desorption or stripping showing $(\tilde{L}/\tilde{G})_{max}$.

For gas absorption the rich (o) end of the tower is at the bottom of the actual tower, whereas for desorption, it is at the top. The equations for the number of transfer units can be used with no change except the sign of the appropriate driving force. Figure 35-6 shows that the pinch point does not necessarily occur at one end of a tower if the operating or equilibrium line is not straight.

Liquid-liquid extraction In this operation a liquid solvent B is used to extract a solute A from a second liquid C in which it is dissolved. We consider in this chapter only systems for which liquids B and C are immiscible. Liquid extraction is useful for separating A from C when A and C have similar volatilities, so that separation by distillation is difficult. The solute A is extracted by solvent B, which is chosen so that A and B are easily separable by distillation or some other means. The rich solvent stream is called the extract, and the stripped C stream is called the *raffinate*.

It is seen that extraction is very similar to desorption, or stripping, so we adopt the convention that the solvent is the y or G phase and the raffinate is the x or L phase. It is common to use mass units rather than molar units. As shown in Fig. 35-7, the operating line is below the equilibrium line. The minimum amount of solvent corresponds to the maximum possible slope L/G for the extraction of a given feed with a solvent of given composition.

The equations of Chaps. 31 and 35 given in terms of mole fractions are valid in terms of mass fractions if the superscripts are removed.

$$n_G = \int_{y_{A1}}^{y_{Ao}} \frac{(1 - y_A)_{lm}\, dy_A}{(1 - y_A)(y_{As} - y_A)} \tag{35-25}$$

This equation is to be used with H_G or k_y defined by Eqs. (31-52) and (31-53). An alternative, more cumbersome procedure is to use H_G defined by Eq. (31-49) and find

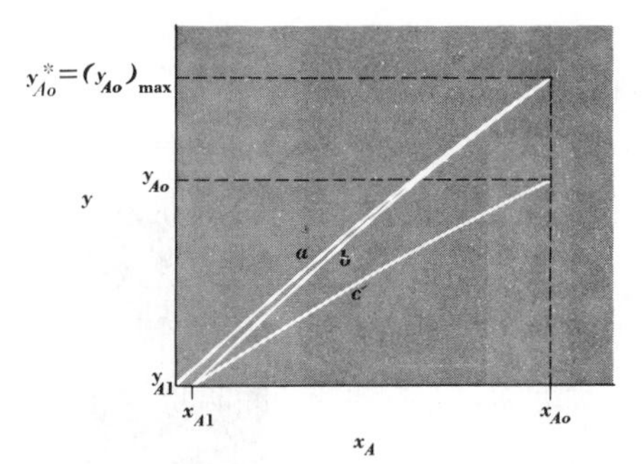

FIGURE 35-7
Graph for liquid-liquid extraction showing conditions for minimum solvent (concentrated solutions).
a Equilibrium line.
b Operating line for minimum solvent.
c Actual operating line.

n_G by replacing the mole fractions in Eq. (35-12) or (35-15) by mass fractions according to Eq. (30-2). Although this procedure is slightly more rigorous, the low precision of available mass-transfer data would probably not justify the use of this approach.

For dilute solutions, Eq. (35-21) becomes

$$n_{OL} = \frac{x_{Ao} - x_{A1}}{(x_A - x_A^*)_{lm}} \qquad (35\text{-}26)$$

The equilibrium line in Fig. 35-7 is sometimes referred to as a distribution line. The constant slope m is called a *distribution coefficient*, because it indicates the equilibrium distribution of A between the liquids B and C.

For liquid-liquid extraction, the rich end of a countercurrent contactor can be either the top or the bottom of the actual tower, depending on whether the extract is lighter or heavier than the raffinate.

Figure 35-8 shows two ways in which a liquid-extraction tower may be operated. The phase which forms the drops is called the *dispersed*, or *discontinuous*, phase; the other phase is the continuous phase. There are no internals in the spray tower shown; for a more efficient contact between the phases, the tower might be filled with packing or fitted with baffles or agitators.

Operation with concurrent flow Although countercurrent flow has many advantages over concurrent flow, the latter may be used in some situations. For instance, the flow in an individual stage of some stagewise contactors may be concurrent. Also, some towers are operated with concurrent flow in order to achieve high velocities and the resultant high mass-transfer coefficients. The position of the operating line for concurrent gas absorption is shown in Fig. 35-9. The method of calculating the tower height is not changed from the methods applied to countercurrent flow.

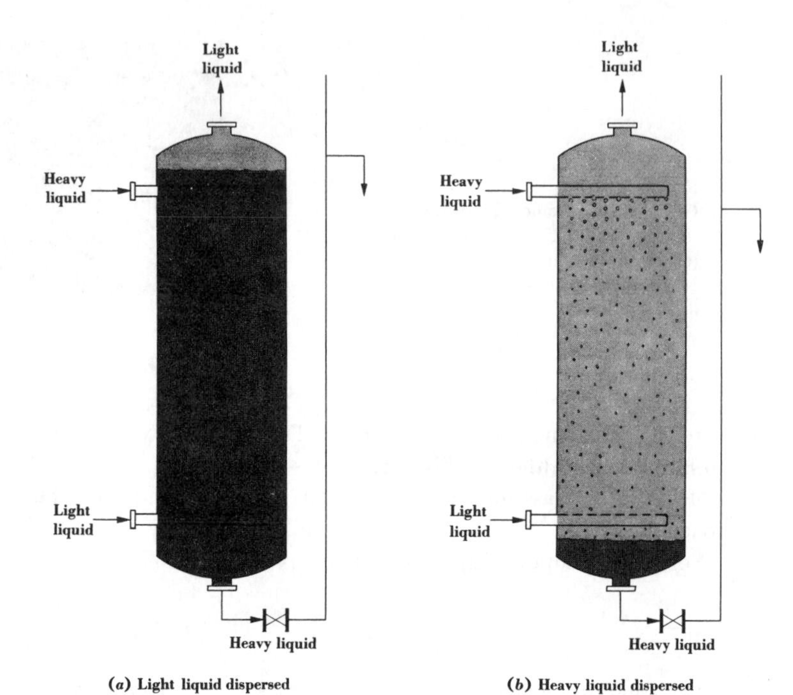

FIGURE 35-8
Alternate ways of operating a spray tower.

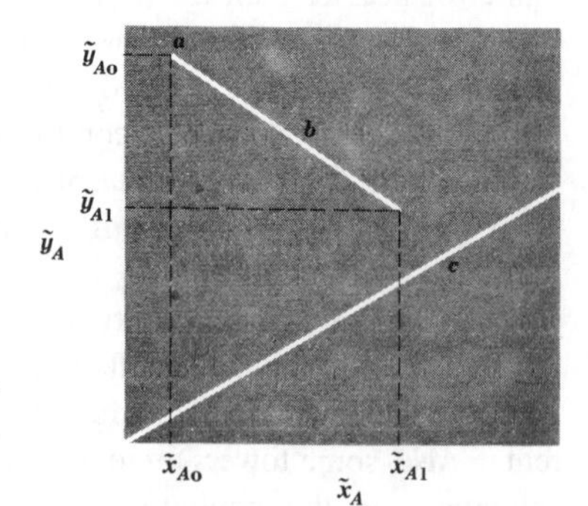

FIGURE 35-9
Gas absorption with concurrent flow.
a Top of actual tower.
b Operating line (slope = $-\tilde{L}/\tilde{G}$).
c Equilibrium line.

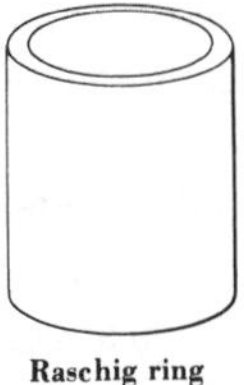

FIGURE 35-10
A few examples of tower packings.

Notes on Equipment Design

Gas-liquid operations The most common kind of continuous contacting device is the packed tower. A few types of packing are shown in Fig. 35-10; the properties of these packings are given in Table 35-1. Packing is designed to promote a large area of contact between phases with a minimum resistance to the flow of the two phases.

We have seen that the height of a packed tower is determined by the rate of mass transfer; the diameter of the tower is set by the flow rate of material to be treated. If the gas and liquid rates to a tower are doubled, the tower cross-sectional area must be doubled in order to keep the design mass velocities the same. The height of the tower and the number of transfer units are unchanged.

In Chap. 14 we studied the flow of a single fluid through a packed bed; we now wish to consider the countercurrent flow of a gas and a liquid. If liquid flows down over the packing, a certain amount is held in the various interstices and cavities of the rings or saddles; there is a minimum liquid holdup, no matter how low the

Table 35-1 TOWER PACKING CHARACTERISTICS

Nominal size, in	Ceramic Raschig rings			Plastic Pall rings			Ceramic Intalox saddles			Ceramic Berl saddles		
	ε	c_f	a_p	ε	c_f	a_p	ε	c_f	a_p	ε	c_f	a_p
$\frac{1}{4}$	0.73	1000	240				0.75	600	300	0.60	900	274
$\frac{1}{2}$	0.63	640	111				0.78	265	190	0.63	380	142
$\frac{3}{4}$	0.73	255	80				0.77	130	102	0.66	170	82
1	0.73	160	58	0.90	52	63	0.78	98	78	0.69	110	70
$1\frac{1}{2}$	0.71	95	38	0.91	32	39	0.81	52	60	0.75	65	44

SOURCE: R. E. Treybal, "Mass Transfer Operations," 3d ed., pp. 196–199, McGraw-Hill Book Company, New York, 1980.
ε = void fraction when dry
c_f = characterization factor for the correlation of Fig. 35-12
a_p = surface area, ft^2/ft^3, when dry
The data are for 16- and 30-in-ID towers filled by dumping the packing into the tower when filled with water. See also, "Packed Distillation Column Design," R. F. Strigle, Jr. and F. Rukovena, *Chem. Eng. Progr.*, **75**(3):86 (1979).

liquid-flow rate is. As L is increased, the holdup is increased. Even if there is no upward gas flow, the rate of liquid flow downward can be increased until the tower is flooded, i.e., completely filled with liquid.

When liquid flows down over the packing at an intermediate rate, so that there is not an excessive holdup, gas can be blown up through the tower. The pressure drop is larger than with unirrigated packing, for the cross section available to flow is reduced by the liquid holdup. As G is increased with a fixed L, the interphase friction increases and the holdup increases. Finally, at a certain value of G, the holdup is so high that the tower starts to fill with liquid. The tower cannot be operated above this flooding velocity, which is a function of the liquid velocity, the fluid properties, and the characteristics of the packing.

A typical plot showing schematically the relation between the pressure drop over a tower and the liquid and gas flow rates is given in Fig. 35-11. In Fig. 35-12 a correlation is given for the flooding velocity in packed columns. The data from Table 35-1 can be used to estimate c_f, which appears in the ordinate of Fig. 35-12, but actual data should be used if possible; the void fraction is very sensitive to the manner in which the packing is dumped in a particular tower. It is customary to fix the diameter of a gas-absorption tower so that the gas rate is 50 to 70 percent of the flooding rate. Extensive data on pressure drop and flooding are given in Perry's Handbook.[1]

Liquid-liquid operations The situation in a packed liquid-extraction tower is basically the same as in a packed tower for gas-liquid contacting. Since the densities of the two phases are not widely different, the limiting velocities in countercurrent liquid-liquid operations are less than those in gas-liquid operations. A correlation for flooding velocities in packed towers is given in Fig. 35-13.

Unpacked spray towers, as shown in Fig. 35-8, are sometimes used for liquid-liquid operations. The flooding velocities can be estimated from the minimum fluidization velocity of the solid particle of the same diameter as the dispersed-phase droplets.[2] The diameter of the droplets is governed by the size of the holes in the distribution nozzles. Correlations giving the pertinent relations are found in Treybal.[3]

Example 35-1 It is desired to absorb 95 percent of the acetone in a 2.00 mol percent mixture of acetone in air in a continuous, countercurrent absorption tower using 20 percent more than the minimum water rate. Pure water is introduced at the top of the tower, and the gas mixture is blown into the

[1] Perry, pp. 18-21 to 18-29. More details are given by R. E. Treybal, "Mass Transfer Operations," 3d ed., McGraw-Hill Book Company, New York, 1980.

[2] R. E. C. Weaver, L. Lapidus, and J. C. Elgin, *AIChE J.*, **5**:533 (1959).

[3] Op. cit.

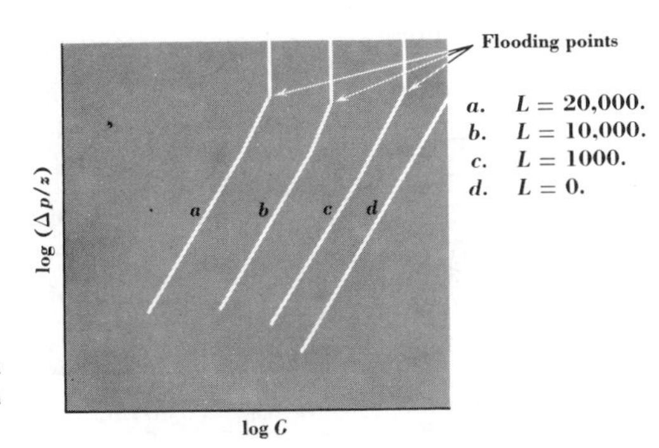

FIGURE 35-11
Example of the effect of fluid velocities on pressure drop in a packed column (countercurrent flow).

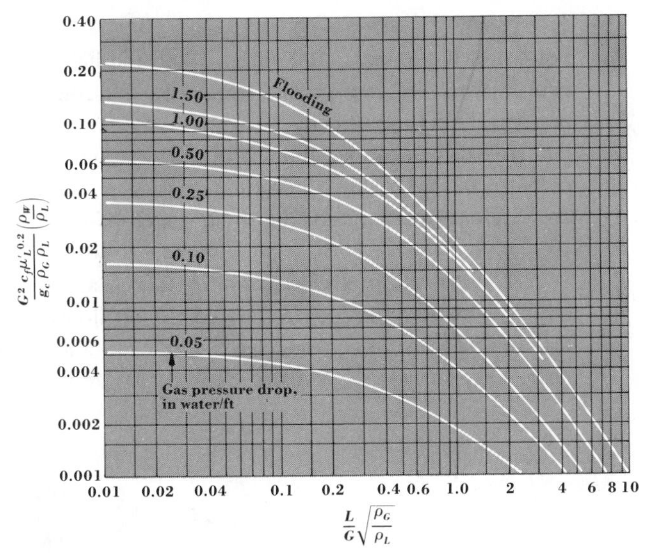

FIGURE 35-12
Flooding velocities for random packings, gas absorption. Units: lb, h, ft, except μ'_L which is expressed in cP. ρ_w refers to water. [*From R. E. Treybal, "Mass Transfer Operations," 2d ed., p. 160, McGraw-Hill Book Company, New York, 1968.*]

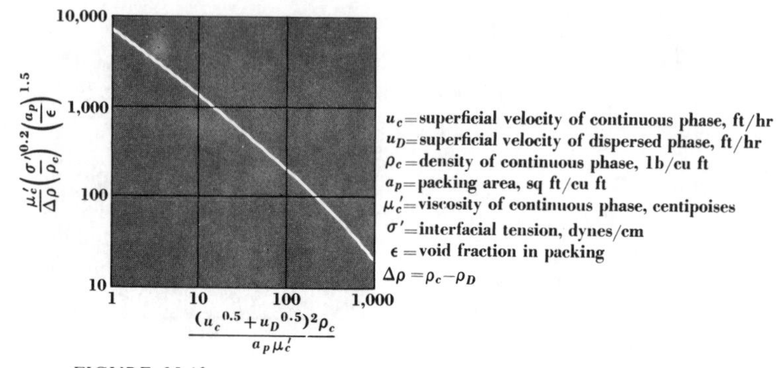

FIGURE 35-13
Flooding velocities in packed extraction towers. [*From J. W. Crawford and C. R. Wilke, Chem. Eng. Progr.*, **47**:*423 (1951)*.]

bottom of the tower at 1000 lb/h. Find the height and diameter of a tower packed with 1-in, wet-packed, stoneware Raschig rings and operated at 40 percent of flooding velocity.

It will be assumed that the tower operates isothermally at atmospheric pressure and 80°F; the equilibrium relation is $\tilde{y}_A = 2.53\tilde{x}_A$. The diffusivities are acetone-air, $D_{AB} = 0.368$ ft^2/h; acetone-water (liquid), $D_{AC} = 4.81 \times 10^{-5}$ ft^2/h.

The streams are labeled as in Fig. 35-1, and the conditions of the problem give $\tilde{y}_{Ao} = 0.02$ and $\tilde{x}_{A1} = 0$. The rate of air flow at both ends of the tower is the same, so that it is convenient to find the exit-air composition by using molar ratios, as follows: $\tilde{Y}_{Ao} = 0.02/0.98 = 0.0204$ mol acetone/mol air. At the

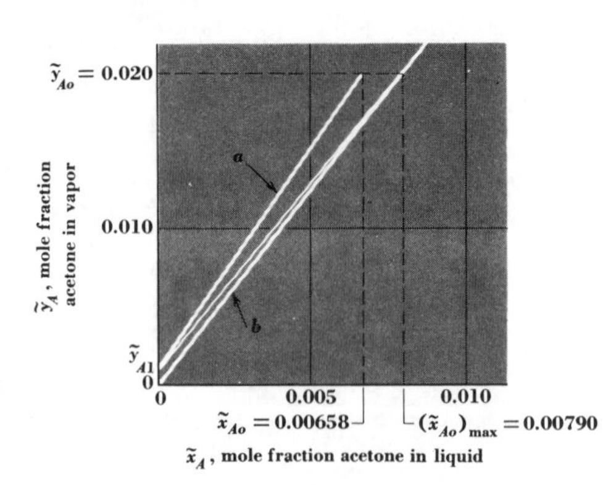

FIGURE 35-14
Operating diagram for Example 35-1.
a Actual operating line.
b Equilibrium line.

dilute end of the tower, for 95 percent absorption, $\tilde{Y}_{A1} = (0.05) \times (0.0204) = 0.00102 \cong y_{A1}$.

Since both the operating and equilibrium lines are almost straight, the minimum water rate corresponds to equilibrium at the rich end of the tower, so that $(\tilde{x}_{Ao})_{\max} = 0.02/2.53 = 0.00790$. We then compute $(\tilde{L}/\tilde{G})_{\min}$.

$$\left(\frac{\tilde{L}}{\tilde{G}}\right)_{\min} = \frac{0.0200 - 0.00102}{0.00790 - 0} = 2.41$$

Figure 35-14 illustrates this procedure. The actual slope of the operating line is then $1.2(2.41) = 2.89$, and $\tilde{x}_{Ao} = 0.00658$.

The number of transfer units can be calculated from Eq. (35-20)

$$(\tilde{y}_A - \tilde{y}_A^*)_o = 0.02 - 0.00658(2.53) = 0.00336$$

$$(\tilde{y}_A - \tilde{y}_A^*)_1 = 0.00102 - 0 = 0.00102$$

$$(\tilde{y}_A - \tilde{y}_A^*)_{lm} = 0.001967$$

$$n_{OG} = \frac{0.0200 - 0.00102}{0.001967} = 9.65$$

Alternatively, Eq. (35-22) can be used with $\mathscr{S}$ equal to 0.875.

The value of H_{OG} is needed to calculate the tower height, but the liquid and vapor mass velocities must be found before the HTU can be determined. The flooding velocity is found from Fig. 35-12 as follows:

$$\frac{L}{G}\sqrt{\frac{\rho_G}{\rho_L}} = \frac{2.89(18)}{29}\sqrt{\frac{0.0737}{62.3}} = 0.0617$$

$$0.17 = \frac{G^2(c_f)(\mu_L')^{0.2}}{g_c \rho_G \rho_L}$$

$$G^2 = \frac{(0.17)(4.17 \times 10^8)(0.0737)(62.3)}{(160)(0.860)^{0.2}}$$

$$G = 1615 \text{ lb/(h)(ft}^2\text{) at flooding}$$

The tower is to operate at 40 percent of this mass velocity, say 650 lb/(h)(ft^2). The required tower diameter for a gas flow of 1000 lb/h is

$$D = \sqrt{\frac{(4)(1000)}{(3.14)(650)}} = 1.40 \text{ ft}$$

The liquid rate is $(650)(2.89)(\frac{18}{29}) = 1167$ lb/(h)(ft^2).

The height of the gas-phase transfer unit is found from Eq. (34-11) and Table 34-1.

$$\mathbf{Sc} = \frac{\mu}{\rho D_{AB}} = \frac{(0.018)(2.42)}{(0.0737)(0.368)} = 1.607$$

$$\begin{aligned} H_G &= 6.41G^{0.32}L^{-0.51}\ \mathbf{Sc}^{0.5} \\ &= (6.41)(650)^{0.32}(1167)^{-0.51}(1.607)^{0.5} \\ &= 1.78 \text{ ft} \end{aligned}$$

For the liquid phase we have

$$\mathbf{Sc} = \frac{(0.860)(2.42)}{(62.3)(4.81 \times 10^{-5})} = 694$$

Equation (34-12) and Table 34-2 yield

$$\begin{aligned} H_L &= 0.01\left[\frac{1167}{(0.860)(2.42)}\right]^{0.22}(694)^{0.5} \\ &= 1.06 \text{ ft} \end{aligned}$$

The overall gas-phase HTU is obtained from Eq. (31-56):

$$H_{OG} = 1.78 + (0.875)(1.06) = 2.71 \text{ ft}$$

The tower height required is

$$z = (9.65)(2.71) = 26.2 \text{ ft}$$

The pressure drop obtained from a plot like Fig. 35-11 is 0.568 psi, which corresponds to a power consumption of approximately

$$\frac{(0.568)(144)(1000)}{(0.0737)(3600)(550)(0.70)} = 0.802 \text{ hp}$$

The compression efficiency is assumed to be 70 percent. ////

Example 35-2 A solution of acetic acid in water containing 0.15 mass fraction acetic acid is to be extracted using pure methyl isobutyl ketone as a solvent. The extract is to contain 0.075 mass fraction acetic acid, and the raffinate 0.010 mass fraction acid. Estimate the height of a tower to perform this separation.

The equilibrium data are plotted in Fig. 35-15; the line is slightly curved. The solutions are too concentrated to assume that the operating line is straight, so L/G must be considered a variable.

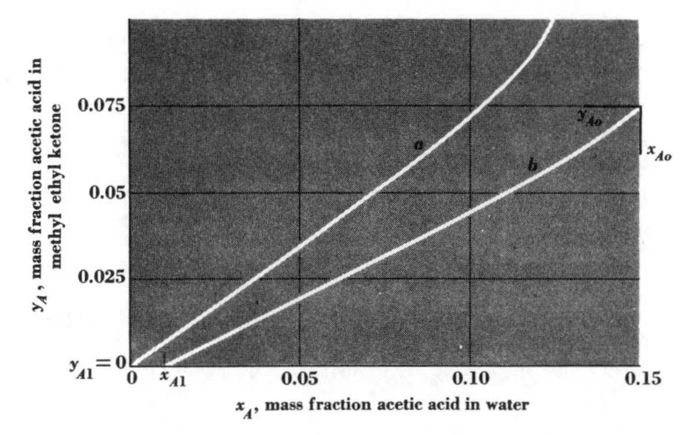

FIGURE 35-15
Operating diagram for Example 35-2.
a Equilibrium line.
b Operating line.

We shall solve this problem by assuming that the phases are immiscible; the following relations then hold:

$$G = \frac{(1 - y_{Ao})G_o}{1 - y_A}$$

$$L = \frac{(1 - x_{Ao})L_o}{1 - x_A}$$

The computations are simplified by expressing the material balance in terms of the flows of water and ketone and using concentrations based on mass ratios:

$$Y_A = \frac{y_A}{1 - y_A}$$

$$X_A = \frac{x_A}{1 - x_A}$$

$$y_A = \frac{Y_A}{1 + Y_A}$$

$$x_A = \frac{X_A}{1 + X_A}$$

The quantity Y_A is the ratio of A to B in the G stream (solvent), so that $G_B Y_A$ is the flux of A carried along. It is easy to show that $G_B Y_A = G y_A$ for immiscible phases. For the raffinate (L) phase we have $L_C X_A = L x_A$. A mass balance over the tower then gives, where L_C/G_B is a constant,

$$\frac{L_C}{G_B} = \frac{Y_{Ao} - Y_{A1}}{X_{Ao} - X_{A1}} \tag{1}$$

where C refers to water and B to ketone. For the present problem we have

$$y_{Ao} = 0.075 \qquad Y_{Ao} = 0.0811$$
$$x_{Ao} = 0.150 \qquad X_{Ao} = 0.1768$$
$$y_{A1} = 0 \qquad Y_{A1} = 0$$
$$x_{A1} = 0.0100 \qquad X_{A1} = 0.0101$$

Thus we have

$$\frac{L_C}{G_B} = \frac{0.0811}{0.1768 - 0.0101} = 0.487$$

We write Eq. (1) for a general point as

$$Y_A = 0.487X_A - 0.0050 \tag{2}$$

and construct the following table:

X_A	Y_A	x_A	y_A
0.0101	0	0.0100	0
0.0527	0.0206	0.0500	0.0202
0.1111	0.0491	0.1000	0.0468
0.1768	0.0811	0.1500	0.0750

The operating line is plotted in Fig. 35-15; it is also slightly curved.

It is normal to calculate n_{OL} or n_L for extraction (stripping) operations. Since the equilibrium and operating lines are similarly curved, Eq. (35-26) can be used to a good approximation. Strictly, the individual transfer units should be used, but the only data available for the system are in terms of overall transfer units.

We proceed to apply Eq. (35-26):

$$x_{A1} - x_{A1}^* = 0.0100 - 0 = 0.0100$$

$$x_{Ao} - x_{Ao}^* = 0.1500 - 0.1030 = 0.0470$$

$$(x_A - x_A^*)_{lm} = \frac{0.0370}{\ln 4.70} = 0.0207$$

$$n_{OL} = \frac{0.1500 - 0.0100}{0.0207} = 6.77$$

The same result is obtained by graphical integration.

Values of the HTU for a spray tower have been reported in the literature.[1] If water is the dispersed phase, as shown in Fig. 35-8*b*, it is found that the HTU for the continuous phase, that is, H_{OG}, is 6 ft. From Eq. (31-56) we have

$$H_{OL} = \frac{L}{Gm} H_{OG}$$

$$m \cong 0.700$$

$$\frac{L}{G} = 0.487$$

$$H_{OL} = (0.696)(6) = 4.17 \text{ ft}$$

$$z = (6.77)(4.17) = 28.3 \text{ ft}$$

Alternatively, n_{OG} can be calculated directly from Fig. 35-15, using, however, mass fractions rather than mole fractions. A much longer portion of the equilibrium curve must be used in this case so that the effects of the curvatures of the two lines do not appear to cancel out. A graphical integration is done, starting from the following table:

y_A	$y_A^* - y_A$	$\frac{1}{y_A^* - y_A}$
0	0.0065	154.0
0.0125	0.0110	90.9
0.0250	0.0160	62.5
0.0375	0.0215	48.8
0.0500	0.0285	35.1
0.0625	0.0350	28.6
0.0750	0.0415	24.1

$$n_{OG} = 4.44$$

$$z = (4.44)(6) = 26.7$$

This height is a good check with that previously obtained. Note that a term similar to the second term on the right-hand side of Eq. (35-15) has not been included because of the approximate nature of the calculations.

The effect of using a packed tower can be found.[2] If the ketone is dispersed as in Fig. 35-8*a* and $\frac{1}{2}$-in C rings are used as packing, H_{OG} is 1.43 ft. Thus z is (1.43)(4.44), or only 6.37 ft. The increase in extraction rate

[1] H. F. Johnson and R. H. Bliss, *Trans. Am. Inst. Chem. Eng.*, **42**:331 (1946).

[2] T. K. Sherwood, J. E. Evans, and J. V. A. Longrov, *Ind. Eng. Chem.*, **31**:1144 (1939).

is caused by the increased turbulence and interfacial area given by the packing. The HTU values are strongly influenced by the type and size of distributing nozzle, by the direction of mass transfer, and by the choice of which phase is dispersed. These matters are best considered empirically at present; a detailed discussion is given by Treybal.[1] ////

PROBLEMS

35-1 Ammonia is to be absorbed from an air stream in a tower packed with 1-in Raschig rings. The inlet gas is 2 mol percent ammonia, and 90 percent of the ammonia is to be recovered in the exit liquid using pure water as the absorbent. Calculate the height of packing required if a liquid rate 25 percent more than the minimum is used with a gas mass velocity of 400 lb/(h)(ft^2). Assume that the temperature is constant at 30°C and the total pressure is 1 atm throughout. Data are given in Perry.

35-2 The benzoic acid in a toluene solution is to be extracted using pure water as the solvent in a continuous, countercurrent packed tower. The feed solution contains 0.010 lb mole benzoic acid/ft^3, and 12.5 volumes of water per volume of toluene is used. Ninety-five percent of the acid is to be recovered in the water phase.

The equilibrium data are:

Concentration of benzoic acid in water phase, lb mol/ft^3	Concentration of benzoic acid in toluene phase, lb mol/ft^3
1.0×10^{-4}	0.7×10^{-3}
2.0×10^{-4}	1.2×10^{-3}
4.0×10^{-4}	3.0×10^{-3}
6.0×10^{-4}	5.6×10^{-3}
8.0×10^{-4}	9.8×10^{-3}
9.0×10^{-4}	12.7×10^{-3}

The two phases can be considered immiscible.

Find the height of packing needed if the individual transfer unit based on the toluene phase (H_L) is 3 ft and the mass-transfer coefficient $k_{\tilde{\rho}}$ in the water phase is three times that in the toluene phase.

Although the equations necessary for the solution of this problem have been derived using mole or mass fractions, it is a simple matter to express them in terms of concentrations, so that the data given here need not be converted to mass fractions.

35-3 It is desired to absorb the ammonia from an air stream at 1 atm in a tower wet-packed with 1-in Raschig rings operating at 30°C. The inlet gas is 10 mol percent ammonia, and 99.9 percent of the ammonia is to be recovered in the solution leaving the tower. Pure water is fed to the top of the tower at a rate 20 percent greater than the minimum rate. Find the height and diameter of a tower designed to operate at one-half the flooding velocity needed to treat 1360 kg/h of gas.

[1] R. E. Treybal, "Liquid Extraction," 2d ed., McGraw-Hill Book Company, New York, 1963.

35-4 Four hundred pounds per hour of a 20 mass percent solution of dioxane in water is to be treated with pure benzene at 25°C in a tower packed with $\frac{3}{8}$-in Raschig rings to reduce the dioxane content in the raffinate to 3.0 mass percent. The two phases may be considered immiscible; the equilibrium relation is $y_A = 1.30x_A$. Estimate the height and diameter of a tower operating at a solvent rate 40 percent above the minimum. The height of the overall transfer unit based on a continuous water phase (H_{OL}) is 4 ft. Assume operation at 70 percent of flooding.

35-5 Individual mass-transfer coefficients are sometimes found by the Wilson method, which has been described for heat transfer in Chap. 27. Describe the experiments which would be necessary and how the data would be interpreted in order to find individual mass-transfer coefficients by this method.

35-6 R. P. Whitney and J. E. Vivian [*Chem. Eng. Progr.*, **45**:323 (1949)] have measured rates of absorption of SO_2 in water and found the following expressions for 1-in Raschig rings at 70°F:

$$k_L a = 0.044L^{0.82}$$

$$k_G a = 0.028G^{0.7}L^{0.25}$$

Wilson's method of analysis was used.

Use these equations to specify the height and diameter of a tower to treat 40,000 ft^3/h at 1 atm and 70°F of air containing 20 mole percent SO_2. It is planned to recover 99.5 percent of the SO_2 by using a liquid rate 25 percent greater than the minimum. A gas rate of 50 percent of flooding may be used. Note that the driving forces for $k_L a$ and $k_G a$ are, respectively, in lb mol/ft^3 and atmospheres. Check the values of the HTUs obtained by equations of Whitney and Vivian against those found from the general correlations given in Chap. 34. Can you explain any discrepancies?

35-7 A packed absorption tower is being used to scrub 0.38 kg/s of 2 mol percent A in air. The absorbing liquid is pure water flowing at a rate which is 1.2 times the minimum possible. The air leaving the top of the tower contains 0.2 mol percent A. The equilibrium line is straight, and $m = 100$. The diameter of the tower is such that it operates at 80 percent of the flooding gas velocity.

It now occurs that 0.76 kg/s of the same gas must be treated, i.e., twice the former rate. How must the liquid flow rate be changed, and what happens to the concentration of A in the scrubbed gas ($\tilde{y}_{A1}$)?

35-8 The waste product from a process contains 1 mol percent hydrogen sulfide in air. The H_2S concentration is to be reduced to 100 ppm (parts per million) by volume in an absorption tower using a dilute alkaline solution in water as a solvent at a rate of 1.2 times the minimum possible. A gas rate of 200 lb/(h)(ft^2) will be used with 1-in Raschig rings. Find the tower height.

Data:

$H_A = 4.83$ atm/mol fraction (Henry's constant)
$H_L = 0.5$ ft (HTU for the liquid-side resistance)
Sc $= 1$ for the gas (H_2S in air)
The tower operates at 1 atm abs and 20°C.

35-9 A stream in a pilot plant contains 1 mol percent benzoic acid dissolved in benzene. The concentration of acid is to be reduced to 0.10 mol percent by washing with water in a countercurrent tower packed with $\frac{1}{2}$-in Raschig rings. The feed rate is 300 lb/h and the water rate is adjusted so that 1.2 times the minimum quantity is needed. Calculate the height and diameter of the tower to be used at 70 percent of flooding velocity.

For this system, when the benzene is in the dispersed phase, the H_{OG} based on the continuous phase is 6 ft. The distribution coefficient can be found in Appendix Table A-24.

35-10 Benzene is to be used as a solvent at 25°C to extract phenol from an aqueous waste stream in which it is present in a concentration of 2.0 parts phenol/100 parts water. Design a tower to recover 90 percent of the phenol in a feed of 300 lb/h of this solution. The tower is to be packed with $\frac{1}{2}$-inch Raschig rings and operated at 70 percent of the flooding velocity and 1.3 times the minimum benzene flow rate.

When water is the continuous phase, the overall HTU based on this phase is 8 ft. The phases are immiscible and the distribution coefficient is $K = 2.3$ where K is defined as in Prob. 35-9.

35-11 A stream of water containing 2×10^{-4} mol fraction of hydrogen sulfide (200 ppm on a molar basis) is to be stripped by a stream of air in a packed tower containing 1-in Raschig rings. To remove 99 percent of the H_2S from the water leaving the tower, 40 percent more than the minimum amount of air is used. Water is to be treated at 1000 lb/h and 50 percent of the flooding velocity is to be used. Design a suitable tower.

35-12 Derive Eqs. (35-22) and (35-24).

35-13 Air containing 2 mol percent A is being treated at a certain rate in a countercurrent absorption tower fed by pure water to reduce the A content in the air leaving the top of the tower to 0.2 mol percent. The slope of the equilibrium line is 20, and 1.2 times the minimum water rate is being used.

A new regulation requires that the exit A concentration now be reduced to 0.02 mol percent. If the same tower is used, at what fraction of the original rate is it possible to treat the gas now? Assume that the tower is operated at 50 percent of the flooding velocity in each case, and neglect any difference in the H_{OG} for the two situations.

Calculate also the ratio of the new liquid rate to the old one.

35-14 An absorption tower is designed to scrub a gas A at 25°C and 1 atm from air fed at 2300 kg/h containing 2 mol percent A. Pure water is the solvent, and the flow rates are chosen so that $\mathscr{S}$ is 0.75; the concentration of A in the scrubbed gas is then 20 ppm (molar basis). The stage efficiency is 1.0. How many stages are needed?

After the tower is built, a foaming problem requires the reduction of the liquid flow rate of three-quarters of its design value. The gas flow rate stays the same. What is the concentration of A in the scrubbed gas for these altered conditions?

35-15 A dilute mixture of carbon dioxide in air is to be scrubbed at 40°C and 100 atm in a tower packed with 1-in Berl saddles. The diameter is such that $G = 200$ lb/(h)(ft^2), and the absorption factor is set at 1.4. The concentration of CO_2 in the scrubbed air is to be 1 percent of that in the feed gas; pure water is fed to the top of the tower. What height of packing and tower diameter are required?

35-16 A 1-mol-percent solution of A in water is being desorbed from the liquid by pure air fed to the bottom of a continuous countercurrent packed tower operating at 1 atm and 30°C. The concentration of A in the water leaving the bottom of the tower is 10 ppm (moles) when $m = 2$ and $\tilde{L}/\tilde{G} = 1.4$.

We now wish to increase the liquid fed into the tower, leaving the gas flow rate constant; the new water flow is 1.49 times the original rate. Find the new concentration of A in the water leaving the bottom of the tower. Assume H_G and H_L remain constant.

35-17 The solubility of A in water for an air-water system at 1 atm and 30°C is given by the equation $\tilde{y}_A = 1.5\tilde{x}_A(1 - \tilde{x}_A)$. A plate tower is to be used to absorb the A from a feed of 28.6 mol percent A in air, recovering 95 percent of entering A in the water leaving the bottom of the countercurrent contacting system. Pure water is fed to the top of the tower. Find the concentration of A in the water leaving the bottom of the tower if a liquid rate 1.2 times the minimum is used. Find also the number of ideal stages required.

35-18 A tower is being designed to reduce the concentration of nitric oxide (NO) in an air stream from 1 to 0.01 mol percent by absorption in a countercurrent packed tower fed with pure water at the top. The water rate used will be 50 percent more than the minimum possible.

Two engineers are discussing whether to design the tower to run at 40 or at 5°C. The lower temperature will increase the solubility of the gas but may reduce the rates of mass transfer.

Data:

	5°C	40°C
m	1.93×10^4	3.52×10^4
H_G		1.0 ft
H_L		1.0 ft

The tower is to be packed with 1-in Raschig rings; L will be greater than 500 lb/(h)(ft^2). Calculate the tower height needed at 40 and at 5°C.

36

SIMULTANEOUS MOMENTUM, HEAT, AND MASS TRANSFER

In studying mass-transfer coefficients for binary mixtures in some of the previous chapters, we learned that the mass flux has an effect on the mass-transfer coefficient. For the laminar boundary layer on a flat plate, the effect was evaluated quantitatively in Chap. 32. For a packed tower or for turbulent flow, the effect of the net mass flux was taken into account approximately by the use of the factor $(1 - \tilde{x}_A)_{lm}$ for the common case of $\tilde{N}_B$ equal to zero.

Because of the analogy between mass and energy transfer, it is logical that the net mass flux affect the value of the heat-transfer coefficient as well as the mass-transfer coefficient; this phenomenon has been discussed in Chap. 32. Notice that we are speaking here of the effect of the mass flux on the energy flux for a given Δt and given flow parameters such as u_o, μ, etc. The effect arises as a result of the influence of the normal velocity at the surface on the velocity distribution near the interface, i.e., on the boundary layer. This velocity distribution in turn controls the magnitude of the mass- and heat-transfer coefficients. Momentum transfer (as reflected in the value of f or C_D) may also be affected. In many common situations, however, the mass flux is not sufficiently large to have an appreciable effect on these coefficients.

This chapter will also deal with the effect of the mass flux on the energy flux as distinguished from its effect on the heat-transfer coefficient. For instance, the energy required to evaporate a liquid drop suspended in a gas, if there is no radiation from solids, must be transferred from the bulk of the gas phase. This heat transfer occurs only if there is a temperature-difference driving force, so the result is that the mass flux of evaporated vapor is coupled to the energy flux necessary to furnish the latent heat of vaporization at the liquid surface. Mass transfer from the region of high concentration at the liquid surface to the bulk of the gas is accompanied by simultaneous heat transfer in the opposite direction to furnish the necessary heat of vaporization. In order to transfer this heat, the drop must be cooler than the gas. The heat-transfer coefficient is the same as it would be without mass transfer unless the mass flux is high enough to bring into play the factors mentioned in the previous paragraph.

In liquid extraction or gas absorption in some dilute solutions, the temperature effects which arise from the mass transfer are negligible. We shall, however, consider a case of gas absorption for which the temperature changes are appreciable. The unit operations of humidification, dehumidification, and drying are concerned with the evaporation of a pure liquid, and the temperature difference between phases is often considerable. Some examples of these operations will be presented in the last part of this chapter to show some of the practical consequences of the simultaneous transfer of heat and mass.

Laminar Boundary Layer on a Flat Plate

In Chap. 32 mass transfer for a binary mixture in the laminar boundary layer on a flat plate was studied. The effect on the mass-transfer coefficient of a net mass flux in the direction perpendicular to the plate was evaluated quantitatively. The results are represented in Fig. 32-1. It was mentioned that this mass flux had an effect on the heat-transfer coefficient. We shall see that it also affects the drag coefficient, or friction factor.

In the present discussion, we extend the results of Chap. 32 to heat and momentum transfer. This extension is obtained by analogy; the detailed derivations are not given. We begin by writing the following equations, derived in previous chapters, for momentum, heat, and mass transfer in the laminar boundary layer on a flat plate. From Eqs. (11-17) and (11-19) we have

$$\frac{d^2(u_x/u_o)}{d\eta^2} + \frac{f(\eta)}{2}\frac{d(u_x/u_o)}{d\eta} = 0 \qquad (36\text{-}1)$$

from Eq. (22-3):

$$\frac{d^2[(t_s - t)/(t_s - t_o)]}{d\eta^2} + \frac{f(\eta)}{2}\mathbf{Pr}\frac{d[(t_s - t)/(t_s - t_o)]}{d\eta} = 0 \qquad (36\text{-}2)$$

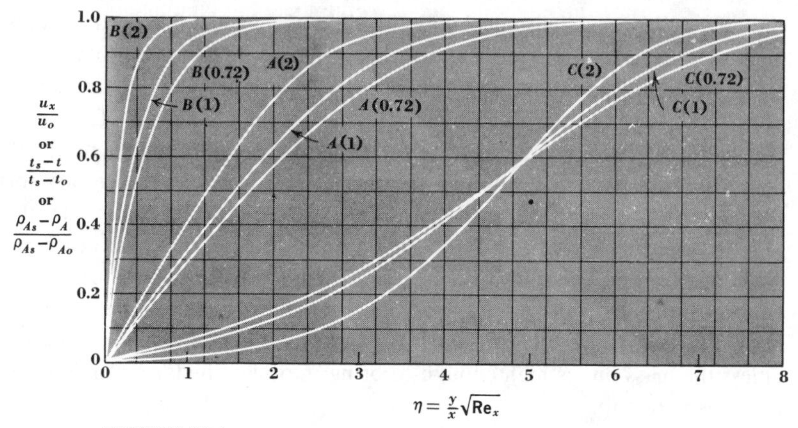

FIGURE 36-1
Profiles in the laminar boundary layer on a flat plate. (*From H. S. Mickley, R. C. Ross, A. L. Squyers, and W. E. Stewart, Natl. Adv. Comm. Aeronaut. Tech. Note 3208, 1954.*)

and from Eq. (32-4):

$$\frac{d^2[(\rho_{As}-\rho_A)/(\rho_{As}-\rho_{Ao})]}{d\eta^2}+\frac{f(\eta)}{2}\,\mathbf{Sc}\,\frac{d[(\rho_{As}-\rho_A)/(\rho_{As}-\rho_{Ao})]}{d\eta}=0 \tag{36-3}$$

The definitions of η and $f(\eta)$ are given in Chap. 11 as

$$\eta = y\sqrt{\frac{u_o}{\nu x}} = \frac{y}{x}\sqrt{\mathbf{Re}_x} \tag{11-15}$$

$$f(\eta) = \frac{\psi}{\sqrt{xu_o\nu}} \tag{11-16}$$

In Eq. (11-16) ψ is the stream function. The analogy among momentum, energy, and mass transfer is well illustrated by Eqs. (36-1) to (36-3). For the case in which $\mathbf{Pr} = \mathbf{Sc} = 1.0$, the solutions for these three equations give u_x/u_o, $(t_s - t)/(t_s - t_o)$, and $(\rho_{As} - \rho_A)/(\rho_{As} - \rho_{Ao})$ as the same function of the position variable η. Thus, in Fig. 36-1, the curve $A(1)$ gives the variation of the velocity, temperature, and concentration as a function of $\mathbf{Re}_x$ and position in the boundary layer. It is natural for the curve representing these three quantities to be identical for Prandtl and Schmidt numbers of unity because the Prandtl number is the ratio of the diffusivity of momentum (kinematic viscosity) to the thermal diffusivity, and the Schmidt number is the ratio of the diffusivity of momentum to the diffusivity of mass (molecular diffusivity).

For the velocity distribution given by curve $A(1)$ there are other possible temperature and concentration distributions, depending on the values of **Pr** and **Sc**. Equations (36-2) and (36-3) can be solved for values of **Pr** and **Sc** other than 1.0; two of the resulting functions are plotted as curve $A(0.72)$ (meaning **Pr** or **Sc** $= 0.72$) and curve $A(2)$ (meaning **Pr** or **Sc** $= 2.0$) in Fig. 36-1. For instance, for pentane vapor diffusing from a heated flat plate through a laminar boundary layer of air, we have **Pr** $\cong 0.72$ and **Sc** $\cong 2.0$; u_x/u_o is given by $A(1)$, $(t_s - t)/(t_s - t_o)$ is given by $A(0.72)$, and $(\rho_{As} - \rho_A)/(\rho_{As} - \rho_{Ao})$ is given by $A(2)$. For water vapor diffusing from a heated plate through air, **Pr** is still 0.72 and **Sc** is a little less than 0.72. Thus curve $A(0.72)$ represents approximately both the temperature and concentration profiles. The Lewis number **Sc**/**Pr** for this system is about unity.

In Chap. 32, we learned that the mass flux at the surface of the plate, represented by the dimensionless group $u_{ys}\sqrt{\mathbf{Re}_x}/u_o$, has an important effect on the curves in Fig. 32-1; this effect must now appear in Fig. 36-1. The group $u_{ys}\sqrt{\mathbf{Re}_x}/u_o$ becomes appreciably different from zero either because of a large concentration difference or because of the addition or withdrawal of fluid through pores provided in the flat plate. The curves B in Fig. 36-1 apply when mass is being transferred from the fluid stream to the plate at a rate such that $u_{ys}\sqrt{\mathbf{Re}_x}/u_o = -2.5$. The curves C are for mass transfer from plate to stream, and $u_{ys}\sqrt{\mathbf{Re}_x}/u_o = 0.5$.

It is of interest to give now the effect of the distributions shown in Fig. 36-1 on the transfer coefficients C_D, k'_ρ, and h. These coefficients are proportional to the slopes of the lines in Fig. 36-1 at $y = 0$. The coefficient found from these slopes is a point coefficient. Since $u_{ys}\sqrt{\mathbf{Re}_x}/u_o$ is constant for the length of the plate, Eq. (32-5), the correction to the average coefficient for a length of plate L is the same as it is for the point coefficient.

For momentum transfer, the drag coefficient is given by

$$C_D = \frac{2g_c F_d}{u_o{}^2 \rho A} = \frac{2g_c \tau_s}{u_o{}^2 \rho} \tag{11-28}$$

in which F_d/A should be interpreted as the point value of the shear stress at the surface τ_s. This quantity is given by

$$\tau_s = \frac{\mu}{g_c}\left(\frac{\partial u_x}{\partial y}\right)_{y=0} \tag{11-23}$$

Equations (11-28) and (11-23) are combined to yield

$$C_D = \frac{2\mu}{u_o \rho}\left[\frac{\partial(u_x/u_o)}{\partial y}\right]_{y=0} \tag{36-4}$$

By using the definition of η, Eq. (11-15), Eq. (36-4) is readily converted to

$$\frac{C_D}{2} = \frac{\mu}{u_o \rho}\left[\frac{d(u_x/u_o)}{d\eta}\right]_{\eta=0}\sqrt{\frac{\rho u_o}{\mu x}} \tag{36-5}$$

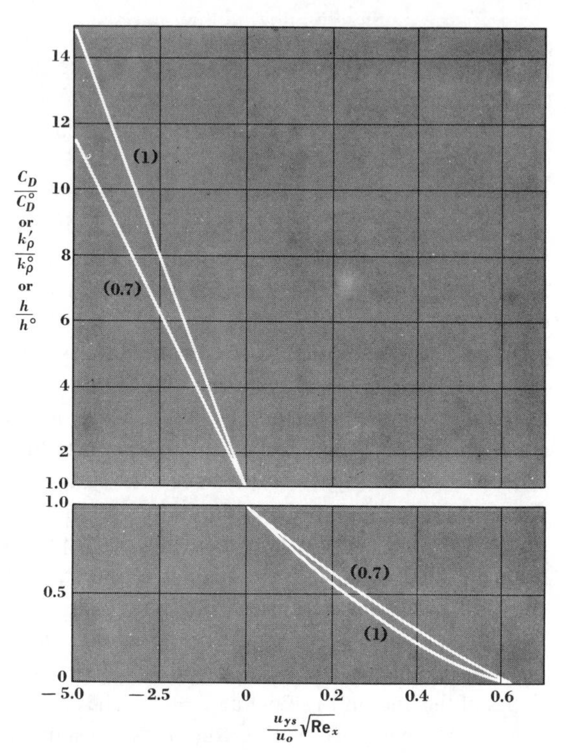

FIGURE 36-2
The effect of mass transfer on transfer coefficients. The parameters on the curves refer to **Pr** or **Sc**. [*From J. P. Hartnett and E. R. G. Eckert, Trans. Am. Soc. Mech. Eng.*, **69**:*247–254* (*1957*).]

or

$$\frac{C_D}{2} = \frac{1}{\sqrt{\mathbf{Re}_x}}\left[\frac{d(u_x/u_o)}{d\eta}\right]_{\eta=0} \tag{36-6}$$

The derivative in Eq. (36-6) is obtained from the slopes at $\eta = 0$ of curves $A(1)$, $B(1)$, and $C(1)$ in Fig. 36-1, and it is evident that the drag coefficient is a strong function of the mass-transfer group $u_{ys}\sqrt{\mathbf{Re}_x}/u_o$. The effect of this group on C_D is shown as curve 1 in Fig. 36-2; $C_D{}^o$ is the value C_D would have if u_{ys} were zero.

The mass-transfer coefficient k'_ρ, which is plotted as $k'_\rho/k_\rho{}^o$ in Fig. 36-2, is defined by an equation equivalent to Eq. (32-6).

$$N_A = k'_\rho(\rho_{As} - \rho_{Ao}) + Nx_{As} \tag{36-7}$$

The fluxes N_A and N are those at $s(y = 0)$. From Eq. (32-7) we obtain the definition of k'_ρ.

$$k'_\rho = D_{AB}\left\{\frac{\partial[(\rho_{As} - \rho_A)/(\rho_{As} - \rho_{Ao})]}{\partial y}\right\}_{y=0} \tag{36-8}$$

This equation is analogous to Eq. (36-4). However, it will be recalled that in Chaps. 31 to 35, we have used a coefficient defined by

$$N_A = k_\rho(\rho_{As} - \rho_{Ao}) \tag{36-9}$$

We can obtain a relation between k'_ρ and k_ρ by combining Eqs. (36-7) and (36-9).

$$k_\rho = \frac{k'_\rho(\rho_{As} - \rho_{Ao}) + Nx_{As}}{\rho_{As} - \rho_{Ao}}$$

In the limit of $N = 0$ this equation shows that k_ρ and k'_ρ are identical. In this situation they are designated as $k_\rho{}^o$. In the common case for which $N_B = 0$ and $N = N_A$, we have, from Eqs. (36-7) and (36-9),

$$k'_\rho = (1 - x_{As})k_\rho \tag{36-10}$$

Equation (36-8) is next changed so that the independent variable in the derivative is η, and we get

$$k'_\rho = D_{AB}\left\{\frac{d[(\rho_{As} - \rho_A)/(\rho_{As} - \rho_{Ao})]}{d\eta}\right\}_{\eta=0}\sqrt{\frac{\rho u_o}{\mu x}} \tag{36-11}$$

In terms of dimensionless groups, this equation becomes

$$\frac{k'_\rho}{u_o} = \frac{1}{\mathbf{Sc}\sqrt{\mathbf{Re}_x}}\left\{\frac{d[(\rho_{As} - \rho_A)/(\rho_{As} - \rho_{Ao})]}{d\eta}\right\}_{\eta=0} \tag{36-12}$$

This equation, with the derivative evaluated from the slopes of the curves in Fig. 36-1, shows the effect of mass transfer on the coefficient k'_ρ; the results are also given in Fig. 36-2. The parameters (0.72) and (1) refer to values of the Schmidt number. The equation given above for mass transfer can also be given in terms of coefficients such as k_x, $k_{\tilde{x}}$, etc. The coefficient $k_\rho{}^o$ is the value k'_ρ would have if u_{ys} were zero; k_ρ is obtained from k'_ρ by Eq. (36-10) for $N_B = 0$.

The heat-transfer coefficient h, which is plotted as h/h^o in Fig. 36-2, is defined by

$$h = k\left\{\frac{\partial[(t_s - t)/(t_s - t_o)]}{\partial y}\right\}_{y=0} \tag{22-11}$$

We proceed as in the previous paragraph and write

$$h = k\left\{\frac{d[(t_s - t)/(t_s - t_o)]}{d\eta}\right\}_{\eta=0}\sqrt{\frac{\rho u_o}{\mu x}} \tag{36-13}$$

This equation can be converted to a relation among dimensionless groups by dividing both sides by $C_p u_o \rho$; the result is

$$\frac{h}{C_p u_o \rho} = \frac{k}{C_p \mu} \left\{ \frac{d[(t_s - t)/(t_s - t_o)]}{d\eta} \right\}_{\eta=0} \sqrt{\frac{\mu}{\rho u_o x}} \tag{36-14}$$

or

$$\mathbf{St} = \frac{1}{\mathbf{Pr}\sqrt{\mathbf{Re}_x}} \left\{ \frac{d[(t_s - t)/(t_s - t_o)]}{d\eta} \right\}_{\eta=0} \tag{36-15}$$

The derivative in Eq. (36-15) is found from the curves of Fig. 36-1; it is a function of both **Pr** and $u_{ys}\sqrt{\mathbf{Re}_x}/u_o$. The curves in Fig. 36-2 are applied to heat transfer by interpreting the parameters 0.72 and 1 on the curves as values of the Prandtl number. The coefficient h^o is the value h would have if u_{ys} were zero.

Simultaneous Heat and Mass Transfer with Turbulent Flow

An exact solution of the boundary-layer equation for heat transfer with simultaneous mass transfer is not available for turbulent flow. An approximate solution will be given which follows the methods used in Chap. 31 to obtain Eq. (31-26) for mass transfer. The present derivation is oriented toward an evaluation of the effect of mass transfer on heat transfer.

A stationary film is postulated adjacent to the surface which is exchanging heat and mass with the fluid. The effective thickness of the layer δ_m is such that the transfer rates through it equal those obtained for the actual flow situation. As before, $\tilde{N}_B$ is zero.

We study the more common case, for which the fluid is hotter than the surface. There is mass transfer of component A from the surface to the fluid and no net flow of B. The variables involved are shown in Fig. 36-3. The flux $\tilde{N}_A$ through the interface arises either from the evaporation of a volatile liquid or solid or from a forced flow of A through a porous plate. There is an enthalpy change $\tilde{H}_{As} - \tilde{H}_{A1}$ in bringing the transferred mass from its condensed state, or its state on the side of the porous plate away from the fluid, to its state in the vapor phase adjacent to the plate. For evaporation or sublimation $\tilde{H}_{As} - \tilde{H}_{A1}$ is simply the heat of vaporization. We assume that all the energy required for the evaporation is supplied by conduction from the fluid through the hypothetical layer of thickness δ_m. In addition, heat must be conducted in to make up for the energy which is carried by the molecules of A as they diffuse away from the surface. The overall energy balance is thus written over the length δ_m as

$$\tilde{N}_A(\tilde{H}_{As} - \tilde{H}_{A1}) + \tilde{N}_A(\tilde{H}_{Ao} - \tilde{H}_{As}) = h'(t_o - t_s) \tag{36-16}$$

It is customary, however, to include in the heat-transfer coefficient the energy carried

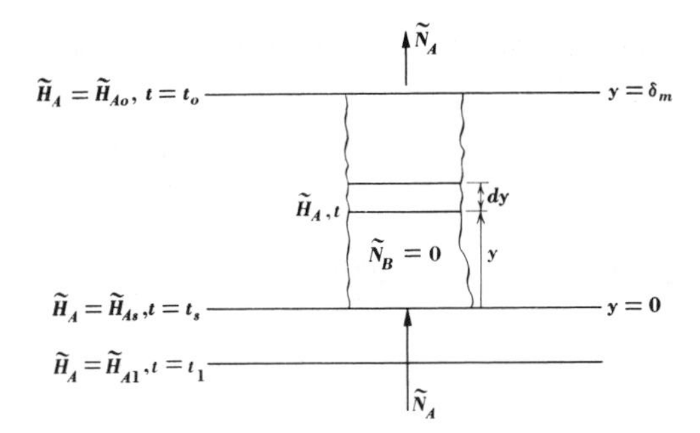

FIGURE 36-3
Film mechanism for simultaneous heat and mass transfer.

by the diffusing molecules of A, so that we have, by definition,

$$h(t_o - t_s) = \tilde{N}_A(\tilde{H}_{As} - \tilde{H}_{A1}) = h'(t_o - t_s) - \tilde{N}_A(\tilde{H}_{Ao} - \tilde{H}_{As}) \qquad (36\text{-}17)$$

The use of the film theory and the application of an integral energy balance over the thickness δ_m have resulted in the definition of the coefficient h'. From the relation

$$\tilde{H}_{Ao} - \tilde{H}_{As} = \tilde{C}_{pA}(t_o - t_s) \qquad (36\text{-}18)$$

and Eq. (36-17), we find

$$h' = h\left[1 + \frac{\tilde{C}_{pA}(t_o - t_s)}{\tilde{H}_{As} - \tilde{H}_{A1}}\right] \qquad (36\text{-}19)$$

The coefficient h' is sometimes used in the analysis of simultaneous heat and mass transfer.

Since we usually apply the present analysis to a gas phase, we use molar units so that $\tilde{N}_A$ is constant for ideal gases for all values of y from zero to δ_m; that is, $\tilde{N}_B$ is everywhere zero. At some point in the film between $y = 0$ and $y = \delta_m$, we write

$$\tilde{N}_A(\tilde{H}_{As} - \tilde{H}_{A1}) + \tilde{N}_A(\tilde{H}_A - \tilde{H}_{As}) = k\frac{dt}{dy} \qquad (36\text{-}20)$$

We next integrate this equation from $y = 0$ to $y = \delta_m$ and from $t = t_s$ to $t = t_o$. Before the integration, $\tilde{H}_{As} - \tilde{H}_{A1}$ is replaced by $\Delta\tilde{H}_A$ and $\tilde{H}_A - \tilde{H}_{As}$ by $\tilde{C}_{pA}(t - t_s)$. Equation (36-20) is then written as

$$\tilde{N}_A \int_0^{\delta_m} dy = k \int_0^{t_o - t_s} \frac{d(t - t_s)}{\Delta\tilde{H}_A + \tilde{C}_{pA}(t - t_s)} \qquad (36\text{-}21)$$

Integration of Eq. (36-21) yields

$$\tilde{N}_A = \frac{k}{\delta_m \tilde{C}_{pA}} \ln\left[1 + \frac{\tilde{C}_{pA}}{\Delta\tilde{H}_A}(t_o - t_s)\right] \qquad (36\text{-}22)$$

The total energy flux at the interface is $\tilde{N}_A \, \Delta\tilde{H}_A$, so that we get, by the definition of h, Eq. (36-17),

$$h = \frac{k \, \Delta\tilde{H}_A \ln [1 + (\tilde{C}_{pA}/\Delta\tilde{H}_A)(t_o - t_s)]}{\delta_m \tilde{C}_{pA}(t_o - t_s)} \qquad (36\text{-}23)$$

If there is no mass transfer, we have simply

$$h^o = \frac{k}{\delta_m} \qquad (36\text{-}24)$$

From Eqs. (36-23) and (36-24) there results

$$h^o = h\left[1 + \frac{\tilde{C}_{pA}}{\Delta\tilde{H}_A}(t_o - t)\right]_{lm} \qquad (36\text{-}25)$$

in which

$$\left[1 + \frac{\tilde{C}_{pA}}{\Delta\tilde{H}_A}(t_o - t)\right]_{lm} = \frac{(\tilde{C}_{pA}/\Delta\tilde{H}_A)(t_o - t_s)}{\ln [1 + (\tilde{C}_{pA}/\Delta\tilde{H}_A)(t_o - t_s)]} \qquad (36\text{-}26)$$

Note that Eqs. (36-17) and (36-20) reduce to the equivalent of Eq. (22-11) at $y = 0 (\tilde{H}_A = \tilde{H}_{As})$.

Equation (36-25) is analogous to Eq. (31-26). It has been derived for heat transfer toward the surface and mass transfer away from the surface, the most common situation. With a suitable adjustment in signs, Eq. (36-25) can be applied to other combinations of heat and mass transfer for which $\tilde{N}_B = 0$.

Although the analysis for diffusion and conduction through a hypothetical stagnant film is a model which is known to be only approximate, Eqs. (31-26) and (36-25) are all that are presently available for turbulent flow. The values of h^o and $k_\rho{}^o$ are obtained from suitable correlations in Chaps. 23 and 33.

Example 36-1 The temperatures reached in rockets are so high that special precautions must be taken to prevent the walls of the combustion chamber from being destroyed. In a liquid-fueled rocket the walls may be cooled by making them porous and forcing liquid oxygen through the walls into the combustion chamber; this is called *transpiration cooling*. Let us study a situation for which the temperature of the gas stream is 3000°R and it is desired to keep the inside surface of the walls at 1160°R by adding liquid oxygen at 162°R. Calculate the necessary rate of oxygen addition if the heat-transfer coefficient for the turbulent flow with no addition is 250 Btu/(h)(ft^2)(°F).

This situation corresponds to the transfer of A through stagnant B, and an approximate result can be obtained from Eq. (36-22); for turbulent flow, only the film theory is available. We replace k/δ_m by h^o, or 250, and get

$$\tilde{N}_A = \frac{h^o}{\tilde{C}_{pA}} \ln \left[1 + \frac{\tilde{C}_{pA}(t_o - t_s)}{\Delta \tilde{H}_A} \right]$$

For oxygen, an average value of $\tilde{C}_{pA}$ is taken as 7.20 Btu/(°F)(lb mol) between 162 and 1160°R, and the latent heat of vaporization is 2930 Btu/lb mol. Thus we get

$$\Delta \tilde{H}_A = 2930 + 7.20(1160 - 162) = 10{,}120$$

Between 1160 and 3000°R, $\tilde{C}_{pA}$ is taken as 8.50; thus we have

$$\tilde{N}_A = \frac{250}{8.50} \ln \left[1 + \frac{8.50(1840)}{10{,}120} \right]$$

$$= 27.5 \text{ lb mol/(h)(ft}^2\text{)}$$

It is of interest to calculate the value of h from Eqs. (36-25) and (36-26):

$$\left[1 + \frac{\tilde{C}_{pA}}{\Delta \tilde{H}_A} (t_o - t) \right]_{lm} = \frac{1.545}{\ln (2.545)} = 1.658$$

$$h = \frac{h^o}{1.658} = \frac{250}{1.658} = 151 \text{ Btu/(h)(ft}^2\text{)(°F)}$$

From Eq. (36-19) we calculate

$$h' = h(2.545) = 384 \text{ Btu/(h)(ft}^2\text{)(°F)}$$

The significance of the three coefficients can be summarized as follows. If there were no mass transfer, the value $h^o = 250$ would prevail; because of the mass transfer, this value is changed to $h = 151$. However, this latter value of h includes the energy carried by the mass transfer of A; the rate of heat transfer by conduction across the fictitious film is given by $h' = 384$. If the flow were laminar, Fig. 36-2 could be used with a trial-and-error procedure; it would be found that h/h^o is of the same order as 1/1.658. ////

Mass Transfer in a Nonisothermal Contactor

Let us now discuss a contacting tower in which an exothermic heat effect accompanies the dissolution of component A which is being transferred from one phase to another. The two phases are immiscible, so that N_B is again zero. In order to speak of a concrete example, we apply our reasoning to gas absorption, but the solutions obtained can be adapted to desorption, liquid-liquid extraction, or other separation operations.

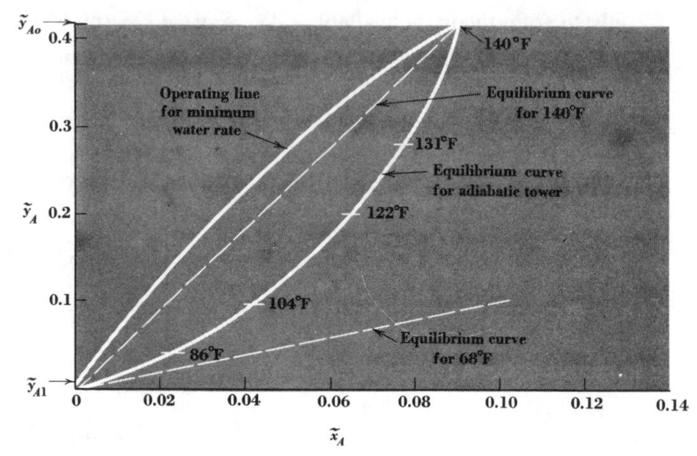

FIGURE 36-4
Operating diagram for the absorption of ammonia in water. The effect of the heat of absorption on the liquid temperature is shown. Heat transfer between phases is neglected. (*From T. K. Sherwood and R. L. Pigford, "Absorption and Extraction," 2d ed., p. 162, McGraw-Hill Book Company, New York, 1952.*)

If the absorbing liquid introduced at the top of a gas-absorption tower is at the same temperature as the rich gas introduced at the bottom, the liquid temperature increases as it flows down the column. The exothermic heat of absorption is converted to sensible heat in the liquid; the tower operation is taken as adiabatic. In rough calculations the heat transfer between phases is sometimes assumed to be zero. The gas-phase temperature is then constant throughout the tower, and the liquid temperature increases to a maximum at the bottom of the tower. Since the mass-transfer coefficient or HTU is only slightly affected by temperature, the only effect of the variation of liquid temperature is to increase the value of m as $\tilde{x}_A$ and $\tilde{y}_A$ increase. The operating line is unchanged, and the equilibrium line is concave upward; the situation is illustrated in Fig. 36-4.

The analysis we have just described may lead to incorrect results. From the analogy between heat and mass transfer it is evident that if mass is transferred between the phases at an appreciable rate, heat must be similarly transferred. This statement would be inaccurate only if the Lewis number were much greater than 1. For most gas-liquid systems, **Le** is of the order of unity, i.e., between 0.5 and 3.0.

A typical system in which the temperatures vary is shown in Fig. 36-5. The liquid entering the tower is in contact with gas which has been heated by the hot liquid lower in the tower; heat flows from gas to liquid, so that the liquid temperature

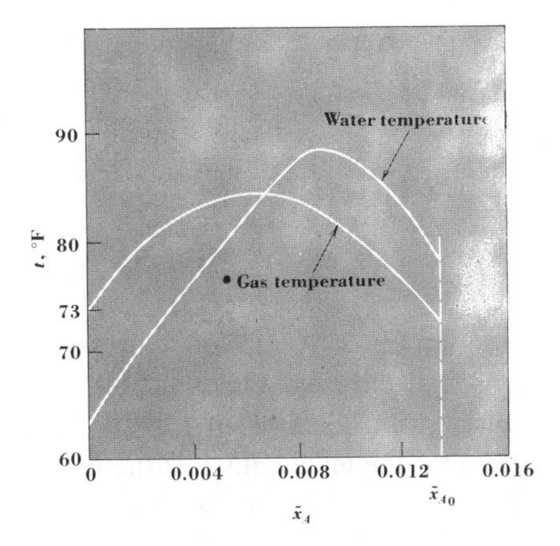

FIGURE 36-5
Estimated temperature profiles for the adiabatic absorption of acetone. (*From T. K. Sherwood and R. L. Pigford, "Absorption and Extraction," 2d ed., p. 165, McGraw-Hill Book Company, New York, 1952.*)

increases more rapidly than it would have increased without interphase heat transfer. Near the bottom of the tower the liquid loses heat to the cold incoming gas. Its temperature thus increases less rapidly or decreases (as in Fig. 36-5) as the liquid flows downward.

Since the gas temperature is governed only by sensible heat effects, it is heated by the hot liquid near the bottom of the tower and cooled by the cold liquid near the top of the tower. The temperature curves of the two phases thus cross at a point within the tower. The gas temperature reaches a maximum at an interior point in the tower; the liquid temperature either increases to maximum at the bottom of the tower or reaches a maximum within the tower, depending on the particular situation.

We are simplifying our discussion by studying a case for which B and C are immiscible. Actual fluids may be partially miscible; for instance, water used as an absorbing liquid vaporizes if the gas phase is not saturated with water vapor.

In order to calculate the height of a nonisothermal contacting tower, we must in general resort to a numerical integration of the differential equations which apply at a point within the tower. We proceed to write these equations for a gas-absorption tower. The rate of mass transfer between phases is, from Eqs. (35-10) and (31-49),

$$-\frac{d\tilde{y}_A}{dz} = \frac{\tilde{y}_A - \tilde{y}_{As}}{H_G}\frac{1 - \tilde{y}_A}{(1 - \tilde{y}_A)_{lm}} \tag{36-27}$$

or, for dilute solutions,

$$-\frac{d\tilde{y}_A}{dz} = \frac{\tilde{y}_A - \tilde{y}_{As}}{H_G} \tag{36-28}$$

The overall mass balance applied to the differential height dz gives

$$\frac{d(\tilde{G}\tilde{y}_A)}{d(\tilde{L}\tilde{x}_A)} = 1.0 \qquad (36\text{-}29)$$

and for dilute solutions

$$\frac{d\tilde{y}_A}{d\tilde{x}_A} = \frac{\tilde{L}}{\tilde{G}} \qquad (36\text{-}30)$$

Equation (36-29) or (36-30) can be integrated independently from the other equations to give the operating line, Eq. (35-6). In other words, for given terminal mole fractions and for a given $\tilde{L}/\tilde{G}$, the operating line is not affected by temperature. However, the temperature of the liquid determines $\tilde{y}_{As}$ and thus has an important effect on Eq. (36-27) or (36-28).

The rate of heat transfer between phases is written in a form analogous to Eq. (35-7) as

$$-dz = \frac{d(\tilde{C}_{py}\tilde{G}t_y)}{h_y{}^o a(t_y - t_s)} \qquad (36\text{-}31)$$

If $\tilde{C}_{py}$ and $\tilde{G}$ are assumed constant, Eq. (36-31) can be written as

$$-\frac{dt_y}{dz} = \frac{t_y - t_s}{H_{ty}} \qquad (36\text{-}32)$$

in which H_{ty}, the height of a transfer unit for heat transfer for the y phase, is defined by

$$H_{ty} = \frac{\tilde{G}\tilde{C}_{py}}{h_y{}^o a} \qquad (36\text{-}33)$$

It is interesting to observe that the Chilton-Colburn analogy in the form of Eqs. (23-27) and (33-27) can be combined with the definitions of the HTUs to give

$$\frac{H_G}{H_{ty}} = \frac{h_y{}^o}{k_{\tilde{y}}{}^o\tilde{C}_{py}} = \mathbf{Le}^{2/3} \qquad (36\text{-}34)$$

Since **Le** is close to unity for most systems, the two HTUs are of the same order of magnitude.

An overall energy balance applied to a height of tower dz, assuming steady adiabatic operation, yields

$$d(\tilde{L}\tilde{H}_x) = d(\tilde{G}\tilde{H}_y) \qquad (36\text{-}35)$$

The enthalpies are defined by

$$\tilde{H}_x = \tilde{C}_{px}(t_x - t_r) \qquad (36\text{-}36)$$

and

$$\tilde{H}_y = \tilde{C}_{py}(t_y - t_r) + Q_r\tilde{y}_A \qquad (36\text{-}37)$$

The equation for the gas (y) phase applies to dilute mixtures for which $\tilde{y}_A$ is not large enough to effect $\tilde{C}_{py}$ appreciably. If the vapor phase approximates ideal-gas behavior, for which $(\partial \tilde{H}_y/\partial p)_T = 0$, it is convenient to refer both $\tilde{H}_x$ and $\tilde{H}_y$ to zero for the dilute liquid solution at t_r, so that the above simple equations for the enthalpy are obtained. The quantity Q_r is the exothermic molar heat of solution of component A in the absorbing liquid at the reference temperature t_r. If it is assumed that $\tilde{L}$, $\tilde{C}_{px}$, $\tilde{G}$, and $\tilde{C}_{py}$ are constant, Eq. (36-35) can be put into the form

$$\tilde{L}\tilde{C}_{px}\frac{dt_x}{dz} = \tilde{G}\tilde{C}_{py}\frac{dt_y}{dz} + Q_r\tilde{G}\frac{d\tilde{y}_A}{dz} \qquad (36\text{-}38)$$

The values of the interfacial mole fraction and temperature, $\tilde{y}_{As}$ and t_s, are found as described in Chap. 31 from

$$-\frac{k_{\tilde{x}}}{k_{\tilde{y}}} = \frac{\tilde{y}_A - \tilde{y}_{As}}{\tilde{x}_A - \tilde{x}_{As}} \qquad (31\text{-}22)$$

and from the analogous equation for heat transfer,

$$-\frac{h_x}{h_y} = \frac{t_y - t_s}{t_x - t_s} \qquad (36\text{-}39)$$

If it is possible to use overall transfer coefficients, Eqs. (31-22) and (36-39) are not necessary and Eqs. (36-28) and (36-32) become

$$-\frac{d\tilde{y}_A}{dz} = \frac{\tilde{y}_A - \tilde{y}_A^*}{H_{OG}} \qquad (36\text{-}40)$$

and

$$-\frac{dt_y}{dz} = \frac{t_y - t_x}{H_{toy}} \qquad (36\text{-}41)$$

The equations given above cannot be integrated analytically; the calculation of the height of a nonisothermal tower must be done by a stepwise numerical method. The procedure is as follows. Assuming that the temperatures and compositions of both phases are known at the bottom of the tower ($z = 0$), the derivatives $d\tilde{y}_A/dz$ and dt_y/dz can be calculated from Eqs. (36-28) and (36-32) or (36-40) and (36-41). These derivatives also permit the calculation of dt_x/dz from Eq. (36-38). A short increment of tower height Δz is chosen over which it is assumed that the derivatives are constant. In this way the values of t_x, t_y, and $\tilde{y}_A$ can be calculated at the end of the first interval Δz_a. From the operating line and Eqs. (31-22) and (36-39), $\tilde{y}_{As}$ and t_s can be calculated at the end of the interval Δz_a. These values permit the calculation of new values of $d\tilde{y}_A/dz$, dt_y/dz, and dt_x/dz at the point $z = \Delta z_a$. A simple procedure is then to assume that these new values of the derivatives are constant over a second interval Δz_b; the calculations are continued until the mole fraction is reached corresponding to the top of the tower.

The integration procedure described above becomes more accurate as the magnitude of Δz is decreased; it is often not accurate for conveniently large values of the increment. With a reasonably large increment, improved accuracy can be obtained if the values of the derivatives at the two ends of an interval are averaged. Systematic procedures, such as the methods of Euler and of Runge and Kutta, are described in the literature.[1] These repetitive calculations are well suited for performance by digital computer. The estimated temperatures in a tower absorbing acetone in water are shown in Fig. 36-5.

The integration of simultaneous nonlinear ordinary differential equations, such as those considered here, can be performed rapidly by an analog computer. However, numerical integration by digital computer is more accurate and has a more convenient output form. Such programs (software) are now readily available under such names as CSMP, ACSL, or CSSL, so that the user can solve the differential equations without preparing his own numerical integration procedure. Since these programs facilitate the study of various operating conditions and equipment sizes, the result is a simulation on the computer of the actual behavior of various towers. The assumption of dilute solutions has been made in the foregoing discussion in order to bring out clearly the principles of the procedure involved. However, it is clear that if the computations are done by computer it is quite possible to consider the more complicated forms which are obtained if $\tilde{C}_{py}$, $\tilde{G}$, $\tilde{L}$, or H_G, for example, vary from one end of the tower to the other, i.e., the solutions are concentrated.

The equations and procedures given in this section can also be used if a volatile liquid is being vaporized or condensed in a tower. In general, the only simplifications for this case are that $k_{\tilde{x}}$ is infinite (there is no mass-transfer resistance in the pure liquid) and that $\tilde{y}_{As}$ is a function of t_s only, since $\tilde{x}_{As}$ and $\tilde{x}_A$ are always unity. A stepwise calculation is still necessary. However, added simplifications are possible if the liquid is water; also, a special nomenclature exists for air-water mixtures. These matters are discussed in the next section.

Air-Water Operations

Definitions The *humidity* Y_A is defined as the pounds of water (component A) carried by one pound of dry air (component B). A *molar humidity* $\tilde{Y}_A$ can be

[1] For a discussion from an engineering point of view, see H. S. Mickley, T. K. Sherwood, and C. E. Reed, "Applied Mathematics in Chemical Engineering," 2d ed., chap. 5, McGraw-Hill Book Company, New York, 1957. A good treatment of the basic mathematics is given in F. B. Hildebrand, "Introduction to Numerical Analysis," McGraw-Hill Book Company, New York, 1956. Computer-oriented discussion is given by B. Carnahan, H. A. Luther, and J. O. Wilkes, "Applied Numerical Methods," John Wiley & Sons, Inc., New York, 1969. For a more advanced treatment see L. Lapidus, "Digital Computation for Chemical Engineers," McGraw-Hill Book Company, New York, 1962.

defined, but most of the accumulated literature on humidification and drying is in terms of Y_A. The two quantities are related by

$$Y_A = \tfrac{18}{29}\tilde{Y}_A \qquad (36\text{-}42)$$

The humidity is then related to the mole fraction by

$$Y_A = \frac{18}{29}\frac{\tilde{y}_A}{1 - \tilde{y}_A} \qquad (36\text{-}43)$$

Air is said to be saturated with water vapor at a given pressure and temperature when it holds the maximum possible amount of water vapor. This condition is achieved when the air is in equilibrium with liquid water. The *saturation humidity* Y_{As} is the value of Y_A corresponding to a partial pressure $\bar{p}_A$ equal to the vapor pressure of water p_A at the given temperature. The *relative humidity* or *relative saturation* is defined by

$$\text{Relative humidity} = 100\,\frac{\bar{p}_A}{p_A} \qquad (36\text{-}44)$$

The *percent humidity* or *percent saturation* is given by

$$\text{Percent saturation} = 100\,\frac{\tilde{Y}_A}{\tilde{Y}_{As}} = \frac{\bar{p}_A}{p_A}\frac{1 - p_A}{1 - \bar{p}_A}(100) \qquad (36\text{-}45)$$

where p_A and $\bar{p}_A$ are expressed in atmospheres and the total pressure is one atmosphere.

The *humid heat* C_{pH} is defined by

$$C_{pH} = C_{pB} + Y_A C_{pA} \qquad (36\text{-}46)$$

or

$$C_{pH} = 0.24 + 0.46\,Y_A \qquad (36\text{-}47)$$

or

$$C_{pH} = C_{py}(1 + Y_A) \qquad (36\text{-}48)$$

and measured in Btu/(°F)(lb of dry air). The quantity C_{pB} is the specific heat of dry air alone, and C_{py} is that of the air plus water vapor mixture, measured in Btu/(°F)(lb of mixture).

The *humid volume* is defined by

$$V_H = V_B + Y_A V_A \qquad (36\text{-}49)$$

and measured in ft^3/lb of dry air.

The enthalpy of moist air is defined by Eq. (36-37) in units consistent with the present discussion:

$$H_y = C_{pH}(t_y - t_r) + \lambda_r Y_A \qquad (36\text{-}50)$$

The units are Btu/lb of dry air; λ_r is the heat of vaporization at the reference temperature t_r. Table A-23 gives values of the enthalpy of saturated air.

It will sometimes be convenient to use the latent heat λ at a temperature different from t_r. Since the enthalpy is a state function, it can be referred to the reference conditions (dry air plus liquid at t_r) in two ways:

$$C_{pH}(t_y - t_r) + \lambda_r Y_A = C_{pB}(t_y - t_r) + Y_A C_{px}(t_y - t_r) + \lambda Y_A \qquad (36\text{-}51)$$

where C_{px} refers to the liquid and λ is at t_y. The humid heat C_{pH} is a function of Y_A according to Eq. (36-46) so that we obtain

$$Y_A C_{pA}(t_y - t_r) + \lambda_r Y_A = Y_A C_{px}(t_y - t_r) + \lambda Y_A \qquad (36\text{-}52)$$

or

$$\lambda = \lambda_r + C_{pA}(t_y - t_r) - C_{px}(t_y - t_r) \qquad (36\text{-}53)$$

The above relations for the air-water system are not limited to dilute solutions, although the gas phase is assumed to be ideal.

The *adiabatic saturation temperature* of moist air is the temperature which this air reaches when enough water is evaporated into it to cause saturation; the water added is defined as added at the final temperature (t_{as}). The whole process is carried out adiabatically; the heat required to vaporize the water is obtained by cooling the moist air. This process can be visualized as occurring in a very tall air-water contacting tower. The air leaving the top of the tower is saturated at the temperature of the entering water. The water from the bottom of the tower is led to the top and recirculated through the tower, the entire apparatus being insulated from the surroundings. As air at a given state is blown through the tower, it is found that the recirculated water achieves an equilibrium temperature, which is t_{as}. The energy balance ($\Delta H = 0$), in a form similar to Eq. (36-35), can be applied to the tower to give

$$L_o H_{xo} - L_1 H_{x1} = G_B(H_{yo} - H_{y1}) \qquad (36\text{-}54)$$

The rich (o) end of the tower corresponds to t_{as}, and the dilute (1) end to the moist feed, which we designate simply as air at t (so $H_{y1} = H_y$, $H_{yo} = H_{as}$, $Y_{A1} = Y_A$, $Y_{Ao} = Y_{as}$). For the liquid, $H_{xo} = H_{x1} = H_{xas}$. Then we have

$$(L_o - L_1)H_{xas} = G_B(H_{as} - H_y) \qquad (36\text{-}55)$$

A mass balance yields

$$(L_o - L_1) = G_B(Y_{as} - Y_A) \qquad (36\text{-}56)$$

and $(H_{as} - H_y)$ can be expressed in terms of the humid heat from Eq. (36-50) so that Eq. (36-55) becomes

$$(Y_{as} - Y_A)C_{px}(t_{as} - t_r) = C_{pH_{as}}(t_{as} - t_r) + \lambda_r Y_{as} - C_{pH}(t_y - t_r) - \lambda_r Y_A \qquad (36\text{-}57)$$

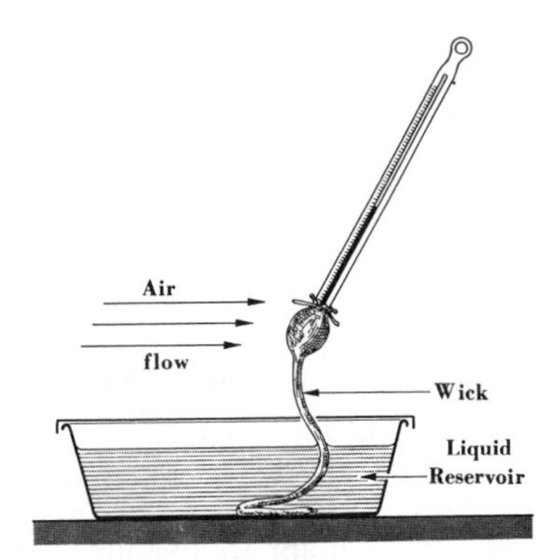

FIGURE 36-6
Sketch of a wet-bulb thermometer.

The definition of the humid heat (Eq. 36-46) permits Eq. (36-57) to be written as

$$\lambda_r(Y_{as} - Y_A) - C_{px}(t_{as} - t_r)(Y_{as} - Y_A) + C_{pA}(t_{as} - t_r)Y_{as} - C_{pA}(t_y - t_r)Y_A + C_{pB}(t_{as} - t_y) = 0 \qquad (36\text{-}58)$$

This equation can be simplified by subtracting $C_{pA}(t_{as} - t_r)Y_A$ after the first three terms and adding it to the fourth term so that we have

$$[\lambda_r - C_{px}(t_{as} - t_r) + C_{pA}(t_{as} - t_r)](Y_{as} - Y_A) + C_{pA}(t_{as} - t_y)Y_A + C_{pB}(t_{as} - t_y) = 0 \qquad (36\text{-}59)$$

The quantity in the brackets in the first term is λ_{as}, by Eq. (36-53), so that the resulting equation for t_{as} is

$$t_{as} = t_y - \frac{\lambda_{as}}{C_{pH}}(Y_{as} - Y_A) \qquad (36\text{-}60)$$

This equation is represented on the humidity chart (Fig. 36-7) as almost straight lines with a negative slope (adiabatic-saturation lines). The lines are slightly curved because C_{pH} is a function of Y_A.

The *wet-bulb temperature* t_{wb} is the temperature attained by a small reservoir of water in contact with a large amount of air flowing past it. A device for measuring this temperature is shown in Fig. 36-6. All the energy required for the vaporization of the water flows to the water by sensible heat transfer from the air. This situation is the same as that leading to the derivation of Eq. (36-17). In our present nomenclature, we write

$$N_A \lambda_{wb} = k_Y(Y_{wb} - Y_A)\lambda_{wb} = h_y(t_y - t_{wb}) \qquad (36\text{-}61)$$

or

$$t_{wb} = t_y - \frac{k_Y \lambda_{wb}}{h_y} (Y_{wb} - Y_A) \qquad (36\text{-}62)$$

In terms of the units of the present discussion, Eq. (36-34) becomes, for dilute solutions ($k_Y = k_Y{}^o$, $h_y = h_y{}^o$),

$$\frac{h_y}{k_Y C_{py}} = \mathbf{Le}^{2/3} \qquad (36\text{-}63)$$

so that Eq. (36-62) can be written as

$$t_{wb} = t_y - \frac{\lambda_{wb}}{C_{py}\, \mathbf{Le}^{2/3}} (Y_{wb} - Y_A) \qquad (36\text{-}64)$$

This equation applies to the vaporization of any liquid; for water in air, $\mathbf{Le} = 1.0$, and for sufficiently dilute gaseous solutions, $C_{pH} \cong C_{py}$, so that

$$t_{wb} = t_y - \frac{\lambda_{wb}}{C_{pH}} (Y_{wb} - Y_A) \qquad (36\text{-}65)$$

If, in addition, any difference between λ_{wb} and λ_{as} is neglected, it is seen that t_{wb} and t_{as} are the same for the water-air system at temperatures low enough to form a dilute gas-phase solution (say $t = 120°F$). For other liquids, the adiabatic saturation and wet-bulb temperatures are not equal.

The *dew-point temperature* is the temperature at which a given sample of moist air becomes saturated as it is cooled at constant pressure and humidity. The *dew-point pressure* is the total pressure to which moist air must be compressed at constant temperature and humidity to bring it to saturation.

Humidity chart The relations among some of the foregoing variables are conveniently given by the humidity chart of Fig. 36-7. Its use is illustrated by the following example.

Example 36-2 The air used in a certain process at 760 mmHg is found by measurement to have a dry-bulb temperature of 110°F and a wet-bulb temperature of 85°F.

From Fig. 36-7 the humidity is $Y_A = 0.0205$. If the air under these conditions is cooled to saturation, dew appears at 77.5°F. Air at 110°F could hold 0.0595 lb water/lb dry air; saturation is thus 34.5 percent, as can also be read from the curved lines on the chart. The mole fraction water vapor in the moist air is found from Eq. (36-43), rearranged as

$$\tilde{y}_A = \frac{Y_A}{\frac{18}{29} + Y_A}$$

$$\tilde{y}_A = 0.0320$$

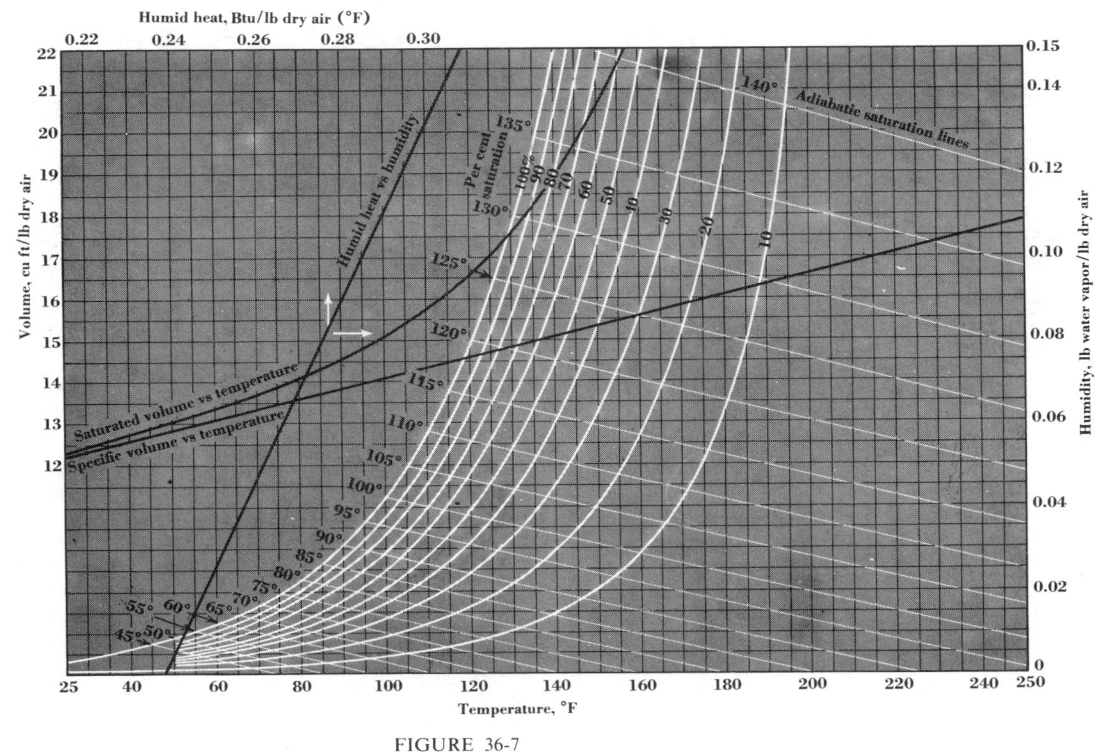

FIGURE 36-7
Humidity chart for the air-water system at 1 atm. (*From G. G. Brown et al., "Unit Operations," p. 545, fig. 499, John Wiley & Sons, Inc., New York, 1950.*)

The partial pressure of water vapor is

$$\bar{p}_A = (0.0320)(760) = 24.3 \text{ mmHg}$$

From Perry, p. 12-7, we have

$$p_A = 2.597 \text{ in Hg} = 65.9 \text{ mmHg}$$

The relative humidity is then 24.3/65.9, or 36.9 percent.

The enthalpy of the moist air at 110°F is found from Eq. (36-50) as

$$\begin{aligned} H_y &= (0.249)(110) + (1075)(0.0205) \\ &= 49.5 \text{ Btu/lb dry air, for dry air at } 0°\text{F and liquid water at } 32°\text{F} \end{aligned}$$

The value of the latent heat of vaporization used above is for the vaporization of liquid water at 32°F. Because the wet-bulb and adiabatic saturation temperatures for the air-water system have been assumed identical, the enthalpy of the moist air at 110°F should be the same as that of saturated air at 85°F. From the table in Perry, p. 12-8,[1] it is found that $H_{as} = 49.4$ Btu/lb at 85°F (h_s in Perry's notation). ////

Cooling towers[2] As warm water flows down a tower countercurrent to rising unsaturated air, the water is cooled by furnishing part of the latent heat required to vaporize some of the water into the air stream. The air is thus humidified as it rises. The calculation of the tower height for the vaporization of liquids other than water must be done by the stepwise procedure using Eqs. (36-27) to (36-41). For water, the fact that **Le** is unity leads to an important simplification. We first rewrite Eq. (36-38) in the units used with air-water operations:

$$LC_{px}\frac{dt_x}{dz} = G_B C_{pH}\frac{dt_y}{dz} + G_B \lambda_r \frac{dY_A}{dz} \tag{36-66}$$

This equation is limited to dilute solutions ($t < 120°$F), for which C_{pH} and L can be assumed constant. The derivative dt_y/dz is replaced by using Eq. (36-32), and dY_A/dz is replaced by using

$$-\frac{dY_A}{dz} = \frac{Y_A - Y_{As}}{H_G} \tag{36-67}$$

which is analogous to Eq. (36-28). The result is

$$-LC_{px}\frac{dt_x}{dz} = G_B C_{pH}\frac{t_y - t_s}{H_{ty}} + G_B \lambda_r \frac{Y_A - Y_{As}}{H_G} \tag{36-68}$$

[1] See Table A-23.

[2] Practical aspects are covered in "Cooling Towers," American Institute of Chemical Engineers, New York, 1972.

For the air-water system, H_{ty} approximately equals H_G, so we write

$$-LC_{px}\frac{dt_x}{dz} = \frac{G_B}{H_G}[C_{pH}(t_y - t_s) + \lambda_r(Y_A - Y_{As})] \qquad (36\text{-}69)$$

If this equation is compared with Eq. (36-50), the definition of the enthalpy of the gas phase, it can be changed to

$$-LC_{px}\frac{dt_x}{dz} = \frac{G_B}{H_G}(H_y - H_{ys}) \qquad (36\text{-}70)$$

The left-hand side of the equation is also equal to $-G_B(dH_y/dz)$, Eq. (36-35). Therefore we have

$$-\frac{dH_y}{dz} = \frac{H_y - H_{ys}}{H_G} \qquad (36\text{-}71)$$

In addition, we combine Eqs. (36-70) and (36-71) to give

$$LC_{px}\frac{dt_x}{dz} = G_B\frac{dH_y}{dz} \qquad (36\text{-}72)$$

and thus

$$\frac{dH_y}{dt_x} = C_{px}\frac{L}{G_B} \qquad (36\text{-}73)$$

Since the right-hand side of Eq. (36-73) is constant (it is assumed that the amount of water evaporated in the tower does not greatly change L), this equation represents a straight operating line on a graph of H_y versus t (Fig. 36-8). Equation (36-73) can be integrated to give the relation between H_y and t_x at any point in the tower.

$$H_y = \frac{C_{px}L}{G_B}t_x + H_{yo} - \frac{C_{px}L}{G_B}t_{xo} \qquad (36\text{-}74)$$

From data on the vapor pressure of water and the enthalpies of air and water, the curve shown in Fig. 36-8 for H_{ys} versus t_s is obtained. If there is negligible resistance to heat transfer in the liquid phase ($h_y/h_x = 0$), t_s and t_x are identical. Once the operating and equilibrium lines are plotted on a graph like Fig. 36-8, it is possible to find $H_{ys} - H_y$ as a function of H_y and evaluate the integral for the tower height,

$$z = H_G\int_{H_{y1}}^{H_{yo}}\frac{dH_y}{H_{ys} - H_y} \qquad (36\text{-}75)$$

The above procedure does not give the gas temperature t_y or the humidity Y_A as a function of H_y or t_x. This relation must be found by a stepwise numerical

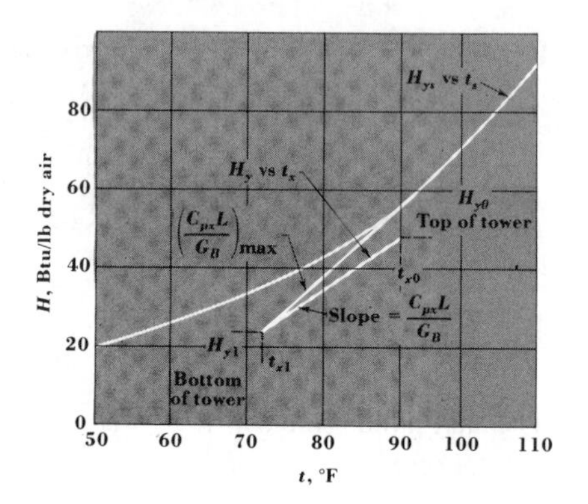

FIGURE 36-8
Operating diagram for a cooling tower.

calculation[1] using the following equation, which is a combination of Eqs. (36-32) and (36-71):

$$\frac{dH_y}{dt_y} = \frac{H_y - H_{ys}}{t_y - t_s} \tag{36-76}$$

It is evident that the preceding equation can also be solved easily by one of the analog simulation computer programs referred to previously (CSMP, etc.).

For the successful operation of a tower, the air must not become saturated, for this leads to fog formation. If the curve of H_y versus t_y crosses the equilibrium curve, fog forms in the air.

For the cooling or humidification tower which we have just discussed, the operating line lies below the equilibrium line. There is a minimum air rate which can be used to accomplish a specified water-cooling duty, as shown in Fig. 36-8 $[(C_{px} L/G_B)_{\max}]$.

In a dehumidification tower cool water is used to reduce the humidity (and temperature) of air blown in at the foot of the tower. For this case the operating line is above the equilibrium line and there exists a minimum amount of water which can be used to dehumidify a given quantity of air.

Example 36-3 Water is to be cooled from 110 to 80°F in a forced draft tower under conditions such that H_G is 8 ft. Air enters the bottom of the tower at 75°F and a wet-bulb temperature of 70°F. Find the tower height if 1.33 times the minimum air rate is used. Neglect the heat-transfer resistance of the liquid phase.

The equilibrium curve shown in Fig. 36-9 is plotted from data in Perry on pp. 12-7, 12-8. The enthalpy of the air entering the bottom of the tower is the

[1] This procedure has been described by H. S. Mickley, *Chem. Eng. Progr.*, **45**:739 (1949).

same as the enthalpy of saturated air at 70°F, the adiabatic saturation temperature of the unsaturated inlet air. The point on Fig. 36-9 representing the inlet air having an adiabatic saturation temperature of 70°F is on a horizontal line to the right of the equilibrium curve, at an enthalpy of 34.1 Btu/lb. The end of the operating line is then fixed by the outlet water temperature, $t_{x1} = 80°F$.

The maximum slope of the operating line is shown in the diagram; the actual operating line, having a slope of 1/1.33 of the limiting line, results in a value of H_{yo} of 75.6 Btu/lb.

The number of transfer units is found from Eq. (36-75); the following table is made from Fig. 36-9.

t_x, °F	H_y	$1/(H_{ys} - H_y)$
80	34.1	0.104
90	47.9	0.125
100	61.8	0.101
110	75.6	0.060

A graphical or numerical integration can be performed; the result is

$$n_G = \int_{34.1}^{75.6} \frac{dH_y}{H_{ys} - H_y} = 4.1$$

and so we obtain

$$z = H_G n_G = (8)(4.1) = 32.8 \text{ ft}$$

Before accepting this result it is well to check the air temperature in the tower. At the bottom of the tower, Eq. (36-76) is

$$\frac{dH_y}{dt_y} = \frac{H_{ys} - H_y}{t_{ys} - t_y} = \frac{43.7 - 34.1}{80 - 75} = 1.92 \text{ Btu/(°F)(lb)}$$

If an interval $\Delta t_y = 5°$ is chosen, H_y at $t_y = 80°F$ is $34.1 + 5(1.92) = 43.7$ Btu/lb. The slope at this point is

$$\frac{dH_y}{dt_y} = \frac{51.8 - 43.7}{87 - 80} = 1.16$$

The average slope over the 5° interval is $(\frac{1}{2})(1.92 + 1.16)$, or 1.54, so the revised value of H_y at $t_y = 80°$ is $34.1 + (5)(1.54) = 41.8$. This point is labeled 2 on the line of H_y versus t_y in Fig. 36-9. A horizontal line from 2 to the operating line locates the water temperature, $t_x = t_{ys} = 86°F$. In a similar way the curve of H_y versus t_y is extended to points 3 and 4, as shown in Fig. 36-9. Since this

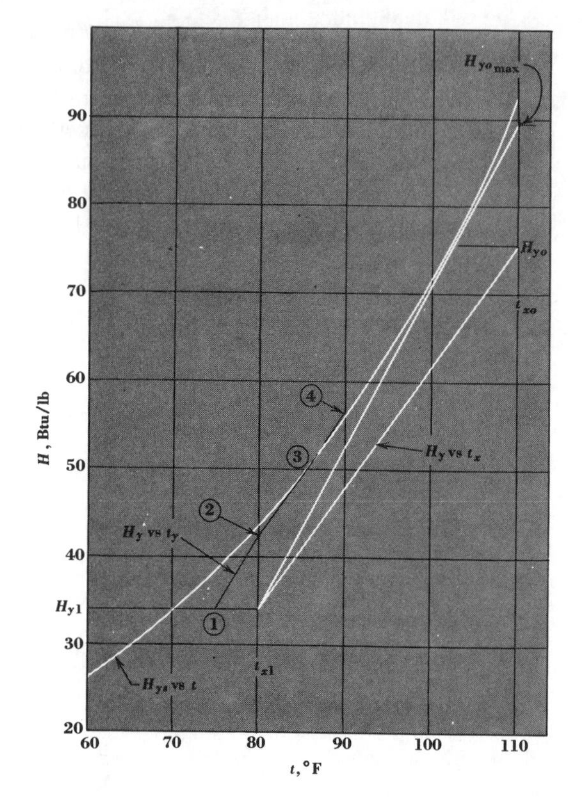

FIGURE 36-9
Operating diagram for Example 36-3. (Enthalpy data from Appendix 3, Table A-23.)

curve crosses the equilibrium line, there would be fog formation in the tower. It would be necessary to operate with a higher air rate in order to accomplish the desired cooling of the water without fog formation. ////

Drying

In the unit operation of drying, water or another liquid is removed from a solid material. The removal of moisture from gases and liquids is also called drying, but the present discussion concerns only the drying of solids.

In the most common type of drying equipment water is evaporated and carried off by a stream of air. However, another unsaturated phase such as superheated steam can also be used as a carrier gas. Alternatively, rather than blow an unsaturated gas past the solid, the pressure can be reduced by a vacuum pump so that the liquid boils at the prevailing temperature.

The heat required for the vaporization of the liquid in the solid is in most cases largely furnished from the inert carrier gas, which enters the drier hot. In some driers the solid is in contact with heated metal surfaces, so part of the heat of vaporization flows to the solid by conduction. In vacuum drying, for which there is no carrier gas, the heat of vaporization must be furnished by conduction or radiation; the capacity of the drier is largely influenced by the heat-transfer surface available.

There are many types of drying equipment;[1] only a few of the most important will be discussed here. The simplest drier is a tray drier, in which the material to be dried is placed on trays contained in an ovenlike enclosure. Heated air may be forced past the solid, or the enclosure may be evacuated. The material is dried in batches, and the trays are loaded and unloaded intermittently.

Continuous operation is usually desirable; a continuous belt drier is shown in Fig. 36-10. Wet material is fed to a moving belt and exposed to hot air as it moves through the tunnel-like drier. The dry material is discharged continuously from the far end of the drier. The rotary steam-tube drier shown in Fig. 36-11 is used for drying granular solids. As the body of the drier turns, the solids are lifted and fall through the central air space. Hot air is blown into one end of the drier or in through louvers in the shell. The heated surfaces in the drier increase the heat-transfer rate to the solid and thus promote rapid drying.

The moisture content of a solid being dried in a typical batch situation is plotted in Fig. 36-12 against time. From the slope of the line in Fig. 36-12 the drying-rate curve of Fig. 36-13 is obtained. In this discussion we assume that a large excess of air of constant wet- and dry-bulb temperatures is being used so that the humidity of the air does not change with time or change appreciably as it passes through the drier. The drying rate is then a function only of the moisture content of the solid.

The shape of the curve shown in Fig. 36-13 is typical for many solids. At first the surface of the solid is flooded with liquid, so that water is carried away from the surface just as it would be from a liquid surface. For constant air conditions, the drying rate is constant; the solid temperature is the wet-bulb temperature of the air unless the solid is heated by conduction or radiation from other solid surfaces. The principles previously developed for air-water contacting permit the necessary design calculation for the constant-rate drying period.

During the constant-rate period water is supplied from the solid near the surface fast enough to keep the surface entirely wet. If the solid shrinks on drying,

[1] See Perry, pp. 20-22 to 20-65.

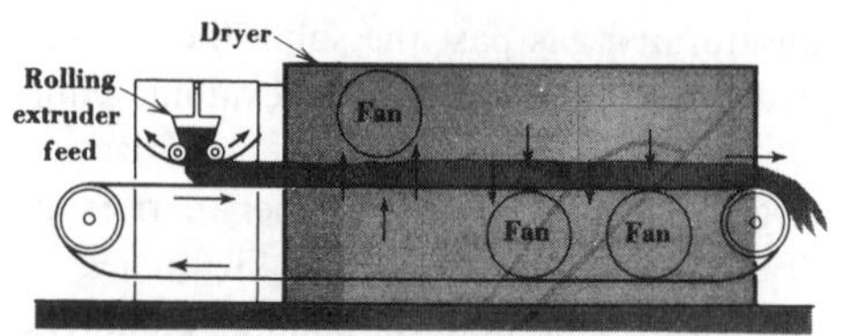

(a) Path of travel of permeable bed through a 3 unit through-circulation dryer.

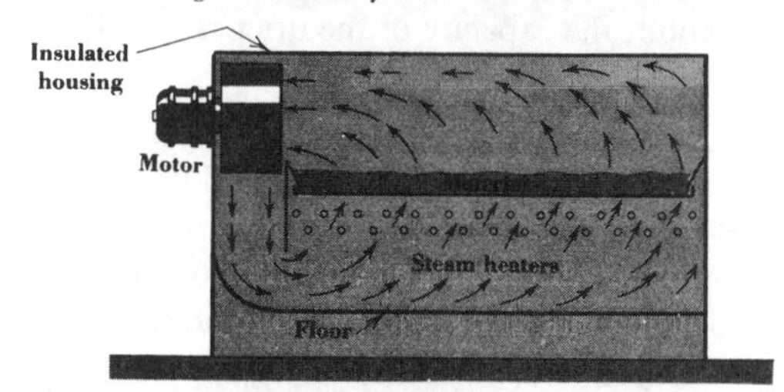

(b) Air flow in wet end

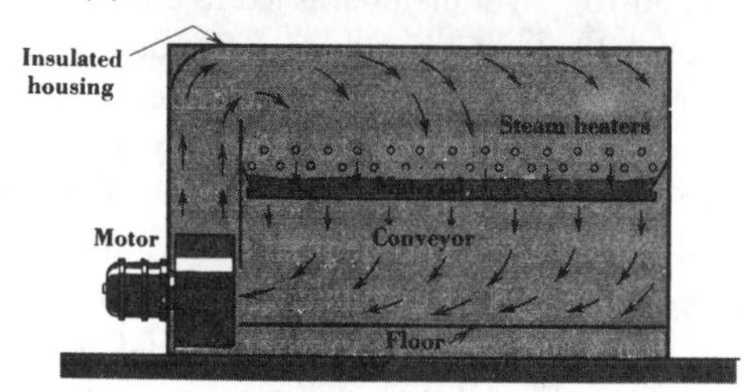

(c) Air flow in dry end

FIGURE 36-10
Continuous belt drier. (*Proctor and Schwartz, Inc.*)

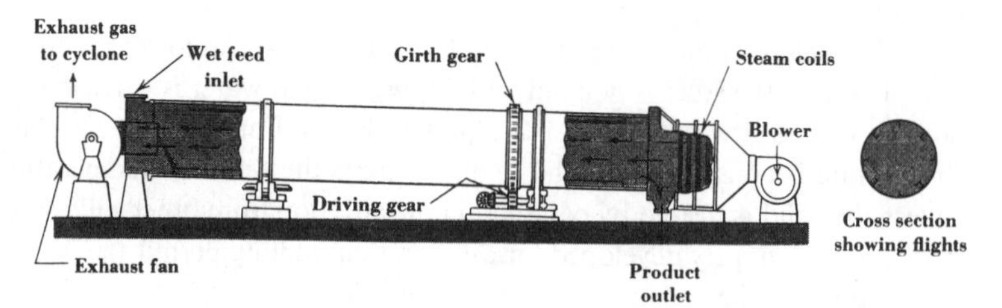

FIGURE 36-11
Continuous rotary drier.

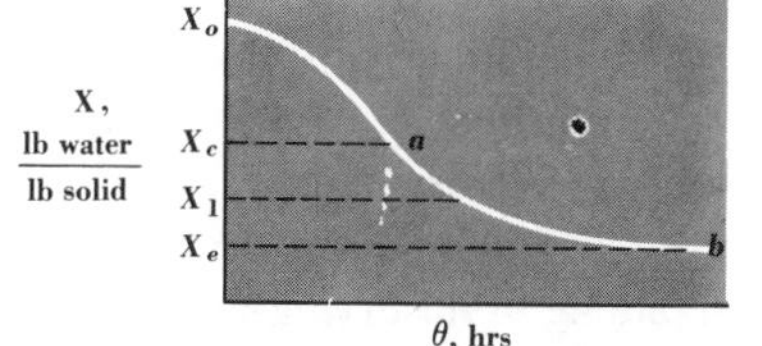

FIGURE 36-12
A typical curve of moisture content as a function of drying time for fixed air conditions.
a Critical moisture content.
b Equilibrium moisture content.

water is squeezed out from the solid, and there may be a long constant-rate period. At a certain *critical moisture content* dry patches begin to exist on the surface and the drying rate begins to fall. The liquid surface retreats into the interior of the solid, so that the evaporated moisture must diffuse through the pores of the solid. Because of the effect of capillarity, the liquid surface rises higher in small pores than in larger ones. Eventually the entire irregular liquid front moves to the interior of the solid and leaves the surface entirely dry. The time during which the surface is drying up is sometimes called the first falling-rate period. Often these periods are indistinguishable, but sometimes the rate falls more rapidly in the second period than in the first.

The diffusion of water vapor in the pores of a solid can be described by the differential mass balance, Eq. (7-19). The diffusivity based on the solid volume is less than that through air, for the cross section is partially filled with solid and the path for diffusion is lengthened by the tortuous nature of the pores. These effects reduce the effective diffusivity to something on the order of one-tenth that for air-water alone. Although the effective diffusivity within the solid is less than in the

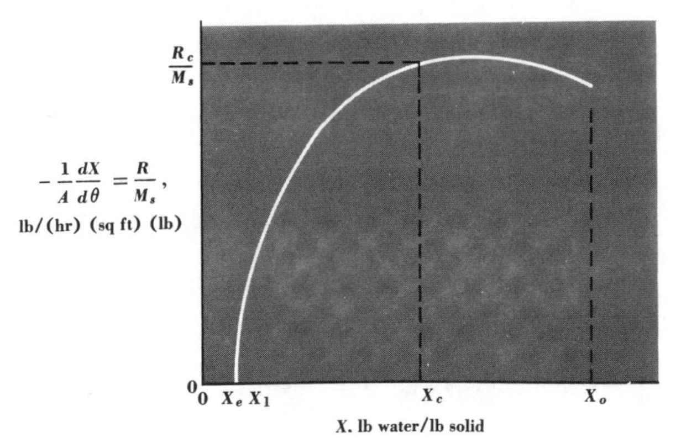

FIGURE 36-13
Drying-rate curve for fixed air conditions.

boundary layer, the effective thermal conductivity is usually higher than it is in the boundary layer. Heat flows through the dried solid by conduction. The result is that the temperature at the vapor-liquid interface rises above the wet-bulb temperature of the drying air.

The solution of the differential mass balance for the boundary values corresponding to the drying situation described above is difficult, but it is analogous to the freezing of a liquid by heat removal through the frozen layer. The unfrozen liquid is assumed to be well stirred. This heat-transfer problem has been discussed extensively.[1]

As the liquid surface retreats from the edge of the solid, islands of liquid may be left in the solid. In addition, the liquid may travel toward the surface by capillary movement through the pores or by diffusion in the fine structure of the solid itself. For these reasons the simple mechanism just described may not apply, and the rate of drying in many cases must be determined entirely empirically.

The rate of drying, as shown in Figs. 36-12 and 36-13, may reach zero before the solid is completely dry. The resulting *equilibrium moisture content* for a given solid is a function of temperature and of the condition of the drying air, as shown in Fig. 36-14.

The drying rate R for a mass M_s of solid being dried is defined as

$$R = -\frac{M_s}{A}\frac{dX}{d\theta} \qquad (36\text{-}77)$$

The units of R are lb water/(h)(ft^2). For a batch-drying operation Eq. (36-77) is integrated to

$$\theta = -M_s \int_{X_o}^{X_1} \frac{dX}{AR} \qquad (36\text{-}78)$$

In the simplest situation, A and R are constant, and Eq. (36-78) is integrated to

$$\theta = \frac{M_s}{AR}(X_o - X_1) \qquad (36\text{-}79)$$

If the variation of A and R is available in the form of a graph such as Fig. 36-13, Eq. (36-78) can be integrated graphically. Sometimes the curve of R versus X is simply a horizontal line followed by a straight line connecting the critical moisture content X_c, where $R = R_c$, to the equilibrium moisture content X_e, where $R = 0$. The integrated expression is then

$$\theta = \frac{M_s}{AR_c}\left[(X_o - X_c) + (X_c - X_e)\ln\frac{X_c - X_e}{X_1 - X_e}\right] \qquad (36\text{-}80)$$

[1] H. S. Carslaw and J. C. Jaeger, "Conduction of Heat in Solids," 2d ed., chap. 11, Clarendon Press, Oxford, 1959.

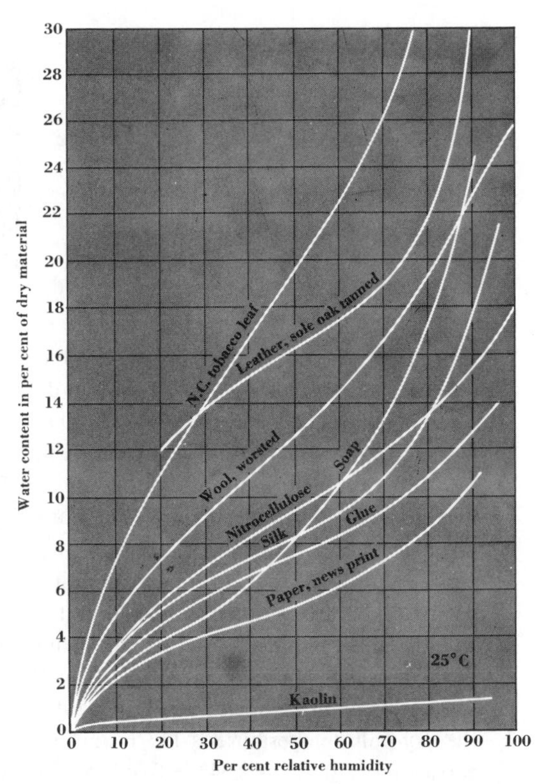

FIGURE 36-14
Equilibrium moisture content for various materials. (*From "International Critical Tables," vol. 2, pp. 322–325, McGraw-Hill Book Company, New York, 1928.*)

For a continuous drier, with air flow, for example, countercurrent to the movement of solid, the drying rate is a function both of moisture content of the solid and of the humidity and temperature of the air. The equations developed for a cooling tower apply to the constant-rate period. For the falling-rate period, the calculations must be based on experimental data obtained under conditions which simulate those obtained in the actual countercurrent apparatus.[1]

Example 36-4 A granular solid containing 93 percent water is to be dried in a fluidized drier to produce 100 lb/h of dry material. Atmospheric air is passed through a steam-heated exchanger in which it is heated to 170°F and then passed through the bed of solid, from which it emerges 90 percent

[1] D. B. Broughton and H. S. Mickley, *Chem. Eng. Progr.*, **49**:319 (1953).

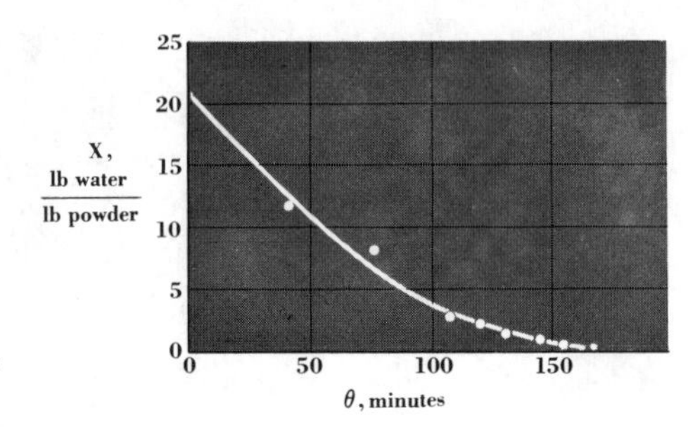

FIGURE 36-15
Experimental drying data for Example 36-4.

saturated. The hot-air temperature is limited to 170°F because of the heat-sensitive nature of the solid. Base the design on the worst probable atmospheric conditions: $t = 100$°F, $t_{wb} = 100$°F. Estimate the quantity of air needed and the duty of the heat exchanger.

This problem is solved with the aid of Fig. 36-7. The humidity of the inlet air is 0.043 lb water/lb air. The humidity is unchanged as the air passes through the heat exchanger. At the entrance to the drier, where $Y_A = 0.043$ and $t = 170$°F, t_{wb} is 110°F. As the air flows through the fluidized bed of drying solid, the operation is adiabatic; it is assumed that no heat is exchanged with the surroundings. The average temperature of the solid particles and their associated water is constant. Thus the air follows an adiabatic cooling line; its wet-bulb temperature is constant at 110°F. At 90 percent saturation the dry-bulb temperature is 113°F and $Y_A = 0.057$. Each pound of air thus carries away $0.057 - 0.043 = 0.014$ lb of water. The total water to be evaporated is $100/0.07 - 100 = 1330$ lb, so the air needed to produce 100 lb/h of dry material is $1330/0.014 = 95{,}000$ lb/h.

The heat required is computed from the enthalpies in Appendix Table A-23. At $t_{wb} = 110$°F, $H = 92.3$ Btu/lb, and at $t_{wb} = 100$°F, $H = 71.7$ Btu/lb, so the exchanger duty must be $(95{,}000)(92.3 - 71.7) = 1.96 \times 10^6$ Btu/h. This figure neglects the heat capacity of the solid and the heat needed to heat the charge to about 110°F.

It is possible to reduce the amount of air required by reheating the air that has passed through the bed of solid and using it to dry a second bed. The air at $t = 113$, $t_{wb} = 110$ is heated to 170° at $Y_A = 0.057$; at this condition $t_{wb} = 116$°F. After passing through the wet solid the temperature at 90 percent saturation is 119° and the humidity is 0.068. The total water carried is now $0.068 - 0.043$, or 0.025 lb/lb air, so the air required is reduced to 53,200 lb/h.

The rate of drying is best determined by a small-scale experiment; the results of such a run are shown in Fig. 36-15 for single-stage drying. The drier could be designed for a batch of 400 lb with 3 h for drying and 1 h between runs. During the falling-rate period the temperature of the solid would rise above the wet-bulb temperature, so that some safety factor should be allowed in the design. ////

PROBLEMS

36-1 Calculate the ratio h/h^o for the gas phase in a cooling tower at a point where $t_x = t_s =$ 120°F, $t_y = 140$°F, and $t_{ywb} = 105$°F.

36-2 Nitrogen at 1000°R and 1 atm is flowing past a flat plate. At a point 1 ft from the leading edge of the plate the temperature of the plate is 800°R and $\mathbf{Re}_x$ is 100,000. What is the heat flux at this point? Calculate the rate at which liquid oxygen at about 1 atm must be added for transpiration cooling of the wall at the same position if the new wall temperature is 500°R. The material of the wall is assumed to be a perfect insulator in the second case.

36-3 An air-water mixture has a dry-bulb temperature of 150°F and a wet-bulb temperature of 130° at 1 atm.

(*a*) What is the percent saturation of this mixture?
(*b*) What is the absolute humidity?
(*c*) What is the volume in cubic feet per pound of dry air?
(*d*) What is the volume in cubic feet per pound of mixture?
(*e*) What is the dew point?
(*f*) The mixture is cooled to 105°F. What percent of the moisture condenses out?
(*g*) The gaseous mixture from (*f*) left after condensation is then heated to 155°F. What is the percent saturation of this mixture?
(*h*) What is the humid heat, and how many Btu's are required to heat the air from 105 to 155°F?

36-4 Calculate the wet-bulb temperature of an ethanol-air mixture at 120°F and a percent saturation of 20 percent.

36-5 Find the height and diameter of a cooling tower to cool 5000 gal/h of water from 130 to 90°F using 2.5 times the minimum air rate. The air has a dry-bulb temperature of 80°F and a wet-bulb temperature of 65°F. The HTU is 20 ft, and the tower will be operated at $G = 700$ lb/(h)(ft^2). Neglect any resistance to heat transfer in the liquid phase.

36-6 A tray drier is to be used to dry sheets of material from 70 to 5 percent moisture content. The sheets are 4 ft by 5 ft by 2 in. From experiments on this material with air at the same condition as that to be used in the full-scale drier, it is found that the rate during the constant-rate period is 1.5 lb water/(h)(ft^2). The critical moisture content is 30 percent, and the equilibrium moisture content is negligible. If the material is dried from both sides and has a bone-dry density of 25 lb/ft^3, calculate the time required for drying, assuming the falling-rate period to be linear.

36-7 Three thousand pounds (bone dry) of a granular solid is to be dried under constant drying conditions from a moisture content of 0.20 lb/lb to final moisture content of

0.02 lb/lb. The material has an effective area of 0.30 ft^2/lb. Under the same conditions, the following rates were previously known. Calculate the time required for the drying.

Moisture content	Rate, lb/(h)(ft^2)
0.300	0.35
0.200	0.35
0.140	0.35
0.114	0.30
0.096	0.265
0.056	0.180
0.042	0.150
0.026	0.110
0.016	0.075

36-8 Nine hundred pounds per hour of sand (bone dry) is to be dried from a moisture content of 1 lb/lb to a moisture content of 0.2 by means of a continuous countercurrent tunnel drier. The air enters the drier with a humidity of 0.013 and a temperature of 195°F at a rate of 40,000 lb dry air/h. The drier will operate adiabatically, and the sand can be considered to enter and leave at the wet-bulb temperature of the air. From laboratory experiments under constant drying conditions with the air at the initial condition, it was found that the sand had a critical moisture content of 0.5 lb/lb. It was found that the rate was constant at 1.0 lb/(h)(ft^2) above the critical point and fell off linearly to zero for bone-dry sand during the falling-rate period.

Calculate the length of drier required if the sand has an equivalent area of 7 ft^2/ft of drier, assuming that the rate is a linear function of humidity difference.

36-9 Solve Prob. 35-1 for an adiabatic tower, taking into account the heat effects. Data on the ammonia-water system can be found in Perry.

36-10 It is proposed to recover 99.0 percent of the CO_2 in a gas mixture by scrubbing with a 1 *N* solution of diethanolamine in a tower packed with $\frac{3}{4}$-in Raschig rings. The inlet gas is 25 mol percent CO_2 and 75 mol percent H_2 and is to be treated at 5000 ft^3/h at 25°C and 1 atm. Consider the effect of the temperature rise of the absorbing solution, but the sensible heat of the gas phase may be neglected. The absorbing solution entering the top of the column contains 0.30 mol CO_2/mol amine and is to be supplied at a rate $2\frac{1}{2}$ times the true minimum rate (not the isothermal minimum rate). The necessary data can be found in an article by Cryder and Maloney, *Trans. Am. Inst. Chem. Eng.*, **37**:827 (1941). Estimate the height and diameter of the tower required.

36-11 A wet solid is to be treated in a two-stage drying system in such a way that the solid always remains at the wet-bulb temperature of the air entering the particular stage.

Saturated air at 60°F is preheated to 135°F and then passed through the first stage. The contact is such that the air leaves at 80 percent saturation. This air is then reheated to 164°F and passed through the second stage, again reaching 80 percent saturation. Calculate the heat needed in the two heat exchangers, Btu/lb of total water evaporated in the two stages.

37

SEPARATION BY EQUILIBRIUM STAGES; IMMISCIBLE PHASES

The Concept of the Ideal Stage

In Chap. 35 we studied separation operations in which two fluid phases flow past one another countercurrently or concurrently in such a way that they are in continuous contact. The solute tends to migrate from one phase to the other throughout the length of the contacting device. It was seen that an infinite interphase area is required to achieve equilibrium between the phases at one end of the tower, and it was assumed that there is negligible mixing in the direction of flow (longitudinal mixing).

A large proportion of industrial separations are performed in equipment in which the interphase contacting is done countercurrently in *stages*. The two phases are intimately mixed and then separated and led independently to the adjacent stages. The contact between the phases as they move through the device is intermittent rather than continuous. Typical stages used in liquid-liquid extraction are shown in Fig. 37-1.

In an ideal, or equilibrium, stage, the phases are mixed long enough so that the streams leaving the stage are in equilibrium, and the phases are mixed so thoroughly

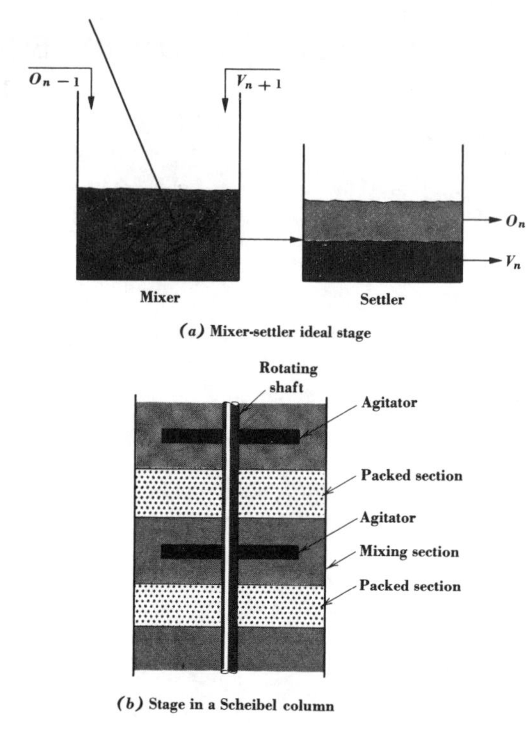

FIGURE 37-1
Contactors for liquid-liquid extraction.

that there are no gradients of concentration in them. By contrast, in a real stage, there may not be sufficient contact between the phases to bring the streams into equilibrium. Such a stage accomplishes less interphase transfer than an ideal stage; it is said to be less efficient than an ideal stage. At the end of this chapter we shall study the factors which affect stage efficiency, but first we shall consider the relation between the number of ideal stages and the separation achieved in a cascade of these stages.

Separation by a Single Equilibrium Contact

In gas absorption a single equilibrium contact can be obtained by introducing a gas containing solute A through a bubbler at the bottom of a vessel containing the absorbing solvent. The gas is added at a constant rate V, and the liquid enters the

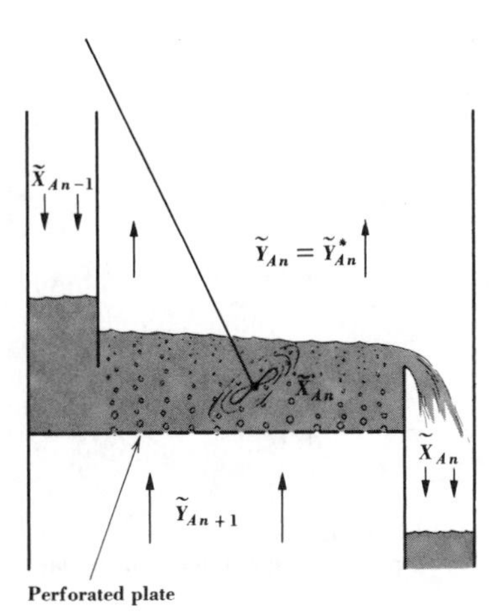

FIGURE 37-2
Sketch of a hypothetical ideal contacting stage.

contactor at a constant rate O.[1] If the bubbles are small and the liquid is well mixed and deep enough to give sufficient time of contact, the gas leaving is in equilibrium with the liquid. The device is equivalent to one ideal stage.

In calculations for separations by stages the only need to consider mass-transfer resistance is in the case of real stages, for which the efficiency can be related to the mass-transfer rate between the phases in contact. It is simply assumed that in an ideal stage there is enough contact so that the phases leave the stage at equilibrium.

The material balance for the stage is conveniently expressed in terms of mole ratios and the flow rates of the immiscible inert streams. The units of composition used in Chaps. 35 and 36 were mole or mass fractions, which are the customary measure of the driving force for mass transfer. In mass-balance or operating-line equations, however, it is often convenient to use mole or mass ratio units ($\tilde{Y}_A$ or Y_A), as was illustrated in Examples 35-1 and 35-2, particularly if the phases are immiscible.

A single stage for gas absorption is shown schematically in Fig. 37-2. A mass balance on component A can be made to obtain the equation

$$\frac{\tilde{Y}_{An} - \tilde{Y}_{An+1}}{\tilde{X}_{An} - \tilde{X}_{An-1}} = -\frac{\tilde{O}_C}{\tilde{V}_B} \tag{37-1}$$

The symbol $\tilde{O}_C$ is the molar flow rate of inert constituent C in the liquid phase, while

[1] The quantities O and V are measured in pounds per hour for steady operation or simply in pounds for batch operation. The ratios O/V and L/G are equal, but in stagewise operations it is not so natural to refer both items to the same cross-sectional area. Especially when dealing with one stream alone, we use the symbols O and V rather than the mass velocities.

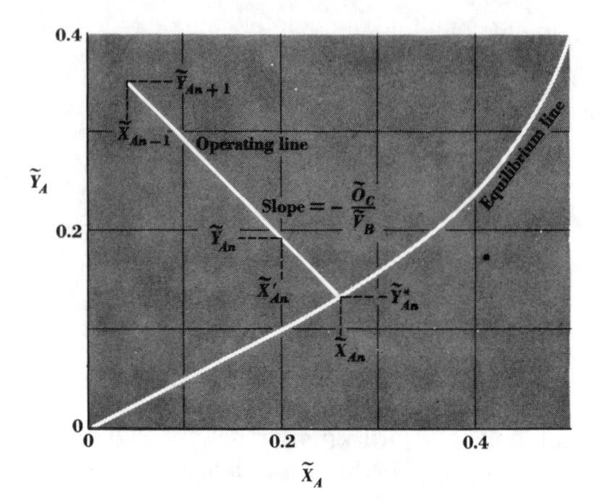

FIGURE 37-3
Operating diagram for a single stage.

$\tilde{V}_B$ is the molar flow rate of the insoluble constituent B in the gas. For immiscible fluids, $\tilde{O}_C/\tilde{V}_B$ is a constant; that is, $\tilde{O}_{Cn} = \tilde{O}_{Cn-1}$, and $\tilde{V}_{Bn} = \tilde{V}_{Bn+1}$. If mole fractions were used, the analogous equation would of course be valid, but $\tilde{O}/\tilde{V}$ would be a function of the composition of a phase. The balance could also be expressed in terms of mass ratios or fractions.

The operating diagram for a single-stage contactor is shown in Fig. 37-3. Since the streams leaving are in equilibrium, we have $\tilde{Y}_{An} = \tilde{Y}^*_{An}$; $\tilde{Y}^*_{An}$ and $\tilde{X}_{An}$ are on the equilibrium line. The four compositions pertinent to the contactor are shown in Fig. 37-3. The position of the operating line is similar to that given in Fig. 35-9. However, for the well-mixed phases involved in the equilibrium contact, intermediate points on the operating line do not represent adjacent phases as in true concurrent flow. As applied to the equilibrium stage, only the extremities of the operating line shown in Fig. 37-3 have meaning.

If the time of contact or interfacial area in the stage is not sufficient to bring the phases leaving the contactor into equilibrium, the compositions of the streams are given by $\tilde{X}'_{An}$ and $\tilde{Y}_{An}$, shown in Fig. 37-3. The efficiency of the stage can be defined by the equation

$$E_G = \frac{\tilde{Y}_{An+1} - \tilde{Y}_{An}}{\tilde{Y}_{An+1} - \tilde{Y}^*_{An}} \tag{37-2}$$

in which E_G is the Murphree V-phase efficiency of the stage.

In the device shown in Fig. 37-2 the liquid is well mixed and each phase may be presumed to leave the contactor with a uniform composition. However, interphase contacting is often done by a bubble-cap tray such as the one shown in Fig. 37-4. The liquid flow is normal to the vapor flow, so that there must be a gradient of

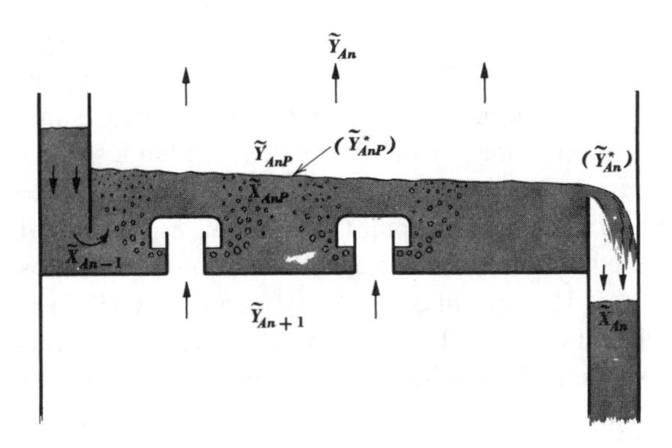

FIGURE 37-4
A bubble-cap tray.

composition in the liquid phase within the stage. As applied to this contacting device, the symbol $\tilde{Y}^*_{An}$ in Eq. (37-2) refers to the vapor composition which would be in equilibrium with the actual liquid leaving the tray, that is, $\tilde{X}_{An}$. As the liquid travels from inlet to outlet, its composition varies from $\tilde{X}_{An-1}$ to $\tilde{X}_{An}$. The vapor entering the tray is assumed to have a uniform composition $\tilde{Y}_{An+1}$, and $\tilde{Y}_{An}$ is the bulk, or mixing-cup, composition of the vapor leaving the tray.

For a contactor with incompletely mixed phases we define a Murphree point efficiency,

$$E_{GP} = \frac{\tilde{Y}_{An+1} - \tilde{Y}_{AnP}}{\tilde{Y}_{An+1} - \tilde{Y}^*_{AnP}} \tag{37-3}$$

In this equation $\tilde{Y}_{AnP}$ and $\tilde{Y}^*_{AnP}$ represent the actual and equilibrium compositions at a point on the tray at the upper liquid surface, as shown in Fig. 37-4. The quantity E_{GP} is a measure of the efficiency of contacting in a small column of liquid located at the point P, in which the rising Y phase contacts an X phase of composition $\tilde{X}_{AnP}$. The Murphree point efficiency can be related to the mass-transfer rate in this hypothetical small column, as characterized, for example, by the quantities n_{OG} and H_{OG}. The maximum value of E_{GP} is unity.

If the contacting on the plate of Fig. 37-4 is good enough so that E_{GP} is everywhere equal to 1.0, it is evident that, for gas absorption, $\tilde{Y}_{An}$ would be less than $\tilde{Y}^*_{An}$, for the latter would be in equilibrium with the richest liquid. According to Eq. (37-2), E_G, the Murphree stage efficiency, would be greater than 1. Since an ideal stage is defined as one for which $E_G = 1.0$, a stage with cross flow such as shown in Fig. 37-4 would not be ideal if E_{GP} were everywhere unity. Note that for the well-mixed stage shown in Fig. 37-2, E_{GP} is everywhere equal to E_G.

The value of E_G can be obtained by the integration of E_{GP} over the plate. The quantity E_G/E_{GP} is a function of the flow pattern in the stage and the concentration distribution in the $\tilde{Y}_{An+1}$ stream, as well as the values of m and $\tilde{O}_C/\tilde{V}_B$. The possible variations of the flow pattern within a single stage can be compared with the various flow patterns which are possible in a heat exchanger. The integration procedure used to obtain the ratio E_G/E_{GP} is somewhat analogous to the integration used to obtain the correction factor Y for cross-flow or multipass heat exchangers.

The units of composition used in Eqs. (37-2) and (37-3) can be changed as may be convenient. In liquid-liquid extraction, mass ratios or fractions may be used, and we shall find in Chap. 39 that mole fractions are used in the analysis of distillation, so that we can define another form as

$$E_G = \frac{\tilde{y}_{An+1} - \tilde{y}_{An}}{\tilde{y}_{An+1} - \tilde{y}^*_{An}} \qquad (37\text{-}4)$$

Before further discussion of stage efficiencies, it is necessary to discuss the operation of multistage separation equipment. We shall return to the subject of stage efficiency at the end of this chapter.

Example 37-1 The acetic acid in a solution in water containing 0.15 mass fraction acid is to be extracted in a single equilibrium stage using pure methyl isobutyl ketone as the solvent. Solvent is used in the amount of 1.745 lb for every pound of rich solution treated. Find the compositions of the streams leaving the stage.

The compositions of the entering streams are the same as those of Example 35-2. It is convenient to use mass-ratio units, so the equilibrium data are converted from mass fractions as shown in Fig. 35-15 to mass ratios as shown in Fig. 37-5. The equations given in Example 35-2 have been used for this calculation.

The equation for the operating line can be deduced from Eq. (37-1) to be

$$\frac{Y_{An} - Y_{An+1}}{X_{An} - X_{An-1}} = -\frac{O_C}{V_B} \qquad (1)$$

For this problem, we have

$$X_{An-1} = 0.15/0.85 = 0.1768$$

$$Y_{An+1} = 0$$

$$\frac{O_C}{V_B} = 0.85/1.745 = 0.487$$

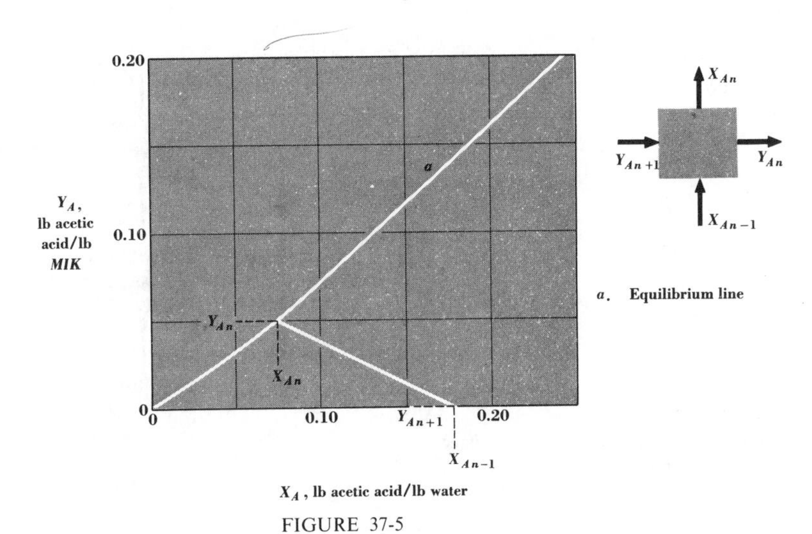

FIGURE 37-5
Operating diagram for Example 37-1; single-stage extraction.

In Fig. 37-5, one end of the operating line is at X_{An-1}, Y_{An+1}, and its slope is $-O_C/V_B$, or -0.487. The intersection of the operating line with the equilibrium line is at X_{An}, Y_{An}, so that we obtain

$$X_{An} = 0.074$$
$$Y_{An} = 0.050$$

or, expressed in mass fractions,

$$x_{An} = 0.0688$$
$$y_{An} = 0.0476$$

These values show that much less extraction has been obtained in one equilibrium stage than was obtained in the continuous countercurrent tower of Example 35-2. Note that the ratio of solvent to feed is the same for both examples. ////

Multiple Contact with Fresh Absorbent (Solvent) at Each Stage

Let us reconsider the gas absorption discussed in the previous section. Instead of contacting the phases in a single stage, it is possible to use a number of stages. In Fig. 37-6 the absorbent is split into three equal parts and added to three stages as

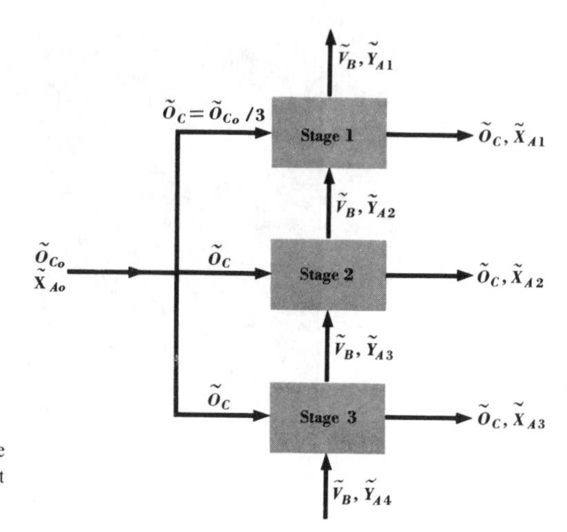

FIGURE 37-6
Schematic diagram of a three-stage contactor with fresh absorbent or solvent at each stage.

shown. The total amounts and the compositions of the two phases entering the system are the same as in Figs. 37-2 and 37-3.

The operation of the cascade of stages is analyzed in Fig. 37-7. We start with stage 3, in which the phases of composition $\tilde{Y}_{A4}$ and $\tilde{X}_{Ao}$ are brought into contact. Equation (37-1) applies, with $\tilde{O}_C = \tilde{O}_{Co}/3$. All rates $\tilde{V}_{Bn}$ are equal because component B is insoluble in the O phases, so that these streams are referred to as $\tilde{V}_B$. The same construction is used as in Fig. 37-3, and the points $\tilde{Y}_{A3}$ and $\tilde{X}_{A3}$ are found on Fig. 37-7. In stage 2, $\tilde{Y}_{A3}$ is contacted with $\tilde{X}_{Ao}$, and the resulting phase compositions are $\tilde{Y}_{A2}$ and $\tilde{X}_{A2}$. The first stage reduces the concentration in the V

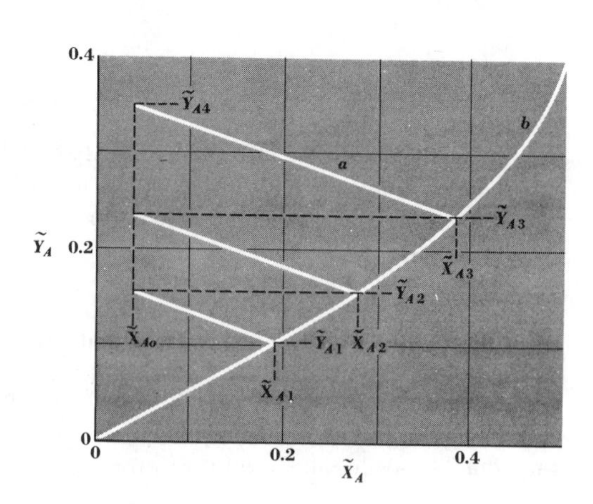

FIGURE 37-7
Operating diagram for multiple-stage contact with fresh absorbent at each stage.
a Slope $= -\tilde{O}_C/\tilde{V}_B$.
b Equilibrium line.

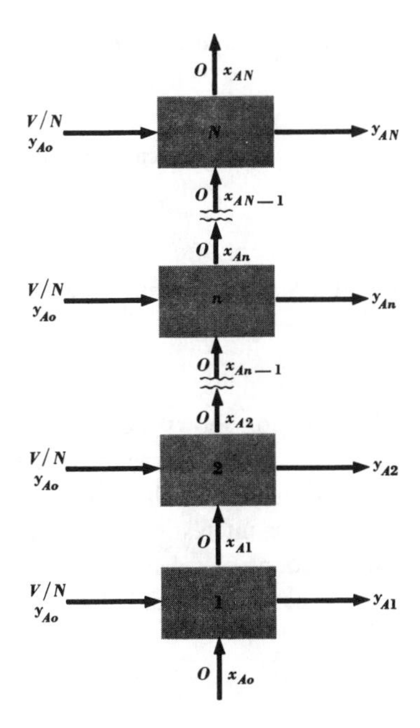

FIGURE 37-8
Multistage extraction with solvent at each stage.

phase to $\tilde{Y}_{A1}$, which equals 0.102, as against 0.135 when all the contacting was done in one stage as shown in Fig. 37-3.

The method of analysis which has been outlined can be applied to liquid-liquid extraction, as shown in Fig. 37-8. If the concentration of A in both phases is small, it is convenient to use mass fractions, and rates O and V can be considered constant through the system. For extraction or stripping, the operating lines on a figure like Fig. 37-7 would be on the opposite side of the equilibrium line.

Let us analyze a simple extraction problem with dilute solutions for which the equilibrium line is given by $y_A = mx_A$ and for which an equal amount of pure solvent ($y_A = 0$) is added to each stage. For this situation, with m, O, and V approximately constant, it is convenient to do the analysis algebraically. For the first stage, we write the mass balance for A as

$$O(x_{Ao} - x_{A1}) = \frac{V}{N} y_{A1} \tag{37-5}$$

in which N is the total number of stages.

We next replace y_{A1} by mx_{A1} and solve for x_{A1} to get

$$x_{A1} = \left(\frac{O}{mV/N + O}\right) x_{Ao} \tag{37-6}$$

A balance around the second stage yields

$$x_{A2} = \left(\frac{O}{mV/N + O}\right) x_{A1} \tag{37-7}$$

These two equations can be combined to give

$$x_{A2} = \left(\frac{O}{mV/N + O}\right)^2 x_{Ao} \tag{37-8}$$

For the N stages we have, by induction,

$$\frac{x_{AN}}{x_{Ao}} = \left(\frac{1}{mV/NO + 1}\right)^N \tag{37-9}$$

This equation shows the extraction which is achieved in the N-stage system of Fig. 37-8. This type of extraction is often used in organic chemistry laboratory techniques.

We have seen that increasing the number of stages increases the fraction extracted; it is interesting to find the most extraction which would be possible with a fixed amount of solvent. This quantity is found by taking the limit of the ratio x_{AN}/x_{Ao} as N increases indefinitely. We proceed by writing Eq. (37-9) as

$$\ln \frac{x_{AN}}{x_{Ao}} = \frac{\ln [1/(mV/NO + 1)]}{1/N} \tag{37-10}$$

This expression becomes indeterminate as N increases, so L'Hôpital's rule is used to find

$$\lim_{N \to \infty} \ln \frac{x_{AN}}{x_{Ao}} = -\frac{mV}{O} \tag{37-11}$$

or

$$\lim_{N \to \infty} \frac{x_{AN}}{x_{Ao}} = e^{-mV/O} \tag{37-12}$$

Multiple-Stage Countercurrent Contact

We have seen in the previous sections of this chapter that the fraction of absorption or extraction obtained by a given amount of absorbent or solvent can be increased by dividing it among a number of stages. Referring to Fig. 37-8, it is evident that the extract emerging from the Nth stage contains little of component A and could be used as solvent in the richer stages of the system. It is therefore logical to arrange the stages so that the flow is countercurrent, as is illustrated for gas absorption in Fig. 37-9.

A mass balance around the stage n of Fig. 37-9 can be written as

$$\frac{\tilde{Y}_{An+1} - \tilde{Y}_{An}}{\tilde{X}_{An} - \tilde{X}_{An-1}} = \frac{\tilde{O}_C}{\tilde{V}_B} \tag{37-13}$$

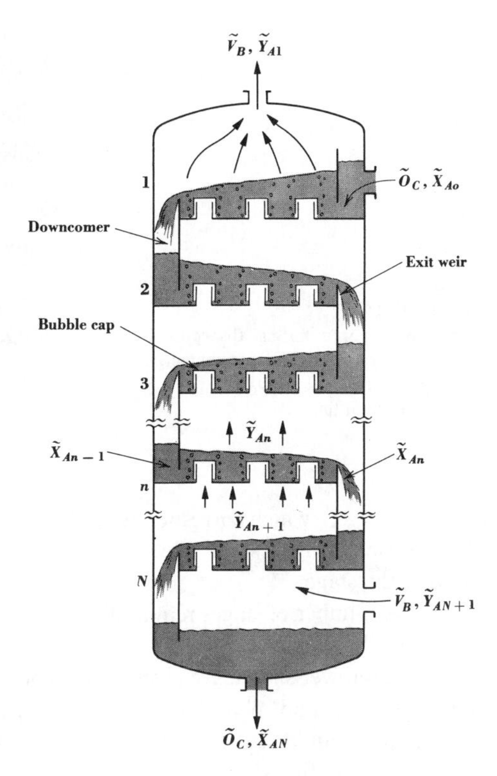

FIGURE 37-9
Countercurrent gas-absorption tower.

Alternatively, the balance can be written over the first n stages as

$$\frac{\tilde{Y}_{An+1} - \tilde{Y}_{A1}}{\tilde{X}_{An} - \tilde{X}_{Ao}} = \frac{\tilde{O}_C}{\tilde{V}_B} \tag{37-14}$$

or over all N stages as

$$\frac{\tilde{Y}_{AN+1} - \tilde{Y}_{A1}}{\tilde{X}_{AN} - \tilde{X}_{Ao}} = \frac{\tilde{O}_C}{\tilde{V}_B} \tag{37-15}$$

Although Eq. (37-13) is mathematically identical to Eq. (37-1), it is used in a different way. In Eq. (37-1) $\tilde{Y}_{An}$ and $\tilde{X}_{An}$ were considered as the coordinates of a point on the operating line, as in concurrent flow. In using Eqs. (37-13) and (37-14) we consider rather $\tilde{Y}_{An+1}$ and $\tilde{X}_{An}$ to be the coordinates of a point on the operating line. These quantities are often spoken of as the compositions of adjacent or passing streams in a contactor. Figure 37-10 shows the operating and equilibrium lines for the gas-absorption tower of Fig. 37-9. A point on the operating line, Eq. (37-14), represents

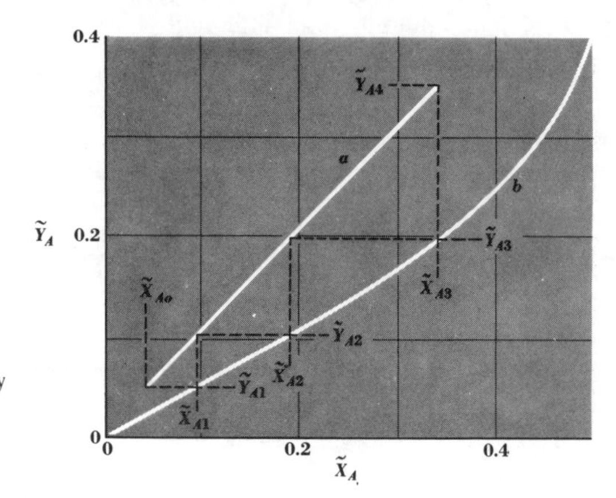

FIGURE 37-10
Operating diagram for gas absorption, by multiple-stage countercurrent contact.
a Operating line slope $\tilde{O}_C/\tilde{V}_B$.
b Equilibrium line.

a value of $\tilde{Y}_{An+1}$ which corresponds to a value of $\tilde{X}_{An}$, that is, the composition of the V phase entering a stage which corresponds to the composition of the O phase leaving the stage.

The number of stages required for a given separation is found by a procedure which is illustrated by Figs. 37-9 and 37-10. The conditions at the bottom of the absorption tower shown are given by the coordinates at the upper end of the operating line—$\tilde{Y}_{A4}$ and $\tilde{X}_{A3}$ for the three-stage countercurrent contactor of Fig. 37-10. The liquid leaving the third equilibrium stage is in equilibrium with the vapor leaving the stage, so $\tilde{Y}_{A3}$ is found from the intersection of $\tilde{X}_{A3}$ and the equilibrium line. With $\tilde{Y}_{A3}$ known, the composition of the liquid leaving the second stage $\tilde{X}_{A2}$ is found from the intersection of $\tilde{Y}_{A3}$ with the operating line. The result of the analysis which we are describing is a set of steps drawn between the operating and equilibrium lines, as shown in Fig. 37-10. The number of stages is given by the number of times the steps must touch the equilibrium line in order to reach the compositions $\tilde{Y}_{A1}$ and $\tilde{X}_{Ao}$ at the other extremity of the operating line.

The procedure of counting stages could obviously have been started alternatively at the dilute end of the tower. In Fig. 37-10 an integral number of stages is obtained, but such is not usually the case. A fraction of a stage has no physical meaning; however, if the overall stage efficiency is 50 percent and $3\frac{1}{2}$ equilibrium stages are obtained by the stepwise construction, there would be seven actual stages. Thus it is customary to speak of a fraction of a stage rather than merely say that, for example, four stages are obtained by the graphical construction.

The ratio $\tilde{O}_C/\tilde{V}_B$ and the compositions $\tilde{X}_{Ao}$ and $\tilde{Y}_{AN+1} = \tilde{Y}_{A4}$ have been chosen in Fig. 37-10 to be the same as in Figs. 37-7 and 37-3. The operating line in

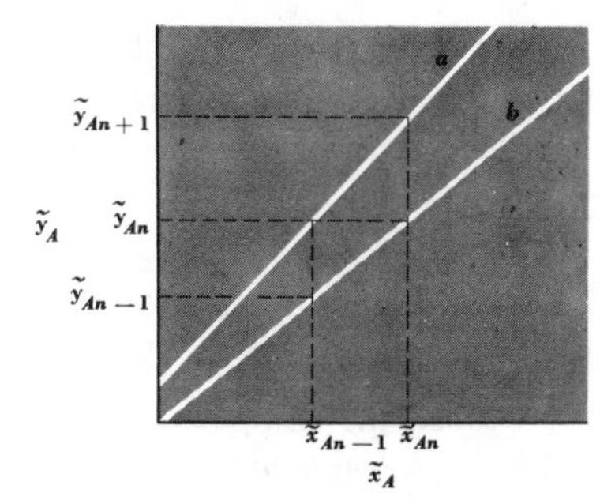

FIGURE 37-11
Height of a packed section equivalent to a theoretical stage for gas absorption.
a Operating line.
b Equilibrium line.

Fig. 37-10 has been placed so that three equilibrium stages are needed, just as in Fig. 37-7. A comparison of the three methods of operation shows that for a rich gas composition ($\tilde{Y}_{AN+1} = 0.35$) the exit-gas composition is 0.135 by a single-stage-equilibrium contact, 0.102 by a three-stage system with fresh absorbent to each stage, and 0.05 by a three-stage system with countercurrent flow.

For given flow rates and terminal compositions, the operating and equilibrium lines are the same on an $\tilde{x}_A$-$\tilde{y}_A$ or on an $\tilde{X}_A$-$\tilde{Y}_A$ diagram for countercurrent stagewise contact as for countercurrent continuous contact in a packed tower. As the operating and equilibrium lines become closer together, the number of stages goes up just as the quantity NTU goes up. All the remarks in Chap. 35 concerning the minimum absorbent or solvent rates apply equally well for stagewise contact. If there is a pinch point, the number of stages becomes infinite. We have used mole ratios in Fig. 37-10 in order to obtain a straight operating line, but mole fractions could have been used equally well, as in Chap. 35.

Algebraic Relations for the Number of Ideal Stages

From the methods of finding the number of theoretical stages for countercurrent flow illustrated in this chapter, it may be suspected that there is a relation between the number of transfer units, say n_{OG}, and the number of theoretical stages, N. If the operating and equilibrium lines are straight, the analysis is simple and is explained by reference to Fig. 37-11. Consider a section of a packed tower which is just sufficient to accomplish the separation which would be obtained by one theoretical stage. The height of this packed section is given for gas absorption by

$$\begin{aligned} z_{TS} &= H_{OG} n_{OG} \\ &= H_{OG} \frac{(\tilde{y}_{An+1} - \tilde{y}_{An}) \ln [(\tilde{y}_{An+1} - \tilde{y}_{An})/(\tilde{y}_{An} - \tilde{y}_{An-1})]}{(\tilde{y}_{An+1} - \tilde{y}_{An}) - (\tilde{y}_{An} - \tilde{y}_{An-1})} \end{aligned} \tag{37-16}$$

The logarithmic mean V-phase driving force has been used to find n_{OG} for the separation occurring at plate n.

We can also write the identity

$$\frac{\tilde{y}_{An+1} - \tilde{y}_{An}}{\tilde{y}_{An} - \tilde{y}_{An-1}} = \frac{\tilde{y}_{An+1} - \tilde{y}_{An}}{\tilde{x}_{An} - \tilde{x}_{An-1}} \frac{\tilde{x}_{An} - \tilde{x}_{An-1}}{\tilde{y}_{An} - \tilde{y}_{An-1}} \tag{37-17}$$

The right-hand side of this equation is recognized as $\tilde{L}/m\tilde{G}$ or $1/\mathcal{S}$, Eq. (35-23), where $\mathcal{S}$ is called the stripping factor. The height z_{TS}, which is sometimes called the *height equivalent to a theoretical stage*, or HETS, is found by combining Eqs. (37-16) and (37-17) with the definition of $\mathcal{S}$ to give

$$z_{TS} = \text{HETS} = \frac{H_{OG} \ln \mathcal{S}}{\mathcal{S} - 1} \tag{37-18}$$

For an entire tower we can write

$$z = \text{HETS}(N) = H_{OG} n_{OG} \tag{37-19}$$

so that we have

$$N = \left(\frac{\mathcal{S} - 1}{\ln \mathcal{S}} \right) n_{OG} \tag{37-20}$$

The combination of Eqs. (37-20) and (35-22) then gives the following formula for the number of theoretical stages:

$$N = \frac{\ln \{(1 - \mathcal{S})[\tilde{y}_{AN+1} - m\tilde{x}_{Ao})/(\tilde{y}_{A1} - m\tilde{x}_{Ao})] + \mathcal{S}\}}{\ln (1/\mathcal{S})} \tag{37-21$\mathcal{A}$}$$

The subscripts on the compositions, given here as mole fractions, correspond for gas absorption to those given in Fig. 37-9. If the operating and equilibrium lines are parallel, Eq. (37-20) shows that the number of transfer units and the number of theoretical stages are identical.

Although the height of a packed tower can be found from the HETS and N, it is preferable to use HTUs and the methods of Chap. 35. However, Eq. (37-21) is convenient for stagewise contactors if the operating and equilibrium lines are straight.

If the number of theoretical stages is given, it is desirable to obtain an expression for the degree of separation as an explicit function of N. The first step is to rearrange Eq. (37-21) to yield

$$\frac{\tilde{y}_{A1} - m\tilde{x}_{Ao}}{\tilde{y}_{AN+1} - m\tilde{x}_{Ao}} = \frac{\mathcal{A} - 1}{\mathcal{A}^{N+1} - 1} \tag{37-22}$$

where $1/\mathcal{S}$ has been replaced by the absorption factor $\mathcal{A}$ (that is, $\tilde{L}/m\tilde{G}$). We write next the identity

$$\frac{\tilde{y}_{A1} - \tilde{y}_{Ao}}{\tilde{y}_{AN+1} - \tilde{y}_{Ao}} = 1 - \frac{\tilde{y}_{AN+1} - \tilde{y}_{A1}}{\tilde{y}_{AN+1} - \tilde{y}_{Ao}} \tag{37-23}$$

in which $\tilde{y}_{Ao}$ has replaced $m\tilde{x}_{Ao}$. Equation (37-22) then becomes

$$\frac{\tilde{y}_{AN+1} - \tilde{y}_{A1}}{\tilde{y}_{AN+1} - \tilde{y}_{Ao}} = \frac{\mathcal{A}^{N+1} - \mathcal{A}}{\mathcal{A}^{N+1} - 1} \tag{37-24$\mathcal{A}$}$$

This equation is convenient because its left-hand side is the ratio of the actual absorption accomplished by N stages to the maximum possible absorption which could be obtained. This maximum occurs when $N = \infty$ and $\tilde{y}_{A1} = \tilde{y}_{Ao}$ for the more common case of $\mathcal{A} > 1.0$. It was mentioned in Chap. 35 that $\mathcal{A}$ (or $1/\mathcal{S}$) often has an economic value of about 1.4 for absorption in packed towers; the same is true for stagewise contactors such as the bubble-cap tower shown in Fig. 37-9. Equation (37-24) was first derived by Kremser[1] and by Souders and Brown[2] by a method different from the one we have used.

Equations (37-21$\mathcal{A}$) and (37-24$\mathcal{A}$) apply to absorption (operating line above equilibrium line); for stripping or extraction (operating line below equilibrium line) the proper relations are

$$N = \frac{\ln\left\{(1 - \mathcal{A})\left[\left(\tilde{x}_{Ao} - \frac{\tilde{y}_{AN+1}}{m}\right)\Big/\left(\tilde{x}_{AN} - \frac{\tilde{y}_{AN+1}}{m}\right)\right] + \mathcal{A}\right\}}{\ln(1/\mathcal{A})} \tag{37-21$\mathcal{S}$}$$

and

$$\frac{\tilde{x}_{Ao} - \tilde{x}_{AN}}{\tilde{x}_{Ao} - \tilde{x}_{AN+1}} = \frac{\mathcal{S}^{N+1} - \mathcal{S}}{\mathcal{S}^{N+1} - 1} \tag{37-24$\mathcal{S}$}$$

The algebraic relations for N given in this section assume $\mathcal{A}$ (or $\mathcal{S}$) is constant through the tower. If there is a slight variation, the value of $\mathcal{A}$ to use is that in the region where the operating lines and equilibrium line are close together. For the usual situation this region is at the dilute end of the contactor ($\mathcal{A} > 1.0$ for absorption, $\mathcal{S} > 1.0$ for stripping or extraction).

Example 37-2 It is desired to absorb 95 percent of the acetone in a 2.00 mol percent mixture of acetone in air in a countercurrent bubble-cap tower using 20 percent more than the minimum water rate. Pure water is introduced at the top of the tower, and the gas mixture is blown into the bottom at 1000 lb/

[1] A. Kremser, *Natl. Pet. News*, **22**(21):42 (1930).

[2] M. Souders and G. G. Brown, *Ind. Eng. Chem.*, **24**:519 (1932).

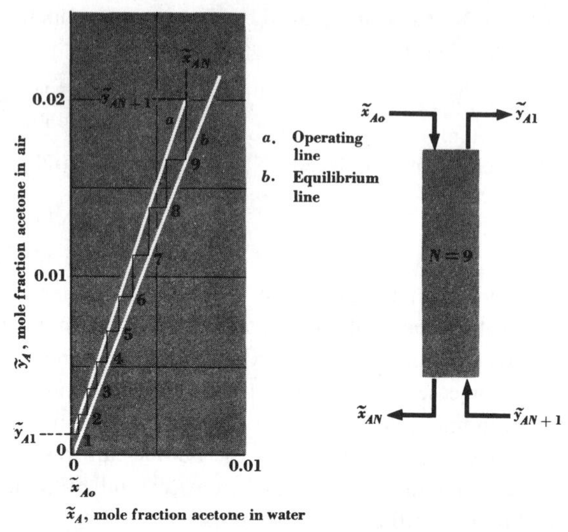

FIGURE 37-12
Operating diagram for Example 37-2; countercurrent absorption.

h. Find the number of equilibrium stages required for this separation. It is assumed that the tower operates isothermally at 80°F and 1 atm; the equilibrium relation is $\tilde{y}_A = 2.53\tilde{x}_A$.

The separation to be performed is the same as that of Example 35-1, and the equilibrium and operating lines of Fig. 37-12 are the same as those of Fig. 35-14. The terminal compositions are

$$\tilde{x}_{Ao} = 0$$

$$\tilde{x}_{AN} = 0.00658$$

$$\tilde{y}_{AN+1} = 0.0200$$

$$\tilde{y}_{A1} = 0.00102$$

The slope of the operating line is $\tilde{L}/\tilde{G} = 2.89$.

The stages are counted as shown on Fig. 37-12, starting at the first stage, i.e., the top of the tower. The points on the equilibrium line represent the compositions of streams leaving a given stage. Points on the operating line represent the compositions of the streams passing between stages, such as $\tilde{y}_{A5}$ and $\tilde{x}_{A4}$. The number of stages required is 9.0.

If Eq. (37-21) is used with $\mathscr{S} = 0.875$ and $\tilde{y}_{AN+1}/\tilde{y}_{A1} = 20$ $(\tilde{x}_{Ao} = 0)$, it is found that $N = 9.0$. Since the operating and equilibrium lines are almost parallel, the number of stages differs little from the number of transfer units found in Example 35-1 as 9.65. ////

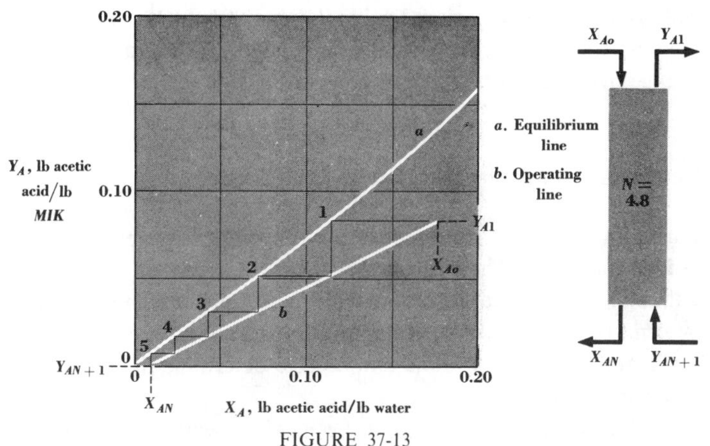

FIGURE 37-13
Countercurrent extraction by stages, Example 37-3.

Example 37-3 The acetic acid in a solution in water containing 0.15 mass fraction acid is to be extracted using pure methyl isobutyl ketone as a solvent. The extract is to contain 0.075 mass fraction acetic acid and the raffinate 0.010 mass fraction acid. Estimate the number of equilibrium stages required for this separation if countercurrent multiple contact is used.

The terminal compositions are the same as those of Example 35-2 and give the following mass ratios:

$$Y_{A1} = 0.0811$$

$$X_{Ao} = 0.1768$$

$$Y_{AN+1} = 0$$

$$X_{AN} = 0.0101$$

The operating line is straight and connects these terminal points as shown on Fig. 37-13.

The stages are counted starting from the rich end of the tower. We need about 4.8 stages for the desired separation. ////

In the discussion and in the examples which have been presented, it has been assumed that the temperature is constant throughout the tower. However, the same heat effects occur in a stagewise contactor as have been described for the continuous contactors of Chap. 36. Thus an energy balance as well as a mass balance must be made around a stage when heat effects are important. These temperature effects will displace the equilibrium line, as shown in Chap. 36. Sometimes cooling coils or intercoolers are placed between certain stages in order to keep the net temperature change over a contactor to a minimum.

Determination of the Number of Real Stages

Earlier in this chapter we defined the Murphree stage efficiency and the Murphree point efficiency. These quantities have been based on y-phase compositions; although similar equations exist in terms of x-phase compositions, we use only those in terms of y-phase compositions in this discussion. Let us now see how the values of E_G and E_{GP} may be found and used to determine the number of real stages in a separation device.

The point efficiency E_{GP} has been defined with respect to the performance of a small column of fluid at a point on a bubble-cap tray or similar contacting device. We again base our explanation on the example of gas absorption, but the generality of the equations should be recalled; they apply to liquid-liquid extraction, to distillation, and to other operations, provided suitable units are used. By analogy with Eq. (37-3), we have

$$E_{GP} = \frac{\tilde{y}_{An+1} - \tilde{y}_{AnP}}{\tilde{y}_{An+1} - \tilde{y}^*_{AnP}} \tag{37-25}$$

Mole fractions are the appropriate units for gas absorption involving dilute solutions, and they will also be used for distillation.

The operation of the hypothetical column at a point on the tray shown in Fig. 37-4 can also be expressed in terms of the equations of Chap. 35. For example, we have, from Eq. (35-18),

$$n_{OG} = \int_{\tilde{y}_{AnP}}^{\tilde{y}_{An+1}} \frac{d\tilde{y}_A}{\tilde{y}_A - \tilde{y}^*_A} \tag{37-26}$$

The liquid composition at a point on the tray is taken as constant over the height of the little column; $\tilde{y}^*_A$ is thus constant and equal to $\tilde{y}^*_{AnP}$. Equation (37-26) is integrated to give

$$n_{OG} = \ln \frac{\tilde{y}_{An+1} - \tilde{y}^*_{AnP}}{\tilde{y}_{AnP} - \tilde{y}^*_{AnP}} \tag{37-27}$$

This equation is rearranged to give

$$\frac{\tilde{y}_{AnP} - \tilde{y}^*_{AnP}}{\tilde{y}_{An+1} - \tilde{y}^*_{AnP}} = e^{-n_{OG}} \tag{37-28}$$

The left-hand side of this equation can be changed to give

$$\frac{\tilde{y}_{An+1} - \tilde{y}^*_{AnP}}{\tilde{y}_{An+1} - \tilde{y}^*_{AnP}} - \frac{\tilde{y}_{An+1} - \tilde{y}_{AnP}}{\tilde{y}_{An+1} - \tilde{y}^*_{AnP}} = e^{-n_{OG}} \qquad (37\text{-}29)$$

or, from Eq. (37-25),

$$1 - E_{GP} = e^{-n_{OG}} \qquad (37\text{-}30)$$

The NTU is equal to z/H_{OG}, where z is the height of the little column in which the rising bubbles are in contact with the liquid on the tray. Equation (31-56) can be written in terms of NTU as

$$\frac{1}{n_{OG}} = \frac{1}{n_G} + \frac{m\tilde{G}}{\tilde{L}}\frac{1}{n_L} \qquad (37\text{-}31)$$

The number of individual transfer units is in turn equal to z/H_G or z/H_L, and we know from Chap. 34 that H_G and H_L are functions of the Reynolds number and the Schmidt number for the appropriate phase. A number of empirical correlations exist for finding n_G and n_L from the Schmidt number, the tray geometry, and the flow rates.[1] These correlations are based on a considerable amount of data for gas-liquid contactors; little work has been done for other systems.

Strictly speaking, the equations given require a constant value of m, but it is permissible, within the precision of the method, to use the average slope of the equilibrium line (not $\tilde{y}/\tilde{x}$) for the concentration range on the tray.

The next step in finding the number of real trays is to obtain E_G by integrating E_{GP} over the surface area of the tray. The process has been described qualitatively earlier in this chapter. The result depends on a number of factors, such as the extent of mixing in the phase; if both phases are well mixed, $E_G = E_{GP}$. Another possible assumption is that the gas phase is well mixed but that the liquid flows across the tray with no mixing in the direction of flow. A third assumption might be that the gas phase retains a concentration gradient in the horizontal direction as it rises from one plate to the next. A number of these situations have been evaluated.[2] Values of E_G/E_{GP} considerably more than unity are common in practice, and the ratio can increase to the order of 10s or 100s in special situations.

As the rate at which gas is bubbled through the liquid on the tray is increased, a point is reached where a considerable quantity of liquid droplets is carried up to the tray above. This entrainment partially nullifies the countercurrent flow and decreases the efficiency of the tray. An equation can be derived[3] which gives the

[1] For an excellent reference on this subject see Bubble-tray Design Manual, American Institute of Chemical Engineers, New York, 1958.

[2] For example, see W. K. Lewis, *Ind. Eng. Chem.*, **28**:399 (1936).

[3] A. P. Colburn, *Ind. Eng. Chem.*, **28**:526 (1936).

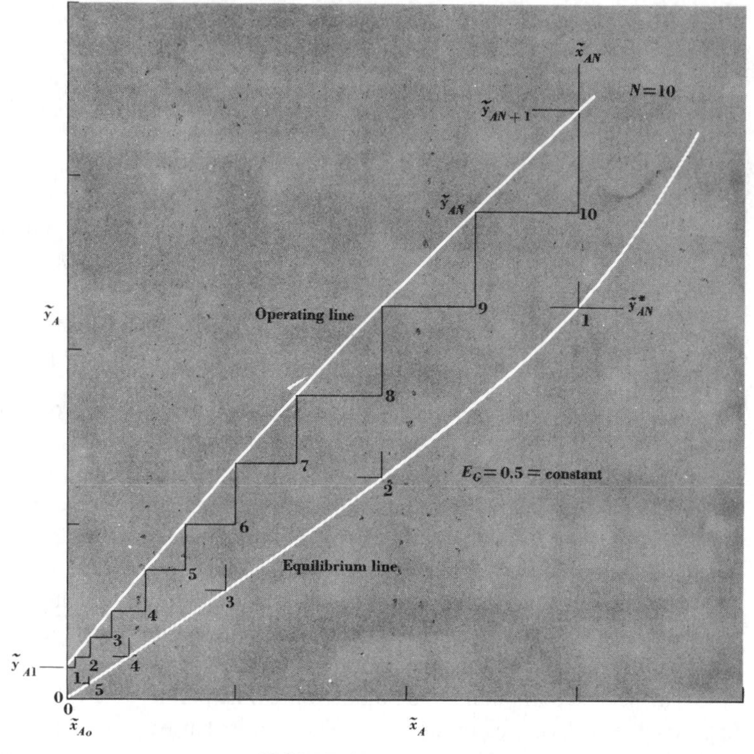

FIGURE 37-14
Calculation of the number of real trays for gas absorption.

effect on the efficiency of the entrainment of a known fraction of the liquid flowing across a plate. Entrainment is kept at a minimum in a properly designed tower.

The way in which the number of real stages for $E_G = 0.5$ is found is shown in Fig. 37-14. Starting from the rich end of the tower, the liquid leaving the bottom plate has the composition $\tilde{x}_{AN}$. The gas passing this liquid is $\tilde{y}_{AN+1}$, and the gas which would be in equilibrium with the liquid is $\tilde{y}^*_{AN}$. Since the value of E_G is 0.5, we proceed only one-half the way from $\tilde{y}_{AN+1}$ to $\tilde{y}^*_{AN}$ in order to locate $\tilde{y}_{AN}$, as shown in Fig. 37-14. The succeeding steps are counted as shown; 10 trays are needed for the separation shown.

The overall efficiency E_o is defined as the ratio of the number of ideal stages to the number of real stages. If this ratio is known, it is a simple matter to obtain the number of real stages after calculating the number of ideal stages. However, the value of E_G may vary with position in a contactor, and even if it is constant, as shown in Fig. 37-14, E_o does not equal E_G unless the operating and equilibrium lines are straight and parallel. The inequality of E_G and E_o may be verified by counting off the theoretical stages on Fig. 37-14; only about 4.5 are required, so E_o is 0.45. There is an analytical relation[1] between E_G and E_o if the operating and equilibrium lines are straight and E_G is constant.

The method of approach just outlined, which is that of the Am. Inst. Chem. Eng. Manual, is the most reliable procedure now available. However, a rough estimate of E_o can be obtained from empirical correlations.[2]

Notes on Tower Design

It has already been mentioned that entrainment is undesirable in a gas-liquid contactor. The allowable gas velocity in a tower is limited to a value which does not cause appreciable entrainment. A liquid drop will be entrained when the upward vapor velocity exceeds the terminal velocity of the drop. The drops are usually large enough so that the flow is turbulent, and the following equation can be derived.

$$u_t\left(\frac{\rho_G}{\rho_L - \rho_G}\right)^{1/2} = (3Dg)^{1/2} \qquad (37\text{-}32)$$

The diameters of the drops which may be entrained are influenced by the depth of liquid over the slots in the bubble caps, and the allowable approach to the terminal velocity is governed by the distance between plates. The correlation shown in Fig. 37-15 is based on these facts.

Equation (37-32) may also be written as

$$G_a = C_{sb}\sqrt{\rho_G(\rho_L - \rho_G)} \qquad (37\text{-}33)$$

in which G_a is the allowable mass velocity in lb/(s)(ft^2). Values of C_{sb} are given in Perry, p. 18-6. All velocities are based on the empty cross section of the tower.

The pressure drop from one tray to the next and the liquid head needed to cause the liquid to flow across the tray are discussed in Perry, sec. 18. The methods of calculation are applications of the principles of fluid dynamics presented in the first part of this book. The resistance to liquid flow determines the height to which the liquid backs up into the downspouts and thus the necessary plate spacing. If the liquid forms a foam in the tower, the height of liquid plus foam in the downcomer necessary to overcome a given resistance of flow increases, and flooding may result.

[1] Ibid.

[2] H. G. Drickamer and J. R. Bradford, *Trans. Am. Inst. Chem. Eng.*, **39**:319 (1943); H. E. O'Connell, *Trans. Am. Inst. Chem. Eng.*, **42**:741 (1946).

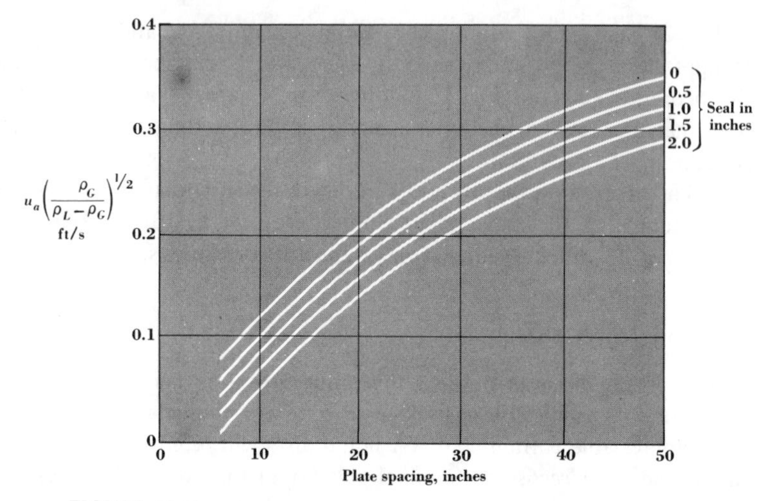

FIGURE 37-15
Allowable gas velocity u_a in gas-liquid tray towers. (*From C. S. Robinson and E. R. Gilliland, "Elements of Fractional Distillation," 4th ed., p. 430, McGraw-Hill Book Company, New York, 1950.*)

The liquid flowing across a tray is deeper on the upstream side near the exit of the downcomer from the tray above. Thus the gas tends to flow preferentially through the bubble caps or other gas openings near the exit weir on the tray. The pressure drop through the various rows of caps may be adjusted to equalize the gas flow, but the tray cannot operate at maximum efficiency with large variations in liquid or gas rates. Some of the recently developed valve-type plates almost eliminate the effects of the hydraulic gradient over a tray and have a somewhat wider operating range.

PROBLEMS

37-1 Find the number of ideal stages required for the separation of Prob. 35-1. Determine the allowable gas mass velocity if bubble-cap trays are used with a slot submergence or liquid seal of 2 in and a tray spacing of 24 in.

37-2 Estimate the number of real bubble-cap trays needed in Prob. 37-1 by using correlations found in Perry or the Am. Inst. Chem. Eng. Manual.[1]

37-3 Calculate the number of ideal stages needed for the separation of Prob. 35-2.

37-4 Estimate the height and diameter of a plate-type column with a 2-ft plate spacing and $E_o = 0.60$ to perform the separation of Prob. 35-3. The liquid seal is 2 in.

[1] Loc. cit.

37-5 Find the number of theoretical stages required for the separation of Prob. 35-4.

37-6 Derive the equation

$$E_o = \frac{\ln [1 + E_G(\mathcal{S} - 1)]}{\ln \mathcal{S}} \qquad (37\text{-}34)$$

in which $\mathcal{S}$ and E_G are constant over the tower. One approach is to find the height of a packed tower equivalent to a real stage and then obtain an equation which can be compared with Eq. (37-21) to find the desired relation.

37-7 The solubility of a substance A in water is given by the equation

$$\tilde{y}_A = 1.4\tilde{x}_A - 20\tilde{x}_A^{\,2}$$

(*a*) What is the minimum ratio of water rate to air rate which can be used in a tower with infinite plates to absorb all the entering solute using pure water as the solvent? The inlet gas contains 0.02 mol fraction A.

(*b*) What is the concentration of the strong liquor in part (*a*)?

(*c*) If the tower is used to desorb A from water, what is the minimum ratio of air rate to water rate which could be used if all the solute is to be desorbed using pure air as the desorbent? The inlet liquid contains 0.02 mol fraction A.

This problem is best solved analytically.

38

CONTACTING OF PARTIALLY MISCIBLE PHASES

Phase-Equilibrium Relations

In the study of separation processes up to this point we have been concerned with cases for which the two phases in contact are immiscible. The distributed component A (the solute) is present in both phases, but there has been no C in the B phase (which we refer to as the extract) and no B in the C phase, which we call the raffinate. This condition is approximately fulfilled for most gas-absorption problems and for certain problems in liquid-liquid extraction for low concentrations of solute. However, in many extraction problems the phases are partly miscible; i.e., there is considerable B in the raffinate phase and considerable C in the extract phase. In this chapter our explanation will be given in terms of liquid-liquid extraction, but the principles and equations can also be applied to other separation operations such as gas absorption or leaching.

Phase relations for ternary systems are shown graphically by triangular diagrams such as Fig. 38-1. The mass fraction of A is plotted in the horizontal direction, and lines of constant x_A or y_A are parallel to the BC side of the triangle. The mass fraction of B is plotted vertically, and lines of constant x_B and y_B are parallel to the AC side of the triangle. The mass fraction of C is easily found from the equation

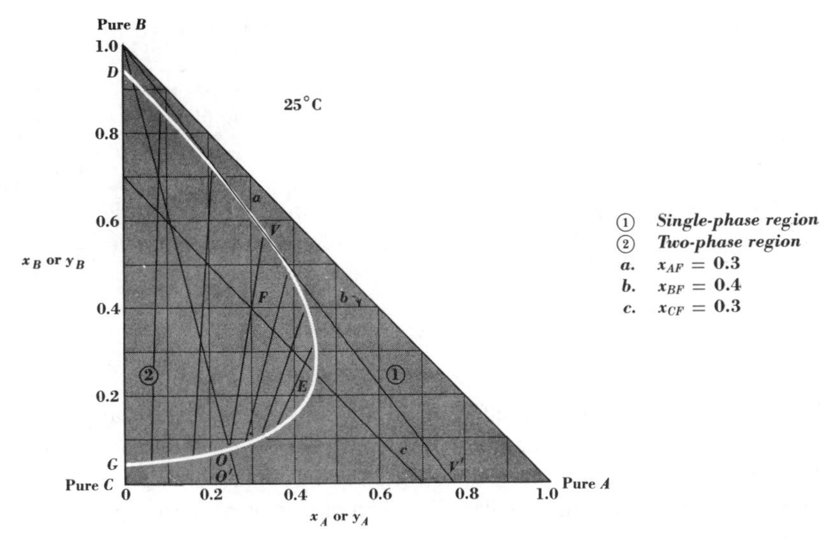

FIGURE 38-1
Equilibrium in a liquid-liquid system.

$x_C = 1 - (x_A + x_B)$; lines of constant x_C or y_C are parallel to the AB side of the triangle. These principles are illustrated in Fig. 38-1 by the point F, which has the composition $x_{AF} = 0.3$, $x_{BF} = 0.4$, $x_{CF} = 0.3$.

The line DEG in Fig. 38-1 is the solubility envelope of a liquid system. At equilibrium, two liquid phases are present within the solubility curve; one phase is present in the rest of the triangle. A mixture which has an overall composition lying within the two-phase region separates into two layers. The compositions of these layers, which are in equilibrium, are given by the slanting lines connecting points on the DE branch of the solubility curve to points on the EG branch. These lines are called *tie lines*. For example, if the composition of the raffinate phase is at O, the composition of the extract phase in equilibrium with it is at V, the other end of the tie line starting at O. The composition x_A is associated with the point O, and y_A with the point V. From the set of tie lines on Fig. 38-1 it is possible to construct the y-x, or distribution, diagram of Fig. 38-2. The point E in Figs. 38-1 and 38-2 is called the *plait point*; here the compositions of the two phases become identical.

At low concentrations of A the so-called distribution law, $y_A = mx_A$, is followed as shown on Fig. 38-2. In this range of concentration there is only a small amount of C in the extract phase and B in the raffinate phase. The operating line can be found here to a good approximation by the methods of Chaps. 35 and 37. At

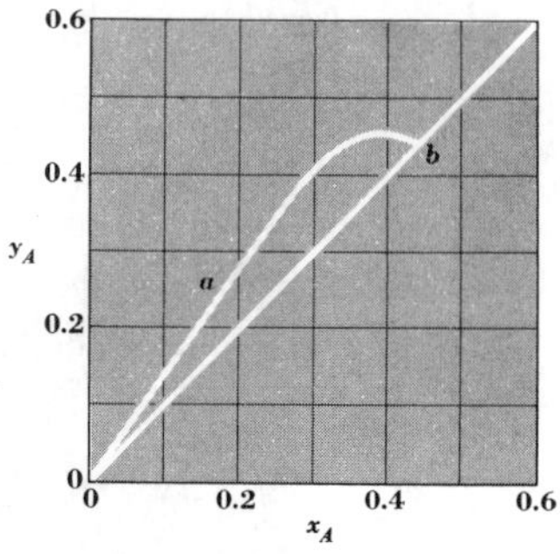

FIGURE 38-2
Distribution diagram which corresponds to Fig. 38-1.
a $y_A = mx_A$.
b E, plait point.

higher concentrations of A the effect of the miscibility of the phases may be considerable in locating the operating line. However, once the operating and equilibrium lines are plotted on a diagram such as Fig. 38-2, the number of stages can be stepped off graphically as was done in Chap. 37. This chapter is concerned principally with graphical methods of applying the overall mass balance to find the curved operating line which we see plotted in Fig. 38-12, for example.

Although Fig. 38-1 shows a common type of solubility curve, other configurations are often encountered; some of these are shown in Fig. 38-3. All the phase diagrams shown so far are for systems at one temperature. As the temperature is increased, miscibility almost always increases, as shown in Fig. 38-4.

We use right triangles to represent data on ternary systems, but equilateral triangles are often used in the literature. The principles set forth in this chapter apply to triangular diagrams of any shape or proportions. A right triangle is preferred because it is easy to construct on ordinary graph paper, and its proportions can be adjusted to expedite quantitative graphical procedures.

Mass-balance calculations are readily done graphically on a triangular diagram, and a number of similar diagrams[1] by use of the so-called *lever-arm principle*. When two streams of different composition are mixed, this principle requires that the point representing the composition of the resulting mixture lie on a straight line between the points. Conversely, if a mixture is separated into two streams, the overall mass balance requires that the points representing the compositions of the three streams involved lie on a straight line through the points. We now proceed to prove these assertions and show that certain quantitative relations exist among the various line segments.

Figure 38-5 can represent any physical separation process, such as distillation, in which a ternary feed stream is separated into two product streams. For the

[1] Extraction calculations are also often done on a Janecke diagram: see R. E. Treybal, "Mass Transfer Operations," 2d ed., chap. 10, McGraw-Hill Book Company, New York, 1968.

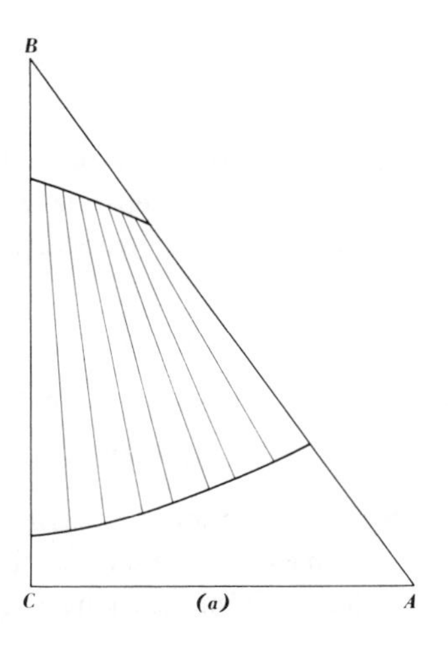

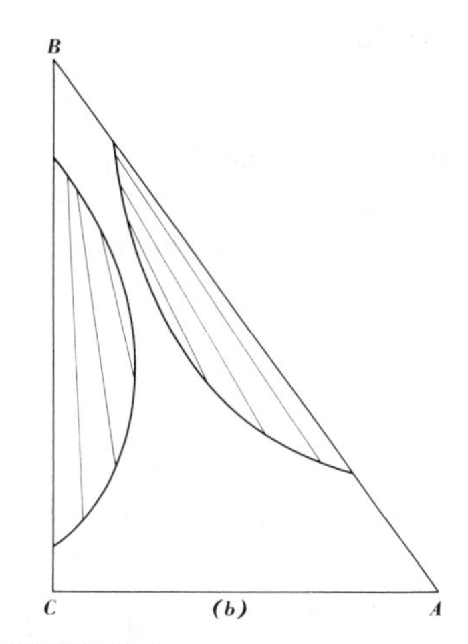

FIGURE 38-3
Two types of liquid equilibrium systems.

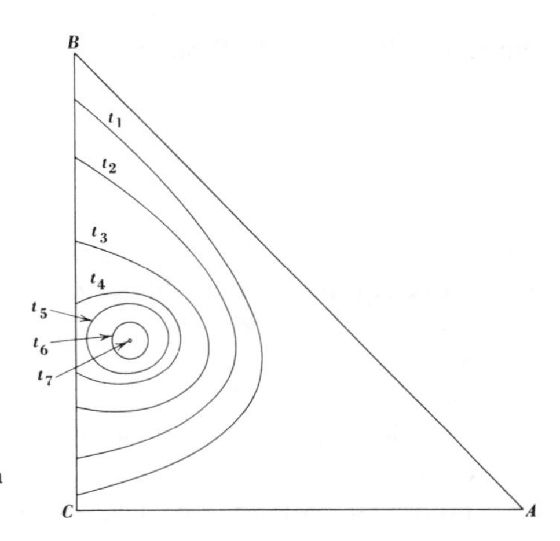

FIGURE 38-4
The effect of temperature on solubility in a typical liquid system; $t_{k+1} > t_k$.

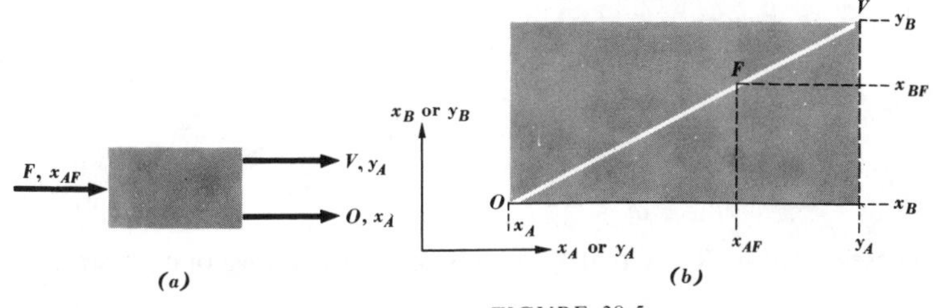

FIGURE 38-5
Demonstration of the lever-arm principle.

present example the feed stream is taken as a well-mixed, two-phase fluid which is introduced into a settling tank. In this tank the mixture separates into two layers which can be drawn off independently. The following mass balances can be written:[1]

$$F = O + V \tag{38-1}$$

$$Fx_{AF} = Ox_A + Vy_A \tag{38-2}$$

$$Fx_{BF} = Ox_B + Vy_B \tag{38-3}$$

The mass balance on C is not an independent equation. Referring now to Fig. 38-5*b*, we wish to prove that the slope of the line $\overline{OF}$ is the same as the slope of the line $\overline{FV}$; if this is so, then $\overline{OFV}$ is a straight line. From the triangle in the figure we have

$$\text{Slope of } \overline{OF} = \frac{x_{BF} - x_B}{x_{AF} - x_A} \tag{38-4}$$

Equations (38-1) and (38-3) can be solved for x_A and x_B to yield

$$x_A = \frac{(O + V)x_{AF} - Vy_A}{O} \tag{38-5}$$

$$x_B = \frac{(O + V)x_{BF} - Vy_B}{O} \tag{38-6}$$

When these equations are substituted in Eq. (38-4), we obtain

$$\text{Slope of } \overline{OF} = \frac{y_B - x_{BF}}{y_A - x_{AF}} \tag{38-7}$$

This equation is also the slope of the line $\overline{FV}$, so the assertion that $\overline{OFV}$ is a straight line is proved.

From Eqs. (38-1) and (38-2) we can also obtain

$$\frac{O}{V} = \frac{y_A - x_{AF}}{x_{AF} - x_A} \tag{38-8}$$

$$\frac{O}{F} = \frac{y_A - x_{AF}}{y_A - x_A} \tag{38-9}$$

$$\frac{V}{F} = \frac{x_{AF} - x_A}{y_A - x_A} \tag{38-10}$$

These equations show that the ratios of the quantities of each stream can easily be

[1] The symbols G and L represent mass velocities. V and O are flow rates in a steady process in pounds per hour; L/G and O/V are identical.

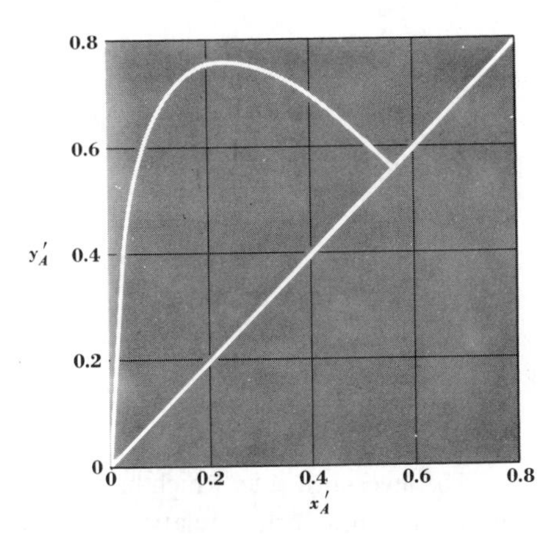

FIGURE 38-6
Selectivity diagram which corresponds to Fig. 38-1.

found graphically. By the principle of similar triangles it can be seen that the ratios can be found from the vertical distances or from the ratios of the line segments $\overline{OF}$, $\overline{FV}$, and $\overline{OV}$, themselves. For quantitative work using a right-triangular diagram, the horizontal or vertical distances are easily read from the graph.

The graphical principles we have just described can be applied to the system shown in Fig. 38-1. If a two-phase mixture having a composition given by point F is fed to a separator, represented schematically by Fig. 38-5*a*, the two phases into which the feed separates must lie both on a tie line and on a line through F, that is, on a tie line through F. The compositions of the two streams are given by the points O and V in Fig. 38-1. For a unit mass of F, we have, using B compositions,

$$O = \frac{0.57 - 0.40}{0.57 - 0.08} = \frac{0.17}{0.49} = 0.347$$

The amount of V is $1 - 0.347$, or 0.653.

The lever-arm principle can also be used for subtraction. For instance, if the stream V is put through a distillation column which removes all the B as one pure product, the remaining mixture of A and C must be on the AC side of the triangle and on a line through B and V, as shown by the point V' on Fig. 38-1. Here y_A equals 0.763. We refer to the composition given by point V' as the solvent-free composition of the mixture at V; we symbolize it by y'_A. The solvent-free composition of point O is found by drawing a line through B and O to $x'_A = 0.26$.

The solvent-free compositions of points O and V are of particular interest, for they represent the separation achieved between A and C after the extraneous solvent B has been removed. The values of x'_A and y'_A found in this fashion from Fig. 38-1 can be plotted as in Fig. 38-6. Such a graph is called a *selectivity diagram.*

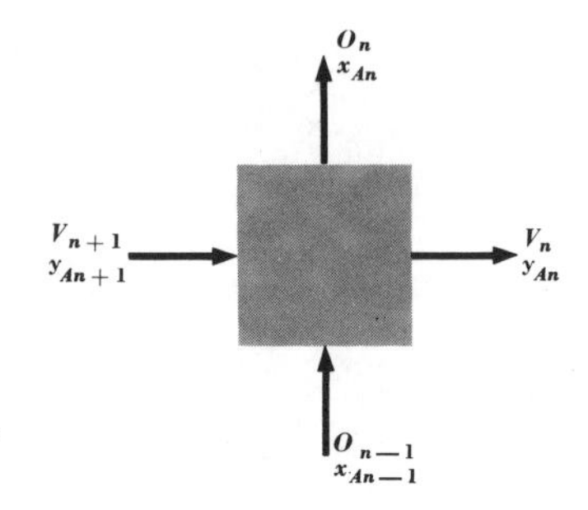

FIGURE 38-7
Schematic diagram of a single-stage extraction.

The lever-arm rule, which has been illustrated in connection with this description of phase-equilibrium relations in liquid-liquid systems, will be used extensively in the application of mass and energy balances to calculate operating lines for problems in extraction and distillation.

Separation by a Single Equilibrium Contact

Let us now study the separation of A from a mixture of A and C by extraction with a solvent B. If one equilibrium stage is used, the process can be schematically represented by the diagram in Fig. 38-7. The operation is the same as that illustrated by Fig. 37-5. The stream O_{n-1}, for which x_{An-1} is 0.25, is treated with pure solvent stream V_{n+1}; the liquid layers leaving the contactor are O_n and V_n; the mass fractions of solute A are shown on the diagram also.

The problem is solved on the triangular diagram of Fig. 38-8. When streams V_{n+1} and O_{n-1} are mixed, a stream of composition given by the point M is obtained. The position of M on the line between V_{n+1} and O_{n-1} is found from the relative quantities of the two streams and the application of the lever-arm principle. If 0.408 lb of solvent is used per pound of feed O_{n-1}, the ratio of lines $\overline{O_{n-1}M}/\overline{V_{n+1}M}$ must equal 0.408. The ratio $\overline{O_{n-1}M}/\overline{O_{n-1}V_{n+1}}$ is then 0.408/1.408, or 0.290. Rather than measure 0.29 of the distance from O_{n-1} to V_{n+1} along the line $\overline{O_{n-1}V_{n+1}}$, it is much simpler to use the ratios of the vertical coordinates of the points. The point M is thus simply located on an ordinate of 0.290, as shown. The point M represents the net composition of the fluid entering the contactor, and it must also be the net (overall) composition of the fluid leaving the contactor. This stream divides into two layers which are in equilibrium; thus points O_n and V_n are on the ends of a tie line through M. The mass fraction of A in the raffinate stream has been reduced to $x_A = 0.16$, or, on a solvent-free basis, $x'_A = 0.17$.

Note that in the problems treated in this chapter, O, V, and similar symbols can

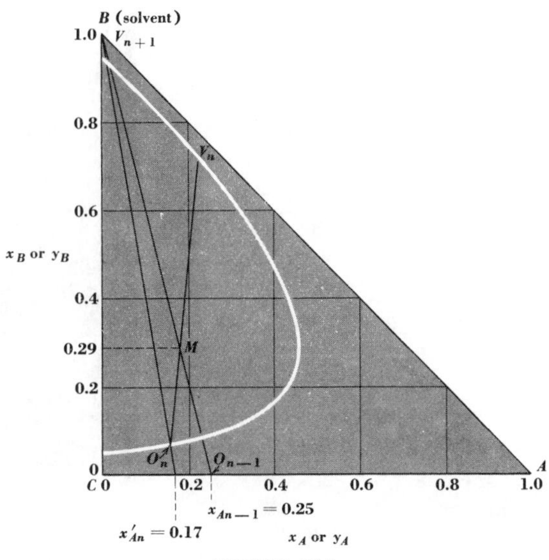

FIGURE 38-8
Extraction by a single equilibrium stage.

represent either flow rates in pounds per hour for a steady operation or quantities in pounds for a batch extraction. Although these symbols will be referred to as flow rates, the second interpretation is sometimes useful.

Multiple Contact with Fresh Solvent at Each Stage

This operation has been described in Chap. 37, and we refer to the sketch of the flows shown in Fig. 37-8.

The operation of a three-stage system is explained by reference to Fig. 38-9. To the first stage is added 0.408 lb of solvent for every pound of feed O_o. The point M_1 is the same as the point M on Fig. 38-8, and O_1 and V_1 correspond to O_n and V_n. The stream O_1 is now introduced to the second stage, to which is added another 0.408 lb solvent/lb feed O_o. To find M_2, we must first know the quantity O_1. The lever-arm rule gives $O_1/M_1 = \overline{V_1 M_1}/\overline{V_1 O_1}$; from the ordinates of the three points on $O_1 M_1 V_1$ we find $O_1/M_1 = 0.435/0.658 = 0.661$. Since we know that $M_1 = 0.408 + 1.00$, we have $O_1 = (0.661)(1.408) = 0.931$.

When 0.408 lb of solvent is mixed in stage 2 with 0.931 lb of raffinate from stage 1, the result is the mixture M_2; this point is located on the line $\overline{O_1 M_2 V_{N+1}}$ by the

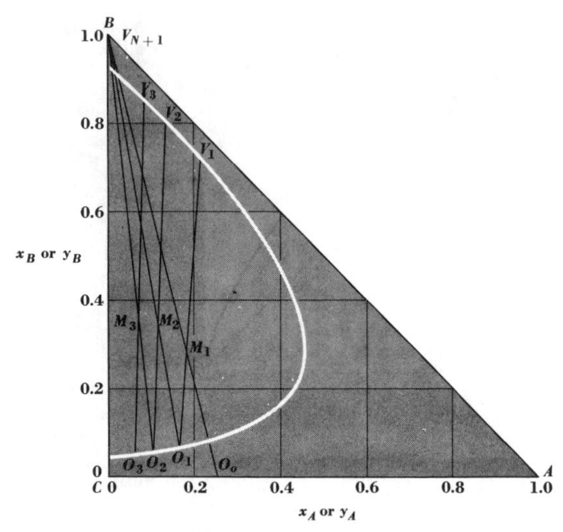

FIGURE 38-9
Extraction by multiple contact with fresh solvent at each stage.

lever-arm rule. The mixture M_2 separates into two layers having compositions given by the ends of the tie line through M_2. This line is easily found by trial and error with the aid of Fig. 38-2 plus Fig. 38-9 and a straightedge. The compositions of O_2 and V_2 are shown. For the example given, we obtain $M_2 = 1.339$ lb, $O_2 = 0.815$ lb, and $V_2 = 0.524$ lb. The computations are continued to a third stage, for which $M_3 = 1.223$ lb, $O_3 = 0.740$ lb, and $V_3 = 0.483$ lb. The mass fraction of A in the raffinate from the third stage has been reduced to $x_{A3} = 0.060$ or $x'_{A3} = 0.061$ on a solvent-free basis. This result compares with $x'_{An} = 0.17$ for the same feed with a single stage, but three times as much solvent has been used in this example. Note that the quantity O_n varies from stage to stage, and O_{Cn} also varies by about 10 percent; this computation could thus not be done conveniently on a diagram such as Fig. 37-7 since O_{Cn}/V_{Bn+1} (or L_C/G_B) is not constant.

Multiple-Stage Countercurrent Contact

The method of countercurrent contacting with multiple stages is shown in Fig. 38-10. The rectangle which symbolizes each stage can represent a compartment or tray in a vertical tower or one of a train of stages each composed of a mixer and a settler. The symbols V and O refer to the extract and raffinate streams; for a vertical tower, the flows shown in Fig. 38-10 are obtained when the extract phase is lighter

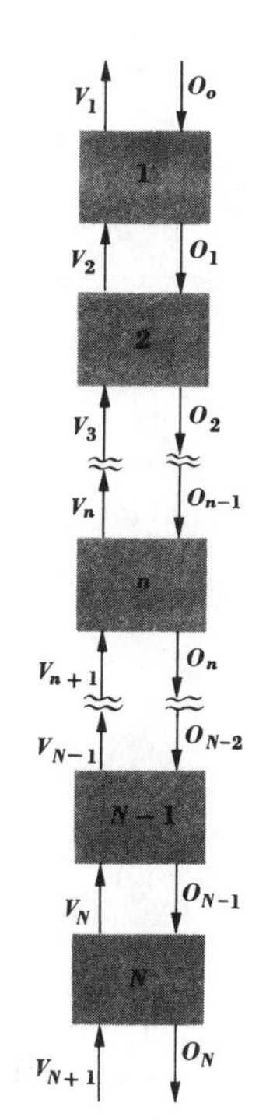

FIGURE 38-10
Schematic diagram of a countercurrent extraction operation using multiple stages.

than the raffinate phase. If the reverse is true, the raffinate phase flows upward and the extract phase flows downward. In any case the stages are usually counted from the A-rich end of the system.

Let us suppose that the composition of the four terminal streams of Fig. 38-10 are those shown in Fig. 38-11; that is, $x_{A_o} = 0.25$, $x_{AN} = 0.06$, $y_{A1} = 0.325$, and $y_{AN+1} = 0$. From the equation

$$V_{N+1} + O_o = V_1 + O_N = M \qquad (38\text{-}11)$$

and the lever-arm rule, it is seen from Fig. 38-11 that V_{N+1}/O_o is 0.408, as in the discussions in the previous sections. Equation (38-11) and the corresponding mass balances on components A and B are symbolized by the lines $\overline{O_N M V_1}$ and $\overline{V_{N+1} M O_o}$ and represent the overall mass balance for the multiple-stage system.

The number of stages required to achieve the separation described above can be stepped off directly on Fig. 38-11, or the stages can be stepped off on the $x - y$ diagram of Fig. 38-2 once the operating line has been plotted. A mass balance around the first stage is written as

$$V_1 - O_o = V_2 - O_1 = \Delta \qquad (38\text{-}12)$$

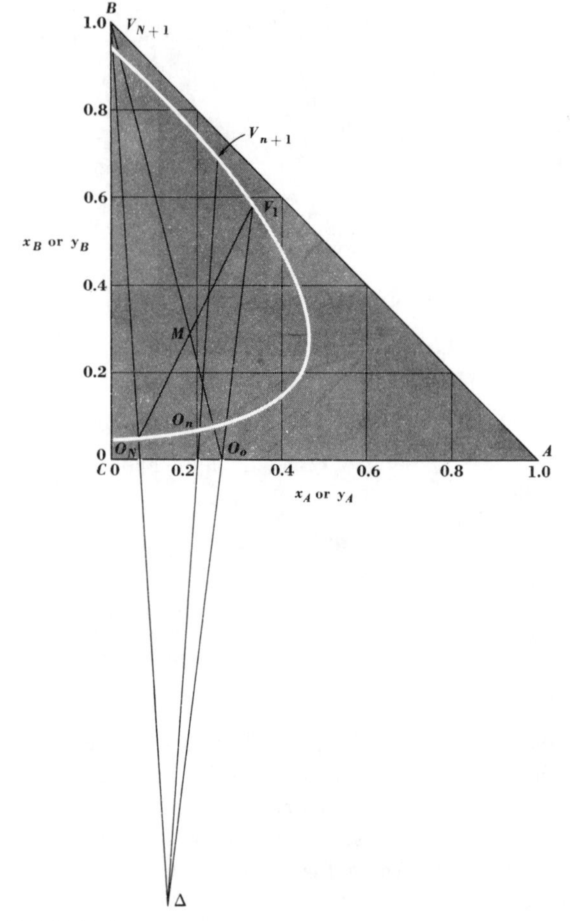

FIGURE 38-11
Graph for countercurrent multiple-contact extraction computations.

with similar equations for components A and B. The quantity Δ is the net flow toward the rich end of the cascade. For the first two stages, we have

$$V_1 - O_o = V_3 - O_2 = \Delta \qquad (38\text{-}13)$$

for the first n stages,

$$V_1 - O_o = V_{n+1} - O_n = \Delta \qquad (38\text{-}14)$$

and for all N stages,

$$V_1 - O_o = V_{N+1} - O_N = \Delta \qquad (38\text{-}15)$$

These equations show that the net flow Δ is constant throughout the cascade. From the four known terminal compositions and the requirement according to Eqs. (38-12) and (38-15) that $\overline{V_1 O_o \Delta}$ and $\overline{V_{N+1} O_N \Delta}$ be straight lines, the position of Δ is found as shown on Fig. 38-11.

From Fig. 38-11, O_o is the sum of Δ and V_1. Thus O_o is larger than V_1 and Δ is negative; this means that the net flow is toward the lean (N) end of the cascade. The lever-arm rule can be used to find the quantities of the streams. For instance, $O_o/V_1 = \overline{V_1 \Delta}/\overline{O_o \Delta}$; from the ordinates of the three points we obtain $O_o/V_1 = 1.60/1.02 = 1.57$. If O_o is 1.0, V_1 is 0.637 and $\Delta = -0.363$.

If the separation is changed so that V_1 travels along the V-phase solubility line toward smaller values of y_A, and V_{N+1}, O_N, and O_o remain unchanged, the ratio O_o/V_1 decreases. When the lines $\overline{V_1 O_o \Delta}$ and $\overline{V_{N+1} O_N \Delta}$ become parallel, O_o/V_1 is unity and Δ is zero. Further displacement of V_1 to the left means that Δ is now located above the triangle; Δ is positive, and V_1 is greater than O_o. There is a net flow toward the rich (1) end of the cascade.

An arbitrary line can be drawn through Δ cutting the solubility envelope between the terminal compositions. We know that for ideal stages the compositions of all streams in the cascade are found on the solubility envelope. Thus, if the composition of a stream O_n is as shown on Fig. 38-11, the composition of the stream V_{n+1} is found by the intersection of a line through O_n and Δ and the upper boundary of the solubility envelope, which is the locus of V-phase compositions. An operating line can be drawn on an x-y diagram by choosing arbitrary values of x_{An} on the O-phase solubility curve and using the Δ point to find the corresponding values of y_{An+1} for the passing streams. By this technique the operating line shown in Fig. 38-12 has been drawn.

Once the operating and equilibrium lines are plotted in Fig. 38-12, the stages are stepped off just as they were in similar diagrams in Chap. 37. From the known value of y_{A1} found from Fig. 38-11, x_{A1} is found from the equilibrium curve. The value of y_{A2} is found from the intersection of x_{A1} with the operating line. From y_{A2} and the equilibrium curve, x_{A2} is found. This construction is continued until x_{AN} is reached. The number of stages is counted as shown in the figure.

The stages can be counted directly on Fig. 38-11, although it is usually preferable to use a diagram such as Fig. 38-12 for this purpose so that the triangular diagram is not confounded with many lines. The composition of O_1 is at the intersection of a tie line through V_1 and the O-phase solubility curve. The composition of the passing phase V_2 is found from the intersection of a line through O_1 and Δ and the V-phase solubility curve. In this way stages are counted until x_{AN} is reached.

Just as for the contacting of immiscible phases, there is a minimum ratio V_{N+1}/O_o which gives an operable system. As V_{N+1}/O_o is decreased, M travels toward O_o and V_1 travels along the solubility curve to the right (higher y_{A1}). The upper end of the operating line in Fig. 38-12 approaches the equilibrium line; the

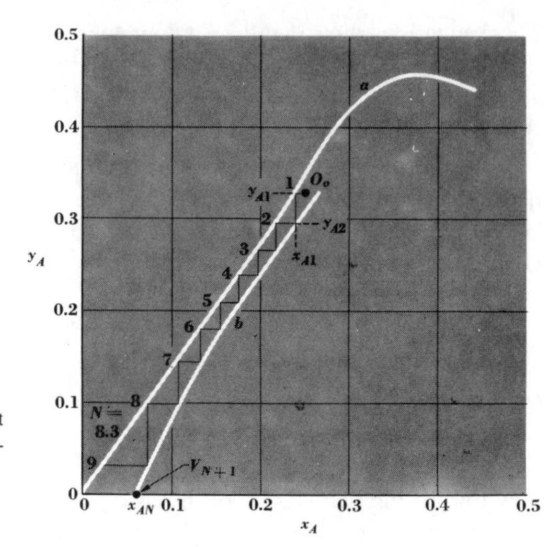

FIGURE 38-12
Stages for countercurrent multiple-contact extraction counted on a distribution diagram.
a Equilibrium line.
b Operating line.

minimum solvent/feed ratio occurs when these lines touch and create a pinch point. Although the pinch point often occurs at one of the ends of the cascade, if the operating and equilibrium lines are curved, the lines may become tangent at an intermediate point.

For an operable system the slope of any possible line $\overline{V_{n+1}\, O_n\, \Delta}$ must be greater than the slope of the tie line through the corresponding point O_n. Minimum solvent corresponds to a condition for which these lines coincide at any stage within the cascade.

Countercurrent Contact with Reflux

The countercurrent system described in the previous section can strip most of component A from component C. If pure solvent is used, x_{AN} can be reduced to an arbitrarily small quantity by the use of a suitable ratio O_o/V_{N+1} and a sufficiently large number of stages. However, the concentration of A in the rich extract y_{A1} is limited; if an infinite number of stages are used, y_{A1} can be increased only to a value in equilibrium with x_{Ao}. It is to enrich the solvent stream to a concentration greater than y^*_{Ao} that a recycle, or reflux, stream is used at the extract end of the system.

A cascade of stages with reflux "above" the feed is shown in Fig. 38-13. An enriching section has been added; the solvent stream is enriched by contact with a raffinate stream obtained by returning to the enriching section a portion of the extract product D from the system. The returned stream is called *reflux*, and the

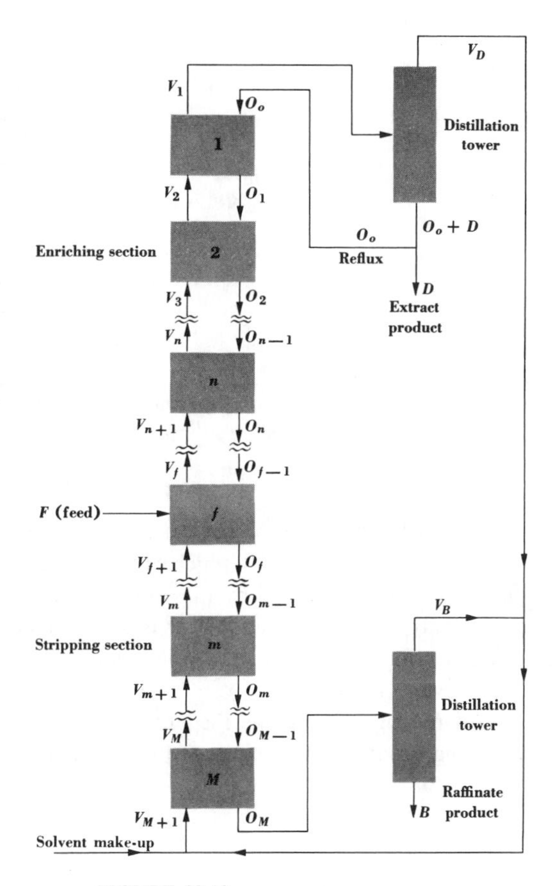

FIGURE 38-13
Countercurrent multiple contact with extract reflux.

ratio O_o/D is the *reflux ratio*. The extract product and the reflux are obtained from the extract V_1 by removing the solvent in a distillation tower, as shown in Fig. 38-13.

The calculations for this type of separation process can be made on a triangular diagram. However, it is usually found that the Δ points (two are needed when there is reflux) are crowded into the region near the B apex of the triangle, and the constructions required become quite inaccurate. For this reason we shall do the graphical work on a Janecke diagram.[1]

[1] See R. E. Treybal, "Mass Transfer Operations," 3d ed., chap. 10, McGraw-Hill Book Company, New York, 1980.

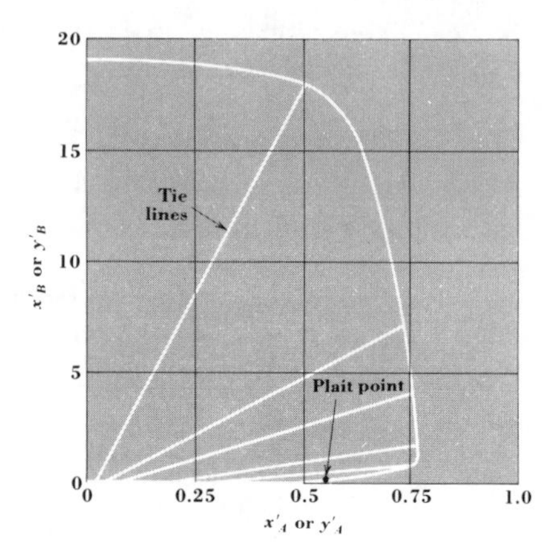

FIGURE 38-14
Solvent-free basis diagram which corresponds to Fig. 38-1.

In such a diagram the solvent concentrations are expressed on a solvent-free basis (y'_B = lb B/lb A + lb C), which causes a lengthening of the vertical axis. In fact, pure solvent is represented by $y'_B = \infty$. To be consistent in the mass balances, the concentrations of the V and O phases are expressed also on solvent-free bases: y'_A and x'_A (Fig. 38-6). The flows to be used are also solvent free: V_{AC} and O_{AC}. For pure solvent, $V_{AC} = 0$.

A mixing process like the one shown in Fig. 38-5 can be expressed by the following equations, analogous to Eqs. (38-1) to (38-3):

$$F_{AC} = O_{AC} + V_{AC} \tag{38-16}$$

$$F_{AC}x'_{AF} = O_{AC}x'_A + V_{AC}y'_A \tag{38-17}$$

$$F_{AC}x'_{BF} = O_{AC}x'_B + V_{AC}y'_B \tag{38-18}$$

As on the triangular diagram, mixing O and V gives F on a straight line between them; the lever-arm principle applies also to the Janecke diagram.

Figure 38-14 shows the equilibrium relations for a system which behaves like that of Fig. 38-1. If one attempts to do a graphical solution such as that shown in Fig. 38-11, it is found that the shape of the envelope and the positions of the tie lines make the constructions inconvenient; the triangular diagram is preferable. However, extraction with reflux is usually practical with systems such as that shown in Fig. 38-3(a); such a system is plotted in Fig. 38-15. This type of diagram does prove to be convenient for systems with reflux, and we now shall see how to do this analysis.

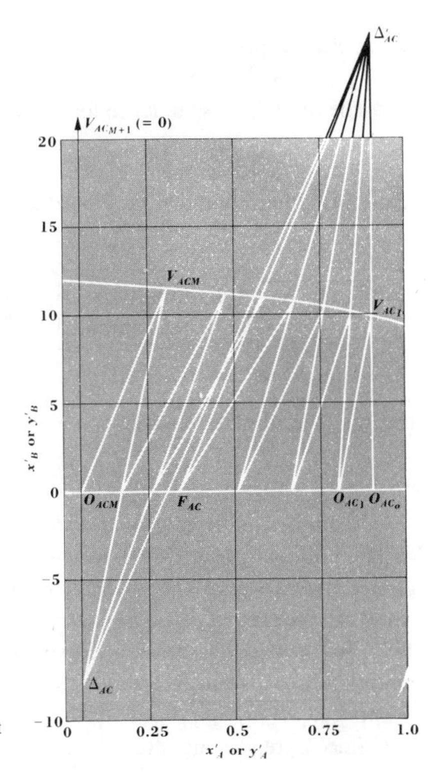

FIGURE 38-15
Solvent-free basis graph for countercurrent multiple contact with extract reflux.

In the enriching section of the cascade we have the mass balances

$$V_{AC1} - O_{AC0} = V_{AC2} - O_{AC1} = \Delta'_{AC} \tag{38-19}$$

$$V_{ACn+1} - O_{ACn} = \Delta'_{AC} \tag{38-20}$$

$$V_{ACf} - O_{ACf-1} = \Delta'_{AC} \tag{38-21}$$

The point Δ'_{AC} is on a straight line through O_{AC0} and V_{AC1} and will serve to locate the operating line for the enriching section. For the rest of the column, called the *stripping section*, the mass balances are

$$V_{ACf+1} - O_{ACf} = \Delta_{AC} \tag{38-22}$$

$$V_{ACm+1} - O_{ACm} = \Delta_{AC} \tag{38-23}$$

$$V_{ACM+1} - O_{ACM} = \Delta_{AC} \tag{38-24}$$

The operating point Δ_{AC} for the stripping section is on a line through V_{ACM+1} and O_{ACM}.

The overall mass balance for the cascade is written

$$F_{AC} + V_{ACM+1} - O_{ACM} = V_{AC1} - O_{AC0} \tag{38-25}$$

From this equation and Eqs. (38-19) and (38-24) we obtain

$$F_{AC} + \Delta_{AC} = \Delta'_{AC} \tag{38-26}$$

and $\overline{F_{AC}\Delta'_{AC}\Delta_{AC}}$ is a straight line.

If F_{AC} and O_{AC0} are combined mathematically to give J_{AC}, as follows,

$$F_{AC} + O_{AC0} = J_{AC} \tag{38-27}$$

then the overall mass balance can also be written as

$$J_{AC} + V_{ACM+1} = O_{ACM} + V_{AC1} = M_{AC} \tag{38-28}$$

A simple example of countercurrent extraction with reflux is shown in Fig. 38-15. Following Fig. 38-13, we assume that the two distillation towers shown are capable of removing all the solvent from the feeds which enter them. On the solvent-free basis, we have $x'_{A0} = x'_D = x_D = y'_{A1}$, set at 0.9 in the figure. The composition of V_1 is on the y-saturation curve, at V_{AC1} in Fig. 38-15. Equation (38-19) corresponds to a vertical line at $x'_A = 0.9$. The reflux ratio on the solvent-free basis is simply $\overline{\Delta'_{AC}V_{AC1}}/\overline{V_{AC1}O_{AC0}}$.

Since pure solvent enters the cascade at the bottom of Fig. 38-13, V_{ACM+1} has the coordinates y'_{AM}, $y'_B = \infty$, where $y'_{AM} = x'_{AM}$ has been set at 0.05. The point Δ_{AC} is on the vertical line at 0.05 (Eq. 38-24), and on the line through Δ'_{AC} (known from the reflux ratio) and F_{AC}, the known feed point (Eq. 38-26).

The ideal stages can now be obtained from Fig. 38-15. Knowing V_{AC1}, we follow a tie line to O_{AC1}, as shown. Then V_{AC2} is at the intersection of the y-saturation curve and a line through O_{AC1} and Δ'_{AC}. To obtain the minimum stages for the given reflux ratio and other conditions, the feed is added after the fifth stage ($f = 5$). To find V_{AC6} from O_{AC5} the lower operating point Δ_{AC} now is used, and the stepping process is continued until $x'_{AM} \leq 0.05$ is found. Seven stages are needed in all.

The reflux ratio can vary between a certain minimum operable value and total reflux, where no product D is removed. In the latter case F and B are also zero. With total reflux, the minimum possible number of stages is required. Note that under these conditions, Δ' is at infinity, and O_m/V_{m+1} is a minimum.

As the reflux ratio is progressively decreased, the operating lines on a graph such as Fig. 38-12 approach the equilibrium line and more and more stages are required for the separation. At the minimum reflux ratio, the lines touch at a pinch point and an infinite number of stages would be required for the separation. The pinch point can occur anywhere in the cascade, but it usually occurs at the intersection of the two operating lines. On a Janecke diagram, minimum reflux is reached as soon as any line through one of the Δ points coincides with a tie line.

There is an economic optimum value of the reflux ratio. Operation near minimum reflux reduces solvent losses and the size of the individual stages but results in a large number of stages. Operation at high reflux ratios reduces the number of stages but increases the size of the individual stages because of the high circulation rates.

There is a maximum solvent-free mass fraction of A that can be obtained in the extract product with a system such as that of Fig. 38-1. This y'_{A1} is given by a line through pure B, tangent to the upper solubility envelope. For the system of Fig. 38-1, this maximum is given by the line $\overline{BVV'}$ in Fig. 38-1; $y'_{A1,\max}$ is 0.77.

For a ternary system such as that of Fig. 38-3*a*, a cascade with reflux such as shown in Fig. 38-13 can produce *both* almost pure C and almost pure A. For instance, a mixture of hexane and methylcyclopentane can be separated into almost pure fractions by a system like that of Fig. 38-13, which uses aniline as a solvent.

PROBLEMS

38-1 A countercurrent liquid-liquid extraction battery operates as shown in Fig. 38-16. Find the number of stages necessary to give the desired separation.

38-2 It is proposed to separate a mixture F containing 33 mass percent *n*-heptane and 67 mass percent methylcyclohexane (MCH) into a raffinate stream O_M containing 18 mass percent MCH and a product stream D containing 92 mass percent MCH and 8 mass percent *n*-heptane. A stream of pure solvent (aniline) is supplied to the extraction battery, and a lesser amount of the same solvent (also pure) is produced from the reflux unit. The system operates at a reflux ratio O_0/D of 6. Assuming the feed to be added at the optimum location in the system, find the number of stages required to produce the desired separation when the system is operated as shown in Fig. 38-13. Equilibrium data for the ternary system have been reported at 25°C by K. A. Vartaressian and M. R. Fenske, *Ind. Eng. Chem.*, **29**:270 (1937), and are as follows:

Hydrocarbon layer		Solvent layer	
MCH	Heptane	MCH	Heptane
0	92.6	0	6.2
9.2	83.1	0.8	6.0
18.6	73.4	2.7	5.3
33.8	57.6	4.6	4.5
46.0	45.0	7.4	3.6
59.7	30.7	9.2	2.8
71.6	18.2	12.7	1.6
83.3	5.4	15.6	0.6
88.1	0	16.9	0

All figures are in mass percent.

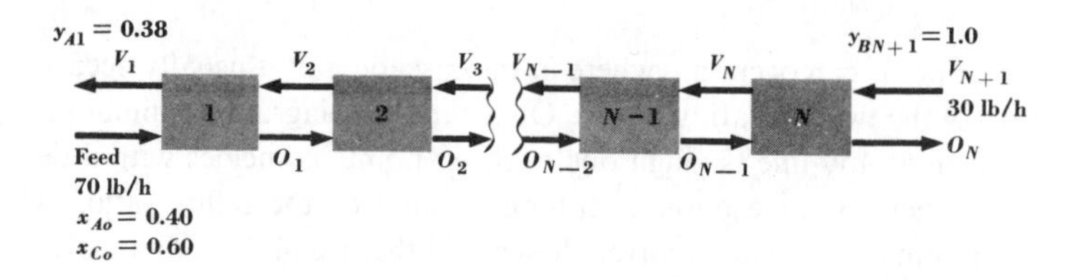

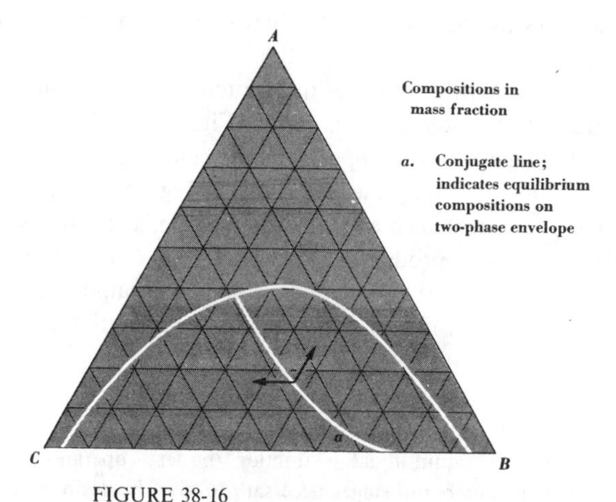

FIGURE 38-16
Countercurrent extraction battery and triangular diagram for Prob. 38-1.

It is convenient to relate x'_A and y'_A by the empirical equation given by the authors:

$$y'_A = \frac{\beta x'_A}{1 + (\beta - 1)x'_A}$$

where $\beta = 1.90$. Verify that this equation fits the data.

38-3 An extraction battery operates as shown in Fig. 38-17. Find the points on the triangular diagram which represent the composition of streams V_4 and O_3. Stream V_4 is not pure solvent but is saturated ternary mixture from a stage not shown.

38-4 Fish oil is to be extracted from pulverized fish heads containing 70 mass percent insoluble solid and 30 mass percent oil. The extraction is to be accomplished in a multiple-contact extraction battery consisting of two equilibrium stages with fresh solvent added to each stage. The feed rate is 1000 lb/h of fish heads, and the solvent rate is 500 lb/h to each stage. The underflow from each stage contains 0.30 lb solution/lb solids. Find the fraction of oil initially present in the fish heads that is recovered in the extract stream.

38-5 Sixty tons per day of oil sand (25 mass percent oil and 75 mass percent sand) is to be extracted with 40 tons/day of naphtha in a continuous, countercurrent extraction bat-

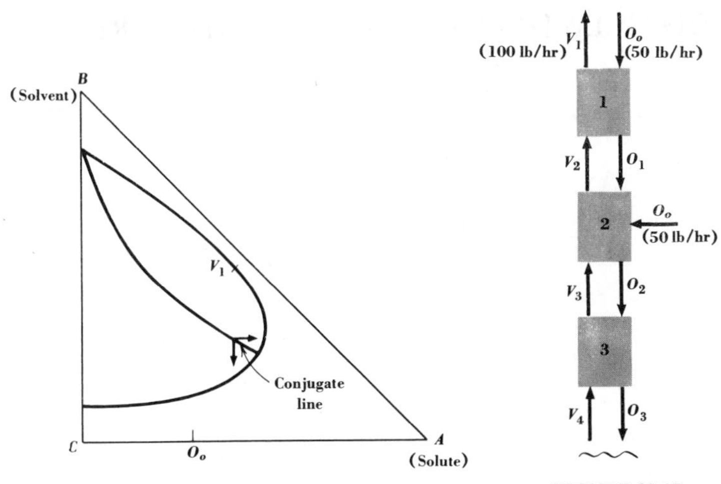

FIGURE 38-17
Data for Prob. 38-3.

tery. The final extract from the battery is to contain 40 mass percent oil and 60 mass percent naphtha, and the underflow from each unit is expected to consist of 35 mass percent solution and 65 mass percent sand. If the overall efficiency of the battery is 50 percent, how many stages will be required?

38-6 Consider a liquid-liquid system having equilibrium conditions given by Figs. 38-1 and 38-2. A mixture of 100 kg of 0.40 mass fraction A in C is to be treated with pure solvent B so that the A content of the raffinate product is reduced. In a first stage 100 kg of pure solvent is used. The equilibrated phases are separated and the raffinate layer is treated in a second stage with 100 kg of pure solvent again. Estimate the raffinate quantities and compositions leaving the two stages.

38-7 The feed to a multistage countercurrent extraction system is 0.20 mass fraction A in C; the equilibrium relations are given by Figs. 38-1 and 38-2. We want to reduce the A content in the raffinate to $x_{AN} = 0.05$ by using pure solvent at a rate of 1.3 times the minimum. Find the extract composition (y_{A1} and y_{B1}) and the number of ideal stages needed.

39

DISTILLATION OF BINARY MIXTURES

In this chapter we shall consider the separation of binary mixtures by the technique of distillation. The importance of understanding the principles of binary distillation lies only partly in their industrial application; these principles are equally important because of the assistance they provide in understanding the principles of multicomponent distillation, in which most of the same concepts exist but are encumbered by a large amount of algebra necessary to describe the behavior of all the components.

The batch distillation of what are essentially binary mixtures of ethanol and water has been practiced by brewers and distillers for centuries. It is only in the present century, however, that distillation has been developed on a large scale as a steady-state operation. The major contributions in this field have come from engineers in the petroleum industry, where large numbers of multistage distillation towers operate on a continuous basis.

Although the term distillation is occasionally used to describe the removal of volatile materials from solids, we shall be concerned primarily with the separation of volatile components found in liquid solutions. If a homogeneous liquid mixture is boiled, the vapor bubbles leaving are richer in the more volatile components than is the liquid, whereas the liquid remaining is richer in the less volatile components. If a batch of liquid is treated in this way until, say, half of it has been vaporized, the liquid remaining will have a different composition from the original mixture. The overhead vapor, if removed from the system continuously and condensed, would have a composition also different from that of the original mixture. Thus a separation would be achieved.

It should not be thought that either the overhead product or the residual liquid would necessarily be very pure. At any instant the vapor leaving would not be richer than the equilibrium composition. Thus the collected overhead product could be regarded as the sum of a number of increments of vapor, each of which was in equilibrium with the liquid at the time it was produced. For example, a liquid with an initial composition of 50 percent A and 50 percent B might give a cumulative overhead product containing 70 percent A and 30 percent B, with a residual liquid containing 20 percent A and 80 percent B.

Unless an azeotrope is formed in the system, a greater separation may be achieved than in the example chosen. In batch distillation, however, as the liquid becomes richer in the less volatile component, the amount of liquid remaining becomes smaller. In the example chosen, if the vapor were continually removed and boiling continued until the liquid was nearly all gone, the liquid composition would approach 100 percent B. The collected overhead product would, of course, have almost the same composition as the original charge of liquid, so no useful result would have been achieved.

In continuous, multistage distillation the separation of components into nearly pure streams requires large numbers of stages compared with the number of stages required, say, for the production of streams of 90 percent purity. The difficulty of preparing very pure products by distillation is reflected by thermodynamic calculations[1] for the minimum work of separation, which show that the minimum work required increases sharply as the purity of the products approaches 100 percent.

The simplest distillation operation, which we call batch distillation, can be carried out using a vessel called a *still*, equipped with a steam jacket or heating coil. The vapor bubbles produced at the heating surface rise through the liquid and pass out the top of the vessel to a condenser, where they are condensed back into liquid form and collected in a receiver. As was seen in Chap. 34, mass transfer to and from vapor bubbles rising through a liquid proceeds very rapidly, so the vapor

[1] B. F. Dodge, in "Chemical Engineering Thermodynamics," p. 597, McGraw-Hill Book Company, New York, 1944, shows the following results for the calculation of the minimum work for separating a mixture of 50 mol percent benzene in toluene into the following products at 25°C:

1 Both products 95 mol percent pure, −526 Btu/lb mol feed
2 Both products 99 mol percent pure, −555 Btu/lb mol feed
3 Both products 100 percent pure, −738 Btu/lb mol feed

composition is seldom far from the equilibrium value when the bubbles break the upper surface of the liquid in the still.

A more complex piece of distillation apparatus is the fractionating column. This is a cylindrical vessel containing various types of internal construction devised for the purpose of bringing the vapor and liquid into successive contact as the two phases move generally countercurrent in the column. In this device liquid reflux must be added at the top of the tower; this liquid is usually supplied by condensing part of the overhead vapor. The internal construction may consist of nothing more than a fixed bed of Berl saddles, Raschig rings, or some other kind of packing commonly used in gas absorption. More often, however, the internal construction consists of a series of plates or stages. The plates may be perforated with holes or rectangular slots, or they may be of the more elaborate bubble-cap or valve-plate construction. A bubble-cap tray has been shown in Fig. 37-4. Most of our analysis of multistage distillation will be made with the bubble-cap plate column as an implicit model. As usual, it will be assumed that for an ideal stage the gas and liquid phases *leaving* each plate are in equilibrium. The discussion of stage efficiency given in Chap. 37 of course applies also to distillation.

The batch still as we have described it consists of a vessel which provides one equilibrium stage. The fractionating column, on the other hand, usually operates under steady conditions and contains a number of plates, each of which is approximately an equilibrium stage. The two pieces of apparatus can be combined by having the vapors from the batch still pass up into the fractionating column so that the batch distillation is performed with a number of stages operating simultaneously. The composition at any point in the fractionating column is then a function of time; the system is operated only for a sufficient period to produce cumulative overhead and residual products of the desired composition.

Phase Equilibrium

Extensive use is made of the principles of phase equilibrium in distillation calculations. In Chap. 31 we discussed the use of the phase rule in determining the number of independent variables that could be specified in a system at equilibrium. In a two-phase binary system we saw from Eq. (31-1) that two variables could be arbitrarily specified. Because pressure is usually fixed by design considerations, the specification of only one additional variable for any particular binary system is sufficient to fix all other variables in the system. For example, if a liquid is boiling in a batch still at 1 atm pressure and the liquid composition is 65 mol percent ethanol and 35 mol percent water, then both the temperature of the liquid-vapor system and the composition of the vapor are fixed. As another example, if the pressure in a fractionating column is 1 atm and the temperature of the liquid leaving the top plate is known, then for a particular binary system the compositions of both the vapor and liquid streams leaving the plate are fixed if they are in equilibrium.

The equilibrium compositions are found from experimental measurements and by the use of various equations. In Chap. 31 we saw how Raoult's law could be used to predict equilibrium compositions in a vapor-liquid system if the liquid phase were an ideal solution and the gas phase an ideal gas.

One useful concept is relative volatility. The volatility, which has the units of pressure, is written for component A as

$$v_A = \frac{\bar{p}_A}{\tilde{x}_A} \qquad (39\text{-}1)$$

For a pure substance the volatility is identical with the vapor pressure. The relative volatility is written as

$$\alpha_{AB} = \frac{v_A}{v_B} = \frac{\bar{p}_A \tilde{x}_B}{\tilde{x}_A \bar{p}_B} \qquad (39\text{-}2)$$

The choice of components for A and B is usually made so that α_{AB} is greater than 1. In this sense component A is the more volatile of the two. From the definition of partial pressure, Eq. (31-2), we can write

$$\alpha_{AB} = \frac{\tilde{x}_B}{\tilde{x}_A} \frac{\tilde{y}_A}{\tilde{y}_B} \qquad (39\text{-}3)$$

or if Raoult's law applies to the system, we can substitute Eq. (31-3) in Eq. (39-2) to give

$$\alpha_{AB} = \frac{p_A}{p_B} \qquad (39\text{-}4)$$

The discussion in this chapter is restricted to binary systems, so we shall usually simplify the nomenclature from this point on by dropping the subscripts which designate components A and B. Henceforth, the mole fractions of the more volatile component A in the vapor and liquid will be written as $\tilde{y}$ and $\tilde{x}$, respectively, and the mole fractions of the less volatile component B as $1 - \tilde{y}$ and $1 - \tilde{x}$. Using this convention, we write Eq. (39-3) as

$$\alpha = \frac{1 - \tilde{x}}{\tilde{x}} \frac{\tilde{y}}{1 - \tilde{y}} \qquad (39\text{-}5)$$

which may be solved for $\tilde{y}$ to give

$$\tilde{y} = \frac{\alpha \tilde{x}}{1 + (\alpha - 1)\tilde{x}} \qquad (39\text{-}6)$$

This relationship is particularly useful in binary distillation when the temperature range between the boiling points of the pure components is small. Under these conditions the relative volatility is often approximately constant. This makes it possible to construct a plot of $\tilde{y}$ versus $\tilde{x}$ over the composition range of 0 to 1 by using a single mean value of α. If the binary system obeys Raoult's law, α can be calculated from the ratio of vapor pressures as shown in Eq. (39-4). Although the vapor pressures may change considerably over the range of temperatures involved, the ratio of vapor pressures will change much less.

If the liquid and gas phases do not behave as an ideal solution and an ideal gas, respectively, the relative volatility can often be calculated with the aid of thermodynamics. However, the relative volatility is used for little more than calculating the phase compositions, so the thermodynamic calculations are usually made instead for a quantity which we shall write as K. For this quantity, known as the *vapor-liquid equilibrium ratio*,[1] we have the simple definition

$$K = \frac{\tilde{y}}{\tilde{x}} \tag{39-7}$$

which enables us to calculate phase compositions.

It is not our purpose here to discuss the thermodynamic calculations from which K values are determined. This topic is found in most textbooks on chemical engineering thermodynamics. It will be sufficient here to remind the student that K is a function of temperature, pressure, and generally also of composition. Although K values are available from a variety of sources, we shall use only the values given by Perry, 3d ed. These values depend only on pressure and temperature; hence they apply only to liquids and gases which behave as ideal solutions. If more accurate values are required, the student is referred to a series of 276 charts known as the Kellogg charts,[2] which are applicable to nonideal solutions.

Example 39-1 Construct a diagram of $\tilde{y}$ versus $\tilde{x}$ for mixtures of n-butane and n-hexane under the following conditions:

(*a*) Total pressure is 20 psia. Assume that Raoult's law applies.

(*b*) Total pressure is 115 psia. Use K values from Perry.

(*a*) At 20 psia the boiling points of n-butane and n-hexane are 50 and 175°F, respectively (from Perry, p. 4-49). Hence mixtures of these two com-

[1] This quantity is known by a variety of names. S. B. Adler and D. F. Palazzo, in *Chem. Eng.*, June 29, 1959, p. 95, point out that it is also known as the *vaporization constant*, the *vaporization ratio*, the *volatility equilibrium distribution ratio*, and simply as the *equilibrium constant*. They give a concise summary of equations for calculating values of K.

[2] The M. W. Kellogg Company, "Liquid-Vapor Equilibria in Mixtures of Light Hydrocarbons," New York, 1950. The data in these charts are summarized in articles by C. L. DePriester, *Chem. Eng. Progr. Symp. Ser.*, no. 7, p. 1 (1953), and W. C. Edminster and C. L. Ruby, *Chem. Eng. Progr.*, **51**:95 (1955). Also see appendix, Figs. A-7 and A-8.

ponents can exist as a vapor-liquid system at equilibrium at 20 psia only between these two temperatures. The relative volatility at 50°F is found using Eq. (39-4).

$$\alpha = 20/1.50 = 13.3$$

At 175°F the relative volatility is

$$\alpha = 150/20 = 7.5$$

The variation of relative volatility in this example is such that it would be unwise to attempt to use an average value of α in the relation between $\tilde{y}$ and $\tilde{x}$, Eq. (39-6). As an alternative, we select a number of temperatures between 50 and 175°F and calculate $\tilde{x}$ and $\tilde{y}$ as shown below. The partial pressures are written using Raoult's law and added:

$$\tilde{x}p_A + (1 - \tilde{x})p_B = 20 \tag{1}$$

Solving for $\tilde{x}$, we obtain

$$\tilde{x} = \frac{20 - p_B}{p_A - p_B} \tag{2}$$

At 100°F the vapor pressure of n-butane is 52 psia, and the vapor pressure of n-hexane is 5 psia. Substituting these numbers in Eq. (2), we obtain

$$\tilde{x} = \frac{20 - 5}{52 - 5}$$

$$= 0.32 \tag{3}$$

The vapor composition is found by dividing the partial pressure of n-butane by the total pressure:

$$\tilde{y} = \frac{(0.32)(52)}{20}$$

$$= 0.83 \tag{4}$$

We proceed in this fashion to find values of $\tilde{y}$ versus $\tilde{x}$, which are plotted on Fig. 39-1.

(*b*) At 115 psia we see from the K chart (Perry 3d ed., p. 569) that the K values for n-butane and n-hexane are unity at 155 and 310°F, respectively. These represent, therefore, the boiling points of the pure components at this pressure. We designate n-butane as component A and n-hexane as component B and write

$$\tilde{x}_A + \tilde{x}_B = 1 \tag{1}$$

$$\tilde{y}_A + \tilde{y}_B = 1 \tag{2}$$

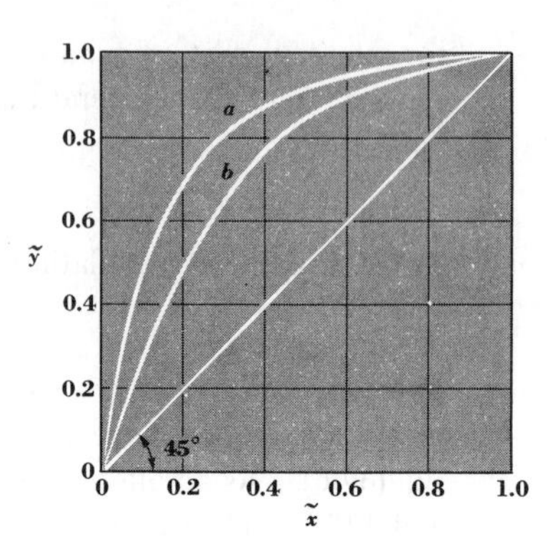

FIGURE 39-1
Mole fraction *n*-butane in the vapor versus mole fraction *n*-butane in the liquid for systems containing *n*-butane and *n*-hexane at total pressures of 20 and 115 psia (calculated as shown in Example 39-1).
a 20 psia.
b 115 psia.

Using the definition of K, we substitute, in Eq. (2),

$$K_A \tilde{x}_A + K_B \tilde{x}_B = 1 \tag{3}$$

We then substitute Eq. (1) in Eq. (3),

$$K_A \tilde{x}_A + K_B(1 - \tilde{x}_A) = 1 \tag{4}$$

and solve for $\tilde{x}_A$ from Eq. (4),

$$\tilde{x}_A = \frac{1 - K_B}{K_A - K_B} \tag{5}$$

At 200°F $K_A = 1.60$ and $K_B = 0.30$. We substitute these values in Eq. (5) and find $\tilde{x}_A = 0.54$. The vapor composition is

$$\begin{aligned} \tilde{y}_A &= (1.60)(0.54) \\ &= 0.86 \end{aligned}$$

Values of $\tilde{y}_A$ versus $\tilde{x}_A$ are calculated in this manner at a number of temperatures between 155 and 310°F and plotted on Fig. 39-1. ////

It will be noticed on Fig. 39-1 that the curve of $\tilde{y}$ versus $\tilde{x}$ for a pressure of 115 psia is closer to a line drawn at 45° through the origin than is the curve for 20 psia. The 45° line represents the relation $\tilde{y} = \tilde{x}$; a vapor-liquid equilibrium curve coinciding with this line would represent a condition at which no separation could be achieved by distillation. On the other hand, a vapor-liquid equilibrium curve at a great distance from the 45° line represents a system in which a substantial separation is achieved by distillation. Thus the separation of mixtures of *n*-butane and *n*-hexane could be more easily achieved at 20 psia than at 115 psia. By this we mean

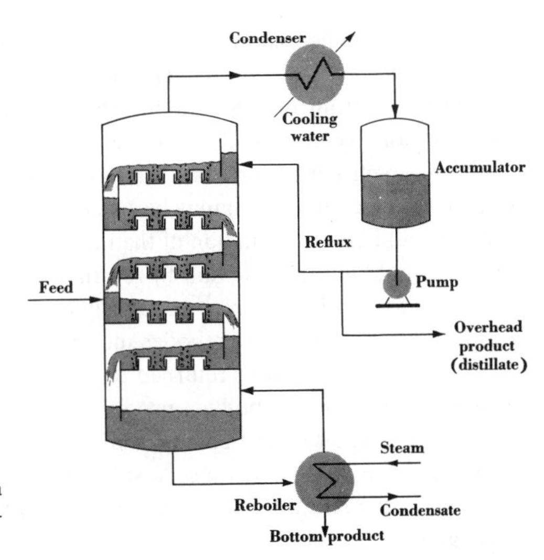

FIGURE 39-2
Schematic diagram of operation of a fractionating column containing bubble-cap plates.

that fewer stages would be required to achieve a specified separation in a fractionating column, or a greater difference in vapor and liquid composition would occur with a single-stage batch distillation.

When an azeotrope occurs in a binary system, the curve of $\tilde{y}$ versus $\tilde{x}$ crosses the 45° line. Thus, if the system is ethanol and water at 1 atm and the mole fraction of ethanol in the vapor is plotted against the mole fraction of ethanol in the liquid, the curve of $\tilde{y}$ versus $\tilde{x}$ is above the 45° line from $\tilde{x} = 0$ up to the azeotrope composition of 0.894 mole fraction ethanol. At this point the curve crosses the 45° line and continues beneath the line up to $\tilde{x} = 1$.

The Operation of a Fractionating Column

A fractionating column operating at steady state is shown schematically in Fig. 39-2. A feed enters the column somewhere near the middle; overhead and bottom products are withdrawn as shown. The column contains a number of bubble-cap plates. Vapor from the top plate passes to a condenser, where it is condensed to a saturated liquid which is collected in a vessel known as an *accumulator*. As we shall see later, the condenser may also be operated to produce either a subcooled liquid or a mixture of liquid and vapor; for the present purpose, however, we shall consider the product to be a liquid at its saturation temperature. Some of the liquid in the accumulator is returned as reflux continuously to the top plate of the column because it is necessary to have liquid on this plate and on succeeding plates to achieve any mass transfer. The remainder of the liquid is withdrawn continuously and is referred to as the *overhead-product stream*, or *distillate.*

At the bottom of the column the liquid collects in the reboiler, where it is heated, usually by steam coils. The reboiler may be an integral part of the cylindrical vessel which constitutes the column, or it may be a vessel of different shape attached to the column, or it may be a vessel external to the column (as shown in Fig. 39-2). In all cases the function of the reboiler is to receive the liquid overflow from the lowest bubble-cap plate and to return a portion of this as a vapor stream while the remainder is withdrawn continuously as a liquid bottom product. The conditions for mass transfer in the reboiler such as degree of mixing and time of contact are usually such that the vapor returned to the column is in equilibrium with the liquid leaving the reboiler as a bottom product.

The feed entering the column may be a saturated liquid, a saturated vapor, a mixture of both liquid and vapor, a subcooled liquid, or a superheated vapor. The condition of the feed affects the performance of the column, as we shall see later in this chapter.

The Sorel Method of Analysis

The bases of most methods of analysis of fractionating columns lie in the equations of Sorel.[1] These equations are fundamental to the solution of multicomponent distillation problems, which are usually solved by an electronic computer. The equations are generally not used for binary systems because, although the general approach is rather simple, the solution requires numerous trial-and-error calculations. Simpler methods are available: the McCabe-Thiele graphical method, in which certain simplifying assumptions are made, and the Ponchon-Savarit method, which is the graphical equivalent of the Sorel method but does not require a trial-and-error solution. We shall examine first the Sorel method.

A portion of a fractionating column is shown in Fig. 39-3. The vapor and liquid streams are shown schematically for the top few plates. The designations not only identify the streams, but also represent the amounts in pound moles per hour. A material balance around the top plate gives

$$\tilde{O}_R + \tilde{V}_2 = \tilde{O}_1 + \tilde{V}_1 \qquad (39\text{-}8)$$

and an enthalpy balance around the top plate yields[2]

$$\tilde{O}_R \tilde{h}_R + \tilde{V}_2 \tilde{H}_2 = \tilde{O}_1 \tilde{h}_1 + \tilde{V}_1 \tilde{H}_1 \qquad (39\text{-}9)$$

[1] E. Sorel, "La Rectification de l'alcool," Paris, 1893.

[2] The symbols $\tilde{h}$ and $\tilde{H}$ refer to the enthalpies in Btu/lb mol of liquid and vapor phases, respectively.

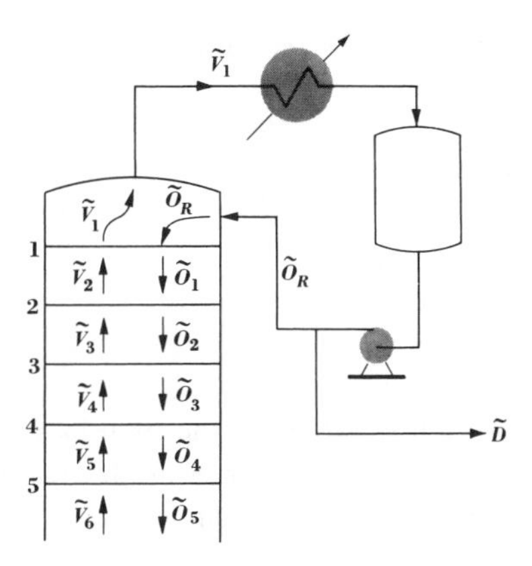

FIGURE 39-3
Upper section of a fractionating column with bubble-cap plates.

These two equations contain eight variables, so it is apparent that six variables must be known if the equations are to be solved. In the initial stages of design the column pressure is usually specified, as we shall see near the end of this chapter. Because all streams are binary mixtures and are assumed to be at saturation conditions, we need to specify only one variable for each stream, in addition to pressure, to fix the properties of the stream. For example, the composition of the distillate stream $\tilde{x}_D$ is usually specified arbitrarily. All the vapor entering the condenser leaves as condensate of the same composition, so $\tilde{x}_D = \tilde{x}_R = \tilde{y}_1$. Furthermore, stream O_1 is in equilibrium with V_1, so we can find $\tilde{x}_1$ from a value of $\tilde{y}_1$ and a plot of $\tilde{y}$ versus $\tilde{x}$. Therefore we can determine the compositions of all streams arriving at or leaving plate 1 except V_2. Knowing the three stream compositions and the total pressure, we find the stream temperatures and enthalpies $\tilde{H}_1$, $\tilde{h}_R$, and $\tilde{h}_1$ from tabulated information in handbooks or in the literature.

Two other variables in Eqs. (39-8) and (39-9) can be found directly. The amount of overhead product $\tilde{D}$ is usually specified at the beginning of the problem or determined from an overall material balance. In addition, the reflux ratio, which we define as $\tilde{O}_R/\tilde{D}$, is more or less arbitrarily chosen. These two pieces of information make it possible to calculate the reflux rate $\tilde{O}_R$, and hence the amount of vapor leaving the top plate $\tilde{V}_1$.

The three variables in Eqs. (39-8) and (39-9) which we do not know are the rates $\tilde{O}_1$ and $\tilde{V}_2$ and the enthalpy $\tilde{H}_2$. To obtain a solution we assume a value for one of the variables. For example, as a first trial we might say that $\tilde{H}_2$ is the same as $\tilde{H}_1$ (which is known). This permits us to solve the two equations for the remaining variables $\tilde{O}_1$ and $\tilde{V}_2$. To check the assumed value of $\tilde{H}_2$, we write a material balance around plate 1 for the more volatile component. This equation is

$$\tilde{O}_R \tilde{x}_R + \tilde{V}_2 \tilde{y}_2 = \tilde{O}_1 \tilde{x}_1 + \tilde{V}_1 \tilde{y}_1 \qquad (39\text{-}10)$$

which may be rearranged to give

$$\tilde{y}_2 = \frac{\tilde{O}_1}{\tilde{V}_2}\tilde{x}_1 + \frac{\tilde{V}_1\tilde{y}_1 - \tilde{O}_R\tilde{x}_R}{\tilde{V}_2} \tag{39-11}$$

The numerator of the last term in Eq. (39-11) is simply $\tilde{D}\tilde{x}_D$, so we write the expression as

$$\tilde{y}_2 = \frac{\tilde{O}_1}{\tilde{V}_2}\tilde{x}_1 + \frac{\tilde{D}\tilde{x}_D}{\tilde{V}_2} \tag{39-12}$$

All the variables on the right-hand side of this equation are known or have been found in the calculations described above, so we solve Eq. (39-12) directly for $\tilde{y}_2$. Knowing the composition of this stream and the temperature, we can find its enthalpy $\tilde{H}_2$ from tables of enthalpy data, and in this way we check the value assumed for $\tilde{H}_2$ earlier in the calculations.

After the calculations for conditions around plate 1 have been repeated until they give consistent results, the entire procedure is repeated with balances around plate 2. As before, the material and enthalpy balances contain eight variables, of which five are known and three are unknown ($\tilde{O}_2$, $\tilde{V}_3$, $\tilde{H}_3$). $\tilde{H}_3$ is assumed (as a first guess we should assume $\tilde{H}_3 = \tilde{H}_2$), and the balances solved for $\tilde{O}_2$ and $\tilde{V}_3$. A balance is then written around the plate for the volatile component. This yields

$$\tilde{y}_3 = \frac{\tilde{O}_2}{\tilde{V}_3}\tilde{x}_2 + \frac{\tilde{D}\tilde{x}_D}{\tilde{V}_3} \tag{39-13}$$

which is solved for $\tilde{y}_3$ and permits the independent determination of $\tilde{H}_3$ as a check on the assumed value of $\tilde{H}_3$.

The procedure outlined for plates 1 and 2 is repeated for successive plates to the point where the feed is added. If the feed is to be added at the optimum location, it will usually be admitted to the column at some point where it has approximately the same composition as some stream in the column (this matter will be discussed later). At the feed plate the contributions of the feed stream must be considered in the material, enthalpy, and component balances. When the calculations apply to plates beneath the feed, they proceed just as they did for the column above the feed plate, except that if Eq. (39-13) is used, it must contain a term to account for the feed stream.

The calculations continue for successive plates until a vapor stream is found which has a composition that indicates that it is in equilibrium with the bottom product $\tilde{x}_B$. This vapor stream will be the vapor from the reboiler and marks the end of the calculations. The principal object, of course, has been to determine the number of stages required to give a specified separation of a certain feed into overhead and bottom products of predetermined compositions $\tilde{x}_D$ and $\tilde{x}_B$. In the course of making the calculations, the compositions and enthalpies of all streams have been determined, so the determination of the heat-transfer rate in the condenser and reboiler is a matter simply of making enthalpy balances on the fluid streams entering and leaving these auxiliary pieces of equipment. In addition, the feed plate has been located and the rates of all streams have been determined. The temperatures will have been found (or can readily be found from a knowledge of total pressure and stream composition), so that the volumetric flow rates of all streams can be found directly. This information is necessary so that the tower diameter and the dimensions of such mechanical equipment in the column as the downcomers and bubble caps can be determined.

The McCabe-Thiele Method of Analysis

In the Sorel analysis the principal difficulty was in the trial-and-error determinations of the enthalpies of the vapor streams $\tilde{H}_2$, $\tilde{H}_3$, etc. As a first approximation, it was suggested that these enthalpies might be considered equal. This assumption made it possible to solve for the vapor compositions and then to calculate the vapor enthalpies. Lewis[1] made a similar assumption when he suggested that the vapor rates $\tilde{V}_1$, $\tilde{V}_2$, $\tilde{V}_3$, etc., might be considered to be equal. In sections of the column where no feed is added or side streams withdrawn, the assumption of equal vapor rates implies equal liquid rates, as can be seen from Eq. (39-8).

For many systems in which the components have similar molecular structures, the assumption of Lewis is sufficiently accurate so that the trial-and-error part of the Sorel solution is eliminated. The overall material and enthalpy balances, Eqs. (39-8) and (39-9), become superfluous, and only the balance on the more volatile component, Eq. (39-12), is needed. If all vapor streams in the column above the feed are flowing at the same rate, we can designate this rate by a common symbol $\tilde{V}$; similarly, we designate the rate of all liquid streams as $\tilde{O}$. Thus Eq. (39-12) is written as

$$\tilde{y}_2 = \frac{\tilde{O}}{\tilde{V}} \tilde{x}_1 + \frac{\tilde{D}\tilde{x}_D}{\tilde{V}} \qquad (39\text{-}14)$$

This equation applies equally well as a relation between $\tilde{y}_3$ and $\tilde{x}_2$, or, for that matter, between any pair of passing streams in Fig. 39-3. Thus the Lewis modification reduces the Sorel method to a series of alternate equilibrium and component-balance calculations without trial and error.

With reference to Fig. 39-3, the first calculation would be to find $\tilde{x}_1$ knowing $\tilde{y}_1$ and the equilibrium data. Then $\tilde{y}_2$ would be found from Eq. (39-14). Next $\tilde{x}_2$

[1] W. K. Lewis, *Ind. Eng. Chem.*, **14**:492 (1922).

would be found as the composition of a stream in equilibrium with $\tilde{y}_2$. After this, $\tilde{y}_3$ would be found by substituting $\tilde{y}_3$ and $\tilde{x}_2$ in Eq. (39-14) for $\tilde{y}_2$ and $\tilde{x}_1$. In this manner the calculations would proceed down the column. At and below the feed plate, the addition of the feed would have to be considered in the material-balance equation; otherwise the method of calculation would be unaltered.

The method of McCabe and Thiele[1] is simply a graphical solution of the Sorel equations, with the simplification of constant vapor and liquid rates as suggested by Lewis. The advantage of the McCabe-Thiele method depends on the linearity of Eq. (39-14). McCabe and Thiele suggested that this function be plotted as an operating line on the same graph as the equilibrium curve of $\tilde{y}$ versus $\tilde{x}$, so that the number of equilibrium stages can be determined by a graphical construction. The stages are counted between the equilibrium and operating lines just as in Chaps. 37 and 38.

Before proceeding with the details of the McCabe-Thiele method we shall examine the assumption of constant vapor and liquid rates (often called *constant molar overflow*). The principal source of enthalpy on a bubble-cap plate is the saturated vapor approaching the plate from below. This enthalpy, some of which is given up by condensation in the liquid on the plate, causes the vaporization of some of the liquid on the plate. The vapor produced joins the uncondensed vapor rising from the plate and represents the principal enthalpy stream leaving the plate. In addition, there may be a heat loss through the walls of the column, though this is usually slight, and there is also a small difference in the enthalpies of the liquid stream arriving at the plate and the liquid stream leaving. It can be seen, therefore, that the two vapor streams are the principal "carriers" of enthalpy. If the two components have similar molecular structure and behavior, the molar latent heat of vaporization divided by the absolute temperature is approximately constant. This generalization is known as *Trouton's rule.* Since the streams of vapor leaving adjacent plates are at nearly the same temperature, the molar latent heats are nearly the same if the system follows Trouton's rule. Thus, if heat losses are negligible, a mole of vapor condensing in the liquid on a plate will cause the generation of another mole of vapor from the liquid. In these circumstances, therefore, the condition of equal molar vapor rates between plates is achieved. It is important to note that Trouton's rule refers only to molar latent heats. The latent heats written as Btu/lb are far from constant. Hence an equation similar to Eq. (39-14) in which the liquid and vapor rates are expressed in lb/h would not be linear because the rates O and V would not be constant.

Rectifying columns Fractionating columns in which the feed enters the bottom of the column as a vapor are known as *rectifying*, or *enriching*, columns. An

[1] W. L. McCabe and E. W. Thiele, *Ind. Eng. Chem.*, **17**:605 (1925).

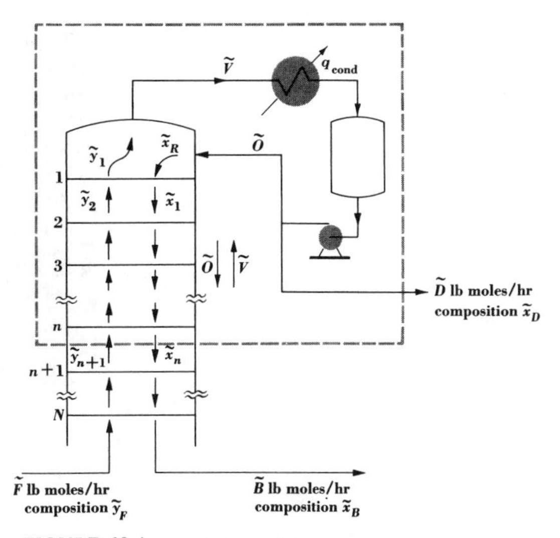

FIGURE 39-4
Operation of a rectifying column with constant molar overflow.

overhead product is produced in the same manner as in a complete fractionating column; this product may be quite rich in the more volatile component. The bottom product, however, which is the stream of liquid from the bottom plate, is seldom much leaner than the feed, so that only one of the two product streams can be said to be a fairly pure product. In the sense that the less volatile component is preferentially removed from the vapor ascending the column, the operation of a rectifying column can be considered somewhat analogous to that of a gas-absorption tower. It also resembles the enriching section of the extraction system with reflux, described in Chap. 38.

The rectifying column is somewhat simpler to analyze than a complete fractionating column, so we shall consider the rectifying column first. We shall refer to Fig. 39-4, in which the system is supplied with a saturated vapor feed at the rate of $\tilde{F}$ lb mol/h and produces saturated-liquid products at the rate of $\tilde{D}$ and $\tilde{B}$ lb mol/h. The condenser at the top produces a saturated liquid of composition $\tilde{x}_D$, part of which is removed as the overhead product. The remainder of the condensate is returned to the column on the top plate as a saturated-liquid reflux.

A material balance is written for the more volatile component for the portion of the column above plate $n + 1$, as shown in Fig. 39-4.

$$\tilde{V}\tilde{y}_{n+1} = \tilde{O}\tilde{x}_n + \tilde{D}\tilde{x}_D \qquad (39\text{-}15)$$

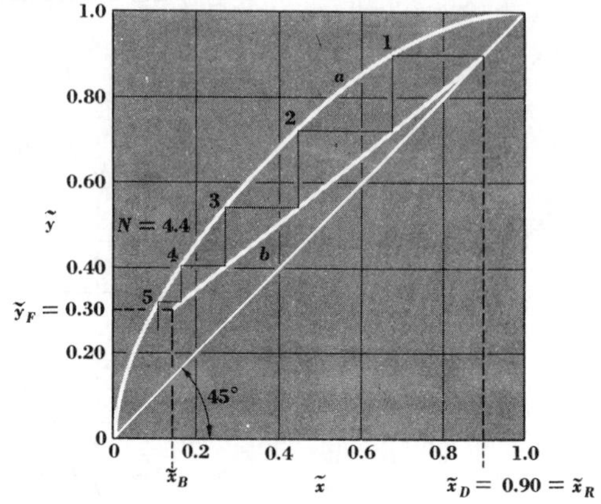

FIGURE 39-5
McCabe-Thiele diagram for a rectifying column (Example 39-2).
a Equilibrium line.
b Enriching line, slope $\tilde{O}/\tilde{V} = 0.80$.

This may be written explicitly for $\tilde{y}_{n+1}$ as

$$\tilde{y}_{n+1} = \frac{\tilde{O}}{\tilde{V}}\tilde{x}_n + \frac{\tilde{D}\tilde{x}_D}{\tilde{V}} \tag{39-16}$$

Equation (39-16) is a linear relation if $\tilde{O}$ and $\tilde{V}$ are constant. When plotted on a graph of $\tilde{y}$ versus $\tilde{x}$, it gives a straight line with a slope $\tilde{O}/\tilde{V}$. The intercept if extended to the y axis is $\tilde{D}\tilde{x}_D/\tilde{V}$. The line is shown in Fig. 39-5, which also shows the equilibrium curve and a line drawn at 45° from the origin. The line representing Eq. (39-16) is an operating line, or more specifically, an enriching line. It can be located in a variety of ways. Points on the line represent passing streams; because $\tilde{y}_1$, $\tilde{x}_D$, and $\tilde{x}_R$ are identical and $\tilde{y}_1$ and $\tilde{x}_R$ are passing streams, a point on the 45° line at $\tilde{x}_D$ represents these three streams and is a point on the operating line. Another point can be found on the operating line at any section where the compositions of the passing streams are known. For example, the balance which gave Eqs. (39-15) and (39-16) could have been written to include all the plates in the column. Thus $\tilde{y}_{N+1}$ would be $\tilde{y}_F$ and $\tilde{x}_N$ would be $\tilde{x}_B$. If we know the feed and bottom-product compositions, we can plot this point on the operating line as representing passing streams at the lowest point in the column. This is also the lowest point on the enriching line and is shown in Fig. 39-5.

For a rectifying column $\tilde{O} = \tilde{B}$ and $\tilde{V} = \tilde{F}$, so that if the flow rates of these external streams are known, $\tilde{O}/\tilde{V}$ is readily determined. Alternatively, $\tilde{O}/\tilde{V}$ can be found from the reflux ratio $\tilde{O}/\tilde{D}$ by writing

$$\frac{\tilde{O}}{\tilde{V}} = \frac{\tilde{O}}{\tilde{O} + \tilde{D}} \tag{39-17}$$

Both numerator and denominator of the right side are divided by $\tilde{D}$ to give

$$\frac{\tilde{O}}{\tilde{V}} = \frac{\tilde{O}/\tilde{D}}{\tilde{O}/\tilde{D} + 1} \tag{39-18}$$

Hence, if $\tilde{O}/\tilde{D}$ is specified, the slope of the operating line $\tilde{O}/\tilde{V}$ can be found. Sometimes $\tilde{O}/\tilde{V}$ is referred to as the internal reflux ratio and $\tilde{O}/\tilde{D}$ as the external reflux ratio.

The number of equilibrium stages required to give a specified separation is found as in previous chapters. By successive horizontal and vertical steps between the equilibrium and operating lines, we find the number of stages required to give the specified separation between the overhead and bottom streams. Each stage is represented by an intersection, or step, which touches the equilibrium curve.

Example 39-2 A rectifying column is supplied with a saturated-vapor feed containing 30 mol percent A and 70 mol percent B and is to produce a saturated-liquid distillate containing 90 mol percent A and 10 mol percent B. The reflux ratio $\tilde{O}_R/\tilde{D}$ is 4. Find the number of theoretical stages required to give the separation, assuming constant molar overflow. Vapor-liquid equilibrium data are given on a plot of $\tilde{y}$ versus $\tilde{x}$ shown in Fig. 39-5. Component A is more volatile than component B.

The slope of the operating line $\tilde{O}/\tilde{V}$ is found from Eq. (39-18) to be 0.80. The line starts at the 45° line at $\tilde{x}_D = \tilde{x}_R = \tilde{y}_1 = 0.90$ and is drawn with a slope of 0.80. The line terminates at the lower end at $\tilde{y}_F = 0.30$. The value on the $\tilde{x}$ axis at this point represents the bottom-product composition $\tilde{x}_B$.

Equilibrium stages are stepped off as described above. It is seen that 4.4 stages are required to give the desired separation. ////

Stripping A stripping column is usually characterized by having the feed supplied to the top of the column, as shown in Fig. 39-6. In this way it resembles the portion of a complete fractionating column from the feed plate to the reboiler. The feed is usually supplied as a saturated liquid, and the overhead product is the vapor rising from the top plate. This vapor is somewhat richer in terms of the more volatile component than is the feed. However, it is usually not a stream of very great purity. The only stream leaving the stripping column that has a very high concentration is the saturated-liquid bottom product from the reboiler. This stream can be quite rich with respect to the less volatile component. In this sense the stripping column in distillation is analogous to the stripping column in gas absorption in which a liquid stream passing down the column is stripped of a volatile component by a stream of vapor passing upward.

The stripping column shown in Fig. 39-6 is analyzed by the McCabe-Thiele

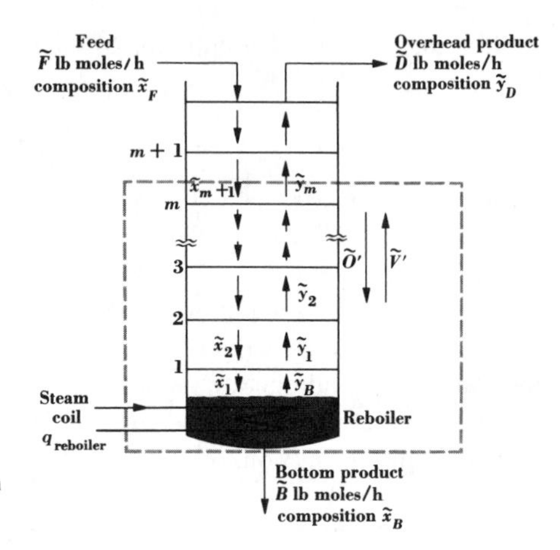

FIGURE 39-6
Operation of a stripping column with constant molar overflow.

method in Fig. 39-7. Constant molar rates of $\tilde{O}'$ and $\tilde{V}'$ are assumed for all streams in the column. A material balance is written for the more volatile component around that section of the column below plate $m + 1$.

$$\tilde{O}'\tilde{x}_{m+1} = \tilde{V}'\tilde{y}_m + \tilde{B}\tilde{x}_B \qquad (39\text{-}19)$$

which is rearranged to give

$$\tilde{y}_m = \frac{\tilde{O}'}{\tilde{V}'}\tilde{x}_{m+1} - \frac{\tilde{B}\tilde{x}_B}{\tilde{V}'} \qquad (39\text{-}20)$$

If $\tilde{O}'$ and $\tilde{V}'$ are constant, then Eq. (39-20) is linear. On Fig. 39-7 the function is designated as the stripping, or operating, line and represents compositions of passing streams at any horizontal cross section in the column. The stripping line can be found by a variety of methods. At the top of the column we may know the compositions of the feed entering and the vapor leaving. If the feed is a saturated liquid, the point $\tilde{x}_F$, $\tilde{y}_D$ represents the upper end of the line. The lower end can be found by determining the intersection with the 45° line. At the 45° line $\tilde{y} = \tilde{x}$, and making this substitution in Eq. (39-19), we get

$$\tilde{O}'\tilde{x} = \tilde{V}'\tilde{x} + \tilde{B}\tilde{x}_B$$

which gives the $\tilde{x}$ value at the intersection as

$$\tilde{x} = \frac{\tilde{B}\tilde{x}_B}{\tilde{O}' - \tilde{V}'}$$

$$= \tilde{x}_B \qquad (39\text{-}21)$$

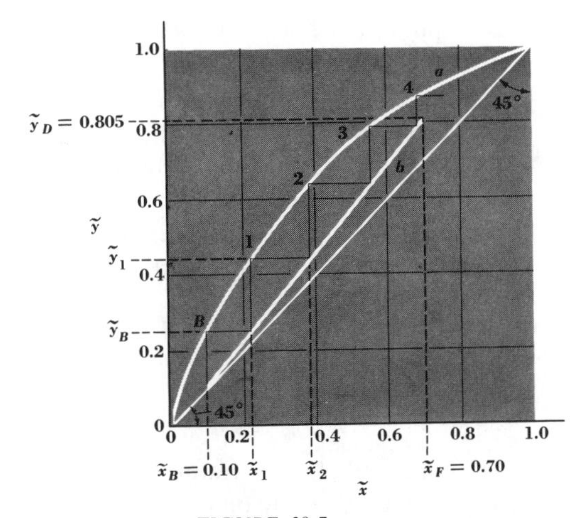

FIGURE 39-7
MCabe-Thiele diagram for a stripping column (Example 39-3).
a Equilibrium curve.
b Stripping-line slope = 100/85.

An additional device for locating the line is to use the slope. If the feed is a saturated liquid, $\tilde{O}' = \tilde{F}$ and $\tilde{V}' = \tilde{D}$, so that $\tilde{O}'/\tilde{V}'$ can be found merely by knowing the flow rates of the terminal streams.

The determination of the number of equilibrium stages in a stripping column is similar to the procedure used for an enriching column. We can start at either end of the operating line; we arbitrarily choose the bottom. The composition of the vapor leaving the reboiler $\tilde{y}_B$ is found on the equilibrium curve at $\tilde{x}_B$. This is a vertical step from the 45° line at $\tilde{x}_B$. The composition of the liquid flowing down from the plate to the reboiler $\tilde{x}_1$ is found by solving Eq. (39-20) graphically. This requires the construction of a horizontal step at $\tilde{y}_B$ to the operating line, giving $\tilde{x}_1$ at the intersection. Composition $\tilde{y}_1$ is found by a vertical step at $\tilde{x}_1$ to the equilibrium line, and composition $\tilde{x}_2$ by a horizontal step to the operating line. This procedure is continued until the upper end of the operating line is reached or passed. The number of equilibrium stages is found by counting the intersections on the equilibrium curve. Each intersection represents a pair of streams in equilibrium: a vapor rising from a stage and a liquid descending from the same stage. Hence each intersection represents an equilibrium stage.

Example 39-3 A stripping column is supplied with 100 lb mol/h of saturated-liquid feed containing 70 mol percent *A* and 30 mol percent *B*. It is to produce a bottom product containing 10 mol percent A and 90 mol percent B at the rate of 15 lb mol/h. Find the number of theoretical stages required to give this separation, assuming constant molar overflow. Vapor-liquid equilibrium data are plotted on Fig. 39-7 in terms of the mole fraction of component A, which is the more volatile of the two components.

By an overall material balance we see that the overhead product is produced at the rate of 85 lb mol/h. A material balance for component A gives

$$(100)(0.70) = (15)(0.10) + (85)\tilde{y}_D$$

from which we find

$$\tilde{y}_D = 0.805$$

The upper end of the operating line is located at $\tilde{y}_D = 0.805$, $\tilde{x}_F = 0.70$, and the lower end at $\tilde{x}_B = 0.10$ on the 45° line. Thus the position of the operating line is determined. The slope is $\frac{100}{85}$, though this piece of information need not be used.

The number of stages may be seen on Fig. 39-7 to be 4.3, including the stage which is the reboiler. ////

Complete fractionating columns When a column contains bubble-cap plates both above and below the feed injection point, a possibility exists of producing both rich overhead and bottom products. In this sense, as well as in schematic appearance, the complete fractionating column combines the properties of both enriching and stripping columns. The complete column has both a reflux condenser and a reboiler, as shown in Fig. 39-2.

The operating lines in the McCabe-Thiele analysis of a complete column are illustrated in Fig. 39-8. Two operating lines are shown; they are described by the same equations, (39-16) and (39-20), as the enriching and stripping lines of the respective columns. Furthermore, the lines may be located by some of the same techniques as were used for the separate columns.

One minor change which applies to the analysis of complete fractionating columns is in the treatment of the feed stream. A number of possibilities exist, depending on the condition of the feed. The feed may be a subcooled liquid (below its saturation temperature), a saturated liquid, a mixture of saturated liquid and saturated vapor, a saturated vapor, or a superheated vapor. The quantitative relations between the streams $\tilde{V}$, $\tilde{V}'$, $\tilde{O}$, and $\tilde{O}'$ depend on which of the above conditions describes the feed. For convenience, we shall introduce a quantity designated q, which is defined as the number of moles of saturated liquid produced on the feed plate by each mole of feed added to the column. With this definition we write the

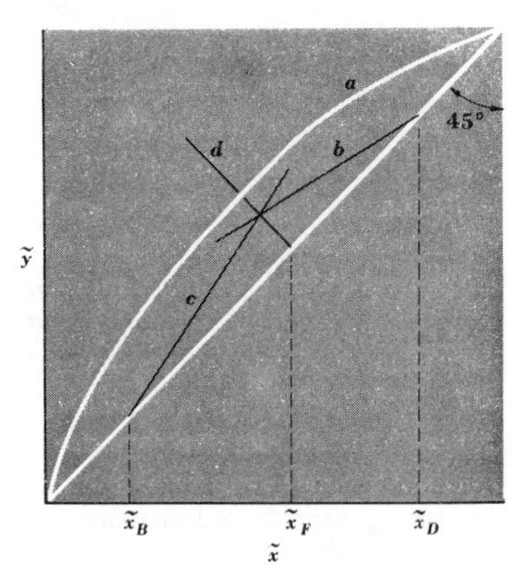

FIGURE 39-8
Operating lines for a complete fractionating column with a feed consisting of a mixture of vapor and liquid.
a Equilibrium curve.
b Enriching line.
c Stripping line.
d *q* line.

following relationships between the streams above and below the feed plate:

$$\tilde{O}' = \tilde{O} + q\tilde{F} \tag{39-22}$$

$$\tilde{V}' = \tilde{V} + (q - 1)\tilde{F} \tag{39-23}$$

These operating conditions are represented in Fig. 39-9 for a feed containing saturated liquid, saturated vapor, or a mixture of the two.

Because the amount and quality of the feed affect the relative amounts of the vapor and liquid streams in the column, it can be seen that the feed condition also affects the slopes of the two operating lines $\tilde{O}/\tilde{V}$ and O'/V'. On the McCabe-Thiele diagram, the quality of the feed determines the locus of intersection of the two operating lines. At the point of intersection, the material balances, Eqs. (39-15) and (39-19), which are represented graphically by the operating lines, can be subtracted

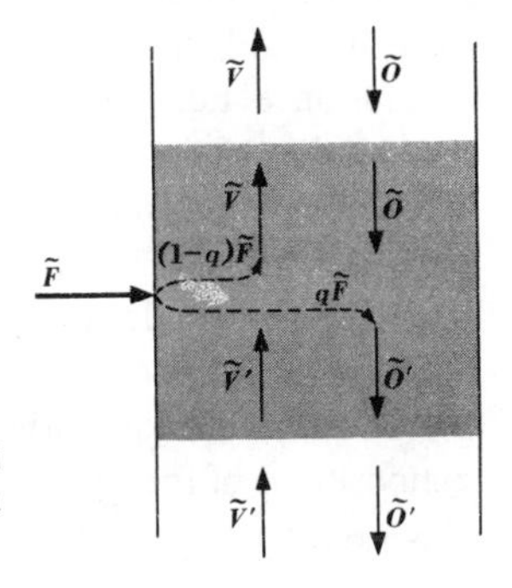

FIGURE 39-9
Schematic diagram showing effect of feed quality when feed is a saturated liquid or a saturated vapor.

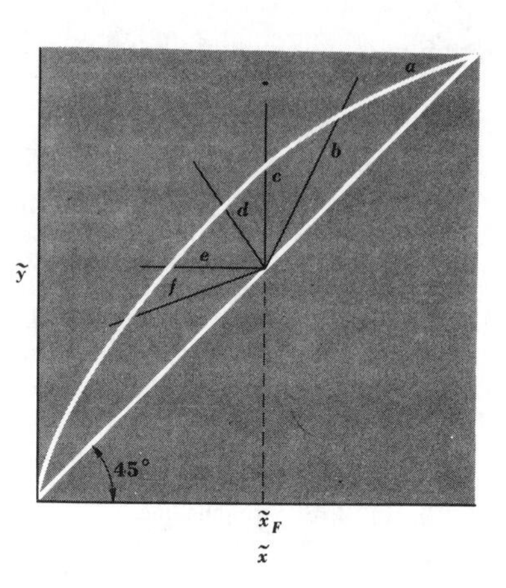

FIGURE 39-10
Location of q line as a function of feed quality. ($\tilde{x}_F$ is overall feed composition regardless of quality.)
a Equilibrium line.
b Subcooled liquid q line.
c Saturated liquid q line.
d Liquid and vapor q line.
e Saturated vapor q line.
f Superheated vapor q line.

to give a locus of intersection on the plot of $\tilde{y}$ versus $\tilde{x}$:

$$\tilde{y}(\tilde{V}' - \tilde{V}) = \tilde{x}(\tilde{O}' - \tilde{O}) - \tilde{B}\tilde{x}_B - \tilde{D}\tilde{x}_D \tag{39-24}$$

In this equation we substitute for $\tilde{V}' - \tilde{V}$ from Eq. (39-23) and for $\tilde{O}' - \tilde{O}$ from Eq. (39-22). In addition, the sum of the last two terms on the right-hand side of Eq. (39-24) is equal to $\tilde{F}\tilde{x}_F$, in which $\tilde{x}_F$ represents the overall composition of the feed, regardless of quality. Making these substitutions, the locus of intersection of the two operating lines becomes

$$\tilde{y}(q - 1)\tilde{F} = \tilde{x}q\tilde{F} - \tilde{F}\tilde{x}_F \tag{39-25}$$

which is rearranged to give

$$\tilde{y} = \frac{q}{q - 1}\tilde{x} - \frac{\tilde{x}_F}{q - 1} \tag{39-26}$$

The straight line described by Eq. (39-26) is known as the q line. The intersection of the q line and the 45° line is found by equating $\tilde{y}$ and $\tilde{x}$ in Eq. (39-26). In this circumstance Eq. (39-26) reduces to $\tilde{x} = \tilde{x}_F$, in which $\tilde{x}_F$ designates the overall composition of the feed rather than just the composition of the liquid phase. The other piece of information used to locate the q line is its slope. The relation between slope and feed quality is indicated in Table 39-1.

The possible locations of the q line are shown in Fig. 39-10. For feed which contains a mixture of saturated liquid and saturated vapor it can be shown that the q line intersects the equilibrium curve at values of $\tilde{y}$ and $\tilde{x}$ which are identical with the compositions of the vapor and liquid phases in the feed. That this is true for a feed which is all saturated vapor or all saturated liquid may be seen in Fig. 39-10. For subcooled liquid or superheated vapor feed, the intersection of the q line and equilibrium curve has no such significance.

The determination of stream compositions in a complete fractionating column is very similar to the procedure used for stripping and rectifying columns. A situation for which the feed is added at the optimum plate is illustrated in Fig. 39-11*a*. It can be seen that passing-stream compositions are found on the top operating line above the point of intersection of the two operating lines; below this intersection, passing-stream compositions are found using the bottom operating line.

For the system illustrated in Fig. 39-11*a*, the second plate is the optimum feed plate. This means that if the feed is a saturated or a subcooled liquid, it is added to the liquid flowing onto plate 2 from the plate above. If the feed is a saturated or superheated vapor, the construction of Fig. 39-11*a* implies that the feed is added beneath plate 2 and joins the vapor rising from the plate below. The final possibility is that the feed is a mixture of vapor and liquid.[1] The construction of Fig. 39-11*a* implies that the vapor and the liquid of the feed are separated and the liquid is added directly above the feed plate while the vapor is injected underneath the feed plate. Although this may not be done in practice, such procedure, nevertheless, is the implication of the commonly used analysis shown in Fig. 39-11*a*.

The feed plate shown in Fig. 39-11*a* is the optimum feed plate, but not the only possible choice. The feed plate for the same operating lines may be the first plate as shown in Fig. 39-11*b* or the third plate as in Fig. 39-11*c*. It can be seen on the

Table 39-1 EFFECT OF FEED QUALITY ON SLOPE OF q LINE

Feed quality	q	Slope of q line, $q/(q-1)$
Subcooled liquid†	>1	>1
Saturated liquid	1	∞
Liquid and vapor	0–1	Negative
Saturated vapor	0	0
Superheated vapor	<0	0–1

† For example, suppose a liquid of molar heat capacity 18 Btu/(lb mol)(°F) and molar latent heat 18,000 Btu/lb mol and subcooled 20°F is fed to a tower at tower pressure. Then we have $q = 1 + 18(20)/18{,}000 = 1.02$. An extra 0.02 mol of saturated liquid is produced by condensing some of the vapor in the tower as the liquid is heated by 20°.

[1] From the slope of the q line in Fig. 39-11*a*, we see that this is the situation actually represented here.

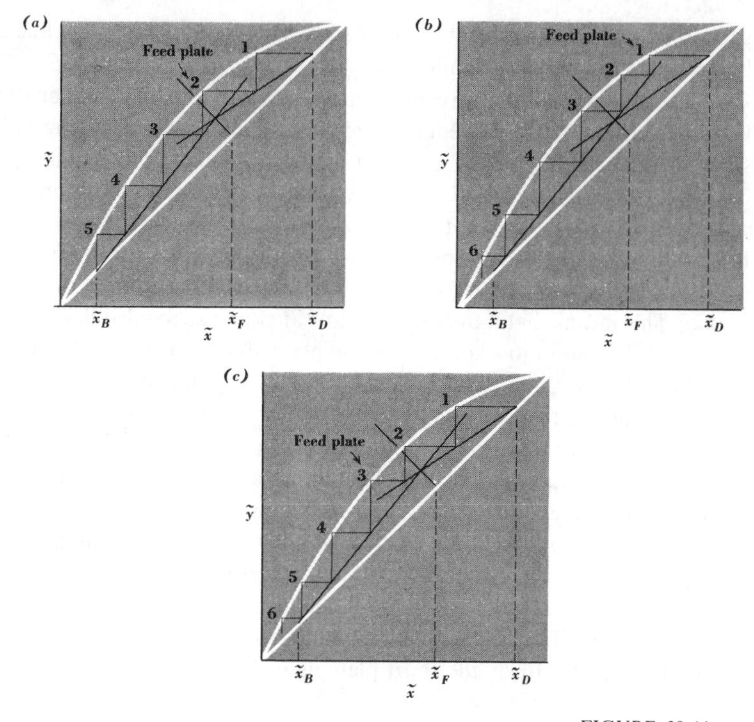

FIGURE 39-11
Feed-plate location.

diagrams that, for these latter possibilities, a greater number of plates is required to give the same specified separation between overhead and bottom products. Thus the location of the feed plate at the intersection of the operating lines is the optimum choice. From the constructions of Fig. 39-11, it is evident that the feed plate may be located anywhere between the points where the two operating lines intersect the equilibrium curve.

If a column operates with more than one feed stream, each feed is represented by its own q line, which is found by locating a point on the 45° line representing the overall composition of that feed and drawing a line with a slope of $q/(q - 1)$, as shown by Eq. (39-26). Each q line is the locus of intersection of the operating lines representing passing streams immediately above and below the feed in question. For example, a column with two feeds would have three operating lines, one for each section of the column in which molar liquid and vapor rates are assumed to be constant. For a given set of operating lines, each feed stream would have an

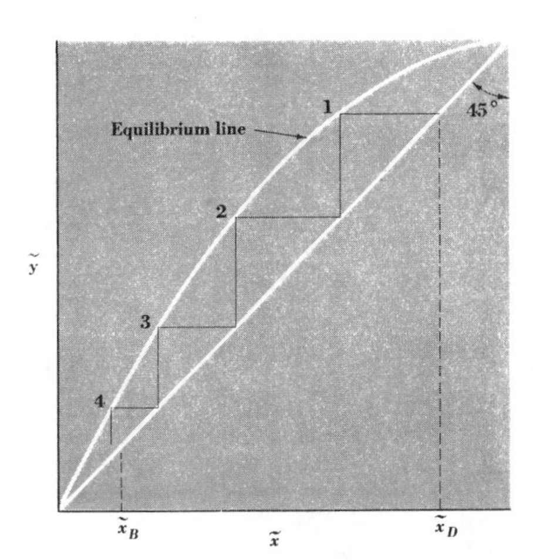

FIGURE 39-12
McCabe-Thiele analysis for total reflux and minimum plates.

optimum injection point, which would give the specified separation with a minimum number of equilibrium stages.

Limiting reflux conditions In Chap. 38 we considered for extraction two limiting reflux conditions which are also of interest in distillation. One is the condition of total reflux. This means that all vapor from the top plate is condensed and returned as liquid reflux. Thus $\tilde{O} = \tilde{V}$, and the slope of the top operating line is unity. If the feed stream and bottom-product stream are also shut off, then the liquid and vapor rates $\tilde{O}'$ and $\tilde{V}'$ beneath the feed are also equal and are equal to the rates above the feed plate. Thus both operating lines have slopes of unity and coincide with the 45° line. The situation is illustrated in Fig. 39-12, in which the equilibrium stages are stepped off. It can be seen that for this system the number of stages for a separation between specified values of $\tilde{x}_B$ and $\tilde{x}_D$ is a minimum.

A second limiting reflux condition is minimum reflux. If the overhead-product composition $\tilde{x}_D$ is specified, the location of the top operating line is dependent on the slope $\tilde{O}/\tilde{V}$. For a given vapor rate $\tilde{V}$, any increase in the rate of product withdrawal $\tilde{D}$ means a decrease in the rate of reflux being returned to the column, and hence a decrease in the liquid rate $\tilde{O}$. As $\tilde{O}/\tilde{V}$ decreases, the intersection between the top and bottom operating lines at the q line moves farther from the 45° line in the direction of the equilibrium curve. This is accompanied by an increase in the number of stages necessary to give a fixed product separation $\tilde{x}_B$ and $\tilde{x}_D$. At the point where the intersection of the operating lines touches the equilibrium curve, an infinite number of stages will be required. This is illustrated in Fig. 39-13, in which the zone of zero separation (the pinch zone) occurs on both sides of the feed

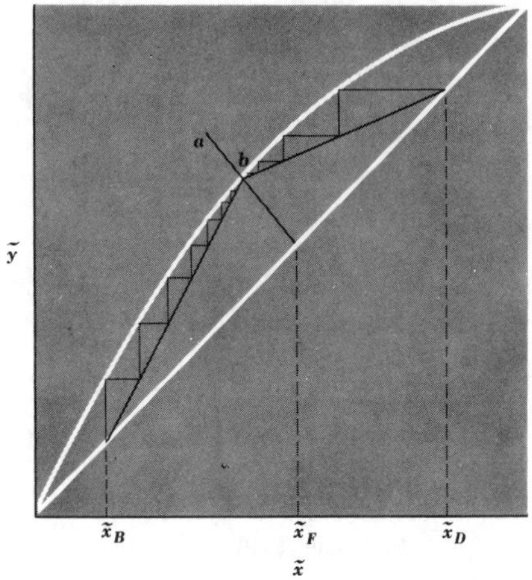

FIGURE 39-13
McCabe-Thiele analysis for minimum reflux and an infinite number of plates.
a q line.
b Pinch zone.

plate. If a still lower value is chosen for $\tilde{O}/\tilde{V}$, the intersection of the operating lines would occur outside the equilibrium curve and two zones of infinite plates would occur. By definition, however, we shall restrict the term minimum reflux to apply to the maximum reflux rate at which infinite plates occur.

For an azeotropic system the zone of infinite plates does not necessarily occur at the feed plate. This is shown in Fig. 39-14, in which the zone of infinite plates is located by drawing the operating line tangent to the equilibrium curve. The reflux

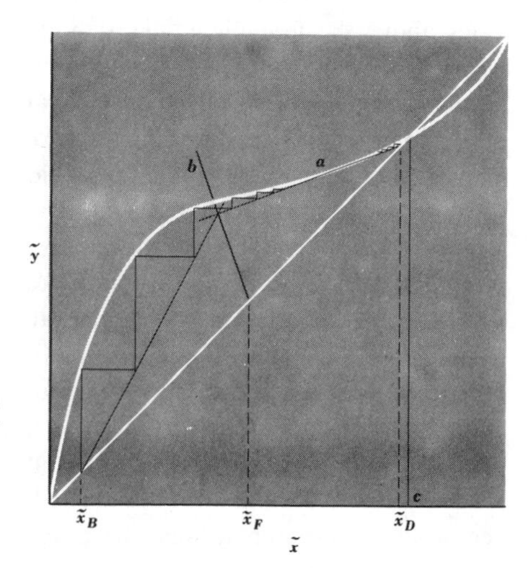

FIGURE 39-14
McCabe-Thiele analysis for minimum reflux and an infinite number of plates for an azeotropic system.
a Pinch zone.
b q line.
c Azeotropic composition.

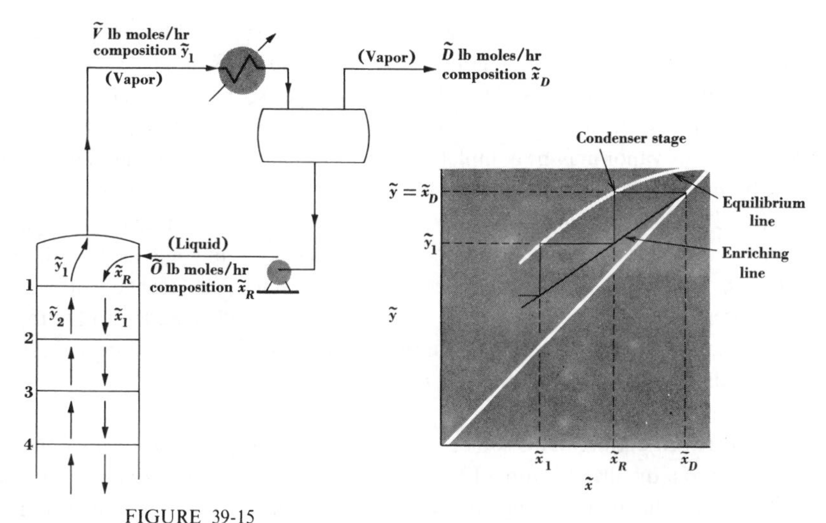

FIGURE 39-15
Analysis for a partial condenser from which the vapor and liquid phases leave in equilibrium.

rate which would give this operating line is the minimum reflux rate. For the system shown in Fig. 39-14, the choice of a sufficiently low value for $\tilde{x}_D$ would mean that the zone of infinite plates would occur just at the feed, as it does in nonazeotropic systems.

Partial condensers The overhead product may be produced as a saturated vapor if the vapor from the top plate is only partially condensed. The liquid condensate is returned to the column as reflux. The reason for this method of operation may be that an excessive column pressure would be required to cause total condensation with cooling water at atmospheric temperature. Alternatively, total condensation might be achieved with a lower column pressure, but with some form of refrigeration to remove heat from the condenser. Both these methods of operation are expensive, however, and if the overhead product is to be used as fuel or as a gaseous reactant in some other part of the process, it will usually be taken as a vapor from the overhead condenser.

In some partial condensers the vapor and liquid phases are in equilibrium. For this case the McCabe-Thiele analysis, shown in Fig. 39-15, is the same as for a total condenser. The equation for the enriching line is the same as Eq. (39-16); $\tilde{x}_D$ represents the composition of the gaseous product, and $\tilde{D}$ is the rate at which it is produced in pound moles per hour. The enriching line still crosses the 45° line at

$\tilde{x} = \tilde{x}_D$, as may be shown by substituting $\tilde{x} = \tilde{x}_n = \tilde{y}_{n+1}$ in Eq. (39-16). The first step on the McCabe-Thiele diagram represents conditions in the condenser, so that the condenser provides one of the theoretical stages required for the separation.

Other possible methods of condenser operation were discussed in Chap. 25. A portion of the vapor may be completely condensed, and the remainder unchanged, so that both the gaseous overhead product and the liquid reflux have the same composition, and no separation is achieved in the condenser. Alternatively, a separation equivalent to several theoretical stages may be achieved in condensers with vertical tubes in which the vapor and liquid flow countercurrently to each other. Such condensers often subcool the liquid. Subcooled reflux when added to the top plate returns to the saturation temperature and produces an amount of internal reflux, $\tilde{O}$, greater than the amount of external reflux added, $\tilde{O}_R$.

Steam injection It is sometimes convenient to supply heat by injecting steam directly into a distillation apparatus instead of having the steam condense in a closed coil. When the liquid being distilled is immiscible with water, the technique is known as *steam distillation.* The presence of the steam in the vapor phase reduces the partial pressure of the other components at a fixed total pressure and thus lowers the saturation temperature of the liquid to be distilled. Steam distillation is commonly used to distill organic liquids which might decompose if heated to temperatures high enough to cause them to boil without the presence of the steam. Gases other than steam may also be used, but steam is usually preferred because of its inertness and availability.

Steam is also injected into binary-distillation systems in which water is the less volatile of two components being separated. In this case the closed steam coil in the reboiler is replaced by a sparger, from which the steam is injected as small bubbles into the liquid in the reboiler. Mass transfer occurs between the liquid and the bubbles, and if sufficient contact time is available, the bubbles are in equilibrium with the liquid when they leave the reboiler.

The analysis of a system in which open steam is added to the reboiler may be carried out using the customary stripping-line equation, Eq. (39-20). This expression is valid because it represents a mass balance for the more volatile component, none of which has been added in the steam. The stripping line no longer crosses the 45° line at $\tilde{x} = \tilde{x}_B$, however. In the analysis which leads to Eq. (39-21), we now substitute $\tilde{B} - \tilde{S}$ for $\tilde{O}' - \tilde{V}'$, so that the intersection with the 45° line occurs at $\tilde{x} = \tilde{B}\tilde{x}_B/(\tilde{B} - \tilde{S})$, in which $\tilde{S}$ is the rate of steam addition in pound moles per hour. A more useful piece of information is the intersection of the stripping line with the $\tilde{x}$ axis. When $\tilde{y}_m$ in Eq. (39-19) is zero, we get

$$\tilde{x}_{m+1} = \frac{\tilde{B}\tilde{x}_B}{\tilde{O}'} \qquad (39\text{-}27)$$

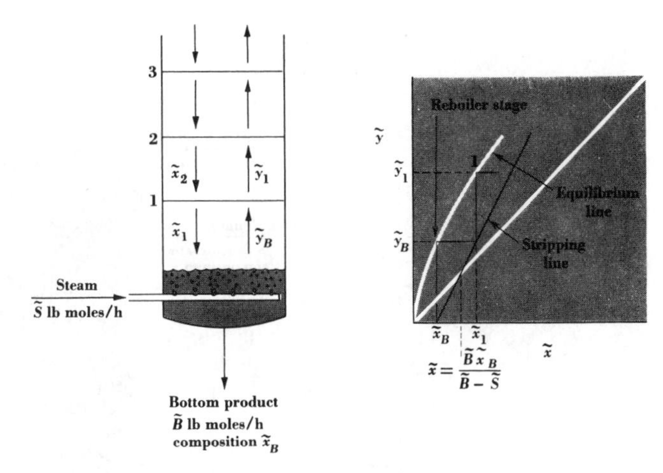

FIGURE 39-16
McCabe-Thiele analysis for a miscible binary system with open steam.

If constant molar overflow exists, $\tilde{S} = \tilde{V}'$, so that $\tilde{B} = \tilde{O}'$ and we see that $\tilde{x}_{m+1} = \tilde{x}_B$. The analysis is illustrated in Fig. 39-16. The counting of steps begins at $\tilde{x}_B$ as usual.

For systems identical in all other respects, the use of open rather than closed steam dilutes the bottom product. In addition, an extra fraction of a stage is required, as may be seen in Fig. 39-16. The advantage of open steam lies in the simpler construction of the heating coil.

In binary systems in which water is the more volatile of the two components, the use of open steam in the reboiler would not be suitable. In such systems the overhead product is the water-rich stream; pure water may be used as reflux to increase the concentration of the more volatile component on the plates at the top of the column. As in the use of open steam, this causes the useful portion of the operating line (this time the enriching line) to extend below the 45° line.

Side streams Product streams are occasionally taken from sections of the column between the top and bottom; these are known as side streams. The side stream may be either vapor or liquid. Occasionally a vapor side stream is withdrawn, condensed, and returned to the column as liquid. The purpose is to reduce the vapor velocity in the tower and provide higher plate efficiency. Side streams may be withdrawn, however, merely because a product of that composition is desired.

The McCabe-Thiele analysis for a column with a liquid side stream is il-

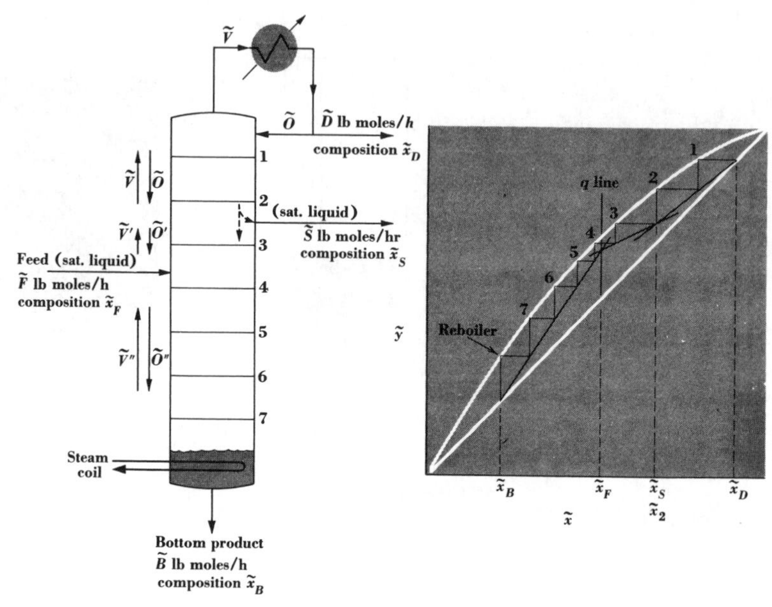

FIGURE 39-17
McCabe-Thiele analysis for a system with a side stream.

lustrated in Fig. 39-17. The principal effect of a side stream on the analysis is to alter either the vapor or the liquid rate, so that a third operating line must be introduced to describe passing streams in the column. The top and bottom operating lines are found in the usual way, and the analysis which gives the q line is unaffected by the side stream. The operating line for the middle portion of the column is found by writing a material balance around either the top or bottom of the column, including the plate from which the side stream is taken. In Eq. (39-28) a material balance is written for the more volatile component around the top:

$$\tilde{y}_4 \tilde{V}' = \tilde{x}_3 \tilde{O}' + \tilde{S}\tilde{x}_s + \tilde{D}\tilde{x}_D \qquad (39\text{-}28)$$

which is solved for $\tilde{y}_4$ to give

$$\tilde{y}_4 = \frac{\tilde{O}'}{\tilde{V}'}\tilde{x}_3 + \frac{\tilde{S}\tilde{x}_s + \tilde{D}\tilde{x}_D}{\tilde{V}'} \qquad (39\text{-}29)$$

This is the equation for the middle operating line. Like the other operating lines, it has a slope equal to the mole ratio of the liquid and vapor rates. This operating line may be located by the q line which determines the intersection of the operating lines immediately above and below the feed, or it may be fixed by the arbitrary specification of the side-stream composition $\tilde{x}_s$. The point representing the side stream is on both the upper and middle operating lines, because the withdrawal of a side stream changes only the amount of the stream from which it is taken and not its composition. If a step on the McCabe-Thiele diagram does not actually land at the intersection, this indicates that, for the conditions chosen, an impossible side-stream composition has been specified. This is usually remedied by changing the reflux ratio slightly so that the composition on all stages is altered.

Dilute solutions For dilute solutions the graphical analysis of McCabe and Thiele may best be carried out on log-log graph paper. The equilibrium curve reduces to a straight line with a slope of 1, because in the dilute range Henry's law usually applies. The operating line, however, is curved and must be plotted by calculating a number of points, using the operating-line equation and locating these on the graph. The steps representing equilibrium stages are marked off with vertical and horizontal lines, just as they would be on rectangular-coordinate graph paper.

The Ponchon-Savarit Method of Analysis

The analysis of binary-distillation systems by the method of Ponchon[1] and Savarit[2] represents a graphical solution of the equations of Sorel. The principal difference from the McCabe-Thiele method is that the liquid and vapor rates in moles per hour are not assumed to be constant for flows between adjacent plates. The only assumption we make in the Ponchon-Savarit method is that there are no heat losses through the walls of the column.

The Ponchon-Savarit solution is usually performed on an enthalpy-composition diagram similar to Fig. 39-18. The units used in this figure are Btu/lb mole and mole fraction. Some writers use diagrams in which the units are Btu/lb and mass fraction. The method is valid for any consistent set of units.

The enthalpy-composition diagram represents data at a fixed total pressure. The upper line represents the enthalpy of saturated vapor, and the lower line the enthalpy of saturated liquid. Superimposed on the diagram in Fig. 39-18 is a $\tilde{y}$-$\tilde{x}$ diagram as used in a McCabe-Thiele analysis. The $\tilde{y}$ scale is on the right-hand vertical axis, while the $\tilde{x}$ scale is the same horizontal composition scale used for values of both $\tilde{x}$ and $\tilde{y}$ on the enthalpy-composition diagram. The $\tilde{y}$-$\tilde{x}$ diagram is used to locate tie lines on the enthalpy-composition diagram. The technique is illustrated in the examples shown in Fig. 39-19.

[1] M. Ponchon, *Tech. mod.*, **13**:20 (1921).

[2] R. Savarit, *Arts et métiers*, pp. 65, 142, 178, 241, 266, 307 (1922).

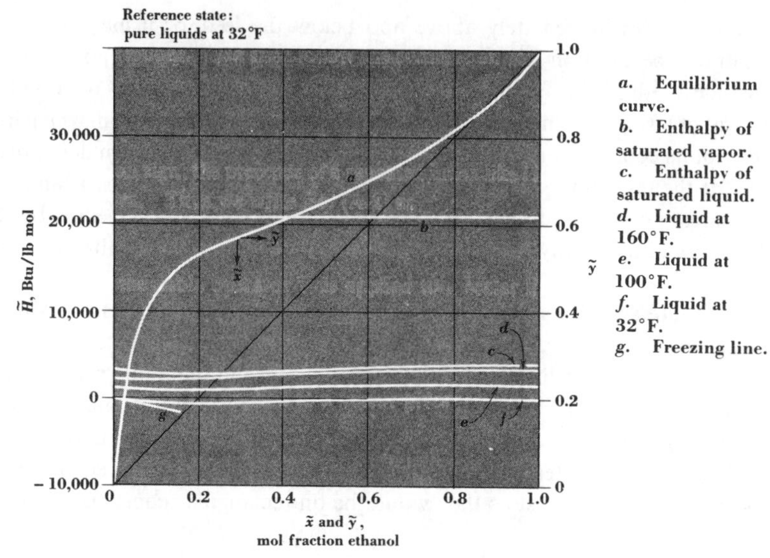

FIGURE 39-18
Enthalpy-composition diagram for ethanol-water system at 1 atm; equilibrium diagram ($\tilde{y}$ versus $\tilde{x}$) superimposed.

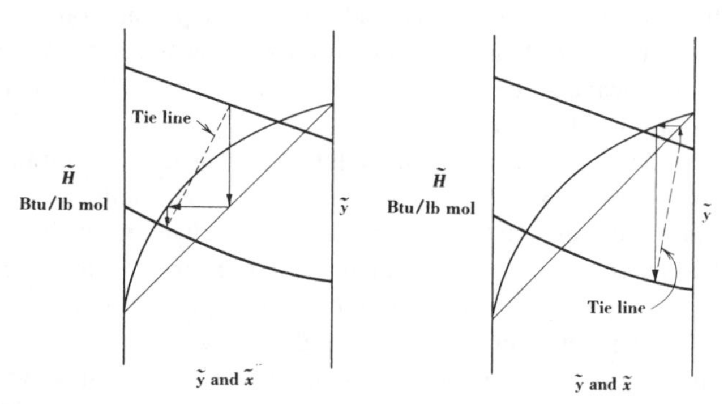

FIGURE 39-19
Use of $\tilde{y}$ versus $\tilde{x}$ diagram in locating tie lines on an enthalpy-composition diagram.

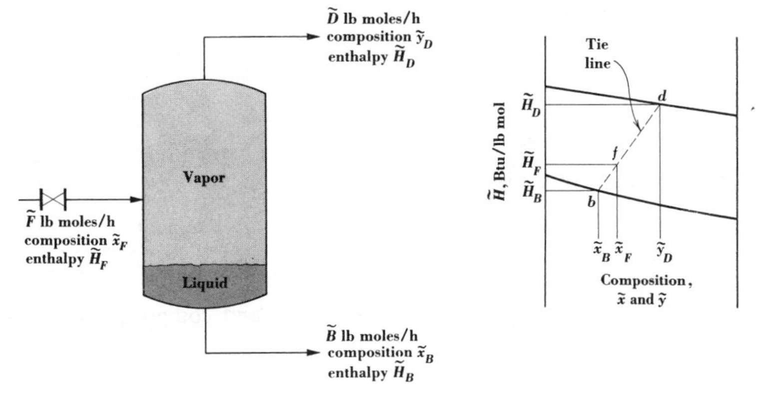

FIGURE 39-20
Example showing use of lever-arm principle.

Lever-arm principle The Ponchon-Savarit method of analysis depends upon the graphical addition and subtraction of properties of streams of fluid by the lever-arm principle. This has already been considered in Chap. 38; it will be reviewed here briefly with reference to some of the special problems encountered in distillation.

In the system shown in Fig. 39-20 a stream of fluid (a binary mixture) expands continuously through a valve into a separator at a pressure such that the fluid forms an equilibrium mixture of vapor and liquid. The rates $\tilde{F}$, $\tilde{D}$, and $\tilde{B}$ lb mol/h are constant. The separator is insulated, so there are no heat leaks. An overall mass balance and a mass balance on the more volatile component may be written as follows:

$$\tilde{F} = \tilde{D} + \tilde{B} \tag{39-30}$$

$$\tilde{x}_F \tilde{F} = \tilde{y}_D \tilde{D} + \tilde{x}_B \tilde{B} \tag{39-31}$$

If we substitute for $\tilde{F}$ from Eq. (39-30) in Eq. (39-31) and rearrange, we obtain

$$\frac{\tilde{D}}{\tilde{B}} = \frac{\tilde{x}_F - \tilde{x}_B}{\tilde{y}_D - \tilde{x}_F} \tag{39-32}$$

This equation provides a relation between the rates of $\tilde{D}$ and $\tilde{B}$ and the compositions of the three streams.

We now write an enthalpy balance for the separator.

$$\tilde{H}_F \tilde{F} = \tilde{H}_D \tilde{D} + \tilde{H}_B \tilde{B} \tag{39-33}$$

If we substitute for $\tilde{F}$ from Eq. (39-30) in Eq. (39-33) and rearrange, we get

$$\frac{\tilde{D}}{\tilde{B}} = \frac{\tilde{H}_F - \tilde{H}_B}{\tilde{H}_D - \tilde{H}_F} \tag{39-34}$$

We combine Eqs. (39-32) and (39-34) to obtain a relation between rates, compositions, and enthalpies.

$$\frac{\tilde{D}}{\tilde{B}} = \frac{\tilde{x}_F - \tilde{x}_B}{\tilde{y}_D - \tilde{x}_F} = \frac{\tilde{H}_F - \tilde{H}_B}{\tilde{H}_D - \tilde{H}_F} \tag{39-35}$$

The proportionality between enthalpies and compositions expressed in this equation can also be rearranged to give

$$\frac{\tilde{y}_D - \tilde{x}_F}{\tilde{H}_D - \tilde{H}_F} = \frac{\tilde{x}_F - \tilde{x}_B}{\tilde{H}_F - \tilde{H}_B}$$

which indicates that points on the enthalpy-composition diagram representing streams B, F, and D must be on a straight line. Streams B and D are in equilibrium and must be represented at the ends of a tie line, so stream F must be on the tie line.

If we know the composition and enthalpy of the feed, $\tilde{x}_F$ and $\tilde{H}_F$, the point on the enthalpy-composition diagram which represents the feed can be located immediately (point f on Fig. 39-20). Since only one tie line passes through this point, it is located by trial and error and fixes the locations of points d and b at the ends of the tie line.

The rates $\tilde{D}$ and $\tilde{B}$ are fixed by the ratios of composition differences and enthalpy differences as expressed in Eq. (39-35). If $\overline{bd}$ is considered to be the hypotenuse of a right triangle, it is seen by using the properties of similar triangles that the proportionalities expressed in Eq. (39-35) can be extended to include the ratio

$$\frac{\tilde{D}}{\tilde{B}} = \frac{\overline{bf}}{\overline{fd}} \tag{39-36}$$

The foregoing derivation of the lever-arm principle is similar to the development given in Chap. 38. The three mass balances, Eqs. (38-1) to (38-3), were written for the total streams, for component A, and for component B. For the enthalpy-composition diagram the three balances are written for the overall streams, for component A, and for the enthalpies of the streams.

If we had substituted for $\tilde{D}$ from Eq. (39-30) in Eqs. (39-31) and (39-33), we should have obtained

$$\frac{\tilde{B}}{\tilde{F}} = \frac{\tilde{y}_D - \tilde{x}_F}{\tilde{y}_D - \tilde{x}_B} = \frac{\tilde{H}_D - \tilde{H}_F}{\tilde{H}_D - \tilde{H}_B} \tag{39-37}$$

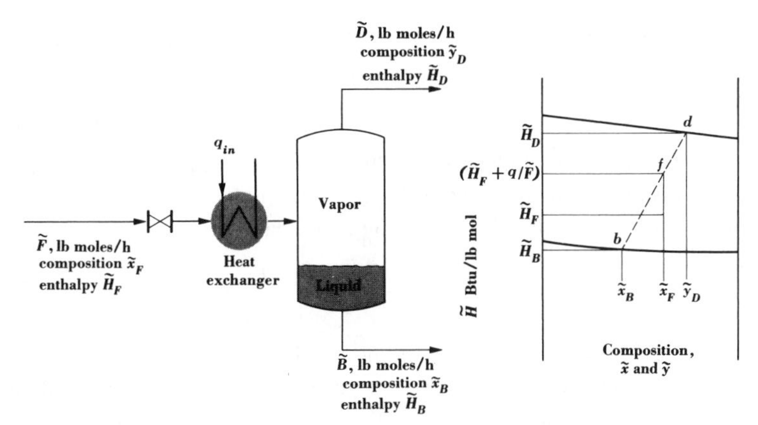

FIGURE 39-21
Example showing use of lever-arm principle for streams of mass and enthalpy.

Thus we see that the mass rates $\tilde{B}$ and $\tilde{F}$ can be expressed by the proportionality

$$\frac{\tilde{B}}{\tilde{F}} = \frac{\overline{fd}}{\overline{bd}} \tag{39-38}$$

By a similar procedure we can derive the expression

$$\frac{\tilde{D}}{\tilde{F}} = \frac{\overline{fb}}{\overline{bd}} \tag{39-39}$$

In addition to relating graphically streams containing both mass and enthalpy, we can include the effect of a stream of enthalpy, i.e., a heat input. In Fig. 39-21 we have a system in which heat is added to the feed stream at the rate q Btu/hr. Otherwise it is similar to the system shown in Fig. 39-20.

The overall material balance and the balance on the more volatile component are represented by Eqs. (39-30) and (39-31) of the previous discussion. The enthalpy balance, however, has become

$$\tilde{H}_F \tilde{F} + q = \tilde{H}_D \tilde{D} + \tilde{H}_B \tilde{B} \tag{39-40}$$

A stream of enthalpy cannot be represented on an enthalpy-composition diagram by itself, but must be associated with a stream of mass. Therefore we write Eq. (39-40) as

$$\left(\tilde{H}_F + \frac{q}{\tilde{F}}\right)\tilde{F} = \tilde{H}_D \tilde{D} + \tilde{H}_B \tilde{B} \tag{39-41}$$

If we now substitute for $\tilde{F}$, as was done to obtain Eq. (39-35), we get

$$\frac{\tilde{D}}{\tilde{B}} = \frac{\tilde{x}_F - \tilde{x}_B}{\tilde{y}_D - \tilde{x}_F} = \frac{(\tilde{H}_F + q/\tilde{F}) - \tilde{H}_B}{\tilde{H}_D - (\tilde{H}_F + q/\tilde{F})} \tag{39-42}$$

This result is similar to Eq. (39-35) except for the enthalpy of the feed. If the point f is located at composition $\tilde{x}_F$ and enthalpy $\tilde{H}_F + q/\tilde{F}$, the analogy is exact, and we see that for the line $\overline{bfd}$ in Fig. 39-21 we can write

$$\frac{\tilde{D}}{\tilde{B}} = \frac{\overline{bf}}{\overline{fd}} \qquad \frac{\tilde{B}}{\tilde{F}} = \frac{\overline{fd}}{\overline{bd}} \qquad \frac{\tilde{D}}{\tilde{F}} = \frac{\overline{fb}}{\overline{bd}} \tag{39-43}$$

just as we did with the corresponding line in Fig. 39-20.

To complete the picture we recognize that the enthalpy of the fluid leaving the heat exchanger is actually $\tilde{H}_F + q/\tilde{F}$. If the properties of the feed stream had been considered at this point rather than before the exchanger, the entire analysis would have been identical with that in Fig. 39-20.

As an alternative to the addition of enthalpy to the feed stream in Fig. 39-21, we shall consider the subtraction of enthalpy from one of the product streams, say the bottom product. This is done for the system shown in Fig. 39-22. We write an overall enthalpy balance just as we did in Eq. (39-41), except that we arbitrarily place q on the right-hand side of the equation.

$$\tilde{H}_F \tilde{F} = \tilde{H}_D \tilde{D} + \left(\tilde{H}_B - \frac{q}{\tilde{B}}\right)\tilde{B} \tag{39-44}$$

If we substitute for $\tilde{F}$ from the material and component balances, we get

$$\frac{\tilde{D}}{\tilde{B}} = \frac{\tilde{x}_F - \tilde{x}_B}{\tilde{y}_D - \tilde{x}_F} = \frac{\tilde{H}_F - (\tilde{H}_B - q/\tilde{B})}{\tilde{H}_D - \tilde{H}_F} \tag{39-45}$$

Thus, if we locate point b at composition $\tilde{x}_B$ and enthalpy $\tilde{H}_B - q/\tilde{B}$, we have a proportionality much as before, and the lever-arm principle, as represented in Eqs. (39-43), applies as usual.

The reasoning associated with Fig. 39-22 is somewhat more abstract than that for Fig. 39-21. In the one problem the addition of enthalpy occurred and produced a stream (the fluid leaving the heat exchanger) which actually had an enthalpy equal to $\tilde{H}_F + q/\tilde{F}$. In the other case, however, the stream with enthalpy $\tilde{H}_B - q/\tilde{B}$ is purely imaginary. No such stream exists in the system. The concept is, nevertheless, of great value, as will be shown in our subsequent analysis.

The enthalpy stream q could just as well have been subtracted from the overhead-product stream $\tilde{D}$ as from $\tilde{B}$. In fact, the point $(\tilde{y}_D, \tilde{H}_D - q/\tilde{D})$ can be found quite readily on Fig. 39-22 by drawing a straight line from $(\tilde{x}_B, \tilde{H}_B)$ through $(\tilde{x}_F, \tilde{H}_F)$ to a vertical line through $\tilde{y}_D$. This line would represent the material and enthalpy balances just as completely as the line $\overline{dfb}$ shown.

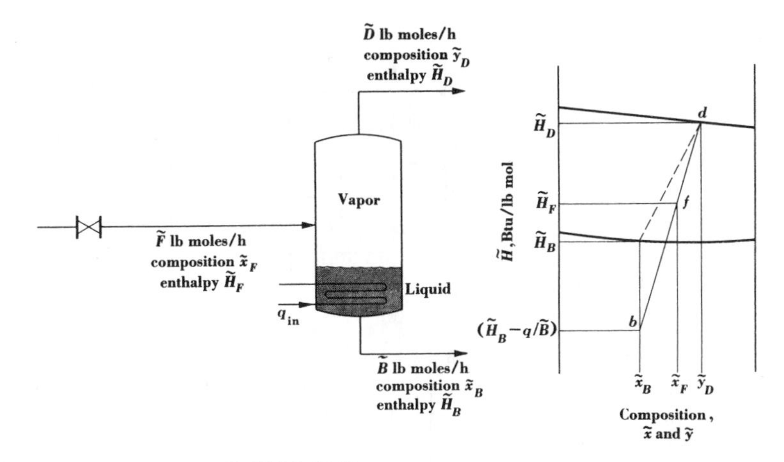

FIGURE 39-22
Example showing subtraction of enthalpy from a stream of mass.

Analysis of distillation columns The use of the lever-arm principle in the analysis of a section of a distillation column is shown in Fig. 39-23. The bubble-cap plates are assumed to be equilibrium stages, and we shall assume that we know the amounts, compositions, and enthalpies of streams O_3 and V_4. The points representing these streams are shown in the enthalpy-composition diagrams in Fig. 39-23.

The composition of stream O_4 is found from vapor-liquid equilibrium data and plotted at the opposite end of the tie line from V_4, with which it is in equilibrium. This is shown in Fig. 39-23*a*. To find the point representing V_5, we write a material balance around plate 4,

$$\tilde{O}_3 + \tilde{V}_5 = \tilde{O}_4 + \tilde{V}_4 \qquad (39\text{-}46)$$

and rearrange this to provide an equation expressing the difference between passing streams,

$$\tilde{V}_4 - \tilde{O}_3 = \tilde{V}_5 - \tilde{O}_4 \qquad (39\text{-}47)$$

If $V_4 - O_3$ is treated as a single hypothetical stream of mole rate $\tilde{\Delta}$, we can write

$$\tilde{V}_4 - \tilde{O}_3 = \tilde{\Delta} = \tilde{V}_5 - \tilde{O}_4 \qquad (39\text{-}48)$$

The point representing stream Δ on the diagram can be found readily since we know the amounts of V_4 and O_3. If $\tilde{V}_4 > \tilde{O}_3$, a straight line is drawn from point O_3 through V_4 and extrapolated a distance such that the following ratio applies. This extrapolation is shown in Fig. 39-23*b*.

$$\frac{\tilde{V}_4}{\tilde{O}_3} = \frac{\overline{\Delta O_3}}{\overline{\Delta V_4}} \qquad (39\text{-}49)$$

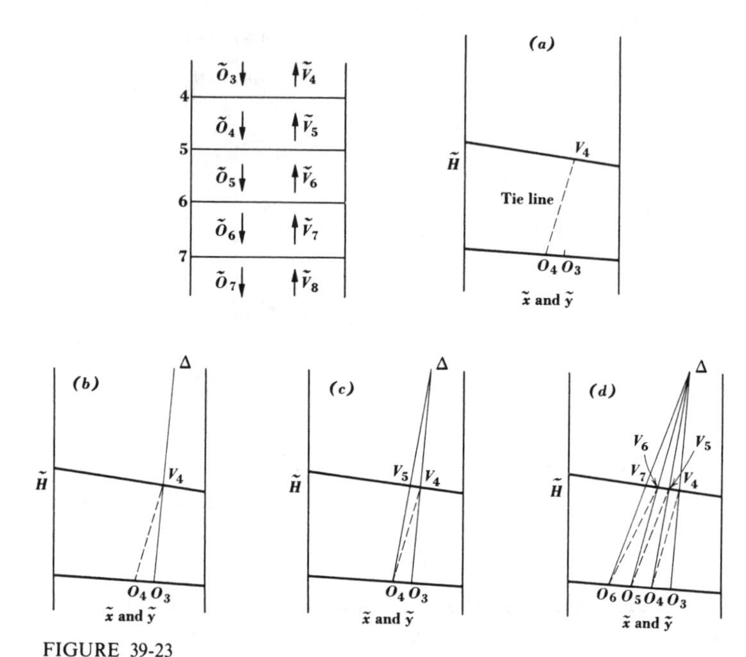

FIGURE 39-23
Use of an enthalpy-composition diagram for finding passing streams in a distillation column.

This basis of this step is perhaps more apparent if we write the first equality of Eq. (39-48) as

$$\tilde{V}_4 = \tilde{O}_3 + \tilde{\Delta} \qquad (39\text{-}50)$$

and apply the lever-arm rule to streams O_3 and Δ. From this equation it is seen that the following additional proportionalities between mole rates and the lengths of line segments on Fig. 39-23 apply.

$$\frac{\tilde{V}_4}{\tilde{\Delta}} = \frac{\overline{\Delta O_3}}{\overline{V_4 O_3}} \qquad \frac{\tilde{\Delta}}{\tilde{O}_3} = \frac{\overline{V_4 O_3}}{\overline{\Delta V_4}} \qquad (39\text{-}51)$$

Either of these equations would be easier to use in finding the point Δ than Eq. (39-49), because each contains only one segment of unknown length. The rate $\tilde{\Delta}$ is, of course, simply an arithmetic difference as defined in Eq. (39-48).

Having found the point Δ, the rest of the analysis follows directly. A relation between $\tilde{O}_4$, $\tilde{V}_5$, and $\tilde{\Delta}$ is given in Eq. (39-48). We have already found the point O_4, and Eq. (39-48) tells us that

$$\tilde{V}_5 = \tilde{O}_4 + \tilde{\Delta} \qquad (39\text{-}52)$$

Thus we know that point V_5 lies on a straight line joining points O_4 and Δ. In addition, V_5 is a saturated vapor, so we find point V_5 at the intersection of the saturated-vapor line and the line joining Δ to O_4. This is illustrated in Fig. 39-23*c*.

To find the points representing the rest of the streams we repeat the above procedures. For example, O_5 is found on a tie line at the opposite end from V_5. Point V_6 is found by writing a material balance around plate 5,

$$\tilde{O}_4 + \tilde{V}_6 = \tilde{O}_5 + \tilde{V}_5 \qquad (39\text{-}53)$$

which may be rearranged to give

$$\tilde{V}_5 - \tilde{O}_4 = \tilde{V}_6 - \tilde{O}_5 = \tilde{\Delta} \qquad (39\text{-}54)$$

Since we know points O_5 and Δ, we locate V_6 at the intersection of the saturated-vapor line with the line joining O_5 and Δ. This is shown in Fig. 39-23*d*.

The procedure we have followed is the graphical equivalent of a series of equilibrium calculations alternating with material and enthalpy balances. The use of a tie line is similar to an equilibrium calculation, and the use of the Δ point in finding a vapor stream is equivalent to writing equations for material and enthalpy balances around a plate and solving for the properties of the one unknown stream. The value of $\tilde{\Delta}$ between all plates is identical if there is no heat loss, no feed stream added, and no side stream taken off; it is a difference which can be considered as a hypothetical stream. This hypothetical stream has a finite rate ($\tilde{V} - \tilde{O}$ for the case shown), a finite composition, and a finite enthalpy. If $\tilde{O}$ had been greater than $\tilde{V}$, the point Δ would have been below the saturated-liquid line and, in fact, could have a large negative enthalpy. In some systems, such as columns with two feed streams, the point Δ can lie outside the composition range of 0 to 1. The only place it does not appear is between the saturated-vapor and saturated-liquid lines. In this region it would represent the addition of a vapor and a liquid stream rather than the difference between them.

Rectifying columns The general principles of operation of rectifying columns have already been discussed in the section on the McCabe-Thiele method of analysis. We shall show here the analysis according to the Ponchon-Savarit method.

The operation of a typical rectifying column is shown in Fig. 39-4. For the present discussion we shall assume that we know the compositions and enthalpies of streams F, B, and D. Each plate will be assumed to be an equilibrium stage, but constant molar overflow will not be assumed. As the first step, an enthalpy balance is written for the entire column.

$$\tilde{F}\tilde{H}_F = \tilde{B}\tilde{H}_B + \tilde{D}\tilde{H}_D + q \qquad (39\text{-}55)$$

As seen earlier, q cannot be represented on an enthalpy-concentration diagram unless it is expressed in terms of its effect on a stream of mass. Therefore we rewrite Eq. (39-55) as

$$\tilde{F}\tilde{H}_F = \tilde{B}\tilde{H}_B + \tilde{D}\left(\tilde{H}_D + \frac{q}{\tilde{D}}\right) \qquad (39\text{-}56)$$

In this equation we have simply an addition of three streams, two with their actual enthalpies and the third, D, with a hypothetical enthalpy $\tilde{H}_D + q/\tilde{D}$. If the points F, B, and D are given (as on Fig. 39-24), a line through F and B extrapolated to a composition $\tilde{x}_D$ (a vertical line through D) terminates at the point $(\tilde{x}_D, \tilde{H}_D + q/\tilde{D})$. This point, as we shall see presently, also represents a difference between passing streams, so it will be designated Δ on Fig. 39-24. From Eq. (39-56) we see that the rate $\tilde{\Delta}$ equals the rate $\tilde{D}$. The stream of overhead vapor V_1 has the same composition as the product stream D but lies on the saturated-vapor line. The reflux stream O_R is identical in all properties with the product stream D but differs in amount.

The liquid overflow from the first plate O_1 is in equilibrium with V_1, so O_1 lies on a tie line at the opposite end from V_1. To find V_2 we write a balance on a volume which encloses the condenser and the top plate. In terms of enthalpy we get

$$\tilde{V}_2\tilde{H}_2 = \tilde{O}_1\tilde{H}_1 + \tilde{D}\tilde{H}_D + q \qquad (39\text{-}57)$$

which may also be written

$$\tilde{V}_2\tilde{H}_2 - \tilde{O}_1\tilde{H}_1 = \tilde{D}\left(\tilde{H}_D + \frac{q}{\tilde{D}}\right) \qquad (39\text{-}58)$$

We recognize the right-hand side of Eq. (39-58) as designating the stream represented by the point Δ on Fig. 39-24. We have previously found the location of O_1, so the point representing stream V_2 can be found on a line joining points Δ and O_1. We also know that stream V_2 is a saturated vapor, so that it must lie at the intersection of the saturated-vapor locus with the line joining Δ to O_1.

The analysis continues with the location of O_2 on a tie line opposite V_2. The vapor stream V_3 which is passing O_2 is found by an enthalpy balance including the condenser and the first two plates.

$$\tilde{V}_3\tilde{H}_3 - \tilde{O}_2\tilde{H}_2 = \tilde{D}\left(\tilde{H}_D + \frac{q}{\tilde{D}}\right) \qquad (39\text{-}59)$$

We join the point O_2 and the point Δ, which represents the right-hand side of Eq. (39-59), and get V_3 at the intersection with the saturated-vapor line.

Further steps are similar to those indicated above. Liquid compositions are found by drawing tie lines from the vapor points, whereas vapor points are found

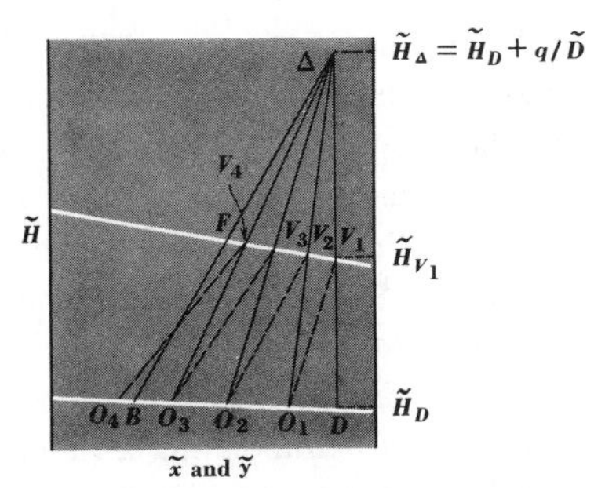

FIGURE 39-24
Ponchon-Savarit analysis of a rectifying column.

using Δ, which represents the difference between passing streams in both amount and enthalpy. When a tie line hits point B (or passes it on the left), then the correct number of steps will have been taken. Each tie line represents an equilibrium stage. As in the McCabe-Thiele analysis, it is seldom that a whole number of stages will give the specified bottom-product composition.

The reflux ratio $\tilde{O}_R/\tilde{D}$ is equal to the ratio of the segments $\overline{\Delta V_1}/\overline{V_1 D}$ on Fig. 39-24. This may be proved as follows: The line segment $\overline{\Delta D}$ represents an enthalpy difference $\tilde{H}_\Delta - \tilde{H}_D$, which is equal to $q/\tilde{D}$. The line segment $\overline{V_1 D}$ represents an enthalpy difference $\tilde{H}_{V1} - \tilde{H}_D$ equal to $q/\tilde{V}_1$. If we take the difference between these quantities, we see that the segment $\overline{\Delta V_1}$ must be represented by the difference in enthalpies $\tilde{H}_\Delta - \tilde{H}_{V1}$ or $q(1/\tilde{D} - 1/\tilde{V}_1)$. We then write the following proportionality:

$$\frac{\overline{\Delta V_1}}{\overline{V_1 D}} = \frac{q(1/\tilde{D} - 1/\tilde{V}_1)}{q/\tilde{V}_1} = \frac{\tilde{V}_1}{\tilde{D}} - 1 = \frac{\tilde{O}_R}{\tilde{D}} \tag{39-60}$$

At total reflux, $\tilde{D}$ is zero, so the ratio $\overline{\Delta V_1}/\overline{V_1 D}$ is infinite. To represent this, the point Δ must be an infinite distance above the diagram. Thus vertical lines are drawn from the liquid locus when the Δ point is used for finding the passing vapor streams at total reflux.

Minimum reflux occurs when the tie lines and the difference lines (those drawn to Δ) coincide at any stage. Starting from V_1, the difference lines decrease in slope as we move to the left on the diagram. The difference line with the lowest slope would be one passing from B through F. If Δ is sufficiently low, the line from B through F

to Δ will coincide with a tie line. This condition indicates that even if an infinite number of steps were taken, at this point the composition on succeeding plates would not change. Thus we have the zone of infinite plates at the feed point, which we discussed earlier in the section on the McCabe-Thiele analysis. The value of $\Delta V_1/V_1 D$ when this occurs is the minimum reflux ratio and is analogous to the situation shown for a complete column in Fig. 39-13.

If the equilibrium behavior of the mixture is such that an inflection point occurs in the $\tilde{y}$-$\tilde{x}$ diagram (as in Fig. 39-14), the minimum reflux ratio is found by the procedure of extrapolating a number of tie lines to the vertical line from D and V_1. The highest intersection of tie lines (considering only those to the right of F) locates the Δ point which corresponds to the minimum reflux ratio.

Stripping columns The operation of a stripping column is shown in Fig. 39-6, and the Ponchon-Savarit analysis in Fig. 39-25. We shall assume that we know the compositions and enthalpies of streams F, D, and B. First, an enthalpy balance around the entire column is written.

$$\tilde{F}\tilde{H}_F + q = \tilde{D}\tilde{H}_D + \tilde{B}\tilde{H}_B \qquad (39\text{-}61)$$

The heat q which enters the system from the steam coils in the reboiler is combined mathematically with the bottom product, and Eq. (39-61) is written as

$$\tilde{F}\tilde{H}_F = \tilde{D}\tilde{H}_D + \tilde{B}\left(\tilde{H}_B - \frac{q}{\tilde{B}}\right) \qquad (39\text{-}62)$$

By the usual method of adding streams graphically, we see that the point $(\tilde{x}_F, \tilde{H}_F)$ must lie on a line joining point $(\tilde{y}_D, \tilde{H}_D)$ to point $(\tilde{x}_B, \tilde{H}_B - q/\tilde{B})$. Thus we find this latter point on Fig. 39-25 by the intersection of a vertical line through B with the extrapolation of the straight line through D and F.

The point $(\tilde{x}_B, \tilde{H}_B - q/\tilde{B})$ is designated as Δ because it represents a difference between passing streams, just as the Δ point did in Fig. 39-24 for a rectifying column. From the considerations which led to its location, we see that the Δ point represents the material and enthalpy difference between streams F and D. By writing an overall balance for any section of the stripping column (as shown by the dashed line in Fig. 39-6), we see also that the Δ point represents the material and enthalpy difference between any pair of passing streams. This difference is equal in amount to the amount of bottom product withdrawn per unit of time, and the difference in enthalpy is equal to $\tilde{B}\tilde{H}_B - q$. Hence, if we know the location of Δ and also the point on the enthalpy-concentration diagram representing any vapor or liquid stream in the column, we can find the point representing the passing stream at that section.

The procedure for locating the streams in the stripping column is shown on Fig. 39-25. First we find the point representing the vapor V_B from the reboiler on a tie

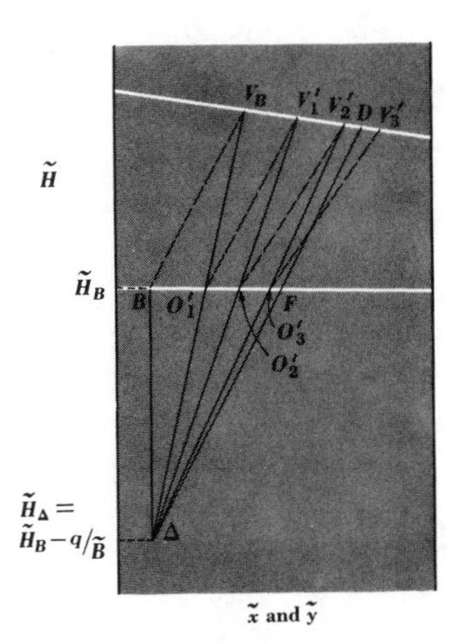

FIGURE 39-25
Ponchon-Savarit analysis of a stripping column.

line with B. The point representing O'_1 is found by joining V_B to Δ. The vapor stream V'_1 is found on a tie line from O'_1, and the liquid O'_2 on a line joining V'_1 with Δ. The construction proceeds in this fashion until a tie line reaches or passes to the right of point D. Each tie line represents an equilibrium stage.

Complete fractionating columns The Ponchon-Savarit analysis of a complete fractionating column is illustrated in Fig. 39-26. The column shown in Fig. 39-2 is the column we analyze here.

From the McCabe-Thiele analysis it will be recalled that for a complete fractionating column there are two operating-line equations. In the Ponchon-Savarit analysis these relations are expressed as Δ points. One Δ point will describe the difference between passing streams in the column above the feed plate, and the other below the feed plate. The analysis is first directed to finding these two Δ points. The information specified will be the composition and enthalpy of the feed, overhead distillate and bottom product, and the reflux ratio. Thus we can fix the points F, B, D, and Δ_1 on Fig. 39-26. The location of Δ_2 is found at the intersection of a vertical line through B with a line extrapolated from Δ_1 through F. The basis for this procedure may be shown by writing an enthalpy balance for the entire column.

$$\tilde{F}\tilde{H}_F + q_{\text{reb}} = \tilde{B}\tilde{H}_B + \tilde{D}\tilde{H}_D + q_{\text{cond}} \qquad (39\text{-}63)$$

This is rearranged in the following form:

$$\tilde{F}\tilde{H}_F = \tilde{B}\left(\tilde{H}_B - \frac{q_{\text{reb}}}{\tilde{B}}\right) + \tilde{D}\left(\tilde{H}_D + \frac{q_{\text{cond}}}{\tilde{D}}\right) \qquad (39\text{-}64)$$

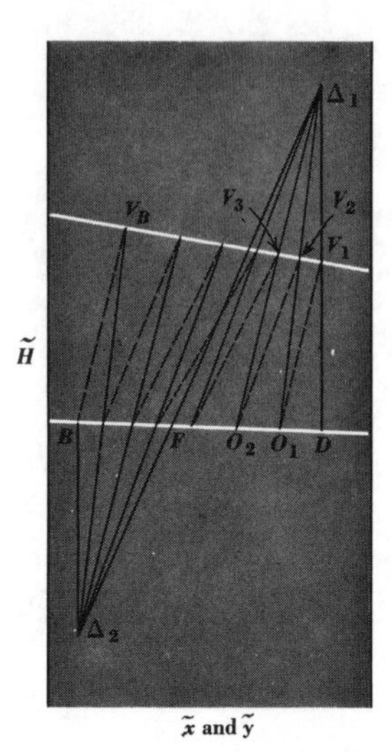

FIGURE 39-26
Ponchon-Savarit analysis of a complete fractionating column.

The two terms on the right-hand side of Eq. (39-64) represent Δ_2 and Δ_1, respectively, and because of the additive relation between F, Δ_2, and Δ_1 shown in the equation, the three points are on the same straight line.

The equilibrium stages may be constructed starting from either B or V_1. The first step is to construct a tie line; this represents an equilibrium calculation. A line is then drawn to the appropriate Δ point to determine the point representing the passing stream; this represents a material and enthalpy balance. Successive tie lines and difference lines are constructed until the desired range of composition is covered. The choice of which difference point to use is usually based on the desire to achieve a maximum separation with each stage. With this criterion, point Δ_1 is used for all difference lines constructed to the right of $\overline{\Delta_1 F \Delta_2}$, and Δ_2 is used for all difference lines to the left of $\overline{\Delta_1 F \Delta_2}$. If the criterion of maximum separation is not used, then either Δ point can be used within certain limits. The significance of switching from one Δ point to the other is that this step indicates the feed-plate location. Since each tie line is an equilibrium stage, the tie line which touches difference lines going to separate Δ points is the tie line representing the feed plate. The situation illustrated in Fig. 39-26 is one in which the optimum feed-plate location is used. If Δ_1 is used for the first difference line to the left of $\overline{\Delta_1 F \Delta_2}$, a smaller separation would be achieved than by using Δ_2. Similarly, if Δ_2 were used for difference lines to the right of $\overline{\Delta_1 F \Delta_2}$, a poorer separation would result than that shown.

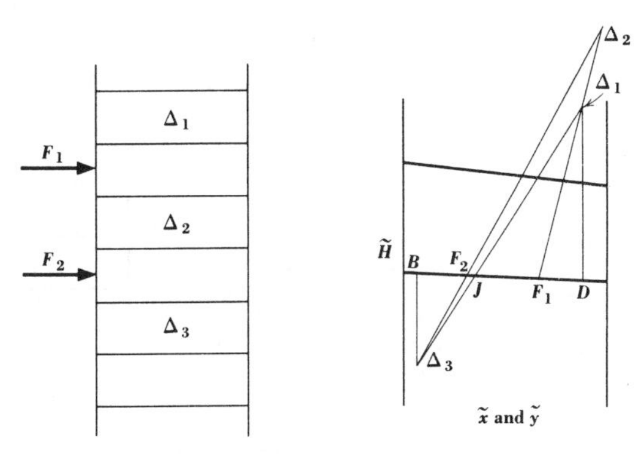

FIGURE 39-27
Ponchon-Savarit analysis of a column with two feeds.

Other systems A number of special situations exist which illustrate the versatility of the Ponchon-Savarit method. A column with two feeds is analyzed in Fig. 39-27. As usual, the Δ points are located first. It is assumed that B, F_1, F_2, and D are defined and can be located immediately and that the reflux ratio is known. This enables us to find Δ_1. There will be three Δ points for the column (just as there would be three operating lines in a McCabe-Thiele analysis), and we first find the Δ point for the portion of the column beneath the lower feed stream F_2. This point we call Δ_3. The procedure is to add streams F_1 and F_2 graphically by joining them with a straight line and locating the addition point J on the basis of the lever-arm principle.

$$\frac{\tilde{F}_1}{\tilde{F}_2} = \frac{\overline{JF_2}}{\overline{JF_1}} \tag{39-65}$$

The overall enthalpy and material balances are the same for a system with separate feeds F_1 and F_2 as for a system in which F_1 and F_2 are physically combined as stream J before being added to the column. The point Δ_3 is found by drawing a line from Δ_1 through J and extrapolating this to the intersection of a vertical line through B.

We next find Δ_2, which represents the difference between passing streams in that section of the column between the injection points for F_1 and F_2. We do this by recognizing that material and enthalpy balances can be written for streams F_2, Δ_2, and Δ_3. Thus these three points must be on a single straight line. Likewise, a material and enthalpy balance can be written containing only terms for streams F_1, Δ_1, and Δ_2, so that these three points must also be on a single straight line. Thus we

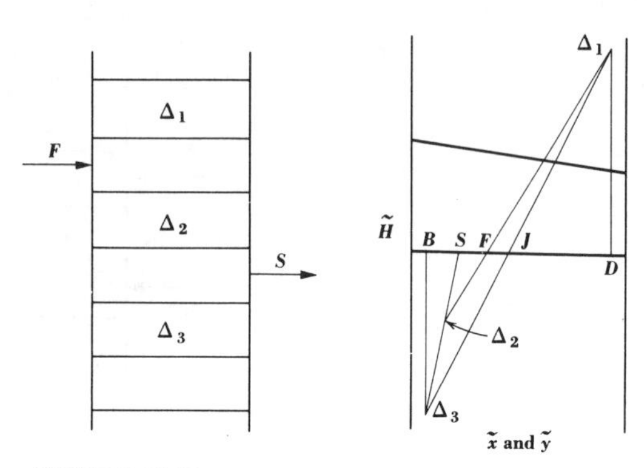

FIGURE 39-28
Ponchon-Savarit analysis of a column with a side stream.

find Δ_2 at the intersection of one line from Δ_3 through F_2 with another line from F_1 through Δ_1. The equilibrium stages are marked off in the customary fashion by alternately drawing tie lines and difference lines. The criterion for each feed-plate location is usually that the difference lines be drawn to the Δ point which will give the maximum separation for that step.

Side streams can be treated in a similar fashion. The procedure is illustrated in Fig. 39-28. The feed stream and the side stream are subtracted because they flow in opposite directions relative to the column. The subtraction is done graphically using the lever-arm rule

$$\frac{\tilde{F}}{\tilde{S}} = \frac{\overline{JS}}{\overline{JF}} \tag{39-66}$$

or more conveniently as

$$\frac{\tilde{F} - \tilde{S}}{\tilde{S}} = \frac{\tilde{J}}{\tilde{S}} = \frac{\overline{SF}}{\overline{JF}} \tag{39-67}$$

Stream J, which is the equivalent of the feed stream minus the side stream, is used in the same manner as the combined feeds in a two-feed column. A line is drawn from Δ_1 through J to the intersection of a vertical line through B. This locates the Δ point for the lowest portion of the column, Δ_3. The intermediate Δ point, Δ_2, is found at the intersection of a line through Δ_3 and S with another line through Δ_1 and F. The location of the equilibrium stages is done by methods similar to those already shown. In this case, however, a tie line must pass through S, and different Δ points *must* be used on either side of this tie line.

The injection of saturated steam into the reboiler is illustrated in Fig. 39-29. The only unusual thing about this system is that the lower Δ point represents

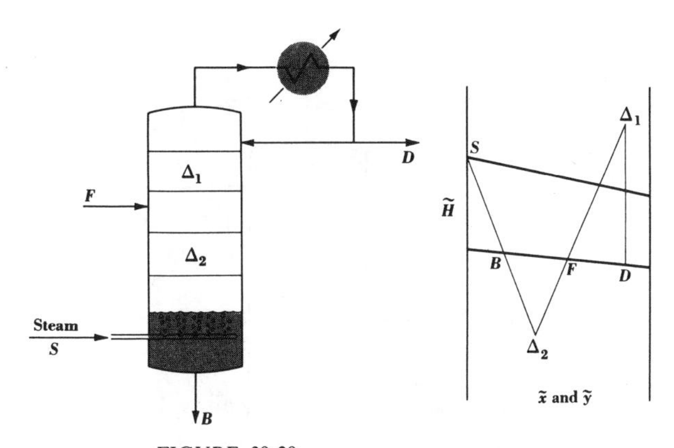

FIGURE 39-29
Ponchon-Savarit analysis of a column with open steam.

the difference between the bottom product withdrawn and the steam added, as can be shown by writing the appropriate material and enthalpy balances. Consequently, Δ_2 lies on a line through S and B. As usual, it also lies on a line from Δ_1 through F, so that the intersection determines Δ_2. Equilibrium stages are marked off in the usual fashion.

One other system of interest is the partial condenser. This is shown in Fig. 39-30, which is self-explanatory. The reflux ratio $\tilde{O}_R/\tilde{D}$ is the ratio of lengths $\overline{\Delta_1 V_1}/\overline{V_1 O_R}$.

Comparison of the Ponchon-Savarit and McCabe-Thiele methods of analysis The Ponchon-Savarit method, as we have said, is more rigorous than the McCabe-Thiele method. However, in many systems the difference is not great. It can be seen from an enthalpy-composition diagram that when the saturated vapor and liquid loci are parallel, straight lines, the ratio of liquid to vapor rates represented by the following equation will be constant.

$$\frac{\tilde{O}}{\tilde{V}} = \frac{\overline{\Delta_1 V}}{\overline{\Delta_1 O}} \tag{39-68}$$

If the ratio $\tilde{O}/\tilde{V}$ is constant, then the operating-line equations for the McCabe-Thiele analysis will be straight lines, and the two methods of analysis will give identical results. The extent of deviation of the vapor and liquid loci from being straight, parallel lines is therefore a measure of the inaccuracy of the McCabe-Thiele method. It is important to note that the loci must be straight and parallel on a plot of enthalpy per unit mole versus mole fraction. Ponchon-Savarit analyses are

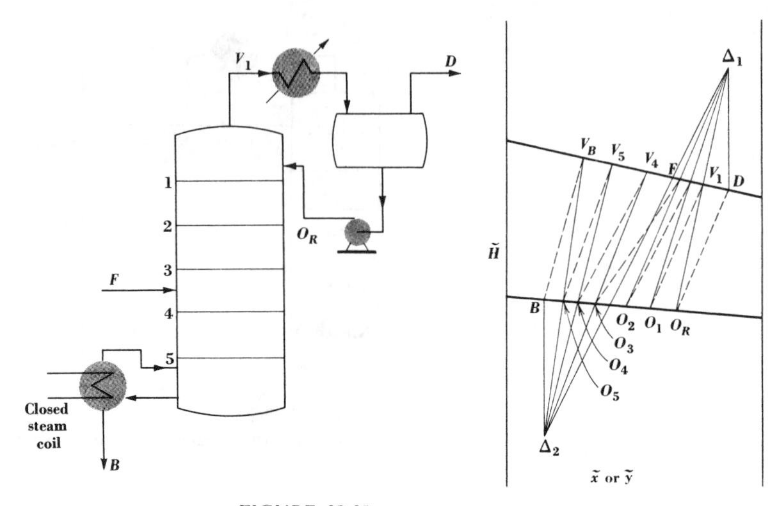

FIGURE 39-30
Ponchon-Savarit analysis of a column with a partial condenser.

frequently made (with perfect rigor) on diagrams of enthalpy per unit mass versus mass fraction. The deviation of the vapor and liquid loci from being straight, parallel lines is usually much greater on this type of diagram than on a diagram using mole units. This kind of diagram, however, does not give any direct indication of the accuracy of the McCabe-Thiele analysis for the system represented.

One criticism often made of the usefulness of the Ponchon-Savarit method is that enthalpy-composition data are available for only a very few systems. The enthalpy-composition diagram can be constructed approximately, however, if data on the latent heats of vaporization of the pure substances are available. The procedure is simply to join the points representing enthalpies of the pure substances with straight lines. This will usually be quite accurate for the gas phase but less accurate for the liquid phase. Nevertheless, the use of an enthalpy-composition diagram constructed in this way will generally give more accurate results than the McCabe-Thiele analysis.

Example 39-4 A fractionating column operating at 1 atm pressure is supplied at the optimum location with a saturated-liquid feed containing 40 mol percent ethanol and 60 mol percent water. The column produces a saturated-liquid overhead product containing 80 mol percent ethanol and a saturated-liquid bottom product containing 20 mol percent ethanol. The reflux ratio is 2.0. Find:

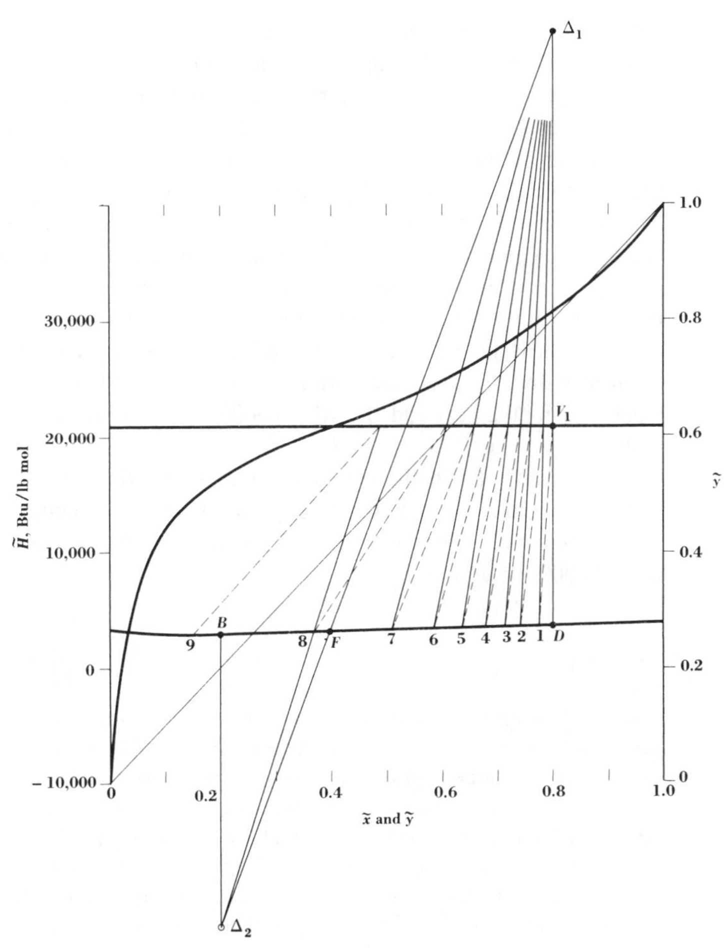

FIGURE 39-31
Ponchon-Savarit analysis of the problem in Example 39-4.

(*a*) The number of theoretical stages required to give the separation

(*b*) The optimum feed-plate location assuming 100 percent plate efficiency

(*c*) The heat load in the reboiler in Btu/lb mol of bottom product

(*d*) The heat load in the condenser in Btu/lb mol of overhead product

The solution is illustrated in Fig. 39-31. Points *F*, *D*, and *B* can be plotted immediately. The upper difference point Δ_1 is plotted using Eq.

(39-60). The lower difference point Δ_2 is found at the intersection of a line extrapolated from Δ_1 through F with a vertical line through B. The equilibrium stages are marked off starting at V_1 and ending with the ninth stage, which would give a liquid composition leaner than that specified for B. Thus the number of stages required for the specified separation is 8.8. The feed plate is the eighth plate because the tie line representing this plate crosses the $\Delta_1 F \Delta_2$ line. It can be seen that a line extrapolated from Δ_2 through the lower end of this tie line gives a leaner vapor composition than is obtained with a line joining the bottom of the tie line with Δ_1. Since the feed is a saturated liquid, it is added to the liquid flowing onto the eighth plate.

The heat load in the reboiler is represented by the vertical distance between B and Δ_2, since the bottom product and the hypothetical stream Δ_2 differ only by the heat added to the reboiler. This heat load is 25,500 Btu/lb mol of B.

The heat load in the condenser is represented by the vertical distance between Δ_1 and D, since Δ_1 can be regarded as the equivalent of stream D plus the heat removed in the condenser. This heat load can be seen on the diagram to be 51,300 Btu/lb mol of D. ////

PROBLEMS

39-1 A rectifying column operating at 1 atm is supplied with a saturated-vapor feed containing 40 mol percent ethanol and 60 mol percent water. The bottom product and the overhead product are saturated liquids containing 15 mol percent and 76 mol percent ethanol, respectively. Assuming constant-molar-flow rates in the column, find the internal reflux ratio $\tilde{O}/\tilde{V}$ and the external reflux ratio $\tilde{O}_R/\tilde{D}$. In addition, find the number of equilibrium stages required to give the specified separation.

39-2 A fractionating column operating at 1 atm pressure is supplied with 100 lb mol/h of a feed stream which is 25 mol percent saturated vapor and 75 mol percent saturated liquid. This feed has an overall composition of 40 mol percent ethanol and 60 mol percent water and is added at the optimum location. The overhead product leaves the condenser as a saturated liquid containing 80 mol percent ethanol. The bottom product is also a saturated liquid and contains 5 mol percent ethanol. The column operates with constant molar overflow and at a reflux ratio $\tilde{O}_R/\tilde{D}$ equal to $1.5(\tilde{O}_R/\tilde{D})_{\min}$. Find:

(*a*) The minimum reflux ratio

(*b*) The vapor and liquid rates both above and below the feed plate at the specified operating reflux ratio

(*c*) The heat loads on the reboiler and condenser in Btu/h

(*d*) The number of equilibrium stages required to perform the separation

(*e*) The minimum number of equilibrium stages required to perform the separation

39-3 A feedstock containing 32.1 mol percent *n*-butane and 67.9 mol percent *n*-pentane is supplied as a saturated liquid at the optimum location of a fractionating column operating at 50 psia. The feed rate is 100 lb mol/h, and a distillate product containing 88.5 mol percent *n*-butane is removed at the rate of 30 lb mol/h. The condenser at the top of the column removes 1 million Btu/h of heat and produces saturated-liquid product and reflux streams. Assuming constant molar overflow, determine:

(*a*) The reflux ratio

(*b*) The vapor and liquid rates in lb mol/h both above and below the feed plate

(*c*) The number of theoretical stages required to produce the separation

(*d*) The composition of the bottom product

The following equilibrium data are available, expressed as mole fraction of *n*-butane:

$\tilde{x}$	1	0.725	0.568	0.439	0.321	0.207	0.108	0
$\tilde{y}$	1	0.885	0.785	0.681	0.553	0.397	0.232	0

The latent heat of vaporization of the feed may be taken as 8500 Btu/lb mol.

39-4 A rectifying column containing 16 actual plates (overall efficiency 50 percent) is fed a saturated vapor containing 38 mol percent *A* and 62 mol percent *B*. The saturated-liquid distillate produced from the column contains 80 mol percent *A*, and the bottom product contains 30 mol percent *A*. A liquid side stream is also taken off the column. If the external reflux ratio $\tilde{O}_R/\tilde{D}$ is 10.9 and the column is achieving a maximum separation under the conditions, find:

(*a*) The composition of the side stream

(*b*) The internal reflux ratio above the side-stream plate

(*c*) The internal reflux ratio beneath the side-stream plate

(*d*) The actual plate from which the side stream is taken

The equilibrium relationship between vapor and liquid mixtures of *A* and *B* may be satisfactorily represented by a relative volatility of 1.50. Assume constant molar overflow.

39-5 A stream of 1000 lb mol/h of saturated liquid containing 30 mol percent methanol in water is fed at the optimum location of a distillation column. The column operates with constant molar overflow at a reflux ratio $\tilde{O}_R/\tilde{D}$, which is twice the theoretical minimum reflux ratio. A saturated-liquid distillate which contains 90 mol percent methanol is produced from the overhead condenser. The bottom product from the reboiler contains 10 mol percent methanol.

(*a*) Find the number of theoretical stages provided by the column and the vapor and liquid rates above and below the feed plate.

(*b*) After several months of successful operation in the manner described above, a water leak develops in the overhead condenser so that 900 lb/h of water is leaking into the condensate. The condenser is operated in such a manner that the reflux and product streams are still saturated liquids and the reflux ratio $\tilde{O}_R/\tilde{D}$ is the same as it was in part (*a*). Furthermore, the boil-up rate from the reboiler remains the same as before. However, internal reflux ratios and terminal compositions will obviously be changed. Find the new values for the internal vapor and liquid rates and the overhead- and bottom-product compositions.

Equilibrium data (mole fraction methanol):

$\tilde{x}$	0.046	0.094	0.157	0.217	0.321	0.425	0.534	0.632	0.727
$\tilde{y}$	0.267	0.402	0.533	0.602	0.680	0.745	0.791	0.829	0.883

39-6 A rectifying column is fed a saturated vapor containing 20 mol percent A and 80 mol percent B. An overhead distillate is produced at the rate of 200 lb mol/h containing 90 mol percent A. The bottom product contains 15 mol percent A.

To prevent an excessive vapor rate in the column, a vapor stream is taken off between the third and fourth plates from the bottom of the column and completely condensed. It comes off at the rate of 400 lb mol/h. After condensation it returns to the column as a saturated liquid and is fed onto the fifth plate from the bottom of the column. Assuming 100 percent plate efficiency and constant molar overflow, calculate:

(*a*) The feed and bottom-product rates
(*b*) The vapor and liquid rates in all sections of the column
(*c*) The side-stream composition
(*d*) The number of plates required to perform the separation

Equilibrium data (mole fraction A):

$\tilde{x}$	0.1	0.2	0.3	0.4	0.5	0.6	0.7	0.8	0.9
$\tilde{y}$	0.163	0.305	0.428	0.538	0.636	0.723	0.803	0.875	0.940

39-7 A stripping column operating at 1 atm pressure is fed a saturated liquid containing 38 mol percent ethanol. The bottom product is a saturated liquid containing 5 mol percent ethanol. The heat added to the reboiler is 30,000 Btu/lb mol of bottom product. Using an enthalpy-composition diagram, find by graphical construction:

(*a*) The number of theoretical stages required to perform the separation
(*b*) The heat input to the reboiler per pound mole of feed
(*c*) The heat input to the reboiler per pound mole of overhead product

39-8 A fractionating column is supplied with a saturated-liquid feed containing 21 mol percent ethanol and 79 mol percent water. The overhead distillate product is a saturated liquid which contains 83 mol percent ethanol; the bottom product is a saturated liquid which contains 5 mol percent ethanol. The column operates at 1 atm pressure with a reflux ratio twice the minimum value. Use the Ponchon-Savarit method to find:

(*a*) The number of theoretical stages required
(*b*) The optimum feed-plate location
(*c*) The heat load in the reboiler (Btu/lb mol of bottom product)
(*d*) The heat load in the condenser (Btu/lb mol of distillate product)

39-9 A conventional fractionating column, operating continuously at 1 atm pressure, is equipped with a reboiler and an overhead partial condenser, both of which act as equilibrium stages, and six bubble plates. The feed, which is added at the optimum location, contains 30 mol percent ethanol and 70 mol percent water and has an enthalpy of 10,000 Btu/lb mol. The partial condenser produces a saturated liquid and a saturated vapor at equal mole rates. All the liquid from the condenser is returned to the column as reflux, and all the vapor is taken off as a product. The bottom product is a saturated liquid containing 5 mol percent ethanol and 95 mol percent water. The heat added to the reboiler amounts to 10,000 Btu/lb mol bottom product leaving the reboiler. Find the overall plate efficiency.

39-10 An equilibrium vapor-liquid mixture of 0.40 mol fraction acetone and 0.60 mol fraction ethanol which is 50 percent liquid is fed to a distillation tower at the optimal plate. The pressure is 1 atm. The product is removed as a vapor of 0.90 mol fraction acetone, and the bottoms is 0.10 mol fraction acetone. Heating is by open steam introduced into the still pot at 1.19 mol per mol of total feed; a partial condensor is to be used.

Find the reflux ratio and the number of trays in the column itself if the overall plate efficiency is 50 percent. Estimate also the tower diameter.

39-11 A stripping tower is to be used to reduce the A content of a saturated liquid feed from $\tilde{x}_F = 0.0100$ to $\tilde{x}_B = 10^{-6}$ (1 ppm). The tower is run so that $y_D = 0.0105$. Find the number of ideal stages needed. The relative volatility is 1.105; use an analytical method.

39-12 A distillation tower has a total condenser and a thermosyphon reboiler: the feed is a saturated liquid ($\tilde{z}_f = 0.40$) mixture of acetone and ethanol fed to the optimal stage. The pressure is 1 atm. The tower is designed so that a reflux ratio of 1.4 times the minimum gives $\tilde{x}_D = 0.80$ and $\tilde{x}_B = 0.10$. Find the number of ideal stages in the tower.

After the tower is built, we want to increase the reflux ratio so that $\tilde{x}_B = 0.030$; $\tilde{x}_D$ stays at 0.80. Assuming that the tower capacity is limited by upward vapor flow rate, find the ratio of the new capacity to the old.

39-13 Essentially pure oxygen is to be made in a stripping column fed with liquid air at its boiling point. The column operates at 760 torr total pressure. At $-196°C$, which is the normal boiling point of nitrogen, the vapor pressure of oxygen is 162 torr. The two substances can be assumed to form an ideal solution, and it can also be assumed that the relative volatility is constant over the range of temperatures in the tower. Pure oxygen boils at $-183°C$.

What is the maximum moles of oxygen which can be obtained per mole of air feed?

39-14 A 41.7 mol percent saturated vapor mixture of acetone (A) in ethanol is to be separated in a continuous fractionating column operating at 1 atm. The top product is to be 90 percent A and the bottom product 10 percent A. Calculate the theoretical minimum steam consumption (mol/mol of feed) for this system. Assume that the latent heats of vaporization of acetone, ethanol, and water are all equal to 10,000 cal/mol.

40

MULTICOMPONENT SEPARATIONS

The separation operations which we have studied in the preceding chapters have been restricted to systems of two or three components. In the discussion of gas absorption in Chaps. 35 to 37, the solute transferred from the gas to the liquid was a single component. In distillation (Chap. 39), a binary system was considered in which the aim of the operation was to alter the proportions of the two constituents. In the study of extraction in Chap. 35, we considered the transfer of one component between a pair of insoluble liquids. Later in Chap. 38 we examined extraction operations for systems in which the liquid solvents were partially miscible, but we were interested in the transfer of only one component. In all these operations, the possibility exists that a multitude of components may be present and transferred between phases.

Whether in nature or industry, systems reacting chemically usually produce a number of products. One aim of the engineer in industry is, of course, to diminish the number of side reactions by choosing the most favorable conditions for a reaction. It is often not possible, however, to eliminate the production of by-products. Hence the engineer must frequently design equipment to separate the desired product, not only from the remainder of the unreacted feed, but also from the by-products of the reaction. A good example from nature is crude petroleum, in which mixtures of hundreds of chemical compounds have been produced by naturally occurring processes. The principal efforts of the petroleum-refining industry in the past have been to separate the crude petroleum into fractions, each consisting of a mixture of closely related compounds of similar properties. These fractions we know as gasoline, kerosene, lubricating oil, etc.

The methods of analyzing the separation of the binary and ternary systems discussed in the previous chapters can be extended to multicomponent systems rather simply for some operations, such as gas absorption with dilute solutions. For other operations, such as multicomponent distillation, the analysis becomes rather complex. The difference can be attributed to the effects of the additional components on the two basic steps of any mass-transfer analysis: the equilibrium and the material-balance calculations. If these are affected significantly for any one solute by the presence of the other solutes, then the interactions must be considered in the analysis. If, however, the transfer of any one component is unaffected by the simultaneous transfer of the other components, the analysis with respect to each component can be considered independently, except for the restriction of common solvent rates, temperature, pressure, and tower height or number of stages.

Absorption of a Dilute Multicomponent Mixture

To illustrate the remarks made above we consider the absorption by a liquid solvent of several solutes present in small amounts in a gas. The operation may be done either in a packed tower or in a tower containing a number of bubble plates. Because the initial gas concentration is low, the removal of some or all of any solute does not affect the gas rate significantly. Likewise the absorption of some or all of any solute into the liquid phase does not increase the liquid rate significantly. Thus it is apparent that a mass balance may be solved for the mole fraction of each component in the gas and liquid phases at any cross section of the tower from a knowledge of the gas and liquid rates and the terminal compositions at one end of the tower. Such equations were presented in Chaps. 35 and 37 and are referred to as operating-line equations.

The other important point to consider is the effect of the additional solutes on the equilibrium relations. If the solutions behave as ideal solutions, the equilibrium concentrations are not affected by the presence of the other components, whether solutes or solvent. For dilute solutions this condition is often achieved, so that the Henry's law constant for each solute is unaffected by the presence of additional solutes.

As a consequence we see that the design of a gas-absorption system for treating lean multicomponent mixtures is performed exactly as outlined in previous chapters for a single solute. Since only a single set of gas and liquid rates and one tower height exist for each system, we are not free to specify all terminal compositions independently. Rather, a specified separation may be chosen for one solute and the required tower dimensions and flow rates determined which best give this separation. The separations obtained for the other components with these tower dimensions and flow rates can then be calculated.

In the specific case of a packed tower, the first calculation might be a determination of the number of transfer units, using Eq. (35-20), (35-21), or (35-22), and the height of a transfer unit for one solute. These would then be multiplied to give the height of the tower. Then, for a second solute, the height of the tower would be divided by the height of a transfer unit for that component to get the number of transfer units. The equation for the number of transfer units would then be solved for the terminal composition of the second solute. This procedure could be repeated for any number of solutes.

The situation described above is often encountered in gas absorption, stripping, and extraction processes and is so straightforward that the methods and equations given for the systems with only one solute will not be rewritten here.

If the analysis were for the purpose of designing a stage contactor, as described in Chap. 37, similar procedures would be employed. The operating lines such as the one relating $\tilde{Y}_{An+1}$ to $\tilde{X}_{An}$ would remain straight and parallel regardless of concentration, because their slopes are merely the molar ratio of inert liquid to insoluble gas. For dilute mixtures the equilibrium lines would also be straight, so that the determination of the number of theoretical plates could be performed for each component independently.

The operation of a plate column for the absorption of a gas containing three components is illustrated in Fig. 40-1. The three components differ in volatility, A being most volatile and C least volatile. They are present in equal amounts in the inlet gas; hence all operating lines terminate at the same ordinate value ($\tilde{Y}_{N+1} = \tilde{Y}_A = \tilde{Y}_B = \tilde{Y}_C$). The solvent is pure, so the lower ends of all operating lines terminate at the same value of the abscissa ($\tilde{X}_{Ao} = \tilde{X}_{Bo} = \tilde{X}_{Co} = 0$). The ratio of the solvent rate $\tilde{L}_s$ to the insoluble-gas rate $\tilde{G}_I$ is the common slope of all operating lines.

If the exit concentration of one component is specified, the number of plates is fixed, so that all other exit concentrations are likewise fixed. To illustrate this assertion let us arbitrarily say that half of component B is to be absorbed in the column. This choice of $\tilde{Y}_{B1}$ enables us to locate the lower end of the operating line for B. We also know the slope $\tilde{L}_s/\tilde{G}_I$, so we draw the operating line to end at the specified value of $\tilde{Y}_B$, which is designated simply as $\tilde{Y}_{N+1}$ in Fig. 40-1. We can immediately determine the number of stages in the column either by the graphical construction of steps, as was shown in Chap. 37, or by solving Eq. (37-21).

Because the number of stages has now been fixed, we cannot arbitrarily specify the exit-gas concentration for any other component. This quantity must be

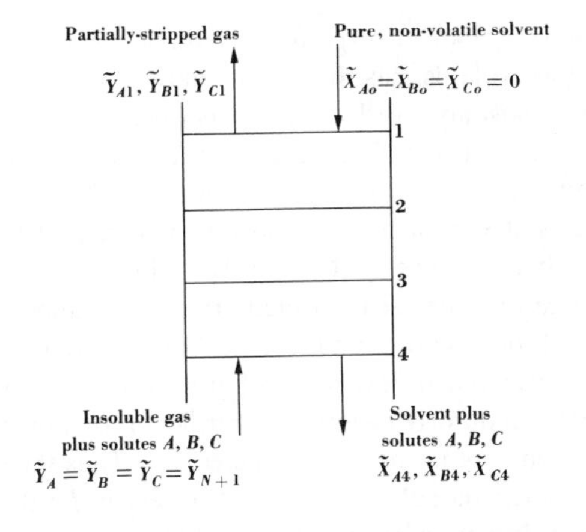

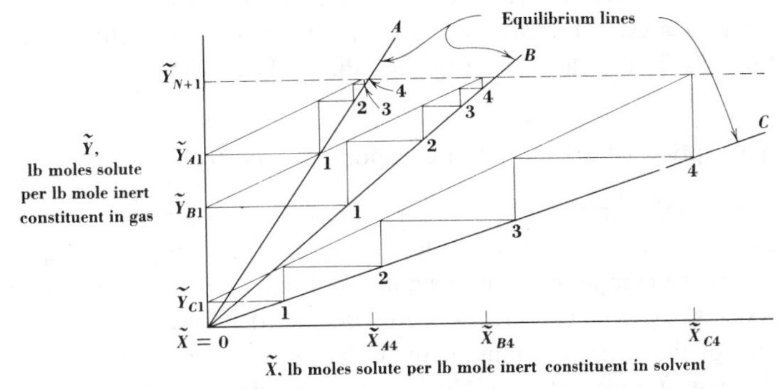

FIGURE 40-1
Gas absorption of a multicomponent mixture in a plate column.

determined either by graphical trial-and-error procedures or by solving Eq. (37-24) for the terminal concentration. The graphical procedure can be illustrated with reference to Fig. 40-1. Considering for the moment component A, we draw above the equilibrium line for A a straight line which we intend to test as a possible operating line for A. It has the same slope as the operating line for B and ends at $\tilde{X}=0$ and $\tilde{Y}=\tilde{Y}_{N+1}$ just as the line for B does. The steps representing plates are then constructed. If the number coincides with the number already found for the specified separation of component B, we have chosen the correct location for the operating line for A. We locate the operating line for C by an identical procedure.

Examination of Fig. 40-1 shows that, in spite of the fact that the concentrations of all components in the inlet gas are the same, the concentration in the exit gas of the most volatile component (A) is the highest, since it is least absorbed. Component C, which is least volatile, has the lowest concentration in the exit gas because it is the component most absorbed. The solvent becomes almost saturated with A shortly after entering the column and leaves in this state of near saturation with respect to A at the bottom. Large quantities of C are absorbed by the solvent near the bottom, and the closest approach to equilibrium with respect to C occurs at the top, where the gas is almost stripped of C. The behavior of any additional soluble components in the gas can be analyzed in exactly the same manner.

The absorption of a multicomponent mixture in a packed tower proceeds in a manner similar to that described in Fig. 40-1 for absorption in a plate column. In a packed column the driving force for transfer of component A decreases quickly as the concentration of A in the solvent increases. Thus the greatest driving force occurs at the top, where the solvent is pure. Conversely, for the least volatile component C, the greatest driving force occurs at the bottom, where the concentration of C in the gas is greatest. The driving force diminishes only at the top, where the concentration of C in the gas phase is greatly diminished.

Absorption of a Rich Multicomponent Mixture

The design of an absorption tower for a rich multicomponent mixture may be a fairly complex and tedious task. One of the difficulties is obtaining equilibrium data. Where the compounds are of similar molecular structure (as in many hydrocarbon systems), it may be permissible to assume that ideal solutions are formed, even though the concentrations of the diffusing components are appreciable. Even in this case, however, the equilibrium lines if plotted as $\tilde{Y}$ versus $\tilde{X}$ are curved, so that analytical solutions assuming linear equilibrium behavior are ruled out for this type of representation. In general, ideal-solution behavior cannot be assumed, and experimental data or some empirical correlation derived from experimental data on similar systems are essential.

The second major difficulty in analyzing the absorption of rich mixtures is in determining the location of the operating line for each component. As mentioned earlier, the operating line, if dependent on total gas and liquid rates, will not be straight because the absorption of substantial amounts of solute will change the gas and liquid rates in the column. This is a major difficulty when analyzing multicomponent systems because, as we have already seen, the various solutes are absorbed at different rates in each section of the column. Since the rates of absorption depend on the gas and liquid flow rates and the flow rates depend on the absorption rates, we have a problem which can be solved only by a trial-and-error procedure.

The problems encountered in the design of packed towers and plate towers differ somewhat because of the nature of the information sought. We shall first consider the design of a packed tower. A system is chosen arbitrarily in which three solutes A, B, and C of known concentration are carried into the tower by an insoluble carrier gas I and absorbed by a nonvolatile liquid S. The problem is to find the height of packing necessary to give a specified exit-gas composition $\tilde{y}_{A1}$. For each of the three solutes we can write a mass balance around a section of the tower extending from some arbitrarily chosen cross section to the bottom, where all compositions are known. These equations are similar to Eq. (35-6). We can also write similar but simpler material-balance equations for the insoluble carrier gas I and the nonvolatile component S in the solvent stream. These five equations contain 10 variables, $\tilde{y}_A$, $\tilde{y}_B$, $\tilde{y}_C$, $\tilde{y}_I$, $\tilde{x}_A$, $\tilde{x}_B$, $\tilde{x}_C$, $\tilde{x}_S$, $\tilde{L}$, and $\tilde{G}$. Two additional equations which can be written are $\sum \tilde{y} = 1$ and $\sum \tilde{x} = 1$, so that three more equations are needed to provide the total of 10 necessary to solve for the 10 variables. These three equations are the integrated rate equations for the three diffusing components:

$$z = (H_G n_G)_A = (H_G n_G)_B = (H_G n_G)_C \tag{40-1}$$

We have, however, introduced an eleventh variable, z, the height of packing from the bottom of the tower to the cross section being considered. Thus we have 11 variables and 10 equations, and one variable can be chosen arbitrarily. If this is one of the composition variables at the cross section within the tower, say $\tilde{y}_A$, the 10 equations can be solved for all 10 remaining variables, including $\tilde{x}_A$. Thus a point $(\tilde{x}_A, \tilde{y}_A)$ can be established on the operating line. The procedure is repeated for values of $\tilde{y}_A$, ranging from $\tilde{y}_{Ao}$ to $\tilde{y}_{A1}$. In this way the complete operating line can be established for component A (and for B and C also, in the course of these calculations).

One major difficulty in this technique is that analytical expressions are not available for the numbers of transfer units n_{GA}, n_{GB}, or n_{GC} for concentrated solutions. These must be found by graphical integration, Eq. (35-15), and the possibility of a direct algebraic solution of the 10 simultaneous equations is thus eliminated. Hence the solution must be found by some stepwise procedure. One which is fairly straightforward and can be made as accurate as desired is to analyze the column as a series of short sections. Over each section, gas and liquid rates are assumed to be constant and the methods described earlier for dilute solutions applied. The fact that the solutions are not dilute is of little consequence. The significant thing is that only small changes of composition occur over each segment of the column being analyzed. As a result, the operating lines for any one segment can be assumed to be straight and the analysis proposed earlier for dilute solutions applied to that segment. The repetition of this procedure makes it possible ultimately to determine all exit-gas compositions and the tower height if only one exit-gas composition is specified.

As the size of the segments is reduced, we approach a differential segment as a limit. The differential equations for the mass and energy balances and for interphase transfer for a point in a tower have been written as Eqs. (36-27) *et seq.* In Chap. 36 the only solute considered was the component A; if the other solutes, say D and E, are present, we merely add to the list of differential equations expressions such as Eqs. (36-28) and (36-30) written for D, E, etc. The set of differential equations must be solved simultaneously by a numerical method such as the Euler or Runge-Kutta procedure. Such calculations are now rarely done by hand, but are easily solved by computer.

We next examine the absorption of a rich mixture using a plate tower. If $\tilde{Y}$ is plotted against $\tilde{X}$, the operating lines-remain straight, just as for a dilute solution. The slopes of the operating lines represent the ratio of solute-free liquid to solute-free gas, which is constant regardless of the amount of solute present in either phase. The difficulties in locating the equilibrium line are, however, considerable. Even when ideal solutions are assumed, the equilibrium values of $\tilde{Y}$ and $\tilde{X}$, by virtue of their definition, depend on the concentrations of all solutes in each phase. Therefore, before any point $(\tilde{X}_A, \tilde{Y}_A)$ can be plotted on the equilibrium line, the concentrations of B and C in the solvent and in the gas phase must be known. This requirement is shown by writing the definitions of $\tilde{Y}_A$ and $\tilde{X}_A$:

$$\tilde{y}_A = \frac{\tilde{Y}_A}{1 + \tilde{Y}_A + \tilde{Y}_B + \tilde{Y}_C} \tag{40-2}$$

and

$$\tilde{x}_A = \frac{\tilde{X}_A}{1 + \tilde{X}_A + \tilde{X}_B + \tilde{X}_C} \tag{40-3}$$

Thus, even if the equilibrium relation between $\tilde{y}_A$ and $\tilde{x}_A$ is known, it is also necessary to know $\tilde{Y}_B, \tilde{Y}_C, \tilde{X}_B$, and $\tilde{X}_C$ for each point where $\tilde{Y}_A$ and $\tilde{X}_A$ are to be calculated from $\tilde{y}_A$ and $\tilde{x}_A$. Trial-and-error procedures can be used by guessing the rates of separation and then using this information to locate the equilibrium curves. If several solutes of widely differing volatilities are present in large amounts, this procedure can become tedious. A simpler method is probably that of constructing a diagram in which both equilibrium and operating lines are plotted in terms of $\tilde{y}_A$ versus $\tilde{x}_A$, as described earlier for the analysis of packed towers. The equilibrium stages can be counted on this diagram just as well as on a $\tilde{Y}_A$ versus $\tilde{X}_A$ diagram. Another trial-and-error procedure is that proposed by Sherwood and Pigford,[1] involving a modified gas-composition variable differing from both $\tilde{Y}_A$ and $\tilde{y}_A$.

Absorption calculations for multicomponent mixtures are done by hand as described above only for preliminary estimates. For a reliable design, plate-to-plate energy and mass balances are performed by a computer according to the method of Sorel, as described below for multicomponent distillation. This method applies

[1] T. K. Sherwood, R. L. Pigford, and C. R. Wilke, "Mass Transfer," McGraw-Hill Book Company, New York, 1975.

even when there is no insoluble vapor-phase component or nonvolatile liquid-phase component. The actual heat effects are also considered, and these may be far from negligible.

Multicomponent Distillation[1]

The analysis of multicomponent distillation offers many of the same problems encountered in the analysis of gas-absorption systems. Distillation is seldom done in packed towers, so we need consider only plate columns. Several new problems, however, are added. First, in distillation we usually are not dealing with systems in which there is a nonsoluble constituent in the gas and a nonvolatile constituent in the liquid, so we must be concerned with the transfer of all components in both phases. Second, the magnitude of the change in composition for most constituents in a distillation column is so great that we are concerned with "rich mixtures" in most problems.

Other problems primarily associated with distillation systems include thermal effects, the use of reflux, the withdrawal of side streams, and the introduction of feed into the column at some point other than the top or bottom. Because of these complications we seldom attempt graphical or analytical solutions for problems of multicomponent distillation. Instead we perform plate-to-plate calculations similar to those already described for binary systems in Chap. 39. All methods involve a succession of equilibrium calculations alternating with material-balance calculations. The methods differ chiefly in the assumptions used in making these calculations.

We shall first apply the Sorel method. If we consider a system containing only the top plate and the condenser of a distillation column (as shown in Fig. 39-3), we see that the first step is to obtain the composition of the liquid leaving the plate (assuming a total condenser and saturated reflux). If the composition of the vapor leaving the top plate is known (that is, the composition of the product stream) and the pressure in the column is fixed, according to the phase rule the composition and temperature of the liquid leaving the plate are fixed, regardless of the number of components present.

To obtain the composition of the vapor arriving at the top plate from the plate below, we write mass balances around the top plate for all components. These are of the same form as Eq. (39-12), derived for the more volatile component of a binary system. Writing Eq. (39-12) for component A, we have

$$\tilde{y}_{A2} = \frac{\tilde{O}_1}{\tilde{V}_2}\tilde{x}_{A1} + \frac{\tilde{D}\tilde{x}_{AD}}{\tilde{V}_2} \qquad (40\text{-}4)$$

[1] C. D. Holland, in "Multicomponent Distillation," Prentice-Hall, Inc., Englewood Cliffs, N.J., 1963, presents a general discussion of multicomponent distillation with a number of illustrative examples.

The liquid composition $\tilde{x}_{A1}$ has been found by the equilibrium calculation, so Eq. (40-4) contains only three unknowns, $\tilde{y}_{A2}$, $\tilde{O}_1$, and $\tilde{V}_2$. The product rate $\tilde{D}$ is usually specified in advance. Similar equations are written for all components, so for a system of i components we obtain i equations containing $i + 2$ unknowns.

Two additional equations which may be written are the total mass and enthalpy balances.

$$\tilde{O}_R + \tilde{V}_2 = \tilde{O}_1 + \tilde{V}_1 \qquad (39\text{-}8)$$

$$\tilde{O}_R \tilde{h}_R + \tilde{V}_2 \tilde{H}_2 = \tilde{O}_1 \tilde{h}_1 + \tilde{V}_1 \tilde{H}_1 \qquad (39\text{-}9)$$

The rates $\tilde{O}_R$ and $\tilde{V}_1$ are fixed by the product rate $\tilde{D}$ and the reflux ratio $\tilde{O}_R/\tilde{D}$, so that if all the enthalpies were known, we should now have $i + 2$ equations and $i + 2$ unknowns and the equations could be solved simultaneously. Three of the enthalpies, $\tilde{h}_R$, $\tilde{h}_1$, and $\tilde{H}_1$, are known (that is, they can be computed from the known compositions and temperatures of their respective saturated streams). The fourth enthalpy, $\tilde{H}_2$, is not known, however, and cannot be calculated until the unknown stream compositions ($\tilde{y}_{A2}$, $\tilde{y}_{B2}$, etc.) are determined.

The solution is obtained by assuming a value for one of the unknown variables, say $\tilde{H}_2$. The rates $\tilde{O}_1$ and $\tilde{V}_2$ are found from Eqs. (39-8) and (39-9), and then the mole fractions of all components in stream V_2 are determined, using component balances like Eq. (40-4). From the compositions obtained and the appropriate enthalpy data on the components, we calculate $\tilde{H}_2$ to check the initial assumption.

When the calculations have been repeated a sufficient number of times to provide a satisfactory check on $\tilde{H}_2$, we proceed to the calculations for the next plate. The first step is an equilibrium calculation to determine the composition of stream O_2. Next we write the mass and enthalpy balances around plate 2, which are similar to the balances around plate 1. The solution of these balances is done by a trial-and-error procedure, similar to that employed for the top plate.

For a rectifying column the plate-to-plate calculations are continued downward until the vapor composition obtained by the trial-and-error calculation matches the specified feed composition. This is not likely to occur at any one plate for all components because the initial choice of the overhead-product composition, product rate, and reflux ratio will have been made arbitrarily. To obtain a vapor composition beneath the bottom plate which matches the feed composition, alterations must be made in the variables assumed for the top of the column. Thus the entire set of plate-to-plate calculations must be repeated until the compositions check at the bottom with the desired accuracy.

For a stripping column, calculations are made starting with the streams entering and leaving the reboiler. The rate and composition of the bottom product and the rate of vapor production are arbitrarily chosen and plate-to-plate calculations made in an upward direction in the manner described for the rectifying column. When the liquid composition matches that of the feed stream, calculations are discontinued. As in the design of the rectifying column, the bottom conditions chosen arbitrarily are altered and the calculations repeated until the liquid composition at the top of the stripping column checks that of the feed.

The design of a complete fractionating column containing both rectifying and stripping sections requires that both sets of calculations be made in such a way that stream compositions match at the feed section. In view of the large amount of work required and the repetitive nature of the calculations, it is not surprising that plate-to-plate calculations are now done largely by digital computers. If the calculations are done manually, it is customary to use some short-cut procedure.[1]

In Chap. 39 we saw that the plate-to-plate calculations for binary distillation were greatly simplified when it was assumed that constant-molar-flow rates existed for each stream; that is, $\tilde{O}_R = \tilde{O}_1 = \tilde{O}_2$, etc., and $\tilde{V}_1 = \tilde{V}_2 = \tilde{V}_3$, etc. This eliminated the need for overall mass and enthalpy balances. If we use this assumption in analyzing multicomponent systems, Eq. (40-4) can be solved for the mole fraction of any component in stream V_2 without the necessity of making trial-and-error calculations. This simplification greatly reduces the total labor required, but does not affect the trial-and-error equilibrium calculations which must still be made at each plate. The assumption of constant-molar-flow rates can be made only for systems obeying Trouton's rule. Now that digital computers are available for the design of equipment for multicomponent separations, the necessity for making this assumption has greatly diminished.

Even more simplified procedures may be applied to the determination of the number of stages required for multicomponent distillation if only a preliminary design is required. Perry, pp. 13-26 to 13-30, presents an empirical method which involves:

1 The calculation of the number of stages at total reflux, $N_{\min}$

2 The calculation of the minimum reflux ratio $(\tilde{O}/\tilde{D})_{\min}$

3 The use of an empirical correlation relating $N_{\min}$ and $(\tilde{O}/\tilde{D})_{\min}$ to the actual number of theoretical stages N required at a finite reflux ratio $(\tilde{O}/\tilde{D})$

This method is useful when making calculations to determine the economic optimum reflux ratio for a tower. The final design, however, is better made by plate-to-plate calculations.

Example 40-1 Ethylene is being manufactured by passing a saturated hydrocarbon feed, such as ethane, through a heater in which the gas is cracked at

[1] An analysis of multicomponent distillation and a review of various design procedures for use with digital computers are presented by B. D. Smith in "Equilibrium Stage Processes," McGraw-Hill Book Company, New York, 1963. See also D. N. Hanson, J. H. Duffin, and G. F. Somerville, "Computation of Multistage Separation Processes," Reinhold Publishing Corporation, New York, 1962.

about 1500°F. The mixture leaving the cracking tubes is rapidly cooled and then separated into a number of streams in a series of fractionating towers. After the lightest and heaviest components have been removed, the remaining mixture is fed to a tower called a deethanizer. Ethane and ethylene are removed from the top, and the bottoms consist of almost pure propylene. The feed to such a tower is as given in the following table.

Component	Lb mol/h	Mole fraction
CH_4	0.09	0.0002
C_2H_4	302.29	0.5300
C_2H_6	122.18	0.2140
C_3H_6	141.96	0.2482
C_3H_8	4.28	0.0075
C_4H_{10}	0.08	0.0001
Total	570.88	1.000

It is required to find the number of theoretical stages necessary in a tower in which the propylene mole fraction in the distillate is reduced to 0.0054 and the ethane mole fraction in the bottoms to 0.0002. The ethane and the propylene are the key, or distributed, components. Lighter components appear in the bottoms only as traces; these are the nondistributed components. A material balance over the tower can now be built up as shown in the following table. The tower operates at about 200 psia; the distillate is produced as a vapor from a partial condenser.

	Flow, lb mol/h		
Component	Feed	Distillate	Bottoms
CH_4	0.09	0.09	
C_2H_4	302.29	302.29	
C_2H_6	122.18	122.15	0.03
C_3H_6	141.96	2.31	139.65
C_3H_8	4.28	......	4.28
C_4H_{10}	0.08	......	0.08
Total	570.88	426.84	144.04

Since the total quantity of the bottoms stream is little affected by the ethane in this stream, the mole fraction ethane can be used to estimate the quantity of ethane in the bottoms as 0.03 mol/h. The total bottoms are now subtracted from the feed to find the total distillate, and the amount of propylene in the distillate is found by difference.

It is helpful to do some approximate calculations before starting the plate-to-plate procedure. We begin by using the Fenske-Underwood equation to find the minimum plates required to obtain the desired distribution of the key components. This minimum occurs at total reflux, so the operating-line equation for a component reduces to $\tilde{y}_{An+1} = \tilde{x}_{An}$.

The Fenske-Underwood equation can be derived by the following analysis. At the top of the tower we have

$$\tilde{y}_{A1} = K_{A1}\tilde{x}_{A1} \tag{1}$$

$$\tilde{y}_{B1} = K_{B1}\tilde{x}_{B1} \tag{2}$$

and a similar equation for each component. We let A be the light key and B the heavy key, and we define the relative volatility α_{AB} by

$$\alpha_{AB} = \frac{K_A}{K_B} \tag{3}$$

so that we have

$$\frac{\tilde{y}_{A1}}{\tilde{y}_{B1}} = \alpha_{AB}\frac{\tilde{x}_{A1}}{\tilde{x}_{B1}} \tag{4}$$

The operating-line equation at total reflux then permits us to write

$$\frac{\tilde{y}_{A1}}{\tilde{y}_{B1}} = \alpha_{AB}\frac{\tilde{y}_{A2}}{\tilde{y}_{B2}} \tag{5}$$

Similarly, at the second plate down, we find

$$\frac{\tilde{y}_{A2}}{\tilde{y}_{B2}} = \alpha_{AB}\frac{\tilde{y}_{A3}}{\tilde{y}_{B3}} \tag{6}$$

An assumption of the method is to take α_{AB} as constant at a suitable average value in the tower; note that α_{AB} varies much less than K_A or K_B.

Equations (5) and (6) are combined to give

$$\frac{\tilde{y}_{A1}}{\tilde{y}_{B1}} = \alpha_{AB}{}^2\frac{\tilde{y}_{A3}}{\tilde{y}_{B3}} \tag{7}$$

For all N stages we have

$$\frac{\tilde{y}_{A1}}{\tilde{y}_{B1}} = \alpha_{AB}{}^{N-1}\frac{\tilde{y}_{AN}}{\tilde{y}_{BN}} \tag{8}$$

or

$$\frac{\tilde{y}_{AD}}{\tilde{y}_{BD}} = \alpha_{AB}{}^{N}\frac{\tilde{x}_{AB}}{\tilde{x}_{BB}} \tag{9}$$

where N includes the reboiler and partial condenser, if any. A similar equation holds for any other component; for example,

$$\frac{\tilde{y}_{CD}}{\tilde{y}_{BD}} = \alpha_{CB}{}^{N}\frac{\tilde{x}_{CB}}{\tilde{x}_{BB}} \tag{10}$$

When the terminal compositions for the light and heavy keys are substituted in Eq. (9), with α_{AB} corresponding to an average tower temperature of 2.2°F, it is found that $N_{\min}$ is 8.8 stages. The estimation of the average temperature is based on empirical procedures and is not given here.

The next step in the work is to find the approximate value of the minimum reflux ratio. The method of Underwood (Perry, p. 13-29), which assumes a constant α, gives $(\tilde{O}/\tilde{D})_{\min}$ equal to 0.447. The number of stages at intermediate reflux ratios is then found by the Erbor-Maddox correlation (Perry, p. 13-28), which gives N as a function of $N_{\min}$, $(\tilde{O}/\tilde{D})_{\min}$, and $\tilde{O}/\tilde{D}$. The calculations described up to this point can be programmed for a digital computer; the results of such a computation are as given in the following table.

$\tilde{O}/\tilde{D}$	N
0.447	∞
0.492	25.4
0.655	20.2
0.792	17.8
1.055	15.2
1.544	12.8
2.055	11.7
3.010	10.7
∞	8.8

From these data it is possible to make an economic analysis of operating and capital costs. It is found that the optimum $\tilde{O}/\tilde{D}$ is about 1.5 times the minimum, or about 0.66. We are now ready to perform the plate-to-plate calculations to find the temperature and composition profiles in the tower and the exact number of theoretical stages.

The method of Sorel is used, and the first step is the calculation of the temperature, quantity, and composition of the liquid leaving the first stage (the partial condenser) based on the vapor quantity and composition leaving the stage, that is, $\tilde{V}_1$, $\tilde{y}_{AD}$, $\tilde{y}_{BD}$, $\tilde{y}_{CD}$, etc. However, the distillate composition given in the table lacks values for the heavy nondistributed components. These substances must actually be present in trace quantities, so that they will appear on the lower trays of the column. We estimate the necessary compositions by using Eq. (10) with N equal to 8.8; all the mole fractions except $\tilde{y}_{CD}$ are known. For butane, for instance, the computation gives $\tilde{y}_{\text{butane } D} = 1.68 \times 10^{-12}$. From the complete composition of stream V_1 (or D) thus estimated, the temperature and composition of the dew-point liquid are calculated by the trial-and-error calculation described in the text. From this composition and temperature and the known reflux ratio, the values of $\tilde{V}_2$, $\tilde{y}_{A2}$, $\tilde{y}_{B2}$, $\tilde{y}_{C2}$, etc., are calculated by trial and error from the mass and energy balances around the first stage in the manner already described in the text.

The plate-to-plate calculations are best done on a digital computer. One of the simplest procedures is to calculate downward until the liquid leaving a stage is about what is expected at the feed tray. Then a similar set of calculations is started from the bottom of the tower and continued upward until the liquid leaving a tray is as close as possible in composition to that found on the last tray calculated from the top of the tower. The lack of matching of compositions at the middle of the tower requires the adjustment of the compositions at the ends of the tower; the overall material balance and the mole fraction of the key components must of course be respected. A suitable computer program for making the adjustments necessary to obtain a match at the feed tray can be devised, and iterations are performed on the tray-to-tray calculations until the match is good enough. Usually two or three iterations will suffice.

The plate-to-plate calculations described above have been programmed for a digital computer, and the results for the problem given are shown in Fig. 40-2. The temperature and composition profiles are shown; 24 theoretical stages are required. The vapor rate varies from $\tilde{V}_1 = 708.6$ lb mol/h to $\tilde{V}_8 = 644.1$ lb mol/h in the top section, so that the importance of not assuming a constant molar overflow is evident. ////

The preceding discussion and example have been presented in terms of a plate-to-plate calculation which determines the number of plates needed to reach a specified separation. Early computer programs[1] followed this procedure.

It has since been found that it is preferable to solve such problems by simulation; the number of plates is fixed, and the program calculates the separation obtained. Other solutions can also be done for different numbers of plates. A clear statement of this method is given by Wang and Henke.[2]

In the simulation method, it is necessary to assume the profiles of temperature and vapor (or liquid) flow rate at every plate. It then becomes possible to express material balances for all the plates as a set of linear equations. These can be arranged so that their constant coefficients (for the assumed profiles) form a tridiagonal matrix, easy to invert by a computer program, leading to a solution for all the compositons in the tower. These compositions are then used to recalculate all the tray temperatures which are consistent with the equilibrium relation, and all the vapor rates which are consistent with the enthalpy balances. The profiles are then changed to their updated forms and the compositions are solved for again. The calculations are iterated until certain convergence criteria are met.

[1] J. S. Bonner, *Petrol. Process*, **11**(6):64 (1956).
[2] J. C. Wang and G. E. Henke, *Hydrocarbon Processing*, **45**(8):155 (1966).

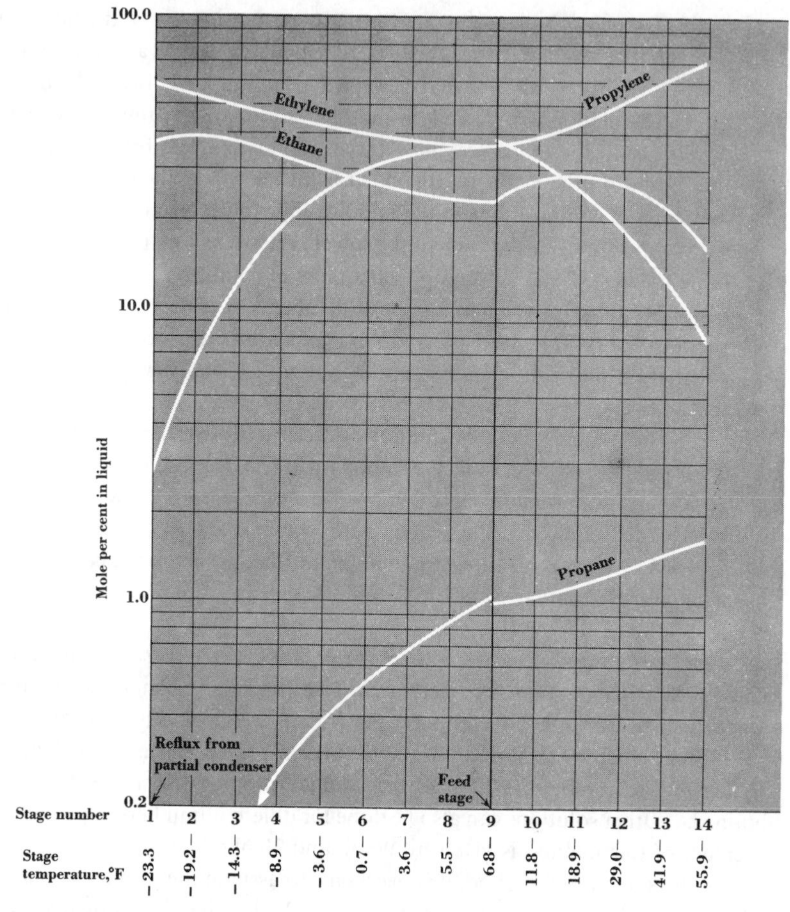

FIGURE 40-2
Compositions in a multicomponent distillation column; Example 40-1.

The above description gives an idea of the procedure, often called a *method of successive approximations*. Convergence methods and the way they are used (for example, simultaneously or sequentially) vary among the authors of various computer programs. For more details refer to King,[1] to previously cited works, and to Perry, pp. 13-30 to 13-35.

[1] C. J. King, "Separation Processes", McGraw-Hill Book Company, New York, 1971.

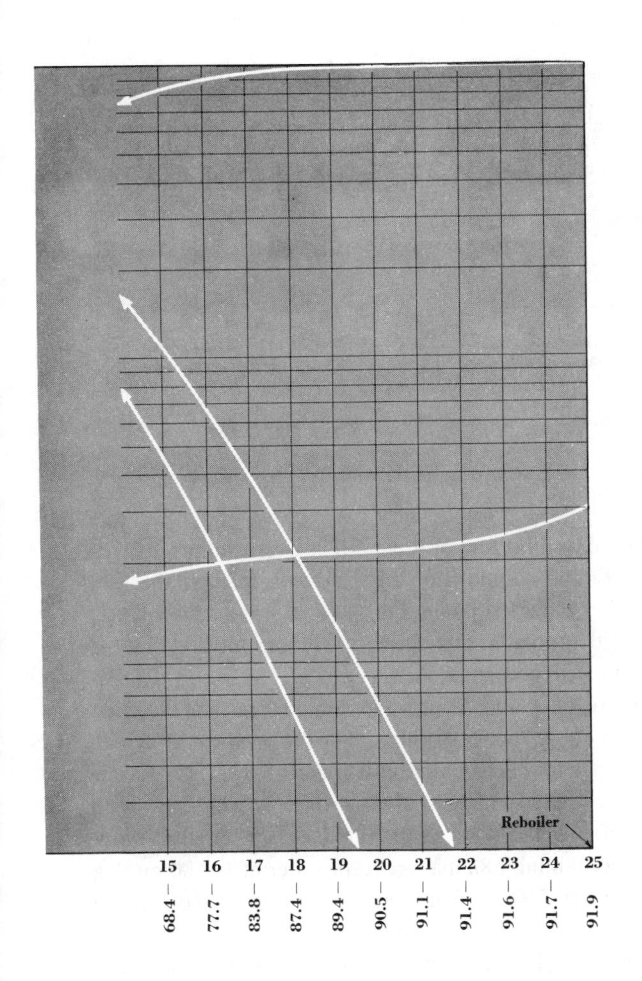

Azeotropic and Extractive Distillation

The separation by distillation of components in a binary system is sometimes hindered by the formation of a constant boiling mixture known as an azeotrope. A system containing an azeotrope is shown in Fig. 40-3. If a mixture rich in component *B* is distilled in a fractionating column, the limiting product compositions, regardless of the number of stages, are pure *B* and a mixture of azeotropic composition. Because the stream approaching pure *B* in composition has the

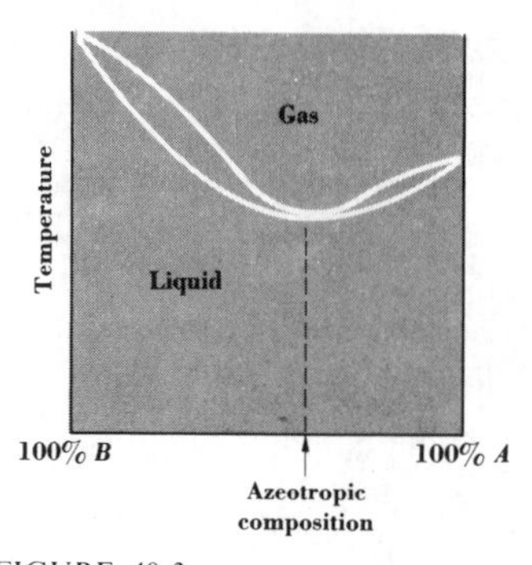

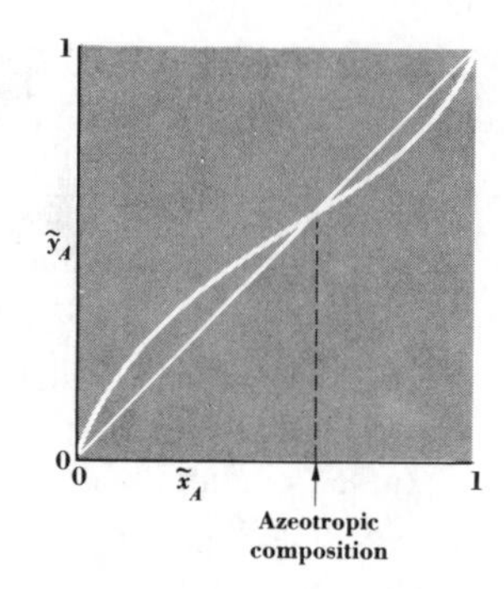

FIGURE 40-3
Phase behavior at constant pressure of a binary system containing a homogeneous azeotrope.

higher boiling point, it is produced as a bottom product. The overhead product, or lower-boiling stream, approaches the composition of the azeotrope. If the feed composition is to the right of the azeotrope in Fig. 40-3, a different situation exists. The overhead product still approaches the azeotrope in composition, but the bottom product approaches pure A in composition.

The separation of a mixture of A and B into two streams, each of which is nearly pure, cannot be done by a single binary distillation at the pressure for which the data in Fig. 40-3 are shown. A number of alternatives exist, however. One is that the pressure in the column be altered to some value at which an azeotrope is no longer formed. An example of this behavior is provided by the ethanol-water system, which forms azeotropes containing 7.88 mass percent water at 14,500 mmHg pressure, 4.4 mass percent water at 760 mmHg pressure, and no azeotrope at 70 mmHg pressure.

A second method of performing the separation is to use two distillation columns which operate at different pressures. For the ethanol-water system, one column operating at atmospheric pressure might produce a bottom product rich in water and an overhead product of nearly azeotropic composition. If this overhead product contained less than 7.88 mass percent water, it could be fractionated in a column operating at 14,500 mmHg pressure to produce nearly pure ethanol as the bottom product.

A third method of separating the two components is to add another component, which alters the relative volatility of the two original components. If the proper additive is chosen, the formation of an azeotrope may be entirely eliminated and the original components separated to any desired extent. This method of altering the relative volatility has also been used in systems in which no azeotrope exists but in which the closeness of the relative volatility to unity makes a separation very difficult. Because additional devices must be supplied to remove the additive from the product streams, an economic analysis does not always justify the use of the additive with its recovery system in place of the straightforward, but difficult, binary separation.

The technique of adding a third component to alter the relative volatility is successful mainly for binary mixtures which do not obey Raoult's law. For example, the original mixture may consist of two substances of nearly the same volatility but of greatly different polarity. If the added constituent has a greater polarity than either of the original two components, it generally increases the volatility of the less polar constituent relative to the more polar constituent. An example of this behavior is shown in Table 40-1, in which phenol is added to toluene and a mixture of paraffin hydrocarbons which have boiling points in the vicinity of the toluene boiling point. Another example would be the addition of water to a mixture of acetone and methanol. Water is more polar than methanol, and methanol is more polar than acetone. The addition of water increases the volatility of acetone relative to the methanol. If an additive less polar than either of the original pair were added, it would increase the volatility of the more polar substance. As an example, the addition of a hydrocarbon to acetone-methanol mixtures would make the methanol more volatile than the acetone. The thermodynamic effects of additives are discussed by Van Winkle.[1]

The terms *extractive* and *azeotropic* distinguish the classes of compounds used as additives. Extractive distillation usually refers to systems in which the added compound has a substantially lower vapor pressure than either compound in the original binary mixture. An example of extractive distillation is the use of acetone in the separation of a mixture of *n*-butane and 2-butene. As would be expected, the

Table 40-1 EFFECT OF PHENOL ON THE RELATIVE VOLATILITY OF PARAFFIN-TOLUENE MIXTURES

Composition, mole percent			Relative volatility (paraffin-toluene)
Paraffin	Toluene	Phenol	
50	50	0	1
30	30	40	1.5
20	20	60	2.0
0	0	100	3.7 (as a limit)

[1] M. Van Winkle, "Distillation," McGraw-Hill Book Company, New York, 1967.

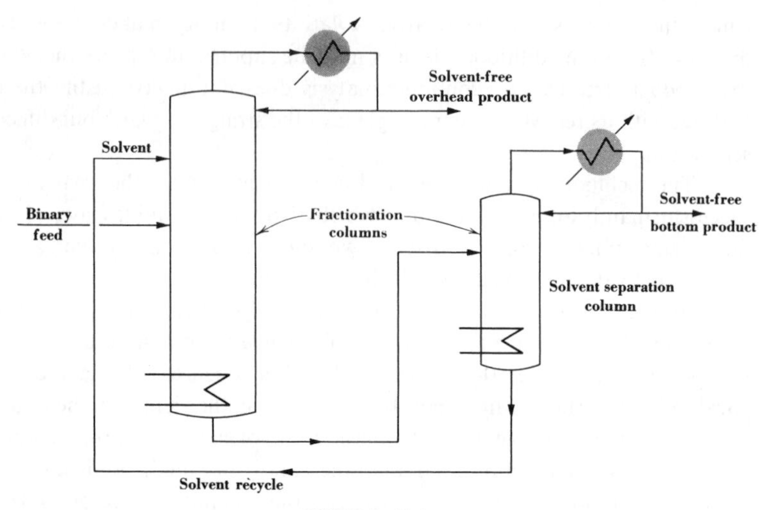

FIGURE 40-4
Typical flow sheet for an extractive distillation process.

volatility of n-butane is increased relative to that of the 2-butene. Acetone boils at a temperature 60°C greater than either component, so that it does not appear in substantial amounts in the vapor phase. For this reason it must be added near the top of the column as a liquid.

A typical extractive-distillation process is shown in Fig. 40-4. One of the advantages of extractive distillation is that the separation of the additive from both components of the binary mixture is generally quite easy. Only a few plates are needed above the solvent injection plate to fractionate the solvent out of the vapor stream. In addition, a simple auxiliary distillation column serves to remove the additive from the bottom product. An example of the design of an extractive-distillation system is shown in Perry, 13-45. The principal difference from the design of an ordinary ternary distillation column is in the nonideality of the liquid solution. Reliable equilibrium data are often difficult to find for new systems.

Azeotropic distillation is the term descriptive of processes in which the additive has a boiling point of approximately the same magnitude as those of the original components in the binary mixture being separated. As a result, the additive appears in significant concentrations on all plates in the system. This would ordinarily mean that separation of the additive from each of the original components would be difficult. Usually, however, the additive is chosen so that it forms a heterogeneous azeotrope with one or both of the original components. A heterogeneous azeotrope

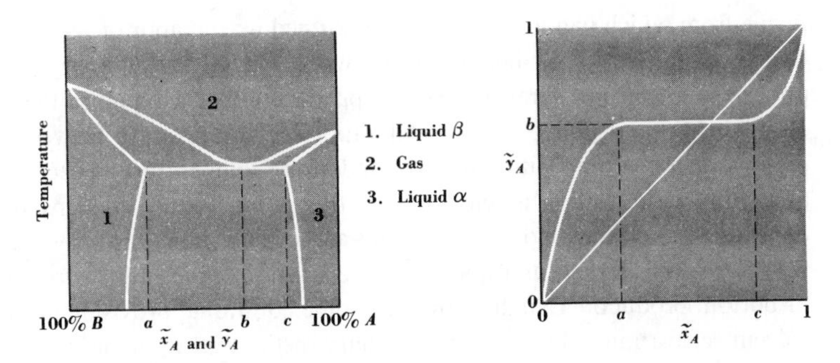

FIGURE 40-5
Phase behavior at constant pressure of a binary system containing a heterogeneous azeotrope.

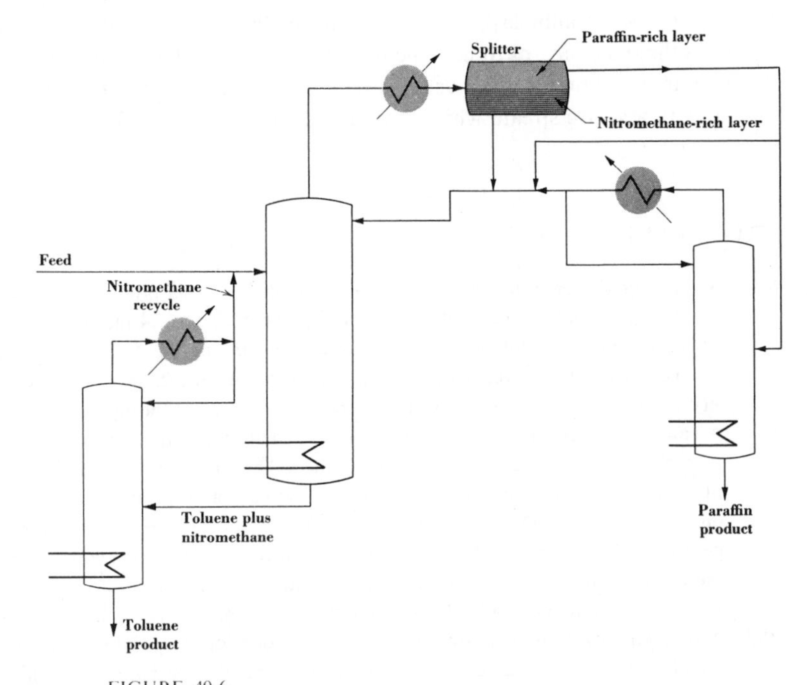

FIGURE 40-6
Flow sheet for azeotropic distillation of a paraffin-toluene mixture with nitromethane as an added agent.

is one from which two liquid phases are formed when vapor of azeotropic composition is condensed. This behavior is shown in Fig. 40-5. A vapor at a temperature above the azeotropic temperature can produce either a liquid in the α region or a liquid in the β region or both. All vapors of composition between the α and β regions alter in composition as the condensing temperature is decreased until the azeotropic composition is reached. At this point, further condensation occurs at constant temperature and α and β liquids are formed simultaneously. This property may be of considerable assistance in the separation of the additive from the overhead or bottom product. Usually, however, some additional provision must be made for solvent separation. This may be auxiliary distillation equipment or a liquid-liquid extraction unit.

An example of azeotropic distillation is the paraffin-toluene system with nitromethane as the additive. A flow sheet is shown in Fig. 40-6. The paraffin-toluene mixture boils at 110°C at atmospheric pressure and is treated with nitromethane, which boils at 101°C. The overhead product from the main tower consists of a paraffin-nitromethane mixture which, when condensed, forms two liquid phases. The nitromethane-rich phase is returned directly to the top plate of the tower, and the paraffin layer is sent to an auxiliary fractionating column which separates the remaining nitromethane from the paraffin. The bottom product from the main fractionating tower must also be fractionated to produce the toluene product and an additive stream which is returned to the tower.

PROBLEMS

40-1 A vapor-liquid separator operating adiabatically at 215 psia produces a vapor stream containing 35 mol percent methane, 23 mol percent ethane, 24 mol percent propane, 12 mol percent *n*-butane, 4 mol percent *n*-pentane, and 2 mol percent *n*-hexane. The vapor stream leaving the separator constitutes 60 mole percent of the feed entering. Find the composition of the feed stream and the liquid-product stream.

40-2 A natural gas stream contains 88 mol percent methane, 6 mol percent ethane, 4 mol percent propane, and 2 mol percent *n*-butane. The gas is to be treated in a bubble-plate tower with a nonvolatile absorber oil flowing at the rate of 4 mol absorber oil/mole inlet gas. Determine the number of theoretical stages required to recover 80 percent of the propane, and find the fractional recovery of the other components under these conditions. The tower operates at 3 atm pressure and 60°F. Assume that Raoult's law applies for all solutes except methane, which can be considered insoluble.

40-3 A fractionating column is supplied with 500 mol/h of a feed which is a saturated liquid at 65 psia. The feed contains 1.2 mol percent propane, 23.0 mol percent *n*-butane, and 75.8 mol percent *i*-butane. The column is equipped with a total condenser which produces a saturated-liquid reflux at a pressure of 65 psia. The overhead-product composition is to be 5 mol percent propane, 5 mol percent *n*-butane, and 90 mol percent *i*-butane. The bottom-product composition is not specified, but it is apparent that only a negligible amount of propane will be present.

(*a*) Find the overhead- and bottom-product rates and the bottom-product composition.

(*b*) Find by plate-to-plate calculations the minimum number of plates required to perform the specified separation. Check this with the results of the Fenske-Underwood equation [Eq. (9), Example 40-1].

40-4 A hydrocarbon stream containing 37 mol percent *n*-butane, 32 mol percent *i*-pentane, 21 mol percent *n*-pentane, and 10 mol percent *n*-hexane is fed as a saturated liquid at column pressure to a debutanizing column at the rate of 4750 lb mol/d. The debutanizer is to produce a feedstock for an isomerization process at the rate of 1250 lb mol/d with a minimum of 95 mol percent *n*-butane and the remainder chiefly *i*-pentane. The column has a total condenser and is to produce a saturated-liquid overhead product at 100°F. The pressure drop in the column will be neglected. Find:

(*a*) The pressure in the column

(*b*) The temperature of the feed

(*c*) The temperature of the bottom product

(*d*) The number of plates required to give the specified separation at a reflux ratio $(\tilde{O}_R/\tilde{D})$ of 2

40-5 In extractive distillation the added component has a much lower vapor pressure than the components in the original mixture. In azeotropic distillation the added component has about the same vapor pressure as the original mixture. Discuss the possible advantages and disadvantages of a type of distillation in which the third component has a much higher vapor pressure than the original mixture.

501)

Ch21.1. $q = C_p W \cdot \Delta t = 149074\ Btu/hr$
물의 출구 온도 $= 99.78°F$

$C_{pw} = 4184 \frac{J}{kg} = 1 Btu/lb°F$

$\rho_{water} = \frac{1000kg}{m^3} \cdot \frac{1lb}{0.4536kg} \cdot \frac{2.832 \times 10^{-2} m^3}{ft^3} = 62.4 lb/ft^3$

(a) concurrent.

$\Delta t_1 = 180°F$ $q = U_o A_o \cdot \frac{\Delta t_2 - \Delta t_1}{\ln \Delta t_2 / \Delta t_1}$, $A_o = 19.43 ft^2$. $/ 2.36 ft^2 = 8.24$ (9개)
$\Delta t_2 = 20.22°F$ — 2.36 ft² = one section

(b) counter current.

$\Delta t_1 = 50°F$ $A_o = \frac{15.6 ft^2}{2.36 ft^2} = 6.61$ 개 (7개)
$\Delta t_2 = 150°F$.

Ch21.19. $t_{bi} = 70°F$, $t_{bo} = 0°F$, Δx

$k = 0.10 Btu/hr \cdot ft \cdot °F$, $t_{si} = 50°F$, $h = 0.18 \cdot (\Delta t)^{1/3}$.

$h_i \cdot (t_{bi} - t_{si}) = -k \cdot \frac{t_{so} - t_{si}}{L} = h_o (t_{so} - t_{bo})$

$\therefore t_{so} = 20°F$, $L = 0.31 ft$

Ch.21.10, $h \cdot A \cdot (200 - t) = V \cdot \rho \cdot c_p \cdot \frac{dt}{d\theta}$, $\ln \frac{200-t}{200-25} = -0.268\theta$, $\theta = 16.7 s$

ch 21.15 $t_1 (37°C)$, $t_2 (43°C)$, Δx_1, Δx, t_3 (water)

$\frac{q}{A} = \frac{t_3 - t_1}{\frac{\Delta x}{k} + \frac{1}{h}} = \frac{t_2 - t_1}{\frac{\Delta x_1}{k}} = 2.4 \times 10^3$, $t_3 = 53.2°C$

ch21.17 $k = 377.36 W/m \cdot K$. $q = \frac{100-t}{\Sigma R} = V \rho \cdot c_p \cdot \frac{dt}{d\theta} = m \cdot c_p \cdot \frac{dt}{d\theta}$ → $4.187 \times 10^3 = a$

$\frac{1}{h_i A_i}$: 1.02×10^{-4}; $\frac{\Delta r}{k A_{lm}}$: 5.77×10^{-6}; $\frac{L_1}{h_o A_o}$: 2.64×10^{-3}; $a(1.32 \times 10^{-3})$.

$\therefore \ln \frac{100-t}{100-t_0} = -\frac{\theta}{aR}$

→ $h_o = 3.95 \times 10^2 W/m^2 \cdot K$.
— a) $h_o \times 2$, $\theta = 1.245 \times 10^3 s = 20.75 min$
b) $t_s = 150$, $\theta = 890 s = 14.8 min$

Ch18.1). $10°F$ | wood | cork board | | $70°F$; $t_1 (12.6°F)$, $t_2 = 67.4°F$

$R_{wood} = 0.67$, $R_{cork} = 13.9$, $q = \frac{\Delta T}{2R_1 + R_2} = 3.9 Btu/hr$ — moisture ⇒ 4″, frost ⇒ 1.93″

18-4) $k_{glass} = 0.5$ — $R_1 = 1.04 \times 10^{-3}$; $k_{air} = 0.016$ — $R_2 = 3.26 \times 10^{-2}$; $q = \frac{\Delta T}{2R_1 + R_2} = 865 Btu/hr$

$q_{single} = 1.92 \times 10^4$, $q_{si} - q_{th} = 1.83 \times 10^4$ ‖ (198.7일)

2) B120 = 6ton
$6000kg \times \frac{1lb}{0.4536kg} \cdot \frac{13,200 Btu}{1lb} \times 0.5$ eff. $= 8.73 \times 10^7 Btu$

18-5) $k_{glass} = 0.80$, $k_{steel} = 448$

glass/steel t_1, steel/air t_2.

$R_{gl} = 3.33 \times 10^{-5}$, $q = \frac{50 - t_1}{R_1}$, $t_1 = 32.8°C$

$R_{st} = 4.94 \times 10^{-7}$.

$R_{air} = R_g + R_s = 3.38 \times 10^{-5}$, $q = \frac{t_2 - 32.8}{R_2}$, $t_2 = 32.5°C$

$q = \frac{\Delta T}{\Sigma R} = 5.18 \times 10^5 J/s$

18.7). $R = \rho_\Omega \frac{60}{(\frac{0.051}{12})(\frac{0.50}{12})} = \rho_\Omega \cdot 3.39 \times 10^5 [\Omega]$

$P = \frac{V^2}{R} = \frac{3.57 \times 10^{-2}}{\rho_\Omega} [W]$

$S = \frac{P}{V} = \frac{3.57 \times 10^{-2}}{\rho_\Omega} \frac{J}{s} \times \frac{Btu}{1055J} \times \frac{3600s}{1hr} \times \frac{1}{60 \times \frac{0.051}{12} \times \frac{0.50}{12}} = \frac{11.47}{\Omega} \left[\frac{Btu}{h \cdot ft^3}\right]$

$k \cdot \frac{d^2t}{dx^2} = -s \Rightarrow t = -\frac{s}{k}x^2 + c_2$ B.C. $x = \pm\frac{1}{2}(0.051'')$ $t = 1400°F$ — ①; $x = 0$, $\frac{dt}{dx} = 0$.

APPENDIXES

$\frac{s}{k} = \frac{11.47}{\rho a k} = \frac{11.47}{25.7 \times 10^{6}} = 4.46 \times 10^{5} \frac{°F}{ft^2}$

$t = -4.46 \times 10^{5} \cdot x^2 + c_1 \Rightarrow c_1 = 1402°F$, $x = 0 \nearrow t_{max}$.

18-8) $\frac{\partial}{\partial r}\left(k \cdot r \frac{\partial t}{\partial r}\right) = -rs$ $\Downarrow$ $k \cdot \frac{dt}{dr} = -\frac{r}{2}s$, $k = K(1+\alpha t)$

$\int_{t_i}^{t_o} K(1+\alpha t)\,dt = \int_{r_i}^{r_o} -\frac{s}{2} r\,dr$

$\rightarrow K[(t_o - t_i) + \frac{\alpha}{2}(t_o^2 - t_i^2)] = -\frac{s}{4}(r_o^2 - r_i^2)$

$t_o = 100°C \nearrow$ $8.0 \times 10^{-2} t_i^2 - 200 t_i + 6.96 \times 10^4 = 0$

$t_i = 417.8°C$ (not saftey)

22-6) $\frac{t - t_s}{t_m - t_s} = 1 - \frac{r^2}{r_i^2}$, $q = -k \frac{dt}{dr}\Big|_{r_i} = -h(t_s - t_m) \Rightarrow h = \frac{k}{t_s - t_m} \cdot \frac{dt}{dr}\Big|_{r_i}$

$\frac{dt}{dr}\Big|_{r_i} = -\frac{2}{r_i}(t_m - t_s)$ $h = \frac{2k}{r_i} \Rightarrow Nu = h \times \frac{D}{k} = \frac{2k}{r_i} \times \frac{2r_i}{k} = ④$

fig 22-4). d) ⇒ 3.66 Nu, ○ ~ 4 fully developed; ④ 아래서 fully developed 는 아니다.

22-4) $\frac{\partial^2 t}{\partial r^2} + \frac{1}{r}\frac{\partial t}{\partial r} = \frac{2u_b}{\alpha}\left[1 - \left(\frac{r}{r_i}\right)^2\right]\frac{\partial t}{\partial x}$, $\frac{\partial t}{\partial x} = const$, $\frac{2u_b}{\alpha} \cdot \frac{\partial t}{\partial x} = \frac{2u_b}{\alpha}\frac{dt}{dx} = a$

$h_x = \frac{w \cdot c_p}{\pi D}\left(\frac{dt}{dx}\right)\frac{1}{t_s - t_b} = \frac{w \cdot c_p}{\pi D} \cdot \left(\frac{a\alpha}{2u_b}\right) \cdot \frac{1}{t_s - t_b}$, $\left(w = \rho u_b \left(\frac{\pi}{4}D^2\right)\right) = \frac{a \times D}{8(t_s - t_b)}$

$Nu_x = h_x \times \frac{D}{k} = \frac{aD^2}{8(t_s - t_b)}$, $r\frac{\partial}{\partial r}\left(r\frac{\partial t}{\partial r}\right) = a\left(1 - \left(\frac{r}{r_i}\right)^2\right) \Rightarrow t = a\left(\frac{r^2}{4} - \frac{r^4}{16 r_i^2} - \frac{3}{16}r_i^2\right) + t_s$

$t_{bulk} = \frac{2\pi \int_0^{r_i} u\, t\, r\, dr}{u_b \pi r_i^2} = \frac{2}{u_b r_i^2}\int_0^{r_i} 2u_b\left(1 - \left(\frac{r}{r_i}\right)^2\right)\left[a\left(\frac{r^2}{4} - \frac{r^4}{16 r_i^2} - \frac{3}{16}r_i^2\right) + t_s\right] r\,dr$

$\frac{r^2}{r_i^2} = s \Rightarrow \frac{1}{r_i^2} 2r\,dr = ds = 2\int_0^1 (1-s)\left[a\left(\frac{s}{4}r_i^2 - \frac{1}{16}s^2 r_i^2 - \frac{3}{16}r_i^2\right) + t_s\right]ds$

$= 2\int_0^1 a\left(\frac{s}{4}r_i^2 - \frac{1}{16}s^2 r_i^2 - \frac{3}{16}r_i^2\right) - a\left(\frac{s^2}{4}r_i^2 - \frac{1}{16}s^3 - \frac{3}{16}r_i^2 s\right) + t_s - s t_s$

Appendix 1 The Basic Balances in Compact Notation 777
Appendix 2 The Differential Balances Expressed in Several Coordinate Systems 788
Appendix 3 Some Useful Data 792
Table A-5 Properties of Ferrous Pipe 793
Table A-5A Properties of Ferrous Pipe—Metric 794
Table A-6 Tubing Characteristics 795
Table A-6A Tubing Characteristics—Metric 796
Table A-7 Approximate Specific Gravities and Densities of Miscellaneous Solids and Liquids 797
Table A-8 Viscosities of Gases 800
Table A-9 Viscosities of Liquids 801
Table A-10 Viscosity of Water 802
Table A-11 Thermal Conductivities of Metals 802
Table A-12 Thermal Conductivities of Liquids 803
Table A-13 Thermal Conductivities of Gases and Vapors 804
Table A-14 Thermal Conductivities of Some Building and Insulating Materials 805

if) $\frac{\partial^2 t}{\partial r^2} + \frac{1}{r}\frac{\partial t}{\partial r} = 0$; $Nu = \frac{48}{5} = 9.6$; $t_b = t_o + \frac{s u_b^2}{k} \cdot \frac{5}{6}$

$= t_s + 2r_i^2 a\left(\frac{1}{8} - \frac{1}{48} - \frac{3}{16} - \frac{1}{12} + \frac{1}{64} + \frac{3}{32}\right) = t_s - \frac{11}{96}r_i^2 a$

$h = \frac{aD^2}{8(t_s - t_b)} = \frac{a(2r_i)^2}{8 \cdot \frac{11}{96} \cdot a r_i^2} = \left(\frac{48}{11}\right)$

Table A-15	Thermal Conductivities of Insulating Materials at High Temperatures	806
Table A-16	Saturated Steam Temperature Table	806
Table A-17	Specific Heats of Miscellaneous Liquids and Solids	807
Table A-18	Diffusion Coefficients of Gases and Vapors in Air at 25°C, 1 atm	811
Table A-19	Diffusion Coefficients in Liquids at 20°C	812
Table A-20	Ammonia (NH_3) Solubility in Water	813
Table A-21	Carbon Dioxide (CO_2) Solubility in Water	813
Table A-22	Chlorine (Cl_2) Solubility in Water	813
Table A-23	Thermodynamic Properties of Moist Air	814
Table A-24	Selected Distribution Coefficients for Aqueous Systems	816
Table A-25	Equilibrium Data for Monoethanolamine Solutions	817
Table A-26	Constant-pressure Liquid-vapor Equilibrium Data for Binary Mixtures	818
Figure A-2	Viscosities of Gases at 1 atm	798
Figure A-3	Viscosities of Liquids at 1 atm	799
Figure A-4	Specific Heats of Liquids	808
Figure A-5	Specific Heats (C_p) of Gases at 1 atm Pressure	809
Figure A-6	Latent Heat of Vaporization	810
Figure A-7	Equilibrium Constants in Light-hydrocarbon Systems. Low-temperature Range	819
Figure A-8	Equilibrium Constants in Light-hydrocarbon Systems. High-temperature Range	820

APPENDIX 1

THE BASIC BALANCES IN COMPACT NOTATION

In Part 1 of this book we have developed the mass, energy, and momentum balances in the cartesian coordinate system. Similar equations for the balances in cylindrical and spherical coordinates can be obtained from a consideration of volume elements expressed in the appropriate coordinate systems, and such equations appear in the various applications treated later in the text. Since the balances express the same ideas independently of the coordinate system, it is common to use dyadic vector-tensor notation to write a given balance in an invariant form. A knowledge of the meaning of the symbols in this compact notation permits one to expand them into a suitable coordinate system; this step is usually necessary for the application of the balances to a specific problem and for the integration of the differential balances.

In dyadic notation we write a vector, such as the velocity, as a boldface symbol **u**. In cartesian coordinates this vector is the sum of the three velocity components in the three coordinate directions:

$$\mathbf{u} = u_1\,\boldsymbol{\delta}_1 + u_2\,\boldsymbol{\delta}_2 + u_3\,\boldsymbol{\delta}_3 \qquad \text{(A-1)}$$

Here the subscripts 1, 2, and 3 represent the axes x, y, and z of the text. The $\boldsymbol{\delta}_i$ are unit vectors in the appropriate directions. We shall follow the summation convention. In the equation

$$\mathbf{u} = \sum_{i=1}^{3} u_i \, \boldsymbol{\delta}_i \tag{A-2}$$

we omit the summation sign; the presence of a repeated index like i in a single term means that we sum over that index. Thus we write Eq. (A-2) as

$$\mathbf{u} = u_i \, \boldsymbol{\delta}_i \tag{A-3}$$

As another example, the nabla operator ∇ as applied to a scalar like the density gives

$$\nabla \rho = \boldsymbol{\delta}_i \frac{\partial \rho}{\partial x_i} \tag{A-4}$$

and the divergence operator $\nabla \cdot$ applied to a vector gives

$$\nabla \cdot \mathbf{u} = \frac{\partial u_i}{\partial x_i} \tag{A-5}$$

Standard references can be consulted for the meaning of these and other dyadic vector symbols in cylindrical and spherical coordinates.

For the actual mathematical manipulation of equations, it is convenient to retain the cartesian form by using cartesian tensor notation. The unit vectors are omitted, and we write the vector $\mathbf{u}$ as u_i, where

$$u_i = (u_1, u_2, u_3) \tag{A-6}$$

We symbolize the equivalence of the two notations by an arrow $\Rightarrow$:

$$u_i \Rightarrow \mathbf{u} \tag{A-7}$$

$$\frac{\partial \rho}{\partial x_i} \Rightarrow \nabla \rho \tag{A-8}$$

$$\frac{\partial u_i}{\partial x_i} = \nabla \cdot \mathbf{u} \tag{A-9}$$

In other words, in the cartesian tensor notation we write only the rectangular components of a vector (or tensor); the unit vectors are omitted. The index notation used here can be generalized to apply to any orthogonal system of coordinates (e.g., rectangular, cylindrical, or spherical) or even further, to nonorthogonal systems. The latter notation is useful in an advanced treatment of nonnewtonian fluids but is not needed for the subjects treated in our book.

It should be clear that u_i *does not equal* $\mathbf{u}$; the two are related by Eq. (A-3). This idea is expressed by the symbol $\Rightarrow$. Of course, when the result of an operation gives a scalar, the equality sign is correct, as in Eq. (A-9).

We shall proceed with the essentially simple transcription of the balances of Part 1 to cartesian tensor notation and to dyadic notation. From the latter, the equations can be written in the other orthogonal systems; the results are given in Tables A-1 to A-4.

MASS BALANCES

Single component systems:

$$\frac{\partial(u_i\rho)}{\partial x_i} + \frac{\partial \rho}{\partial \theta} = 0 \qquad \text{(7-7C)†}$$

$$\Rightarrow \qquad \nabla \cdot \mathbf{u}\rho + \frac{\partial \rho}{\partial \theta} = 0 \qquad \text{(7-7D)}$$

$$\rho\left(\frac{\partial u_i}{\partial x_i}\right) + u_i \frac{\partial \rho}{\partial x_i} + \frac{\partial \rho}{\partial \theta} = 0 \qquad \text{(7-8C)}$$

$$\Rightarrow \rho\nabla \cdot \mathbf{u} + \mathbf{u} \cdot \nabla\rho + \frac{\partial \rho}{\partial \theta} = 0 \qquad \text{(7-8D)}$$

The definition of the substantial derivative applied to the density yields

$$\frac{d\rho}{d\theta} = \frac{\partial \rho}{\partial x_i}\frac{dx_i}{d\theta} + \frac{\partial \rho}{\partial \theta} \qquad \text{(A-10)}$$

$$\frac{d\rho}{d\theta} = \frac{D\rho}{D\theta} = u_i \frac{\partial \rho}{\partial x_i} + \frac{\partial \rho}{\partial \theta} \qquad \text{(A-11C)}$$

$$\Rightarrow \qquad \frac{D\rho}{D\theta} = \mathbf{u} \cdot \nabla\rho + \frac{\partial \rho}{\partial \theta} \qquad \text{(A-11D)}$$

The equation of continuity is then

$$\frac{\partial u_i}{\partial x_i} + \frac{1}{\rho}\frac{D\rho}{D\theta} = 0 \qquad \text{(7-9C)}$$

$$\Rightarrow \qquad \nabla \cdot \mathbf{u} + \frac{1}{\rho}\frac{D\rho}{D\theta} = 0 \qquad \text{(7-9D)}$$

For an incompressible fluid we thus have

$$\frac{\partial u_i}{\partial x_i} = \nabla \cdot \mathbf{u} = 0 \qquad \text{(7-12C or D)}$$

Two-component systems:

$$\frac{\partial(u_i\rho_A)}{\partial x_i} + \frac{\partial \rho_A}{\partial \theta} + \frac{\partial J_{Ai}}{\partial x_i} - r_A = 0 \qquad \text{(7-14C)}$$

$$\Rightarrow \qquad \nabla \cdot \mathbf{u}\rho_A + \frac{\partial \rho_A}{\partial \theta} + \nabla \cdot \mathbf{J}_A - r_A = 0 \qquad \text{(7-14D)}$$

The flux J_{Ai} is given by

$$J_{Ai} = -D_{AB}\frac{\partial \rho_A}{\partial x_i} \qquad \text{(7-16C)}$$

$$\Rightarrow \qquad \mathbf{J}_A = -D_{AB}\,\nabla\rho_A \qquad \text{(7-16D)}$$

† The equation numbers refer to the equivalent equations in the text. The suffix C means "in cartesian tensor notation," and suffix D means "in dyadic notation."

For a constant D_{AB}, Eq. (7-14) then becomes

$$\frac{\partial(u_i \rho_A)}{\partial x_i} + \frac{\partial \rho_A}{\partial \theta} = D_{AB} \frac{\partial^2 \rho_A}{\partial x_i \, \partial x_i} + r_A \qquad \text{(7-17C)}$$

$$\Rightarrow \qquad \nabla \cdot \mathbf{u} \rho_A + \frac{\partial \rho_A}{\partial \theta} = D_{AB} \nabla^2 \rho_A + r_A \qquad \text{(7-17D)}$$

We notice that

$$\frac{D\rho_A}{D\theta} = u_i \frac{\partial \rho_A}{\partial x_i} + \frac{\partial \rho_A}{\partial \theta} \qquad \text{(A-12C)}$$

$$\Rightarrow \qquad \frac{D\rho_A}{D\theta} = \mathbf{u} \cdot \nabla \rho_A + \frac{\partial \rho_A}{\partial \theta} \qquad \text{(A-12D)}$$

We can then transform Eq. (7-17) to give

$$\rho_A \frac{\partial u_i}{\partial x_i} + \frac{D\rho_A}{D\theta} = D_{AB} \frac{\partial^2 \rho_A}{\partial x_i \, \partial x_i} + r_A \qquad \text{(7-18C)}$$

$$\Rightarrow \qquad \rho_A \nabla \cdot \mathbf{u} + \frac{D\rho_A}{D\theta} = D_{AB} \nabla^2 \rho_A + r_A \qquad \text{(7-18D)}$$

and, for a constant ρ,

$$\frac{D\rho_A}{D\theta} = D_{AB} \nabla^2 \rho_A + r_A \qquad \text{(7-19D)}$$

It is often convenient to express the species balance in terms of the mass fraction x_A or x_B, where

$$x_A = \frac{\rho_A}{\rho} \qquad \text{(A-13)†}$$

Then Eq. (7-14) becomes

$$x_A \frac{\partial(u_i \rho)}{\partial x_i} + u_i \rho \frac{\partial x_A}{\partial x_i} + x_A \frac{\partial \rho}{\partial \theta} + \rho \frac{\partial x_A}{\partial \theta} = - \frac{\partial J_{A_i}}{\partial x_i} + r_A \qquad \text{(A-14)}$$

Equation (7-7C), the equation of continuity, is used to simplify this expression to yield

$$u_i \rho \frac{\partial x_A}{\partial x_i} + \rho \frac{\partial x_A}{\partial \theta} = - \frac{\partial J_{A_i}}{\partial x_i} + r_A \qquad \text{(A-15)}$$

or

$$\rho \frac{Dx_A}{D\theta} = - \frac{\partial J_{A_i}}{\partial x_i} + r_A \qquad \text{(A-16)}$$

Compositions expressed as mass fractions are most convenient when the fluid has a constant density; we can then combine Eqs. (7-16C), (A-13), and (A-16) to obtain

† We use an upper-case subscript on x to denote a mass fraction and a lower-case subscript to denote a coordinate.

$$\rho \frac{Dx_A}{D\theta} = \rho D_{AB} \frac{\partial^2 x_A}{\partial x_i \, \partial x_i} + r_A \qquad \text{(A-17C)}$$

$$\Rightarrow \qquad \rho \frac{Dx_A}{D\theta} = \rho D_{AB} \, \nabla^2 x_A + r_A \qquad \text{(A-17D)}$$

OVERALL BALANCES

The overall mass balance of Chap. 3 can be obtained by writing Eq. (7-7C) as

$$\frac{\partial(u_i \rho)}{\partial x_i} \, d\check{V} + \frac{\partial \rho}{\partial \theta} d\check{V} = 0$$

and integrating over the control volume:

$$\iiint_{\check{V}} \frac{\partial(u_i \rho)}{\partial x_i} \, d\check{V} + \iiint_{\check{V}} \frac{\partial \rho}{\partial \theta} d\check{V} = 0$$

$$\Rightarrow \qquad \iiint_{\check{V}} \nabla \cdot \mathbf{u}\rho \, d\check{V} + \iiint_{\check{V}} \frac{\partial \rho}{\partial \theta} d\check{V} = 0$$

The Gauss divergence theorem[1] is then used to transform the first integral to a surface integral, and the formula of Leibnitz[2] for the differentiation of a definite integral is used to transform the second integral. The limits on the integrals, represented by $\check{V}$, are constants. Thus we get

$$\iint_A u_i \rho \, dA_i + \frac{\partial}{\partial \theta} \iiint_{\check{V}} \rho \, d\check{V} = 0 \qquad \text{(3-12C)}$$

$$\Rightarrow \qquad \iint_A \mathbf{u}\rho \cdot d\mathbf{A} + \frac{\partial}{\partial \theta} \iiint_{\check{V}} \rho \, d\check{V} = 0 \qquad \text{(3-12D)}$$

The direction of **A** is described by its outward-directed normal, so that the scalar product in the first integral becomes

$$\mathbf{u}\rho \cdot d\mathbf{A} = u\rho \cos \alpha \, dA$$

where u and A are the (scalar) magnitudes of the vectors **u** and **A**, and α is defined in Fig. 3-3.

The equality of the left-hand members of Eqs. (A-14) and (A-16) establishes the following relation:

$$\rho \frac{Dx_A}{D\theta} = \frac{\partial(u_i \rho x_A)}{\partial x_i} + \frac{\partial(\rho x_A)}{\partial \theta} \qquad \text{(A-18)}$$

[1] C. R. Wylie, Jr., "Advanced Engineering Mathematics," 3d ed., p. 572, McGraw-Hill Book Company, New York, 1966.

[2] Ibid., p. 274.

This equation applies for other scalar, vector, or tensor properties carried by a fluid; for example, in general:

$$\rho \frac{DZ_j}{D\theta} = \frac{\partial(u_i \rho Z_j)}{\partial x_i} + \frac{\partial(\rho Z_j)}{\partial \theta} \tag{A-19}$$

We can thus transform a Lagrange derivative of a property moving with a fluid to Euler derivatives referred to fixed axes. The right-hand side of Eq. (A-19) is in a form which is convenient to integrate over a fixed control volume, with the aid of the divergence theorem:

$$\iiint_{\check{V}} \rho \frac{DZ_j}{D\theta}\, d\check{V} = \iint_A u_i \rho Z_j\, dA_i + \iiint_{\check{V}} \frac{\partial(\rho Z_j)}{\partial \theta}\, d\check{V}$$

or

$$\iiint_{\check{V}} \rho \frac{DZ_j}{D\theta}\, d\check{V} = \iint_A u_i \rho Z_j dA_i + \frac{\partial \check{Z}_j}{\partial \theta} \tag{A-20}$$

Since $\rho\, dV = dM$, it is clear that the left-hand side of Eq. (A-20) is simply $D\check{Z}_j/D\theta$, so that

$$\frac{D\check{Z}_j}{D\theta} = \iint_A u_i \rho Z_j\, dA_i + \frac{\partial \check{Z}_j}{\partial \theta} \tag{A-21C}$$

This equation is a form of the Reynolds transport theorem and is equivalent to Eq. (4-9).

ENERGY BALANCES

The derivations of Chap. 8 lead to the following expressions:

$$\rho \frac{DH}{D\theta} - \frac{Dp}{D\theta} = -\frac{\partial(q/A)_i}{\partial x_i} + \Phi \tag{8-8C}$$

$$\left(\frac{q}{A}\right)_i = -k \frac{\partial t}{\partial x_i} \tag{8-9C}$$

$$\rho \frac{DH}{D\theta} - \frac{Dp}{D\theta} = k\, \nabla^2 t + \Phi \tag{8-10D}$$

$$\frac{Dt}{D\theta} = \frac{k}{\rho C_p} \nabla^2 t + \Phi \tag{8-14D}$$

The overall energy balance is ($u_s = 0$)

$$\iint_A u_i \rho \left(\frac{u^2}{2g_c} + \frac{g}{g_c} Z + H\right) dA_i + \frac{\partial \check{E}}{\partial \theta} = q - W_s \tag{4-12C}$$

MOMENTUM BALANCE

The equations of Chap. 5 can be immediately expressed in compact notation:

$$\check{F}_i = \frac{1}{g_c}\frac{D(Mu_i)}{D\theta} = \frac{1}{g_c}\frac{D\check{P}_i}{D\theta} \tag{5-1C}$$

$$\check{F}_i = \frac{1}{g_c}\left[\iint_A u_i \rho u_j \, dA_j + \frac{\partial}{\partial\theta}\iiint_{\check{V}} u_i \rho \, d\check{V}\right] \tag{5-2C}$$

or

$$\check{F}_i = \iint_A \frac{u_i \rho u_j}{g_c} dA_j + \frac{1}{g_c}\frac{\partial \check{P}_i}{\partial\theta} \tag{5-3C}$$

From here on, in the development of the differential momentum balance, we shall adopt a different procedure from that used in Chap. 9. Tensor notation will now permit a considerable simplification of the derivation of the Navier-Stokes equation.

If the control volume over which Eq. (5-2C) applies does not cut any solid wall, we can divide F_i into just two kinds of forces: body forces X_i (lb_f/lb) like gravity or electrostatic forces; and surface forces τ_i (lb_f/ft^2) like viscous force (drag force) and pressure force.

We express this concept by the following equation:

$$\check{F}_i = \iiint_{\check{V}} X_i \rho \, d\check{V} + \iint_A \tau_i \, dA \tag{A-22}$$

Note that the surface force is written in terms of the vector τ_i and the scalar dA. The value of τ_i depends on both the state of stress at a point and the orientation of the vector dA_j at the point.

The reader familiar with vector and tensor notation will recognize that the stresses such as τ_{xy} discussed on pages 100 to 103 are components of the stress tensor.

On each face of an infinitesimal cubic element of fluid there acts a force which has three components. Thus we define the state of stress at a point by the tensor τ_{ji}, which has nine components.

$$\tau_{ji} = \begin{pmatrix} \tau_{11} & \tau_{12} & \tau_{13} \\ \tau_{21} & \tau_{22} & \tau_{23} \\ \tau_{31} & \tau_{32} & \tau_{33} \end{pmatrix} \Rightarrow \boldsymbol{\tau} \tag{A-23}$$†

There are only nine components because for an infinitesimal element (a point) the stress components on opposite faces become identical. In Fig. (9-1) the small quantities dx, dy, $dz \Rightarrow dx_i$ approach zero as a limit.

In the integral

$$\iint_A \tau_i \, dA \tag{A-24}$$

† $\boldsymbol{\tau}$ represents the stress tensor.

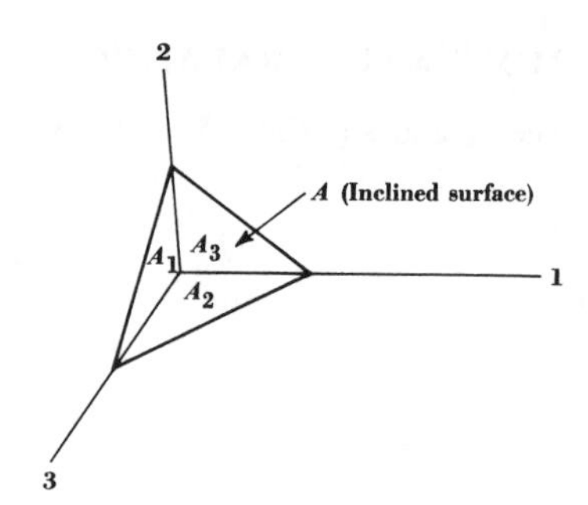

FIGURE A-1
Element for analysis of stresses.

we wish to express the stress τ_i on an arbitrarily oriented differential area in terms of the state of stress at the point in question, i.e., the stress tensor. Consider an arbitrarily positioned area A as shown in Fig. A-1. The limits of the area are determined by the three coordinate planes. We now make a force balance on the four-sided fluid element thus defined. We adopt the sign convention that the material outside the element A exerts a positive force in a given direction when the corresponding component of the outward normal of the area element points in the positive direction. For the element in Fig. A-1 there results, in the 1 direction,

$$\bar{\tau}_1 A - \bar{\tau}_{11} A_1 - \bar{\tau}_{21} A_2 - \bar{\tau}_{31} A_3 + \bar{f}_1 \rho \check{V} = 0 \qquad \text{(A-25)}$$

where the superbars indicate average values.

Here $\bar{f}_1$ is the component of the body and inertia forces acting on the element of volume $\check{V}$, which is given by $\frac{1}{3}hA$, h being the altitude perpendicular to A. For any component i, we get

$$\bar{\tau}_i A - \bar{\tau}_{ji} A_j - \bar{f}_i \rho \tfrac{1}{3}hA = 0 \qquad \text{(A-26)}$$

and the limit as h goes to zero is

$$\tau_i A = \tau_{ji} A_j \qquad \text{(A-27)}$$

or

$$\tau_i \, dA = \tau_{ji} \, dA_j \qquad \text{(A-28)}$$

We notice that A_j, which equals $n_j A$, where n_j is the unit outward normal to A, has as its components the three areas in Fig. A-1, given the same symbols.

Thus, the stress vector on a differential element of area cutting the material is related to the orientation of the area n_j and the state of stress τ_{ji} by

$$\tau_i = \tau_{ji} n_j \qquad \text{(A-29)}$$

This expression is now used to convert the surface force to

$$\iint_A \tau_{ji} \, dA_j$$

and the divergence theorem then gives

$$\iint_A \tau_{ji}\, dA_j = \iiint_{\check{V}} \frac{\partial \tau_{ji}}{\partial x_j} d\check{V} \qquad \text{(A-30)}$$

The contraction of τ_{ji} to $(\partial \tau_{ji}/\partial x_j)$ by differentiation (a divergence) yields a vector

$$\frac{\partial \tau_{1i}}{\partial x_1} + \frac{\partial \tau_{2i}}{\partial x_2} + \frac{\partial \tau_{3i}}{\partial x_3} \Rightarrow \nabla \cdot \tau \qquad \text{(A-31)}$$

Equation (A-30) is now used with Eq. (A-22), Eq. (5-2C), and the divergence theorem to give

$$\frac{1}{g_c} \iiint_{\check{V}} \frac{\partial (u_i \rho u_j)}{\partial x_j} d\check{V} + \frac{1}{g_c} \iiint_{\check{V}} \frac{\partial (u_i \rho)}{\partial \theta} d\check{V}$$

$$= \iiint_{\check{V}} X_i \rho\, d\check{V} + \iiint_{\check{V}} \frac{\partial \tau_{ji}}{\partial x_j} d\check{V} \qquad \text{(A-32)}$$

Application of this equation to a vanishingly small volume $\check{V}$ yields Cauchy's equation of motion:

$$\frac{1}{g_c} \frac{\partial (u_i \rho u_j)}{\partial x_j} + \frac{1}{g_c} \frac{\partial (u_i \rho)}{\partial \theta} = X_i \rho + \frac{\partial \tau_{ji}}{\partial x_j} \qquad \text{(A-33)}$$

This equation may be written as

$$\frac{\rho}{g_c} \frac{Du_i}{D\theta} = \rho X_i + \frac{\partial \tau_{ji}}{\partial x_j} \qquad \text{(9-11C)}$$

$$\Rightarrow \qquad \frac{\rho}{g_c} \frac{D\mathbf{u}}{D\theta} = \rho \mathbf{X} + \nabla \cdot \tau \qquad \text{(9-11D)}$$

As shown in Chap. 9, we also know that τ_{ji} is symmetric, that is,

$$\tau_{ji} = \tau_{ij} \qquad \text{(A-34)}$$

It will be convenient, as in Chap. 9, to divide the stress tensor into the pressure and the viscous stress tensor by the equation

$$\tau_{ij} = -p\, \delta_{ij} + \sigma_{ij} \qquad \text{(9-22C)}$$

when δ_{ij} is the Kronecker delta

$$\delta_{ij} = \begin{pmatrix} 1 & 0 & 0 \\ 0 & 1 & 0 \\ 0 & 0 & 1 \end{pmatrix}$$

and is the only isotropic tensor of the second order. Thus its value is the same for all possible orientations of the axes, as is characteristic of the pressure. For other tensors, such as τ_{ij}, the nine components depend on the orientation of the axes.

For example, if τ_{ij} is known with respect to a given set of cartesian coordinates, the same state of stress at the point is represented by a new set of components τ_{mn} if the axes are given an arbitrary rotation. From the principles of tensor analysis we have

$$\tau'_{mn} = \ell_{mi}\,\ell_{nj}\,\tau_{ij} \tag{A-35}$$

where ℓ_{rs} is the cosine of the angle between the primed r axis and the unprimed s axis.

Equation (9-22C) is next substituted into Eq. (9-11C) to give

$$\frac{\rho}{g_c}\frac{Du_i}{D\theta} = \rho X_i + \frac{\partial \sigma_{ij}}{\partial x_j} - \frac{\partial p}{\partial x_i} \tag{A-36}$$

where we have used the relation

$$\delta_{ij}\frac{\partial p}{\partial x_j} = \frac{\partial p}{\partial x_i}$$

which can easily be verified. The symbol δ_{ij} is thus also called a substitution operator because of the general relation

$$\delta_{ij} B_j = B_i$$

$$\delta_{ij} B_i = B_j$$

In order to continue our derivation of the Navier-Stokes equations, we need an expression for σ_{ij} as a function of the velocity gradient, just as we replaced J_{Ai} in the mass balance by $-D_{AB}(\partial \rho_A/\partial x_i)$. In fluid mechanics this relation is called the *constitutive equation.* For a newtonian fluid we use the linear relation

$$g_c \sigma_{ij} = c_{ijk\ell}\, e_{k\ell} \tag{A-37}$$

where $c_{ijk\ell}$ is a constant isotropic tensor of the fourth order. The symmetric tensor e_{ij} is called the *rate of deformation tensor* and is given in terms of the velocity-gradient tensor by

$$e_{ij} = \frac{1}{2}\left(\frac{\partial u_i}{\partial x_j} + \frac{\partial u_j}{\partial x_i}\right) \tag{A-38}$$

The viscosity tensor must not depend on the arbitrary orientation of the axes, so it is a fourth-order isotropic tensor. We know from tensor analysis that a fourth-order isotropic tensor with the required symmetry must have the form

$$c_{ijk\ell} = \lambda \delta_{ij}\delta_{k\ell} + \mu(\delta_{ik}\delta_{j\ell} + \delta_{i\ell}\delta_{jk}) \tag{A-39}$$

where λ and μ are two scalar constants which describe the fluid. Substitution of this equation into Eq. (A-37) yields

$$g_c \sigma_{ij} = \lambda\, \delta_{ij} e_{kk} + 2\mu e_{ij} \tag{A-40}$$

We note from Eq. (A-38) that e_{kk} is the divergence $\partial u_k/\partial x_k$. When Eq. (A-40) is substituted into Eq. (A-36) there results

$$\rho\frac{Du_i}{D\theta} = \rho X_i - \frac{\partial p}{\partial x_i} + \frac{\lambda}{g_c}\frac{\partial^2 u_k}{\partial x_k\,\partial x_i} + \frac{\mu}{g_c}\frac{\partial^2 u_i}{\partial x_j\,\partial x_j} + \frac{\mu}{g_c}\frac{\partial^2 u_j}{\partial x_i\,\partial x_j} \tag{A-41}$$

It is customary to define a bulk viscosity κ by

$$\lambda = \kappa - \tfrac{2}{3}\mu \qquad \text{(A-42)}$$

The bulk viscosity is a measure of the dissipation (friction) experienced by a fluid undergoing a volumetric deformation without shear. We then substitute this expression for λ into Eq. (A-41), noting that $(\partial^2 u_k/\partial x_k\, \partial x_i)$ is the same as $(\partial^2 u_j/\partial x_i\, \partial x_j)$,

$$\frac{\rho}{g_c}\frac{Du_i}{D\theta} = \rho X_i - \frac{\partial p}{\partial x_i} + \frac{\kappa}{g_c}\frac{\partial^2 u_j}{\partial x_i\, \partial x_j} + \frac{\mu}{g_c}\frac{\partial^2 u_i}{\partial x_j\, \partial x_j} + \frac{1}{3}\frac{\mu}{g_c}\frac{\partial^2 u_j}{\partial x_i\, \partial x_j} \qquad \text{(A-43)}$$

For an ideal monatomic gas, κ is zero, and we have

$$\frac{\rho}{g_c}\frac{Du_i}{D\theta} = \rho X_i - \frac{\partial p}{\partial x_i} + \frac{\mu}{g_c}\frac{\partial^2 u_i}{\partial x_j\, \partial x_j} + \frac{1}{3}\frac{\mu}{g_c}\frac{\partial}{\partial x_i}\left(\frac{\partial u_j}{\partial x_j}\right) \qquad \text{(9-50C)}$$

$$\Rightarrow \qquad \frac{\rho}{g_c}\frac{D\mathbf{u}}{D\theta} = \rho\mathbf{X} - \nabla p + \frac{\mu}{g_c}\nabla^2\mathbf{u} + \frac{1}{3}\frac{\mu}{g_c}\nabla(\nabla\cdot\mathbf{u}) \qquad \text{(9-50D)}$$

For an incompressible fluid the Navier-Stokes equation becomes

$$u_j\frac{\partial u_i}{\partial x_j} + \frac{\partial u_i}{\partial \theta} = g_c X_i - \frac{g_c}{\rho}\frac{\partial p}{\partial x_i} + \nu\frac{\partial^2 u_i}{\partial x_j\, \partial x_j} \qquad \text{(9-52C)}$$

$$\Rightarrow \qquad \mathbf{u}\cdot\nabla\mathbf{u} + \frac{\partial\mathbf{u}}{\partial\theta} = g_c\mathbf{X} - \frac{g_c}{\rho}\nabla p + \nu\nabla^2\mathbf{u} \qquad \text{(9-52D)}$$

The various differential balances are written out in the common coordinate systems in Tables A-1 to A-4 for constant properties (μ, ρ, C_p, k, D_{AB}), and r_A and Φ zero.

APPENDIX 2

THE DIFFERENTIAL BALANCES EXPRESSED IN SEVERAL COORDINATE SYSTEMS

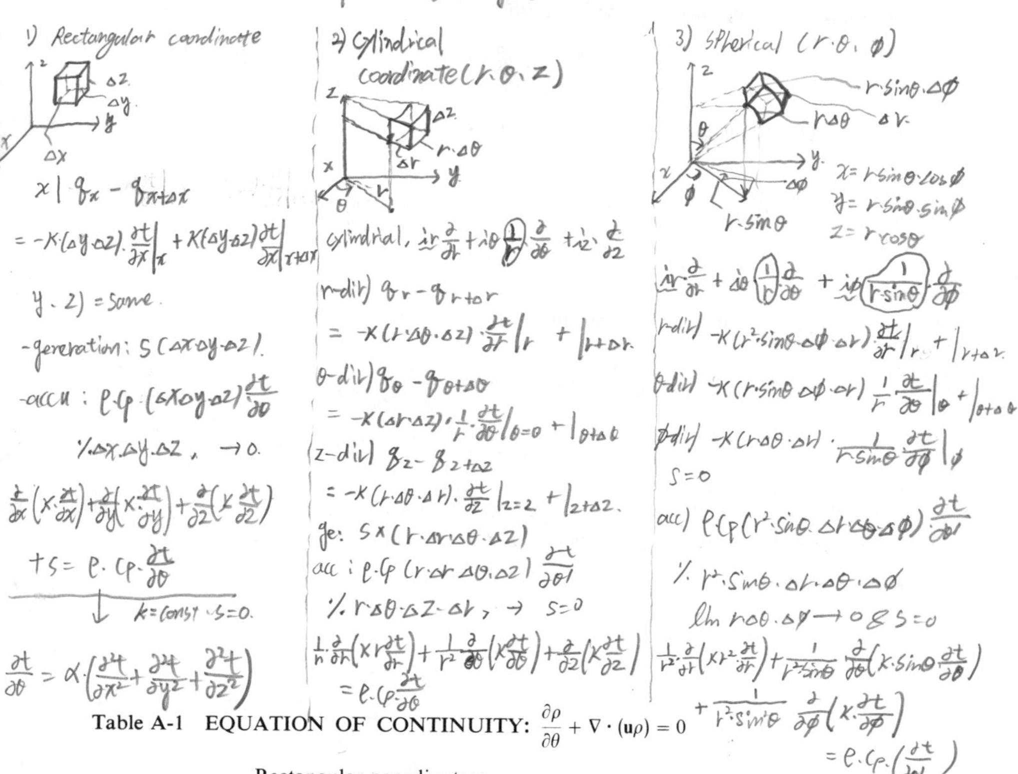

Table A-1 EQUATION OF CONTINUITY: $\dfrac{\partial \rho}{\partial \theta} + \nabla \cdot (\mathbf{u}\rho) = 0$

Rectangular coordinates:

$$\frac{\partial \rho}{\partial \theta} + \frac{\partial (u_x \rho)}{\partial x} + \frac{\partial (u_y \rho)}{\partial y} + \frac{\partial (u_z \rho)}{\partial z} = 0$$

Cylindrical coordinates[1]:

$$\frac{\partial \rho}{\partial \theta'} + \frac{1}{r}\frac{\partial}{\partial r}(\rho r u_r) + \frac{1}{r}\frac{\partial}{\partial \theta}(\rho u_\theta) + \frac{\partial}{\partial z}(\rho u_z) = 0$$

Spherical coordinates:

$$\frac{\partial \rho}{\partial \theta'} + \frac{1}{r^2}\frac{\partial}{\partial r}(\rho r^2 u_r) + \frac{1}{r \sin \theta}\frac{\partial}{\partial \theta}(\rho u_\theta \sin \theta) + \frac{1}{r \sin \theta}\frac{\partial}{\partial \phi}(\rho u_\phi) = 0$$

[1] In the equations in cylindrical and spherical coordinates the notation for time is θ'; θ is the angular coordinate.

Table A-2 DIFFERENTIAL MASS BALANCE FOR COMPONENT A

For a binary mixture, ρ and D_{AB} constant, and r_A zero.

$$\frac{D\rho_A}{D\theta} = D_{AB}\,\nabla^2\rho_A$$

Rectangular coordinates:

$$\frac{\partial \rho_A}{\partial \theta'} + u_x\frac{\partial \rho_A}{\partial x} + u_y\frac{\partial \rho_A}{\partial y} + u_z\frac{\partial \rho_A}{\partial z} = D_{AB}\left(\frac{\partial^2 \rho_A}{\partial x^2} + \frac{\partial^2 \rho_A}{\partial y^2} + \frac{\partial^2 \rho_A}{\partial z^2}\right)$$

Cylindrical coordinates:

$$\frac{\partial \rho_A}{\partial \theta'} + u_r\frac{\partial \rho_A}{\partial r} + \frac{u_\theta}{r}\frac{\partial \rho_A}{\partial \theta} + u_z\frac{\partial \rho_A}{\partial z}$$

$$= D_{AB}\left[\frac{1}{r}\frac{\partial}{\partial r}\left(r\frac{\partial \rho_A}{\partial r}\right) + \frac{1}{r^2}\frac{\partial^2 \rho_A}{\partial \theta^2} + \frac{\partial^2 \rho_A}{\partial z^2}\right]$$

Spherical coordinates:

$$\frac{\partial \rho_A}{\partial \theta'} + u_r\frac{\partial \rho_A}{\partial r} + \frac{u_\theta}{r}\frac{\partial \rho_A}{\partial \theta} + \frac{u_\phi}{r\sin\theta}\frac{\partial \rho_A}{\partial \phi}$$

$$= D_{AB}\left[\frac{1}{r^2}\frac{\partial}{\partial r}\left(r^2\frac{\partial \rho_A}{\partial r}\right) + \frac{1}{r^2\sin\theta}\frac{\partial}{\partial \theta}\left(\sin\theta\frac{\partial \rho_A}{\partial \theta}\right) + \frac{1}{r^2\sin^2\theta}\frac{\partial^2 \rho_A}{\partial \phi^2}\right]$$

Table A-3 DIFFERENTIAL ENERGY BALANCE

For ρ, C_p, k constant and Φ zero.

$$\frac{Dt}{D\theta} = \alpha\,\nabla^2 t \qquad \alpha = \frac{k}{C_p\rho}$$

Rectangular coordinates:

$$\frac{\partial t}{\partial \theta} + u_x\frac{\partial t}{\partial x} + u_y\frac{\partial t}{\partial y} + u_z\frac{\partial t}{\partial z} = \alpha\left(\frac{\partial^2 t}{\partial x^2} + \frac{\partial^2 t}{\partial y^2} + \frac{\partial^2 t}{\partial z^2}\right)$$

Cylindrical coordinates:

$$\frac{\partial t}{\partial \theta'} + u_r\frac{\partial t}{\partial r} + \frac{u_\theta}{r}\frac{\partial t}{\partial \theta} + u_z\frac{\partial t}{\partial z} = \alpha\left[\frac{1}{r}\frac{\partial}{\partial r}\left(r\frac{\partial t}{\partial r}\right) + \frac{1}{r^2}\frac{\partial^2 t}{\partial \theta^2} + \frac{\partial^2 t}{\partial z^2}\right]$$

Spherical coordinates:

$$\frac{\partial t}{\partial \theta'} + u_r\frac{\partial t}{\partial r} + \frac{u_\theta}{r}\frac{\partial t}{\partial \theta} + \frac{u_\phi}{r\sin\theta}\frac{\partial t}{\partial \phi}$$

$$= \alpha\left[\frac{1}{r^2}\frac{\partial}{\partial r}\left(r^2\frac{\partial t}{\partial r}\right) + \frac{1}{r^2\sin\theta}\frac{\partial}{\partial \theta}\left(\sin\theta\frac{\partial t}{\partial \theta}\right) + \frac{1}{r^2\sin^2\theta}\frac{\partial^2 t}{\partial \phi^2}\right]$$

Table A-4 DIFFERENTIAL MOMENTUM BALANCE

Equation of motion for μ and ρ constant.

$$\frac{D\mathbf{u}}{D\theta} = g_c\mathbf{X} - \frac{g_c}{\rho}\nabla p + \nu\nabla^2\mathbf{u} \qquad \nu = \frac{\mu}{\rho}$$

Rectangular coordinates:

x component
$$\frac{\partial u_x}{\partial \theta} + u_x\frac{\partial u_x}{\partial x} + u_y\frac{\partial u_x}{\partial y} + u_z\frac{\partial u_x}{\partial z} = g_c X - \frac{g_c}{\rho}\frac{\partial p}{\partial x} + \nu\left(\frac{\partial^2 u_x}{\partial x^2} + \frac{\partial^2 u_x}{\partial y^2} + \frac{\partial^2 u_x}{\partial z^2}\right)$$

y component
$$\frac{\partial u_y}{\partial \theta} + u_x\frac{\partial u_y}{\partial x} + u_y\frac{\partial u_y}{\partial y} + u_z\frac{\partial u_y}{\partial z} = g_c Y - \frac{g_c}{\rho}\frac{\partial p}{\partial y} + \nu\left(\frac{\partial^2 u_y}{\partial x^2} + \frac{\partial^2 u_y}{\partial y^2} + \frac{\partial^2 u_y}{\partial z^2}\right)$$

z component
$$\frac{\partial u_z}{\partial \theta} + u_x\frac{\partial u_z}{\partial x} + u_y\frac{\partial u_z}{\partial y} + u_z\frac{\partial u_z}{\partial z} = g_c Z - \frac{g_c}{\rho}\frac{\partial p}{\partial z} + \nu\left(\frac{\partial^2 u_z}{\partial x^2} + \frac{\partial^2 u_z}{\partial y^2} + \frac{\partial^2 u_z}{\partial z^2}\right)$$

Cylindrical coordinates:

r component
$$\frac{\partial u_r}{\partial \theta'} + u_r\frac{\partial u_r}{\partial r} + \frac{u_\theta}{r}\frac{\partial u_r}{\partial \theta} - \frac{u_\theta{}^2}{r} + u_z\frac{\partial u_r}{\partial z}$$
$$= g_c X_r - \frac{g_c}{\rho}\frac{\partial p}{\partial r} + \nu\left\{\frac{\partial}{\partial r}\left[\frac{1}{r}\frac{\partial}{\partial r}(ru_r)\right] + \frac{1}{r^2}\frac{\partial^2 u_r}{\partial \theta^2} - \frac{2}{r^2}\frac{\partial u_\theta}{\partial \theta} + \frac{\partial^2 u_r}{\partial z^2}\right\}$$

θ component
$$\frac{\partial u_\theta}{\partial \theta'} + u_r\frac{\partial u_\theta}{\partial r} + \frac{u_\theta}{r}\frac{\partial u_\theta}{\partial r} + \frac{u_r u_\theta}{r} + u_z\frac{\partial u_\theta}{\partial z}$$
$$= g_c X_\theta - \frac{g_c}{\rho}\frac{1}{r}\frac{\partial p}{\partial \theta} + \nu\left\{\frac{\partial}{\partial r}\left[\frac{1}{r}\frac{\partial}{\partial r}(ru_\theta)\right] + \frac{1}{r^2}\frac{\partial^2 u_\theta}{\partial \theta^2} + \frac{2}{r^2}\frac{\partial u_r}{\partial \theta} + \frac{\partial^2 u_\theta}{\partial z^2}\right\}$$

z component
$$\frac{\partial u_z}{\partial \theta'} + u_r\frac{\partial u_z}{\partial r} + \frac{u_\theta}{r}\frac{\partial u_z}{\partial \theta} + u_z\frac{\partial u_z}{\partial z}$$
$$= g_c X_z - \frac{g_c}{\rho}\frac{\partial p}{\partial z} + \nu\left[\frac{1}{r}\frac{\partial}{\partial r}\left(r\frac{\partial u_z}{\partial r}\right) + \frac{1}{r^2}\frac{\partial^2 u_z}{\partial \theta^2} + \frac{\partial^2 u_z}{\partial z^2}\right]$$

Spherical coordinates:

r component
$$\frac{\partial u_r}{\partial \theta'} + u_r\frac{\partial u_r}{\partial r} + \frac{u_\theta}{r}\frac{\partial u_r}{\partial \theta} + \frac{u_\phi}{r\sin\theta}\frac{\partial u_r}{\partial \phi} - \frac{u_\theta{}^2 + u_\phi{}^2}{r}$$
$$= g_c X_r - \frac{g_c}{\rho}\frac{\partial p}{\partial r}$$

(continued on p. 791)

$$+ \nu\left[\frac{1}{r^2}\frac{\partial}{\partial r}\left(r^2\frac{\partial u_r}{\partial r}\right) + \frac{1}{r^2 \sin\theta}\frac{\partial}{\partial\theta}\left(\sin\theta\frac{\partial u_r}{\partial\theta}\right) + \frac{1}{r^2\sin^2\theta}\frac{\partial^2 u_r}{\partial\phi^2} - \frac{2}{r^2}u_r - \frac{2}{r^2}\frac{\partial u_\theta}{\partial\theta} - \frac{2}{r^2}u_\theta\cot\theta - \frac{2}{r^2\sin\theta}\frac{\partial u_\phi}{\partial\phi}\right]$$

θ component

$$\frac{\partial u_\theta}{\partial\theta'} + u_r\frac{\partial u_\theta}{\partial r} + \frac{u_\theta}{r}\frac{\partial u_\theta}{\partial\theta} + \frac{u_\phi}{r\sin\theta}\frac{\partial u_\theta}{\partial\phi} + \frac{u_r u_\theta}{r} - \frac{u_\phi^2\cot\theta}{r}$$

$$= g_c X_\theta - \frac{g_c}{\rho}\frac{1}{r}\frac{\partial p}{\partial\theta}$$

$$+ \nu\left[\frac{1}{r^2}\frac{\partial}{\partial r}\left(r^2\frac{\partial u_\theta}{\partial r}\right) + \frac{1}{r^2\sin\theta}\frac{\partial}{\partial\theta}\left(\sin\theta\frac{\partial u_\theta}{\partial\theta}\right) + \frac{1}{r^2\sin^2\theta}\frac{\partial^2 u_\theta}{\partial\phi^2} + \frac{2}{r^2}\frac{\partial u_r}{\partial\theta} - \frac{u_\theta}{r^2\sin^2\theta} - \frac{2\cos\theta}{r^2\sin^2\theta}\frac{\partial u_\phi}{\partial\phi}\right]$$

ϕ component

$$\frac{\partial u_\phi}{\partial\theta'} + u_r\frac{\partial u_\phi}{\partial r} + \frac{u_\theta}{r}\frac{\partial u_\phi}{\partial\theta} + \frac{u_\phi}{r\sin\theta}\frac{\partial u_\phi}{\partial\phi} + \frac{u_\phi u_r}{r} + \frac{u_\theta u_\phi}{r}\cot\theta$$

$$= g_c X_\phi - \frac{g_c}{\rho}\frac{1}{r\sin\theta}\frac{\partial p}{\partial\phi}$$

$$+ \nu\left[\frac{1}{r^2}\frac{\partial}{\partial r}\left(r^2\frac{\partial u_\phi}{\partial r}\right) + \frac{1}{r^2\sin\theta}\frac{\partial}{\partial\theta}\left(\sin\theta\frac{\partial u_\phi}{\partial\theta}\right) + \frac{1}{r^2\sin^2\theta}\frac{\partial^2 u_\phi}{\partial\phi^2} - \frac{u_\phi}{r^2\sin^2\theta} + \frac{2}{r^2\sin\theta}\frac{\partial u_r}{\partial\phi} + \frac{2\cos\theta}{r^2\sin^2\theta}\frac{\partial u_\theta}{\partial\phi}\right]$$

APPENDIX 3

SOME USEFUL DATA

Table A-5 PROPERTIES OF FERROUS PIPE

Nominal pipe size, in	OD, in	Schedule Number*	Wall thickness, in	ID, in	Cross-sectional area: Metal, in²	Cross-sectional area: Flow, ft²	Circumference, ft, or surface, ft²/ft of length: Outside	Circumference, ft, or surface, ft²/ft of length: Inside	Capacity at 1 ft/s velocity: U.S. gal/min	Capacity at 1 ft/s velocity: Lb/h water	Weight of plain-end pipe, lb/ft
1/8	0.405	10S	0.049	0.307	0.055	0.00051	0.106	0.0804	0.231	115.5	0.19
		40ST, 40S	.068	.269	.072	.00040	.106	.0705	.179	89.5	.24
		80XS, 80S	.095	.215	.093	.00025	.106	.0563	.113	56.5	.31
1/4	0.540	10S	.065	.410	.097	.00092	.141	.107	.412	206.5	.33
		40ST, 40S	.088	.364	.125	.00072	.141	.095	.323	161.5	.42
		80XS, 80S	.119	.302	.157	.00050	.141	.079	.224	112.0	.54
3/8	0.675	10S	.065	.545	.125	.00162	.177	.143	.727	363.5	.42
		40ST, 40S	.091	.493	.167	.00133	.177	.129	.596	298.0	.57
		80XS, 80S	.126	.423	.217	.00098	.177	.111	.440	220.0	.74
1/2	0.840	5S	.065	.710	.158	.00275	.220	.186	1.234	617.0	.54
		10S	.083	.674	.197	.00248	.220	.176	1.112	556.0	.67
		40ST, 40S	.109	.622	.250	.00211	.220	.163	0.945	472.0	.85
		80XS, 80S	.147	.546	.320	.00163	.220	.143	0.730	365.0	1.09
		160	.188	.464	.385	.00117	.220	.122	0.527	263.5	1.31
		XX	.294	.252	.504	.00035	.220	.066	0.155	77.5	1.71
3/4	1.050	5S	.065	.920	.201	.00461	.275	.241	2.072	1036.0	0.69
		10S	.083	.884	.252	.00426	.275	.231	1.903	951.5	0.86
		40ST, 40S	.113	.824	.333	.00371	.275	.216	1.665	832.5	1.13
		80XS, 80S	.154	.742	.433	.00300	.275	.194	1.345	672.5	1.47
		160	.219	.612	.572	.00204	.275	.160	0.917	458.5	1.94
		XX	.308	.434	.718	.00103	.275	.114	0.461	230.5	2.44
1	1.315	5S	.065	1.185	.255	.00768	.344	.310	3.449	1725	0.87
		10S	.109	1.097	.413	.00656	.344	.287	2.946	1473	1.40
		40ST, 40S	.133	1.049	.494	.00600	.344	.275	2.690	1345	1.68
		80XS, 80S	.179	0.957	.639	.00499	.344	.250	2.240	1120	2.17
		160	.250	0.815	.836	.00362	.344	.213	1.625	812.5	2.84
		XX	.358	0.599	1.076	.00196	.344	.157	0.878	439.0	3.66
1 1/4	1.660	5S	.065	1.530	0.326	.01277	.435	.401	5.73	2865	1.11
		10S	.109	1.442	0.531	.01134	.435	.378	5.09	2545	1.81
		40ST, 40S	.140	1.380	0.668	.01040	.435	.361	4.57	2285	2.27
		80XS, 80S	.191	1.278	0.881	.00891	.435	.335	3.99	1995	3.00
		160	.250	1.160	1.107	.00734	.435	.304	3.29	1645	3.76
		XX	.382	0.896	1.534	.00438	.435	.235	1.97	985	5.21
1 1/2	1.900	5S	.065	1.770	0.375	.01709	.497	.463	7.67	3835	1.28
		10S	.109	1.682	0.614	.01543	.497	.440	6.94	3465	2.09
		40ST, 40S	.145	1.610	0.800	.01414	.497	.421	6.34	3170	2.72
		80XS, 80S	.200	1.500	1.069	.01225	.497	.393	5.49	2745	3.63
		160	.281	1.338	1.429	.00976	.497	.350	4.38	2190	4.86
		XX	.400	1.100	1.885	.00660	.497	.288	2.96	1480	6.41
2	2.375	5S	.065	2.245	0.472	.02749	.622	.588	12.34	6170	1.61
		10S	.109	2.157	0.776	.02538	.622	.565	11.39	5695	2.64
		40ST, 40S	.154	2.067	1.075	.02330	.622	.541	10.45	5225	3.65
		80ST, 80S	.218	1.939	1.477	.02050	.622	.508	9.20	4600	5.02
		160	.344	1.687	2.195	.01552	.622	.436	6.97	3485	7.46
		XX	.436	1.503	2.656	.01232	.622	.393	5.53	2765	9.03
2 1/2	2.875	5S	.083	2.709	0.728	.04003	.753	.709	17.97	8985	2.48
		10S	.120	2.635	1.039	.03787	.753	.690	17.00	8500	3.53
		40ST, 40S	.203	2.469	1.704	.03322	.735	.647	14.92	7460	5.79
		80XS, 80S	.276	2.323	2.254	.02942	.753	.608	13.20	6600	7.66
		160	.375	2.125	2.945	.02463	.753	.556	11.07	5535	10.01
		XX	.552	1.771	4.028	.01711	.753	.464	7.68	3840	13.70
3	3.500	5S	.083	3.334	0.891	.06063	.916	.873	27.21	13,605	3.03
		10S	.120	3.260	1.274	.05796	.916	.853	26.02	13,010	4.33
		40ST, 40S	.216	3.068	2.228	.05130	.916	.803	23.00	11,500	7.58
		80XS, 80S	.300	2.900	3.016	.04587	.916	.759	20.55	10,275	10.25
		160	.438	2.624	4.213	.03755	.916	.687	16.86	8430	14.31
		XX	.600	2.300	5.466	.02885	.916	.602	12.95	6475	18.58
3 1/2	4.0	5S	.083	3.834	1.021	.08017	1.047	1.004	35.98	17,990	3.48
		10S	.120	3.760	1.463	.07711	1.047	0.984	34.61	17,305	4.97
		40ST, 40S	.226	3.548	2.680	.06870	1.047	0.929	30.80	15,400	9.11
		80XS, 80S	.318	3.364	3.678	.06170	1.047	0.881	27.70	13,850	12.51
4	4.5	5S	.083	4.334	1.152	.10245	1.178	1.135	46.0	23,000	3.92
		10S	.120	4.260	1.651	.09898	1.178	1.115	44.4	22,200	5.61
		40ST, 40S	.237	4.026	3.17	.08840	1.178	1.054	39.6	19,800	10.79
		80XS, 80S	.337	3.826	4.41	.07986	1.178	1.002	35.8	17,900	14.98
		120	.438	3.624	5.58	.07170	1.178	0.949	32.2	16,100	19.01
		160	.531	3.438	6.62	.06647	1.178	0.900	28.9	14,450	22.52
		XX	.674	3.152	8.10	.05419	1.178	0.825	24.3	12,150	27.54
5	5.563	5S	.109	5.345	1.87	.1558	1.456	1.399	69.9	34,950	6.36
		10S	.134	5.295	2.29	.1529	1.456	1.386	68.6	34,300	7.77
		40ST, 40S	.258	5.047	4.30	.1390	1.456	1.321	62.3	31,150	14.62
		80XS, 80S	.375	4.813	6.11	.1263	1.456	1.260	57.7	28,850	20.78
		120	.500	4.563	7.95	.1136	1.456	1.195	51.0	25,500	27.04
		160	.625	4.313	9.70	.1015	1.456	1.129	45.5	22,750	32.96
		XX	.750	4.063	11.34	.0900	1.456	1.064	40.4	20,200	38.55

* 5S, 10S, 40S, and 80S refer to stainless-steel pipe; all other numbers refer to wrought-steel pipe.

Table A-5A PROPERTIES OF FERROUS PIPE—METRIC

Nominal pipe size, in	OD, cm	Schedule Number*	Wall thickness, cm	ID, cm	Cross-sectional area: Metal, cm^2	Cross-sectional area: Flow, cm^2	Circumference, cm, or surface, cm^2/cm of length: Outside	Circumference, cm, or surface, cm^2/cm of length: Inside	Capacity at 30.48 cm/s velocity: L/min	Capacity at 30.48 cm/s velocity: kg/h water	Weight of plain-end pipe, kg/m
$\frac{1}{8}$	1.029	10S	0.124	0.780	0.355	0.474	3.231	2.451	0.874	52.4	0.283
		40ST, 40S	0.173	0.683	0.465	0.372	3.231	2.149	0.677	40.6	0.357
		80XS, 80S	0.241	0.546	0.600	0.232	3.231	1.716	0.428	35.7	0.462
$\frac{1}{4}$	1.372	10S	0.165	1.041	0.626	0.855	4.298	3.261	1.559	93.8	0.492
		40ST, 40S	0.224	0.925	0.807	0.669	4.298	2.896	1.222	73.3	0.626
		80XS, 80S	0.302	0.767	1.013	0.465	4.298	2.408	1.848	50.8	0.804
$\frac{3}{8}$	1.715	10S	0.165	1.384	0.807	1.505	5.395	4.359	2.751	165.0	0.626
		40ST, 40S	0.231	1.252	1.077	1.236	5.395	3.932	2.255	135.3	0.849
		80XS, 80S	0.320	1.074	1.400	0.910	5.395	3.383	1.665	99.9	1.102
$\frac{1}{2}$	2.134	5S	0.165	1.803	1.019	2.555	6.706	5.669	4.669	280.1	0.804
		10S	0.211	1.712	1.271	2.304	6.706	5.364	4.208	252.4	0.998
		40ST, 40S	0.277	1.580	1.613	1.960	6.706	4.968	3.576	214.3	1.266
		80XS, 80S	0.373	1.387	2.065	1.514	6.706	4.359	2.762	165.7	1.624
		160	0.478	1.179	2.484	1.087	6.706	3.719	1.994	119.6	1.951
		XX	0.747	0.640	3.252	0.325	6.706	2.012	0.587	35.2	2.547
$\frac{3}{4}$	2.667	5S	0.165	2.337	1.297	4.283	8.382	7.346	7.840	470.3	1.028
		10S	0.211	2.245	1.626	3.958	8.382	7.041	7.201	431.9	1.281
		40ST, 40S	0.287	2.093	2.149	3.447	8.382	6.584	6.300	377.9	1.683
		80XS, 80S	0.391	1.885	2.794	2.787	8.382	5.913	5.089	305.3	2.190
		160	0.556	1.554	3.691	1.895	8.382	4.877	3.470	208.2	2.890
		XX	0.782	1.102	4.633	0.957	8.382	3.475	1.744	104.6	3.634
1	3.340	5S	0.165	3.010	1.645	7.135	10.485	9.449	13.051	783.2	1.296
		10S	0.277	2.786	2.665	6.094	10.485	8.748	11.148	668.5	2.085
		40ST, 40S	0.338	2.664	3.187	5.574	10.485	8.382	10.179	610.6	2.502
		80XS, 80S	0.455	2.431	4.123	4.636	10.485	7.620	8.476	508.5	3.232
		160	0.635	2.070	5.394	3.363	10.485	6.492	6.149	368.9	4.230
		XX	0.909	1.521	6.942	1.821	10.485	4.785	3.322	199.3	5.452
$1\frac{1}{4}$	4.216	5S	0.165	3.886	2.103	11.863	13.259	12.222	21.682	1300.7	1.653
		10S	0.277	3.663	3.426	10.535	13.259	11.521	19.261	1155.4	2.696
		40ST, 40S	0.356	3.505	4.310	9.662	13.259	11.003	17.293	1037.4	3.381
		80XS, 80S	0.485	3.246	5.684	8.277	13.259	10.211	15.098	905.7	4.469
		160	0.635	2.946	7.142	6.819	13.259	9.266	12.449	746.8	5.601
		XX	0.970	2.276	9.897	4.069	13.259	7.163	7.454	477.2	7.760
$1\frac{1}{2}$	4.826	5S	0.165	4.496	2.420	15.877	15.149	14.112	29.023	1741.1	1.907
		10S	0.277	4.272	3.962	14.334	15.149	13.411	26.261	1573.1	3.113
		40ST, 40S	0.368	4.089	5.162	13.136	15.149	12.832	23.991	1439.2	4.051
		80XS, 80S	0.508	3.810	6.897	11.380	15.149	11.979	20.774	1246.2	5.407
		160	0.714	3.399	9.220	9.067	15.149	10.668	16.574	994.3	7.239
		XX	1.016	2.794	12.162	6.131	15.149	8.778	11.201	671.9	9.548
2	6.033	5S	0.165	5.702	3.045	25.538	18.959	17.922	46.695	2801.2	2.398
		10S	0.277	5.479	5.007	23.578	18.959	17.221	42.100	2585.5	3.932
		40ST, 40S	0.391	5.250	6.936	21.646	18.959	16.490	39.543	2372.2	5.437
		80ST, 80S	0.554	4.925	9.530	19.045	18.959	15.484	34.813	2088.4	7.477
		160	0.874	4.285	14.162	14.418	18.959	13.289	26.374	1582.2	11.112
		XX	1.107	3.818	17.137	11.445	18.959	11.979	20.926	1255.3	13.450
$2\frac{1}{2}$	7.303	5S	0.211	6.881	4.697	37.188	22.951	21.610	67.998	4079.2	3.694
		10S	0.305	6.693	6.704	35.181	22.951	21.031	64.328	3859.0	5.258
		40ST, 40S	0.516	6.271	10.994	30.861	22.951	19.721	56.457	3386.8	8.624
		80XS, 80S	0.701	5.900	14.543	27.331	22.951	18.532	49.949	2996.4	11.410
		160	0.953	5.398	19.001	22.881	22.951	16.947	41.889	2512.9	14.910
		XX	1.402	4.498	25.989	15.895	22.951	14.143	29.061	1743.4	20.406
3	8.890	5S	0.211	8.468	5.749	56.325	27.920	26.609	102.963	6176.7	4.513
		10S	0.305	8.280	8.220	53.845	27.920	25.999	98.460	5906.5	6.450
		40ST, 40S	0.549	7.793	14.375	47.658	27.920	24.475	87.032	5221.0	11.290
		80XS, 80S	0.762	7.366	19.459	42.613	27.920	23.134	77.761	4664.9	15.267
		160	1.113	6.665	27.182	34.884	27.920	20.940	63.798	3827.2	21.315
		XX	1.524	5.842	35.267	26.802	27.920	18.349	49.003	2939.7	27.675
$3\frac{1}{2}$	10.160	5S	0.211	9.738	6.587	74.478	31.913	30.602	136.148	8167.5	5.183
		10S	0.305	9.550	9.439	71.635	31.913	29.992	130.964	7856.5	7.403
		40ST, 40S	0.574	9.012	17.291	63.822	31.913	28.316	116.547	6991.6	13.569
		80XS, 80S	0.808	8.545	23.730	57.319	31.913	26.853	104.817	6287.9	18.634
4	11.430	5S	0.211	11.008	7.433	95.176	35.905	34.595	174.064	10442.0	5.839
		10S	0.305	10.820	10.652	91.952	35.905	33.985	168.010	10078.8	8.356
		40ST, 40S	0.602	10.226	20.453	82.124	35.905	32.126	149.846	8989.2	16.072
		80XS, 80S	0.856	9.718	28.453	74.190	35.905	30.541	135.467	8126.6	22.313
		120	1.113	9.205	36.002	66.609	35.905	28.926	121.845	7309.4	28.315
		160	1.349	8.733	42.712	61.751	35.905	27.432	109.358	6560.3	33.544
		XX	1.712	8.006	52.261	50.343	35.905	25.146	91.951	5516.1	41.021
5	14.130	5S	0.277	13.576	12.065	144.738	44.379	42.642	264.502	15867.3	9.473
		10S	0.340	13.449	14.775	142.044	44.379	42.245	259.582	15572.2	11.573
		40ST, 40S	0.655	12.819	27.744	129.131	44.379	40.264	235.743	14142.1	21.777
		80XS, 80S	0.953	12.225	39.422	117.333	44.379	38.405	218.337	13097.9	30.952
		120	1.270	11.590	51.293	105.534	44.379	36.424	192.984	11577.0	40.276
		160	1.588	10.955	62.584	94.294	44.379	34.412	172.172	10328.5	49.094
		XX	1.905	10.320	73.166	83.610	44.379	32.431	152.874	9170.8	57.420

* 5S, 10S, 40S, and 80S refer to stainless-steel pipe; all other numbers refer to wrought-steel pipe.

Table A-6 TUBING CHARACTERISTICS*

OD of tubing	BWG gauge	Thick-ness, in	Internal area, in^2	Ft^2 external surface per ft length	Ft^2 internal surface per ft length	Weight per ft length steel, lb †	ID tubing, in	Moment of inertia, in^4	Section modulus, in^3	Radius of gyration, in	Constant C‡	OD / ID	Metal area (transverse metal area), in^2
1/4	22	0.028	0.0295	0.0655	0.0508	0.066	0.194	0.00012	0.00098	0.0792	46	1.289	0.0195
1/4	24	.022	.0333	.0655	.0539	.054	.206	.00011	.00083	.0810	52	1.214	.0159
1/4	26	.018	.0360	.0655	.0560	.045	.214	.00009	.00071	.0824	56	1.168	.0131
3/8	18	.049	.0603	.0982	.0725	.171	.277	.00068	.0036	.1164	94	1.354	.0502
3/8	20	.035	.0731	.0982	.0798	.127	.305	.00055	.0029	.1213	114	1.233	.0374
3/8	22	.028	.0799	.0982	.0835	.104	.319	.00046	.0025	.1227	125	1.176	.0305
3/8	24	.022	.0860	.0982	.0867	.083	.331	.00038	.0020	.1248	134	1.133	.0244
1/2	16	.065	.1075	.1309	.0969	.302	.370	.0022	.0086	.1556	168	1.351	.0888
1/2	18	.049	.1269	.1309	.1052	.236	.402	.0018	.0072	.1606	198	1.244	.0694
1/2	20	.035	.1452	.1309	.1126	.174	.430	.0014	.0056	.1649	227	1.163	.0511
1/2	22	.028	.1548	.1309	.1162	.141	.444	.0012	.0046	.1671	241	1.126	.0415
5/8	12	.109	.1301	.1636	.1066	6.02	.407	.0061	.0197	.1864	203	1.536	.177
5/8	13	.095	.1486	.1636	.1139	.537	.435	.0057	.0183	.1903	232	1.437	.158
5/8	14	.083	.1655	.1636	.1202	.479	.459	.0053	.0170	.1938	258	1.362	.141
5/8	15	.072	.1817	.1636	.1259	.425	.481	.0049	.0156	.1971	283	1.299	.125
5/8	16	.065	.1924	.1636	.1296	.388	.495	.0045	.0145	.1993	300	1.263	.114
5/8	17	.058	.2035	.1636	.1333	.350	.509	.0042	.0134	.2016	317	1.228	.103
5/8	18	.049	.2181	.1636	.1380	.303	.527	.0037	.0118	.2043	340	1.186	.089
5/8	19	.042	.2298	.1636	.1416	.262	.541	.0033	.0105	.2068	358	1.155	.077
5/8	20	.035	.2419	.1636	.1453	.221	.555	.0028	.0091	.2089	377	1.126	.065
3/4	10	.134	.1825	.1963	.1262	.884	.482	.0129	.0344	.2229	285	1.556	.260
3/4	11	.120	.2043	.1963	.1335	.809	.510	.0122	.0326	.2267	319	1.471	.238
3/4	12	.109	.2223	.1963	.1393	.748	.532	.0116	.0309	.2299	347	1.410	.220
3/4	13	.095	.2463	.1963	.1466	.666	.560	.0107	.0285	.2340	384	1.339	.196
3/4	14	.083	.2679	.1963	.1529	.592	.584	.0098	.0262	.2376	418	1.284	.174
3/4	15	.072	.2884	.1963	.1587	.520	.606	.0089	.0238	.2410	450	1.238	.153
3/4	16	.065	.3019	.1963	.1623	.476	.620	.0083	.0221	.2433	471	1.210	.140
3/4	17	.058	.3157	.1963	.1660	.428	.634	.0076	.0203	.2455	492	1.183	.126
3/4	18	.049	.3339	.1963	.1707	.367	.652	.0067	.0178	.2484	521	1.150	.108
3/4	20	.035	.3632	.1963	.1780	.269	.680	.0050	.0134	.2532	567	1.103	.079
7/8	10	.134	.2892	.2291	.1589	1.061	.607	.0221	.0505	.2662	451	1.441	.312
7/8	11	.120	.3166	.2291	.1662	.969	.635	.0208	.0475	.2703	494	1.378	.285
7/8	12	.109	.3390	.2291	.1720	.891	.657	.0196	.0449	.2736	529	1.332	.262
7/8	13	.095	.3685	.2291	.1793	.792	.685	.0180	.0411	.2778	575	1.277	.233
7/8	14	.083	.3984	.2291	.1856	.704	.709	.0164	.0374	.2815	616	1.234	.207
7/8	16	.065	.4359	.2291	.1950	.561	.745	.0137	.0312	.2873	680	1.174	.165
7/8	18	.049	.4742	.2291	.2034	.432	.777	.0109	.0249	.2925	740	1.126	.127
7/8	20	.035	.5090	.2291	.2107	.313	.805	.0082	.0187	.2972	794	1.087	.092
1	8	.165	.3526	.2618	.1754	1.462	.670	.0392	.0784	.3009	550	1.493	.430
1	10	.134	.4208	.2618	.1916	1.237	.732	.0350	.0700	.3098	656	1.366	.364
1	11	.120	.4536	.2618	.1990	1.129	.760	.0327	.0654	.3140	708	1.316	.332
1	12	.109	.4803	.2618	.2047	1.037	.782	.0307	.0615	.3174	749	1.279	.305
1	13	.095	.5153	.2618	.2121	.918	.810	.0280	.0559	.3217	804	1.235	.270
1	14	.083	.5463	.2618	.2183	.813	.834	.0253	.0507	.3255	852	1.199	.239
1	15	.072	.5755	.2818	.2241	.714	.856	.0227	.0455	.3291	898	1.167	.210
1	16	.065	.5945	.2618	.2278	.649	.870	.0210	.0419	.3314	927	1.149	.191
1	18	.049	.6390	.2618	.2361	.496	.902	.0166	.0332	.3366	997	1.109	.146
1	20	.035	.6793	.2618	.2435	.360	.930	.0124	.0247	.3414	1060	1.075	.106
1 1/4	7	.180	.6221	.3272	.2330	2.057	.890	.0890	.1425	.3836	970	1.404	.605
1 1/4	8	.165	.6648	.3272	.2409	1.921	.920	.0847	.1355	.3880	1037	1.359	.565
1 1/4	10	.134	.7574	.3272	.2571	1.598	.982	.0741	.1186	.3974	1182	1.273	.470
1 1/4	11	.120	.8012	.3272	.2644	1.448	1.010	.0688	.1100	.4018	1250	1.238	.426
1 1/4	12	.109	.8365	.3272	.2702	1.329	1.032	.0642	.1027	.4052	1305	1.211	.391
1 1/4	13	.095	.8825	.3272	.2775	1.173	1.060	.0579	.0926	.4097	1377	1.179	.345
1 1/4	14	.083	.9229	.3272	.2838	1.033	1.084	.0521	.0833	.4136	1440	1.153	.304
1 1/4	16	.065	.9852	.3272	.2932	.823	1.120	.0426	.0682	.4196	1537	1.116	.242
1 1/4	18	.049	1.042	.3272	.3016	.629	1.152	.0334	.0534	.4250	1626	1.085	.185
1 1/4	20	.035	1.094	.3272	.3089	.456	1.180	.0247	.0395	.4297	1707	1.059	.134
1 1/2	10	.134	1.192	.3927	.3225	1.955	1.232	.1354	.1806	.4853	1860	1.218	.575
1 1/2	12	.109	1.291	.3927	.3356	1.618	1.282	.1159	.1546	.4933	2014	1.170	.476
1 1/2	14	.083	1.398	.3927	.3492	1.258	1.334	.0931	.1241	.5018	2181	1.124	.370
1 1/2	16	.065	1.474	.3927	.3587	.996	1.370	.0756	.1008	.5079	2299	1.095	.293
2	11	.120	2.433	.5236	.4608	2.410	1.760	.3144	.3144	.6660	3795	1.136	.709
2	13	.095	2.573	.5236	.4739	1.934	1.810	.2586	.2586	.6744	4014	1.105	.569
2 1/2	9	.148	3.815	.6540	.5770	3.719	2.204	.7592	.6074	.8332	5951	1.134	1.094

* Standards of Tubular Exchanger Manufacturers Association, 4th ed., 1960.

† Weights are based on low-carbon steel with a density of 0.2833 lb/in^3. For other metals multiply by the following factors:

Aluminum	0.35	Nickel-chrome-iron	1.07
A.I.S.I. 400 series stainless steels	0.99	Admiralty	1.09
A.I.S.I. 300 series stainless steels	1.02	Nickel and nickel-copper	1.13
Aluminum bronze	1.04	Copper and cupronickels	1.14
Aluminum brass	1.06		

‡ Liquid velocity $= \dfrac{\text{lb per tube per h}}{C \times \text{specific gravity of liquid}}$ in ft/s (specific gravity of water at 60°F = 1.0).

Table A-6A Tubing Characteristics—Metric

OD of tubing	BWG gauge	Thickness, cm	Internal area cm^2	cm^2 external surface per cm length	cm^2 internal surface per cm length	Weight per m length steel, kg	ID tubing, cm	Moment of inertia, cm^4	Section modulus, cm^3	Radius of gyration, cm	OD/ID	Metal area (transverse metal area), cm^2
$\frac{1}{4}$	22	0.071	0.190	1.996	1.548	0.098	0.493	0.0050	0.0161	0.201	1.289	0.126
$\frac{1}{4}$	24	0.056	0.215	1.996	1.643	0.080	0.523	0.0046	0.0136	0.206	1.214	0.103
$\frac{1}{4}$	26	0.046	0.232	1.996	1.707	0.067	0.544	0.0038	0.0116	0.209	1.168	0.085
$\frac{3}{8}$	18	0.124	0.389	2.993	2.210	0.255	0.704	0.0283	0.0590	0.296	1.354	0.324
$\frac{3}{8}$	20	0.089	0.472	2.993	2.432	0.189	0.775	0.0229	0.0475	0.308	1.233	0.241
$\frac{3}{8}$	22	0.071	0.516	2.993	2.545	0.155	0.810	0.0192	0.0410	0.312	1.176	0.197
$\frac{3}{8}$	24	0.056	0.555	2.993	2.643	0.124	0.841	0.0159	0.0328	0.317	1.133	0.157
$\frac{1}{2}$	16	0.165	0.694	3.990	2.954	0.450	0.940	0.0916	0.1409	0.395	1.351	0.573
$\frac{1}{2}$	18	0.124	0.819	3.990	3.206	0.352	1.021	0.0749	0.1180	0.408	1.244	0.448
$\frac{1}{2}$	20	0.089	0.937	3.990	3.432	0.259	1.092	0.0583	0.0918	0.419	1.163	0.330
$\frac{1}{2}$	22	0.071	0.999	3.990	3.542	0.210	1.128	0.0500	0.0754	0.424	1.126	0.268
$\frac{5}{8}$	12	0.277	0.839	4.987	3.249	0.897	1.034	0.2539	0.3228	0.473	1.536	1.142
$\frac{5}{8}$	13	0.241	0.959	4.987	3.472	0.800	1.105	0.2373	0.2999	0.483	1.437	1.019
$\frac{5}{8}$	14	0.211	1.068	4.987	3.664	0.713	1.166	0.2206	0.2785	0.492	1.362	0.910
$\frac{5}{8}$	15	0.183	1.172	4.987	3.837	0.633	1.222	0.2040	0.2556	0.501	1.299	0.807
$\frac{5}{8}$	16	0.165	1.241	4.987	3.950	0.578	1.257	0.1873	0.2376	0.506	1.263	0.736
$\frac{5}{8}$	17	0.147	1.313	4.987	4.063	0.521	1.293	0.1748	0.2196	0.512	1.228	0.665
$\frac{5}{8}$	18	0.124	1.407	4.987	4.206	0.451	1.339	0.1540	0.1934	0.519	1.186	0.574
$\frac{5}{8}$	19	0.107	1.483	4.987	4.316	0.390	1.374	0.1374	0.1721	0.525	1.155	0.497
$\frac{5}{8}$	20	0.089	1.561	4.987	4.429	0.329	1.410	0.1166	0.1491	0.531	1.126	0.419
$\frac{3}{4}$	10	0.340	1.177	5.983	3.847	1.317	1.224	0.5370	0.5637	0.566	1.556	1.678
$\frac{3}{4}$	11	0.305	1.318	5.983	4.069	1.205	1.295	0.5079	0.5342	0.576	1.471	1.536
$\frac{3}{4}$	12	0.277	1.434	5.983	4.246	1.114	1.351	0.4829	0.5064	0.584	1.410	1.419
$\frac{3}{4}$	13	0.241	1.589	5.983	4.468	0.992	1.422	0.4454	0.4670	0.594	1.339	1.265
$\frac{3}{4}$	14	0.211	1.728	5.983	4.660	0.882	1.483	0.4080	0.4293	0.604	1.284	1.123
$\frac{3}{4}$	15	0.183	1.861	5.983	4.837	0.775	1.539	0.3705	0.3900	0.612	1.238	0.987
$\frac{3}{4}$	16	0.165	1.948	5.983	4.947	0.709	1.575	0.3455	0.3622	0.618	1.210	0.903
$\frac{3}{4}$	17	0.147	2.037	5.983	5.060	0.638	1.610	0.3164	0.3327	0.624	1.183	0.813
$\frac{3}{4}$	18	0.124	2.154	5.983	5.203	0.547	1.656	0.2789	0.2917	0.631	1.150	0.697
$\frac{3}{4}$	20	0.089	2.343	5.983	5.425	0.401	1.727	0.2081	0.2196	0.643	1.103	0.510
$\frac{7}{8}$	10	0.340	1.866	6.983	4.843	1.580	1.542	0.9200	0.8275	0.676	1.441	2.013
$\frac{7}{8}$	11	0.305	2.043	6.983	5.066	1.443	1.613	0.8659	0.7784	0.687	1.378	1.839
$\frac{7}{8}$	12	0.277	2.187	6.983	5.243	1.327	1.669	0.8159	0.7358	0.695	1.332	1.690
$\frac{7}{8}$	13	0.241	2.378	6.983	5.465	1.180	1.740	0.7493	0.6735	0.706	1.277	1.503
$\frac{7}{8}$	14	0.211	2.570	6.983	5.657	1.049	1.801	0.6827	0.6129	0.715	1.234	1.336
$\frac{7}{8}$	16	0.165	2.812	6.983	5.944	0.836	1.892	0.5703	0.5113	0.730	1.174	1.065
$\frac{7}{8}$	18	0.124	3.060	6.983	6.200	0.643	1.974	0.4537	0.4080	0.743	1.126	0.819
$\frac{7}{8}$	20	0.089	3.284	6.983	6.422	0.466	2.045	0.3413	0.3064	0.755	1.087	0.594
1	8	0.419	2.275	7.980	5.346	2.178	1.702	1.6318	1.2847	0.764	1.493	2.774
1	10	0.340	2.715	7.980	5.840	1.843	1.859	1.4570	1.1471	0.787	1.366	2.349
1	11	0.305	2.927	7.980	6.066	1.682	1.930	1.3612	1.0717	0.798	1.316	2.142
1	12	0.277	3.099	7.980	6.239	1.545	1.986	1.2780	1.0078	0.806	1.279	1.968
1	13	0.241	3.325	7.980	6.465	1.367	2.057	1.1656	0.9160	0.817	1.235	1.742
1	14	0.211	3.525	7.980	6.654	1.211	2.118	1.0532	0.8308	0.827	1.199	1.542
1	15	0.183	3.713	7.980	6.831	1.064	2.174	0.9450	0.7456	0.836	1.167	1.355
1	16	0.165	3.836	7.980	6.943	0.967	2.210	0.8742	0.6866	0.842	1.149	1.232
1	18	0.124	4.123	7.980	7.196	0.736	2.291	0.6909	0.5440	0.855	1.109	0.942
1	20	0.089	4.383	7.980	7.422	0.536	2.362	0.5162	0.4048	0.867	1.075	0.684
$1\frac{1}{4}$	7	0.457	4.014	9.973	7.102	3.064	2.261	3.7049	2.3351	0.974	1.404	3.903
$1\frac{1}{4}$	8	0.419	4.289	9.973	7.343	2.861	2.337	3.5259	2.2204	0.986	1.359	3.645
$1\frac{1}{4}$	10	0.340	4.887	9.973	7.836	2.380	2.494	3.0846	1.9435	1.009	1.273	3.032
$1\frac{1}{4}$	11	0.305	5.169	9.973	8.059	2.157	2.565	2.8640	1.8026	1.021	1.238	2.749
$1\frac{1}{4}$	12	0.277	5.397	9.973	8.236	1.980	2.621	2.6725	1.6829	1.029	1.211	2.523
$1\frac{1}{4}$	13	0.241	5.694	9.973	8.458	1.747	2.692	2.4103	1.5174	1.041	1.179	2.226
$1\frac{1}{4}$	14	0.211	5.955	9.973	8.650	1.539	2.753	2.1688	1.3650	1.051	1.153	1.961
$1\frac{1}{4}$	16	0.165	6.357	9.973	8.937	1.226	2.845	1.7734	1.1176	1.066	1.116	1.561
$1\frac{1}{4}$	18	0.124	6.723	9.973	9.193	0.937	2.926	1.3904	0.8751	1.080	1.085	1.194
$1\frac{1}{4}$	20	0.089	7.058	9.973	9.415	0.679	2.997	1.0282	0.6473	1.091	1.059	0.865
$1\frac{1}{2}$	10	0.340	7.691	11.969	9.830	2.912	3.129	5.6364	2.9595	1.233	1.218	3.710
$1\frac{1}{2}$	12	0.277	8.330	11.969	10.229	2.410	3.256	4.8247	2.5334	1.253	1.170	3.071
$1\frac{1}{2}$	14	0.211	9.020	11.969	10.644	1.874	3.388	3.8756	2.0336	1.275	1.124	2.387
$1\frac{1}{2}$	16	0.165	9.510	11.969	10.933	1.484	3.480	3.1471	1.6518	1.290	1.095	1.890
2	11	0.305	15.698	15.959	14.045	3.590	4.470	13.0878	5.1521	1.692	1.136	4.574
2	13	0.241	16.601	15.959	14.444	2.881	4.597	10.7650	4.2377	1.713	1.105	3.671
$2\frac{1}{2}$	9	0.376	24.614	19.934	17.587	5.539	5.598	31.6040	9.9535	2.116	1.134	7.058

Table A-7 APPROXIMATE SPECIFIC GRAVITIES AND DENSITIES OF MISCELLANEOUS SOLIDS AND LIQUIDS* (water at 4°C and normal atmospheric pressure taken as unity)

Substance	Specific gravity	Aver. weight lb/ft³
Metals, Alloys, Ores		
Aluminum cast-hammered	2.55–2.80	165
bronze	7.7	481
Brass, cast-rolled	8.4–8.7	534
Bronze, 7.9 to 14% Sn	7.4–8.9	509
phosphor	8.88	554
Copper, cast-rolled	8.8–8.95	556
ore, pyrites	4.1–4.3	262
German silver	8.58	536
Gold, cast-hammered	19.25–19.35	1205
coin (U.S.)	17.18–17.2	1073
Iridium	21.78–22.42	1383
Iron, gray cast	7.03–7.13	442
cast, pig	7.2	450
wrought	7.6–7.9	485
spiegeleisen	7.5	468
ferro-silicon	6.7–7.3	437
ore, hematite	5.2	325
ore, limonite	3.6–4.0	237
ore, magnetite	4.9–5.2	315
slag	2.5–3.0	172
Lead	11.34	710
ore, galena	7.3–7.6	465
Manganese	7.42	475
ore, pyrolusite	3.7–4.6	259
Mercury	13.6	849
Monel metal, rolled	8.97	555
Nickel	8.9	537
Platinum, cast-hammered	21.5	1330
Silver, cast-hammered	10.4–10.6	656
Steel, cold-drawn	7.83	489
machine	7.80	487
tool	7.70–7.73	481
Tin, cast-hammered	7.2–7.5	459
cassiterite	6.4–7.0	418
Tungsten	19.22	1200
Zinc, cast-rolled	6.9–7.2	440
blende	3.9–4.2	253
Various Solids		
Cereals, oats, bulk	0.51	26
barley, bulk	0.62	39
corn, rye, bulk	0.73	45
wheat, bulk	0.77	48
Cork	0.22–0.26	15
Cotton, flax, hemp	1.47–1.50	93
Fats	0.90–0.97	58
Flour, loose	0.40–0.50	28
pressed	0.70–0.80	47
Glass, common	2.40–2.80	162
plate or crown	2.45–2.72	161
crystal	2.90–3.00	184
flint	3.2–4.7	247
Hay and straw, bales	0.32	20
Leather	0.86–1.02	59
Paper	0.70–1.15	58
Potatoes, piled	0.67	44
Rubber, caoutchouc	0.92–0.96	59
goods	1.0–2.0	94
Salt, granulated, piled	0.77	48
Saltpeter	1.07	67
Starch	1.53	96
Sulfur	1.93–2.07	125
Wool	1.32	82
Timber, Air-dry		
Apple	0.66–0.74	44
Ash, black	0.55	34
white	0.64–0.71	42
Birch, sweet, yellow	0.71–0.72	44
Cedar, white, red	0.35	22
Cherry, wild red	0.43	27
Chestnut	0.48	30
Cypress	0.45–0.48	29
Elm, white	0.56	35
Fir, Douglas	0.48–0.55	32
balsam	0.40	25
Hemlock	0.45–0.50	29
Hickory	0.74–0.80	48
Locust	0.67–0.77	45
Mahogany	0.56–0.85	44
Maple, sugar	0.68	43
white	0.53	33
Oak, chestnut	0.74	46
live	0.87	54
red, black	0.64–0.71	42
white	0.77	48
Pine, Norway	0.55	34
Oregon	0.51	32
red	0.48	30
Southern	0.61–0.67	38–42
white	0.43	27
Poplar	0.43	27
Redwood, California	0.42	26
Spruce, white, red	0.45	28
Teak, African	0.99	62
Indian	0.66–0.88	48
Walnut, black	0.59	37
Willow	0.42–0.50	28
Various Liquids		
Alcohol, ethyl (100%)	0.789	49
methyl (100%)	0.796	50
Acid, muriatic, 40%	1.20	75
nitric, 91%	1.50	94
sulfuric, 87%	1.80	112
Chloroform	1.500	95
Ether	0.736	46
Lye, soda, 66%	1.70	106
Oils, vegetable	0.91–0.94	58
mineral, lubricants	0.88–0.94	57
Turpentine	0.861–0.867	54
Water, 4°C max. density	1.0	62.428
100°C	0.9584	59.830
ice	0.88–0.92	56
snow, fresh fallen	0.125	8
sea water	1.02–1.03	64
Ashlar Masonry		
Bluestone	2.3–2.6	153
Granite, syenite, gneiss	2.4–2.7	159
Limestone	2.1–2.8	153
Marble	2.4–2.8	162
Sandstone	2.0–2.6	143
Rubble Masonry		
Bluestone	2.2–2.5	147
Granite, syenite, gneiss	2.3–2.6	153
Limestone	2.0–2.7	147
Marble	2.3–2.7	156
Sandstone	1.9–2.5	137
Dry Rubble Masonry		
Granite, syenite, gneiss	1.9–2.3	130
Limestone, marble	1.9–2.1	125
Sandstone, bluestone	1.8–1.9	110
Brick Masonry		
Hard brick	1.8–2.3	128
Medium brick	1.6–2.0	112
Soft brick	1.4–1.9	103
Sand-lime brick	1.4–2.2	112
Concrete Masonry		
Cement, stone, sand	2.2–2.4	144
slag, etc.	1.9–2.3	130
cinder, etc.	1.5–1.7	100
Various Building Materials		
Ashes, cinders	0.64–0.72	40–45
Cement, Portland, loose	1.5	94
Lime-gypsum, loose	0.85–1.00	53–64
Mortar, lime, set	1.4–1.9	103
Portland cement	2.08–2.25	94–135
Portland cement	3.1–3.2	196
Slags, bank slag	1.1–1.2	67–72
bank screenings	1.5–1.9	98–17
machine slag	1.5	96
slag sand	0.8–0.9	49–55
Earth, etc., Excavated		
Clay, dry	1.0	63
damp plastic	1.76	110
and gravel, dry	1.6	100
Earth, dry, loose	1.2	76
dry, packed	1.5	95
moist, loose	1.3	78
moist, packed	1.6	96
mud, flowing	1.7	108
mud, packed	1.8	115
Riprap, limestone	1.3–1.4	80–85
Riprap, sandstone	1.4	90
Riprap, shale	1.7	105
Sand, gravel, dry, loose	1.4–1.7	90–105
gravel, dry, packed	1.6–1.9	100–120
gravel, wet	1.89–2.16	126
Excavations in Water		
Clay	1.28	80
River mud	1.44	90
Sand or gravel	0.96	60
and clay	1.00	65
Soil	1.12	70
Stone riprap	1.00	65
Minerals		
Asbestos	2.1–2.8	153
Barytes	4.50	281
Basalt	2.7–3.2	184
Bauxite	2.55	159
Bluestone	2.5–2.6	159
Borax	1.7–1.8	109
Chalk	1.8–2.8	143
Clay, marl	1.8–2.6	137
Dolomite	2.9	181
Feldspar, orthoclase	2.5–2.7	162
Gneiss	2.7–2.9	175
Granite	2.6–2.7	165
Greenstone, trap	2.8–3.2	187
Gypsum, alabaster	2.3–2.8	159
Hornblende	3.0	187
Limestone	2.1–2.86	155
Marble	2.6–2.86	170
Magnesite	3.0	187
Phosphate rock, apatite	3.2	200
Porphyry	2.6–2.9	172

* From Marks, "Mechanical Engineers' Handbook," McGraw-Hill, New York, 1941.

Note: For SI units, multiply density in lb/ft³ by 16.02 to get kg/m³.

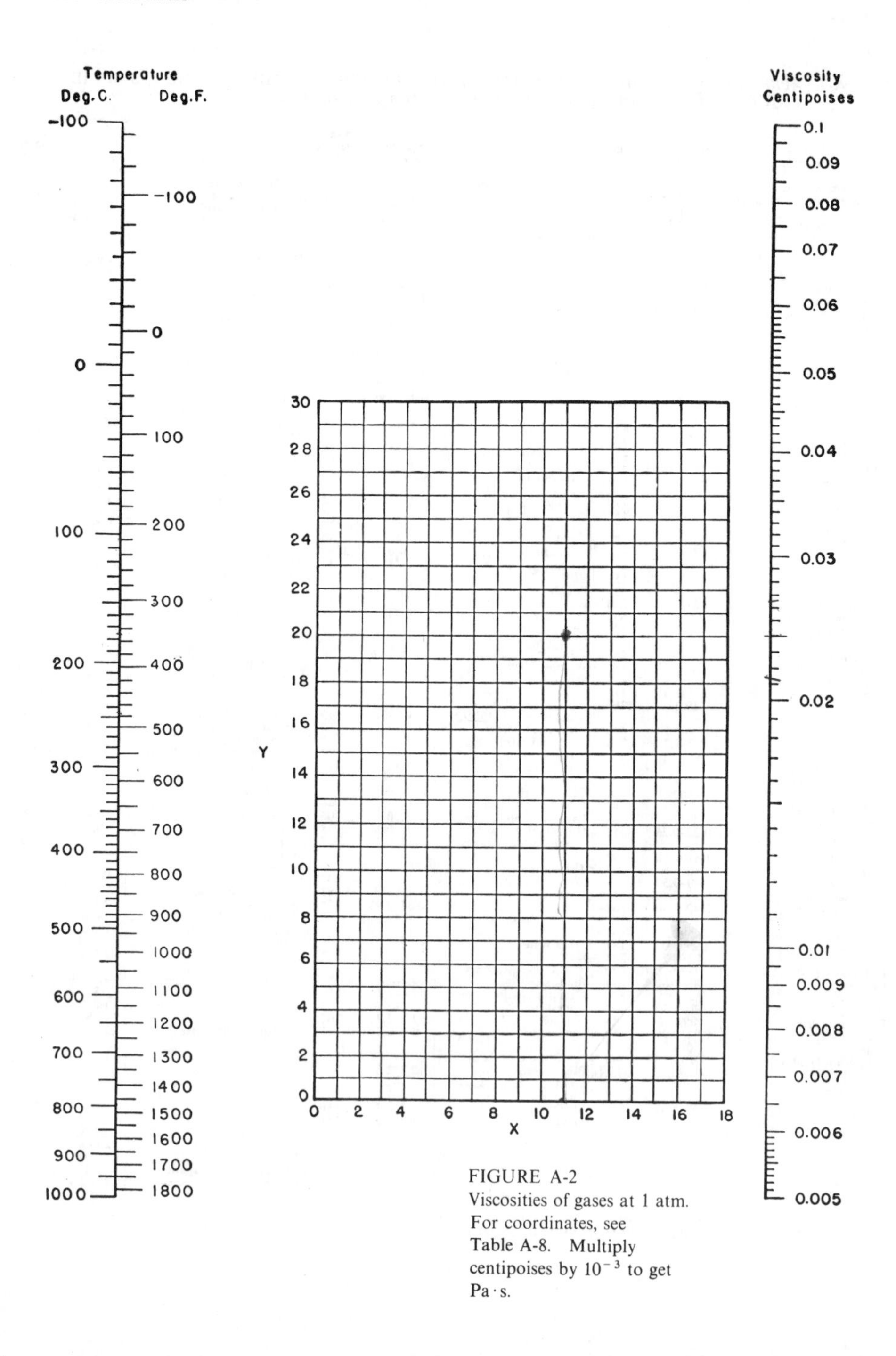

FIGURE A-2
Viscosities of gases at 1 atm. For coordinates, see Table A-8. Multiply centipoises by 10^{-3} to get Pa · s.

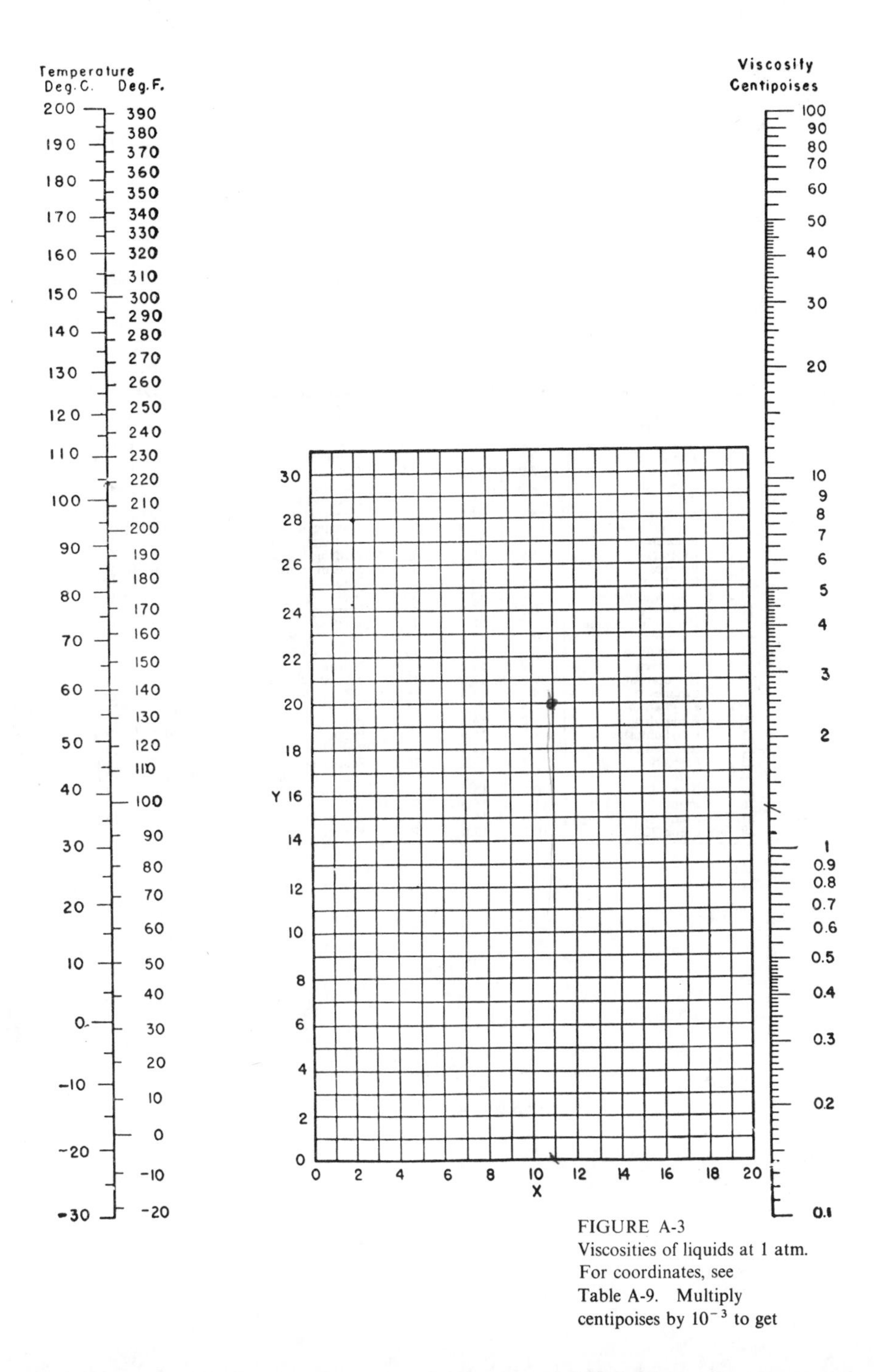

FIGURE A-3
Viscosities of liquids at 1 atm. For coordinates, see Table A-9. Multiply centipoises by 10^{-3} to get

Table A-8 VISCOSITIES OF GASES (coordinates for use with Fig. A-2)

No.	Gas	*X*	*Y*	No.	Gas	*X*	*Y*
1	Acetic acid	7.7	14.3	29	Freon-113	11.3	14.0
2	Acetone	8.9	13.0	30	Helium	10.9	20.5
3	Acetylene	9.8	14.9	31	Hexane	8.6	11.8
4	Air	11.0	20.0	32	Hydrogen	11.2	12.4
5	Ammonia	8.4	16.0	33	$3H_2 + 1N_2$	11.2	17.2
6	Argon	10.5	22.4	34	Hydrogen bromide	8.8	20.9
7	Benzene	8.5	13.2	35	Hydrogen chloride	8.8	18.7
8	Bromine	8.9	19.2	36	Hydrogen cyanide	9.8	14.9
9	Butene	9.2	13.7	37	Hydrogen iodide	9.0	21.3
10	Butylene	8.9	13.0	38	Hydrogen sulfide	8.6	18.0
11	Carbon dioxide	9.5	18.7	39	Iodine	9.0	18.4
12	Carbon disulfide	8.0	16.0	40	Mercury	5.3	22.9
13	Carbon monoxide	11.0	20.0	41	Methane	9.9	15.5
14	Chlorine	9.0	18.4	42	Methyl alcohol	8.5	15.6
15	Chloroform	8.9	15.7	43	Nitric oxide	10.9	20.5
16	Cyanogen	9.2	15.2	44	Nitrogen	10.6	20.0
17	Cyclohexane	9.2	12.0	45	Nitrosyl chloride	8.0	17.6
18	Ethane	9.1	14.5	46	Nitrous oxide	8.8	19.0
19	Ethyl acetate	8.5	13.2	47	Oxygen	11.0	21.3
20	Ethyl alcohol	9.2	14.2	48	Pentane	7.0	12.8
21	Ethyl chloride	8.5	15.6	49	Propane	9.7	12.9
22	Ethyl ether	8.9	13.0	50	Propyl alcohol	8.4	13.4
23	Ethylene	9.5	15.1	51	Propylene	9.0	13.8
24	Fluorine	7.3	23.8	52	Sulfur dioxide	9.6	17.0
25	Freon-11	10.6	15.1	53	Toluene	8.6	12.4
26	Freon-12	11.1	16.0	54	2, 3, 3-Trimethylbutane	9.5	10.5
27	Freon-21	10.8	15.3	55	Water	8.0	16.0
28	Freon-22	10.1	17.0	56	Xenon	9.3	23.0

Table A-9 VISCOSITIES OF LIQUIDS (coordinates for Fig. A-3)

Liquid	X	Y	Liquid	X	Y
Acetaldehyde	15.2	14.8	Freon-113	12.5	11.4
Acetic acid, 100%	12.1	14.2	Glycerol, 100%	2.0	30.0
Acetic acid, 70%	9.5	17.0	Glycerol, 50%	6.9	19.6
Acetic anhydride	12.7	12.8	Heptane	14.1	8.4
Acetone, 100%	14.5	7.2	Hexane	14.7	7.0
Acetone, 35%	7.9	15.0	Hydrochloric acid, 31.5%	13.0	6.6
Acetonitrile	14.4	7.4	Iodobenzene	12.8	15.9
Acrylic acid	12.3	13.9	Isobutyl alcohol	7.1	18.0
Allyl alcohol	10.2	14.3	Isobutyric acid	12.2	14.4
Allyl bromide	14.4	9.6	Isopropyl alcohol	8.2	16.0
Allyl iodide	14.0	11.7	Isopropyl bromide	14.1	9.2
Ammonia, 100%	12.6	2.0	Isopropyl chloride	13.9	7.1
Ammonia, 26%	10.1	13.9	Isopropyl iodide	13.7	11.2
Amyl acetate	11.8	12.5	Kerosene	10.2	16.9
Amyl alcohol	7.5	18.4	Linseed oil, raw	7.5	27.2
Aniline	8.1	18.7	Mercury	18.4	16.4
Anisole	12.3	13.5	Methanol, 100%	12.4	10.5
Arsenic trichloride	13.9	14.5	Methanol, 90%	12.3	11.8
Benzene	12.5	10.9	Methanol, 40%	7.8	15.5
Brine, $CaCL_2$, 25%	6.6	15.9	Methyl acetate	14.2	8.2
Brine, NaCl, 25%	10.2	16.6	Methyl acrylate	13.0	9.5
Bromine	14.2	13.2	Methyl *i*-butyrate	12.3	9.7
Bromotoluene	20.0	15.9	Methyl *n*-butyrate	13.2	10.3
Butyl acetate	12.3	11.0	Methyl chloride	15.0	3.8
Butyl acrylate	11.5	12.6	Methyl ethyl ketone	13.9	8.6
Butyl alcohol	8.6	17.2	Methyl formate	14.2	7.5
Butyric acid	12.1	15.3	Methyl iodide	14.3	9.3
Carbon dioxide	11.6	0.3	Methyl propionate	13.5	9.0
Carbon disulfide	16.1	7.5	Methyl propyl ketone	14.3	9.5
Carbon tetrachloride	12.7	13.1	Methyl sulfide	15.3	6.4
Chlorobenzene	12.3	12.4	Naphthalene	7.9	18.1
Chloroform	14.4	10.2	Nitric acid, 95%	12.8	13.8
Chlorosulfonic acid	11.2	18.1	Nitric acid, 60%	10.8	17.0
Chlorotoluene, ortho	13.0	13.3	Nitrobenzene	10.6	16.2
Chlorotoluene, meta	13.3	12.5	Nitrogen dioxide	12.9	8.6
Chlorotoluene, para	13.3	12.5	Nitrotoluene	11.0	17.0
Cresol, meta	2.5	20.8	Octane	13.7	10.0
Cyclohexanol	2.9	24.3	Octyl alcohol	6.6	21.1
Cyclohexane	9.8	12.9	Pentachloroethane	10.9	17.3
Dibromomethane	12.7	15.8	Pentane	14.9	5.2
Dichloroethane	13.2	12.2	Phenol	6.9	20.8
Dichloromethane	14.6	8.9	Phosphorus tribromide	13.8	16.7
Diethyl ketone	13.5	9.2	Phosphorus trichloride	16.2	10.9
Diethyl oxalate	11.0	16.4	Propionic acid	12.8	13.8
Diethylene glycol	5.0	24.7	Propyl acetate	13.1	10.3
Diphenyl	12.0	18.3	Propyl alcohol	9.1	16.5
Dipropyl ether	13.2	8.6	Propyl bromide	14.5	9.6
Dipropyl oxalate	10.3	17.7	Propyl chloride	14.4	7.5
Ethyl acetate	13.7	9.1	Propyl formate	13.1	9.7
Ethyl acrylate	12.7	10.4	Propyl iodide	14.1	11.6
Ethyl alcohol, 100%	10.5	13.8	Sodium	16.4	13.9
Ethyl alcohol, 95%	9.8	14.3	Sodium hydroxide, 50%	3.2	25.8
Ethyl alcohol, 40%	6.5	16.6	Stannic chloride	13.5	12.8
Ethyl benzene	13.2	11.5	Succinonitrile	10.1	20.8
Ethyl bromide	14.5	8.1	Sulfur dioxide	15.2	7.1
2-Ethyl butyl acrylate	11.2	14.0	Sulfuric acid, 110%	7.2	27.4
Ethyl chloride	14.8	6.0	Sulfuric acid, 100%	8.0	25.1
Ethyl ether	14.5	5.3	Sulfuric acid, 98%	7.0	24.8
Ethyl formate	14.2	8.4	Sulfuric acid, 60%	10.2	21.3
2-Ethyl hexyl acrylate	9.0	15.0	Sulfuryl chloride	15.2	12.4
Ethyl iodide	14.7	10.3	Tetrachloroethane	11.9	15.7
Ethyl propionate	13.2	9.9	Thiophene	13.2	11.0
Ethyl propyl ether	14.0	7.0	Titanium tetrachloride	14.4	12.3
Ethyl sulfide	13.8	8.9	Toluene	13.7	10.4
Ethylene bromide	11.9	15.7	Trichloroethylene	14.8	10.5
Ethylene chloride	12.7	12.2	Triethylene glycol	4.7	24.8
Ethylene glycol	6.0	23.6	Turpentine	11.5	14.9
Ethylidene chloride	14.1	8.7	Vinyl acetate	14.0	8.8
Fluorobenzene	13.7	10.4	Vinyl toluene	13.4	12.0
Formic acid	10.7	15.8	Water	10.2	13.0
Freon-11	14.4	9.0	Xylene, ortho	13.5	12.1
Freon-12	16.8	15.6	Xylene, meta	13.9	10.6
Freon-21	15.7	7.5	Xylene, para	13.9	10.9
Freon-22	17.2	4.7			

Values of Seiner, *Chem. Eng.*, September, 1958, have been included in this listing.

Table A-10 VISCOSITY OF WATER*

(Multiply centipoises by 10^{-3} to get Pa·s)

Temp., °C	Viscosity, centipoises	Temp., °C	Viscosity, centipoises	Temp., °C	Viscosity, centipoises
0	1.7921	33	0.7523	67	0.4233
1	1.7313	34	0.7371	68	0.4174
2	1.6728	35	0.7225	69	0.4117
3	1.6191	36	0.7085	70	0.4061
4	1.5674	37	0.6947	71	0.4006
5	1.5188	38	0.6814	72	0.3952
6	1.4728	39	0.6685	73	0.3900
7	1.4284	40	0.6560	74	0.3849
8	1.3860	41	0.6439	75	0.3799
9	1.3462	42	0.6321	76	0.3750
10	1.3077	43	0.6207	77	0.3702
11	1.2713	44	0.6097	78	0.3655
12	1.2363	45	0.5988	79	0.3610
13	1.2028	46	0.5883	80	0.3565
14	1.1709	47	0.5782	81	0.3521
15	1.1404	48	0.5683	82	0.3478
16	1.1111	49	0.5588	83	0.3436
17	1.0828	50	0.5494	84	0.3395
18	1.0559	51	0.5404	85	0.3355
19	1.0299	52	0.5315	86	0.3315
20	1.0050	53	0.5229	87	0.3276
20.20	*1.0000*	54	0.5146	88	0.3239
21	0.9810	55	0.5064	89	0.3202
22	0.9579	56	0.4985	90	0.3165
23	0.9358	57	0.4907	91	0.3130
24	0.9142	58	0.4832	92	0.3095
25	0.8937	59	0.4759	93	0.3060
26	0.8737	60	0.4688	94	0.3027
27	0.8545	61	0.4618	95	0.2994
28	0.8360	62	0.4550	96	0.2962
29	0.8180	63	0.4483	97	0.2930
30	0.8007	64	0.4418	98	0.2899
31	0.7840	65	0.4355	99	0.2868
32	0.7679	66	0.4293	100	0.2838

* Calculated by the formula:

$$1/\mu = 2.1482[(t - 8.435) + \sqrt{8078.4 + (t - 8.435)^2}] - 120.$$

From Bingham, "Fluidity and Plasticity," p. 340, McGraw-Hill, New York, 1922.

Table A-11 THERMAL CONDUCTIVITIES OF METALS*

(k = Btu/(h)(ft²)(°F/ft) (Multiply by 1.731 to get W/m·K)

Substance	t, °F	k
Metals		
Antimony	32	10.6
Antimony	212	9.7
Bismuth	64	4.7
Bismuth	212	3.9
Cadmium	64	53.7
Cadmium	212	52.2
Gold	64	169.0
Gold	212	170.0
Iron, pure	64	39.0
Iron, pure	212	36.6
Iron, wrought	64	34.9
Iron, wrought	212	34.6
Iron, cast	129	27.6
Iron, cast	216	26.8
Steel (1% C)	64	26.2
Steel (1% C)	212	25.9
Magnesium	32–212	92.0
Mercury	32	4.8
Nickel alloy (62 Ni, 12 Cr, 26 Fe)	68	7.8
Platinum	64	40.2
Platinum	212	41.9
Alloys		
Constantan (60 Cu, 40 Ni)	64	13.1
Constantan (60 Cu, 40 Ni)	212	15.5
Nickel silver	32	16.9
Nickel silver	212	21.5
Manganin (84 Cu, 4 Ni, 12 Mn)	64	12.8
	212	15.2
Platinoid (54 Cu, 25 Ni, 20 Zn)	64	14.5

* Marks, "Mechanical Engineers' Handbook," 4th ed., McGraw-Hill, New York, 1941. For A.I.S.I. 403 steel, Refractalloy 26, S816 alloy, porous 301 stainless steel, X40 alloy, Inconel X, N155 (low C) alloy, Nimonic 80, aluminum alloys 145, 245, 355, brass and silver from about 250° to 1100°F, see Evans, *N.A.C.A. Research Mem.* E50L07, 1951. For 23 solids from 2° to 300°K, see Scott, "Cryogenic Engineering," Van Nostrand, Princeton, N.J., 1959. For brass, nylon, beryllium, copper, soft solder, and Woods metal, 2° to 90°K, see Berman, Foster, et al., *Brit. J. Appl. Phys.*, **6**, 181 (1955). For 41 elements and binary alloys, 1° to 3500°K, see Purdue Univ. TPRC "Data Book," vol. I.

Table A-12 THERMAL CONDUCTIVITIES OF LIQUIDS

$k = \text{Btu/(h)(ft}^2\text{)(°F/ft)}$ (Multiply by 1.731 to get W/m·K)

A linear variation with temperature may be assumed. The extreme values given constitute also the temperature limits over which the data are recommended.

Liquid	t, °F	k
Acetic acid 100%[5]	68	0.099
50%[5]	68	.20
Acetone[4]	86	.102
	167	.095
Allyl alcohol[11]	77–86	.104
Ammonia[8]	5–86	.29
Ammonia, aqueous 26%[6]	68	.261
	140	.29
Amyl acetate[7]	50	.083
alcohol (n-)[6]	86	.094
	212	.089
(iso-)[12]	86	.088
	167	.087
Aniline[9]	32–68	.100
Benzene[12]	86	.092
	140	.087
Bromobenzene[12]	86	.074
	212	.070
Butyl acetate (n-)[11]	77–86	.085
alcohol (n-)[4]	86	.097
	167	.095
(iso)[4]	50	.091
Calcium chloride brine 30%[5]	86	.32
15%[6]	86	.34
Carbon disulfide[4]	86	.093
	167	.088
tetrachloride[10]	32	.107
	154	.094
Chlorobenzene[12]	50	.083
Chloroform[10]	86	.080
Cymene (para-)[12]	86	.078
	140	.079
Decane (n-)[12]	86	.085
	140	.083
Dichlorodifluoromethane[12]	20	.057
	60	.053
	100	.048
	140	.043
	180	.038
Dichloroethane[10]	122	.082
Dichloromethane[10]	5	.111
	86	.096
Ethyl acetate[7]	68	.101
alcohol 100%[2]	68	.105
80%	68	.137
60%	68	.176
40%	68	.224
20%	68	.281
100%[2]	122	.087
benzene[12]	86	.086
	140	.082
bromide[4]	68	.070
ether[4]	86	.080
	167	.078
iodide[4,5]	104	.064
	167	.063
Ethylene glycol[7]	32	.153
Gasoline[6,12]	86	.078
Glycerol 100%[1]	68	.164
80%	68	.189
60%	68	.220
40%	68	.259
20%	68	.278
100%[1]	212	.164
Heptane (n-)[12]	86	.081
	140	.079
Hexane (n-)[12]	86	0.080
	140	.078
Heptyl alcohol (n-)[6]	86	.094
	167	.091
Hexyl alcohol (n-)[6]	86	.093
	167	.090
Kerosene[4]	68	.086
	167	.081
Mercury[7]	82	4.83
Methyl alcohol 100%[2]	68	0.124
80%	68	.154
60%	68	.190
40%	68	.234
20%	68	.284
100%	122	.114
chloride[8,10]	5	.111
	86	.089
Nitrobenzene[12]	86	.095
	212	.088
Nitromethane[12]	86	.125
	140	.120
Nonane (n-)[12]	86	.084
	140	.082
Octane (n-)[12]	86	.083
	140	.081
Oils[5,12*]	86	.079
Oils, castor[9]	68	.104
	212	.100
Oils, olive[9]	68	.097
	212	.095
Paraldehyde[12]	86	.084
	212	.078
Pentane (n-)[12]	86	.078
	167	.074
Perchloroethylene[10]	122	.092
Petroleum ether[4]	86	.075
	167	.073
Propyl alcohol (n-)[5]	86	.099
	167	.095
alcohol (iso-[12])	86	.091
	140	.090
Sodium	212	49
	410	46
Sodium chloride brine 25.0%[5]	86	0.33
12.5%[5]	86	.34
Sulfuric acid 90%[5]	86	.21
60	86	.25
30	86	.30
Sulfur dioxide[8]	5	.128
	86	.111
Toluene[1,12]	86	.086
	167	.084
β-Trichloroethane[10]	122	.077
Trichloroethylene[10]	122	.080
Turpentine[7]	59	.074
Vaseline[7]	59	.106
Water[13]	32	.343
	100	.363
	200	.393
	300	.395
	420	.376
	620	.275
Xylene (ortho-)[7]	68	.090
(meta-)[7]	68	.090

* Thermal conductivity data for a number of oils are available from Reference 12. See also Table 3-276; for many oils an average value of 0.079 may be used.

1 Bates, *Ind. Eng. Chem.*, **28**, 494 (1936).
2 Bates, Hazzard, and Palmer, *Ind. Eng. Chem.*, **10**, 314 (1938).
3 Benning, A. F., private communication, 1940.
4 Bridgman, *Proc. Am. Acad. Arts Sci.*, **59**, 141 (1923).
5 Chilton and Genereaux, personal communication, 1939, based on data selected from the literature.
6 Daniloff, *J. Am. Chem. Soc.*, **54**, 1328 (1932).
7 "International Critical Tables," McGraw-Hill, New York, 1929.
8 Kardos, *Z. Ver. deut. Ing.*,**77**, 1158 (1933); *Z. ges. Kälte-Ind.*, **41**, 1, 29 (1934).
9 Kaye and Higgins, *Proc. Roy. Soc. (London)*, **A117**, 459 (1928).
10 DuPont Chlorinated Hydrocarbons, *Tech. Bull.*, Electrochemicals Dept., du Pont, Buffalo, N.Y., 1938.
11 Shiba, *Sci. Papers Inst. Phys. Chem. Research (Tokyo)*, **16**, 205 (1931).
12 Smith, *Trans. Am. Soc. Mech. Engrs.*, **58**, 719 (1936).
13 Timrot and Vargaftik, *J. Tech. Phys. (U.S.S.R.)*, **10**, 1063 (1940).

Table A-13 THERMAL CONDUCTIVITIES OF GASES AND VAPORS

$k = \mathrm{Btu/(h)(ft^2)(°F/ft)}$ (Multiply by 1.731 to get W/m·K)

The extreme temperature values given constitute the experimental range. For extrapolation to other temperatures, it is suggested that the data given be plotted as log k vs. log T, or that use be made of the asssumption that the ratio $c_p \mu/k$ is practically independent of temperature (or of pressure, within moderate limits).

Substance	t, °F	k
Acetone[11]	32	0.0057
	115	.0074
	212	.0099
	363	.0147
Acetylene[3]	− 103	.0068
	32	.0108
	122	.0140
	212	.0172
Air[1,11]	− 148	.0095
	32	.0140
	212	.0183
	392	.0226
	572	.0265
Ammonia[3]	− 76	.0095
	32	.0128
	122	.0157
	212	.0185
Benzene[11]	32	.0052
	115	.0073
	212	.0103
	363	.0152
	413	.0176
Butane (n-)[8]	32	.0078
	212	.0135
(iso-)[9]	32	.0080
	212	.0139
Carbon dioxide[12]	− 58	.0068
	32	.0085
	212	.0133
	392	.0181
	572	.0228
disulfide[3]	32	.0040
	45	.0042
monoxide[1,3]	− 312	.0041
	− 294	.0046
	32	.0135
tetrachloride[11]	115	.0041
	212	.0052
	363	.0065
Chlorine[8]	32	.0043
Chloroform[11]	32	.0038
	115	.0046
	212	.0058
	363	.0077
Cyclohexane	216	.0095
Dichlorodifluoromethane	32	.0048
	122	.0064
	212	.0080
	302	.0097
Ethane[1,2]	− 94	.0066
	− 29	.0086
	32	.0106
	212	.0175
Ethyl acetate[11]	115	.0072
	212	.0096
	363	.0141
alcohol[11]	68	.0089
	212	.0124
chloride[11]	32	.0055
	212	.0095
	363	.0135
	413	.0152
ether[11]	32	.0077
	115	.0099
	212	.0131
	363	.0189
	413	.0209
Ethylene[3]	− 96	.0064
	32	.0101
	122	.0131
	212	.0161
Heptane (n-)[11]	392	.0112
	212	.0103
Hexane (n-)[9]	32	.0072
	68	.0080

Substance	t, °F	k
Hexene[11]	32	0.0061
	212	.0109
Hydrogen	− 148	.065
	− 58	.083
	32	.100
	122	.115
	212	.129
	572	.178
Hydrogen and carbon dioxide[7]	32	
0% H_2	......	.0083
20%	......	.0165
40%	......	.0270
60%	......	.0410
80%	......	.0620
100%	......	.10
Hydrogen and nitrogen[7]	32	
0% H_2	......	.0133
20%	......	.0212
40%	......	.0313
60%	......	.0438
80%	......	.0635
Hydrogen and nitrous oxide[7]	32	
0% H_2	......	.0092
20%	......	.0170
40%	......	.0270
60%	......	.0410
80%	......	.0650
Hydrogen sulfide[3]	32	.0076
Mercury[8]	392	.0197
Methane[1,3,9]	− 148	.0100
	− 58	.0145
	32	.0175
	122	.0215
Methyl alcohol[11]	32	.0083
	212	.0128
acetate[11]	32	.0059
	68	.0068
Methyl chloride[11]	32	.0053
	115	.0072
	212	.0094
	363	.0130
	413	.0148
Methylene chloride[11]	32	.0039
	115	.0049
	212	.0063
	413	.0095
Nitric oxide[3]	− 94	.0103
	32	.0138
Nitrogen[2,3]	− 148	.0095
	32	.0140
	122	.0160
	212	.0180
Nitrous oxide[2,3]	− 98	.0067
	32	.0087
	212	.0128
Oxygen[1,2,6]	− 148	.0095
	− 58	.0119
	32	.0142
	122	.0164
	212	.0185
Pentane (n-)[9,11]	32	.0074
	68	.0083
(iso-)[11]	32	.0072
	212	.0127
Propane[9]	32	.0087
	212	.0151
Sulfur dioxide[2]	32	.0050
	212	.0069
Water vapor, zero pressure[10,14]*	32	.0132
	200	.0159
	400	.0199
	600	.0256
	800	.0306
	1000	.0495

[1] Chilton and Genereaux, private communication, 1940.
[2] Dickens, *Proc. Roy. Soc.* (*London*), **A143**, 517 (1934).
[3] Eucken, *Physik. Z.*, **12**, 1101 (1911); **14**, 324 (1913) (see footnote 17).
[4] Gregory, *Proc. Roy. Soc.* (*London*), **A149**, 324 (1935).
[5] Gregory and Archer, *Proc. Roy. Soc.* (*London*), **A110**, 119 (1926).
[6] Gregory and Marshall, *Proc. Roy. Soc.* (*London*), **A118**, 594 (1928).
[7] Ibbs and Hirst, *Proc. Roy. Soc.* (*London*), **A123**, 134 (1929).
[8] "International Critical Tables," McGraw-Hill, New York, 1929.
[9] Mann and Dickens, *Proc. Roy. Soc.* (*London*), **A134**, 77 (1931).
[10] Keenan and Keyes, "Thermodynamic Properties of Steam," Wiley, New York, 1944 (tenth impression).
* For saturated vapor (reference 10):

[11] Moser, *Dissertation*, Berlin, 1913 (see footnote 17).
[12] Sherrat and Griffiths, *Phil. Mag.*, **27**, **68** (1939).
[13] Spence and Dock, *Phil. Mag.*, **25**, 129 (1938).
[14] Varhaftik and Timrot, *J. Tech. Phys.* (*U.S.S.R.*), **963** (1939).
[15] Varhaftik and Parquenov, *J. Expt. Theoret. Phys.* (*U.S.S.R.*), **8**, 189 (1938).
[16] Wüllner, *Ann. Physik*, **4**, 321 (1878).
[17] Data from Eucken and Moser are measurements relative to air. Data in this table from these sources are based on the thermal conductivity of air at 32°F of 0.0140 Btu/(h)(ft²)(°F/ft).

psia	250	500	1000	1500	2000
t, °F	401	467	545	596	636
k	0.0748	0.0299	0.0395	0.0486	0.0578

Table A-14 THERMAL CONDUCTIVITIES OF SOME BUILDING AND INSULATING MATERIALS*

$k = \text{Btu/(h)(ft}^2\text{)(°F/ft)}$ (Multiply by 1.731 to get W/m·K)

Material	Apparent density p, lb/ft^3 at room temperature	t, °C	k
Aerogel, silica, opacified	8.5	120	0.013
		290	.026
Asbestos-cement boards	120	20	.43
Asbestos sheets	55.5	51	.096
Asbestos slate	112	0	.087
	112	60	.114
Asbestos	29.3	−200	.043
	29.3	0	.090
	36	0	.087
	36	100	.111
	36	200	.120
	36	400	.129
	43.5	−200	.090
	43.5	0	.135
Aluminum foil (7 air spaces per 2.5 in)	0.2	38	.025
		177	.038
Ashes, wood	....	0–100	.041
Asphalt	132	20	.43
Boiler scale (Note 1)			
Bricks:			
Alumina (92–99% Al_2O_3 by wt) fused	....	427	1.8
Alumina (64–65% Al_2O_3 by wt)	....	1315	2.7
(See also Bricks, fireclay)	115	800	0.62
	115	1100	.63
Building brickwork	....	20	.4
Carbon	96.7	....	3.0
Chrome brick (32% Cr_2O_3 by wt)	200	200	.67
	200	650	.85
	200	1315	1.0
Diatomaceous earth, natural, across strata (Note 2)	27.7	204	0.051
	27.7	871	.077
Diatomaceous, natural, parallel to strata (Note 2)	27.7	204	.081
	27.7	871	.106
Diatomaceous earth, molded and fired (Note 2)	38	204	.14
	38	871	.18
Diatomaceous earth and clay, molded and fired (Note 2)	42.3	204	.14
	42.3	871	.19
Diatomaceous earth, high burn, large pores (Note 3)	37	200	.13
	37	1000	.34
Fire clay (Missouri)	....	200	.58
		600	.85
		1000	.95
		1400	1.02
Kaolin insulating brick (Note 3)	27	500	0.15
	27	1150	.26
Kaolin insulating firebrick (Note 4)	19	200	.050
	19	760	.113
Magnesite (86.8% MgO, 6.3% Fe_2O_3, 3% CaO, 2.6% SiO_2 by wt)	158	204	2.2
	158	650	1.6
	158	1200	1.1
Silicon carbide brick, recrystallized (Note 3)	129	600	10.7
	129	800	9.2
	129	1000	8.0
	129	1200	7.0
	129	1400	6.3
Calcium carbonate, natural	162	30	1.3
White marble	....	....	1.7
Chalk	96	....	0.4
Calcium sulfate ($4H_2O$), artificial	84.6	40	.22
plaster (artificial)	132	75	.43
(building)	77.9	25	.25
Cambric (varnished)	....	38	.091
Carbon, gas	....	0–100	2.0
Carbon stock	94	−184	0.55
		0	3.6
Cardboard, corrugated	....	....	0.037
Celluloid	87.3	30	.12
Charcoal flakes	11.9	80	.043
	15	80	.051
Clinker (granular)	....	0–700	.27
Coke, petroleum	....	100	3.4
		500	2.9
Coke, petroleum (20–100 mesh)	62	400	0.55
Coke (powdered)	....	0–100	.11
Concrete (cinder)	....	....	.20
(stone)	....	....	.54
(1:4 dry)	....	....	.44
Cotton wool	5	30	0.024
Cork board	10	30	.025
Cork (regranulated)	8.1	30	.026
(ground)	9.4	30	.025
Diatomaceous earth powder, coarse (Note 2)	20.0	38	.036
	20.0	871	.082
fine (Note 2)	17.2	204	.040
	17.2	871	.074
molded pipe covering (Note 2)	26.0	204	.051
	26.0	871	.088
4 vol. calcined earth and 1 vol. cement, poured and fired (Note 2)	61.8	204	.16
	61.8	871	.23
Dolomite	167	50	1.0
Ebonite	....	....	0.10
Enamel, silicate	38	....	0.5–0.75
Felt, wool	20.6	30	0.03
Fiber insulating board	14.8	21	.028
Fiber, red	80.5	20	.27
(with binder, baked)	....	20–97	.097
Gas carbon	....	0–100	2.0
Glass	....	....	0.2–0.73
Borosilicate type	139	30–75	0.63
Window glass	....	....	0.3–0.61
Soda glass	....	....	0.3–0.44
Granite	....	....	1.0–2.3
Graphite, longitudinal	....	20	95.
powdered, through 100 mesh	30	40	0.104
Gypsum (molded and dry)	78	20	.25
Hair felt (perpendicular to fibers)	17	30	.021
Ice	57.5	0	1.3
Infusorial earth, see diatomaceous earth			
Kapok	0.88	20	0.020
Lampblack	10	40	.038
Lava	....	....	.49
Leather, sole	62.4	....	.092
Limestone (15.3 vol % H_2O)	103	24	.54
Linen	....	30	.05
Magnesia (powdered)	49.7	47	.35
Magnesia (light carbonate)	13	21	0.034
Magnesium oxide (compressed)	49.9	20	.32
Marble	....	....	1.2–1.7
Mica (perpendicular to planes)	....	50	0.25
Mill shavings	....	....	0.033–0.05
Mineral wool	9.4	30	0.0225
	19.7	30	.024
Paper	....	....	.075
Paraffin wax	....	0	.14
Petroleum coke	....	100	3.4
	....	500	2.9
Porcelain	....	200	.088
Portland cement, see concrete	....	90	.17
Pumice stone	....	21–66	.14
Pyroxylin plastics	....	....	.075
Rubber (hard)	74.8	0	.087
(para)	....	21	.109
(soft)	....	21	0.075–0.092
Sand (dry)	94.6	20	0.19
Sandstone	140	40	1.06
Sawdust	12	21	0.03
Scale (Note 1)			
Silk	6.3	....	.026
varnished	....	38	.096
Slag, blast furnace	....	24–127	.064
Slag wool	12	30	.022
Slate	....	94	.86
Snow	34.7	0	.27
Sulfur (monoclinic)	....	100	0.09–0.097
(rhombic)	....	21	0.16
Wall board, insulating type	14.8	21	.028
Wall board, stiff paste board	43	30	.04
Wood shavings	8.8	30	.034
Wood (across grain):			
Balsa	7–8	30	0.025–0.03
Oak	51.5	15	0.12
Maple	44.7	50	.11
Pine, white	34.0	15	.087
Teak	40.0	15	.10
White fir	28.1	60	.062
Wood (parallel to grain):			
Pine	34.4	21	.20
Wool, animal	6.9	30	.021

* Marks, "Mechanical Engineer's, Handbook," 4th ed., McGraw-Hill, New York, 1941. "International Critical Tables," McGraw-Hill, 1929, and other sources. For additional data, see pp. 458–459.

Note 1: B. Kamp [*Z. tech. Physik*, **12**, 30 (1931)] shows the effect of increased porosity in decreasing thermal conductivity of boiler scale. Partridge [University of Michigan, *Eng. Research Bull.* **15**, 1930] has published a 170-page treatise on Formation and Properties of Boiler Scale.

Note 2: Townshend and Williams, *Chem. & Met.*, **39**, 219 (1932).

Note 3: Norton, "Refractories," 2nd ed., McGraw-Hill, New York, 1942.

Note 4: Norton, private communication.

Table A-15 THERMAL CONDUCTIVITIES OF INSULATING MATERIALS AT HIGH TEMPERATURES*

k = Btu/(h)(ft²)(°F/ft) (Multiply by 1.731 to get W/m·K)

Material	For temperatures, °F up to	Mean temperatures, °F 100	200	300	400	500	600	800	1000	1500	2000
Laminated asbestos felt (approx. 40 laminations per in)	700	0.033	0.037	0.040	0.044	0.048					
Laminated asbestos felt (approx. 20 laminations per in)	500	.045	.050	.055	.060	.065					
Corrugated asbestos (4 plies per in)	300	.050	.058	.069							
85% magnesia (density, 13 lb/ft³)	600	.034	.036	.038	.040						
Diatomaceous earth, asbestos and bonding material	1600	.045	.047	.049	.050	.053	0.055	0.060	0.065		
Diatomaceous earth brick	1600	.054	.056	.058	.060	.063	.065	.069	.073		
Diatomaceous earth brick	2000	.127	.130	.133	.137	.140	.143	.150	.158	0.176	
Diatomaceous earth brick	2500	.128	.131	.135	.139	.143	.148	.155	.163	.183	0.203
Diatomaceous earth powder (density, 18 lb/ft³)	....	.039	.042	.044	.048	.051	.054	.061	.068		
Rock wool	....	.030	.034	.039	.044	.050	.057				

Asbestos cement, 1.2; 85% magnesia cement, 0.05; asbestos and rock wool cement, 0.075 approx.
* Marks, "Mechanical Engineers' Handbook," 4th ed., McGraw-Hill, New York, 1941.

Table A-16 SATURATED STEAM TEMPERATURE TABLE*

Temp., °F t	Abs. pressure, lb/in² p	Volume, ft³/lb Liquid v_f	Volume, ft³/lb Vapor v_g	Enthalpy, Btu/lb Liquid h_f	Enthalpy, Btu/lb Vapor h_g	Entropy, Btu/(lb)(°R) Liquid s_f	Entropy, Btu/(lb)(°R) Vapor s_g
32	0.08854	0.01602	3306	0.00	1075.8	0.0000	2.1877
35	.09995	.01602	2947	3.02	1077.1	.0061	2.1770
40	.12170	.01602	2444	8.05	1079.3	.0162	2.1597
45	.14752	.01602	2036.4	13.06	1081.5	.0262	2.1429
50	.17811	.01603	1703.2	18.07	1083.7	.0361	2.1264
60	.2563	.01604	1206.7	28.06	1088.0	.0555	2.0948
70	.3631	.01606	867.9	38.04	1092.3	.0745	2.0647
80	.5069	.01608	633.1	48.02	1096.6	.0932	2.0360
90	.6982	.01610	468.0	57.99	1100.9	.1115	2.0087
100	.9492	.01613	350.4	67.97	1105.2	.1295	1.9826
110	1.2748	.01617	265.4	77.94	1109.5	.1471	1.9577
120	1.6924	.01620	203.27	87.92	1113.7	.1645	1.9339
130	2.2225	.01625	157.34	97.90	1117.9	.1816	1.9112
140	2.8886	.01629	123.01	107.89	1122.0	.1984	1.8894
150	3.718	.01634	97.07	117.89	1126.1	.2149	1.8685
160	4.741	.01639	77.29	127.89	1130.2	.2311	1.8485
170	5.992	.01645	62.06	137.90	1134.2	.2472	1.8293
180	7.510	.01651	50.23	147.92	1138.1	.2630	1.8109
190	9.339	.01657	40.96	157.95	1142.0	.2785	1.7932
200	11.526	.01663	33.64	167.99	1145.9	.2938	1.7762
210	14.123	.01670	27.82	178.05	1149.7	.3090	1.7598
212	14.696	.01672	26.80	180.07	1150.4	.3120	1.7566
220	17.186	.01677	23.15	188.13	1153.4	.3239	1.7440
230	20.780	.01684	18.283	198.23	1157.0	.3387	1.7288
240	24.969	.01692	16.323	208.34	1160.5	.3531	1.7140
250	29.825	.01700	13.821	216.48	1164.0	.3675	1.6998
260	35.429	.01709	11.763	228.64	1167.3	.3817	1.6860
270	41.858	.01717	10.061	238.84	1170.6	.3958	1.6727
280	49.203	.01726	8.645	249.06	1173.8	.4096	1.6597
290	57.556	.01735	7.461	259.31	1176.8	.4234	1.6472
300	67.013	.01745	6.466	269.59	1179.7	.4369	1.6350
310	77.68	.01755	5.626	279.92	1182.5	.4504	1.6231
320	89.66	.01765	4.914	290.28	1185.2	.4637	1.6115
330	103.06	.01776	4.307	300.68	1187.7	.4769	1.6002
340	118.01	.01787	3.788	311.13	1190.1	.4900	1.5891
350	134.63	.01799	3.342	321.63	1192.3	.5029	1.5783
360	153.04	.01811	2.957	332.18	1194.4	.5158	1.5677
370	173.37	.01823	2.625	342.79	1196.3	.5286	1.5573
380	195.77	.01836	2.335	353.45	1198.1	.5413	1.5471
390	220.37	.01850	2.0836	364.17	1199.6	.5539	1.5371
400	247.31	.01864	1.8633	374.97	1201.0	.5664	1.5272
410	276.75	.01878	1.6700	385.83	1202.1	.5788	1.5174
420	308.83	.01894	1.5000	396.77	1203.1	.5912	1.5078
430	343.72	.01910	1.3499	407.79	1203.8	.6035	1.4982
440	381.59	.01926	1.2171	418.90	1204.3	.6158	1.4887
450	422.6	.0194	1.0993	430.1	1204.6	.6280	1.4793
460	466.9	.0196	0.9944	441.4	1204.6	.6402	1.4700
470	514.7	.0198	.9009	452.8	1204.3	.6523	1.4606
480	566.1	.0200	.8172	464.4	1203.7	.6645	1.4513
490	621.4	.0202	.7423	476.0	1202.8	.6766	1.4419
500	680.8	.0204	.6749	487.8	1201.7	.6887	1.4325
520	812.4	.0209	.5594	511.9	1198.2	.7130	1.4136
540	962.5	.0215	.4649	536.6	1193.2	.7374	1.3942
560	1133.1	.0221	.3868	562.2	1186.4	.7621	1.3742
580	1325.8	.0228	.3217	588.9	1177.3	.7872	1.3532
600	1542.9	.0236	.2668	617.0	1165.5	.8131	1.3307
620	1786.6	.0247	.2201	646.7	1150.3	.8398	1.3062
640	2059.7	.0260	.1798	678.6	1130.5	.8679	1.2789
660	2365.4	.0278	.1442	714.2	1104.4	.8987	1.2472
680	2708.1	.0305	.1115	757.3	1067.2	.9351	1.2071
700	3093.7	.0369	.0761	823.3	995.4	.9905	1.1389
705.4	3206.2	.0503	.0503	902.7	902.7	1.0580	1.0580

* Abridged from Keenan and Keyes, "Thermodynamic Properties of Steam," Wiley, New York, 1936. Copyright, 1937, by Joseph H. Keenan and Frederick G. Keyes.

Table A-17 SPECIFIC HEATS OF MISCELLANEOUS LIQUIDS AND SOLIDS

Material	Specfic Heat cal/g °C
Alumina	0.2 (100°C); 0.274 (1500°C)
Alundum	0.186 (100°C)
Asbestos	0.25
Asphalt	0.22
Bakelite	0.3 to 0.4
Brickwork	About 0.2
Carbon	0.168 (26° to 76°C)
	0.314 (40° to 892°C)
	0.387 (56° to 1450°C)
(gas retort)	0.204
(See under Graphite)	
Cellulose	0.32
Cement, Portland Clinker	0.186
Charcoal (wood)	0.242
Chrome brick	0.17
Clay	0.224
Coal	0.26 to 0.37
tar oils	0.34 (15° to 90°C)
Coal tars	0.35 (40°C); 0.45 (200°C)
Coke	0.265 (21° to 400°C)
	0.359 (21° to 800°C)
	0.403 (21° to 1300°C)
Concrete	0.156 (70° to 312°F); 0.219 (72° to 1472°F)
Cryolite	0.253 (16° to 55°C)
Diamond	0.147
Fireclay brick	0.198 (100°C); 0.298 (1500°C)
Fluorspar	0.21 (30°C)
Gasoline	0.53
Glass (crown)	0.16 to 0.20
(flint)	0.117
(pyrex)	0.20
(silicate)	0.188 to 0.204 (0 to 100°C)
	0.24 to 0.26 (0 to 700°C)
wool	0.157
Granite	0.20 (20° to 100°C)
Graphite	0.165 (26° to 76°C); 0.390 (56° to 1450°C)
Gypsum	0.259 (16° to 46°C)
Kerosene	0.47
Limestone	0.217
Litharge	0.055
Magnesia	0.234 (100°C); 0.188 (1500°C)
Magnesite brick	0.222 (100°C); 0.195 (1500°C)
Marble	0.21 (18°C)
Pyrites (copper)	0.131 (19° to 50°C)
(iron)	0.136 (15° to 98°C)
Quartz	0.17 (0°C); 0.28 (350°C)
Sand	0.191
Silica	0.316
Steel	0.12
Stone	About 0.2
Turpentine	0.42 (18°C)
Wood (oak)	0.570
Most woods vary between	0.45 and 0.65

Note: For SI units, multiply cal/g°C by 4187 to get J/kg·K.

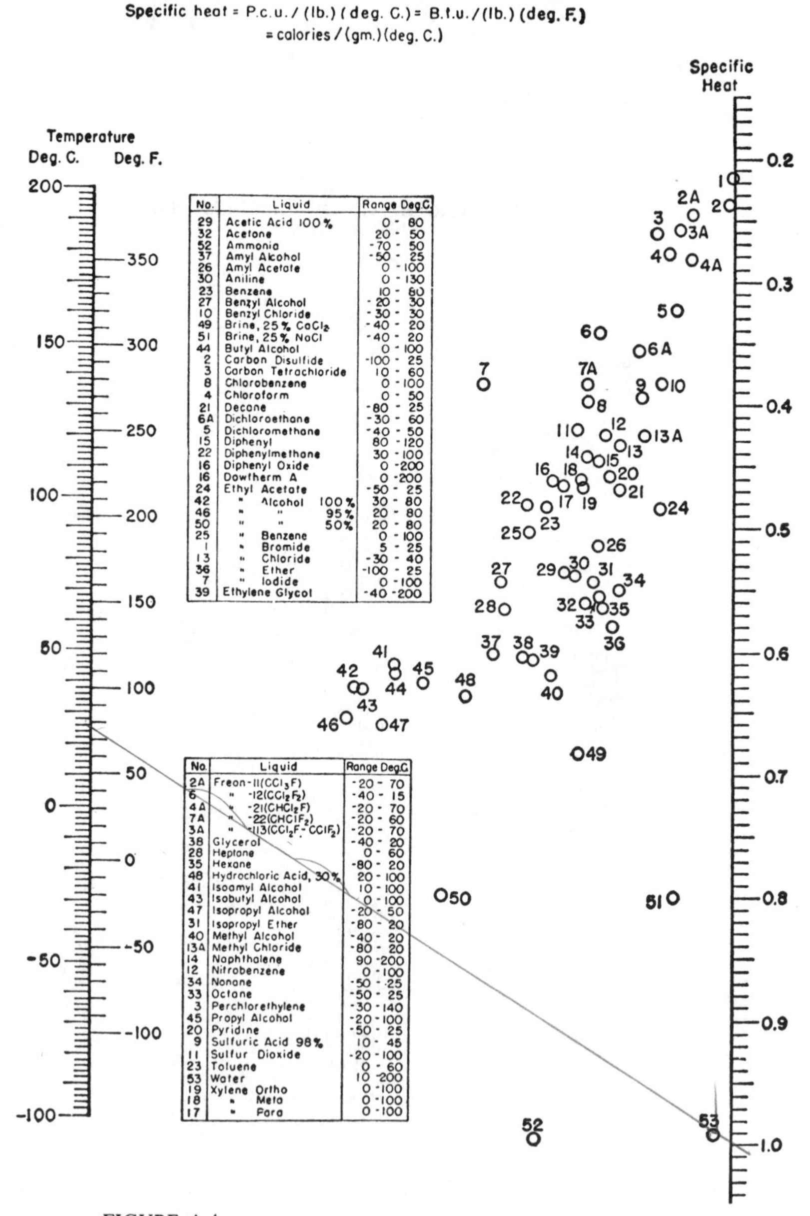

No.	Liquid	Range Deg.C.
29	Acetic Acid 100%	0 - 80
32	Acetone	20 - 50
52	Ammonia	-70 - 50
37	Amyl Alcohol	-50 - 25
26	Amyl Acetate	0 - 100
30	Aniline	0 - 130
23	Benzene	10 - 80
27	Benzyl Alcohol	-20 - 30
10	Benzyl Chloride	-30 - 30
49	Brine, 25% $CaCl_2$	-40 - 20
51	Brine, 25% NaCl	-40 - 20
44	Butyl Alcohol	0 - 100
2	Carbon Disulfide	-100 - 25
3	Carbon Tetrachloride	10 - 60
8	Chlorobenzene	0 - 100
4	Chloroform	0 - 50
21	Decane	-80 - 25
6A	Dichloroethane	-30 - 60
5	Dichloromethane	-40 - 50
15	Diphenyl	80 - 120
22	Diphenylmethane	30 - 100
16	Diphenyl Oxide	0 - 200
16	Dowtherm A	0 - 200
24	Ethyl Acetate	-50 - 25
42	" Alcohol 100%	30 - 80
46	" " 95%	20 - 80
50	" " 50%	20 - 80
25	" Benzene	0 - 100
1	" Bromide	5 - 25
13	" Chloride	-30 - 40
36	" Ether	-100 - 25
7	" Iodide	0 - 100
39	Ethylene Glycol	-40 - 200

No.	Liquid	Range Deg.C.
2A	Freon-11(CCl_3F)	-20 - 70
6	" -12(CCl_2F_2)	-40 - 15
4A	" -21($CHCl_2F$)	-20 - 70
7A	" -22($CHClF_2$)	-20 - 60
3A	" -113(CCl_2F-$CClF_2$)	-20 - 70
38	Glycerol	-40 - 20
28	Heptane	0 - 60
35	Hexane	-80 - 20
48	Hydrochloric Acid, 30%	20 - 100
41	Isoamyl Alcohol	10 - 100
43	Isobutyl Alcohol	0 - 100
47	Isopropyl Alcohol	-20 - 50
31	Isopropyl Ether	-80 - 20
40	Methyl Alcohol	-40 - 20
13A	Methyl Chloride	-80 - 20
14	Naphthalene	90 - 200
12	Nitrobenzene	0 - 100
34	Nonane	-50 - 25
33	Octane	-50 - 25
3	Perchlorethylene	-30 - 140
45	Propyl Alcohol	-20 - 100
20	Pyridine	-50 - 25
9	Sulfuric Acid 98%	10 - 45
11	Sulfur Dioxide	-20 - 100
23	Toluene	0 - 60
53	Water	10 - 200
19	Xylene Ortho	0 - 100
18	" Meta	0 - 100
17	" Para	0 - 100

FIGURE A-4
Specific heats of liquids. Multiply cal/g°C by 4187 to get J/kg·K. (*Based mainly on data from International Critical Tables.*)

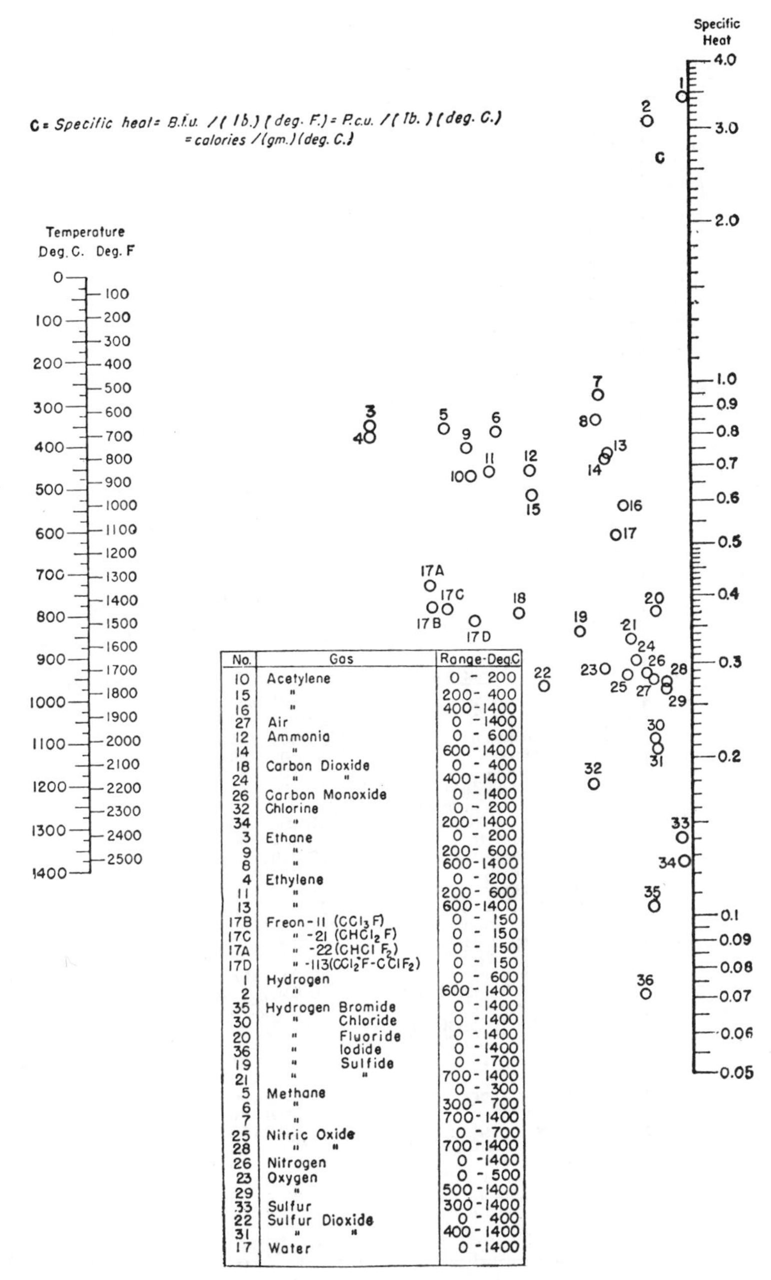

No.	Gas	Range-Deg.C
10	Acetylene	0 - 200
15	"	200 - 400
16	"	400 - 1400
27	Air	0 - 1400
12	Ammonia	0 - 600
14	"	600 - 1400
18	Carbon Dioxide	0 - 400
24	" "	400 - 1400
26	Carbon Monoxide	0 - 1400
32	Chlorine	0 - 200
34	"	200 - 1400
3	Ethane	0 - 200
9	"	200 - 600
8	"	600 - 1400
4	Ethylene	0 - 200
11	"	200 - 600
13	"	600 - 1400
17B	Freon-11 (CCl_3F)	0 - 150
17C	" -21 ($CHCl_2F$)	0 - 150
17A	" -22 ($CHClF_2$)	0 - 150
17D	" -113 (CCl_2F-$CClF_2$)	0 - 150
1	Hydrogen	0 - 600
2	"	600 - 1400
35	Hydrogen Bromide	0 - 1400
30	" Chloride	0 - 1400
20	" Fluoride	0 - 1400
36	" Iodide	0 - 1400
19	" Sulfide	0 - 700
21	" "	700 - 1400
5	Methane	0 - 300
6	"	300 - 700
7	"	700 - 1400
25	Nitric Oxide	0 - 700
28	" "	700 - 1400
26	Nitrogen	0 - 1400
23	Oxygen	0 - 500
29	"	500 - 1400
33	Sulfur	300 - 1400
22	Sulfur Dioxide	0 - 400
31	" "	400 - 1400
17	Water	0 - 1400

FIGURE A-5
Specific heats (C_p) of gases at 1 atm pressure. Multiply cal/g°C by 4187 to get J/kg·K.

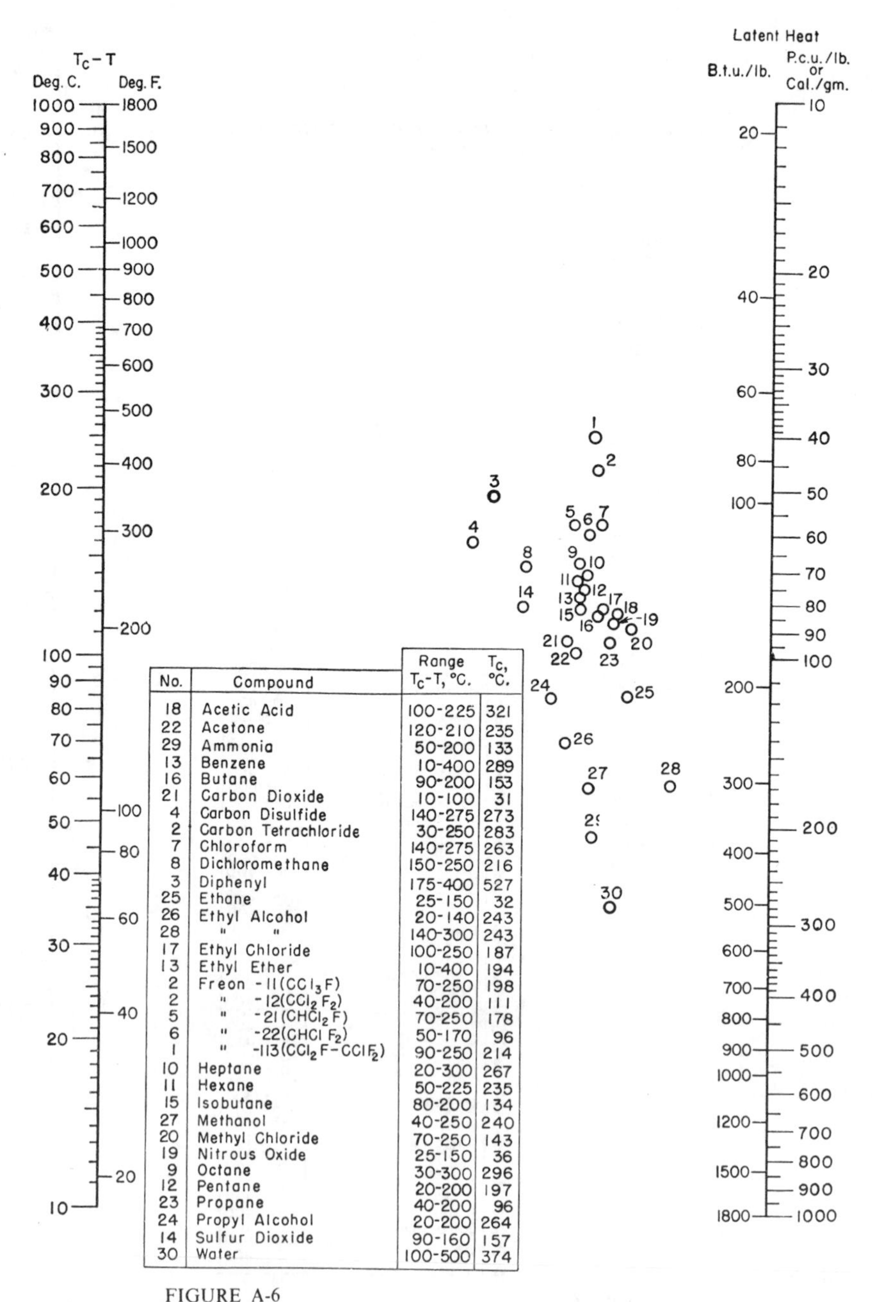

No.	Compound	Range $T_c - T$, °C.	T_c, °C.
18	Acetic Acid	100-225	321
22	Acetone	120-210	235
29	Ammonia	50-200	133
13	Benzene	10-400	289
16	Butane	90-200	153
21	Carbon Dioxide	10-100	31
4	Carbon Disulfide	140-275	273
2	Carbon Tetrachloride	30-250	283
7	Chloroform	140-275	263
8	Dichloromethane	150-250	216
3	Diphenyl	175-400	527
25	Ethane	25-150	32
26	Ethyl Alcohol	20-140	243
28	" "	140-300	243
17	Ethyl Chloride	100-250	187
13	Ethyl Ether	10-400	194
2	Freon -11(CCl_3F)	70-250	198
2	" -12(CCl_2F_2)	40-200	111
5	" -21($CHCl_2F$)	70-250	178
6	" -22($CHClF_2$)	50-170	96
1	" -113($CCl_2F-CClF_2$)	90-250	214
10	Heptane	20-300	267
11	Hexane	50-225	235
15	Isobutane	80-200	134
27	Methanol	40-250	240
20	Methyl Chloride	70-250	143
19	Nitrous Oxide	25-150	36
9	Octane	30-300	296
12	Pentane	20-200	197
23	Propane	40-200	96
24	Propyl Alcohol	20-200	264
14	Sulfur Dioxide	90-160	157
30	Water	100-500	374

FIGURE A-6
Latent heat of vaporization. Multiply cal/g by 4187 to get J/kg. (*Based mainly on data from International Critical Tables.*)

Table A-18 DIFFUSION COEFFICIENTS OF GASES AND VAPORS IN AIR AT 25°C, 1 Atm

Multiply cm^2/s by 10^{-4} to get m^2/s

Substance	D, cm^2/s	$(\mu/\rho D)$
Ammonia	0.28	0.78
Carbon dioxide	.164	.94
Hydrogen	.410	.22
Oxygen	.206	.75
Water	.256	.60
Carbon disulfide	.107	1.45
Ethyl ether	.093	1.66
Methanol	.159	0.97
Ethyl alcohol	.119	1.30
Propyl alcohol	.100	1.55
Butyl alcohol	.090	1.72
Amyl alcohol	.070	2.21
Hexyl alcohol	.059	2.60
Formic acid	.159	0.97
Acetic acid	.133	1.16
Propionic acid	.099	1.56
i-Butyric acid	.081	1.91
Valeric acid	.067	2.31
i-Caproic acid	.060	2.58
Diethyl amine	.105	1.47
Butyl amine	.101	1.53
Aniline	.072	2.14
Chloro benzene	.073	2.12
Chloro toluene	.065	2.38
Propyl bromide	.105	1.47
Propyl iodide	.096	1.61
Benzene	.088	1.76
Toluene	.084	1.84
Xylene	.071	2.18
Ethyl benzene	.077	2.01
Propyl benzene	.059	2.62
Diphenyl	.068	2.28
n-Octane	.060	2.58
Mesitylene	.067	2.31

References: "International Critical Tables," vol. 5, 1928; Landolt-Börnstein, "Physikalische-Chemische Tabellen," 1935.
Note: The group $(\mu/\rho D)$ in the above table is evaluated for mixtures composed largely of air.

Table A-19 DIFFUSION COEFFICIENTS IN LIQUIDS AT 20°C

Multiply cm^2/s by 10^{-4} to get m^2/s

Solute	Solvent	$D \times 10^5$ $(cm^2/s)' \times 10^5$	$\left(\frac{\mu}{\rho D}\right)^*$	Ref.
O_2	Water	1.80	558	1
CO_2	Water	1.77	559	1
N_2O	Water	1.51	665	1
NH_3	Water	1.76	570	1
Cl_2	Water	1.22	824	1
Br_2	Water	1.2	840	2
H_2	Water	5.13	196	1
N_2	Water	1.64	613	1
HCl	Water	2.64†	381	2
H_2S	Water	1.41	712	1
H_2SO_4	Water	1.73	580	2
HNO_3	Water	2.6	390	2
Acetylene	Water	1.56	645	1
Acetic acid	Water	0.88	1140	1
Methanol	Water	1.28	785	1
Ethanol	Water	1.00	1005	1
Propanol	Water	0.87	1150	1
Butanol	Water	.77	1310	1
Allyl alcohol	Water	.93	1080	2
Phenol	Water	.84	1200	2
Glycerol	Water	.72	1400	1
Pyrogallol	Water	.70	1440	2
Hydroquinone	Water	.77	1300	2
Urea	Water	1.06	946	2
Resorcinol	Water	0.80	1260	2
Urethane	Water	.92	1090	2
Lactose	Water	.43	2340	2
Maltose	Water	.43	2340	2
Glucose	Water	.60	....	2
Mannitol	Water	.58	1730	2
Raffinose	Water	.37	2720	2
Sucrose	Water	.45	2230	2
Sodium chloride	Water	1.35	745	2
Sodium hydroxide	Water	1.51	665	2
CO_2	Ethanol	3.4	445	2
Phenol	Ethanol	0.8	1900	2
Chloroform	Ethanol	1.23	1230	2
Phenol	Benzene	1.54	479	1
Chloroform	Benzene	2.11	350	1
Acetic acid	Benzene	1.92	394	1
Ethylene dichloride	Benzene	2.45	301	1

References:
[1] Arnold, *J. Am. Chem. Soc.*, **52**, 3937 (1930).
[2] "International Critical Tables," vol. 5, p. 63.
* Based on $\mu/p = 0.01005$ cm^2/s for water, 0.00737 for benzene, and 0.01511 for ethanol, all at 20°C; applies only for dilute solutes.
† Extrapolated from another temperature, based on rules quoted by Arnold (loc. cit.), i.e., for water solutions D increases 3 percent per °C temperature rise near 20°C, and increases about 2 percent per °C for other liquids.

Table A-20 AMMONIA (NH_3) SOLUBILITY IN WATER

Weight NH_3 per 100 weights H_2O	Partial pressure of NH_3, mmHg							
	0°C	10°C	20°C	25°C	30°C	40°C	50°C	60°C
100	947							
90	785							
80	636	987	1450	....	....	3300		
70	500	780	1170	....	....	2760		
60	380	600	945	....	....	2130		
50	275	439	686	....	....	1520		
40	190	301	470	....	719	1065		
30	119	190	298	....	454	692		
25	89.5	144	227	....	352	534	825	
20	64	103.5	166	....	260	395	596	834
15	42.7	70.1	114	....	179	273	405	583
10	25.1	41.8	69.6	....	110	167	247	361
7.5	17.7	29.9	50.0	....	79.7	120	179	261
5	11.2	19.1	31.7	....	51.0	76.5	115	165
4	....	16.1	24.9	....	40.1	60.8	91.1	129.2
3	....	11.3	18.2	23.5	29.6	45	67.1	94.3
2.5	....	....	15.0	19.4	24.4	(37.6)*	(55.7)	77.0
2	....	....	12.0	15.3	19.3	(30.0)	(44.5)	61.0
1.6	....	....	....	12.0	15.3	(24.1)	(35.5)	48.7
1.2	....	....	....	9.1	11.5	(18.3)	(26.7)	36.3
1.0	....	....	....	7.4	....	(15.4)	(22.2)	30.2
0.5	....	....	....	3.4				

* Extrapolated values.

Table A-21 CARBON DIOXIDE (CO_2) SOLUBILITY IN WATER

Total pressure, atm	Weight of CO_2 per 100 weights of H_2O*								
	12°C	18°C	25°C	31.04°C	35°C	40°C	50°C	75°C	100°C
25	....	3.86	....	2.80	2.56	2.30	1.92	1.35	1.06
50	7.03	6.33	5.38	4.77	4.39	4.02	3.41	2.49	2.01
75	7.18	6.69	6.17	5.80	5.51	5.10	4.45	3.37	2.82
100	7.27	6.72	6.28	5.97	5.76	5.50	5.07	4.07	3.49
150	7.59	7.07	....	6.25	6.03	5.81	5.47	4.86	4.49
200	....	....	....	6.48	6.29	6.28	5.76	5.27	5.08
300	7.86	7.35	....	....	....	....	6.20	5.83	5.84
400	8.12	7.77	7.54	7.27	7.06	6.89	6.58	6.30	6.40
500	....	....	....	7.65	7.51	7.26	....	....	....
700	....	....	....	....	....	....	7.58	7.43	7.61

* In the original, concentration is expressed in cubic centimeters of CO_2 (reduced to 0°C and 1 atm) dissolved in 1 g of water.

Table A-22 CHLORINE (Cl_2) SOLUBILITY IN WATER

Temp., °C	Henry's law coefficient H', lb mol Cl_2/(ft^3)(atm)	Equilibrium constant K_c, (lb mol/ft^3)^2
10	0.00707	7.10
15	.00584	8.55
20	.00469	10.7
25	.00390	12.8

Whitney and Vivian, *Ind. Eng. Chem.*, 33, 741 (1941).

Table A-23 THERMODYNAMIC PROPERTIES OF MOIST AIR (Standard Atmospheric Pressure, 29.921 in Hg)

Temp. t, °F	Saturation humidity $H_s \times 10^8$	Volume, ft³/lb dry air v_a	v_{as}	v_s	Enthalpy, Btu/lb dry air h_a	h_{as}	h_s	Entropy, Btu/(°F)(lb dry air) s_a	s_{as}	s_s	Condensed water: Enthalpy, Btu/lb h_w	Entropy, Btu/(lb)(°F) s_w	Vapor press., in Hg $p_s \times 10^6$	Temp. t, °F
−160	0.2120	7.520	0.000	7.520	−38.504	0.000	−38.504	−0.10300	0.00000	−0.10300	−222.00	−0.4907	0.1009	−160
−155	.3869	7.647	.000	7.647	−37.296	.000	−37.296	−0.09901	.00000	−0.09901	−220.40	−0.4853	.1842	−155
−150	.6932	7.775	.000	7.775	−36.038	.000	−36.088	−0.09508	.00000	−0.09508	−218.77	−0.4800	.3301	−150
−145	1.219	7.902	.000	7.902	−34.881	.000	−34.881	−0.09121	.00000	−0.09121	−217.12	−0.4747	.5807	−145
−140	2.109	8.029	.000	8.029	−33.674	.000	−33.674	−0.08740	.00000	−0.08740	−215.44	−0.4695	1.004	−140
−135	3.586	8.156	.000	8.156	−32.468	.000	−32.468	−0.08365	.00000	−0.08365	−213.75	−0.4642	1.707	−135
−130	6.000	8.283	.000	8.283	−31.262	.000	−31.262	−0.07997	.00000	−0.07997	−212.03	−0.4590	2.858	−130
	$H_s \times 10^7$												$p_s \times 10^5$	
−125	0.9887	8.411	.000	8.411	−30.057	.000	−30.057	−0.07634	.00000	−0.07634	−210.28	−0.4538	0.4710	−125
−120	1.606	8.537	.000	8.537	−28.852	.000	−28.852	−0.07277	.00000	−0.07277	−208.52	−0.4485	.7653	−120
−115	2.571	8.664	.000	8.664	−27.648	.000	−27.648	−0.06924	.00000	−0.06924	−206.73	−0.4433	1.226	−115
−110	4.063	8.792	.000	8.792	−26.444	.000	−26.444	−0.06577	.00000	−0.06577	−204.92	−0.4381	1.939	−110
−105	6.340	8.919	.000	8.919	−25.240	.001	−25.239	−0.06234	.00000	−0.06234	−203.09	−0.4329	3.026	−105
−100	9.772	9.046	.000	9.046	−24.037	.001	−24.036	−0.05897	.00000	−0.05897	−201.23	−0.4277	4.666	−100
	$H_s \times 10^6$												$p_s \times 10^4$	
−95	1.489	9.173	.000	9.173	−22.835	.002	−22.833	−0.05565	.00000	−0.05565	−199.35	−0.4225	0.7111	−95
−90	2.242	9.300	.000	9.300	−21.631	.002	−21.629	−0.05237	.00001	−0.05236	−197.44	−0.4173	1.071	−90
−85	3.342	9.426	.000	9.426	−20.428	.003	−20.425	−0.04913	.00001	−0.04912	−195.51	−0.4121	1.597	−85
−80	4.930	9.553	.000	9.553	−19.225	.005	−19.220	−0.04595	.00001	−0.04594	−193.55	−0.4069	2.356	−80
−75	7.196	9.680	.000	9.680	−18.022	.007	−18.015	−0.04280	.00002	−0.04278	−191.57	−0.4017	3.441	−75
−70	10.40	9.806	.000	9.806	−16.820	.011	−16.809	−0.03969	.00003	−0.03966	−189.56	−0.3965	4.976	−70
−65	14.91	9.932	.000	9.932	−15.617	.015	−15.602	−0.03663	.00005	−0.03658	−187.53	−0.3913	7.130	−65
	$H_s \times 10^5$												$p_s \times 10^3$	
−60	2.118	10.059	.000	10.059	−14.416	.022	−14.394	−0.03360	.00006	−0.03354	−185.47	−0.3861	1.0127	−60
−55	2.982	10.186	.000	10.186	−13.214	.031	−13.183	−0.03061	.00009	−0.03052	−183.39	−0.3810	1.4258	−55
−50	4.163	10.313	.001	10.314	−12.012	.043	−11.969	−0.02766	.00012	−0.02754	−181.29	−0.3758	1.9910	−50
−45	5.766	10.440	.001	10.441	−10.811	.060	−10.751	−0.02474	.00015	−0.02459	−179.16	−0.3707	2.7578	−45
−40	7.925	10.566	.001	10.567	−9.609	.083	−9.526	−0.02186	.00021	−0.02165	−177.01	−0.3655	3.7906	−40
−35	10.81	10.693	.002	10.695	−8.408	.113	−8.295	−0.01902	.00028	−0.01874	−174.84	−0.3604	5.1713	−35
	$H_s \times 10^4$												$p_s \times 10^2$	
−30	1.464	10.820	.002	10.822	−7.207	.154	−7.053	−0.01621	.00038	−0.01583	−172.64	−0.3552	0.70046	−30
−25	1.969	10.946	.004	10.950	−6.005	.207	−5.798	−0.01342	.00051	−0.01291	−170.42	−0.3500	.94212	−25
−20	2.630	11.073	.005	11.078	−4.804	.277	−4.527	−0.01067	.00068	−0.00999	−188.17	−0.3449	1.2587	−20
−15	3.491	11.200	.006	11.206	−3.603	.368	−3.235	−0.00796	.00089	−0.00707	−165.90	−0.3398	1.6706	−15
−10	4.606	11.326	.008	11.334	−2.402	.487	−1.915	−0.00529	.00115	−0.00414	−163.60	−0.3346	2.2035	−10
−5	6.040	11.452	.011	11.463	−1.201	.639	−0.562	−0.00263	.00149	−0.00114	−161.28	−0.3295	2.8886	−5
	$H_s \times 10^3$													
0	0.7872	11.578	.015	11.593	0.000	.835	0.835	0.00000	.00192	0.00192	−158.93	−0.3244	3.7645	0
5	1.020	11.705	.019	11.724	1.201	1.085	2.286	.00260	.00246	.00506	−156.57	−0.3193	4.8779	5
10	1.315	11.831	.025	11.856	2.402	1.401	3 803	.00518	.00314	.00832	−154.17	−0.3141	6.2858	10
15	1.687	11.958	.032	11.990	3.603	1.800	5.403	.00772	.00399	.01171	−151.76	−0.3090	8.0565	15
20	2.152	12.084	.042	12.126	4.804	2.302	7.106	.01023	.00504	.01527	−149.31	−0.3039	10.272	20
25	2.733	12.211	.054	12.265	6.005	2.929	8.934	.01273	.00635	.01908	−146.85	−0.2988	13,032	25
30	3.454	12.338	.068	12.406	7.206	3.709	10.915	.01519	.00796	.02315	−144.36	−0.2936	16.452	30
32	3.788	12.388	.075	12.463	7.686	4.072	11.758	.01617	.00870	.02487	−143.36	−0.2916	18.035	32
32*	3.788	12.388	.075	12.463	7.686	4.072	11.758	.01617	.00870	.02487	0.04	0.0000	18.037	32*
34	4.107	12.438	.082	12.520	8.167	4.418	12.585	.01715	.00940	.02655	2.06	.0041	19.546	34
													p_s	
36	4.450	12.489	.089	12.578	8.647	4.791	13.438	.01812	.01016	.02828	4.07	.0081	0.21166	36
38	4.818	12.540	.097	12.637	9.128	5.191	14.319	.01909	.01097	.03006	6.08	.0122	.22904	38
40	5.213	12.590	.105	12.695	9.608	5.622	15.230	.02005	.01183	.03188	8.09	.0162	.24767	40
42	5.638	12.641	.114	12.755	10.088	6.084	16.172	.02101	.01275	.03376	10.09	.0202	.26763	42
44	6.091	12.691	.124	12.815	10.569	6.580	17.149	.02197	.01373	03570	12.10	.0242	.28899	44
46	6.578	12.742	.134	12.876	11.049	7.112	18.161	.02293	.01478	.03771	14.10	.2082	.31185	46
48	7.100	12.792	.146	12.938	11.530	7.681	19.211	.02387	.01591	.03978	16.11	.0321	.33629	48
50	7.658	12.843	.158	13.001	12.010	8.291	20.301	.02481	.01711	.04192	18.11	.0361	.36240	50
52	8.256	12.894	.170	13.064	12.491	8.945	21.436	.02575	.01839	.04414	20.11	.0400	.39028	52
54	8.894	12.944	.185	13.129	12.971	9.644	22.615	.02669	.01976	.04645	22.12	.0439	.42004	54
56	9.575	12.995	.200	13.195	13.452	10.39	23.84	.02762	.02121	.04883	24.12	.0478	.45176	56
58	10.30	13.045	.216	13.261	13.932	11.19	25.12	.02855	.02276	.05131	26.12	.0517	.48558	58
60	11.08	13.096	.233	13.329	14.413	12.05	26.46	.02948	.02441	.05389	28.12	.0555	.52159	60
62	11.91	13.147	.251	13.398	14.893	12.96	27.85	.03040	.02616	.05656	30.12	.0594	.55994	62
64	12.80	13.197	.271	13.468	15.374	13.94	29.31	.03132	.02803	.05935	32.12	.0632	.60073	64
66	13.74	13.247	.292	13.539	15.855	14.98	30.83	.03223	.03002	.06225	34.11	.0670	.64411	66
68	14.75	13.298	.315	13.613	16.335	16.09	32.42	.03314	.03213	.06527	36.11	.0708	.69019	68

Compiled by John A. Goff and S. Gratch. See also Keenan and Kaye, "Thermodynamic Properties of Air," Wiley, New York, 1945.
Enthalpy of dry air taken as zero at 0°F. Enthalpy of liquid water taken as zero at 32°F.
* Extrapolated to represent metastable equilibrium with undercooled liquid.

Table A-23 (continued)

Temp. t, °F	Saturation humidity $H_s \times 10^2$	Volume, ft³/lb dry air			Enthalpy, Btu/lb dry air			Entropy, Btu/(°F)(lb dry air)			Condensed water			Temp. t, °F
		v_a	v_{as}	v_s	h_a	h_{as}	h_s	s_a	s_{as}	s_s	Enthalpy Btu/lb h_w	Entropy, Btu/(lb)(°F) s_w	Vapor press., inHg p_s	
70	1.582	13.348	.339	13.687	16.816	17.27	34.09	.03405	.03437	.06842	38.11	.0746	.73915	70
72	1.697	13.398	.364	13.762	17.297	18.53	35.83	.03495	.03675	.07170	40.11	.0784	.79112	72
74	1.819	13.449	.392	13.841	17.778	19.88	37.66	.03585	.03928	.07513	42.10	.0821	.84624	74
76	1.948	13.499	.422	13.921	18.259	21.31	39.57	.03675	.04197	.07872	44.10	.0859	.90470	76
78	2.086	13.550	.453	14.003	18.740	22.84	41.58	.03765	.04482	.08247	46.10	.0896	.96665	78
80	2.233	13.601	0.486	14.087	19.221	24.47	43.69	0.03854	.04784	0.08638	48.10	0.0933	1.0323	80
82	2.389	13.651	.523	14.174	19.702	26.20	45.90	.03943	.05105	.09048	50.09	.0970	1.1017	82
84	2.555	13.702	.560	14.262	20.183	28.04	48.22	.04031	.05446	.09477	52.09	.1007	1.7152	84
86	2.731	13.752	.602	14.354	20.663	30.00	50.66	.04119	.05807	.09926	54.08	.1043	1.2529	86
88	2.919	13.803	.645	14.448	21.144	32.09	53.23	.04207	.06189	.10396	56.08	.1080	1.3351	88
90	3.118	13.853	.692	14.545	21.625	34.31	55.93	.04295	.06596	.10890	58.08	.1116	1.4219	90
92	3.330	13.904	.741	14.645	22.106	36.67	58.78	.04382	.07025	.11407	60.07	.1153	1.5135	92
94	3.556	13.954	.795	14.749	22.587	39.18	61.77	.04469	.07480	.11949	62.07	.1188	1.6102	94
96	3.795	14.005	.851	14.856	23.068	41.85	64.92	.04556	.07963	.12519	64.06	.1224	1.7123	96
98	4.049	14.056	.911	14.967	23.548	44.68	68.23	.04643	.08474	.13117	66.06	.1260	1.8199	98
100	4.319	14.106	.975	15.081	24.029	47.70	71.73	.04729	.09016	.13745	68.06	.1296	1.9333	100
102	4.606	14.157	1.043	15.200	24.510	50.91	75.42	.04815	.09591	.14406	70.05	.1332	2.0528	102
104	4.911	14.207	1.117	15.324	24.991	54.32	79.31	.04900	.1020	.1510	72.05	.1367	2.1786	104
	$H_s \times 10$													
106	0.5234	14.258	1.194	15.452	25.472	57.95	83.42	.04985	.1085	.1584	74.04	.1403	2.3109	106
108	.5578	14.308	1.278	15.586	25.953	61.80	87.76	.05070	.1153	.1660	76.04	.1438	2.4502	108
110	.5944	14.359	1.365	15.724	26.434	65.91	92.34	.05155	.1226	.1742	78.03	.1472	2.5966	110
112	.6333	14.409	1.460	15.869	26.915	70.27	97.18	.05239	.1302	.1826	80.03	.1508	2.7505	112
114	.6746	14.460	1.560	16.020	27.397	74.91	102.31	.05323	.1384	.1916	82.03	.1543	2.9123	114
116	.7185	14.510	1.668	16.178	27.878	79.85	107.73	.05407	.1470	.2011	84.02	.1577	3.0821	116
118	.7652	14.561	1.782	16.343	28.359	85.10	113.46	.05490	.1562	.2111	86.02	.1612	3.2603	118
120	.8149	14.611	1.905	16.516	28.841	90.70	119.54	.05573	.1659	.2216	88.01	.1646	3.4474	120
122	.8678	14.662	2.034	16.696	29.322	96.66	125.98	.05656	.1763	.2329	90.01	.1681	3.6436	122
124	.9242	14.712	2.174	16.886	29.804	103.0	132.8	.05739	.1872	.2446	92.01	.1715	3.8493	124
126	.9841	14.763	2.323	17.086	30.285	109.8	140.1	.05821	.1989	.2571	94.01	.1749	4.0649	126
128	1.048	14.813	2.482	17.295	30.766	117.0	147.8	.05903	.2113	.2703	96.00	.1783	4.2907	128
130	1.116	14.864	2.652	17.516	31.248	124.7	155.9	.05985	.2245	.2844	98.00	.1817	4.5272	130
132	1.189	14.915	2.834	17.749	31.729	133.0	164.7	.06067	.2386	.2993	100.00	.1851	4.7747	132
134	1.267	14.965	3.029	17.994	32.211	141.8	174.0	.06148	.2536	.3151	102.00	.1885	5.0337	134
136	1.350	15.016	3.237	18.253	32.692	151.2	183.9	.06229	.2695	.3318	104.00	.1918	5.3046	136
138	1.439	15.066	3.462	18.528	33.174	161.2	194.4	.06310	.2865	.3496	106.00	.1952	5.5878	138
	H_s													
140	0.1534	15.117	3.702	18.819	33.655	172.0	205.7	.06390	.3047	.3686	107.99	.1985	5.8838	140
142	.1636	15.167	3.961	19.128	34.136	183.6	217.7	.06470	.3241	.3888	109.99	.2018	6.1930	142
144	.1745	15.218	4.239	19.457	34.618	196.0	230.6	.06549	.3449	.4104	111.99	.2051	6.5160	144
146	.1862	15.268	4.539	19.807	35.099	209.3	244.4	.06629	.3672	.4335	113.99	.2084	6.8532	146
148	.1989	15.319	4.862	20.181	35.581	223.7	259.3	.06708	.3912	.4583	115.99	.2117	7.2051	148
150	.2125	15.369	5.211	20.580	36.063	239.2	275.3	.06787	.4169	.4848	117.99	.2150	7.5722	150
152	.2271	15.420	5.587	21.007	36.545	255.9	292.4	.06866	.4445	.5132	119.99	.2183	7.9550	152
154	.2430	15.470	5.996	21.466	37.026	273.9	310.9	.06945	.4743	.5438	121.99	.2216	8.3541	154
156	.2602	15.521	6.439	21.960	37.508	293.5	331.0	.07023	.5066	.5768	123.99	.2248	8.7701	156
158	.2788	15.571	6.922	22.493	37.990	314.7	352.7	.07101	.5415	.6125	125.99	.2281	9.2036	158
160	.2990	15.622	7.446	23.068	38.472	337.8	376.3	.07179	.5793	.6511	128.00	.2313	9.6556	160
162	.3211	15.672	8.020	23.692	38.954	363.0	402.0	.07257	.6204	.6930	130.00	.2345	10.125	162
164	.3452	15.723	8.648	24.371	39.436	390.5	429.9	.07334	.6652	.7385	132.00	.2377	10.614	164
166	.3716	15.773	9.339	25.112	39.918	420.8	460.7	.07411	.7142	.7883	134.00	.2409	11.123	166
168	.4007	15.824	10.098	25.922	40.400	454.0	494.4	.07488	.7680	.8429	136.01	.2441	11.652	168
170	.4327	15.874	10.938	26.812	40.882	490.6	531.5	.07565	.8273	.9030	138.01	.2473	12.203	170
172	.4682	15.925	11.870	27.795	41.364	531.3	572.7	.07641	.8927	.9691	140.01	.2505	12.775	172
174	.5078	15.975	12.911	28.886	41.846	576.5	618.3	.07718	.9654	1.0426	142.02	.2537	13.369	174
176	.5519	16.026	14.074	30.100	42.328	627.1	669.4	.07794	1.047	1.125	144.02	.2568	13.987	176
178	.6016	16.076	15.386	31.462	42.810	684.1	726.9	.07870	1.137	1.216	146.03	.2600	14.628	178
180	.6578	16.127	16.870	32.997	43.292	748.5	791.8	.07946	1.240	1.319	148.03	.2631	15.294	180
182	.7218	16.177	18.565	34.742	43.775	821.9	865.7	.08021	1.357	1.437	150.04	.2662	15.985	182
184	.7953	16.228	20.513	36.741	44.257	906.2	950.5	.08096	1.490	1.571	152.04	.2693	16.702	184
186	.8805	16.278	22.775	39.053	44.740	1004	1049	.08171	1.645	1.727	154.05	.2724	17.446	186
188	.9802	16.329	25.427	41.756	45.222	1119	1164	.08245	1.825	1.907	156.06	.2755	18.217	188
190	1.099	16.379	28.580	44.959	45.704	1255	1301	.08320	2.039	2.122	158.07	.2786	19.017	190
192	1.241	16.430	32.375	48.805	46.187	1418	1464	.08394	2.296	2.380	160.07	.2817	19.845	192
194	1.416	16.480	37.036	53.516	46.670	1619	1666	.08468	2.609	2.694	162.08	.2848	20.704	194
196	1.635	16.531	42.885	59.416	47.153	1871	1918	.08542	3.002	3.087	164.09	.2879	21.594	196
198	1.917	16.581	50.426	67.007	47.636	2195	2243	.08616	3.507	3.593	166.10	.2910	22.514	198
200	2.295	16.632	60.510	77.142	48.119	2629	2677	.08689	4.179	4.266	168.11	.2940	23.468	200

Table A-24 SELECTED DISTRIBUTION COEFFICIENTS FOR AQUEOUS SYSTEMS

Solute *B* distributed between nearly immiscible components *A* and *C*: component *C* is water

Components		Temp., °C	Distribution coefficient K*
Solute *B*	*A*		
Chlorine	Carbon tetrachloride	0	5.0
Bromine	Carbon tetrachloride	0	20
		25	27
		40	30
Iodine	Carbon tetrachloride	25	55
Ammonia	Carbon tetrachloride	25	0.0042
Chlorine dioxide	Carbon tetrachloride	0	0.85
		25	0.63
Ammonia	Chloroform	25	0.040
Sulfur dioxide	Chloroform	0	0.71
		20	0.71
Nitric acid	Ethyl ether	25	0.012
Ethanol	Benzene	25	1.1
Isopropanol	Benzene	25	0.50
Isopropanol	Toluene	25	0.21
Phenol	Carbon tetrachloride	25	0.36
Phenol	Chloroform	25	2.8
Phenol	Benzene	25	2.3
Phenol	Xylene	25	1.4
Acetone	Carbon tetrachloride	25	0.44
Acetone	Chloroform	25	5.5
Acetone	Toluene	0	0.48
		20	0.49
		30	0.51
Acetic acid	Carbon tetrachloride	25	0.059
Acetic acid	Bromoform	25	0.083
Acetic acid	Chloroform	20	0.028
Acetic acid	Ethyl ether	0	0.56
		15	0.48
		25	0.45
Acetic acid	Benzene	15	0.012
		25	0.023
Acetic acid	*o*-Xylene	25	0.042
Benzoic acid	Chloroform	10	4.2
		25	2.4
		40	3.2
Benzoic acid	Benzene	6	4.0
		20	4.3
		25	1.8
Diethylamine	Chloroform	25	2.2
Diethylamine	Benzene	25	0.63
Diethylamine	Toluene	18	0.48
		25	0.63
		32	0.90
Diethylamine	Xylene	25	0.20
Aniline	Toluene	25	7.7
Aniline	Xylene	25	3.0

* $K = C_{B(A)}/C_{B(C)}$; C_B = g-formula weight of *B* per liter of solution; values of *K* are extrapolated to zero concentration of *B*.

Reference: "International Critical Tables," vol. III, pp. 418*ff*., 1928.

Table A-25 EQUILIBRIUM DATA FOR MONOETHANOLAMINE SOLUTIONS

Temp., °C	Normality of amine	Partial pressure of CO_2 mmHg	Liquid concentration, moles CO_2 per mole amine
0.0	0.5	745.8	1.110
.0	.5	256.3	0.990
.0	.5	45.3	.817
.0	.5	10.6	.675
25.0	.5	735.7	1.004
25.0	.5	251.8	0.886
25.0	.5	99.6	.795
25.0	.5	44.2	.720
25.0	.5	10.8	.607
50.0	.5	661.3	.880
50.0	.5	228.3	.757
50.0	.5	40.1	.596
75.0	.5	475.8	.685
75.0	.5	130.3	.584
75.0	.5	50.0	.476
0.0	2.0	754.4	.900
.0	2.0	206.1	.776
.0	2.0	79.4	.718
.0	2.0	11.4	.601
25.0	2.0	736.4	.795
25.0	2.0	252.2	.697
25.0	2.0	98.6	.623
25.0	2.0	44.2	.589
25.0	2.0	10.6	.527
50.0	2.0	668.2	.698
50.0	2.0	183.1	.607
50.0	2.0	70.9	.556
50.0	2.0	10.1	.489
75.0	2.0	477.0	.560
75.0	2.0	130.6	.474
75.0	2.0	51.1	.430
0.0	5.0	751.5	.761
.0	5.0	272.2	.679
.0	5.0	206.2	.649
.0	5.0	80.1	.600
.0	5.0	11.5	.600
25.0	5.0	742.9	0.657
25.0	5.0	254.9	.601
25.0	5.0	98.7	.563
25.0	5.0	44.6	.539
25.0	5.0	10.6	.507
50.0	5.0	677.0	.574
50.0	5.0	245.3	.527
50.0	5.0	71.5	.505
50.0	5.0	10.4	.453
75.0	5.0	518.1	.493
75.0	5.0	142.6	.460
75.0	5.0	54.8	.418
0.0	9.5	752.4	.622
.0	9.5	272.2	.592
.0	9.5	79.2	.568
.0	9.5	11.4	.538
25.0	9.5	735.9	.588
25.0	9.5	252.2	.554
25.0	9.5	99.0	.532
25.0	9.5	44.8	.519
25.0	9.5	11.1	.495
50.0	9.5	701.3	.538
50.0	9.5	255.3	.522
50.0	9.5	74.3	.492
50.0	9.5	10.8	.443
75.0	9.5	559.7	.468
75.0	9.5	153.1	.458
75.0	9.5	56.7	.424
25.0	12.5	749.1	.548
25.0	12.5	256.3	.518
25.0	12.5	45.4	.521
50.0	12.5	716.2	.525
50.0	12.5	259.5	.501
50.0	12.5	196.0	.495
50.0	12.5	75.6	.483
50.0	12.5	10.9	.467
75.0	12.5	629.9	.479
75.0	12.5	168.1	.453
75.0	12.5	64.2	.395

Table A-26 CONSTANT-PRESSURE LIQUID-VAPOR EQUILIBRIUM DATA FOR BINARY MIXTURES

Component		Mol % *A* in		Temp., °C	Total pressure, mm
A	*B*	Liquid	Vapor		
Acetone	Ethanol	0.0	0.0	78.3	760
		5.0	15.5	75.4	
		10.0	26.2	73.0	
		15.0	34.8	71.0	
		20.0	41.7	69.0	
		25.0	47.8	67.3	
		30.0	52.4	65.9	
		35.0	56.6	64.7	
		40.0	60.5	63.6	
		50.0	67.4	61.8	
		60.0	73.9	60.4	
		70.0	80.2	59.1	
		80.0	86.5	58.0	
		90.0	92.9	57.0	
		100.0	100.0	56.1	
Butanol (i-)	Water	0.2	4.3	98.9	760
		0.3	6.9	98.1	
		0.4	10.1	97.1	
		0.5	14.7	95.9	
		0.7	16.3	95.1	
		0.9	21.8	93.4	
		1.2	27.0	91.9	
		1.4	28.6	91.5	
		2.0	32.2	89.9	
		2.2	32.7	90.1	
		2.5	32.8	89.5	
		3.2	32.6	89.5	
		4.1	33.0	89.5	
		4.6	33.2	89.5	
		33.1	33.4	89.2	
		33.0	33.1	89.2	
		36.2	32.9	89.4	
		36.5	33.1	89.4	
		39.5	33.3	89.4	
		40.1	33.3	89.5	
		42.4	33.9	89.5	
		43.1	33.9	89.5	
		43.6	34.0	89.5	
		58.7	36.5	90.2	
		60.3	37.4	90.3	
		82.8	55.4	96.0	
		85.0	58.0	97.1	
		86.5	59.9	97.7	
Carbon disulfide	Acetone	0	0	56.2	760
		1.90	8.32	54.0	
		4.76	18.50	51.4	
		13.40	35.10	46.6	
		18.58	44.30	44.0	
		29.12	52.75	41.4	
		37.98	57.40	40.3	
		44.77	59.80	39.8	
		53.60	62.70	39.3	
Carbon disulfide	Acetone	65.30	66.10	39.1	
		78.94	70.50	39.3	
		80.23	72.30	39.6	
		87.99	76.00	40.5	
		96.83	88.60	43.5	
		100.0	100.0	46.3	
Chloroform	Acetone	0	0	56.2	760
		8.55	4.78	57.5	
		14.10	8.35	58.3	
		20.45	13.12	59.4	
		26.12	17.65	60.4	
		33.67	24.95	61.6	
		42.50	35.20	62.8	
		52.29	48.30	63.9	
		73.40	76.30	64.4	
		78.92	82.40	63.8	
		86.25	90.00	63.1	
		88.92	93.50	62.8	
		100.0	100.0	61.3	
Ethanol	Water	0	0	100	760
		1.90	17.00	95.5	
		7.21	38.91	89.0	
		9.66	43.75	86.7	
		12.38	47.04	85.3	
		16.61	50.89	84.1	
		23.37	54.45	82.7	
		26.08	55.80	82.3	
		32.73	58.26	81.5	
		39.65	61.22	80.7	
		50.79	65.64	79.8	
		51.98	65.99	79.7	
		57.32	68.41	79.3	
		67.63	73.85	78.74	
		74.72	78.15	78.41	
		89.43	89.43	78.15	
Methanol	Water	0.0	0.0	100.0	760
		2.0	13.4	96.4	
		4.0	23.0	93.5	
		6.0	30.4	91.2	
		8.0	36.5	89.3	
		10.0	41.8	87.7	
		15.0	51.7	84.4	
		20.0	57.9	81.7	
		30.0	66.5	78.0	
		40.0	72.9	75.3	
		50.0	77.9	73.1	
		60.0	82.5	71.2	
		70.0	87.0	69.3	
		80.0	91.5	67.6	
		90.0	95.8	66.0	
		95.0	97.9	65.0	
		100.0	100.0	64.5	

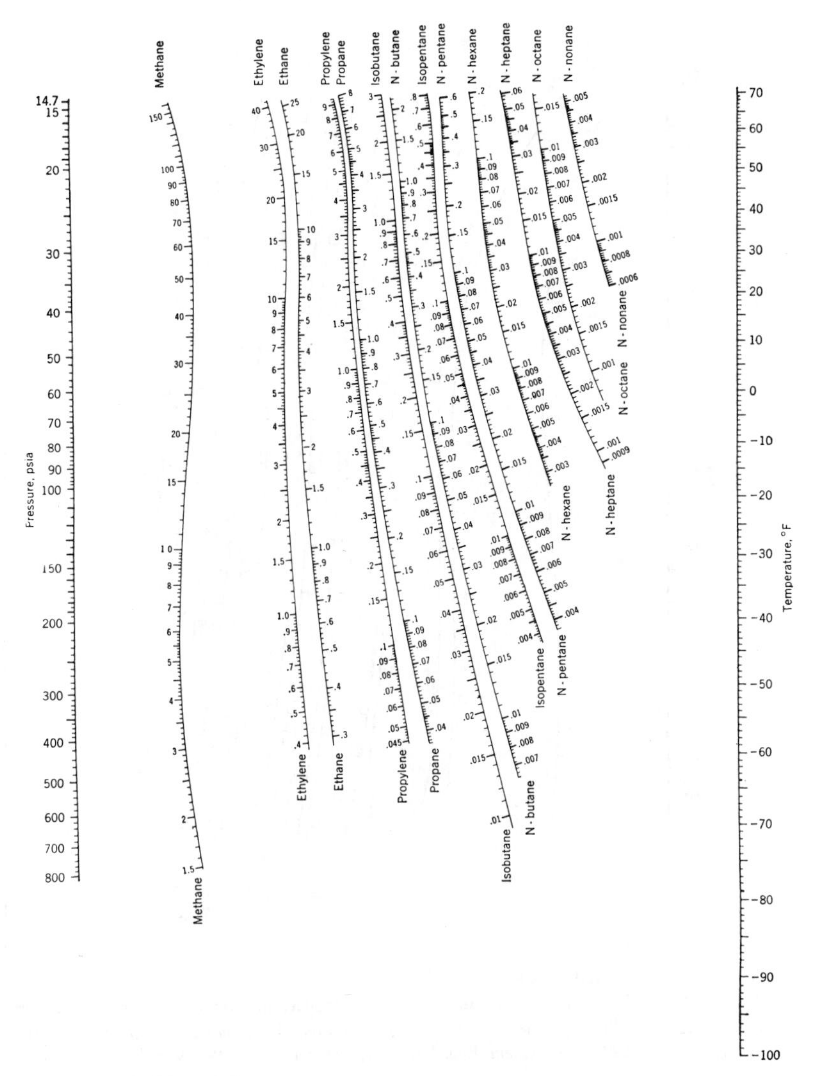

FIGURE A-7
Equilibrium constants in light-hydrocarbon systems. Low-temperature range. Multiply psia by 6895 to get Pa abs. [*Reproduced by permission from C. L. DePriester, Chem. Eng. Progr., Symposium Ser. 7, p. 49 (1953).*]

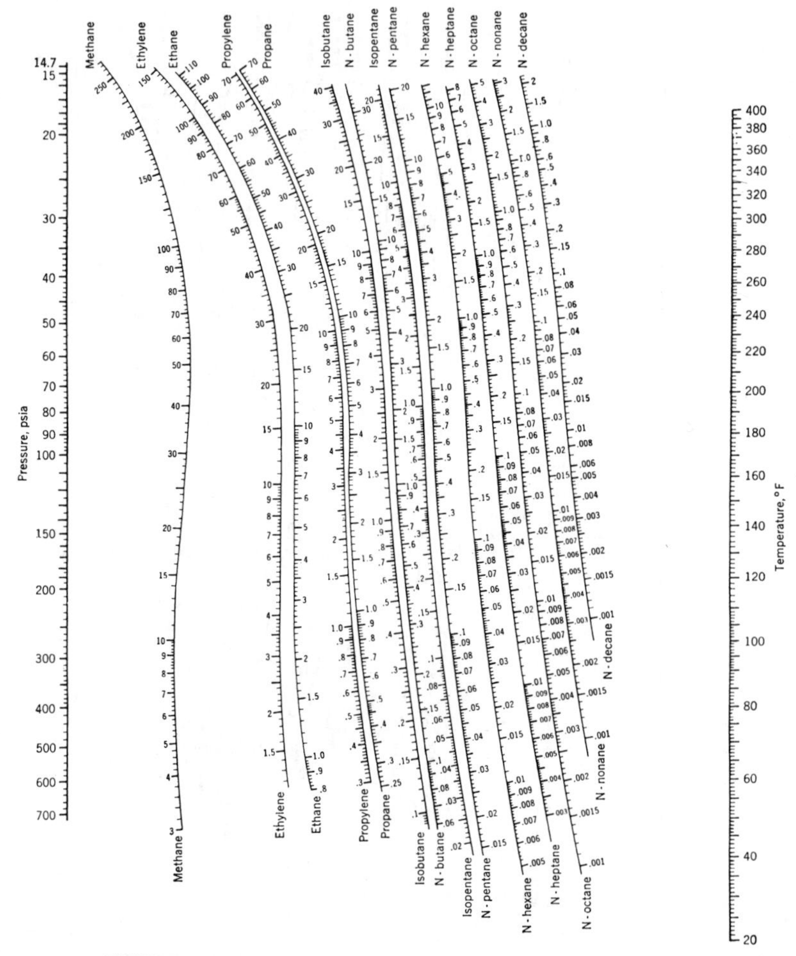

FIGURE A-8
Equilibrium constants in light-hydrocarbon systems. High-temperature range. Multiply psia by 6895 to get Pa abs. [*Reproduced by permission from C. L. DePriester, Chem. Eng. Progr., Symposium Ser. 7, 49 (1953).*]

INDEX

Abramowitz, M., 101
Absorption:
  of gas, 541, 599
  of multicomponent mixtures, 752–759
Absorption factor, 604, 670
Absorptivity, 430
Accumulation, 30
Accuracy of equations, 385
Activity, 497
Activity coefficient, 497
Adiabatic saturation temperature, 640
Adler, S. B., 704
Air-water operations, 638
Aldrich, C. F., 542, 591
Ambrazyavichyus, A. B., 375
American Society of Mechanical Engineers, 22, 80, 86
Ames, W. F., 181
Analogies, 27
  in boundary layer, 625
Analogies (*Cont.*):
  in heat transfer, 358
  in mass transfer, 566
Angular velocity, 106
Angus, F. C., 166
Annulus:
  laminar flow in, 121
  turbulent flow in, 209, 390
Area, heat-transfer, 323, 459
Armstrong, R. C., 23
Arpaci, V., 277, 302
Averaging procedures, 46–47
Azeotropic distillation, 772

Baffles in mixing, 191
Banks of tubes, flow across, 210, 392
Batchelor, Y. K., 155
Beam length, 446
Bénard, H., 375

Benzing, R. J., 193
Bernoulli equation, 53, 130
Bessel functons, 285, 347
Bingham-plastic fluid, 23
Biot number, 286
Bird, R. B., 22, 23, 122
Black body, 430, 435
Blake-Plummer equation, 214
Blasius resistance law, 168
Blasius solution, boundary-layer flow, 141–145
Bliss, R. H., 619
Body force, 104
Boelter, L. M. K., 357
Boiling:
  bubble behavior, 409, 411
  film, 407, 414
  mechanisms, 406
  nucleate, 412
  nuclei curvature, 409
  superheat in, 409
Bonner, J. S., 765
Bosanquet relation, 505
Boundary layer:
  analogies in, 625
  concentration, 551, 577
  definition of, 136
  heat transfer: laminar flow, 335
    turbulent flow, 371
  laminar, 141
  liquid metals, 399
  mass transfer: laminar flow, 551, 625
    turbulent flow, 561
  pressure gradient in, 142
  separation, 137
  thermal, 335
  thickness, 143–145
  turbulent, 137–139
Bradford, J. R., 678
Bradshaw, P., 174
Bradshaw, R. D., 396
Bridgman, P. W., 184, 256
Brink, J. C., 595
Brodkey, R. S., 155
Bromley, L. A., 414
Bubble-cap tray., 660, 676, 702
Bubble-tray Design Manual, American Institute of Chemical Engineers, 676
Buckingham's pi theorum, 184
Buffer layer, 165
Bulk flow in mass transfer, 514, 540
Bulk temperature, 47, 350

Carbon dioxide in the atmosphere, 451
Carrahan, B., 638
Carlsaw, H. S., 277, 523, 652
Cartesian tensors, 778
Cebeci. T., 174
Cellular convection, 375
Centipoise, 19
Cess, R. D., 399
Chang, P., 500
Chemical engineer, 1
Chemical potential, 497
Cheremisinoff, P. M., 452
Christian W. J., 556
Churchill, S. W., 181, 199, 340
Circular pipe:
  average heat-transfer coefficient, 350
  heat transfer: laminar flow, 345, 382
    turbulent flow, 370, 384
  incompressible flow in, 189–207
  laminar flow in, 25, 119
  mass transfer: laminar flow, 557
    turbulent flow, 568–569
  turbulent flow in, 162
Circulating drops, 595
Coefficient of resistance, 168
Colburn, A. P., 384, 569
Colburn entrainment effect, 674
Colburn equation for transfer units, 604
Collision integrals, 503
Compression-permeability technique, 239
Compressive pressure, 229
Computer simulation, 638
Computers for distillation problems, 759–767

Concentration boundary layer, 577
Concurrent flow:
  heat transfer, 327
  mass transfer, 609
Condensation:
  dropwise, 415
  film, 415
  on horizontal tubes, 421
  mechanisms, 413
  of mixtures, 423
  noncondensible gases, 422
  of superheated vapors, 422
  on vertical tubes, 416
Conduction, Fourier equation for, 252, 260, 294
  in heating spheres, 266
  in hollow cylinder, 264–265
  in multilayer wall, 261
  in two-phase system, 258
  unsteady-state, 274
Conservation of mass, 30
Consistency index, 23
Constitutive equation, 786
Continuity, equation of, 92
Continuous belt drier, 650
Control surface, 30
Control volume, 30, 64
Convective heat transfer:
  noncircular conduits, 389
  normal to a cylinder, 389
  in a packed bed, 395
  in pipes, 382
  from a plane surface, 397
  from spheres, 393
Convective mass-transfer coefficients, 493, 535, 580, 586
  (*See also* Mass transfer)
Cooling towers, 644
Cornell, D., 587
Correlation coefficient, 128
Costich, E. W., 189
Crank, J., 523, 594
Creeping flow, 122
Critical moisture content, 651
Cross, H., 207
Crump, J. R., 228
Cullinan, H. T., 498
Curle, N., 132
Curved equilibrium line, 599, 605
Curvilinear squares, method of, 306
Cylindrical coordinates, 106

Daily, J. W., 108
Daives, H. J., 132
Danckwerts, P. V., 573
Darcy's law, 214
Davies, J. T., 151
Deformation of a fluid element, 109
de Groot, S. R., 496
Denbigh, K. A., 524
DePriester, C. L., 704
Design of towers, 541, 599
Desorption or stripping, 591, 607
Dew-point temperature, 625, 642
Diffusion:
  forced, 523
  molecular, 492
  in porous solids, 504
  pressure, 523
  with stagnant B, 514
  thermal, 523
Diffusivities:
  eddy mass, 493, 564
  eddy thermal, 361
  in gases, 502
  in liquids, 486, 500
  in solids, 485, 499
Dilatant fluids, 24
Dimensional analysis:
  in fluid dynamics, 179
  in heat transfer, 378
  in mass transfer, 577
Dimensional homogeneity, 180
Dimensionless group, 179
Distributed properties, 307
Distribution law, 682
Dittus and Boelter equation, 384

Divergence, 93
Dodge, B. F., 701
Donohue, D. A., 211
Double-pipe heat exchangers, 323
Doyle, D., 228
Drag coefficient:
  definition of, 139
  laminar flow, flat plate, 145
  past cylinders, 140
  past spheres and disks, 209
  turbulent flow, flat plate, 175
Drag force, 65
Drake, R. M., Jr., 253, 553
Drew, T. B., 347
Drickamer, H. G., 677
Driving force for mass transfer, 497
Drying, 648
Drying-rate curve, 651
Duffin, J. H., 761
Dühring plot, 477
Dusinberre, G. M., 301
Dwyer, O. E., 399
Dyadics, 777
Dynamic pressure, 118

Eckert, E. R. G., 253, 553
Eddy kinematic viscosity, 160
Eddy mass diffusivity, 493, 564
Eddy thermal diffusivity, 368
Edminster, W. C., 704
Effective particle diameter, 213
Electrolytic analog, 306
Electromagnetic radiation, 427
Elgin, J. C., 614
Emde, F., 285
Emissive power, 434
Emissivity, 431
  of gases, 446
  of solids, 433
Energy flux, 28
Enriching section, 693, 718
  (*See also* Rectifying columns)
Enthalpy, 45, 98
  of moist air, 644
Entrance effects, 147
  in heat transfer: with laminar flow, 312, 343
  with turbulent flow, 356
Equilibrium moisture content, 653
Equilibrium ratio, vapor-liquid, 704
Equimolar counterdiffusion, 520
Equipotential lines, 131
Equivalent diameter, 209
  for annular space, 209
Equivalent length, 201
Erbar-Maddox correlation, 764
Ergun, S., 214
Euler equations, 127
Euler integration method, 637
Euler number, 180
Euler point of view, 93
Evaporation:
  boiling point of solutions, 477
  equipment, 473
  multiple-effect operation, 480
Everett, H. J., 189
Extended surfaces, 465
Extractive distillation, 767

Fair, J. R., 587
Fanning friction factor, 168, 199
Feed quality in distillation, 720
Fenske-Underwood equation, 763
Fick's law, 495
Film theory, 630
Filter cake, 224
  compressible, 231, 239
  incompressible, 231, 242
Filter medium, 224
Filtration:
  constant-pressure, 232
  constant-rate, 233
  variable-rate—variable-pressure, 233

Filtration cycle, 234
Fin efficiencies, 468
First law of thermodynamics, 43
Flint, L. F., 178
Flow-behavior index, 23
Fluctuating velocity, 152
Flugge, W., 285
Fluidization, 225
Flux:
  of energy, 27
  of mass, 27
  of momentum, 21, 27
Fog formation, 646
Force balance, 25
Forced-circulation evaporation, 474
Form drag, 128
Foster, H. K., 412
Fouling coefficients, 321
Fouling factor, 322
Fourier, J. B. J., 252
Fourier equation for conduction (*see* Conduction)
Fractionating column, 707
French, M. J., 166
Friction loss, 54
Friction velocity, 163, 202
Froessling, 394
Froude number, 180
Fuller, E. N., 502

Gas flow in a pipe, 216
Gauss divergence theorem, 781
Gauss error integral, 571
Gay, B., 207
Gebhart, B., 308
General energy balance, 48
Gerritsen, J. K., 259
Giddings, J. C., 502
Gilliland, E. R., 583
Gilliland's correlation, 764
Grace, H. P., 230, 241
Graetz, L., 346
Graetz number, 347, 383
Graetz solutions, 346, 347
Grashof number, 344
Gray surface, 441
Greenhouse effect, 450
Grimison, E. D., 392
Grosh, R. J., 399
Grummer, M., 212
Gurney-Lurie charts, 287–291
Gutenberg, B., 251

Hagen-Poiseuille equation, 116–121
Hanna, O. T., 375
Hanratty, T. J., 593
Hansen, A. G., 181
Hansen, C. M., 376
Hanson, D. N., 761
Harleman, D. R. F., 108
Hartnett, J. P., 7
Harriott, P., 574
Hartland, S., 604
Hassager, O., 23
Hawkins, G. A., 301, 307
Hayward, R. W., 7
Heat exchangers, 323, 460
  double-pipe, 325
  multipass, 463
  shell-and-tube, 460
Heat transfer (*see* Boiling; Condensation; Convection heat transfer; Radiation)
Heat-transfer coefficients, individual, 313–316
Height:
  equivalent to a theoretical stage, 670
  of a transfer unit, 544
  for gases, 588
  for liquids, 589
Heitala, J., 219
Hellums, J. D., 181
Henke, G. E., 765
Henry's law, 533

Higbie, R., 569
Hildebrand, F. B., 638
Himmelblau, D. M., 35
Hindered setting, 225
Hinze, J. O., 155, 566
Hobson, M., 592
Holland, C. D., 759
Holloway, F. A. L., 591
Hooke's law, 108
Horrocks, J. K., 256
Hottel, H. C., 442
Howarth, L., 142
Howe, J. W., 207
Howells, J. R., 430
Huang, Chen, 574
Hughmark, G. A., 593
Humid heat, 639
Humid volume, 639
Humidity, 638
Humidity chart, 643
Hummel, R. L., 166
Hunsaker, J. C., 66
Hydraulic radius, 209
Hydraulically smooth pipe, 167
Hydrodynamics, 125

Ideal fluid, 125
Ideal gas, 497
Ideal solution, 497, 703
Ideal stage(s), 657
  algebraic relations for, 669
Individual heat-transfer coefficients, 313–316
Individual mass-transfer coefficients, 535–536
Infinite plate, 277
Intalox saddles, 611
Integration of mass-transfer-rate equation, 541–542, 598
Intensity:
  of radiation, 436
  of turbulence, 153
Interchange factor, 440
Interfacial area, 591, 599
Interfacial turbulence, 559, 575
Internal energy, 43
Interstitial velocity, 213
Irrotational flow, 127–128
Irvine, T. F., 7
Isotropic solids, 253
Isotropic turbulence, 127
Isotropy, 786

j factor, 369, 584
Jackson, R. M., 82
Jaegar, J. C., 277, 523, 652
Jahnke, E., 285
Jakob, M., 253, 258, 294, 307, 358
Janecke diagram, 693
Jet, flow in, 131
Johnson, A. I., 574
Johnson, H. F., 619
Johnson, R. R., 82
Jordon, A. L., 465
Jorissen, A. L., 78
Juhasz, I. S., 308

Karabelas, A. J., 593
Kármán number, 205
Katz, D. L., 122, 166, 212, 353, 358, 466
Kay, J. M., 108
Kays, W. M., 72, 201, 465
Keenan, J. H., 51
Kellogg charts, 704
Kern, D. Q., 461
Keyes, F. G., 51
Kezios, S. P., 556
Kinetic energy, 43
King, C. J., 766
Kirchhoff's law, 431
Klein, J. S., 346
Klinkenberg, H. A., 181
Knapp, W. G., 587
Knudsen, J. G., 122, 166, 214, 353, 358
Knudsen diffusion, 504

Knudsen gas, 256
Kozeny-Carman equation, 214, 229
Kreider, J. F., 453
Kreith, F., 297, 341, 453
Kremser, A., 671
Kronecker delta, 785
Kronig, R., 595
Kuni, D., 225

Lagrange point of view, 44, 93
Lamb, H., 99, 123
Laminar flow, 19–21
Laminar sublayer, 165, 365
Lamourelle, A. P., 566
Langhaar, H. L., 148, 184
Laplace equation, 128
Lapidus, L., 612, 638
Laplace transforms, 277
Larson, C. A., 82
Leaf-type filter, 227
Le Bas volumes, 501
Lee, C. Y., 502
Leffler, J., 498
Leibnitz formula, 781
Leva, M., 212
Levenspiel, O., 225
Lévêque J., 353
Lever-arm principle, 682, 731
Lewis, W. K., 676, 711
Lewis equation, 567
Lewis number, 554
Lightfoot, E. N., 22, 122
Limiting reflux conditions, 720
Linford, A., 86
Liquid-liquid extraction, 608
  countercurrent contract with reflux, 692
Liquid-liquid operations, 612
Liquid-metals heat transfer, 398
Logarithmic mean:
  area, 265
  mole fraction, 603
  temperature difference, 326
Long-tube natural-circulation evaporator, 474
Loss:
  contraction, 198–202
  expansion, 68, 198–202
  in pumps and turbines, 54
Lost work, 52
Lumley, J. L., 151
Lumped properties, 358
Luther, H. A., 638

McAdams, W. H., 316, 353, 382, 415
McCabe, W. L., 712
McCabe-Thiele method, 711
Mach number, 185
McKelvey, J. M., 23
McLaughlin, E., 256
Magnons, 255
Manometers, 76
Marchello, J. M., 574
Mass average velocity, 508
Mass conservation, 30
Mass flux, 27, 508
Mass fraction, 507
Mass transfer:
  analogies in, 566
  in boundary layer (*see* Boundary layer)
  bulk flow in, 514, 540
  driving force for, 497
  in fixed and fluidized beds, 592
  individual, coefficients, 535–536
  in nonisothermal contactor, 633
  overall coefficients, 537
  from spheres and cylinders, 590
  in spheres, drops, and bubbles, 593
  (*See also* Simultaneous heat and mass transfer)
Mass-transfer-rate equation, integration of, 541–542, 598
Mass velocity, 35
Material derivative, 44
Maximum heat flux, 413

Mazur, P., 496
Mechanical energy balance, 198
Mechanical equivalent of heat, 379
Mellichamp, D. A., 591
Mesler, R. B., 411
Methanol, 34
Metzner, A. B., 23
Mickley, H. S., 144, 277
Middleton, P., 207
Milne-Thomson, L. M., 132
Minimum fluidization velocity, 188–189, 206
Miscibility, 680
Mixing, 189, 191
Mixing-cup (bulk) temperature, 47
Mixing length, 160, 360, 564
Models and similitude, 187
Modulus, 302
Mola average velocity, 508
Mola flux, 495
Mola humidity, 638
Mole fraction, 507
Molecular diffusion, 495
Momentum flux, 21, 27
Monochromatic emissive power, 435
Moody, L. F., 199
Moore, A. D., 299, 306
Moore, F. D., 411
Mooy, H. H., 181
Morrow, D. L., 166
Moyle, M. P., 308
Mullin, J. W., 5
Multicomponent absorption:
  of dilute mixture, 753
  of rich mixture, 756
Multicomponent distillation, 759
Multipass heat exchangers, 463
Multiple contact:
  countercurrent, 666, 688
  with fresh absorbent at each stage, 663
  with fresh solvent at each stage, 665, 689
Murphree efficiency point, 661
Murphree efficiency stage, 660

Nabla operator, 778
Natural convection:
  in heat transfer, 341, 391
  from a vertical plate, 343
Navier-Stokes equations, 114
Nelson, W. L., 321
Newman, A. B., 286
Newton's second law, 64, 103
Nikuradse, J., 166
Nonisothermal contactor, mass transfer in, 633
Nonnewtonian fluids, 23, 400
Normal stress, 105, 110
Notation and units, 4
Nozzle, 58, 67
Nucleation, 405
  (*See also* Boiling)
Nukiyama, Shiro, 407
Number of transfer units, 537
  for dilute solutions, 537
Nusselt, W., 416
Nusselt equation, 416–421
Nusselt number, 338

O'Connell, H. E., 678
Operating line, 538
Operation of a fractionating column, 707
Orifice meter, 79
Origin of turbulence, 152
Overall coefficients:
  heat-transfer, 49, 318
  mass-transfer, 537
Overall efficiency, 677
Overall transfer units, 528, 544
Ozoe, H., 340

Packed bed, flow in, 212
Packed tower:
  design of, 541, 599
  for multicomponent separations, 757
  packing characteristics, 611
Palazzo, D. F., 704

Parabolic velocity distribution, 347
Parallel conduits, 207
Partial condensers, 715, 746
Passive analogs, 308
Peclet number, 347, 580
Penetration theory, 569
  distribution of ages in, 573
Permanent loss, 81
Phase-equilibrium relations:
  liquid systems, 680
  vapor-liquid system, 530, 702
Phase rule, 530
Phonons, 255
Pierce, P. E., 376
Pigford, R. L., 539, 758
Pitot tube, 75
Plait point, 681
Planck's law, 434
Plate-and-frame filter, 226
Pohlausen, E., 171, 335
Pollchik, M., 212
Ponchon, M., 719
Ponchon-Savarit method, 719
Ponchon-Savarit, remarks on utility, 745–746
Poole, J. B., 228
Popovich, A. T., 166
Potential energy, 43
Potential flow, 127
Power number, 189
Prandtl, L., 141
Prandtl mixing length, 162, 360, 564
Prandtl number, 336
Prandtl-Schlichting equation, 177
Prandtl-Taylor analogy, 365
Prausnitz, J. M., 258, 504, 517
Pressure taps for orifices, 79–80
Process development, 1
Proulx, A. Z., 542, 591
Pseudoplastic fluids, 23
Pumps and turbines, efficiency of, 54

Quality of feed in distillation, 720

Raad, T., 411
Radiation, 426
  diffuse, 440
Radiation:
  gamma rays, 427
  from gases, 445
  heat-transfer coefficient, 442
  infrared, 427
  intensity of, 436
  multiple surfaces, 440
  nocturnal, 450
  radio waves, 427
  shield, 444
  solar, 427, 450
  ultraviolet, 427
  visible, 427
  X-rays, 427
Radiation pressure, 429
Radius of gyration, 106
Range of heat-transfer coefficients, 316
Ranz, W. E., 212, 395
Raoult's law, 532, 703
Raschig rings, 611
Rayleigh method, 190
Rayleigh number, 344
Real stages, 674
Recovery factor, 81
Rectifying columns, 712, 737
Reed, C. S., 144, 277
Reflux in extraction, 692
  minimum, 723
  total, 723
Refractory surfaces, 440
Reid, R. C., 258, 504, 512
Regino, T. C., 452
Relative humidity, 639
Relative volatility, 705
Relaxation technique, 294
Residuals, 296
Resistance:
  of filter medium, 231
  coefficient of, 168
  in heat transfer, 261
Resultant of forces, 65

Reynolds, O., 19, 214, 359
Reynolds analogy, 359
Reynolds number, 19, 174
Reynolds stresses, 158
Reynolds transport equation, 45
Rheology, 23
Rheopectic fluid, 25
Rightmire, B. G., 66
Rohsenow, W. M., 412
Roll, J. B., 411
Root-mean-square fluctuating velocity, 154
Rotameter, 83
Rotary vacuum filter, 227
Rotational flow, 129
Rough pipe, flow resistance, 169–171, 200
Roughness, effect on heat transfer, 401
Rouse, H., 207
Ruby, C. L., 704
Rukovena, F., 611
Runge and Kutta method, 638
Rushton, J. H., 189
Ruth, B. F., 239

Sandall, 566, 591
Sarofim, A. F., 442
Satterfield, C. N., 504
Saturation humidity, 639
Savage, D. W., 402
Savarit, R., 729
Scale of turbulence, 154
Scheidegger, A. E., 212
Schettler, P. O., 502
Schlichting, H., 99, 181
Schmidt method in heat conduction, 302
Schmidt number, 552
Schügerl, K., 595
Scriven, L. E., 574, 575
Seban, R. A., 399
Sedimentation, 225
Selectivity, 685
Sellars, J. R., 346
Separation:
  operations for, 494
  point, 138
  of variables, 277
Shaft work, 45
Shear stresses, 105, 108
Shell-and-tube heat exchangers, 460
Sherwood, T. K., 144, 166, 258, 277, 504, 517, 539, 566, 583, 589, 619, 758
Sherwood number, 541, 555
Shewman, P. G., 499
Shimazaki, T. T., 399
Ship models, 187
Shulman, H. L., 542, 591
Side streams, 727, 744
Sieder, E. N., 352, 385
Siegel, R., 430
Similarity transformations, 143, 181, 490, 555
Simultaneous heat and mass transfer:
  with laminar flow, 625
  with turbulent flow, 630
Single equilibrium contact, 686
"Sink" for radiant heat, 440
Skelland, A. H. P., 21, 401
Skin friction, 138
"Slip" in the theory of ideal fluids, 127
Slit, laminar flow in, 117
Smith, B. D., 761
Smith, K. A., 166
Solar constant, 450
Solar heating:
  active systems, 453
  flat-plate collectors, 453
  passive systems, 452
Solar radiation, 427–429, 450
Solubility curve, 680–685
Solvent-free basis, 685, 694
Somerville, G. F., 231
Sonic velocity, 184
Sorel, E., 708
Sorel method in distillation, 708
Souders, M., 671
Source for radiant heat, 440

Southwell, R. V., 294
Spalding, D. B., 178
Specific cake resistance, 230
Specific surface, 213
  mean, 215
Spectrum, 428
Spray towers, 610
Stable solutions, 302
Stage efficiency, 660
Staggered tubes, flow over, 181
Stagnation-flow model, 575
Stagnation pressure, 75
Steady state in turbulent flow, 152
Steam economy, 483
Steam injection in distillation, 726, 745
Stearns, R. R., 82
Stefan-Boltzman law, 434
Stefan experiment, 515
Stegun, I. A., 101
Sternling, C. V., 575
Stewart, W. E., 22, 122
Stokes-Einstein equation, 500
Stokes equation, 124, 500
Storch, H. H., 212
Straightening vanes, 82
Stream function, 124, 143
Streamlines, 124
Streeter, V. L, 132
Streicher, R., 595
Stresses, 104
  apparent, 158
  normal, 110
  Reynolds, 158
  shear, 108
    turbulent, 155
Strigle, R. F., Jr., 611
Stripping columns, 715, 740
Stripping factor, 604
Stripping section in extraction, 693, 695
Stynes, S. K., 471
Substantial derivative, 44
Superficial velocity, 213
Système international (SI), 5

Taborek, J., 461
Tate, G. E., 352, 385
Tensors, 783
Ternary systems, 680
Thermal boundary layer, 311
Thermal conductivities:
  of gases, 256
  of liquids, 255
  of solids, 253
Thermal contact resistance, 267
Thermal diffusion, 523
Thermal diffusivity, 100
Thermocouple errors, 444
Thermodynamics, first law of, 43
Thermophysical properties, 259
Thibessard, G., 80
Thiele, E. W., 712
Thixotropic fluid, 25
Thodos, G., 592
Tie lines, 681, 729
Tiller, F. M., 228
Time constant, 318
Timoshenko, S., 111
Toner, R. K., 35
Toor, H. L., 574
Touloukian, J. K., 259
Touloukian, Y. S., 259
Tower design, 677
Townsend, A. A., 155
Tray drier, 649
Treybal, R. E., 504, 588, 596, 611, 612, 620, 682, 693
Triangular diagrams, 680
Tribus, M., 346
Triple point, 531
Trouton's rule, 712, 761
Turbines and pumps, efficiency of, 54
Turbulent core, 165
Turbulent flow, 19–21
Two feeds in distillation, 743

Uniform heat flux, 349
Unit operations, 2

Units:
American engineering system, 4
British mass system, 4
centimeter-gram-second (cgs) system, 4
meter-kilogram-second (mks) system, 4

van Winkle, M., 771
Vectors, 777
Velocity of approach factor, 78
Velocity profile:
in rough pipe, 170
parabolic velocity distribution, 347
universal, 162
Vena contracta taps, 79
Venturi meter, 77
Vertical-tube evaporator, 473
View factors, 437
Vignes A., 498
Viscoelastic fluid, 25
Viscosity, 17
Viscous-dissipation term, 99
Viscous sublayer, 162–167
Volatility, 703
von Kármán, T., 166
von Kármán, analogy, 368
von Kármán, integral, 371
von Kármán number, 205
Vorticity, 128

Wake, Reynolds numbers and a, 139
Wang, J. C., 765
Weaver, R. E. C., 612
Webb, R. L., 466
Weber number, 185
Wegner, T. W., 593
Weintraub, M., 212
Westwater, J. W., 412
Wet-bulb temperature, 641
Wetted-wall columns, 580
Whitaker, S., 382
White, D. A., 226
Whitehead, J. A., Jr., 376
Whitwell, J. C., 35
Wiedemann-Franz law, 253
Wien's displacement law, 434
Wilke, C. R., 500, 502, 758
Wilkes, J. O., 638
Wilson, E. E., 471
Wislicenus, G. F., 66
Woertz, B. B., 566
Work, (W), 43
Wylie, C. R., 277, 781

Zhukaukas, A. A., 375
Zimmerman, J. O., 542, 591
Zuber, N., 412